Lecture Notes
in Business Information Processing   362

More information about this series at http://www.springer.com/series/7911

Chiara Di Francescomarino · Remco Dijkman ·
Uwe Zdun (Eds.)

Business Process Management Workshops

BPM 2019 International Workshops
Vienna, Austria, September 1–6, 2019
Revised Selected Papers

 Springer

Editors
Chiara Di Francescomarino
Fondazione Bruno Kessler
Trento, Italy

Remco Dijkman
Eindhoven University of Technology
Eindhoven, The Netherlands

Uwe Zdun
University of Vienna
Vienna, Austria

ISSN 1865-1348   ISSN 1865-1356   (electronic)
Lecture Notes in Business Information Processing
ISBN 978-3-030-37452-5   ISBN 978-3-030-37453-2   (eBook)
https://doi.org/10.1007/978-3-030-37453-2

This Springer imprint is published by the registered company Springer Nature Switzerland AG
The registered company address is: Gewerbestrasse 11, 6330 Cham, Switzerland

Preface

The International Conference on Business Process Management (BPM) has established itself over 17 years as the conference where people from academia and industry discuss the latest developments in the area of business process management. This year the conference was organized in Vienna, Austria. Workshops were held on September 2, during one of the conference days and, where the main conference is meant to present finished research, the workshops are meant to discuss research that is still in progress. Each of the workshops focused on particular aspects of business process management, either a particular technical aspect or a particular application domain. These proceedings present the work that was discussed during the workshops.

Workshops were selected based on proposals that were submitted to the workshop co-chairs. The workshops co-chairs evaluated the proposals based on their fit with the conference, the expertise of the workshop organizers, and the expected potential of the workshop to attract high-quality presentations. Subsequently, the workshops themselves called for papers to present. A total of 124 papers was submitted to the workshops. 61 papers were accepted for presentation, leading to a total acceptance rate of 49%. The workshops that attracted sufficient interest from the community created a program, including, for some of them, also keynote talks, as presented in the respective sections of these proceedings. The following workshops were organized:

- The Third International Workshop on Artificial Intelligence for Business Process Management (AI4BPM)
- The Third International Workshop on Business Processes Meet Internet-of-Things (BP-Meet-IoT)
- The 15th International Workshop on Business Process Intelligence (BPI)
- The First International Workshop on Business Process Management in the era of Digital Innovation and Transformation (BPMinDIT)
- The 12th International Workshop on Social and Human Aspects of Business Process Management (BPMS2)
- The 7th International Workshop on Declarative, Decision and Hybrid approaches to processes (DEC2H)
- The Second International Workshop on Methods for Interpretation of Industrial Event Logs (MIEL)
- The First International Workshop on Process Management in Digital Production (PM-DiPro)
- The Second International Workshop on Process-Oriented Data Science for Healthcare (PODS4H)
- The 4th International Workshop on Process Querying (PQ)
- The Second International Workshop on Security and Privacy-enhanced Business Process Management (SPBP)
- The First International Workshop on the Value and Quality of Enterprise Modelling (VEnMo)

As it is customary, the BPM workshop proceedings are post-proceedings meaning that the authors were given the opportunity to revise their papers, based on feedback that they received at the workshops. This book of proceedings contain the revised papers. It also contains a report on a panel that was organized at the conference, during which the major trends and controversies in the field of business process management were discussed.

We are confident that this process of selection, presentation, and revision has led to interesting new insights and research ideas in the field of business process management, and led to compelling discussions and exchange of ideas during the conference. We hope that you – as a reader – will enjoy reading these proceedings and that they inspire your own work.

We would like to take this opportunity to acknowledge the people who helped create these proceedings: Jan Mendling, for carefully managing the conference as a whole, including the workshops; Claudio Di Ciccio, for providing us with useful advice and helpful information about the practical creation of the proceedings; the chairs of the individual workshops and the panel for managing these parts of the program; and of course the people who contributed with their work to these proceedings.

October 2019

Chiara Di Francescomarino
Remco Dijkman
Uwe Zdun

The Panel Discussion at BPM 2019 (Panel Discussion)

Jan Recker[1] and Hajo A. Reijers[2,3]

[1] University of Cologne, Köln, Germany
jan.recker@wiso.uni-koeln.de
[2] Utrecht University, Utrecht, The Netherlands
h.a.reijers@uu.nl
[3] Eindhoven University Technology, Eindhoven, The Netherlands

Abstract. One of the traditional components of the BPM conference is the panel discussion. This report describes the set-up of the 2019 edition, in which we changed procedures somewhat, and summarizes the main insights it generated on basis of the questions discussed. These insights may bear on the development of the discipline as a whole and the conference series specifically.

Keywords: BPM · Panel · Grand challenges · Ethics · Process mining

1 Set-up

A conference focusing on "Business Process Management" (BPM) should look to walk its own talk from time to time. We tried to innovate the processes associated with setting up and running a panel, a traditional component of many academic conferences including the BPM conference series. Because BPM research is situated within information systems scholarship, we looked toward implementing "digital innovations".

For the first time in the history of the conference series, both the composition of the panel and the discussion questions were crowd-sourced. Through mailing lists and postings on social media, we had our community pit nominations for panel members, and nominate topics and questions they would like to see discussed at the panel. In processing this input, we took account of potential biases in the data to compose a representative and knowledgeable panel. The selected panel members were, in alphabetical order:

1. Prof. Avi Gal, Technion – Israel Institute of Technology,
2. Prof. Jan Mendling, Vienna University of Economics and Business,
3. Prof. Stefanie Rinderle Ma, University of Vienna, and
4. Prof. Barbara Weber, University of St. Gallen.

From the wide range of candidate questions we received, we coded the general themes that became evident from the questions and then short-listed the most popular ones. From these, we created a mix of four themes that we considered to be of broad interest to the entire community, to look backward as well as forward in time, and to spark some controversy – hence, excitement among the conference participants. The themes we created ultimately were the following:

Grand Challenges. What are the grand challenges for the BPM community? What direction should our research endeavors take? What is necessary to thrive as a community?

Process Mining. Have we overly invested in research on process mining in the past years at the expense of other topics? Have we missed opportunities because of this focus? Or should we even further deepen and extend our work on process mining?

Success Stories. Are we focused too much on the dissemination of success stories? Should we not be more receptive of learning about failure in BPM research? How could we foster such an attitude?

Ethics. Is there a lack of attention for the ethical implications of the work we generate? How does our work affect the people within the organizations we work with or their wider network? Whose responsibility is it to think through the ethical implications of our work?

The panel discussion was programmed towards the end of the last conference day. It took place on Thursday September 5, 2019, just before the closing session of the conference. After a discussion of the listed themes, the panelists answered questions from the audience. In what follows, we summarize the discussions that ensued about the themes above.

2 Discussion

2.1 Grand Challenges

The panel recognizes that BPM, as a discipline, can connect to many different fields, but that we are often nog clearly recognized by other communities for what we are good at. There is a task for all of us to bring others to our field, while at the same time also reaching out ourselves to other events and outlets for our work. One specific advice for young researchers is also to work research groups with a focus different from BPM, even if only possible for short periods of time.

As to directions for future steps, the panel suggested that we could take a more empirical angle in our work, trying to identify genuine issues that organizations are facing. It would also be good to lessen the tendency of repeatedly fine-tuning existing algorithms and approaches, and instead to be bolder and identify new perspectives on processes. One particular direction noted could be the study the role of people within business processes and how to make them perform better.

2.2 Process Mining

The panel emphasized how process mining has been a big impetus for BPM and has inspired many young, talented people to become active in the discipline. Moreover, the topic also resonates well with industry, which has strengthened our links with practice. This does not mean that there are no other relevant topics. While the human perspective in processes was already mentioned, there seems to be also much potential for exploratory approaches, process design methods, and studies on the link with entrepreneurship.

Another interesting perspective would be "to do new things" by building on the success of process mining. How could process mining become meaningful in IoT (Internet of Things) settings? Is it possible to bring in the management perspective in process mining?

2.3 Success Stories

The panel discussed whether the difference between the engineering/formal sciences on the one hand and the social sciences on the other hand could be behind the apparent difference in appreciation of 'failure' in different communities. While engineers are mostly interested in making things better, social scientists have an interest in explaining the 'how' and 'why' of the way things are and why they are not what they could be. At this point, it seems that in our community scholars are reluctant to send in work that does not show improvement over the state of the art, perhaps because the focus on demonstrating progress is deeply engrained in our authoring and reviewing culture. Yet, the panel was quite clear in that there is much to learn from failure and that failure is, indeed, part of science and theorizing. Even a report on a poorly performing algorithm could make sense if it includes a deep discussion on the reasons that would cause it to perform poorly. Obviously, the panel did not want to give a *carte blanche* to bad science: You have to read related work, design well, report properly, etc. The final insight was that something which is far worse than failing is *faking*: never fake your research.

2.4 Ethics

The community seems to mature with respect to data usage and the appropriate handling of privacy issues, but this is only part of the broader issues we should be aware of. It would be a good idea if we, as researchers, would more consciously reflect on the ethical implications of our work. Of course, researchers themselves cannot completely see through what will be the potential uses of the things they design. A lively discussion took place on the merit (or lack thereof) of a mandatory paragraph in each research paper on the ethical implications of the reported work. A final point made was that we also need to sort out as a community what our ethical standards are for research ethics issues such as acknowledgement and re-use of work or data, for example concerning research deposited in open repositories such as arXiv.

After the discussion of these themes, the panel members interacted with the audience, that raised some additional questions. An idea that was suggested was the further development of process mining into a fully developed research method and to bring it to non-traditional application scenarios. A point of caution that came up was that some of our work may actually be motivated by the convenience of available data sets, rather than other considerations. A final reflection concerned the issue whether BPM should evolve from a data science to a decision science.

A full recording of the panel discussion can be found here: https://www.wu.ac.at/wutv/show/clip/bpm2019-closing.

3 Conclusion

In our opinion, the crowd-sourced approach to organizing the panel worked out well. We felt that the topics that were brought in indeed interested the audience and led people to interact with the panel. The panel members in their turn provided fresh insights, as well as a fair share of amusement. To us, the themes discussed seem to point to a further maturing and growth of the BPM community, both with respect to its coverage of research topics and its critical reflection on the way we carry out our research. In that sense, the panel at BPM 2019 turned out to be a proper vehicle to share points of view within the community not commonly discussed during the research tracks of the conference series.

Contents

Third International Workshop on Artificial Intelligence for Business Process Management (AI4BPM)

Pushing More AI Capabilities into Process Mining to Better Deal
with Low-Quality Logs   5
    Francesco Folino and Luigi Pontieri

Research Challenges for Intelligent Robotic Process Automation   12
    Simone Agostinelli, Andrea Marrella, and Massimo Mecella

Description Logics and Specialization for Structured BPMN   19
    Alexander Borgida, Varvara Kalokyri, and Amélie Marian

Utilizing Ontology-Based Reasoning to Support the Execution
of Knowledge-Intensive Processes   32
    *Carsten Maletzki, Eric Rietzke, Lisa Grumbach, Ralph Bergmann,
    and Norbert Kuhn*

How Cognitive Processes Make Us Smarter   45
    Andrea Zasada

Supporting Complaint Management in the Medical Technology Industry
by Means of Deep Learning   56
    Philip Hake, Jana-Rebecca Rehse, and Peter Fettke

Resource Controllability of Workflows Under Conditional Uncertainty   68
    Matteo Zavatteri, Carlo Combi, and Luca Viganò

Automated Multi-perspective Process Generation
in the Manufacturing Domain   81
    Giray Havur, Alois Haselböck, and Cristina Cabanillas

Data-Driven Workflows for Specifying and Executing Agents
in an Environment of Reasoning and RESTful Systems   93
    Benjamin Jochum, Leonard Nürnberg, Nico Aßfalg, and Tobias Käfer

The Changing Roles of Humans and Algorithms in (Process) Matching   106
    Roee Shraga and Avigdor Gal

Third International Workshop on Business Processes Meet Internet-of-Things (BP-Meet-IoT)

Integrating IoT with BPM to Provide Value to Cattle Farmers in Australia. . .   119
 Owen Keates

Enabling the Discovery of Manual Processes Using a Multi-modal Activity
Recognition Approach.   130
 Adrian Rebmann, Andreas Emrich, and Peter Fettke

15th International Workshop on Business Process Intelligence (BPI)

Impact-Aware Conformance Checking.   147
 Arava Tsoury, Pnina Soffer, and Iris Reinhartz-Berger

Encoding Conformance Checking Artefacts in SAT.   160
 Mathilde Boltenhagen, Thomas Chatain, and Josep Carmona

Performance Mining for Batch Processing Using
the Performance Spectrum.   172
 Eva L. Klijn and Dirk Fahland

LIProMa: Label-Independent Process Matching.   186
 Florian Richter, Ludwig Zellner, Imen Azaiz, David Winkel,
 and Thomas Seidl

A Generic Approach for Process Performance Analysis Using Bipartite
Graph Matching   199
 Chiao-Yun Li, Sebastiaan J. van Zelst, and Wil M. P. van der Aalst

Extracting a Collaboration Model from VCS Logs Based on Process
Mining Techniques   212
 Leen Jooken, Mathijs Creemers, and Mieke Jans

Finding Uniwired Petri Nets Using eST-Miner   224
 Lisa Luise Mannel and Wil M. P. van der Aalst

Discovering Process Models from Uncertain Event Data   238
 Marco Pegoraro, Merih Seran Uysal, and Wil M. P. van der Aalst

Predictive Process Monitoring in Operational Logistics: A Case Study
in Aviation.   250
 Björn Rafn Gunnarsson, Seppe K. L. M. vanden Broucke,
 and Jochen De Weerdt

A Survey of Process Mining Competitions:
The BPI Challenges 2011–2018   263
 Iezalde F. Lopes and Diogo R. Ferreira

**First International Workshop on Business Process Management
in the Era of Digital Innovation and Transformation:
New Capabilities and Perspectives (BPMinDIT)**

The Power of the Ideal Final Result for Identifying Process
Optimization Potential   281
   Ralf Laue

Understanding the Need for New Perspectives on BPM in the Digital Age:
An Empirical Analysis.   288
   Florian Imgrund and Christian Janiesch

The Use of Distance Metrics in Managing Business Process
Transfer - An Exploratory Case Study   301
   Anna-Maria Exler, Jan Mendling, and Alfred Taudes

**12th International Workshop on Social and Human Aspects
of Business Process Management (BPMS2)**

Mining Personal Service Processes: The Social Perspective   317
   Birger Lantow, Julian Schmitt, and Fabienne Lambusch

Supporting ED Process Redesign by Investigating Human Behaviors.   326
   *Alessandro Stefanini, Davide Aloini, Peter Gloor,
   and Federica Pochiero*

The Potential of Workarounds for Improving Processes   338
   Iris Beerepoot, Inge van de Weerd, and Hajo A. Reijers

**7th International Workshop on Declarative, Decision and Hybrid
Approaches to Processes (DEC2H)**

Putting Decisions in Perspective   355
   Marco Montali

DMN for Data Quality Measurement and Assessment   362
   *Álvaro Valencia-Parra, Luisa Parody, Ángel Jesús Varela-Vaca,
   Ismael Caballero, and María Teresa Gómez-López*

Modeling Rolling Stock Maintenance Logistics at Dutch Railways
with Declarative Business Artifacts   375
   Erik Smit and Rik Eshuis

Applying Business Architecture Principles with Domain-Specific Ontology
for ACM Modelling: A Building Construction Project Example   388
   *Antonio Manuel Gutiérrez Fernández, Freddie Van Rijswijk,
   Christoph Ruhsam, Ivan Krofak, Klaus Kogler, Anna Shadrina,
   and Gerhard Zucker*

Checking Compliance in Data-Driven Case Management 400
    Adrian Holfter, Stephan Haarmann, Luise Pufahl, and Mathias Weske

**Second International Workshop on Methods for Interpretation
of Industrial Event Logs (MIEL)**

Graph Summarization for Computational Sensemaking on Complex
Industrial Event Logs 417
    Stefan Bloemheuvel, Benjamin Kloepper, and Martin Atzmueller

Capturing Human-Machine Interaction Events from Radio Sensors
in Industry 4.0 Environments 430
    Stephan Sigg, Sameera Palipana, Stefano Savazzi, and Sanaz Kianoush

**First International Workshop on Process Management in Digital
Production (PMDiPro)**

BPMN and DMN for Easy Customizing of Manufacturing
Execution Systems 441
    René Peinl and Ornella Perak

**Second International Workshop on Process-Oriented Data Science
for Healthcare (PODS4H)**

Analysis and Optimization of a Sepsis Clinical Pathway Using
Process Mining. 459
    *Ricardo Alfredo Quintano Neira, Bart Franciscus Antonius Hompes,
    J. Gert-Jan de Vries, Bruno F. Mazza, Samantha L. Simões de Almeida,
    Erin Stretton, Joos C. A. M. Buijs, and Silvio Hamacher*

Understanding Undesired Procedural Behavior in Surgical Training:
The Instructor Perspective 471
    *Victor Galvez, Cesar Meneses, Gonzalo Fagalde, Jorge Munoz-Gama,
    Marcos Sepúlveda, Ricardo Fuentes, and Rene de la Fuente*

Towards Privacy-Preserving Process Mining in Healthcare 483
    *Anastasiia Pika, Moe T. Wynn, Stephanus Budiono,
    Arthur H. M. ter Hofstede, Wil M. P. van der Aalst, and Hajo A. Reijers*

Comparing Process Models for Patient Populations:
Application in Breast Cancer Care. 496
    *Francesca Marazza, Faiza Allah Bukhsh, Onno Vijlbrief,
    Jeroen Geerdink, Shreyasi Pathak, Maurice van Keulen,
    and Christin Seifert*

Evaluating the Effectiveness of Interactive Process Discovery
in Healthcare: A Case Study.................................. 508
*Elisabetta Benevento, Prabhakar M. Dixit, M. F. Sani, Davide Aloini,
and Wil M. P. van der Aalst*

Developing Process Performance Indicators for Emergency
Room Processes 520
*Minsu Cho, Minseok Song, Seok-Ran Yeom, Il-Jae Wang,
and Byung-Kwan Choi*

Interactive Data Cleaning for Process Mining: A Case Study
of an Outpatient Clinic's Appointment System 532
*Niels Martin, Antonio Martinez-Millana, Bernardo Valdivieso,
and Carlos Fernández-Llatas*

Clinical Guidelines: A Crossroad of Many Research Areas.
Challenges and Opportunities in Process Mining for Healthcare 545
*Roberto Gatta, Mauro Vallati, Carlos Fernandez-Llatas,
Antonio Martinez-Millana, Stefania Orini, Lucia Sacchi,
Jacopo Lenkowicz, Mar Marcos, Jorge Munoz-Gama, Michel Cuendet,
Berardino de Bari, Luis Marco-Ruiz, Alessandro Stefanini,
and Maurizio Castellano*

Predicting Outpatient Process Flows to Minimise the Cost of Handling
Returning Patients: A Case Study 557
Marco Comuzzi, Jonghyeon Ko, and Suhwan Lee

A Data Driven Agent Elicitation Pipeline for Prediction Models 570
*John Bruntse Larsen, Andrea Burattin, Christopher John Davis,
Rasmus Hjardem-Hansen, and Jørgen Villadsen*

A Solution Framework Based on Process Mining, Optimization,
and Discrete-Event Simulation to Improve Queue Performance
in an Emergency Department 583
*Bianca B. P. Antunes, Adrian Manresa, Leonardo S. L. Bastos,
Janaina F. Marchesi, and Silvio Hamacher*

A Multi-level Approach for Identifying Process Change
in Cancer Pathways....................................... 595
*Angelina Prima Kurniati, Ciarán McInerney, Kieran Zucker, Geoff Hall,
David Hogg, and Owen Johnson*

Adopting Standard Clinical Descriptors for Process Mining Case Studies
in Healthcare 608
*Emmanuel Helm, Anna M. Lin, David Baumgartner, Alvin C. Lin,
and Josef Küng*

4th International Workshop on Process Querying (PQ)

Complex Event Processing for Event-Based Process Querying   625
  Han van der Aa

Storing and Querying Multi-dimensional Process Event Logs Using
Graph Databases.   632
  Stefan Esser and Dirk Fahland

**Second International Workshop on Security and Privacy-Enhanced
Business Process Management (SPBP)**

A Legal Interpretation of Choreography Models   651
  Jan Ladleif and Mathias Weske

Provenance Holder: Bringing Provenance, Reproducibility and Trust
to Flexible Scientific Workflows and Choreographies   664
  Ludwig Stage and Dimka Karastoyanova

Mining Roles from Event Logs While Preserving Privacy   676
  Majid Rafiei and Wil M. P. van der Aalst

Extracting Event Logs for Process Mining from Data Stored
on the Blockchain.   690
  *Roman Mühlberger, Stefan Bachhofner, Claudio Di Ciccio,
  Luciano García-Bañuelos, and Orlenys López-Pintado*

A Framework for Supply Chain Traceability Based on Blockchain Tokens. . .   704
  *Thomas K. Dasaklis, Fran Casino, Costas Patsakis,
  and Christos Douligeris*

**First International Workshop on the Value and Quality of Enterprise
Modelling (VEnMo)**

Enterprise Modelling of Digital Innovation in Strategies, Services
and Processes.   721
  Geert Poels

Measuring Business Process Model Reuse in a Process Repository   733
  Ross S. Veitch and Lisa F. Seymour

Anti-patterns for Process Modeling Problems: An Analysis of BPMN
2.0-Based Tools Behavior   745
  *Clemilson Luís de Brito Dias, Vinicius Stein Dani, Jan Mendling,
  and Lucineia Heloisa Thom*

Author Index   759

Third International Workshop on Artificial Intelligence for Business Process Management (AI4BPM)

Third International Workshop on Artificial Intelligence for Business Process Management (AI4BPM)

The field of Artificial Intelligence (AI) continues to grow, with novel methodologies and techniques being applied across numerous areas. In the past few years, there has been a strong interest from both industry and academia in applying AI techniques in the area of Business Process Management (BPM). Indeed, the application of AI is impacting additional areas where BPM perspectives become relevant, including industrial engineering, IoT, and emergency response. The use of AI in BPM has been discussed as the next disruptive technology that will touch almost all business process (BP) activities performed by humans. In some cases, AI will extensively support humans in the execution of tasks, while, in other cases, it will enable full automation of tasks that have traditionally required manual contributions. We believe that over time, AI may lead to entirely new paradigms for BPM in all of its aspects: modeling, analysis, automation, and monitoring. For example, instead of BPM models centered either on process or on case management, we anticipate models that are based fundamentally on goal achievement. Moreover, these models will fully enable continuous improvement and adaptation based on experiential learning with little to no human intervention after the learning phase has been completed.

The goal of this workshop is to establish a forum for researchers and professionals interested in understanding, envisioning, and discussing the challenges and opportunities of moving from current, largely programmatic approaches for BPM, to emerging forms of AI-enabled BPM. The workshop attracted 18 submissions on a large variety of topics including knowledge representation in BPM, deep learning for prediction in BPs, and resource management using AI planning. All submissions were reviewed by at least 3 committee members or their sub-reviewers and eventually 9 papers were accepted (8 full papers and 1 short paper). We believe that the accepted papers provide a novel mix of conceptual and technical contributions that will be of interest for the AI4BPM community. The program of the workshop was completed with a keynote talk given by Luigi Pontieri on extending process mining techniques with AI capabilities to better exploit incomplete/low-level log data, and with an invited talk given by Avidgor Gal on the changing roles of humans and algorithms in BP matching.

Concerning the accepted research contributions, Borgida, Kalokyri, and Marian introduce a new Description Logic, REprocDL, to represent the behavioral semantics of block-structured BPMN models. Maletzki, Rietzke, Grumbach, Bergmann, and Kuhn infer executability of process tasks by designing an inference mechanism to extend a currently researched Ontology-and-Data-Driven Business Process model (ODD-BP model). Agostinelli, Marrella, and Mecella present research challenges for intelligent Robotic Process Automation (RPA), proposing a classification framework to categorize RPA tools on the basis of key dimensions. Hake, Rehse, and Fettke show how data can be used to support complaint management processes in the medical technology industry. Zavatteri, Combi, and Viganó present a technique for automated verification of Access-Controlled Workflows (ACWFs), which are workflow models augmented with information on resource authorizations and constraints. Jochum, Nürnberg, Aßfalg, and Käfer take the Guard-Stage-Milestone (GSM) approach and introduce rules to transfer GSM to a RESTful system with the goal of supporting the READ-WRITE linked data architectures. Zasada conducts a systematic literature in the BPM field in order to highlight the main research directions in cognitive process automation. Havur, Haselboeck, and Cabanillas aim at supporting domain experts in generating manufacturing processes for new products by learning the manufacturing knowledge from existing processes designed for similar products. Finally, Yehezkel, Bialy, Smutko, and Roseberg propose an algorithm for mining low-level event logs. We hope that the mixture of these contributions will ensure that the reader will keep track of the latest advances in the AI4BPM research area.

October 2019

Fabrizio Maria Maggi
Andrea Marrella
Arik Senderovich
Emilio Sulis

Organization

Program Committee

Han van der Aa	Humboldt University of Berlin, Germany
Matteo Baldoni	University of Turin, Italy
Emna Hachicha Belghith	Caen-Normandy University, France
Ralph Bergmann	University of Trier, Germany
Andrea Burattin	DTU, Denmark
Federico Chesani	University of Bologna, Italy
Claudio Di Ciccio	Vienna University of Economics and Business, Austria
Schahram Dustdar	TU Wien, Austria
Peter Fettke	German Research Center for Artificial Intelligence, Germany
Avigdor Gal	Technion, Israel
Krzysztof Kluza	AGH University of Science and Technology, Poland
Henrik Leopold	VU University Amsterdam, The Netherlands
Massimo Mecella	Sapienza University of Rome, Italy
Paola Mello	University of Bologna, Italy
Roberto Micalizio	University of Turin, Italy
Fabio Patrizi	Sapienza University of Rome, Italy
Giulio Petrucci	Google, Switzerland
Manfred Reichert	University of Ulm, Germany
Niek Tax	Booking.com, The Netherlands
Irene Teinemaa	Booking.com, The Netherlands
Daniele Theseider Duprè	University of Eastern Piedmont, Italy
Hagen Voelzer	IBM Zurich Research Lab, Switzerland
Matthias Weidlich	Humboldt University of Berlin, Germany

Pushing More AI Capabilities into Process Mining to Better Deal with Low-Quality Logs

Francesco Folino and Luigi Pontieri[✉]

ICAR-CNR, via P. Bucci 8/9C, 87036 Rende, CS, Italy
{francesco.folino,luigi.pontieri}@icar.cnr.it

Abstract. The ever increasing attention of Process Mining (PM) research to the logs of lowly-structured processes and of non process-aware systems (e.g., ERP, IoT systems) poses several challenges stemming from the lower quality that these logs have, concerning the precision, completeness and abstraction with which they describe the activities performed. In such scenarios, most of the resources spent in a PM project (in terms of time and expertise) are usually devoted to try different ways of selecting and preparing the input data for PM tasks, in order to eventually obtain significant, interpretable and actionable results. Two general AI-based strategies are discussed here that have been partly pursued in the literature to improve the achievements of PM efforts on low-quality logs, and to limit the amount of human intervention needed: (i) using explicit domain knowledge, and (ii) exploiting auxiliary AI tasks. The also provides an overview of trends, open issues and opportunities in the field.

Keywords: Process Mining · Artificial intelligence · Low-quality data

1 Introduction

Process Mining (PM) research has yielded a wide range of data analytics solutions [29] that help turn an event log into process-level knowledge/insight, and provide operational support to running process instances. Specifically, the main classic PM tasks are process discovery, conformance checking and model enhancement. Operational-support tasks range from detecting deviances w.r.t. a given process model, to providing forecasts or recommendations with the help of some kind of predictive model, previously learnt from historical log traces.

However, most PM solutions have problems with lowly structured processes and with fine-grained (possibly imprecise/incomplete) logs, which are a norm in emerging application contexts (e.g., IoT, ERP and service/message -based systems) of PM research. In such cases, even choosing the granularity and scope of the input data for PM tools is difficult, and most part of a PM project's effort (time/expertise) goes into data preparation and intermediate results' evaluation.

C. Di Francescomarino et al. (Eds.): BPM 2019 Workshops, LNBIP 362, pp. 5–11, 2019.
https://doi.org/10.1007/978-3-030-37453-2_1

The paper describes two key log-quality issues and two AI-based strategies for tackling them both (Sect. 2), which rely on using background knowledge and auxiliary AI tasks, respectively. It then overviews recent works that used these strategies to boost, specific PM tasks on low-quality logs (Sect. 3), as well as some related trends, open issues and opportunities (Sect. 4).

2   Log-Quality Issues and AI-based Solution Strategies

2.1   Log-Quality Issues

Among the various kinds of quality problems that may affect a log [3], let us focus below on two issues that both tend to determine a serious gap between the expectancy of PM projects' stakeholders and the nature of the log data available.

Incomplete Data. In many real cases, the information stored in the log does not suffice to produce good process-oriented models/information/decisions. In particular, as concerns control-flow aspects, it may be that: (i) only a subset of work-items' lifecycle transitions were recorded (e.g., complete events only), and/or (ii) the log traces do not meet usual "completeness" assumptions [29] (e.g., two concurrent activities always appear in the same relative order). The scarcity of data labels is another form of incompleteness that may affect certain trace classification settings (e.g., security-breach detection) [7]) where the classifier model must be induced from expert-labeled traces.

Low-Level Data. The log events of non-workflow systems often are not (or cannot be) mapped deterministically to high-level activities, like those analysts are familiar with. Classic PM tools often yield useless results (e.g., incomprehensible process models) when applied directly to such logs, or they cannot be applied at all (e.g., checking log conformance to a model featuring high-level activities). Recent efforts tried to bridge such a gap by devising log abstraction methods [2,4,17,20,26] in two different settings: (a) no predefined activities are known, and the only way to lift the events' representation is employing unsupervised clustering/filtering methods [4,20]; (b) some predefined activities exist (e.g., in process documentation), which provide "supervision" for the abstraction task [2,17,26].

2.2   Two General AI-based Strategies

Using Background Knowledge (BK). A way to deal with low-level/incomplete logs better is to use some base $\mathbb{B}$ of domain-oriented *background knowledge (BK)*, e.g., made of: process taxonomies [10], procedural process models, or declarative models expressed via precedence constraints [17,23] or DECLARE patterns [13].

When $\mathbb{B}$ is both complete and precise, it can be exploited, as a powerful pre-processing tool, to transform a low-quality log L into a more suitable form for the application of PM tools. In particular, behavioural constraints can be used to repair/purge incorrect log data [25], while activity taxonomies [10] can allow for turning the traces of L into sequences of activity-annotated events. However, if the events of L cannot be mapped unambiguously to the concepts in $\mathbb{B}$, the latter data-transformation step can be only done approximately by resorting to heuristics-based log abstraction solutions [2, 26]. A more information-conservative usage of incomplete prior knowledge consists instead in using it in a "deferred" way inside the very discovery, checking and predictive PM tasks.

In general, the attempt to extend Machine Learning (ML) and inference methods with the capability to use background knowledge has a long history in AI. Logics-based representation and reasoning are indeed intrinsic to Inductive Logic Programming (ILP) and Statistical Relational Learning (SRL) frameworks [22], while many constraint-based methods have been applied successfully to several Data Mining tasks, such as itemset/pattern mining and clustering [27].

Exploiting Auxiliary AI Tasks. An alternative/complementary way to alleviate the difficulty of performing some typical PM task T (such as control-flow discovery or predictive monitoring) on low-level/incomplete logs consists in extending current PM algorithms with the capability to extract automatically "auxiliary" knowledge that may help counterbalance the low quality of the data.

In a general data analytics setting, such knowledge could by provided by two different kinds of sub-models: (i) fully-fledged ML models, addressed to correlated/auxiliary tasks and trained synergistically with the primary task T (as in Multi-Task Learning and Learning with Auxiliary Tasks [6] approaches, and in Transfer Learning [30] ones); (ii) "smart" data pre-processing/preparation modules, possibly integrated transparently with the core learning/inference technique —as in the case of the representation learning capabilities that are usually provided by most of the layers of Deep Neural Network models [22]. Both kinds of solutions ground on the expectation that the knowledge or inductive bias coming from correlated (function/representation learning) tasks can offer helpful additional information and skills to an agent that must perform a learning/decision task on the basis of insufficient/unreliable data, and thus lead to better results in terms of generalisation, stability and accuracy.

3   Application to PM Tasks Conducted on Low-Quality Logs

BK-aware Control-Flow Discovery. Behavioural constraints providing a partial description of a process' correct behaviour are exploited in [16, 21, 23] to extract meaningful control-flow models from incomplete/noisy logs. Specifically, this task is stated as a reasoning problem over a given precedence constraints (and solved as a Constraints Satisfaction Problem [27]) in [23], and as an ILP classification problem with a-priori activity dependencies (expressing causality, parallelism, etc.) in [21]. DECLARE models are used instead in [16] to improve an already discovered process model, in a post-processing way; one of the algorithms proposed implements a genetic-programming scheme where candidate models are made evolve according to a fitness score accounting for both the conformance of a model with the log and the fraction of a-constraints satisfied by the model.

BK-aware Conformance Checking. Checking whether low-level traces comply to high-level behavioural models is faced in [1,18] in the challenging setting where the real mapping from the traces' events to the models' activities is uncertain.

Rather than pre-processing the traces with heuristics abstraction methods [2,26], in both proposal the degree of compliance of each trace τ is evaluated probabilistically, either over the set of all possible τ's interpretations [1] or over a Montecarlo-generated sample of the former [18], using prior event-activity mapping probabilities to discard "meaningless" interpretations. The Montecarlo procedure of [17] can also exploit known activity-level constraints for this purpose.

BK-aware Predictive Monitoring. A-priori knowledge encoded via LTL rules are used in [15] to improve the accuracy of an LSTM-based next-activity predictor, when trying to forecast the suffix of an unfinished trace. These a-priori rules are exploited at testing time to prune "invalid" candidate sequences within a beam-search inference scheme, while leaving the underlying LSTM model unchanged.

Auxiliary Tasks: Control-Flow Discovery with Clustering. Trace clustering methods [11] help find homogeneous traces' groups in (heterogenous) logs of complex/flexible processes. Applying process discovery tools to such groups, rather than to an entire log, was often shown to yield: (i) better control-flow models in terms of conformance, generalisation and readability [11]; and (ii) more precise predictors [31]. However, since the clustering task is faced independently of the discovery/prediction task, there is no guarantee that the biases of these two tasks are really aligned.

By contrast, algorithm ActiTrac [12] performs iteratively a joint clustering plus induction scheme, where every cluster is evaluated in terms of its conformance to the control-flow model that is eventually induced from it. Trace clustering is faced synergistically with process discovery also in [8], where an auxiliary conceptual clustering task is introduced to eventually spot subgroups of traces that exhibit deviant behaviors in terms of control-flow and/or performances.

Auxiliary Tasks: Control-Flow Discovery with Abstraction. In order to deal with fine granular and/or heterogenous logs, some control-flow discovery approaches were devised to incorporate automated activity abstraction capabilities, such as: the popular Fuzzy miner plugin [29], and the methods proposed in [5,24] for discovering process models' hierarchies based on automatically induced abstractions patterns. Control-flow discovery from low-level logs is faced in [19] by inducing two models (while possibly using prior knowledge on activity dependencies and event-activity mappings): (i) an HMM roughly describing the process execution flows in terms of (hidden) activities; and (ii) an activity-level control-flow graph, built upon activity-dependency statistics extracted from the HMM model.

Auxiliary Tasks: Prediction with Clustering and Abstraction. The combination of model induction and event abstraction capabilities is also exploited in [20] in a predictive monitoring setting, where the ultimate result is a model for forecasting the remaining time of a partial trace. This task is pursued jointly with two auxiliary tasks, devoted to find two different kinds of clusters (defined via decision rules, for the sake of interpretability), i.e. event clusters and trace clusters — playing as hidden activity types and process variants, respectively. The approach of [20] was extended in [9] to forecast, at run time, the values of an aggregate process-performance indicator, defined over fixed-length time windows; to this end, a further induction task is introduced (and faced jointly with those of [20]): discover a time-series forecasting model, capable to predict, for any new window w, the aggregate performances of the instances of w that have not started yet.

Auxiliary Tasks: Prediction with Deep or Ensemble Models. Deep Neural Nets have become a popular solution for predictive monitoring (cf. [14]), as an effective alternative to feature-based methods [31], thanks to their ability to learn effective representations of the traces jointly with the prediction function.

The ability of ensemble learning to exploit and improve multiple base learners (as a form of meta learning) was leveraged in [7], to discover a (deviance-oriented) trace classifier. This discovery task is accomplished with the help of two kinds of auxiliary tasks: (i) train multiple base classifiers on different pattern-based views of the log; and (ii) train a meta-classifier for combining the base classifiers.

4   Trends, Opportunities and Open Issues

Supervised ensemble models and Deep Neural Nets (DNNs) are powerful solutions for complex classification/prediction tasks, but they both need large amounts of data and time for being trained, and rely on internal data representations and decision logics that are hard to interpret and explain.

The increasing number of solutions that have been proposed for extracting knowledge from complex neural models and for explaining the decisions returned by an ensemble of models could help reduce the latter shortcoming.

On the other hand, the flexibility of DNN architectures makes it easier to define more advanced multi-task learning and transfer learning schemes for tackling the problem of insufficient data, compared to those of Sect. 2.2.

Moreover, the integration of DNN models with background constraints [22] looks a promising direction of research for the analysis of low-quality log data, which has not received attention in the field of Process Mining.

Finally, both Active Learning and Statistical/Deep Reinforcement Learning [28] methods offer under-explored opportunities to PM settings for grasping feedback from the environment and from users, as additional sources of domain knowledge. The latter kind of methods also looks as a natural solution for extending current operational-support PM approaches, and enabling more powerful prescriptive analytics and run-time optimisation mechanisms.

References

1. van der Aa, H., Leopold, H., Reijers, H.A.: Checking process compliance on the basis of uncertain event-to-activity mappings. In: Dubois, E., Pohl, K. (eds.) CAiSE 2017. LNCS, vol. 10253, pp. 79–93. Springer, Cham (2017). https://doi.org/10.1007/978-3-319-59536-8_6

2. Baier, T., Mendling, J., Weske, M.: Bridging abstraction layers in process mining. Inf. Syst. **46**, 123–139 (2014)
3. Bose, R.P.J.C., Mans, R.S., van der Aalst, W.M.P.: Wanna improve process mining results? In: IEEE Symposium on CIDM, pp. 127–134 (2013)
4. Jagadeesh Chandra Bose, R.P., van der Aalst, W.M.P.: Abstractions in process mining: a taxonomy of patterns. In: Dayal, U., Eder, J., Koehler, J., Reijers, H.A. (eds.) BPM 2009. LNCS, vol. 5701, pp. 159–175. Springer, Heidelberg (2009). https://doi.org/10.1007/978-3-642-03848-8_12
5. Bose, R.P.J.C., Verbeek, E.H.M.W., van der Aalst, W.M.P.: Discovering hierarchical process models using ProM. In: Nurcan, S. (ed.) CAiSE Forum 2011. LNBIP, vol. 107, pp. 33–48. Springer, Heidelberg (2012). https://doi.org/10.1007/978-3-642-29749-6_3
6. Caruana, R.: Multitask learning. Mach. Learn. **28**(1), 41–75 (1997)
7. Cuzzocrea, A., Folino, F., Guarascio, M., Pontieri, L.: A robust and versatile multiview learning framework for the detection of deviant business process instances. Int. J. Coop. Inf. Syst. **25**(4), 1–56 (2016)
8. Cuzzocrea, A., Folino, F., Guarascio, M., Pontieri, L.: Deviance-aware discovery of high quality process models. In: ICTAI, pp. 724–731 (2017)
9. Cuzzocrea, A., Folino, F., Guarascio, M., Pontieri, L.: Predictive monitoring of temporally-aggregated performance indicators of business processes against low-level streaming events. Inf. Syst. **81**, 236–266 (2019)
10. De Medeiros, A.A., van der Aalst, W., Pedrinaci, C.: Semantic process mining tools: core building blocks. In: ECIS, pp. 1953–1964 (2008)
11. de Medeiros, A.K.A., et al.: Process mining based on clustering: a quest for precision. In: ter Hofstede, A., Benatallah, B., Paik, H.-Y. (eds.) BPM 2007. LNCS, vol. 4928, pp. 17–29. Springer, Heidelberg (2008). https://doi.org/10.1007/978-3-540-78238-4_4
12. De Weerdt, J., Vanden Broucke, S., Vanthienen, J., Baesens, B.: Active trace clustering for improved process discovery. IEEE TKDE **25**(12), 2708–2720 (2013)
13. van Der Aalst, W.M., Pesic, M., Schonenberg, H.: Declarative workflows: Balancing between flexibility and support. Comput. Sci.-Res. Dev. **23**(2), 99–113 (2009)
14. Di Francescomarino, C., Ghidini, C., Maggi, F.M., Milani, F.: Predictive process monitoring methods: which one suits me best? In: Weske, M., Montali, M., Weber, I., vom Brocke, J. (eds.) BPM 2018. LNCS, vol. 11080, pp. 462–479. Springer, Cham (2018). https://doi.org/10.1007/978-3-319-98648-7_27
15. Di Francescomarino, C., Ghidini, C., Maggi, F.M., Petrucci, G., Yeshchenko, A.: An eye into the future: leveraging a-priori knowledge in predictive business process monitoring. In: Carmona, J., Engels, G., Kumar, A. (eds.) BPM 2017. LNCS, vol. 10445, pp. 252–268. Springer, Cham (2017). https://doi.org/10.1007/978-3-319-65000-5_15
16. Dixit, P.M., Buijs, J.C.A.M., van der Aalst, W.M.P., Hompes, B.F.A., Buurman, J.: Using domain knowledge to enhance process mining results. In: Ceravolo, P., Rinderle-Ma, S. (eds.) SIMPDA 2015. LNBIP, vol. 244, pp. 76–104. Springer, Cham (2017). https://doi.org/10.1007/978-3-319-53435-0_4
17. Fazzinga, B., Flesca, S., Furfaro, F., Masciari, E., Pontieri, L.: Efficiently interpreting traces of low level events in business process logs. Inf. Syst. **73**, 1–24 (2018)
18. Fazzinga, B., Flesca, S., Furfaro, F., Pontieri, L.: Online and offline classification of traces of event logs on the basis of security risks. J. Intell. Inf. Syst. **50**(1), 195–230 (2018)

19. Fazzinga, B., Flesca, S., Furfaro, F., Pontieri, L.: Process discovery from low-level event logs. In: Krogstie, J., Reijers, H.A. (eds.) CAiSE 2018. LNCS, vol. 10816, pp. 257–273. Springer, Cham (2018). https://doi.org/10.1007/978-3-319-91563-0_16

20. Folino, F., Guarascio, M., Pontieri, L.: Mining predictive process models out of low-level multidimensional logs. In: Jarke, M., et al. (eds.) CAiSE 2014. LNCS, vol. 8484, pp. 533–547. Springer, Cham (2014). https://doi.org/10.1007/978-3-319-07881-6_36

21. Goedertier, S., Martens, D., Vanthienen, J., Baesens, B.: Robust process discovery with artificial negative events. J. Mach. Learn. Res. **10**, 1305–1340 (2009)

22. Gori, M.: Machine Learning: A Constraint-Based Approach. Morgan Kaufmann, Burlington (2017)

23. Greco, G., Guzzo, A., Lupia, F., Pontieri, L.: Process discovery under precedence constraints. TKDD **9**(4), 32:1–32:39 (2015)

24. Greco, G., Guzzo, A., Pontieri, L.: Mining taxonomies of process models. Data Knowl. Eng. **67**(1), 74–102 (2008)

25. de Leoni, M., Maggi, F.M., van der Aalst, W.M.: An alignment-based framework to check the conformance of declarative process models and to preprocess event-log data. Inf. Syst. **47**, 258–277 (2015)

26. Mannhardt, F., de Leoni, M., Reijers, H.A., van der Aalst, W.M., Toussaint, P.J.: Guided process discovery-a pattern-based approach. Inf. Syst. **76**, 1–18 (2018)

27. Raedt, L.D., Nijssen, S., O'Sullivan, B., Hentenryck, P.V.: Constraint programming meets machine learning and data mining. Dagstuhl Rep. **1**(5), 61–83 (2011)

28. Sugiyama, M.: Statistical Reinforcement Learning: Modern Machine Learning Approaches. Chapman and Hall/CRC, Boca Raton (2015)

29. Van Der Aalst, W.M.P.: Process Mining: Discovery, Conformance and Enhancement of Business Processes. Springer, Berlin (2011). https://doi.org/10.1007/978-3-642-19345-3

30. Vanschoren, J.: Meta-learning. In: Hutter, F., Kotthoff, L., Vanschoren, J. (eds.) Automated Machine Learning. TSSCML, pp. 35–61. Springer, Cham (2019). https://doi.org/10.1007/978-3-030-05318-5_2

31. Verenich, I., Dumas, M., Rosa, M.L., Maggi, F.M., Teinemaa, I.: Survey and cross-benchmark comparison of remaining time prediction methods in business process monitoring. ACM TIST **10**(4), 34 (2019)

Research Challenges for Intelligent Robotic Process Automation

Simone Agostinelli, Andrea Marrella$^{(\boxtimes)}$, and Massimo Mecella

Sapienza Universitá di Roma, Rome, Italy
{agostinelli,marrella,mecella}@diag.uniroma1.it

Abstract. Robotic Process Automation (RPA) is a fast-emerging automation technology in the field of Artificial Intelligence that allows organizations to automate high volume routines. RPA tools are able to capture the execution of such routines previously performed by a human user on the interface of a computer system, and then emulate their enactment in place of the user. In this paper, after an in-depth experimentation of the RPA tools available in the market, we developed a classification framework to categorize them on the basis of some key dimensions. Then, starting from this analysis, we derived four research challenges necessary to inject intelligence into current RPA technology.

1 Introduction

The recent developments in Artificial Intelligence (AI) force us to continuously revisit the debate on *what should be automated and what should be done by humans*. Robotic Process Automation [4] (RPA) is one of these developments. RPA is a fast-emerging automation approach that uses *software robots* (or simply *SW robots*) to mimic and replicate the execution of highly repetitive tasks performed by humans in their application's user interface (UI). SW robots are mainly used for automating office tasks in operations like accounting, billing and customer service. Typical tasks are: extract semi-structured data from documents, read and write from/to databases, copy and paste data across cells of a spreadsheet, open e-mails and attachments, fill in forms, make calculations, etc. [17].

The Gartner Hype Cycle for AI published in 2018[1] places RPA as one of the technologies at the peak of the hype cycle, meaning that there are nowadays deep expectations of what RPA will be able to deliver to the AI community. In addition, in recent years, a number of case studies have shown that RPA technology can concretely lead to improvements in efficiency for business processes involving routine work in large companies, such as O2 and Vodafone [5,10,12].

Despite this increased interest around RPA, when considering state-of-the-art RPA technology, it becomes apparent that the current generation of RPA tools is driven by predefined rules and manual configurations made by expert users rather than AI [13]. Starting from this statement, in this paper we identify

---

[1] https://www.gartner.com/en/documents/3883863-hype-cycle-for-artificial-intelligence-2018.

C. Di Francescomarino et al. (Eds.): BPM 2019 Workshops, LNBIP 362, pp. 12–18, 2019.
https://doi.org/10.1007/978-3-030-37453-2_2

and test ten RPA tools available in the market and categorize them by means of a classification framework. The results of the classification allow us to derive four research challenges required to evolve RPA towards AI.

2 Classification of RPA Tools

Most of the actual deployments of RPA are industry-specific, e.g., for financial and business services [4]. According to [1], in 2019, the market of RPA solutions includes more than 50 vendors developing tools having different prices and features. Among them, we identified 10 vendors that offer to freely try their RPA tools, i.e., without the need to pay any license. The RPA tools in question are:

- *Automation Anywhere* (https://www.automationanywhere.com/)
- *AssistEdge* (https://www.edgeverve.com/assistedge/)
- *G1ANT* (https://g1ant.com/)
- *Kryon* (https://www.kryonsystems.com/)
- *Rapise* (https://www.inflectra.com/Rapise/)
- *TagUI* (https://github.com/kelaberetiv/TagUI)
- *UiPath* (https://www.uipath.com/)
- *VisualCron* (https://www.visualcron.com/)
- *WinAutomation* (https://www.winautomation.com/)
- *WorkFusion* (https://www.workfusion.com/)

We analyzed any of the above tools leveraging a dedicated case study based on a Purchase-to-Pay process, which includes many standardized and highly repetitive transactions with potential for automation [10]. After selecting the target process to automate, we employed the selected tools to design and train various SW robots, by recording the manual steps of the process. This has allowed us to identify a list of common tasks that must be performed to conduct a RPA project:

1. Determine which process steps (also called *routines*) are good candidates to be automated.
2. Model the selected routines in form of *flowchart diagrams*, which involve the specification of the actions, routing constructs (e.g., parallel and alternative branches), data flow, etc. that define the behaviour of a SW robot.
3. Record the mouse/key events that happen on the UI of the user's computer system. This information is associated to the actions of a routine, enabling it to emulate the recorded human activities by means of a SW robot.
4. Develop each modeled routine by generating the software code required to concretely enact the associated SW robot on a target computer system.
5. Deploy of the SW robots in their environment to perform their actions.
6. Monitor the performance of SW robots to detect bottlenecks and exceptions.
7. Maintenance of the routines, which takes into account each SW robot's performance and error cases. The outcomes of this phase enable a new analysis and design cycle to enhance the SW robots [11].

2.1 Classification Framework

We tested the selected RPA tools with our case study performing the tasks to conduct a RPA project. This has allowed us to realize a *classification framework* for RPA tools, which consists of the following key dimensions:

- **Software (SW) Architecture**: The specific SW architecture adopted by any tool: either *Stand-alone* or *Client-Server*.
- **Coding features**: The behaviour of SW robots can be defined with:
  - *Strong coding*: it is based on the realization of explicit programming scripts, often with the support of a command-line interface (CLI), that instructs the SW robots about the routines to emulate;
  - *Graphical User Interfaces (GUIs)*: user friendly environments providing drag & drop facilities to build the flowchart of the routines to emulate;
  - *Low-code tools*: GUIs that – in addition to drag & drop facilities – provide low-coding functionalities to semi-automatically create software code.
- **Recording facilities**: The actions performed by a human within a software tool can be recorded in many ways:
  - *Web recording*: detection of user actions performed on a web browser;
  - *Desktop recording*: detection of user actions performed on a desktop UI;
  - *Others*: some RPA tools do not support neither web nor desktop recording. Nonetheless, they offer recording tools that work on specific applications only, such as Excel, Acrobat, SAP and Citrix. Some RPA tools provide also traditional screen-scraping recording.
- **Self Learning**: The ability of a RPA tool to automatically understand which user actions belong to which routines (*Intra-routine* learning), and which routines are good candidates for the automation (*Inter-routine* learning).
- **Automation type**: SW robots can either interact with users and/or acting independently. This leads to three different categories of automation:
  - *Attended*: the SW robots constantly require interaction with the users;
  - *Unattended*: the SW robots act like batch processes, i.e., manual intervention is not desired. This is ideal for optimizing back-office work;
  - *Hybrid*: Combination of the two above categories.
- **Routine composition**: The ability of a RPA tool to orchestrate through *manual support* or in *automated way* different (single) routines at run-time associated to different SW robots, when large workflows need to be emulated.
- **Log quality**: The quality of the logs recorded by RPA tools. Since routines consist of collection of activities to be enacted according to certain routing constraints, logs produced by RPA tools resemble *event logs* in process mining. To this end, we measure the quality of such logs using the classification provided in Process Mining Manifesto [2], where five maturity levels are defined, ranging from logs of excellent (★★★★★) to poor quality (★).

Table 1 shows the results of the application of our classification framework to the selected RPA tools. The following aspects become apparent: the majority of the tools provide *(i)* a Client-Server SW architecture, *(ii)* GUIs with drag & drop facilities and low-code functionalities, *(iii)* both web and desktop recording, *(iv)* a hybrid automation type, *(v)* manual-based features to achieve routine composition, *(vi)* logs of poor quality. Interestingly, differently from the other tools, G1ANT and TagUI offer strong-coding functionalities with a basic CLI to support the programming of SW robots. Finally, there is no tool that provides self learning or automated routine composition features.

3   Research Challenges

On the basis of the results discussed in the previous section, we have derived four research challenges (and potential approaches to tackle them) necessary to inject intelligence into current RPA technology.

1. **Intra-routine Self Learning (Segmentation).** Logs recorded by RPA tools are characterized by long sequences of actions and/or events that reflect a number of routine executions. A log can record information about several routines, whose actions and events are mixed in some order that reflects the particular order of their execution by the user [7]. In addition, the same routine can be spread across multiple logs, making the automated identification of routines far from being trivial. One possible approach to tackle this challenge is to rely on log analysis solutions in the Human-Computer Interaction field [9], which focus on identifying frequent user tasks inside logs consisting of actions at different granularity. Alternately, local process mining approaches [15] or sequential pattern mining [8] can be employed to identify sequential patterns of non-consecutive actions that tend to be repeated multiple times across multiple logs [7]. However, to date, no available solution exists that allows to automatically: *(i)* understanding which user actions have to be considered inside the log (separating noise to actions that contribute to routines); *(ii)* interpreting their semantics on the basis of their granularity and *(iii)* identifying to which routines they belong to.

2. **Inter-routine Self Learning (Automated identification of candidate routines to robotize).** To date, current RPA tools provide very limited support to this challenge, which is often performed by means of interviews, walkthroughs, direct observation of workers, and analysis of documentation that may be of poor quality and difficult to understand [11]. This manual approach allows analysts to identify the most obvious routines, while it is not suitable to detect those routines that are not executed on a daily basis or that are performed across multiple business units in different ways [7]. Two potential solutions to this issue are provided respectively by [11] and [7], where the authors propose methods to improve the early stages of the RPA lifecycle using process mining techniques [3].

3. **Automated generation of flowcharts.** In RPA tools, there is a lacking of testing environments. As a consequence, SW robots are developed through a *trial-and-error* approach consisting of three steps that are repeated until success [16]: *(i)* First, a human designer produces a flowchart diagram that includes the actions to be performed by the SW robot on a target system;

Table 1. Results of the application of the classification framework

Tool	SW Arch.		Coding			Recording			Self Learning		Autom. type	Routine comp.		Log quality
	Cl.-Server	St.-alone	Strong	Low	GUI	Web	Desktop	Others	Intra-rout.	Inter-rout.		Manual	Autom.	
Aut. Anyw	✓			✓	✓	✓	✓	✓			Hybrid	✓		★★
AssistEdge		✓		✓	✓	✓		✓			Hybrid	✓		★
G1ANT		✓	✓								Hybrid	✓		★
Kryion	✓			✓	✓	✓	✓	✓			Hybrid	✓		★★
Rapise	✓			✓	✓	✓	✓				Hybrid	✓		★
TagUI		✓	✓		✓	✓	✓				Hybrid	✓		★
UiPath	✓			✓	✓	✓	✓	✓			Hybrid	✓		★★
VisualCron	✓				✓						Attended	✓		★★★
WinAutom	✓			✓	✓	✓	✓				Hybrid	✓		★★
WorkFusion	✓			✓	✓	✓	✓	✓			Hybrid	✓		★★

(ii) Second, SW robots are typically deployed in production environments, where they interact with information systems, with a high risk of errors due to inaccurate modeling of flowcharts; *(iii)* Third, if SW robots are not able to reproduce the behaviour of the users for a specific routine, then the designer adjusts the flowchart diagram to fix the identified gap. While this approach is effective to execute simple rules-based logic in situations where there is no room for interpretation, it becomes time-consuming and error-prone in presence of routines that are less predictable or require some level of human judgment. Indeed, the designer should have a global vision of all possible unfoldings of the routines to define the appropriate behaviours of the SW robot, which becomes complicated when the number of unfoldings increases. In cases where the rule set does not contain a suitable response for a specific situation, robots allow for escalation to a human supervisor. A possible solution to this challenge can be to resort on discovery algorithms from the process mining field [3] and to automatically extract flowcharts in form of Petri nets/BPMN models from RPA logs. Thus, it is necessary to investigate from the literature on process discovery [6] which algorithms suit better to extract the base structure of flowchart diagrams from a RPA log.

4. **Automated routines composition**: In modern contexts, human operators usually enact not just single tasks but complex workflows, consisting of many interrelated routines. However, current RPA technology allows to develop SW robots for executing single, independent routines. Only manual support is provided to orchestrate multiple routines, i.e., the management of more complex workflows is completely delegated to human supervisors. To synthesize complex execution strategies through an intelligent orchestration of the robots' routines, *automated planning* techniques in AI can be employed [14]. The idea is to consider the robots' routines like black boxes, i.e., as planning actions with specific preconditions and effects, and to delegate to a planning system the generation of a proper strategy to automatically compose them in a larger workflow that coordinates their orchestration.

It is worth to notice that, according to Table 1, the logs produced by the tested RPA tools have a poor quality (actions may be missing or not recorded properly), since they are mainly used for debugging purposes. Increasing the quality of RPA logs is a fundamental prerequisite to properly tackle the previous challenges, which leverage a log analysis to discover, identify, model and compose routines in an automated way. To this end, RPA tools should aim at logs at the highest possible quality level.

To conclude, we note that our study has a threat to validity, since we analyzed only a sample of the RPA tools available in the market. As a consequence, our findings can not be generalized beyond the scope of the tested RPA tools. Nonetheless, we consider this work as an important first step towards the realization of intelligent solutions for RPA.

Acknowledgments. This research work has been partly supported by the "Dipartimento di Eccellenza" grant, the H2020 RISE project FIRST (grant #734599), the

Sapienza grants IT-SHIRT, ROCKET and METRICS, the Lazio regional initiative "Centro di eccellenza DTC Lazio" and the project ARCA.

References

1. All 52 RPA Software Tools and Vendors: Sortable List [2019]. https://blog.aimultiple.com/rpa-tools/ (2019)
2. van der Aalst, W., et al.: Process mining manifesto. In: Daniel, F., Barkaoui, K., Dustdar, S. (eds.) BPM 2011, Part I. LNBIP, vol. 99, pp. 169–194. Springer, Heidelberg (2012). https://doi.org/10.1007/978-3-642-28108-2_19
3. van der Aalst, W.M.P.: Process Mining: Data Science in Action, 2nd edn. Springer, Heidelberg (2016). https://doi.org/10.1007/978-3-662-49851-4
4. van der Aalst, W.M.P., Bichler, M., Heinzl, A.: Robotic process automation. BISE **60**(4), 269–272 (2018)
5. Aguirre, S., Rodriguez, A.: Automation of a business process using robotic process automation (RPA): a case study. In: Applied Computer Science in Engineering (2017)
6. Augusto, A., et al.: Automated discovery of process models from event logs: review and benchmark. TKDE **31**(4), 686–705 (2019)
7. Bosco, A., Augusto, A., Dumas, M., La Rosa, M., Fortino, G.: Discovering automatable routines from user interaction logs. In: Hildebrandt, T., van Dongen, B.F., Röglinger, M., Mendling, J. (eds.) BPM 2019. LNBIP, vol. 360, pp. 144–162. Springer, Cham (2019). https://doi.org/10.1007/978-3-030-26643-1_9
8. Dong, G., Pei, J.: Sequence Data Mining, vol. 33. Springer Science & Business Media, Boston (2007)
9. Dumais, S., Jeffries, R., Russell, D.M., Tang, D., Teevan, J.: Understanding user behavior through log data and analysis. In: Olson, J.S., Kellogg, W.A. (eds.) Ways of Knowing in HCI, pp. 349–372. Springer, New York (2014). https://doi.org/10.1007/978-1-4939-0378-8_14
10. Geyer-Klingeberg, J., Nakladal, J., Baldauf, F., Veit, F.: Process mining and robotic process automation: a perfect match. In: BPM 2018 Workshops (2018
11. Jimenez-Ramirez, A., Reijers, H.A., Barba, I., Del Valle, C.: A method to improve the early stages of the robotic process automation lifecycle. In: Giorgini, P., Weber, B. (eds.) CAiSE 2019. LNCS, vol. 11483, pp. 446–461. Springer, Cham (2019). https://doi.org/10.1007/978-3-030-21290-2_28
12. Lacity, M., Willcocks, L.P., Craig, A.: RPA at Telefonica O2. Inst. Repo. for The London School of Economics and Political Science (2015)
13. Lohr, S.: The Beginning of a Wave: A.I. Tiptoes Into the Workplace (2018). https://www.nytimes.com/2018/08/05/technology/workplace-ai.html/
14. Marrella, A.: Automated planning for business process management. J. Data Semant. **8**(2), 79–98 (2019). https://doi.org/10.1007/s13740-018-0096-0
15. Tax, N., Sidorova, N., Haakma, R., van der Aalst, W.M.: Mining local process models. J. Innov. Digit. Ecosyst. **3**(2), 183–196 (2016)
16. Volodymyr, L., Dumas, M., Maggi, F.M., La Rosa, M.: Multi-perspective process model discovery for robotic process automation. In: CAiSE'18 Doct. Cons (2018)
17. Willcocks, L.P., Lacity, M., Craig, A.: The IT function and robotic process automation. Inst. Repo. The London School of Economics and Political Science (2015)

Description Logics and Specialization for Structured BPMN

Alexander Borgida[✉], Varvara Kalokyri, and Amélie Marian

Department of Computer Science, Rutgers University, New Brunswick, USA
{borgida,v.kalokyri,amelie}@cs.rutgers.edu

Abstract. The literature contains arguments for the benefits of representing and reasoning with BPMN processes in (OWL) ontologies, but these proposals are not able to reason about their dynamics. We introduce a new Description Logic, sBPMPROCESSDL, to represent the behavioral semantics of *(block) structured BPMN*. It supports reasoning about process concepts based on their execution traces.

Starting from the traditional notion of subsumption in Description Logics (including sBPMPROCESSDL), we further investigate the notions of specialization and inheritance, as a way to help build and abbreviate large libraries of processes in an ontology, which are needed in many applications.

We also provide *formal* evidence for the intuition that features of structured BPMN diagrams such as AND-gates and sub-processes can provide substantial benefits for their succinctness. The same can be true when moving from a structured to an equivalent unstructured version.

1 Introduction

We begin with some motivation of our work. Process descriptions can be used for *prescriptive* purposes, such as workflows, as well as for *descriptive* purposes, for understanding what actually happens during the execution of a given scenario. Our work [16] uses BPMN for the second purpose: to group and organize "personal digital traces" (byproducts of apps on smart-phones, such as emails, web searches, SMS, check-ins, financial transactions, and GPS), helping to reconstruct the user's "memory of autobiographical events".

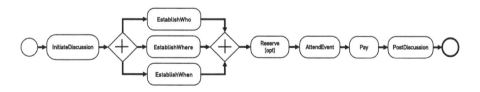

Fig. 1. *Process model for EatingOutAtRestaurant*

An example of such a process model (*"script"*) would be going out to eat at a restaurant, as shown in Fig. 1. Components, such as EstablishWhere, are in

C. Di Francescomarino et al. (Eds.): BPM 2019 Workshops, LNBIP 362, pp. 19–31, 2019.
https://doi.org/10.1007/978-3-030-37453-2_3

turn sub-process descriptions given elsewhere, involving communication loops. This process model can help us organize various digital data into instantiations of restaurant outings. For instance, a thread of messages concerning dinner, an email confirmation of an OpenTable reservation, a credit-card payment at a restaurant, and a Facebook discussion of a meal, all occurring within an appropriate time/place frame, would provide evidence for the InitiateDiscussion, Reserve, Pay and PostDiscussion subprocesses respectively in Fig. 1. Such process instances can then help the user get answers to questions like "When did I go to restaurant X?", or "What restaurant did I got to with Mary in Paris?".

Although this may not look like a business process, arranging a meeting in a company is almost identical: instead of initiating a discussion to go out to eat, one must initiate a discussion for having a meeting; deciding who/when/where the activity will take place happens in both; instead of making a restaurant reservation one might need to reserve a room for the meeting; etc.

In order to make a fully functional system, we need process descriptions for thousands of variants of every-day activities. This will be facilitated by placing the process descriptions in an ontology, and then using specialization and inheritance, so that more specific processes, such as GoingOutToEat, GoingOutToTheater, GoingOutToSports, etc. are obtained by indicating only differences from a common superclass, GoingOutForEntertainment. In turn, GoingOutToEat can be specialized to EatingOutAtRestaurant, EatingFastFood, etc. The main distinction of our work from most prior attempts at using ontologies for processes [25,26], is that we are interested in *reasoning with their behavioural semantics*, rather than just their annotated syntactic structure. For example, we want to be able to infer that "Eating followed by Sleeping" is related to "Eating followed by (Seeping or Working)".

The paper addresses these goals by (i) providing the Description Logic sBPMprocessDL to describe and reason with a subclass of BPMs, and (ii) introduces an improved notion of specialization and inheritance, which allows new actions to be added to subconcepts [1,6] while still using the subsumption algorithm of sBPMprocessDL on the subsumer and a modified subsumee. The algorithms rely on results in the formal language literature.

2   Structured BPM and sBPMprocessDL

We consider a subset of BPMN, called *structured BPM/Workflow* [2,17,18], which has properly nested single-entry/single-exit blocks for (i) sequencing, (ii) branching (deferred choice **xor**), (iii) looping (either BPMN 2 loop marker on the body, or **xor**-join and **xor**-split enclosing the body, with return flows to the beginning), and (iv) concurrent execution (**and**). Although less expressive than its unrestricted version (see [17]), Structured BPMN has the advantages of:

- being more understandable and maintainable by humans (see suggestions for "good BPMN style" in the survey by Corradini et al. [11]);
- facilitating the mining of workflows [18];

- conforming to the flow control structures offered by other process notations such as the OWL-S service model specification language [21];
- being guaranteed not to have anomalies such as deadlock, livelock;
- having clearer semantics (avoiding the complications of global **xor**-join);

while covering a large fraction of the practical applications we have encountered.

It has been noted [2] that a structured workflow block, corresponding to the sub-graph **xor**-split$\{P_1, P_2, ...\}$**xor**-join say, can be represented by a *process tree* with operator **xor** at its root and children $P_1, P_2, ...$, which we'll write as **xor**$(P_1, P_2, ...)$, or the infix-notation $P_1 \sqcup P_2 \sqcup ....$ Similarly, structured workflows have operators **seq**, **and**, and **loop** for sequencing, concurrency and looping.

By treating atomic actions as alphabet symbols, one can see a clear similarity of process trees without **and** to regular expressions (REs) built with $\{ \cdot, | ,^* \}^1$. REs have been extended, among others, with an "interleaving/shuffle" operator # in order to model the notion of concurrency; this will be taken to correspond (in this paper) to **and** in BPMN.

Description Logics (DLs) are knowledge representation languages, which manage information about individuals grouped into classes, possibly related by (binary) relations. DLs are used to define concepts in ontologies, check them for consistency, and especially organize them in hierarchies, ordered by the subsumption relation. The family of DLs is distinguished by the fact the classes in question are   intensional entities that have a compositional structure and are involved in inferences. Almost universally, *concepts* (unary predicates) and *roles* (binary predicates) are built using concept/role *constructors* from atomic identifiers. For example, the concept of "Building one of whose owner is crazy and rich" is represented in *variable-free term notation* as **conjoin**(*Building*, **some**(*owner*,**conjoin**(*Crazy,Rich*))), or, in the more familiar infix notation, as *Building* $\sqcap \exists owner.(Crazy \sqcap Rich)$. This syntax is in contrast with languages using variables and quantifiers (e.g., [27]), or general graphical notations (unless one can find an appropriate syntax for them).

We are now ready to define (see Fig. 2) the DL-like ontology language sBPM-PROCESSDL representing the control flow semantics of structured BPM. Its syntax starts from a finite set of atomic action names $\mathcal{A}$, and a disjoint set $\mathcal{P}$ of (sub)process names. There are some additional useful constant process concepts listed in the top part of the figure; the middle part gives the syntax rules for compound processes P. Please note that sBPMPROCESSDL has the same variable-free term notation as most other members of the family of Description Logics.

We can then express in sBPMPROCESSDL the EatingOutAtRestaurant process as

$$Initiate \cdot (EstablishWho \# EstablishWhen \# EstablishWhere) \cdot Reserve^{opt}$$
$$\cdot AttendEvent \cdot Pay \cdot PostDiscussion$$

where we use an abbreviation for optional tasks, P^{opt}, equivalent to $P \sqcup$ **Null**.

---

[1] This can be used as another argument in favor of Structured BPM: several papers [12,22] have shown how to represent the trace semantics of *declarative BPMs*, such as DECLARE, by mapping to regular expressions.

Name of constructor	Syntax	Term notation	Semantics
$N \in \mathcal{A} \cup \mathcal{P}$	N	N	$N^{\mathcal{I}}$
no-action process	$Null$	**Null**	$\{\lambda\}$
bottom concept	$\bot_P$	**Bottom**$_{Process}$	$\emptyset$
top-process concept	$\top_P$	**Top**$_{Process}$	$Sequence(\Delta_{\mathcal{I}})$
any action	$Action$	**Action**	$\bigcup_{A \in \mathcal{A}} A^{\mathcal{I}}$
sequence	$P_1 \cdot P_2$	**seq**(P_1, P_2)	$\{uw \mid u \in P_1^{\mathcal{I}}, w \in P_2^{\mathcal{I}}\}$
alternation	$P_1 \sqcup P_2$	**xor**(P_1, P_2)	$P_1^{\mathcal{I}} \cup P_2^{\mathcal{I}}$
repetition (base)	P^0	**repeat**(0,P)	$Null^{\mathcal{I}}$
repetition (ind'n)	P^{k+1}	**repeat**(k+1,P)	$P^{\mathcal{I}} \cdot (P^k)^{\mathcal{I}}$
loop	P^*	**loop**(P)	$\bigcup_{i \geq 0} (P^i)^{\mathcal{I}}$
interleaving	$P_1 \# P_2$	**and**(P_1, P_2)	$\{u_1 w_1 ... u_n w_n \mid u_1 ... u_n \in P_1^{\mathcal{I}},$ $w_1 ... w_n \in P_2^{\mathcal{I}}, \text{ for } n \in \mathbb{N}\}$
subprocess N definition	$N \doteq P$		$N^{\mathcal{I}} = P^{\mathcal{I}}$
P_2 subsumes P_1	$P_1 \sqsubseteq P_2$		$P_1^{\mathcal{I}} \subseteq P_2^{\mathcal{I}}$
inconsistency of P	$P \sqsubseteq \bot$		$P^{\mathcal{I}} = \emptyset$
membership of action sequence α in P	$\alpha \in P$		$\alpha \in P^{\mathcal{I}}$

Fig. 2. Syntax and semantics of sBPMPROCESSDL

The semantics of sBPMPROCESSDL[2] is provided by an *interpretation* $\mathcal{I} = (\Delta_{\mathcal{I}}, \cdot^{\mathcal{I}})$, where $\Delta_{\mathcal{I}}$ is intended to be the set of atomic action executions/instances; $A^{\mathcal{I}}$ is a *finite subset of sequences of length <u>one</u>* over $\Delta_{\mathcal{I}}$ for $A \in \mathcal{A}$; in turn, $P^{\mathcal{I}}$ is a finite subset of the set $Sequence(\Delta_{\mathcal{I}})$ of *sequences of individual action executions*. $\mathcal{I}$ is then extended in the natural way to the constants and the process concept constructors in the manner shown in Fig. 2.

It is crucial to note that since the denotation of sBPMPROCESSDL concepts are *sequences of action instances*, this corresponds to the "trace semantics" for BPM/Petri Nets (as used in [12,24,28,29]), rather than bisimulation-based semantics [6]. This means that in our model atomic actions can't be truly parallel; even if we replace A with $StartA \cdot EndA$, the beginnings of two actions must be ordered.

The bottom part of Fig. 2 describes the standard predicates used for inferences and axioms in the field of DLs. It is usual to consider the complexity of these inferences and their implementation. In this case, we rely on results about regular-like expressions in the formal language literature. In particular, let us treat atomic action concept names $A \in \mathcal{A}$ as "characters" in the alphabet Σ usually taken as the base case for regular expressions. Then the above predicates on regular languages will give the same results as the semantics in Fig. 2, since the interpretation $\mathcal{I}$ replaces all occurrences of A by the same set, $A^{\mathcal{I}}$. Therefore, deciding subsumption, inconsistency, and membership in sBPMPROCESSDL correspond exactly to deciding containment, emptiness, and string recognition for this class of extended regular expressions. A summary of results

---

[2] This is a variant of the semantics of one of the plan DLs, $RegExp(\{\cdot, \sqcup, ^*, \#\})$, described in [7,9].

concerning the complexity of these reasoning operations, some based on [19], appear in [7,9]. It is worth mentioning that the preceding references also show how sBPMPROCESSDL can be integrated with standard DLs, like OWL say, using the notion of "concrete domains" [4], thus obtaining more powerful hybrid DLs which contain process descriptions.

3   Specialization and Inheritance for Structured BPM

In software development and AI KR, inheritance is a mechanism for abbreviating descriptions based on the notion of IsA and specialization in class-based formalisms. For example, if *Employee* is a subclass of *Person*, then properties of *Person* such as *name*, *address*, *age*, etc. and constraints on them, are *inherited* by the subclass *Employee*. The purpose of inheritance is reuse (avoiding repetition of work), and effective maintenance: a change on *Person* (e.g., adding property *socialSecurityNr*), results in it being propagated to all subclasses.

When specializing a class, which has associated properties/features common to its instances, several principles have been observed in AI and conceptual modeling [8]:

- **A. Extensional IsA constraint:** the instances of the subclass are instances of the superclass.
- **B. Unmodified inheritance:** properties/features and their constraints are inherited unchanged to the subclass.
- **C. Adding restrictions to existing properties:** an inherited property may have more restrictions placed on it. For example, while the *age* of *Person*s has lower bound 0, the *age* of *Employee*s must be at least 16.
- **D. Adding new properties:** A subclass may have new properties defined over it. For example, *salary* is added for *Employee*.

Recently, another principle has come to dominate in object-oriented programming circles: the "substitution principle" requires that an instance of a subclass must be usable in any situation where an instance of a superclass could appear. This however is not crucial in our situation, since we are not writing software systems (which would use polymorphic typing), but process descriptions, where we try to *describe* variant sub-processes in some principled manner.

Specialization and inheritance are helpful whenever there are many similar classes, as in our case. In fact, already decades ago the Taxis project suggested that the benefits of specialization and inheritance were useful not just for describing entities but also transactions, exceptions, and scripts/workflows [5,8,20]. More recently, Frank [15] has provided an in-depth survey of the field of process specialization.

The question that arises is what we can do during the specialization of process descriptions. A crucial part of this is what counts as "properties/features" of processes. Obvious answers used in the past are the participants in the process, and the mereologic components of the process; these can all be inherited, specialized or added. In our case, it will be the sub-expressions of the sBPMPROCESSDL concept terms that will be viewed as such inheritable features/components.

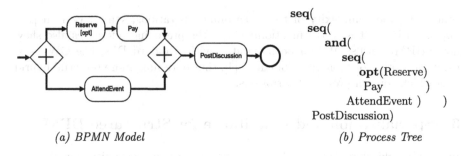

(a) BPMN Model   (b) Process Tree

Fig. 3. GoingOutForEntertainment

The semantics of sBPMprocessDL, as given in Sect. 2, provides a subsumption relationship between $(SitDown \cdot Drink)$ and $SitDown \cdot (Eat \# Drink)$, because every instance scenario (sequence of primitive action instances) of the former is an instance of the latter. This parallels the notion of "extensional IsA constraint" (A) above, and has the immediate side-effect of providing inheritance as in (B). It also corresponds to the notion of process specialization suggested by Wyner and Lee [28]. We therefore start from the assumption that the specialization of class X leads to a class Y, whose extension, as a set of traces, should be contained in that of class X.

Next, we address the issues of how items (C) and (D) above should be handled. Concerning specialization of properties (C), since we want to maintain the "extensional IsA constraint (A)", some natural ways to do this include:

- replace named primitive actions/sub-processes (e.g., *Pay*) by specializations (e.g., *PayByCreditCard* or *PayByCash*). This requires the introduction of a so-called Terminologic Box for named concepts, where IsA hierarchies of them can be kept. (See [9] for a simple way to continue reasoning with sBPMprocessDL in the presence of such taxonomies.)
- replace "control constructs" by ones that allow fewer sequences. For example:
  - eliminate some of the choices in a **xor** or an **and** (and possibly use **xor**(p)≡p, **and**(p)=p to simplify);
  - replace **loop**(p) by the **xor** of zero or more iterations of p;
  - replace **and** by some ordering of its arguments.

The above are easy cases to check syntactically, but unfortunately they do not form a complete set of specialization operations.

So let us return to the idea of allowing process subexpression P to be replaced by Q whenever the extension of Q is contained in that of P (a problem that is algorithmically decidable for sBPMprocessDL as $Q \sqsubseteq P$), but observe that we can do this replacement *at any place*, including the root of the expression.

To illustrate the syntax for specifying the substitution, let us consider in Fig. 3, part of the process GoingOutForEntertainment[3]. If, during specialization for GoingToRockConcert, we want to force Reserve and Pay to occur *before*

---

[3] For brevity, we omit henceforth the beginning, including Initiate, EstablishWho,...

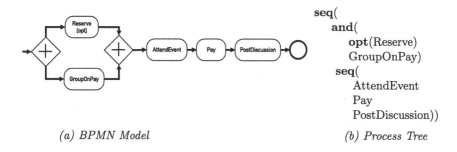

| (a) BPMN Model | (b) Process Tree |

Fig. 4. GroupOn EatingOutAtRestaurant

AttendEvent, we need to replace **and**(?X,?Y) by **seq**(?X,?Y)[4]. To identify the particular node on which we want to match (there are three **seq** nodes in the tree in Fig. 3(b)), we use a notation based on XPath, so that, for example, AttendEvent node is identified by `/.[1]/.[1]/.[2]`. Therefore the above proposed specialization is written as

$$/.[1]/.[1] : \quad \textbf{and}(?X, ?Y) \rightsquigarrow \textbf{seq}(?X, ?Y)$$

In the case of specializing GoingOutForEntertainment to EatingOutAtRestaurant (see Fig. 1) we could use

$$/.[1]/.[1] : \quad \textbf{and}(\textbf{seq}(?X1, ?X2), ?Y) \rightsquigarrow \textbf{seq}(?X1, ?Y, ?X2)$$

to indicate that payment at the restaurant occurs after attending, while the reservation (may) occur before.

Note that both of the above specializations result in properly subsumed concepts, since all the traces in the subclass appear in the extension of the superclass. And this can be detected by the sBPMPROCESSDL subsumption reasoner.

Concerning *adding new properties (D)*, we face a conundrum: in many cases we would like to add new kinds of actions to be carried out as part of the specialized process (like we added *salary* to *Employee*). For example, in creating a specialization GroupOn_EatingOutAtRestaurant, we want to add a step GroupOnPay, for paying for a GroupOn meal (which provides a discount). Let's say GroupOnPay is a new atomic action (in our case signaled by possible digital traces for web search on groupon.com, and a credit card payment to GroupOn). So suppose we carry out the specialization

$$/.[1]/.[1]/.[1] : \textbf{opt}(Reserve) \rightsquigarrow \textbf{and}(\textbf{opt}(Reserve), GroupOnPay)$$

resulting in the process tree shown in Fig. 4, and corresponding BPMN model.

However, this would cause problems because any sequence containing such a new action instance would immediately not be in the extension of the superclass.

---

[4] We use here pattern matching on terms in the style of modern functional languages such as Standard ML, Haskell, etc.

Van der Aalst and Basten [1,6] have thoroughly investigated this question, and suggested four alternative approaches, based on the following two ideas:

Projection inheritance: ignore new action B, by making its instances be invisible. In formal language terms, appropriate for sBPMPROCESSDL, this corresponds to replacing the new atomic concept B by the concept **Null**, *for purposes of computing "proper specialization"*. This substitution can be done very efficiently, so we can use the sBPMPROCESSDL subsumption algorithm to check for projection inheritance.

In the above example, sBPMPROCESSDL subsumption still works, since replacing GroupOnPay by **Null** coincides with the **Null** in **opt**(Reserve), giving only a pre-existing trace. But if reservations were not optional, then we could now have a trace with no reservation (when $GroupOnPay \rightsquigarrow$ **Null**), thus obtaining an invalid specialization under projection inheritance.

Protocol inheritance: eliminate from consideration all action sequences *that contain* the new action B. In formal language terms, this corresponds to reasoning with a more complex specialized process R': the intersection of the original sBPMPROCESSDL R with the "strings" *not* containing B. The latter can be described by the extended regular expression

$$B_{not} = \neg((Action \sqcup B)^* \cdot B \cdot (Action \sqcup B)^*)$$

Therefore, to decide if S is a proper superclass of R according to protocol inheritance, it is sufficient to decide $(R \cap B_{not}) \sqsubseteq S$. Although regular expression reasoning with constructor $\neg$ and $\cap$ is highly complex in general, this particularly simple case can be handled without increase in complexity.

Reconsidering the GroupOn_EatingOutAtRestaurant example, since all traces contain GroupOnPay, they are all eliminated, leaving the extension to be the empty set, whether or not reservation was optional.

In fact, in our case Projection Inheritance implies Protocol Inheritance, since all the sequences where **Null** was inserted would be removed from the extension. (This is due to our use of the trace equivalence instead of branching bisimulation equivalence, as in [6]). Moreover, van der Aalst and Basten suggest two other possible semantics for specialization, involving conjunction and disjunction of the above two. Because of the trace semantics, these collapse in our case.

By adopting one of the semantics for specialization suggested by van der Aalst and Basten, we then achieve a principled work-around the constraints of strict subsumption in sBPMPROCESSDL, which allows the introduction of new activities in the specialized process.

4   On the Succinctness of Some BPMN Models

Based on trace semantics for BPMN, we can use results from the literature on formal languages, to provide *formal proofs* for certain intuitions about constructs in BPMN notation, and how they affect brevity. The proofs of such results almost always involve exhibiting a sequence of languages/process concepts $L_1, L_2, ..., L_n$ such that their description using some feature grows slowly (e.g., linearly) as n

grows, but *any* corresponding description without that feature grows much faster (e.g., exponentially). The following are a few such results:

(a) Allowing sub-process definitions can yield exponentially shorter descriptions of BPMs. In our opinion, this is an important result to remember because definitions of BPMN sometimes omit this notion, possibly considering it trivial. In this one case, we provide a proof outline:

Proof. Letting $\mathcal{A} = \{\ A\ \}$, $\mathcal{P} = \{\ E_1, E_2, ...\ \}$, consider the subprocess definitions:

$$E_1 \doteq A \cdot A \qquad E_2 \doteq E_1 \cdot E_1 \quad ... \quad E_n \doteq E_{n-1} \cdot E_{n-1}$$

We can see that the denotation of E_n is equal to $A^{2^{n+1}}$, and the above description has size $O(n)$. But there is no finite automaton of size less than $O(2^n)$ for this, because *any* finite automaton needs at least k states to distinguish the string A^k from all other strings. There is then no ordinary RE that is smaller in size than the FA describing its language, since there is a well known standard procedure for constructing FAs from REs, which have the same size. It only remains to show that in this case **and** cannot help, which is true because $\mathbf{and}(A^i, A^j) = A^{i+j}$ for any i, j, so nested **and**'s can be inductively eliminated. □

*(b) Eliminating all **and** split/join pairs can result in double-exponentially larger structured BPM.* The basic idea is that, with trace semantics, $\mathbf{and}(A, B)$ on atomic action pairs can be replaced by $\mathbf{xor}(\mathbf{seq}(A, B), \mathbf{seq}(B, A))$, and by iteratively substituting outward nested **and**'s, one can eventually eliminate them. The hard part is finding a family L_n for which one can prove that the result has size at least 2^{2^n} no matter what RE/BPM one uses.

(c) "Structured" BPM comes at a price: there are unstructured BPM such that every structured version is exponentially larger.

5 Related Work

There are two threads of related work. One concerns DLs for representing processes. The second concerns specialization and inheritance of process models.

5.1 Description Logics for Processes

As discussed in Sect. 2, our aim was to develop a DL that can describe the behavioral semantics of (business) processes, and be able to reason with them. Within this class, we focused on a language that had the characteristic DL feature of having *variable-free concept constructors*, through which concepts and axioms can be presented and reasoned with.

One relevant thread of work here is tied to Propositional Dynamic Logic, which has program constructors $\{\ ;\ ,\ |\ ,\ ^*\ \}$ corresponding to our concept constructors $\{\ \cdot\ ,\ \sqcup\ ,\ ^*\ \}$. Among others, [13] developed expressive DLs which use these as *role constructors*, rather than concept constructors. PDL is more expressive (and more complex to reason with) because in addition to complex programs, these are applied to complex formulas describing states. If the state description is just the formula *true*, then PDL expressions and sBPMPROCESSDL, minus #, are in some sense equivalent. However, the above DLs do not involve concurrency. Maier & Stockmayer [19] (among others) have studied extending PDL with interleaving, but it is unclear what concurrency means with DL roles.

Another related domain are temporal logics with modal operators, such as linear temporal logic (LTL). The most relevant work here is the recent formalization of OCBC [3], which has DL concepts such $\bigcirc Eat$, denoting the instances of *Eat* at the next time point. On the one hand, OCBC can express some parallelism, as in $\bigcirc Eat \sqcap \bigcirc Breathe$, which *Eat # Breathe* must linearize. On the other hand, OCBC is based on LTL, which is incapable of representing the sBPMPROCESSDL concept (**Action** $\cdot Breathe)^*$ – breathing every second time point. So the formalisms are not fully comparable.

Finally, there is a large collection of works (e.g., [10,24]) that encode (aspects) of BPMs, and then use Description Logic reasoning to answer particular questions about them. The main difference from these is that they do not have concepts directly denoting process executions, about which one can ask questions like subsumption, and many do not handle concurrency.

5.2 Process Specialization

Our own work in this area in the early 1980's [5,8] added specialization and inheritance to Zisman's pioneering work on workflows [30], by allowing Petri Net edges to be specialized or added. This work was not however formalized.

Van der Aalst and colleagues have carried out influential and extensive related work we have already referenced, including on process trees, and the notions of protocol and projection inheritance. The differences are that they use branching bisimulation as their equivalence semantics, and that for inheritance, they require that on the original "vocabulary" of actions, the specialized process must behave identically to the original one, while we allow sub-process/atomic action specialization.

Wyner and Lee [28] provide a theory of process specialization for processes described by state machines (FAs). Like ours, their approach is based on an extensional semantics of the traces accepted by the FA. Their paper proposes collections of FA editing operators leading to all possible specializations (e.g., deleting nodes, adding edges). This approach is not however fully based on the set of all traces accepted by the FA, because it does not consider alternative FAs that accept the same set of traces. In contrast, by considering the language of REs (possibly built from the automata), and then using algorithms that are provably complete for detecting language containment, sBPMPROCESSDL solves the more general problem.

While the above approaches, like this paper, restrict what can be done during specialization, at the other extreme is work such as [14], which uses ideas of "default inheritance" from Artificial Intelligence to permit arbitrary modifications of web-service method specifications.

An intermediate stance is taken by the subfield of process adaptation/ variability, which uses ideas such as those in [29], to "freeze" and otherwise constrain parts of the process description that can/cannot be modified during specialization. (See Prieto et al. [23] for related literature.)

6 Conclusions and Future Work

The paper is motivated by our desire to reconstruct memories of autobiographical events from digital traces that can be found on smart-phones, for example. We illustrated the use of Structured BPMN to describe every-day "scripts", which support the reconstruction of the memories. We detailed sBPMPROCESSDL, one of the family of process description logics based on regular expressions introduced in [7,9], and showed how it can be used to capture the dynamic semantics of Structured BPMN diagrams. This complements earlier work on ontology languages for static, syntactic aspects of BPMN [26]. The main contribution of the paper is the development of a new, more effective notion of specialization and inheritance for BPMs and workflows, which is needed to gather and organize the large number of scripts required in practice. It is based on the formal notion of subsumption in sBPMPROCESSDL, but allows for new action concepts to appear in specialized process. We use ideas from [1] for this extension, while still relying on the sBPMPROCESSDL subsumption algorithm to check the validity of specialization, albeit on a modified subsumee concept. The paper also discussed some descriptional complexity results for Structured BPMN, made possible by the formalization using regular expressions.

An open problem with specialization is that one cannot eliminate actions that appear in all traces of the superclass, when this class was over-generalized. Such situations are not rare, and we are working on an approach that adjusts definitions in such cases of "default inheritance", while maintaining clear, First Order Logic semantics. We are also investigating how to extend sBPMPROCESSDL with additional constructors that would cover all Petri Net languages. Finally, we plan to explore more refined notions of concurrency.

References

1. van der Aalst, W.M.P., Basten, T.: Life-cycle inheritance. In: Azéma, P., Balbo, G. (eds.) ICATPN 1997. LNCS, vol. 1248, pp. 62–81. Springer, Heidelberg (1997). https://doi.org/10.1007/3-540-63139-9_30
2. van der Aalst, W., Buijs, J., van Dongen, B.: Towards improving the representational bias of process mining. In: Aberer, K., Damiani, E., Dillon, T. (eds.) SIMPDA 2011. LNBIP, vol. 116, pp. 39–54. Springer, Heidelberg (2012). https://doi.org/10.1007/978-3-642-34044-4_3
3. Artale, A., Kovtunova, A., Montali, M., van der Aalst, W.M.: Modeling and reasoning over declarative data-aware processes with object-centric behavioral constraints. In: Proceedings of BPM 2019 (2019, to appear)
4. Baader, F., Hanschke, P.: A scheme for integrating concrete domains into concept languages. In: Proceedings of IJCAI 1991, pp. 452–457 (1991)

30   A. Borgida et al.

5. Barron, J.: Dialogue and process design for interactive information systems using taxis. ACM SIGOA Newsl. **3**, 12–20 (1982). Proceedings of SIGOA 1982
6. Basten, T., van der Aalst, W.: Inheritance of behavior. J. Logic Algebr. Program. **47**(2), 47–145 (2001)
7. Borgida, A.: Initial steps towards a family of regular-like plan description logics. In: Lutz, C., Sattler, U., Tinelli, C., Turhan, A.-Y., Wolter, F. (eds.) Description Logic, Theory Combination, and All That. LNCS, vol. 11560, pp. 90–109. Springer, Cham (2019). https://doi.org/10.1007/978-3-030-22102-7_4
8. Borgida, A., Mylopoulos, J., Wong, H.K.T.: Generalization/specialization as a basis for software specification. In: Brodie, M.L., Mylopoulos, J., Schmidt, J.W. (eds.) On Conceptual Modelling. Topics in Information Systems, pp. 87–117. Springer, New York (1984). https://doi.org/10.1007/978-1-4612-5196-5_4
9. Borgida, A., Toman, D., Weddell, G.: On special description logics for processes and plans. In: Proceedings of Workshop on Description Logics, DL 2019 (2019)
10. Calvanese, D., Montali, M., Patrizi, F., De Giacomo, G.: Description logic based dynamic systems: modeling, verification, and synthesis. In: IJCAI 2015, pp. 4247–4253 (2015)
11. Corradini, F., et al.: A guidelines framework for understandable BPMN models. Data Knowl. Eng. **113**, 129–154 (2018)
12. De Giacomo, G., Dumas, M., Maggi, F.M., Montali, M.: Declarative process modeling in BPMN. In: Zdravkovic, J., Kirikova, M., Johannesson, P. (eds.) CAiSE 2015. LNCS, vol. 9097, pp. 84–100. Springer, Cham (2015). https://doi.org/10.1007/978-3-319-19069-3_6
13. De Giacomo, G., Lenzerini, M.: TBox and ABox reasoning in expressive description logics. In: Proceedings of AAAI 199, pp. 37–48. AAAI Press (1996)
14. Ferndriger, S., Bernstein, A., Dong, J.S., Feng, Y., Li, Y.-F., Hunter, J.: Enhancing semantic web services with inheritance. In: Sheth, A., et al. (eds.) ISWC 2008. LNCS, vol. 5318, pp. 162–177. Springer, Heidelberg (2008). https://doi.org/10.1007/978-3-540-88564-1_11
15. Frank, U.: Specialisation in business process modelling: Motivation, approaches and limitations. Institut für Informatik und Wirtschaftsinformatik (ICB), Universität Duisburg, Essen, Technical report (2012)
16. Kalokyri, V., Borgida, A., Marian, A.: YourDigitalSelf: a personal digital trace integration tool. In: Proceedings of CIKM, pp. 1963–1966. ACM (2018)
17. Kiepuszewski, B., ter Hofstede, A.H.M., Bussler, C.J.: On structured workflow modelling. In: Wangler, B., Bergman, L. (eds.) CAiSE 2000. LNCS, vol. 1789, pp. 431–445. Springer, Heidelberg (2000). https://doi.org/10.1007/3-540-45140-4_29
18. Leemans, S.J.J., Fahland, D., van der Aalst, W.M.P.: Discovering block-structured process models from incomplete event logs. In: Ciardo, G., Kindler, E. (eds.) PETRI NETS 2014. LNCS, vol. 8489, pp. 91–110. Springer, Cham (2014). https://doi.org/10.1007/978-3-319-07734-5_6
19. Mayer, A.J., Stockmeyer, L.J.: The complexity of PDL with interleaving. Theor. Comput. Sci. **161**(1–2), 109–122 (1996)
20. Mylopoulos, J., Bernstein, P.A., Wong, H.K.T.: A language facility for designing database-intensive applications. ACM Trans. Database Syst. **5**(2), 185–207 (1980)
21. OWL-S Coalition: OWL-S 1.1 Release (2004)
22. Prescher, J., Di Ciccio, C., Mendling, J.: From declarative processes to imperative models. Proc. SIMPDA **14**, 162–173 (2014)
23. Prieto, Á.E., Lozano-Tello, A., Rodríguez-Echeverría, R., Preciado, J.C.: A hierarchical adaptation method for administrative workflows. IEEE Access **7**, 11066–11092 (2019)

24. Ren, Y., Gröner, G., Lemcke, J., Rahmani, T., Friesen, A., Zhao, Y., Pan, J.Z., Staab, S.: Validating process refinement with ontologies. In: Proceedings of Workshop on Description Logics, DL 2009 (2009)
25. Riboni, D., Bettini, C.: Owl 2 modeling and reasoning with complex human activities. Pervasive Mob. Comput. **7**(3), 379–395 (2011)
26. Rospocher, M., Ghidini, C., Serafini, L.: An ontology for the business process modelling notation. In: Proceedings of FOIS 2014, pp. 133–146 (2014)
27. Schmiedel, A.: Temporal terminological logic. In: Proceedings of AAAI 1990, pp. 640–645 (1990)
28. Wyner, G.M., Lee, J.: Process specialization: defining specialization for state diagrams. Comput. Math. Organ. Theory **8**(2), 133–155 (2002)
29. Wyner, G.M., Lee, J.: Applying specialization to petri nets: implications for workflow design. In: Bussler, C.J., Haller, A. (eds.) BPM 2005. LNCS, vol. 3812, pp. 432–443. Springer, Heidelberg (2006). https://doi.org/10.1007/11678564_40
30. Zisman, M.: Representation, Specification And Automation of Office Procedures. Ph.D. thesis, University of Pennsylvania (1977)

Utilizing Ontology-Based Reasoning to Support the Execution of Knowledge-Intensive Processes

Carsten Maletzki[1,2], Eric Rietzke[2(✉)], Lisa Grumbach[1,2], Ralph Bergmann[1], and Norbert Kuhn[2]

[1] University of Trier, Trier, Germany
[2] Trier University of Applied Sciences, Trier, Germany
e.rietzke@umwelt-campus.de

Abstract. Supporting knowledge-intensive processes (KiPs) has been widely addressed so far and is still subject of discussion. In this context, little attention was paid to the ontology-driven combination of data-centric and semantic business process modeling, which finds its motivation in supporting KiPs by enabling work sharing between humans and artificial intelligence. Such approaches have characteristics that could allow support for KiPs based on inferencing capabilities of reasoners. We confirm this as we show that reasoners are able to infer the executability of tasks. This is done by designing an inference mechanism to extend a currently researched ontology- and data-driven business process model (ODD-BP model). Further support for KiPs by the proposed inference mechanism results from its ability to infer the relevance of tasks, depending on the extent to which their execution would contribute to process progress. Thereby, it takes into account the dynamic behaviour of KiPs and helps knowledge workers to pursue their process goals.

1 Introduction

Since knowledge-intensive processes (KiPs) have challenging characteristics [7], providing adequate support was and still is subject of research. In general, KiPs are highly dependent on knowledge workers (like managers, researchers and engineers) who perform interconnected knowledge-intensive decision making tasks [7,14]. Since knowledge workers are the greatest determinant of the worth of their companies [6], supporting them in KiPs is important. Various possibilities to provide such support could be opened up by the ontology-driven combination of data-centric and semantic business process modeling. Since semantic business process modeling allows ontology reasoners to infer facts from process models [13], its combination with data-centric business process modeling allows reasoners to access data dependencies of tasks. We expect, that knowledge workers in KiPs can benefit in several ways when these reasoner accessible data dependencies are used to infer process relevant facts. The ability of reasoners to explain

© Springer Nature Switzerland AG 2019
C. Di Francescomarino et al. (Eds.): BPM 2019 Workshops, LNBIP 362, pp. 32–44, 2019.
https://doi.org/10.1007/978-3-030-37453-2_4

inferences could thereby perhaps be used to enhance process support by regarding the background of inferred facts. In correspondence to the observation that progress in KiPs essentially depends on data-availability [11], the most appealing starting point to investigate our expectations is using a reasoner to infer executability of tasks based on available data. Since data-availability in data-centric business process modeling also determines the need for executing tasks [12], inferring executability should also regard the relevance of task execution based on data-availability. The aim of this paper is to develop such an inference mechanism by an extension of our ontology- and data-driven business process model (ODD-BP model). Aligned with our SEMANAS[1] project [11], this paper will use examples of KiPs in the domain of agricultural grant applications.

We will first look at related topics like data-centric and semantic business process modeling and briefly introduce the ODD-BP model. After regarding executability in data-centric business process modeling, we will design an inference mechanism, implement it in the ODD-BP model and demonstrate its capabilities. At last, we will discuss how we believe the proposed inference mechanism could support KiPs and conclude our work.

2   Foundations

Semantic business process modeling uses ontologies to enhance common semi formal representations of processes. During semantic business process modeling, elements of processes (such as tasks) are represented as instances of ontology classes describing the constructs of a process language [13]. Research on semantic business process modeling is often motivated by easing semantic ambiguity in process models, which can e.g. occur when modelers have lacking knowledge about the domain [3,8,13]. In general, research on semantic business process modeling focused on control-flow oriented approaches of business process modeling [3,9,13]. According to Heinrich et al. [9] an essential advantage of semantic business process modeling is accessibility for automated processing. This has been studied in terms of automated creation or adjustment of process models and it was regarded that reasoners can be used to derive facts that are implicitly stated in process models [3,9,13].

In data-centric business process modeling, processes are not described by links between tasks (as in traditional control-flow oriented approaches), but instead tasks are linked to data elements that are required for their execution (input) or the result of it (output). This implies a process logic that focuses rather on data-availability than on predefined sequences of tasks [12]. Several approaches of data-centric business process modeling consider data in context of artifacts that are created, evolved and typically archived during their lifetime in businesses [5,14]. For supporting KiPs, initial research shows that data-centric business process modeling is a promising solution [7].

---

[1] SEMANAS is funded by the Federal Ministry of Education and Research (BMBF), grant no. 13FH013IX6, duration: 2017–2021.

In workflow-management-systems, workflow engines may be used to identify what could be done and who could do it [15]. To the best of our knowledge, it has not been investigated yet whether reasoners are able to take over the job of workflow engines to identify what could be done next in a process.

3   Ontology- and Data-Driven Business Process Model

We now introduce an ontology- and data-driven business process model (ODD-BP model) which combines semantic and data-centric business process modeling with regard to the requirements of KiPs. In our research on the ODD-BP model we aim at providing an ambiguity-free understanding of the process to be supported, so that human and cyber actors can work cooperatively on process enactment. The centerpiece of the ODD-BP model is a base-ontology implemented using the web ontology language 2 (OWL 2).[2] An excerpt of the base-ontology is depicted in Fig. 1, where circles with dashed lines represent classes and directed edges represent object properties between classes.

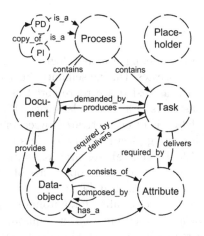

Fig. 1. Base-ontology of the ODD-BP model

Processes in the ODD-BP model can either be process definitions or process instances (Fig. 1: PD, PI). Since process instances represent the ongoings in a business process and are meant to hold data about the process state [15], they are relevant for inferring executability. Central to this are tasks and their input and output relations to documents and attributes (input: demanded_by, required_by; output: delivers, produces; note that the base-ontology also covers the inverse of each property). Documents are abstract data containers that provide attributes. Attributes are key-value pairs that always describe characteristics of entities. The addressed entities are represented by dataobjects that

---

[2] OWL 2 specification: https://www.w3.org/TR/owl2-syntax/.

consist of corresponding attributes. This conceptualization has the advantage of a high expressibility, since attributes can be integrated as wholesome elements in the process and drive its execution. The alternative conceptualization which relies on data-properties attached to representations of entities lacks this ability, since appended literals can not be connected to other elements of a process. Process states in the ODD-BP model are represented by marking unavailable data elements and unexecuted tasks as so-called placeholders. Initially, a process instance only contains elements with placeholder markings that are removed as the process progresses, which leaves only elements behind that describe existing data elements and executed tasks.

Figure 2 depicts an excerpt of a process instance of an application process for agricultural grants which is modeled using the presented base-ontology. The process instance contains a dataobject representing a *Farmer* whose attribute *Date of Birth* is used to perform an *Age Check* to determine if the *Farmer* counts as *Young*. Since young farmers can earn higher grants, this information needs to be filled in the *Application Form*. After the *Application Form* was produced, the *Submission of the Application* is possible. All white elements of the excerpt are marked as placeholder elements, so that only the attribute *Date of Birth* and the dataobject *Farmer* are known. No task was executed and no document produced yet. As this simple excerpt may not represent all characteristics of KiPs, it was chosen due to space limitations and only in order to illustrate how data-driven aspects of KiPs are modeled using the ODD-BP model.

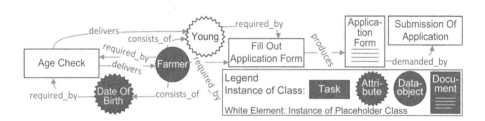

Fig. 2. Example process instance modeled using the base-ontology

4   Inferring Executability and Relevance

This section presents an inference mechanism that aims at supporting knowledge workers by providing execution relevant information about yet unexecuted tasks. For doing so, the inference mechanism regards the states of data dependencies of unexecuted tasks to determine executability and execution relevance. Note that we consider tasks that have to be executed more than once to fulfill their contribution to the process as unexecuted until their final execution is completed.

4.1   Inferring Executability

Among the many reasoning options of OWL 2, most interesting for our purpose is inferring that an individual belongs to a class – in our case, inferring that a task belongs to the class of executable tasks. A class in OWL 2 is a set of individuals that is described by restrictions on the individuals' characteristics [4]. In data-centric business process modeling, executable tasks share the characteristic that all required input data elements are available [12]. In order to infer executability, we need to describe this in OWL 2 by a class. Since OWL 2 relies on the open world assumption (OWA), this is challenging if the result shall be suitable for practice. The OWA makes it impossible to tell if something is inexistent until its inexistence has been stated [10]. If an inference mechanism for executability relies on a class that is restricted by a statement about "everything", considerable effort has to be taken into account in context of KiPs. For example: If a class describes the set of tasks whose input elements are all available, during reasoning, a statement has to tell that a task has no more than the known input data elements. Such is possible by adding statements which restrict the cardinality of relations between individuals [2]. Since KiPs require substantial flexibility at design- and run-time [14], data dependencies often have to be adjusted. This approach would then be impractical, since not only data dependencies but also allowed cardinalities need to be adjusted.

Description logics, which form the base of OWL 2, alternatively allow reasoning with the existential quantifier, which is much easier in the context of the OWA. During reasoning, a statement using the existential quantifier becomes true if the existence of something is declared. It remains false if the existence is not declared. Due to its suitability for practical use, the description of executable tasks should only rely on the existential quantifier. This requires an alternative way of describing executable tasks that can be found by examining unexecuted tasks and data-availability.

Since unexecuted tasks in the ODD-BP model are marked as placeholders, the class of unexecuted tasks corresponds to the intersection between the classes of placeholders and tasks (1). Further, since unexecuted tasks either have all input data elements available or not, the class of unexecuted tasks splits in the two disjoint subclasses of executable and unexecutable tasks (2) (3).

$$Task_{unexecuted} \equiv Task \sqcap Placeholder \tag{1}$$

$$Task_{unexecuted} \equiv Task_{executable} \sqcup Task_{unexecutable} \tag{2}$$

$$Task_{executable} \sqcap Task_{unexecutable} \sqsubseteq \bot \tag{3}$$

Due to placeholder markings, the class of unavailable data elements corresponds to the intersection of the classes of data elements and placeholders (4) (data element is abbreviated in formulas with "DataEl"). Note that dataobjects are not considered by this class because inferring executability on the level of their attributes is more precise. Since unexecutable tasks have at least one unavailable input data element, they can be described by the existential quantifier (7). Note that "input" is the super property of the "required_by" and "demanded_by" properties of the base-ontology (5) (6). The class of executable tasks can then be seen as the intersection between the classes of unexecuted tasks and the inverse of unexecutable tasks (8).

$$DataEl_{unavailable} \equiv (Document \sqcup Attribute) \sqcap Placeholder \tag{4}$$

$$input \sqsupseteq required\_by \tag{5}$$

$$input \sqsupseteq demanded\_by \tag{6}$$

$$Task_{unexecutable} \equiv Task_{unexecuted} \sqcap \exists input^-.DataEl_{unavailable} \tag{7}$$

$$Task_{executable} \equiv Task_{unexecuted} \sqcap \neg Task_{unexecutable} \tag{8}$$

Although this alternative description of executable tasks appears rather cumbersome, it suits very well for practical use since intersections of classes can easily be queried from knowledge bases. Since the exact query routine depends on the usage scenario, it will not be regarded further here. Instead, we focus on the relevance of task execution which depends on the existence of output data and allows further contributions to KiPs.

4.2   Inferring Execution Relevance

In data-centric business process modeling, multiple tasks can lead to the same output data element. For example, one could either make a call or write an email to acquire the same data. When one makes a call and gets the data, writing an email gets irrelevant. Alternatively, if the call did not produce the desired data, one could choose to additionally write an email. Hence, unexecuted tasks whose execution would lead to currently unknown data elements are relevant for execution. The class of such relevant tasks with at least one unavailable output data element can be described by the existential quantifier (11) (relevant is abbreviated in formulas with "Rel"). Note that "output" is the super property of the "delivers" and "produces" properties of the base-ontology (9) (10). Based on our previous findings, we are also able to describe unexecutable relevant tasks (12), which we can use again during querying in order to detect executable relevant tasks (13).

$$output \sqsupseteq delivers \tag{9}$$

$$output \sqsupseteq produces \tag{10}$$

$$Task_{Rel} \equiv Task_{unexecuted} \sqcap \exists output.DataEl_{unavailable} \tag{11}$$

$$Task_{unexecutableRel} \equiv Task_{Rel} \sqcap \exists input^-.DataEl_{unavailable} \tag{12}$$

$$Task_{executableRel} \equiv Task_{Rel} \sqcap \neg Task_{unexecutableRel} \tag{13}$$

An inference mechanism based on the classes above considers tasks as relevant for execution if their output is unavailable. However, it does not regard that an execution can get obsolete when the data it would generate is not required further to achieve a process goal or milestone. This is elucidated by an extension of the example above: Let us assume writing an email or making a call is done to acquire data which is needed by another task to produce a document. The document thereby marks a process goal. If the document already exists, the tasks of writing an email or making a call are obsolete because the goal for which they would be executed for has already been achieved. Since KiPs are goal-oriented as they evolve through the achievement of intermediate goals or milestones [7], they would benefit if goal-relevance was regarded by the inference mechanism.

In the following, we regard goals and milestones as marks on tasks and data elements, denoting that there is a special aim for executing or generating them (for simplicity we will only speak of goals but milestones are considered equally). Figure 3 shows how goal-relevance can be determined in the ODD-BP model. When starting at a goal and iterating backwards though the chain of process elements, all elements marked as placeholders are goal-relevant until the first element without a placeholder marking occurs. Conversely, all elements not included in the loop are goal-irrelevant because their overall contribution has already been fulfilled.

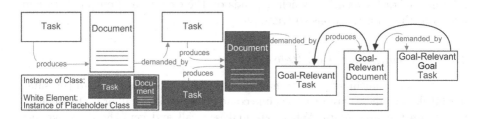

Fig. 3. Determining goal-relevance

In order for the inference mechanism to behave as sketched above, the characteristics of the individuals contained in the loop have to be described by classes. This leads to the classes of goal-relevant data elements (14) and goal-relevant tasks (15) (goal-relevant is abbreviated in formulas with GoalRel). Both classes refer to each other and in case of documents providing attributes the class of goal-relevant data elements refers to itself. This interdependency and self reference forces a reasoner to recursively infer goal-relevance which behaves similar to a loop and matches the above sketched way to determine goal-relevance. Since loops need entry points, both classes include unreached goals (i.e. unexecuted tasks and unavailable data elements that instantiate a class of goals). Since executed tasks and available data elements are not marked as placeholders, they are not contained in both classes, which leads to an end of the recursive reasoning procedure if an element occurs without a placeholder marking. Based on our previous findings, we are also able to describe executable goal-relevant tasks in a way that suits practical use (16) (17).

$$DataEl_{GoalRel} \equiv DataEl_{unavailable} \sqcap$$
$$(Goal \sqcup \exists input.Task_{GoalRel} \sqcup \exists provides.DataEl_{GoalRel}) \qquad (14)$$

$$Task_{GoalRel} \equiv Task_{unexecuted} \sqcap (Goal \sqcup \exists output.DataEl_{GoalRel}) \qquad (15)$$

$$Task_{unexecutableGoalRel} \equiv Task_{GoalRel} \sqcap \exists input^-.DataEl_{unavailable} \qquad (16)$$

$$Task_{executableGoalRel} \equiv Task_{GoalRel} \sqcap \neg Task_{unexecutableGoalRel} \qquad (17)$$

4.3   Implementation

The implementation of an inference mechanism based on the classes above is shown in Fig. 4. The implemented inference mechanism infers executability of goal-relevant tasks, relevant tasks and irrelevant tasks (classes no. 2–7).

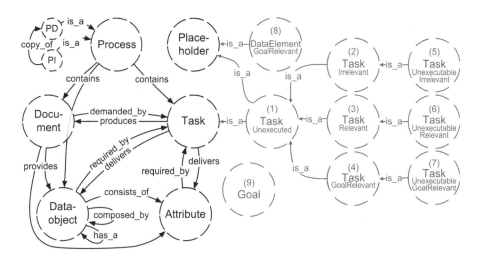

Fig. 4. Extended excerpt of the base-ontology to infer executability

The classes of irrelevant tasks (classes no. 2 & 5) have not been discussed so far because they are equal to the classes of unexecuted and unexecutable tasks which we defined initially in Sect. 4.1. The class of irrelevant tasks is used to infer tasks whose execution would not lead to unknown data and hence are irrelevant for execution. This can be determined through a query routine similar to the query routines required for identifying executable tasks on the other two relevance levels – thus, it only adds minimal complexity. The base-ontology is further extended by a class of goal-relevant data elements (class no. 8), that is used to infer if unavailable data elements are relevant to achieve a goal (for simplicity Fig. 4 does not show that this class further is the subclass of a unified class of documents and attributes). The instantiation of the general placeholder class should be organized by an agent, which can then mark process progress, but also reverse it if required. At last, goal modeling is enabled by a class of goals (class no. 9).

5 Demonstration on Use Cases

The proposed inference mechanism allows using ontologies for classifying unexecuted tasks in data-driven processes in the dimensions executability and relevance while pursuing goals and identifying skippable tasks. Since relevance is determined by the state of output data elements, process participants can choose tasks for execution based on the extent to which their execution would contribute to process progress. Even if the inference mechanism was extensively tested – a formal proof of its correctness in every process situation that can be modeled with the ODD-BP model has not been carried out yet. The capabilities of the inference mechanism will be shown in the following using Protégé 5.2.0[3] and the ontology reasoner Pellet.[4] The demonstration will use the excerpt of an application process for agricultural grants given earlier in this paper (Fig. 2). Figure 5 depicts two cases showing different execution states of the excerpt and marks the intended behaviour of the inference mechanism. In both cases, the task *Submission of Application* is marked as a goal. Since case 1 only consists of placeholders, except the *Date of Birth* attribute and the *Farmer* dataobject, the *Age Check* task should be inferred to be an executable goal-relevant task. All other tasks should have the same relevance level but are unexecutable. Case 2 is similar to case 1, except that the document *Application Form* already exists, which may be the case due to manifold causes. Here, the reasoner should infer that the as goal marked task *Submission of Application* is executable and goal-relevant. In addition, since the output of the execution of *Age Check* would no longer contribute to the achievement of a goal, it should be inferred to be only

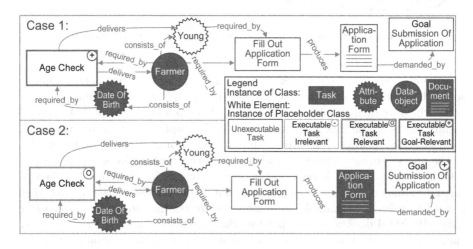

Fig. 5. Intended behaviour of the proposed inference mechanism

---

[3] Protégé is a free-to-use Stanford University ontology editor, available online at: https://protege.stanford.edu/.

[4] Pellet is released under Dual Licensing and available online at: https://github.com/stardog-union/pellet.

relevant and executable. Since the input data elements of *Fill out Application Form* are not available and its execution would not lead to unknown data, it should be irrelevant and unexecutable.

Figure 6 shows the inference results of Pellet. Protégé is used for demonstration because its visualization of inferences mostly matches the required query routine. We used this and labelled the classes no. 1–6 so that Protégé shows the inferred executability and relevance. Since this suffices for displaying inferred executability, it works only partially when showing inferred relevance levels. Hence, for demonstration purposes, only the highest visualized relevance level has to be considered when comparing intended with actual behaviour. However, the feasibility of the query routine was checked and approved. The comparison between the intended and the actual behaviour of the inference mechanism suggests that it works correctly. Also in many different scenarios no errors occurred.

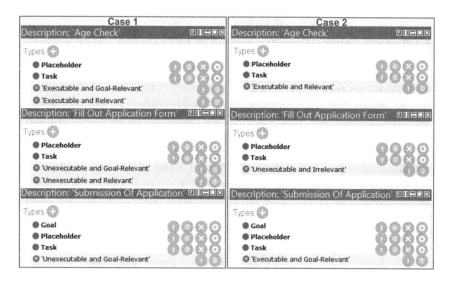

Fig. 6. Inferences about executability and relevance

6 Possible Process Support

The proposed inference mechanism allows inferring whether a task is executable and classifies its execution in terms of relevance. Figure 7 shows a coordinate system that summarizes which statements along the dimensions of executability and relevance are possible. Each field in the coordinate system represents a possible statement and the heading of each field describes what we believe this statement could be used for to support KiPs. Decision makings in KiPs should be supported by e.g. recommendations, choices, advices or contextual information, and knowledge workers should be free to either follow proposed actions or not [1,7,14]. It is obvious that executable and goal-relevant tasks are perfectly suited

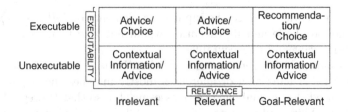

Fig. 7. Possible statements about executability and their use

to be recommended to knowledge workers in order to reach their goals as soon as possible while skipping obsolete tasks. Statements about executable tasks on all relevance levels can be seen as choices for knowledge workers to either follow the inferred ideal execution path to achieve goals or deviate from it. If they choose to deviate, the inference mechanism provides them with advices about which executions could be beneficial if they want to generate data. The advantage of using a reasoner to determine executability of tasks becomes particularly clear if tasks are unexecutable. Reasoners can then explain why they inferred that tasks are unexecutable, which generates lists of unavailable data elements. Since those explanations rely on the triple syntax, they could easily be transformed into human language and displayed as contextual information, so that knowledge workers can figure out how they will generate these data elements to turn unexecutable tasks into executables. This could lead them to ad-hoc planned tasks that exploit existing data elements. E.g. a known email address of some person could be used to generate a missing date of birth attribute of the same person when writing an email and asking for it. Such activities are implicitly contained in process instances and can only be performed when cross relationships between data elements are found. Such cross relationships could probably be detected autonomously by agents, taking explanations of reasoners as an input. Further, they could find appropriate subworkflows which would generate unavailable data elements and advice knowledge workers to start them. This would lead to the ability to make implicit activities explicit and allows proactive support in order to turn unexecutable tasks into executables as soon as possible.

7   Conclusion

In this paper, we propose an inference mechanism to infer executability of tasks in the briefly introduced ODD-BP model to support executions of KiPs. The inference mechanism classifies tasks in terms of execution relevance which detects skippable tasks and allows knowledge workers to weigh their actions based on expected output and goal achievement. Further, we describe various support for KiPs based on the inference mechanism. One of which uses explanations of inferences from reasoners as an input to identify actions that exploit available data to turn unexecutable tasks into executables as soon as possible.

Approaches like the ODD-BP model can easily be extended e.g. by control-flow elements, which would change the process logic. Since the proposed inference mechanism only considers the process logic of data-centric business process modeling, it would then have to be adjusted. Further, the proposed inference mechanism uses recursive reasoning to determine goal-relevance – this could be time consuming in larger knowledge bases. Even if the proposed inference mechanism works correctly in exemplary cases, an evaluation of its correctness in every possible situation was not carried out yet and should be regarded by future work.

References

1. Abecker, A., et al.: FRODO: a framework for distributed organizational memories: DFKI Document D-01-01. DFKI GmbH (2001)
2. Bechhofer, S.: OWL Reasoning Examples. http://owl.man.ac.uk/2003/why/latest/#6
3. Betz, S., Klink, S., Koschmider, A., Oberweis, A.: Automatic user support for business process modeling. In: Proceedings of the Workshop on Semantics for Business Process Management, pp. 1–12 (2006)
4. Bock, C., et al.: OWL 2 Web Ontology Language Structural Specification and Functional-Style Syntax (Second Edition): W3C Recommendation (2012)
5. Cohn, D., Hull, R.: Business artifacts: a data-centric approach to modeling business operations and processes. IEEE Data Eng. Bull. **32**, 3–9 (2009)
6. Davenport, T.H.: Thinking for a Living: How to Get Better Performances And Results from Knowledge Workers. Harvard Business School Press, Boston (2005)
7. Di Ciccio, C., Marrella, A., Russo, A.: Knowledge-intensive processes: characteristics, requirements and analysis of contemporary approaches. J. Data Semant. **4**(1), 29–57 (2015)
8. Fan, S., Hua, Z., Storey, V.C., Zhao, J.L.: A process ontology based approach to easing semantic ambiguity in business process modeling. Data Knowl. Eng. **102**, 57–77 (2016)
9. Heinrich, B., Bewernik, M., Henneberger, M., Krammer, A., Lautenbacher, F.: SEMPA - Ein Ansatz des Semantischen Prozessmanagements zur Planung von Prozessmodellen. WIRTSCHAFTSINFORMATIK **50**(6), 445–460 (2008)
10. Horridge, M.: A Practical Guide To Building OWL Ontologies Using Protégé 4 and CO-ODE Tools Edition 1.3 (2011). http://mowl-power.cs.man.ac.uk/protegeowltutorial/resources/ProtegeOWLTutorialP4_v1_3.pdf
11. Grumbach, L., Rietzke, E., Schwinn, M., Bergmann, R., Kuhn, N.: SEMANAS - semantic support for grant application processes. In: Proceedings of the Conference "Lernen, Wissen, Daten, Analysen", LWDA 2018, Mannheim, Germany, 22–24 August 2018, pp. 126–131. CEUR Workshop Proceedings, CEUR-WS.org (2018)
12. Rietzke, E., Bergmann, R., Kuhn, N.: Semantically-oriented business process visualization for a data and constraint-based workflow approach. In: Teniente, E., Weidlich, M. (eds.) BPM 2017. LNBIP, vol. 308, pp. 142–150. Springer, Cham (2018). https://doi.org/10.1007/978-3-319-74030-0_10
13. Thomas, O., Fellmann, M.: Semantic process modeling - design and implementation of an ontology-based representation of business processes. Bus. Inf. Syst. Eng. **1**(6), 438–451 (2009)

14. Vaculín, R., Hull, R., Heath, T., Cochran, C., Nigam, A., Sukaviriya, P.: Declarative business artifact centric modeling of decision and knowledge intensive business processes. In: 15th IEEE International EDOC Conference Workshops, pp. 151–160. IEEE Computer Society, Los Alamitos, California (2011)
15. Workflow Management Coalition: Workflow Management Coalition Terminology & Glossary, February 1999

How Cognitive Processes Make Us Smarter

Andrea Zasada[(⊠)]

University of Rostock, Rostock, Germany
andrea.zasada@uni-rostock.de

Abstract. Cognitive computing describes the learning effects that computer systems can achieve through training and via interaction with human beings. Developing capabilities like this requires large datasets, user interfaces with cognitive functions, as well as interfaces to other systems so that information can be exchanged and meaningfully linked. Recently, cognitive computing has been applied within business process management (BPM), raising the question about how cognitive computing may change BPM, and even leverage some new cognitive resources. We believe that the answer to this question is linked to the promised learning effects for which we need to explore how cognitive processes enable learning effects in BPM. To this end, we collect and analyze publications on cognitive BPM from research and practice. Based on this information, we describe the principle of cognitive process automation and discuss its practical implications with a focus on technical synergies. The results are used to build a visual research map for cognitive BPM.

Keywords: Cognitive processes · Process automation · Cognitive business process management · Cognitive computing

1 Motivation

Cognitive business process management (BPM) describes the opportunity to develop and deploy process-aware systems that can learn, employ logic and reason, and interact naturally with human beings [14]. This opportunity stems from cognitive computing systems not being explicitly programmed, but rather trained like humans. Systems of this kind gain experience over time and manage both structured and unstructured data using diverse technologies. As such, cognitive computing has the potential to apply business logic [52]. By extending BPM from process logic to business logic, the aim is to establish cognitive processes that make BPM more dynamic and make it easier for us to understand, evaluate, and predict business events. In this way, cognitive processes can indeed make us smarter provided we know how to use the new technical capabilities.

First research efforts in this area have been undertaken to understand the task complexity and requirements of cognitive BPM [23]. The findings indicate that we currently lack a strategy to expand BPM from hard-coded scripts to adaptive and agent-oriented cognitive process automation. Although some research has explored the avenues of cognitive BPM [41], the field is still largely undeveloped, leaving many practical implications of the technologies and the human factor unexplored. In

© Springer Nature Switzerland AG 2019
C. Di Francescomarino et al. (Eds.): BPM 2019 Workshops, LNBIP 362, pp. 45–55, 2019.
https://doi.org/10.1007/978-3-030-37453-2_5

particular, the lack of cognitive approaches to explain how these newly forged inter-actions affect the human user is surprising, since decision making can suffer from any kind of cognitive bias [39].

With this paper, our aim is not only to outline the obvious research perspectives but also to report some missing aspects of cognitive process automation for the continu-ation and growth of research on cognitive BPM. Our paper focuses in particular on the question of *how cognitive processes enable innovations in terms of learning effects*. To this end, we review publications on cognitive BPM, and evaluate how the approaches leverage cognitive processes and learning effects.

The paper starts with an outline of the review procedure and an introduction to cognitive BPM, wherein we discuss the role of cognitive processes and look at the systems and technologies supporting them. Thereafter, we explore the learning effects cognitive processes are associated with. Finally, we envision a research map for cognitive BPM to guide future research activities and point towards current research gaps. With the results we hope to lay the foundation to utilize the whole spectrum of cognitive process automation and facilitate new ideas.

2   Review Procedure

We conducted a systematic literature review following the suggestions of Jones and Gatrell [26]. A main theme of the editorial is that a literature review should be ana-lytical rather than descriptive. As a result we omit most of the quantitative description from this paper and focus on the approach to analyze the literature. First, we deter-mined the search radius. We used Google Scholar to have access to a wide range of scholarly literature [19]. Since we were interested in recent publications on cognitive BPM we set the start time to 2015. The last update was made in May 2019.

Second, we collected alternative search terms that reflect our expectations and understanding of the topic. We checked our ideas in a test run in which we adapted the search terms and identified commonly used phrases. The biggest challenge came from concepts having a different meaning for another discipline. For example, in cognitive science "cognitive processes" refer to mental processes while only recent studies adopt the term in the context of information systems. Another obstacle was the broad con-ception of some terms, such as "collaboration" or "intelligent technology". We finally decided to use the expressions "business process" in combination with "cognitive computing" and "artificial intelligence".

Third and last, we reviewed the literature for relevance to our research question. Beside research papers, we included white papers of several multinational information technology companies (e.g., Accenture, Deloitte, IBM, KPMG), which we retrieved via a Google search using the same search phrases. After we determined the final sample, we analyzed the literature as suggested by [26] with regard to the addressed themes and subthemes, the credibility of sources, missing information and research gaps. In total, we reviewed 386 research papers and 22 industry papers. 76 of them passed the review process. The following sections summarize the outcome structured into the foundations of cognitive BPM (Sect. 3), associated learning effects (Sect. 4), and research per-spectives for cognitive BPM (Sect. 5).

3   Foundations of Cognitive BPM

3.1   Cognitive Processes

Cognitive processes are composed of tasks that are ideally carried out autonomously. This includes the analysis of numbers, words and images, digital tasks and even physical tasks [13]. They are designed to enable more automation while at the same time retaining the flexibility to direct the information flow from one or more learning system [30]. However, while this idea may sound innovative, there is no guarantee that implementing cognitive processes leads to learning effects and improved business performance, even though cognitive technologies, such as linguistic analytics, are readily accessible [24]. To date, traditional BPM has involved a process model that is separate from process execution, which leaves little range for interactions [46]. Thus, the implementation of cognitive processes can be a costly undertaking for BPM, and one that increases the workload through the adoption process, which might negate the expected benefits.

3.2   Learning Systems and Intelligent Technologies

In order to control the adoption risks, we need a systematic implementation process, that is, a guide to cognitive process automation. The next section aims to devise such a guideline. The guideline depends on systems and technologies that are mimicking the cognitive capabilities of the human mind [37]. Concretely, what this really means is conceiving a system that, using unstructured information, can acquire logical insights into the decision-making process [14]. Such systems are highly event-driven goal-oriented, and learn from the exchange of relevant information between systems and interactions with human users [23]. Using predictive and adaptive decision capabilities, the so-called cognitive systems provide not only decision support, but also contribute to the continuous expansion of knowledge by incorporating lessons of prior decisions and reasoning [41].

Table 1. Systems and technologies by task type and level of intelligence (adopted from [13])

Task type	Support for humans	Task automation	Context awareness & learning
Analyze numbers	Business intelligence Data visualization Hypothesis-driven analytics	Operational analytics Model management	Machine learning Deep learning [35] Predictive analytics [15]
Analyze words and images	Character and speech recognition	Image recognition Machine vision	Natural language processing Natural text generation [31]
Perform digital tasks	Digital agents [20] Recommender systems [29]	Rules engines Robotic process automation	Sensor networks [51] Augmented reality [27]
Perform physical tasks	Remote operation of equipment	Industrial and collaborative robotics	Autonomous robots and vehicles

As certain features of these systems are technology-enabled, artificial intelligence (AI) plays an important role in developing these capabilities [32]. Paschek et al. [37] refer to AI as efforts "to pattern human proceedings of problem solving and transfer them to computers in order to invent efficient and new solutions as well as course of actions". Common elements of AI comprise natural language processing (the process through which machines can understand and analyze language), machine learning (algorithms that enable systems to learn), and machine vision (algorithmic inspection and analysis of images) [25]. Further elements are given in Table 1. The sample is based on the work of Davenport and Kirby [13]. It has been tailored to BPM by offering supplementary interpretations with references to respective publications.

3.3   A Guide to Cognitive Process Automation

While the previous sections provided some insights into the general agenda and innovation potentials of cognitive BPM, in the following we introduce a model on the implementation of cognitive processes adopted from Dwarkanhalli et al. [14]. The model describes four phases: process discovery, process definition, process design, and prototype development. Figure 1 illustrates the phases and key elements of cognitive process automation and identifies some characteristic activities for each phase. Cognitive learning is an integral part of this model, as we can see how the cognitive learning strategy evolves and connects basic BPM elements with cognitive tools.

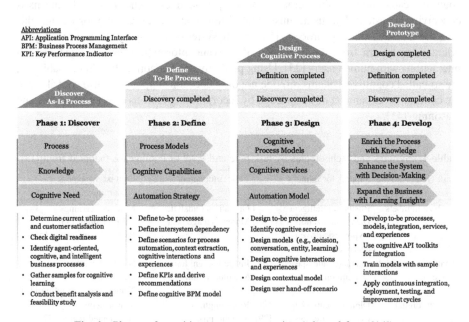

Fig. 1. Phases of cognitive process automation (adopted from [14])

The first phase is essential to assess the organization's digital readiness, and, above all, to identify suitable business processes in which cognitive approaches could leverage learning effects and enhance the user experience. The approach should be selected based on the organizational needs, availability of appropriate data [48], sampling, and current utilization of cognitive systems (cf. Table 1). If the expected benefits outweigh the adoption risks, the cognitive learning strategy can be implemented in the second phase by defining to-be processes, dependencies between systems, as well as use case scenarios and key performance indicators. In the third phase, the cognitive processes and corresponding models are designed with respect to the planned services, interactions, and experiences. The last phase serves the technical implementation and integration of the cognitive learning strategy and comprises testing, deployment, and improvement cycles.

4   Learning from Cognitive Processes

Cognitive processes can be applied across a variety of industries. Accordingly, many different ideas circulate on how cognitive processes can leverage business intelligence. Table 2 summarizes papers on cognitive BPM aggregated by focus and learning effect. In all of these cases, the idea of freeing up cognitive resources stays approximately the same. Instead of following the traditional programmatic approach, cognitive BPM strives for more flexibility by exploiting contextual data with modern tools and analytics to expand the performance spectrum of BPM [23, 33]. The following use cases illustrate which learning effects are typically associated with the implementation of cognitive processes:

Insightful Processes. Creating insightful processes means connecting people, contexts, and content through a cross-channel business model. In healthcare, hospital systems can pull data from social media to monitor the spread of diseases and track outbreaks. By analyzing tweets in real-time, hospitals can take proactive measures to inform patients about precautions and predict future outbreaks.

Scaling Knowledge. Scaling knowledge across the enterprise requires systems that can observe and learn from skilled employees or successfully implemented projects. Systems that are trained this way could be employed in consulting companies in the form of digital assistants supplying employees with information, and thereby improve the advisory quality, as well as the consistency and compliance of given answers.

Informed Decision-Making. Enhancing decision performance means developing automated, adaptive, and predictive decision-making capabilities. Similar to the healthcare scenario, human resource systems could collect and analyze data from social networks to support the selection of potential candidates not only according to formal criteria, but also by character traits.

Increased Productivity. Leveraging cognitive capabilities means establishing agent-oriented processes. Just like the previous examples showed, digital agents can augment individual competences through contextual conversation with learning systems. Service companies could use this to their advantage to bridge the expertise gap between co-workers and achieve learning progress in a relatively short time.

Improved Customer Satisfaction. Improving customer satisfaction means swiftly and adequately responding to customer needs. When customers get approved for a bank loan, the service quality could be enhanced by analyzing communication between the parties, including the customer's sentiment. In this way advisors can not only deliver positive service experiences, but also handle inquiries more efficiently.

Table 2. Papers on cognitive BPM

Focus	#	Learning effect
Connecting people, contexts, and content through a cross-channel business model	11	Insightful processes
Creating the basis for collaborative and interactive learning systems	13	Scaling knowledge
Developing automated, adaptive, and predictive decision-making capabilities	18	Informed decision-making
Establishing agent-oriented processes for contextual conversations	19	Increased productivity
Processing customer inquiries based on customer sentiment analysis	15	Improved customer satisfaction

5   Research Perspectives

Figure 2 structures the field of cognitive BPM according to the topics and questions raised in the previous sections. The four dimensions of the research map are geared towards the requirements of cognitive process automation, the effect of the human-computer interaction, the cognitive technologies used in the context of AI, and finally the utilization of cognitive resources summarized under the umbrella of business processes intelligence.

Most approaches that fell into our review have focused on *cognitive process automation*. A major role in this context is played by AI as the different applications and work areas create new opportunities for data-driven approaches [42]. An important ability to progress their implementation is to provide the right cognitive level for system enactment and automated reasoning [32]. Formalisms from the field of knowledge representation and reasoning may be better suited in this regard to deal with dynamic situations in which data needs to be processed with automatic reasoning capabilities. Given the transdisciplinary of the research field the best results may be achieved through collaboration between researchers from the areas of computer science, linguistics, psychology, cognitive neuroscience, business administration, economics, engineering, law, and others [33, 53]. Furthermore, future research should be conducted in currently less reflected areas, which concerns the development of reference architectures [41] or what is referred to as multi process management [8].

Nevertheless, the biggest research gap seems to lie in the evolution of *human-computer interaction*. In this regard, empirical works would be extremely valuable to access the cognitive strain while performing cognitive processes [28], judging the complexity of process models and procedures [17], or evaluating the decision performance [43]. Cognitive theories as gathered in [22] may also be a lever to design useful interactions with human users and expand cognitive reasoning and learning techniques. However, it should be remembered that the data generated by these interactions poses the risk that private data is being compromised and misused [20, 44]. Thus, cognitive BPM should envision a stringent data policy for the use of interconnected and collaborative systems. Finally, the opportunity to reach into different domains and devise new tasks may require domain knowledge which is still difficult to scale [45]. This is because every interaction requires the input from a human-expert. The reliability of a system-generated answer is therefore limited by the accuracy and reliability of the expert's knowledge and to what extent it can be transferred to new tasks [21]. Thus, we need to device methods and techniques to find and minimize such biases and ensure a good generalization from the data.

Regarding the use of *cognitive technologies*, we discovered that natural language processing (NLP) is often researched in connection to business processes, e.g., to extract process models [2, 16] and requirements [5, 7] from textual descriptions. Accordingly, the application of NLP techniques seems particularly attractive to reduce the effort of modeling [1]. Deep learning, in contrast, extends BPM utilities by predictive capabilities, which is extremely helpful in case models do not exist or are difficult to obtain [15]. Approaches in this area utilize neural networks to monitor the process execution and predict the next event in a business process. Existing methods are, however, tailored to specific prediction tasks and highly sensitive to the dataset [18, 49]. As a neighboring discipline, machine learning is applied to observe and learn from human behavior and assist users in process modeling, e.g., in form of recommendation systems [9, 29] or active learning systems [47]. Problematic in this regard is that process modeling languages usually do not have a formal foundation. In addition,

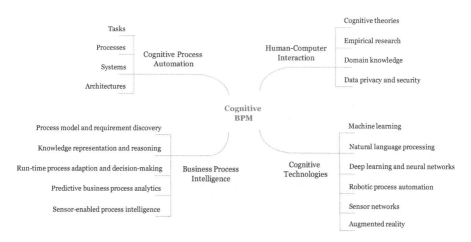

Fig. 2. Research map

certain steps such as clustering and text categorization are still performed in a semi-automated manner. However, it seems that the limits of these techniques have not yet been reached and the combination of different techniques promises further research potential [34]. Approaches surrounding the topic of robotic process automation (RPA) [6, 54] and sensor networks [12, 50] are just as rapidly evolving. RPA aims to automate structured and repetitive tasks to solve them in a time and cost efficient manner [3]. It has to be noted though that the conducted case studies do not always provide supporting evidence. Further research would be needed to clarify these issues and include other factors such as process agility or error reduction [4]. Sensor networks are in contrast mostly used to improve and enrich business processes with relevant context information about the resources deployed in the process [10]. In so far, sensors provide a vivid connection to the actual process execution. Research could drive this development by investigating the correlation between signals or expanding the activity recognition from high-level to fine-grained activities [51]. The probably most surprising use case ideas come from augmented reality (AR). Instead of monitoring the process execution on a stationary computer, AR provides the opportunity to access process executions on a mobile device [27, 38]. Important design decisions pertain the partitioning of processes and possible features such as process guidance (i.e., operative services and analytics). As such, the utilization of AI-related concepts appears to be well underway, although research opportunities remain given the many facets and applications of cognitive processes and technologies for automating the discovery, adaption, and prediction of business processes [23, 36].

The last dimension tackles the new decision-making capabilities which we summarized under the term *business process intelligence*. Interestingly, big data plays an important role in developing these capabilities as it extends the scope to data that used to be too big, complex, fast-paced, or weakly structured to be processed with conventional methods [40]. In this sense, the progress achieved in the field of big data increases the expectations regarding the application of cognitive technologies. However, in order to fully exploit the potential, further research is needed to access the efficiency of distributed execution environments [11], as well as fluctuations in the quality of the data [35], or signs of incomplete training [21].

6 Conclusion

This paper is motivated by the question of how cognitive processes could make us smarter. In order to answer this question, we conducted a systematic literature review. In the course of this review, we collected information about the main constituents and drivers of cognitive process automation, and developed initial ideas to harness cognitive resources and pursue seamless integration into the organizational context. Our findings suggest that cognitive processes could be a vehicle for individual and organizational learning, and thus should not only focus on technical solutions, but cherish human capital to overcome the barriers to innovation.

Limitations come from the scope of research and applied research method. First, the review builds on the search algorithms and metrics of Google Scholar, which seems suitable in the context, but also may have led to some relevant studies being excluded.

Second, we performed a qualitative analysis focusing only on conceptual work and practical applications. Very specific solutions may not be captured by such approach. Third, although we extended the search from research papers to whitepapers, practical insights are yet very limited. Hence, the developed research map can only lay the foundations for future research, although we hope to have added some practical insights to the academic knowledge by integrating ideas and models from both research and industry.

References

1. van der Aa, H., et al.: Challenges and opportunities of applying natural language processing in business process management. In: Proceedings of the 27th International Conference on Computational Linguistics, pp. 2791–2801 (2018)
2. van der Aa, H., Di Ciccio, C., Leopold, H., Reijers, H.A.: Extracting declarative process models from natural language. In: Giorgini, P., Weber, B. (eds.) CAiSE 2019. LNCS, vol. 11483, pp. 365–382. Springer, Cham (2019). https://doi.org/10.1007/978-3-030-21290-2_23
3. van der Aalst, W.M.P., et al.: Robotic process automation. Bus. Inf. Syst. Eng. 60(4), 269–272 (2018)
4. Aguirre, S., Rodriguez, A.: Automation of a business process using robotic process automation (RPA): a case study. In: Figueroa-García, J.C., López-Santana, E.R., Villa-Ramírez, J.L., Ferro-Escobar, R. (eds.) WEA 2017. CCIS, vol. 742, pp. 65–71. Springer, Cham (2017). https://doi.org/10.1007/978-3-319-66963-2_7
5. Arora, C., et al.: Automated checking of conformance to requirements templates using natural language processing. IEEE Trans. Softw. Eng. 41(10), 944–968 (2015)
6. Asatiani, A., Penttinen, E.: Turning robotic process automation into commercial success – case OpusCapita. J. Inf. Technol. Teach. Cases. 6(2), 67–74 (2016)
7. Aysolmaz, B., et al.: A semi-automated approach for generating natural language requirements documents based on business process models. Inf. Softw. Technol. 93, 14–29 (2018)
8. Baesens, B., et al.: Challenges of smart business process management: an introduction to the special issue. Decis. Support Syst. 100, 1–5 (2017)
9. Betz, S., Klink, S.: Automatic user support for business process modeling. In: Proceedings of the Workshop on Semantics for Business Process Management, pp. 1–12 (2006)
10. Caracaş, A., Bernauer, A.: Compiling business process models for sensor networks. In: Proceedings of the International Conference on Distributed Computing in Sensor Systems and Workshops (2011)
11. Chen, M., et al.: Cognitive computing: architecture, technologies and intelligent applications. IEEE Access 6, 19774–19783 (2018)
12. Cheng, Y., et al.: Modeling and deploying IoT-aware business process applications in sensor networks. Sensors 19, 1 (2018)
13. Davenport, T.H., Kirby, J.: Just how smart are smart machines? MIT Sloan Manag. Rev. 57, 3 (2016)
14. Dwarkanhalli, H., et al.: How Cognitive Computing Unlocks Business Process Management's Performance – Enhancing Virtues (2018)
15. Evermann, J., et al.: Predicting process behaviour using deep learning. Decis. Support Syst. 100, 129–140 (2017)

16. Ferreira, R.C.B., et al.: A semi-automatic approach to identify business process elements in natural language texts. In: Proceedings of the International Conference on Enterprise Systems, pp. 250–261 (2017)
17. Figl, K., Laue, R.: Cognitive complexity in business process modeling. In: Mouratidis, H., Rolland, C. (eds.) CAiSE 2011. LNCS, vol. 6741, pp. 452–466. Springer, Heidelberg (2011). https://doi.org/10.1007/978-3-642-21640-4_34
18. Di Francescomarino, C., Ghidini, C., Maggi, F.M., Petrucci, G., Yeshchenko, A.: An eye into the future: leveraging a-priori knowledge in predictive business process monitoring. In: Carmona, J., Engels, G., Kumar, A. (eds.) BPM 2017. LNCS, vol. 10445, pp. 252–268. Springer, Cham (2017). https://doi.org/10.1007/978-3-319-65000-5_15
19. Google Scholar: Content Coverage. https://scholar.google.de/intl/de/scholar/help.html#coverage
20. James, H., Daugherty, P.R.: Collaborative intelligence: humans and AI are joining forces. Harv. Bus. Rev. **96**, 114–123 (2018)
21. Holzinger, A., et al.: Machine learning and knowledge extraction in digital pathology needs an integrative approach. In: Holzinger, A., Goebel, R., Ferri, M., Palade, V. (eds.) Towards Integrative Machine Learning and Knowledge Extraction. LNCS (LNAI), vol. 10344, pp. 13–50. Springer, Cham (2017). https://doi.org/10.1007/978-3-319-69775-8_2
22. Houy, C., et al.: On the theoretical foundations of research into the understandability of business process models. In: Proceedings of the European Conference on Information Systems, pp. 1–38 (2014)
23. Hull, R., Motahari Nezhad, H.R.: Rethinking BPM in a cognitive world: transforming how we learn and perform business processes. In: La Rosa, M., Loos, P., Pastor, O. (eds.) BPM 2016. LNCS, vol. 9850, pp. 3–19. Springer, Cham (2016). https://doi.org/10.1007/978-3-319-45348-4_1
24. IBM: Cognitive Business Operations: Processes and decisions that sense, respond, and learn (2016)
25. Jarrahi, M.H.: Artificial intelligence and the future of work: human-AI symbiosis in organizational decision making. Bus. Horiz. **61**(4), 577–586 (2018)
26. Jones, O., Gatrell, C.: Editorial: The future of writing and reviewing for IJMR (2014). http://doi.wiley.com/10.1111/ijmr.12038
27. Kammerer, K., et al.: Towards context-aware process guidance in cyber-physical systems with augmented reality. In: Proceedings of the 4th International Workshop on Requirements Engineering for Self-Adaptive, Collaborative, and Cyber Physical Systems, pp. 44–51 (2018)
28. Kirschner, P.A.: Cognitive load theory: implications of cognitive load theory on the design of learning. Learn. Instr. **12**(1), 1–10 (2002)
29. Kluza, K., et al.: Overview of recommendation techniques in business process modeling? In: CEUR Workshop Proceedings, pp. 46–57 (2013)
30. KPMG: Embracing the cognitive era. Using automation to break transformation barriers – and make every employee an innovator (2016)
31. Lu, H., et al.: Brain intelligence: go beyond artificial intelligence. Mob. Netw. Appl. **23**(2), 368–375 (2018)
32. Marrella, A., Mecella, M.: Cognitive business process management for adaptive cyber-physical processes. In: Teniente, E., Weidlich, M. (eds.) BPM 2017. LNBIP, vol. 308, pp. 429–439. Springer, Cham (2018). https://doi.org/10.1007/978-3-319-74030-0_33
33. Mendling, J., et al.: How do machine learning, robotic process automation, and blockchains affect the human factor in business process management? Commun. Assoc. Inf. Syst. **43**(1), 297–320 (2018)

34. Metzger, A., Föcker, F.: Predictive business process monitoring considering reliability estimates. In: Dubois, E., Pohl, K. (eds.) CAiSE 2017. LNCS, vol. 10253, pp. 445–460. Springer, Cham (2017). https://doi.org/10.1007/978-3-319-59536-8_28
35. Najafabadi, M.M., et al.: Deep learning applications and challenges in big data analytics. J. Big Data. **2**, 1 (2015)
36. Noor, A.K.: Potential of cognitive computing and cognitive systems. Open Eng. **5**(1), 75–88 (2015)
37. Paschek, D., et al.: Automated business process management–in times of digital transformation using machine learning or artificial intelligence. In: MATEC Web of Conferences (2017)
38. Pryss, R., et al.: Integrating Mobile Tasks with Business Processes: A Self-Healing Approach (2014)
39. Razavian, M., Turetken, O., Vanderfeesten, I.: When cognitive biases lead to business process management issues. In: Dumas, M., Fantinato, M. (eds.) BPM 2016. LNBIP, vol. 281, pp. 147–156. Springer, Cham (2017). https://doi.org/10.1007/978-3-319-58457-7_11
40. Reynolds, H.: Big Data and Cognitive Computing. cognitivecomputingconsortium.com
41. Roeglinger, M., Seyfried, J., Stelzl, S., Muehlen, M.: Cognitive computing: what's in for business process management? an exploration of use case ideas. In: Teniente, E., Weidlich, M. (eds.) BPM 2017. LNBIP, vol. 308, pp. 419–428. Springer, Cham (2018). https://doi.org/10.1007/978-3-319-74030-0_32
42. Samek, W., et al.: Explainable Artificial Intelligence: Understanding, Visualizing and Interpreting Deep Learning Models (2017)
43. Schneider, S., Leyer, M.: Me or information technology? Adoption of artificial intelligence in the delegation of personal strategic decisions. Manag. Decis. Econ. **40**(3), 223–231 (2019)
44. Seth, N., et al.: A conceptual model for quality of service in the supply chain. Int. J. Phys. Distrib. Logist. Manag. **36**(7), 547–575 (2006)
45. Siemens, G.: Learning analytics: the emergence of a discipline. Am. Behav. Sci. **57**(10), 1380–1400 (2013)
46. Smith, H., Fingar, P.: Business Process Management: The Third Wave. Meghan-Kiffer Press, Tampa (2003)
47. Sunkle, S., et al.: Informed active learning to aid domain experts in modeling compliance. In: Proceedings of the IEEE 20th International Enterprise Distributed Object Computing Conference, pp. 129–138 (2016)
48. Tarafdar, M., et al.: Enterprise cognitive computing applications: opportunities and challenges. IT Prof. **19**(4), 21–27 (2017)
49. Tax, N., Verenich, I., La Rosa, M., Dumas, M.: Predictive business process monitoring with LSTM neural networks. In: Dubois, E., Pohl, K. (eds.) CAiSE 2017. LNCS, vol. 10253, pp. 477–492. Springer, Cham (2017). https://doi.org/10.1007/978-3-319-59536-8_30
50. Tranquillini, S., et al.: Process-based design and integration of wireless sensor network applications. In: Barros, A., Gal, A., Kindler, E. (eds.) BPM 2012. LNCS, vol. 7481, pp. 134–149. Springer, Heidelberg (2012). https://doi.org/10.1007/978-3-642-32885-5_10
51. Wang, J., et al.: Deep learning for sensor-based activity recognition: a survey. Pattern Recognit. Lett. **119**, 3–11 (2019)
52. Wang, M., Wang, H.: From process logic to business logic - a cognitive approach to business process management. Inf. Manag. **43**(2), 179–193 (2006)
53. Wang, Y., et al.: Cognitive informatics and cognitive computing in year 10 and beyond. Int. J. Cogn. Informatics Nat. Intell. **5**(4), 1–21 (2012)
54. Willcocks, L., et al.: robotic process automation at Xchanging. In: The Outsourcing Unit Working Research Paper Series, pp. 1–26 (2015)

Supporting Complaint Management in the Medical Technology Industry by Means of Deep Learning

Philip Hake[(✉)], Jana-Rebecca Rehse, and Peter Fettke

Institute for Information Systems (IWi) at the German Research Center
for Artificial Intelligence (DFKI GmbH) and Saarland University,
Campus D3.2, Saarbrücken, Germany
`{philip.hake,jana-rebecca.rehse,peter.fettke}@dfki.de`

Abstract. Complaints about finished products are a major challenge for companies. Particularly for manufacturers of medical technology, where product quality is directly related to public health, defective products can have a significant impact. As part of the increasing digitalization of manufacturing companies ("Industry 4.0"), more process-related data is collected and stored. In this paper, we show how this data can be used to support the complaint management process in the medical technology industry. Working together with a large manufacturer of medical products, we obtained a large dataset containing textual descriptions and assigned error sources for past complaints. We use this dataset to design, implement, and evaluate a novel approach for automatically suggesting a likely error source for future complaints based on the customer-provided textual description. Our results show that deep learning technology holds an interesting potential for supporting complaint management processes, which can be leveraged in practice.

Keywords: Complaint management · Quality management · Process prediction · Machine learning · Deep learning

1 Motivation

For manufacturing companies, complaints about finished products are a major challenge [1]. They not only reduce profits, but also cause additional costs. Complaint management requires time and personnel resources. In addition, regular complaints about quality defects have a lasting negative effect on customer loyalty and therefore on a company's reputation. For manufacturers of medical technology, which must meet special quality requirements in the regulated environment, defective products can be particularly damaging [2]. The new EU regulation on medical devices has further increased the requirements on quality and safety of medical technology [3]. In this industry, lawmakers consider product quality as directly related to public health. Quality defects therefore risk a company's success both by a decline in sales and by official interventions. According to legal requirements, medical technology manufacturers must establish a prompt and consistent approach to the acceptance, assessment, and investigation of complaints and the decision on follow-up measures.

© Springer Nature Switzerland AG 2019
C. Di Francescomarino et al. (Eds.): BPM 2019 Workshops, LNBIP 362, pp. 56–67, 2019.
https://doi.org/10.1007/978-3-030-37453-2_6

In the context of the ongoing digitalization of the economy in general and manu-facturing companies in particular ("Industry 4.0"), more and more process- and production-relevant data is recorded and stored [4]. The considerable amount of real-time sensor, machine, and process data from product lifecycle (PLC), manufacturing execution (MES), and enterprise resource planning (ERP) systems can be further enriched with data from the systems used for complaint and error handling processes as well as customer-related data. This data holds great potential for improved complaint management [5]. For example, it can be used to train a machine learning approach that provides automated support for repetitive, but time-critical process steps.

In this paper, we introduce a new approach to take a first step towards process automation. Working together with a large manufacturer of medical products, we obtained a large dataset containing textual descriptions and assigned error sources for past complaints. We use this dataset to design, implement, and evaluate a novel approach for automatically suggesting a likely error source for future complaints based on the customer-provided textual description. The approach makes use of state-of-the-art deep learning technology, which has already been used for natural language processing (NLP) in other application domains. For this purpose, the paper is organized as follows. In Sect. 2, we report on the foundations of medical technology quality management and related work on machine learning in business process management (BPM). Our suggested solution design and architecture are described in Sect. 3. Section 4 reports on how our suggested approach was realized and evaluated. We discuss implications and challenges in Sect. 5, before concluding the paper in Sect. 6.

2 Preliminaries

2.1 Quality Management in Medical Technology

Both processes and products in the regulated environment are subject to high quality requirements, summarized as GMP (Good Manufacturing Practice). These binding quality requirements result from national and international regulations (such as laws and standards) and must be considered during production [6]. GMP regulations affect central sectors of the economy, such as the pharmaceutical industry, biotechnology, medical technology, chemical industry, and food industry. Compliance with GMP regulations is of fundamental importance to companies in these industries, as they influence their manu-facturing authorization. Core processes of GMP compliance and quality management include process management and document management (Standard Operating Proce-dures, SOPs), improvement management, corrective and preventive actions (CAPA) and controls, risk management, change management, deviations, employee training, as well as internal and external audits for GMP-relevant processes.

A central part of quality management in the regulated environment is complaint management. This is partially regulated by law. For medical device manufacturers, ISO standard 13485 (which largely conforms with ISO 9001) prescribes the use of a quality management system designed to demonstrate consistent compliance with quality standards [7]. Typically, the systems follow the 8D problem-solving process (see Fig. 1) [8]. Originally developed by automotive companies and used across many different industries, this process describes a structured approach to identifying and eliminating problems and their causes and is therefore an integral part of complaint management. The collected data is an important source for internal complaint handling and identifying errors in the production process. Many steps in the 8D process have the potential to be supported by data-driven digital quality management systems.

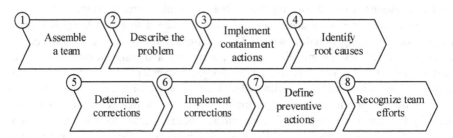

Fig. 1. The 8D reference process for complaint management

2.2   Related Work

Diverse applications for deep neural networks in BPM have recently been presented. In addition to process prediction [9–11], these include, particularly, techniques for anomaly detection [12], representation learning [13] and process modelling [14]. The authors are not aware of an application which uses deep neural networks to support complaint process management. There are several approaches that apply other machine learning techniques in quality management. Coussement et al. introduce a binary classifier that is able to distinguish complaint from non-complaint emails [1]. The approach consists of a rule-based feature extraction and a boosting algorithm for binary classification. Ko et al. deal with the detection of anomalies in engine production [15]. Their approach combines data from production across supply chains with customer data and other quality data to classify the engines' quality. The approach of Weiss et al. is also concerned with the prediction of product quality along a supply chain, but considering microprocessors [16]. In this context, the main challenge is the lengthy production process and the availability of only little measurement data. In addition, there are several approaches that develop models for quality forecasts across multiple production steps. Lieber et al. describe a case of application from the steel industry, in which the quality of interstage products is in focus [17]. Techniques of supervised and unsupervised machine learning, such as clustering or decision trees, are applied to data recorded during production (e.g., by sensors) to identify the most important factors influencing subsequent product quality. The approach of Arif et al., on the other hand, comes from the production of semiconductors, where decision trees are also used to develop a predictive model [18].

3 Conceptual Design

Employees usually file 8D reports after they receive either an internal or external complaint about product quality. At first, filing such a report entails recording a lot of potentially relevant data, but in a second step, the employee also has to assess the claim in terms of its criticality and the potential error source. Both determine the actions to be taken next. The criticality denotes the risk for another customer's health. If, for example, the bacterial load is too high on a previously sterilized product, it must be reported immediately to the responsible authorities in order to avoid public health risks. The potential error source is an internal assessment and the first step towards identifying and fixing the production problem, which has caused the quality complaint. Companies usually have an internal set of pre-defined error codes, which represent potential error sources. These codes vary in terms of specificity, going from a generic (e.g., "packaging error") to a more precise (e.g., "lack of maintenance on machine 5") classification, depending on the information available at the time.

Correctly assessing each filed incident is a difficult and time-consuming task, especially for less experienced employees, who might not have the necessary knowledge. Using a machine learning approach, which is able to automatically analyze all past complaints in order to assist employees in correctly assessing their incidents, may not only reduce the number of wrong assessments, but also accelerate the process, such that the issue can be fixed more quickly. For this purpose, we develop a new approach based on a deep neural network to automatically assign a likely error code to a complaint. As input, the network receives free text as recorded in the 8D report and the error code as assigned by an employee. During training, the network learns which complaint characteristics are decisive for the classification. The trained network can then automatically submit proposals for an assessment to the responsible employee for newly arriving complaints. The general network infrastructure can be transferred to also address other classification issues, such as a criticality assessment.

In order to classify textual descriptions of complaints according to their likely error source, we use a recurrent neural network (RNN) with long short-term memory (LSTM) cells [19]. RNN layer cells feed information back into themselves, evolving their state by "forgetting" or "remembering" previous inputs. Our network consists of one input layer, one or more hidden LSTM layers, and one output layer. The input layer is responsible for generating a numerical representation of the input text, a so-called embedding. We use a pretrained embedding layer of English words [20] and allow the architecture to adapt the word embeddings to the specific context during training. For the hidden layers, we use LSTM cells, because they have been found particularly suitable to manage data with long-term dependencies, such as natural language. The output layer is a fully connected dense layer with a softmax activation, which transforms the activations of the last LSTM layer to the number of potential classes to obtain the probability distribution $\hat{y}$ over the classes.

Overall, our network architecture is a standard one for text classification problems. Our loss function L (Eq. 1), which we use for computing the gradient during training, is given by the categorical cross entropy for the expected output y and the predicted output $\hat{y}$ as well as an additional regularization loss. Given I the number of layers, C_i

the number of cells in layer i and A_c the activation of cell c, the regularization loss $L1$ is defined as the sum over all activations A_c of the hidden layers (Eq. 2). By regularizing the layer activations, we intend to prevent our model from overfitting. Furthermore, we use a dropout probability for the activations of each hidden layer to approach the problem of overfitting [21].

$$L(y, \hat{y}) = -\sum_{i=0}^{C} y_i * \log(\hat{y}_i) + \lambda * L1 \tag{1}$$

$$L1 = \sum_{i=1}^{I-1} \sum_{c=1}^{C_i} |A_c| \tag{2}$$

4 Technical Realization and Evaluation

4.1 Data Characteristics and Data Preparation

To evaluate our solution, we use the complaint management data of a globally operating medical technology company. It contains 15,817 customer complaints about products, including both mass products and products manufactured according to the customer's requirements. The complaints data contains sensitive information about the business processes and products of the manufacturer. Therefore, we cannot make the dataset publicly available. Resolving this issue would require semantically altering the data, resulting in an artificial dataset, which would counteract our goal to provide insights about the performance of machine learning in a real word business process.

Each complaint in our dataset represents a closed case. It consists of a textual description and an error code, which is manually set by the employee handling the complaint. The error code is a numerical representation of the assessment result. The dataset exhibits 186 different error codes. Table 1 compares the characteristics of the codes that occur in at least 500 cases with the remaining codes. The overview reveals that less than 6% of the codes account for more than 46% (7,311) of the cases.

Since a customer may either file a complaint by phone or by letter, the responsible employee summarizes the complaint and submits it to the information system handling the complaint process. The textual description of a complaint exhibits an average length of 122 words with a standard deviation of 124. The following description represents an exemplary complaint: "customer bought the product on 27 May 2019, he claims that the Velcro does not adhere anymore, he also claims that the problem did not occur in previous orders". The dataset contains 1,853,616 tokens, thereof 11,748 distinct words.

Table 1 depicts the number of distinct words that are contained in the textual description of the cases labeled with the same code. In addition, we show the number of distinct words that occur in cases exhibiting the same code but are not contained in any other case. Since machine learning-based classification approaches require sufficient data per class to perform well, we require a class to contain at least 500 samples to be considered for evaluation. Thus, we obtain ten classes that can be directly mapped to error codes and an additional eleventh class containing the samples of the remaining classes. Cases classified with code 1 to 10 are mapped to the respective classes 1 to 10, while the remaining cases exhibiting the codes 11 to 186 are mapped to class 0. Thus, each class is represented by at least 522 samples. To overcome the imbalance in the class distribution we randomly sample 522 cases from each class for the evaluation, resulting in an evaluation dataset of 5,742 samples.

Table 1. Dataset characteristics

Class		Cases	Distinct words	Distinct words compared to other classes
	All codes	15,817	11,748	
Class 1	Code 1	2,499	3,244	337
Class 2	Code 2	1,093	2,154	142
Class 3	Code 3	723	2,201	191
Class 4	Code 4	706	1,940	94
Class 5	Code 5	675	2,668	221
Class 6	Code 6	641	1,947	112
Class 7	Code 7	586	2,300	223
Class 8	Code 8	537	2,347	202
Class 9	Code 9	524	1,536	52
Class 10	Code 10	522	2,385	148
Class 0	Code 11–186	8,506	9,671	4,318

4.2 Evaluation Setup

To evaluate the robustness of a model, we divide the derived dataset into training, validation, and test splits. For each split, we ensure an even distribution of the eleven classes. The test split is composed of 5% (286 samples) of the dataset. From the remaining 5,456 samples, we generate 10 folds containing 4,911 training and 545 validation samples for cross validation. Table 2 shows the hyperparameters of our initial model described in Sect. 3 and the respective search space, whose permutations yield a total of 32 models to evaluate. We use the training splits to optimize the loss function of the models.

Table 2. Hyperparameter search space

Hyperparameter	Search space
LSTM-layers	{1, 2}
Hidden units	{16, 32}
L1 activity regularization (λ)	{0.00, 0.01}
Sequence padding	{100, 200}
Training epochs	{100, 200}
Dropout	{0.1}

The optimization is conducted using a stochastic strategy called Adam [22]. In addition, we perform the incremental optimization on training batches. Each model is trained separately on all ten folds of the training split and evaluated by measuring the accuracy on the respective validation split. Its overall validation accuracy is determined as the average across the ten accuracy values. We select the model with the best validation accuracy and use it to evaluate our approach on the previously unseen test split. Furthermore, we compare our LSTM classifier to a naïve classifier (nc) to assess the achieved performance. The naïve classifier uses a bag of words approach and the Jaccard similarity coefficient (Eq. 3) to map a sample input s to a class $i \in \{0, \ldots, 10\}$. Given a training set, the classifier generates a bag of words b_i for each class based on the words contained in the training samples labeled with class i. A sample s is assigned a class according to the maximum similarity coefficient between b_i and the Bag of Words s_{bow} derived from s (Eq. 4).

$$J(A, B) = \frac{|A \cap B|}{|A \cup B|} \tag{3}$$

$$nc(s) = \min\left(i | J(b_i, s_{bow}) = \max \bigcup_{i=0}^{10} \{J(b_i, s_{bow})\}\right) \tag{4}$$

In the following, we report the mean training, mean validation, and mean test accuracy, as well as the standard deviation for the top model across the 10 folds. Moreover, we compare it to the mean training and test accuracy achieved by the naïve classifier. Furthermore, we report an average confusion matrix of the top model as well as the respective standard deviation.

4.3 Implementation and Results

The presented classifier is implemented in Python 3.6.7 and is online available on GitHub[1]. The LSTM model was implemented, trained and evaluated using TensorFlow[2] version 1.13 and the integrated Keras API. The training of the 32 LSTM models was conducted on a machine with a Intel Xeon W-2175 CPU 2,50 GHz (28 threads), 128 GB RAM and an Nvidia GeForce GTX Titan X GPU.

Training Evaluation. We observed a mean training accuracy of 0.87 and a standard deviation of 0.10 for the 32 LSTM models. Thus, the 32 models are generally capable to fit the training data. Table 3 shows the hyperparameter configuration of the LSTM model that achieved the best validation accuracy. The model achieved a mean training accuracy of 0.94 and standard deviation of 0.02 across the 10 folds. The model achieved a mean validation accuracy of 0.59 and a standard deviation of 0.02.

The low standard deviations show the robustness of the model regarding the evaluation data set. Figure 2 depicts the training history for the 100 epochs. The left curves show the mean training and validation accuracy, while the curves on the right side

---

[1] https://github.com/phakeai/aicomplaint.

[2] https://www.tensorflow.org.

present the mean training and validation loss. While the training loss continuously decreases, the validation loss starts increasing from epoch 20 on. The steep training loss curve shows the capacity of the model to fit the training data. Although we apply dropout as means of regularization, we are not able to prevent the model from over-fitting. Moreover, increasing the degree of regularization by adding L1 activity regularization resulted in a decreased training (0.84) and validation accuracy (0.49).

Table 3. Hyperparameter configuration for the model exhibiting best validation accuracy

Hyperparameter	Search space
LSTM-layers	1
Hidden units	32
L1 activity regularization	0.00
Sequence padding	200
Training epochs	100
Dropout	0.1

Training the naïve classifier achieved a mean training accuracy of 0.41 and a standard deviation of 0.003 on the training folds. The training accuracy shows that the model is not able to fit the training set.

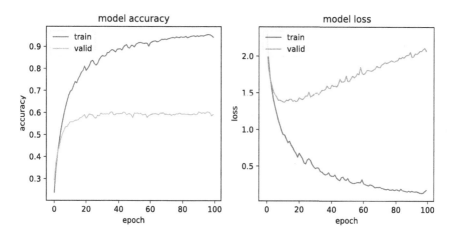

Fig. 2. Accuracy and loss for training and validation

Test Evaluation. The evaluation of the naïve classifier yields an accuracy of 0.21 and a standard deviation of 0.01 on the test data. Although there are distinct words within each class, the bag of words approach appears to be unsuitable for the classification task on our evaluation dataset.

The LSTM model achieved a mean test accuracy of 0.63 and a standard deviation of 0.01 across the 10 models trained for the different folds. The small deviation from the mean accuracy confirms the observed behavior on the validation splits during training.

Figure 3 presents the confusion matrix for the classification task on the test split. For class 7, we observe the highest true positive mean. Our model correctly classifies an average of 22.5 out of 26 samples within class 7, with a standard deviation of 1.3 samples. The lowest mean occurs in class 0, which is the class covering the complaint codes 11 to 186. With a standard deviation of 1.8, only 7.2 samples are classified correctly for this class. The first row of the matrix shows that the trained model is not able to properly distinguish class 0 from the remaining classes.

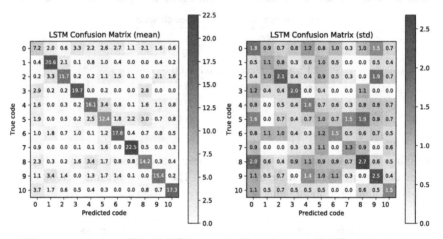

Fig. 3. Confusion matrices displaying the mean true positives and false positives (left) across the 10 folds and the according standard deviation (right)

5   Implications and Challenges

The proposed classifier achieves a mean accuracy of 0.63 on our test data set. Depending on the class predicted, the individual mean accuracy ranges from 0.27 to 0.87. Further models, different hyperparameter configurations as well as different sampling and training strategies could increase the overall accuracy of our approach. In our experiment, the diversity of class 0 (error codes 11 to 186) is covered by 522 samples, resulting in a mean of less than 3 samples per error code. Since the applied regularization yield a decreased trainings and validation accuracy, we assume that further data is required to generalize well. Incorporating further samples from the error codes 11 to 186 during training could increase the overall performance.

To balance our dataset, we applied an under-sampling strategy to the classes containing more than 522 samples. Further investigations should address different sampling strategies, e.g. over-sampling. Moreover, generating synthetic samples should be taken into consideration to augment and balance the dataset. The selection of the error codes for prediction is based on the cutoff of 500 samples. A different selection of error codes or a clustering of error codes are likely to influence the prediction performance. However, the optimal selection or clusters are not only determined by the achieved accuracy, but rather the usefulness in application scenarios. Thus, further constraints, e.g. misclassification costs, need to be considered.

Beside the challenges specific to improving the prediction model, there are challenges in the application of our concept in productive environments. In comparison to other industries and application scenarios, a wrongly classified complaint could have severe legal consequences. Therefore, we consider the prediction result of our model as a recommendation for the employee handling the complaint. Bearing in mind that the error codes represented by classes 1 to 10 account for almost 50% of the incoming complaints, a suggestion of an error code could significantly increase the employee's performance. Moreover, depending on the class predicted, a certainty value or a ranking of k top error codes could be provided to the employee.

There are limitations concerning the application of our approach to other companies. Our model requires a relatively large amount of data to deliver meaningful results. If a company does not have that amount of data, the application described in this work can only be realized to a limited extent. If there are few complaints in a company, it is difficult to correctly classify error codes. This also applies to cases with limited data quality, including incomplete descriptions or incorrectly classified complaints. In this case, our model will not be able to make reliable predictions.

Furthermore, the complaint process in medical technology is dependent on the nature of the individual product. For example, the same legal regulations apply to inexpensive commodity products such as patches and to complex medical devices such as ultrasound machines, but their complaint handling differs considerably. In this respect, our concept might not be applicable to every medical technology company. Therefore, a company needs to individually investigate whether our concept is beneficial in their respective complaint processes. This must be considered particularly given the fact that after the initial training the results of our model should regularly be supervised and re-trained with the appropriate current data so that the quality of the result can be maintained or ideally improved. Thus, the company requires an appropriate infrastructure consisting of computational power, hardware either on premise or as a service, as well as experts maintaining and developing the models.

Finally, we are aware that the work presented does not enable the reader to replicate the evaluation results, since the evaluation data is not publicly available. Nevertheless, by providing the concept implementation and a detailed description of the evaluation setup and experiments conducted, we provide the reader with the necessary information to reproduce the results in similar application scenarios.

6 Conclusion and Outlook

This paper deals with the potential of using machine learning to support complaint management in medical technology companies. Using data from a large manufacturer of medical products, we designed, implemented, and evaluated a deep neural network, which is able to assign the correct error source to a textual complaint description in more than 60% of all cases. We evaluated numerous network configurations to identify the network with the highest accuracy. Our approach was able to correctly classify three times as much samples as the naïve classifier which we used as a baseline comparison. These results show the general potential of machine learning for process automation in medical technology, considering that "classical" approaches such as a keyword search would require a priori knowledge about future customer-specific complaints. Also, ML techniques require very little domain knowledge, such that they can support instead of afflicting the experts. Determining whether full process automation is indeed possible and whether deep learning is in fact the best-suited technology remains as future work.

As we point out in the discussion, our approach still leaves room for optimizations, particularly if the goal is to apply it in a productive complaint management process. However, the domain experts of medical technology company, which provided the data, were already involved in the conceptual design and see much potential in its realization. As a next step, we will further optimize the network and the training data, with the goal to evaluate the approach with its end users in a productive environment.

The technical solution that we suggest can be transferred to assign other attributes to an incoming complaint. Examples are the criticality assessment or the actions necessary for immediate containment. Provided that enough data is available in the correct quality, many tasks in the complaint process can be completely or partially automated, with causality relations (e.g., stage 4) requiring methods for causal inference. Partial automation supports the responsible employees, leaving them with more time to identify and remove the causes of occurring complaints. This is particularly relevant for less experienced employees, since the necessary experience for quickly filing an 8D report can be at least partially replaced by a trained neural network.

However, decisions will not be completely automated for the foreseeable future. For example, the evaluation of the criticality of a complaint can be an extensive decision, which can lead to expensive recalls that may damage the reputation of the company. So, the final assessment will not be automated, but be carried out by an employee instead. This is also relevant for validation purposes. If a neural network made independent decisions within a process related to the manufacturing of medical products instead of just supporting the employees in their decisions, it would be regarded as a production-relevant system. On the one hand, those systems must be formally validated before companies are allowed to use them. On the other hand, neural networks do not behave deterministically, so their functionality can never be validated beyond all doubts.

Acknowledgement. This work was conducted within a project sponsored by the German Ministry for Education and Research (BMBF), project name "Reklamation 4.0", support code "01IS17088B". We also gratefully acknowledge the support of NVIDIA for the donation of a GPU used for this research.

References

1. Coussement, K., van den Poel, D.: Improving customer complaint management by automatic email classification using linguistic style features as predictors. Decis. Support Syst. **44**, 870–882 (2008)
2. Manz, S.: Medical Device Quality Management Systems: Strategy and Techniques for Improving Efficiency and Effectiveness. Elsevier, Amsterdam (2019)
3. European Parliament, Council of the European Union: Regulation (EU) 2017/745 of the European Parliament and of the Council of 5 April 2017 on medical devices

4. Lasi, H., Kemper, H.-G., Fettke, P., Feld, T., Hoffmann, M.: Industry 4.0. Bus. Inf. Syst. Eng. **6**(4), 239–242 (2014)
5. Foidl, H., Felderer, M.: Research challenges of industry 4.0 for quality management. In: Felderer, M., Piazolo, F., Ortner, W., Brehm, L., Hof, H.-J. (eds.) ERP 2015. LNBIP, vol. 245, pp. 121–137. Springer, Cham (2016). https://doi.org/10.1007/978-3-319-32799-0_10
6. European Commission: The rules governing medicinal products in the European Union - EU Guidelines to Good Manufacturing Practice
7. Abuhav, I.: ISO 13485: 2016: A Complete Guide to Quality Management in the Medical Device Industry. CRC Press, Boca Raton (2018)
8. Behrens, B.-A., Wilde, I., Hoffmann, M.: Complaint management using the extended 8D-method along the automotive supply chain. Prod. Eng. **1**, 91–95 (2007)
9. Tax, N., Verenich, I., La Rosa, M., Dumas, M.: Predictive business process monitoring with LSTM neural networks. In: Dubois, E., Pohl, K. (eds.) CAiSE 2017. LNCS, vol. 10253, pp. 477–492. Springer, Cham (2017). https://doi.org/10.1007/978-3-319-59536-8_30
10. Evermann, J., Rehse, J.-R., Fettke, P.: Predicting process behaviour using deep learning. Decis. Support Syst. **100**, 129–140 (2017)
11. Mehdiyev, N., Evermann, J., Fettke, P.: A novel business process prediction model using a deep learning method. Bus. Inf. Syst. Eng. (2018). https://doi.org/10.1007/s12599-018-0551-3
12. Nolle, T., Seeliger, A., Mühlhäuser, M.: BINet: multivariate business process anomaly detection using deep learning. In: Weske, M., Montali, M., Weber, I., vom Brocke, J. (eds.) BPM 2018. LNCS, vol. 11080, pp. 271–287. Springer, Cham (2018). https://doi.org/10.1007/978-3-319-98648-7_16
13. De Koninck, P., vanden Broucke, S., De Weerdt, J.: act2vec, trace2vec, log2vec, and model2vec: representation learning for business processes. In: Weske, M., Montali, M., Weber, I., vom Brocke, J. (eds.) BPM 2018. LNCS, vol. 11080, pp. 305–321. Springer, Cham (2018). https://doi.org/10.1007/978-3-319-98648-7_18
14. Hake, P., Zapp, M., Fettke, P., Loos, P.: Supporting business process modeling using RNNs for label classification. In: Frasincar, F., Ittoo, A., Nguyen, L.M., Métais, E. (eds.) NLDB 2017. LNCS, vol. 10260, pp. 283–286. Springer, Cham (2017). https://doi.org/10.1007/978-3-319-59569-6_35
15. Ko, T., Lee, J.H., Cho, H., Cho, S., Lee, W., Lee, M.: Machine learning-based anomaly detection via integration of manufacturing, inspection and after-sales service data. Ind. Manag. Data Syst. **117**, 927–945 (2017)
16. Weiss, S.M., Dhurandhar, A., Baseman, R.J.: Improving quality control by early prediction of manufacturing outcomes. In: International Conference on Knowledge Discovery and Data Mining, pp. 1258–1266. ACM (2013)
17. Lieber, D., Stolpe, M., Konrad, B., Deuse, J., Morik, K.: Quality prediction in interlinked manufacturing processes based on supervised & unsupervised machine learning. In: Conference on Manufacturing Systems, pp. 193–198 (2013). Procedia CIRP
18. Arif, F., Suryana, N., Hussin, B.: A data mining approach for developing quality prediction model in multi-stage manufacturing. Int. J. Comput. Appl. **69**, 35–40 (2013)
19. Hochreiter, S., Schmidhuber, J.: Long short-term memory. Neural Comput. **9**, 1735–1780 (1997)
20. Mikolov, T., Grave, E., Bojanowski, P., Puhrsch, C., Joulin, A.: Advances in pre-training distributed word representations. In: Proceedings of the International Conference on Language Resources and Evaluation (LREC 2018) (2018)
21. Srivastava, N., Hinton, G., Krizhevsky, A., Sutskever, I., Salakhutdinov, R.: Dropout: a simple way to prevent neural networks from overfitting. J. Mach. Learn. Res. **15**, 1929–1958 (2014)
22. Kingma, D.P., Ba, J.: Adam: a method for stochastic optimization. CoRR. abs/1412.6 (2014)

Resource Controllability of Workflows Under Conditional Uncertainty

Matteo Zavatteri[1]([⊠]), Carlo Combi[1], and Luca Viganò[2]

[1] Department of Computer Science, University of Verona, Verona, Italy
matteo.zavatteri@univr.it
[2] Department of Informatics, King's College London, London, UK

Abstract. An *access controlled workflow (ACWF)* specifies a set of tasks that have to be executed by authorized users with respect to some partial order in a way that all authorization constraints are satisfied. Recent research focused on weak, strong and dynamic controllability of ACWFs under conditional uncertainty showing that directional consistency is a way to generate any consistent assignment of tasks to users efficiently and without backtracking. This means that during execution we never realize that we would have chosen a different user for some previous task to avoid some constraint violation. However, dynamic controllability of ACWFs also depends on how the components of the ACWF are totally ordered. In this paper, we employ *Constraint Networks Under Conditional Uncertainty (CNCUs)* to solve this limitation, and provide an encoding from ACWFs into CNCUs to exploit existing controllability checking algorithms for CNCUs. We also address the execution of a controllable ACWF discussing which (possibly different) users are committed for the same tasks depending on what is going on (online planning).

Keywords: Access controlled workflow · Resource allocation under uncertainty · Online planning · Resource controllability · CNCU · Zeta · AI-based security · Business process compliance under uncertainty

1 Introduction

A *workflow (WF)* is a collection of tasks to be executed in some order to achieve one or more business goals. A *workflow management system (WfMS)* coordinates the execution of tasks and WF instances. An *access controlled workflow (ACWF)* augments a WF by adding users authorized for tasks and authorization constraints saying which combinations of assignments of tasks to users are permitted. When a WF specifies a set of temporal, conditional or authorization constraints and all of its components are under control we simply deal with a *satisfiability problem* asking us to find a fixed solution satisfying all constraints. Instead, when some component is out of control (e.g., task durations or conditional constraints) we deal with a *controllability problem*, where the synthesis of a fixed execution plan is not enough.

© Springer Nature Switzerland AG 2019
C. Di Francescomarino et al. (Eds.): BPM 2019 Workshops, LNBIP 362, pp. 68–80, 2019.
https://doi.org/10.1007/978-3-030-37453-2_7

Controllability implies the existence of a *strategy* to operate (possibly differently) on the part under our control depending on the behavior of some uncontrollable events that we will only be able to observe while executing. This means that, depending on how this uncontrollable part behaves, we might schedule the same tasks at different times or commit different users for the same tasks.

Controllability of *temporal workflows (TWFs)* (i.e., WFs augmented with task duration constraints, delays and deadlines) addresses uncontrollable task durations and conditional uncertainty (i.e., the uncontrollable choice of the WF path to take at runtime). A possible way to check controllability of TWFs is by encoding the TWF into a corresponding temporal network to boil down the controllability of the TWF to that of the temporal network for which controllability checking algorithms exist (see, for example, [2–4,10,12,15,18,19]).

Like TWFs, in ACWFs conditional uncertainty models the uncontrollable choice of the WF path to take during execution. For instance, when a patient comes to the ER, the severity of his condition is not known a priori but it is established by a physician *while* the WF is being executed. Since the result of this condition discriminates what tasks have, or have not, to be executed, and which users are committed for the same tasks, the system must be able to complete the WF by executing all relevant tasks and satisfying all relevant authorization constraints regardless of the result of (any combination) of uncontrollable conditions. When the assignment of tasks to users is generated while the WF executes we must never backtrack. In the real world, this means that we must avoid situations in which, if a patient is urgent, no physician is available because we chose to assign the "wrong" physician to some previous task.

In [17], an approach to address controllability of ACWFs is provided. That approach maps WF paths to *constraint networks (CNs*, [8]), relies on *directional consistency* to guarantee no backtracking when generating any solution to the underlying constraints satisfaction problem [9] and reasons on the intersection of the common parts of the WF paths to achieve a dynamic execution of the ACWFs. However, dynamic controllability of ACWFs also depends on how the components of the ACWF are ordered [20,21]. In [17], the designer encodes manually an ACWF into a CN meaning that it is up to him to choose a suitable total order for each WF path such that these WF paths can be intersected in their common parts to rule out assignments of tasks to users that never satisfy any solution. If the designer chooses the wrong order (even for one WF path only), either the controllability algorithm is not applicable or a controllable ACWF might be classified as uncontrollable.

This paper focuses on the validation and runtime execution of ACWFs under conditional uncertainty. Starting from the observation that dynamic controllability of ACWFs is a matter of order [20,21], our main contribution in this paper is an encoding from ACWFs into *Constraint Networks Under Conditional Uncertainty (CNCUs)* to achieve resource controllability of ACWFs under conditional uncertainty handling ordering issues automatically. As a result, we can validate and execute ACWFs by using ZETA, a tool that was developed for CNCUs.

We proceed as follows. Section 2 introduces a running example we use throughout the paper. Sections 3 and 4 provide background on CNCUs and

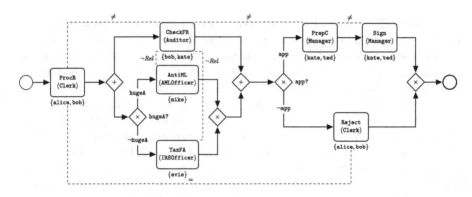

Fig. 1. Example of an ACWF in BPMN for a loan origination process. bob and mike are brothers. kate and evie are sisters.

ACWFs. Section 5 gives the encoding from ACWFs into CNCUs. Section 6 discusses the validation and execution of the running example with ZETA, a tool for CNCUs. Section 7 compares with related work. Section 8 sums up and discusses future work.

2   Running Example

We consider a simplified, synthetic, loan origination process (LOP) depicted in Fig. 1. The WF has 7 tasks, 5 roles (Clerk, Auditor, AMLOfficer, IRSOfficer and Manager) and 6 users (alice, bob, evie, kate, mike and ted). Clerk contains alice and bob, Auditor contains bob and kate, AMLOfficer contains mike only, IRSOfficer contains evie only, Manager contains kate and ted.

A Clerk starts the WF by processing a loan request (ProcR). After that, the flow of execution splits by entering an unconditional parallel block. In this block an Auditor checks the financial records of the customer (CheckFR) and at the same time another further verification takes place depending on if the amount of money requested is huge or not. If hugeA? = ⊤, then an AMLOfficer carries out an anti money laundering assessment (AntiML). If hugeA? = ⊥, an IRSOfficer carries out a simple tax fraud assessment. The Auditor who executes CheckFR must be different from the Clerk who executed ProcR, (as some users, e.g., bob, might belong to both roles) and must also not be a relative of the AMLOfficer who executed AntiML nor of the IRSOfficer who executed TaxFA (different is not necessary since the bank requires AMLOfficers and IRSOfficers to be external disjoint consultants). After that both the internal conditional block and the external parallel one (in this order) complete and the flow of execution enters a new conditional block to carry out the final tasks depending on if the loan has been approved or not. If app? = ⊤, then a Manager prepares the contract (PrepC) and another one, who must be different from the first, signs it (Sign). If app? = ⊥, then the same Clerk who executed the initial ProcR rejects the request (Reject). After that, the WF completes.

Our goal is that of executing the ACWF always satisfying the precedence and the authorization constraints no matter which workflow path will be taken.

3 Constraint Networks Under Conditional Uncertainty

Given a set $\mathcal{P}$ of Boolean propositions, a *label* $\ell = \lambda_1 \ldots \lambda_n$ is any finite conjunction of literals λ_i, where a literal is either a proposition p (positive literal) or its negation $\neg p$ (negative literal). The *empty label* is denoted by $\square$. The *label universe of* $\mathcal{P}$, denoted by $\mathcal{P}^*$, is the set of all possible (consistent) labels drawn from $\mathcal{P}$; e.g., if $\mathcal{P} = \{p, q\}$, then $\mathcal{P}^* = \{\square, p, q, \neg p, \neg q, p \wedge q, p \wedge \neg q, \neg p \wedge q, \neg p \wedge \neg q\}$. Two labels $\ell_1, \ell_2 \in \mathcal{P}^*$ are *consistent* if and only if their conjunction $\ell_1 \wedge \ell_2$ is satisfiable. A label ℓ_1 *entails* a label ℓ_2 (written $\ell_1 \Rightarrow \ell_2$) if and only if all literals in ℓ_2 appear in ℓ_1 too (i.e., if ℓ_1 is more *specific* than ℓ_2). For instance, if $\ell_1 = p \wedge \neg q$ and $\ell_2 = p$, then ℓ_1 and ℓ_2 are consistent since $p \wedge \neg q \wedge p$ is satisfiable, and ℓ_1 entails ℓ_2 since $p \wedge \neg q \Rightarrow p$.

Definition 1. *A* Constraint Network Under Conditional Uncertainty *(CNCU [20,21]) is a tuple* $\mathcal{Z} = \langle \mathcal{V}, \mathcal{D}, D, \mathcal{OV}, \mathcal{P}, O, L, \prec, \mathcal{C} \rangle$, *where:*

- $\mathcal{V} = \{V_1, V_2, \ldots\}$ *is a finite set of variables.*
- $\mathcal{D} = \{D_1, D_2, \ldots\}$ *is a finite set of discrete domains.*
- $D: \mathcal{V} \to \mathcal{D}$ *is a mapping assigning a domain to each variable.*
- $\mathcal{OV} \subseteq \mathcal{V} = \{P?, Q?, \ldots\}$ *is a set of* observation variables.
- $\mathcal{P} = \{p, q, \ldots\}$ *is a finite set of Boolean propositions whose truth values are all initially unknown.*
- $O: \mathcal{P} \to \mathcal{OV}$ *is a bijection assigning a unique observation variable $P?$ to each proposition p. When $P?$ executes, the truth value of p is randomly set (by Nature) and no longer changes.*
- $L: \mathcal{V} \to \mathcal{P}^*$ *is a mapping assigning a label ℓ to each variable V.*
- $\prec \subseteq \mathcal{V} \times \mathcal{V}$ *is a precedence relation on the variables. We write $(V_1, V_2) \in \prec$ (or $V_1 \prec V_2$) to express that V_1 is assigned a value before V_2.*
- $\mathcal{C}$ *is a finite set of labeled relational constraints of the form (R_S, ℓ), where $S \subseteq \mathcal{V}$ and $\ell \in \mathcal{P}^*$. If $S = \{V_1, \ldots, V_n\}$, then $R_S \subseteq D(V_1) \times \cdots \times D(V_n)$.*

$\mathcal{Z}$ *is well defined iff all labels are consistent and the following properties hold.*

(1) A label on a variable V is honest *if for each p or $\neg p$ in $L(V)$ where $p = O(P?)$, $L(V) \Rightarrow L(P?)$ and $P? \prec V$.*

(2) A label on a constraint (R_S, ℓ) is honest *if for each p or $\neg p$ in $L(V)$ where $p = O(P?)$, $\ell \Rightarrow L(P?)$ and it is also* coherent *if for each $V \in S$, $\ell \Rightarrow L(V)$.*

(3) For each pair $V_1, V_2 \in \mathcal{V}$, if $V_1 \prec V_2$, then $L(V_1) \wedge L(V_2)$ is consistent.

(1) and the first part of (2) say that labels on variables must contain the labels of the observation variables associated to each proposition embedded in each contained literal (label honesty). The second part of (2) says that a label on a constraint must be at least as expressive as any label in the scope of the relation (label coherence). (3) says that we cannot impose an order between two variables not taking part together in any execution.

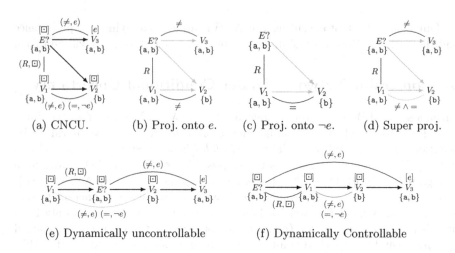

(a) CNCU.   (b) Proj. onto e.   (c) Proj. onto $\neg e$.   (d) Super proj.

(e) Dynamically uncontrollable   (f) Dynamically Controllable

Fig. 2. Weak, strong and dynamic controllability. $R = \{(\mathsf{a},\mathsf{a}),(\mathsf{a},\mathsf{b}),(\mathsf{b},\mathsf{b})\}$. (Color figure online)

A CNCU is *binary* if all constraints (R_S, ℓ) have scope cardinality $|S| \leq 2$. We graphically represent a binary CNCU as a *labeled constraint (multi)graph*, where each variable V is labeled by its label $L(V) \in \mathcal{P}^*$ and its domain $D(V) = \{v_1, \ldots, v_n\}$. Edges are of two kinds: *order edges* (directed unlabeled edges) and *constraint edges* (undirected labeled edges). An order edge $V_1 \to V_2$ models $V_1 \prec V_2$. A constraint edge between V_1 and V_2 models (R_{12}, ℓ). Many constraint edges may possibly be specified between the same pair of variables (or more conveniently, many labels on the same edge), as long as ℓ is different.

Figure 2a shows the graphical representation of a well-defined CNCU specifying 4 variables $E?, V_1, V_2, V_3$, where $D(E?) = D(V_1) = D(V_3) = \{\mathsf{a},\mathsf{b}\}$ and $D(V_2) = \{\mathsf{b}\}$. $E?$ is an observation variable whose associated Boolean proposition is e. Order edges (directed thick edges) say that $E?$ must be assigned a value before V_2 and V_3, whereas V_1 must be assigned a value before V_2. $E?, V_1$ and V_2 are always executed as $L(E?) = L(V_1) = L(V_2) = \square$. V_3 is executed if and only if e is assigned $\top$ as $L(V_3) = e$, ignored otherwise.

The CNCU specifies four constraints represented as labels on constraints edges (undirected thin edges). For example, $(R, \square)$ between $E?$ and V_1 (see caption) says that if $E? = \mathsf{a}$, then V_1 can be any value, whereas if $E? = \mathsf{b}$, then $V_1 = \mathsf{b}$. The constraint holds for any execution as its label is $\square$. Instead, $(\neq, e)$ between $E?$ and V_3 says that if e is assigned true, then $E? \neq V_3$. Likewise, if e is assigned $\top$, then $V_1 \neq V_2$, else $V_1 = V_2$.

In the rest of this section we give the main ideas underlying *weak*, *strong* and *dynamic* controllability of a CNCU and the corresponding checking algorithms.

We require that at least one total order on the variables exists for any CNCU. CNCUs not admitting any are uncontrollable for all three kinds of controllability. To get a total order meeting the restrictions of the partial one, we run topological sort on the graph made of all nodes and order edges only.

A *scenario* $s \colon \mathcal{P} \to \{\bot, \top\}$ is a total assignment of truth values to the propositions in $\mathcal{P}$. A scenario satisfies a label ℓ (in symbols $s \models \ell$) if ℓ valuates true under the interpretation given by s. We execute a variable by assigning it a value. We execute a CNCUs by executing all relevant variables. Variables and constraints are relevant for a scenario s if their labels are satisfied by s. A *projection* of a CNCU onto a scenario s is a classic constraint network (plus the partial order between the survived variables) in which we keep only variables and constraints relevant for s. For instance, Fig. 2b and c show the 2 possible projections of Fig. 2a onto $s(e) = \top$ and $s(e) = \bot$.

A CNCU is *weakly controllable* if any projection admits a total order on the variables and is consistent in the CN-sense. Figure 2a is weakly controllable. If $s(e) = \top$ (Fig. 2b), then $E? = V_1 = $ a and $V_2 = V_3 = $ b. If $s(e) = \bot$ (Fig. 2c), then $E? = $ a and $V_1 = V_2 = $ b. A CNCU is *strongly controllable* if the network obtained by wiping out labels on variables and constraints of the original CNCU admits a total order and is consistent in the CN sense. Figure 2a is not strongly controllable. The super projection (Fig. 2d) contains both $(\neq, e)$ and $(=, \neg e)$ between V_1 and V_2 (of the original CNCU) whose intersection yields an empty relation (red edge). For weak and strong controllability the choice of the total order does not play an important role (that's why we grayed it in Fig. 2b, c and d). Instead, dynamic controllability of CNCUs is a matter of order.

A CNCU is *dynamically controllable* if there exists a total ordering on the variables such that variables are assigned (possibly different) values while executing depending on the scenario being generated. The dynamic controllability checking runs an extension of the adaptive consistency algorithm (ADC, [8]) called LABELEDADC to address labels and iterates on all possible orders stopping as soon as a working one is found. Figure 2a is uncontrollable with respect to the order $V_1 \prec E? \prec V_2 \prec V_3$ mainly because V_1 is executed before having any information on the truth value of e. If $V_1 = E? = $ a and then $s(e) = \bot$ (or $V_1 = E? = $ b and then $s(e) = \top$), there is not valid value for V_2 satisfying $(=, \neg e)$ or $(\neq, e)$, respectively. Instead, the CNCU is dynamically controllable with respect to $E? \prec V_1 \prec V_2 \prec V_3$ (Fig. 2f) because executing $E?$ as first allows us to operate on the other variables with full information on e. A possible *execution strategy* is: $E? = $ a (always). If $s(e) = \top$, then $V_1 = $ a and $V_2 = V_3 = $ b, whereas if $s(e) = \bot$ then $V_1 = V_2 = $ b (when executing we ignore the variables and constraints having labels inconsistent with the scenario being generated). See [20,21] for a discussion on the complexity of the strategy synthesis algorithms.

4 Role-Based Access Controlled Workflows

A *role-based access control model* (*RBAC*, [13]) relies on the concept of *role* that acts as an interface between users and permissions. If a user changes his roles the security officer simply reassigns him to the new ones.

An *RBAC model for ACWFs* is a tuple $\langle$ Roles, Users, Tasks, UA, $TA \rangle$, where Roles $= \{r_1, \ldots, r_n\}$, Users $= \{u_1, \ldots, u_m\}$ and Tasks $= \{t_1, \ldots, t_k\}$ are finite set of roles, users and tasks, respectively, $UA \subseteq $ Users $\times$ Roles and

$TA \subseteq$ Tasks $\times$ Roles are the user-role and task-role assignment relations, respectively. We write $users(t) = \{u \mid \exists r \in$ Roles, $(t,r) \in TA, (u,r) \in UA\}$ for the set of users authorized for t.

Table 1 (left column) shows the fragment of BPMN proposed in [17] to model loop-free ACWFs. Note that loops could be modeled by considering the unfolding of a maximum number of iterations where each iteration is modeled by a choice block (positive condition means iterate, negative one means stop). Tasks are labeled by a finite set of roles $\{r_1, \ldots, r_n\} \subseteq$ Roles meaning that $(t, r_1), \ldots, (t, r_n) \in TA$ (Table 1, 2nd row). Assigning roles to tasks models "who does what". Each mutual exclusive gateway is associated to a unique Boolean variable cond? whose truth value assignment is *out of control* (Table 1, 5th row).

However, classical RBAC models fail to specify security policies at user level such as *separation of duties (SoD)* and *binding of duties (BoD)*.[1] *Authorization constraints* address such an issue, are represented as undirected dashed edges between pairs of non-mutually exclusive tasks t_1, t_2 and are labeled by a finite set of binary relations $\{\rho_1, \neg\rho_2 \ldots\}$ over users ($\rho_i \subseteq$ Users $\times$ Users). Each relation may appear positive (ρ) or negative ($\neg\rho$). If $u_1 \in users(t_1)$ and $u_2 \in users(t_2)$ and (u_1, u_2) satisfies all $(\neg)\rho_i$ in the set, then any execution assigning t_1 to u_1 and t_2 to u_2 *satisfies* the authorization constraint (Table 1, last row).

5   Encoding ACWFs into CNCUs

Table 1 provides an encoding from the fragment of BPMN discussed in [17] into CNCUs, where we assume that wf models the WfMS. Despite such an encoding is quite restrictive with respect to the number of components, it paves the way for future extensions that can consider richer and richer fragments of BPMN.

The start and end of a process are encoded as two variables S and E occurring before and after all other variables, respectively. $L(S) = L(E) = \boxdot$ since the start and the end of a process always occur. wf is the unique authorized user for these variables. No constraint edge involves S and E (Table 1, row 1). A task t having authorized roles $r_1, \ldots, r_n$ is encoded as a homonymous variable whose domain consists of the union of all users belonging to $r_1, \ldots, r_n$ authorized for t; i.e., $users(t)$, whereas its label contains the propositions modeling the Boolean variables associated to the mutual exclusive gateways according to the nesting level of the block in which the task appears (Table 1, row 2). All arcs regulating the control flow are encoded as order edges (Table 1, row 3). Parallel and conditional blocks are straightforwardly encoded mirroring the partial order of the ACWFs in the CNCU. If the block is a parallel, a variable P_S models the parallel gateway, whereas a variable P_E models the corresponding join. $L(P_S) = L(P_E)$ according to the nesting level of the block in the ACWF. All labels of the variables modeling the components inside the block in the ACWF (if any) must entail $L(P_S)$ in the CNCU (Table 1, row 4). If a block is a choice then an observation variable P? having associated proposition p models the mutual

---

[1] SoD is a security policy saying that a subset of tasks must be carried out by different users, whereas BoD says that a subset of tasks must be carried out by the same user.

Table 1. Encoding ACWFs into CNCUs.

ACWF	CNCU
$\bigcirc \!\!\longrightarrow \cdots \longrightarrow\!\! \bigcirc$	$[\Box]$ S {wf} $\longrightarrow \cdots \longrightarrow$ $[\Box]$ E {wf}
$\begin{array}{c} t \\ (r_1,\ldots,r_n) \end{array}$	$[\ell]$ t $\{users(t)\}$
$\cdots \longrightarrow \cdots$	$\cdots \longrightarrow \cdots$
parallel gateway $\diamond{+}\ \cdots \blacktriangleright \cdots \diamond{+}$	$[\ell]$ P_S {wf} $\xrightarrow{[\ell\wedge\ldots]} \cdots \xrightarrow{[\ell\wedge\ldots]}$ $[\ell]$ P_E {wf}
cond $\cdots$; $\diamond\times$ cond? $\diamond\times$; $\neg$cond $\cdots$	$[\ell]$ $P?$ {wf} with $[\ell\wedge p\wedge\ldots]\to\cdots\top$ and $[\ell\wedge\neg p\wedge\ldots]\to\cdots\bot$; $[\ell]$ P_E {wf}
$\begin{array}{c} t_i \\ (r_m,\ldots,r_n) \end{array} \{\rho_1,\neg\rho_2,\ldots\} \begin{array}{c} t_j \\ (r_x,\ldots,r_k) \end{array}$	$\begin{array}{c}[\ell_i]\\ t_i \\ \{users(t_i)\}\end{array} \xrightarrow{(\rho_1\cap\neg\rho_2\ldots,\ \ell_i\wedge\ell_j)} \begin{array}{c}[\ell_j]\\ t_j \\ \{users(t_j)\}\end{array}$

exclusive gateway. P_E still models the corresponding join. Again, $L(P?) = L(P_E)$ but this time as well as entailing $L(P?)$, all labels of the variables modeling the components belonging to the true and false branch in the ACWF (if any) are augmented with p or $\neg p$, respectively in the CNCU (Table 1, row 5). All variables modeling gateways are all executed by wf. An authorization constraint between two non-mutually exclusive tasks is encoded as a constraint edge whose relation is the intersection of all relations appearing on the authorization constraint in the WF block and the label is the conjunction of the labels of the variables modeling tasks (Table 1, row 6). Likewise, despite we do not show it in Table 1, an authorization constraint involving n non-mutually exclusive tasks is encoded into a n-ary relation in the CNCU along with the conjunction of the labels of the corresponding variables.

Figure 3 (above) shows the CNCU encoding the ACWF in Fig. 1. S and E encode the start and end of the process. P_S and P_E encode the parallel gateway and corresponding join of the unconditional parallel block. $H?$ and H_E encode the mutual exclusive gateway and corresponding join of the leftmost conditional block (h models hugeA?). $A?$ and A_E encode the mutual exclusive gateway and corresponding join the rightmost conditional block (a models app?). ProcR, CheckFR, AntiML, TaxFA, PrepC, Sign and Reject model the homonymous tasks.

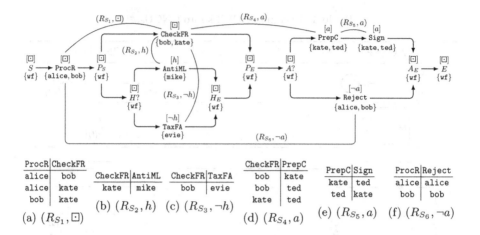

Fig. 3. CNCU encoding the ACWF in Fig. 1 (above). Relational constraints modeling the authorization constraints of the ACWF in Fig. 1 (below).

Figure 3 (below) shows the labeled relational constraints of the CNCU translating the authorization constraints of the ACWF in Fig. 1, where $S_1 = \{\text{ProcR, CheckFR}\}$, $S_2 = \{\text{CheckFR, AntiML}\}$, $S_3 = \{\text{CheckFR, TaxFA}\}$, $S_4 = \{\text{CheckFR, PrepC}\}$, $S_5 = \{\text{PrepC, Sign}\}$ and $S_6 = \{\text{ProcR, Reject}\}$ shorten the scopes.

6   Validation and Online Planning with Zeta

ZETA is a tool for CNCUs that takes in input a specification of a CNCU and acts as a solver for weak, strong and dynamic controllability as well as an executor simulator [20, 21]. We wrote the specification of the CNCU in Fig. 3 into ZETA's input language and run the tool on it to check weak, strong and dynamic controllability. ZETA proved that Fig. 3 is weakly, not strongly but dynamically controllable. Listing 1.1 shows ZETA validating and executing Fig. 3. Due to lack of space, we discuss dynamic controllability only. The strategy synthesized by ZETA for the CNCU in Fig. 3 corresponds to the following strategy for the ACWF in Fig. 1. First, ProcR = alice (always). Then, the WfMS executes the leftmost parallel gateway and the mutual exclusive gateway labeled by hugeA?. If hugeA? = ⊤, then CheckFR = kate and AntiML = mike. After that, the WfMS executes the corresponding joins and also the second mutual exclusive gateway labeled by app?. If app? = ⊤, then PrepC = ted and Sign = kate, whereas if app? = ⊥, then Reject = alice. The process concludes with the WfMS executing the last join. Instead, beside the execution of the gateways (which is the same), if hugeA? = ⊥, then CheckFR = bob and TaxFA = evie. After that, if app? = ⊤, then PrepC = ted and Sign = kate (or vice versa), whereas if app? = ⊥, then Reject = alice. The important thing is to execute the mutual exclusive gateway labeled by hugeA? *before* executing CheckFR (as being in a parallel block, CheckFR could be executed before the gateway). ACWFs like that in Fig. 1 are witnesses of the ordering problem. The implementation of this case study is available at https://github.com/matteozavatteri/zeta.

Listing 1.1. Validation and execution of the CNCU in Fig. 3 with ZETA.

```
1  $ java -jar zeta.jar LOP.cncu --DCchecking LOP.dynamic.ob
2  Dynamically Controllable
3  $ java -jar zeta.jar LOP.cncu --execute LOP.dynamic.ob
4  ================================================================
5  Order: S -> ProcR -> PS -> H -> CheckFR -> AntiML -> TaxFA -> HE -> PE ->
6         A -> PrepC -> Sign -> Reject -> AE -> E
7  ----------------------------------------------------------------
8  S = wf             # The WfMS starts the process
9  ProcR = alice      # Alice processes the request
10 PS = wf            # The WfMS executes the leftmost parallel gateway
11 H = wf, h = false  # The WfMS executes the 1st mutual exclusive gateway
12 CheckFR = bob      # Bob checks the financial records
13 TaxFA = evie       # Evie carries out the tax fraud assessment
14 HE = wf            # The WfMS executes the 1st mutual exclusive join
15 PE = wf            # The WfMS executes the parallel join
16 A = wf, a = true   # The WfMS executes the 2nd mutual exclusive gateway
17 PrepC = ted        # Ted prepares the contract
18 Sign = kate        # Kate signs the contract
19 AE = wf            # The WfMS executes the 1st mutual exclusive join
20 E = wf             # The WfMS ends the process
21 Verifying ... SAT!
22 ================================================================
```

7 Related Work

The problem of verifying WF features related to the assignment of tasks to users is known in literature as WF satisfiability and resiliency [14]. More specifically, the *workflow satisfiability problem (WSP)* is the problem of finding an assignment of tasks to users (i.e., a plan) such that the execution of the WF gets to the end satisfying all authorization constraints. The *workflow resiliency problem* is WSP under the uncertainty that a maximum number of users may become (temporally) absent before or during execution (see [22] for a recent controller synthesis approach). In this work, we dealt with a *dynamic* WSP encoding an ACWF into a CNCU for both checking controllability and executing the WF.

In [1], Cabanillas et al. deal with resource allocation for business processes. They consider an RBAC environment and they do not impose any particular order on activities. They also address loops but their approach does not address *History-Based Allocation* of resources. This work addresses history based allocation of resources exploiting results from dynamic controllability of CNCUs.

In [17], Zavatteri et al., defined and proposed an initial approach to check weak, strong and dynamic controllability for ACWFs under conditional uncertainty (w.r.t. total orders) by mapping WF paths to CNs and reasoning on the intersection of common parts. That work pointed out that dynamic controllability might depend on the chosen total order but did not investigate it further. Later, Zavatteri and Viganò confirmed that hypothesis in [20,21] with the proposal of CNCUs. In this work we encoded ACWFs into CNCUs to exploit existing algorithms and software in order to handle such an issue automatically.

In [7], *Simple temporal networks with uncertainty (STNUs,* [11]) are extended
with security constraints in order to model temporal role-based ACWFs in
which authorization constraints and temporal constraints mutually influence
one another. Dynamic controllability checking is addressed via Timed Game
Automata (TGAs) in [6] with the proposal of Conditional Simple Temporal Networks with Uncertainty and Resources (CSTNURs). Weak and strong controllability are not addressed. *Access-Controlled Temporal Networks (ACTNs* [5])
address ACWFs under conditional and temporal uncertainty simultaneously.
Weak and strong controllability are not addressed. Dynamic controllability is
addressed again via TGAs. This work does not address temporal constraints
(in a quantitative sense) but allows for handling all possible users assignments
(w.r.t. the total order of the components) during execution whereas CSTNURs
and ACTNs don't. A summary of temporal and resource controllability in the
BPM context is in [16].

8   Conclusions and Future Work

We addressed weak, strong and dynamic controllability of ACWFs under conditional uncertainty considering the total ordering problem. We provided an encoding from ACWFs into CNCUs, a recent proposed formalism of CNs addressing
conditional uncertainty. We mapped tasks and gateways to variables. We mapped
mutual exclusive gateways and their associated Boolean variables to observation
variables and propositions. We mapped the partial order to order edges and
authorization constraints to constraint edges. We mapped authorized users to
the domains of the variables. In this way, the encoding ACWFs to CNCUs allows
us to exploit all the algorithms, methods and software available for CNCUs. We
discussed an example affected by the total ordering problem, we encoded it into
a CNCU, we checked the three kinds of controllability and we executed it with
ZETA, a tool for CNCUs. Controllability analysis validates an ACWF once and
executes it anytime. Currently, we are aware that there may exist ACWFs that
are dynamically controllable if and only if the order in which the components
are assigned resources is dynamic. In this more general concept of dynamic controllability precomputing a total ordering on the components leads to an incomplete approach. Therefore, as future work, we plan to investigate such a class of
ACWFs other than addressing a richer fragment of BPMN.

References

1. Cabanillas, C., Resinas, M., del-Río-Ortega, A., Cortés, A.R.: Specification and
   automated design-time analysis of the business process human resource perspective. Inf. Syst. **52**, 55–82 (2015). https://doi.org/10.1016/j.is.2015.03.002
2. Cairo, M., et al.: Incorporating decision nodes into conditional simple temporal
   networks. In: TIME 2017. vol. 90, pp. 9:1–9:17. Schloss Dagstuhl-Leibniz-Zentrum
   fuer Informatik (2017). https://doi.org/10.4230/LIPIcs.TIME.2017.9

3. Combi, C., Gambini, M., Migliorini, S., Posenato, R.: Representing business processes through a temporal data-centric workflow modeling language: an application to the management of clinical pathways. IEEE Trans. Syst. Man Cybern. Syst. **44**(9), 1182–1203 (2014). https://doi.org/10.1109/TSMC.2014.2300055
4. Combi, C., Posenato, R.: Controllability in temporal conceptual workflow schemata. In: Dayal, U., Eder, J., Koehler, J., Reijers, H.A. (eds.) BPM 2009. LNCS, vol. 5701, pp. 64–79. Springer, Heidelberg (2009). https://doi.org/10.1007/978-3-642-03848-8_6
5. Combi, C., Posenato, R., Viganò, L., Zavatteri, M.: Access controlled temporal networks. In: ICAART 2017. ScitePress (2017). https://doi.org/10.5220/0006185701180131
6. Combi, C., Posenato, R., Viganò, L., Zavatteri, M.: Conditional simple temporal networks with uncertainty and resources. J. Artif. Intell. Res. **64**, 931–985 (2019). https://doi.org/10.1613/jair.1.11453
7. Combi, C., Viganò, L., Zavatteri, M.: Security constraints in temporal role-based access-controlled workflows. In: CODASPY 2016. ACM (2016). https://doi.org/10.1145/2857705.2857716
8. Dechter, R.: Constraint Processing. Elsevier, Amsterdam (2003)
9. Dechter, R., Pearl, J.: Network-based heuristics for constraint-satisfaction problems. Artif. Int. **34**(1), 1–38 (1987)
10. Eder, J., Franceschetti, M., Köpke, J.: Controllability of orchestrations with temporal SLA: encoding temporal XOR in CSTNUD. In: IIWAS 2018, pp. 234–242. ACM (2018). https://doi.org/10.1145/3282373.3282398
11. Morris, P.H., Muscettola, N., Vidal, T.: Dynamic control of plans with temporal uncertainty. IJCAI **2001**, 494–502 (2001)
12. Posenato, R., Zerbato, F., Combi, C.: Managing decision tasks and events in time-aware business process models. In: Weske, M., Montali, M., Weber, I., vom Brocke, J. (eds.) BPM 2018. LNCS, vol. 11080, pp. 102–118. Springer, Cham (2018). https://doi.org/10.1007/978-3-319-98648-7_7
13. Sandhu, R.S., Coyne, E.J., Feinstein, H.L., Youman, C.E.: Role-based access control models. Computer **29**(2), 38–47 (1996). https://doi.org/10.1109/2.485845
14. Wang, Q., Li, N.: Satisfiability and resiliency in workflow authorization systems. ACM Trans. Inf. Syst. Secur. **13**(4), 40 (2010)
15. Zavatteri, M.: Conditional simple temporal networks with uncertainty and decisions. In: TIME 2017. vol. 90, pp. 23:1–23:17. Schloss Dagstuhl-Leibniz-Zentrum fuer Informatik (2017). https://doi.org/10.4230/LIPIcs.TIME.2017.23
16. Zavatteri, M.: Temporal and resource controllability of workflows under uncertainty. In: Proceedings of the Dissertation Award, Demonstration, and Industrial Track at BPM 2019, vol. 2420, pp. 9–14. CEUR-WS.org (2019)
17. Zavatteri, M., Combi, C., Posenato, R., Viganò, L.: Weak, strong and dynamic controllability of access-controlled workflows under conditional uncertainty. In: Carmona, J., Engels, G., Kumar, A. (eds.) BPM 2017. LNCS, vol. 10445, pp. 235–251. Springer, Cham (2017). https://doi.org/10.1007/978-3-319-65000-5_14
18. Zavatteri, M., Combi, C., Rizzi, R., Viganò, L.: Hybrid sat-based consistency checking algorithms for simple temporal networks with decisions. In: TIME 2019. vol. 147, pp. 2:1–2:17. Schloss Dagstuhl-Leibniz-Zentrum fuer Informatik (2019). https://doi.org/10.4230/LIPIcs.TIME.2019.2
19. Zavatteri, M., Viganò, L.: Conditional simple temporal networks with uncertainty and decisions. Theoret. Comput. Sci. (2018). https://doi.org/10.1016/j.tcs.2018.09.023

20. Zavatteri, M., Viganò, L.: Constraint networks under conditional uncertainty. In: ICAART 2018. pp. 41–52. SciTePress (2018). https://doi.org/10.5220/0006553400410052
21. Zavatteri, M., Viganò, L.: Conditional uncertainty in constraint networks. In: van den Herik, J., Rocha, A.P. (eds.) ICAART 2018. LNCS (LNAI), vol. 11352, pp. 130–160. Springer, Cham (2019). https://doi.org/10.1007/978-3-030-05453-3_7
22. Zavatteri, M., Viganò, L.: Last man standing: static, decremental and dynamic resiliency via controller synthesis. J. Comput. Secur. 27(3), 343–373 (2019). https://doi.org/10.3233/JCS-181244

Automated Multi-perspective Process Generation in the Manufacturing Domain

Giray Havur[1,2]($\boxtimes$), Alois Haselböck[1], and Cristina Cabanillas[2]

[1] Siemens AG Österreich, Corporate Technology, Vienna, Austria
{giray.havur,alois.haselboeck}@siemens.com
[2] Vienna University of Economics and Business, Vienna, Austria
{giray.havur,cristina.cabanillas}@wu.ac.at

Abstract. Rapid advances in manufacturing technologies have spurred a tremendous focus on automation and flexibility in smart manufacturing ecosystems. The needs of customers require these ecosystems to be capable of handling product variability in a prompt, reliable and cost-effective way that expose a high degree of flexibility. A critical bottleneck in addressing product variability in a factory is the manual design of manufacturing processes for new products that heavily relies on the domain experts. This is not only a tedious and time-consuming task but also error-prone. Our method supports the domain experts by generating manufacturing processes for the new products by learning the manufacturing knowledge from the existent processes that are designed for similar products to the new products. We have successfully applied our approach in the gas turbine manufacturing domain, which resulted in a significant decrease of time and effort and an increase of quality in the final process design.

Keywords: Automated process generation · Graph-based process generation · Resource assignment · Manufacturing process

1 Introduction

With the surging demand for individualized products, the need for flexible production facilities and intelligent manufacturing infrastructures is also increasing. The design of manufacturing processes becomes complex when the products consist of many different parts. The manufacturing industry is still heavily dependent on domain experts to design processes for assembly and production. The search for a feasible process is usually conducted manually by these experts starting with solutions based on analytical calculations and past experiences. The resultant processes are iteratively improved by making changes in the tasks, control-flow and assigned resources. In such a setting, process design becomes a tedious task which is prone to human errors and can lead to sub-optimal

---

Funded by the Austrian Science Fund (FWF) Elise Richter programme under agreement V 569-N31 (PRAIS).

C. Di Francescomarino et al. (Eds.): BPM 2019 Workshops, LNBIP 362, pp. 81–92, 2019.
https://doi.org/10.1007/978-3-030-37453-2_8

processes and a waste of valuable resources. This task also requires a lot of effort when the manufacturing knowledge must be collected from discussions with experts from different disciplines.

Many methods, techniques and tools have been developed to support process generation and decision making depending on the structure, availability, and values of product data [1–5]. However, most of them focus on informational products (where the emphasis is on the collection, processing and aggregation of information), and the automated generation of manufacturing processes for new physical products from the existing processes remains as a challenge.

To assist the domain experts with such a cumbersome task, we have developed a multi-perspective statistical method for generating manufacturing processes from the already existing processes for products which share similar features with the new product. The *functional* perspective is linked with selecting the relevant tasks to the new product, the *behavioral* perspective orders these tasks (i.e. the control flow is implemented), and the *organizational* perspective assigns the necessary resources to the tasks of the generated processes [6]. Therefore, the domain experts are provided with multi-perspective process alternatives that can be reviewed, validated and deployed. We have implemented and tested the approach in a real environment from the gas turbine manufacturing domain. Not only the time and potential cost of the design of the new manufacturing process were saved but also better quality processes happened to be designed following this software-aided approach.

The paper is structured as follows. Section 2 defines concepts needed to understand our approach. Section 3 introduces our process generation method and describes an example scenario that illustrates the method in action. Section 4 briefly describes how the approach has been validated. Section 5 provides a description of the state-of-the-art on related research. Finally, Sect. 6 outlines the conclusions drawn from the work along with its limitations and potential future steps.

2   Background

For the sake of understanding, in the following we briefly define the concepts (stemming from the manufacturing domain) that are involved in our process generation method:

Definition 1 (Product). A product p is a finished good that is manufactured. It consists of parts (materials) that can be described by features. The product parts are usually represented as a Bill of Materials (BoM). Such a product could be the result of a manual product design activity.

Definition 2 (Task). A task t is a discrete step in the manufacturing process of a product p such as a production operation acting on one or a set of input materials (e.g. drilling a hole into a workpiece or inserting a module into a slot of a base plate).

Definition 3 (Resource). A resource r is assigned to the task t for enabling its execution. It can be a production or transportation equipment, like a robot, a computer numerical control (CNC) machine, a 3D printer, or a conveyor belt. Human workers are modeled as resources as well.

Definition 4 (Manufacturing Process). A manufacturing process q is a sequence of tasks for manufacturing a product p. It is a directed path graph[1] where the vertices represent the tasks and the edges represent the control flow for manufacturing a product starting from raw or input materials. $q = (T, E)$, where $t_1, t_2, ..., t_n \in T$ are tasks, and each edge $(t_i, t_j) \in E$ represents a direct precedence relation between two tasks, meaning that task t_j starts after task t_i ends. The function $\gamma : (Q \times T) \to 2^R$ maps each task in a process to a set of possible resources that can execute the task.

Definition 5 (Feature). Each product is described by a set of features. Let F be the set of features of the already manufactured products. In the descriptions below, we use the following mappings:

$\phi_t : T \to 2^F$ maps a task t to the set of features that the task contributes to; $\phi_t(t) \subseteq F$.
$\phi_q : Q \to 2^F$ maps a process $q = (T, E)$ to the set of features that all of its tasks $T = \{t_1, t_2, \ldots t_n\}$ contribute to; $\phi_q(q) = \bigcup_{i=1..n} \phi_t(t_i)$.

A (product) feature is a property of the final product from the customer's point of view. For instance, a metal workpiece could be *hardened, colored,* or *perforated.* We assume that the mapping from tasks to features is labelled manually by domain experts or learned automatically by process mining tools [7] in the manufacturing ecosystem.

The problem space of producing a new product is defined by a set of features, while the manufacturing process with assigned resources represent the solution space. In our context, the relationship between features and tasks is especially important: a task *contributes to* the realization of one or more features. More than one task are typically involved in the realization of a feature.

Therefore, given a new product design *BoM*, a set of features F, a set of tasks T, and a set of resources R extracted from existing processes, the **Process Generation Problem** can be divided into the following sub-problems:

- *Task Selection Problem:* Find a subset T' of the set of tasks T $(T' \subseteq T)$ needed for manufacturing the new product design.
- *Process Sequencing Problem:* Find a total ordering relation on T'.
- *Resource Assignment Problem:* For each task of T', find a resource that is able to perform this task.

A solution to the process generation problem should fulfill several quality criteria. First of all, a solution must be *technically feasible,* satisfying technical

---

[1] A *path graph* is a graph in which the nodes can be listed in an order $t_1, t_2, \ldots, t_n$ such that the edges are (t_i, t_{i+1}) where $i = 1, 2, \ldots, n - 1$.

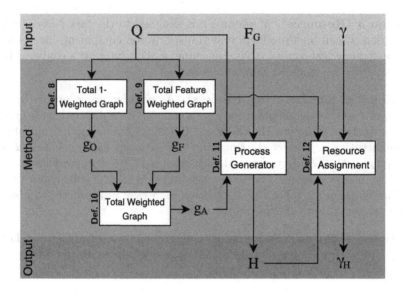

Fig. 1. Technical features diagram.

constraints (e.g. the transport of materials from one resource to the resource of the subsequent process must be possible, and a drilling machine for wood can only drill holes into wooden workpieces, not into metals). Second, a solution must be *effective* (i.e. all features of the new product must be covered by the selected tasks). Lastly, a solution must be *efficient*, which means that it should contain no- or only a minimum- number of tasks that do not contribute to the product features. *Technical feasibility* and *effectiveness* of generated processes depend on the quality of existing processes. The *length-optimality* of generated processes is imposed in the proposed method with an optimization function.

3   Automated Multi-perspective Process Generation

We start explaining our work with an overview of the technical features. We refer the reader to Fig. 1 for an overview of input-output relations of the definitions while reading this section. Our major component, the *Process Generator*, takes a set of manufacturing processes $Q = \{q_1, q_2, ..., q_n\}$ of products already manu-factured, a set of (goal) features F_G of a new product p, and a total weighted graph g_A which contains the task and weights learned from the existing processes and features. A solution to a process generation problem is a set of alternative processes $H = \{h_1, h_2, ..., h_m\}$, each of which is a process candidate for manu-facturing the new product p with features F_G. On the other hand, the *Resource Assignment* component requires Q, H, and a function γ that maps the tasks in the process set Q to their previously assigned resources. The solution to the resource assignment problem is a function γ_H, which learns the preferences of

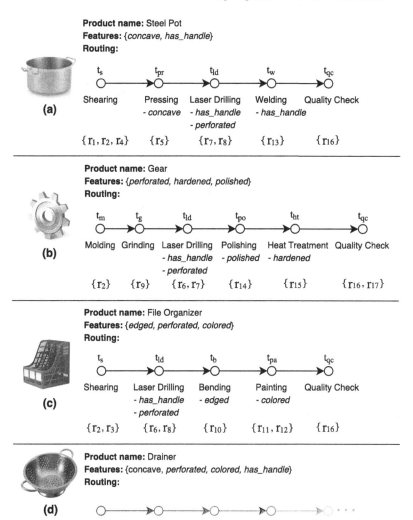

Fig. 2. The products being produced in the metal factory with their corresponding manufacturing processes: (a) Steel Pot with its process q_{sp}, (b) Gear with its process q_g, (c) File Organizer with its process q_{fo}, and the goal product (d) Drainer.

resources from γ and provides a set of resource-probability pairs for each task in H.

We motivate our approach with an example scenario for metalware production derived from the requirements from the gas turbine domain shown in Fig. 2. A metalware factory manufactures three different products: *Steel Pot* (Fig. 2a), *Gear* (Fig. 2b), and *File Organizer* (Fig. 2c). They would like to automatically find out how to produce a *Drainer* (Fig. 2d) by using the existing manufacturing knowledge. A process details this knowledge into a manufacturing process for each product (e.g. the process for producing *Steel Pot* $q_{sp} = (T_{sp}, E_{sp})$).

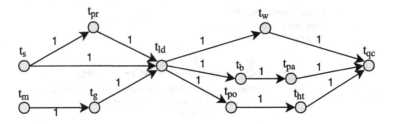

Fig. 3. Total 1-weighted graph.

Note that each task t is remarked by initials of its label (e.g. t_{ld} corresponds to the task *Laser Drilling*). The task set T_{sp} consists of the tasks t_s, t_{pr}, t_{ld}, t_w, and t_{qc}. The edge set E_{sp} consists of the edges (t_s, t_{pr}), (t_{pr}, t_{ld}), (t_{ld}, t_w), and (t_w, t_{qc}). The set of resources that can execute each task is shown below the task, e.g. $\gamma(q_{sp}, t_s) = \{r_1, r_2, r_4\}$. We assume that the tasks in the existing processes are labelled by the features that they contribute to. These features are listed beneath the tasks (e.g. $\phi_t(t_{ld}) = \{has\_handle, perforated\}$). The feature set of the process q_{sp} is $\phi_q(q_{sp}) = \{concave, has\_handle\}$. The feature set of the goal product *Drainer* is $F_G = \{concave, perforated, colored, has\_handle\}$.

Definition 6 (Task Filtering by Features). A function $l_t : 2^T \rightarrow 2^T$ selects all tasks of a set of tasks T that contribute to at least one feature of the goal features F_G:
$$l_t(T) = \{t \in T \mid \phi_t(t) \cap F_G \neq \emptyset\}$$

Definition 7 (Edge Filtering by Features). A function $l_e : 2^E \rightarrow 2^E$ selects all edges of a process (T, E) whose tasks contribute to at least one feature of the goal features F_G:
$$l_e(E) = \{e \in E \mid \phi_t(t_i) \cup \phi_t(t_j) \cap F_G \neq \emptyset \text{ where } e = (t_i, t_j)\}$$

Definition 8 (Total 1-Weighted Graph). A total 1-weighted graph $g_O = (T_O, E_O, w_O)$ is derived from a set of processes $Q = \{q_1, q_2, ..., q_n\}$, where $q_i = (T_i, E_i)$, in the following way:
$$T_O = T_1 \cup T_2 \cup ... \cup T_n$$
$$E_O = E_1 \cup E_2 \cup ... \cup E_n$$
$$w_O : E_O \rightarrow 1 \text{ (all edges have a weight of 1)}$$

Therefore, the total 1-weighted graph is the union of past processes with edge weight of 1. Figure 3 shows the total 1-weighted graph which is the union of processes in Fig. 2: q_{sp}, q_g, and q_{fo}.

Definition 9 (Total Feature Weighted Graph). A total feature weighted graph $g_F = (T_F, E_F, w_F)$ is derived from a set of processes $Q = \{q_1, q_2, ..., q_n\}$, where $q_i = (T_i, E_i)$, in the following way[2]:

---

[2] A *path* in a graph is a sequence of edges which connect a sequence of vertices that are all distinct from one another.

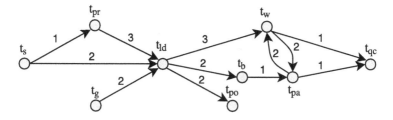

Fig. 4. Total feature weighted graph.

$$T_F = \text{tasks of edges } l_e(E_1) \cup l_e(E_2) \cup \ldots \cup l_e(E_n)$$
$$E_F = l_e(E_1) \cup l_e(E_2) \cup \ldots \cup l_e(E_n) \cup E_+$$
$$\text{where } E_+ = \{(t_i, t_j) \mid t_i, t_j \in l_t(T_F) \wedge \text{ no path between } t_i \text{ and } t_j\}$$

The weight function $w_F : E \to \mathbb{Z}_{\geq 0}$ is defined as

$$w_F(e) = \begin{cases} max(1, |(\phi_t(t_1) \cup \phi_t(t_2)) \cap F_G|) \text{ where } e = (t_1, t_2) \text{ if } e \in E_+ \\ |(\phi_t(t_1) \cup \phi_t(t_2)) \cap F_G| \text{ where } e = (t_1, t_2) \qquad \text{otherwise} \end{cases}$$

Following up on our running example, after deriving the total 1-weighted graph we introduce the total feature weighted graph g_F in Fig. 4, which favours the edges with tasks that have overlapping features with F_G. Let us consider the edge (t_{pr}, t_{ld}). Its weight is $w_F((t_{pr}, t_{ld})) = 3$, because t_{pr} and t_{ld} have 3 features of F_G (*concave*, *perforated*, and *has_handle*). Moreover, the tasks that have overlapping features with F_G, but with no existing path in between, are connected bidirectionally. For example, *Welding* t_w and *Painting* t_{pa} have features *has_handle* and *colored* which are in F_G, however they are not connected in g_O. Therefore, the edges (t_w, t_{pa}) and (t_{pa}, t_w) are introduced in E_+.

Definition 10 (Total Weighted Graph). Let $g_O = (T_O, E_O, w_O)$ and $g_F = (T_F, E_F, w_F)$ be the total 1-weighted graph and the total feature graph corresponding to a set Q of existing processes, respectively. The total weighted graph $g_A = (T_A, E_A, w_A)$ is defined as follows:

$$T_A = T_O$$
$$E_A = E_O \cup E_+$$
$$w_A(e) = \begin{cases} w_O(e) + w_F(e) & \text{if } e \in E_F \\ w_O(e) & \text{if } e \notin E_F \end{cases}$$

The total weighted graph in Fig. 5 is the aggregate graph of g_O and g_F as described above. This graph is the input graph for the process generation.

Definition 11 (Process Sequencing Problem). Let $g_A = (T_A, E_A, w_A)$ be the total weighted graph. Let $T_S \in T_A$ be the set of source tasks, i.e., tasks without incoming edges. Let $T_K \in T_A$ be the set of sink tasks, i.e., tasks without

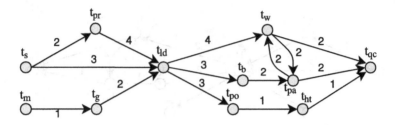

Fig. 5. Total weighted graph.

outgoing edges. The goal of process generator is to compute a set of process graphs $H = \{h_1, h_2, ..., h_n\}$ where any $h = (T, E) \in H$, $T = \{t_i, t_{i+1}, ..., t_j\}$, $E = \{(t_i, t_{i+1}), (t_{i+1}, t_{i+2}), ..., (t_{j-1}, t_j)\}$ has the following properties:

- h is acyclic,
- h starts from a source task $(t_i \in T_S)$ and ends at a sink task $(t_j \in T_K)$, and
- $\phi_q(h) \supseteq F_G$.

The minimization of an objective function $c(E)$ is desirable for ordering the generated process graphs:

$$c(E) = \sum_{e \in E} \frac{1}{w_A(e)}$$

Figure 6 shows three alternative generated processes h_1, h_2, and h_3 for manufacturing *Drainer*. The process with the minimum optimization function value h_1 is highlighted in the total weighted graph g_A. All the generated processes satisfies the properties in Definition 11, namely

- h_1, h_2, and h_3 are path graphs,
- t_s is a source task, and t_{qc} is a sink task, and
- $\phi_q(h_1) \supseteq F_G$, $\phi_q(h_2) \supseteq F_G$, and $\phi_q(h_3) \supseteq F_G$, i.e. the union set of features of each generated process is a superset or equal to F_G.

The processes are ranked with respect to the value of objective function, i.e. $c(E_1) < c(E_2) < c(E_3)$.

Definition 12 (Resource Assignment Problem). The Resource Assignment Problem is defined by:

- a set of processes $Q = \{q_1, q_2, ..., q_n\}$,
- a function $\gamma : (Q \times T) \to 2^R$ that maps a task t in a process q to its set of resources $R_{(q,t)}$,
- a set of generated processes $H = \{h_1, h_2, ..., h_m\}$.

A solution to a Resource Assignment Problem (Q, γ, H) consists of a function $\gamma_H : T \to 2^{(R \times [0..1])}$ that maps a task $t \in T$ to its set of resource-probability pair set $\{(r_i, p(r_i, t)), ..., (r_j, p(r_j, t))\}$, e.g. $p(r_i, t)$ reflects how likely the resource r_i is used for executing the task t.

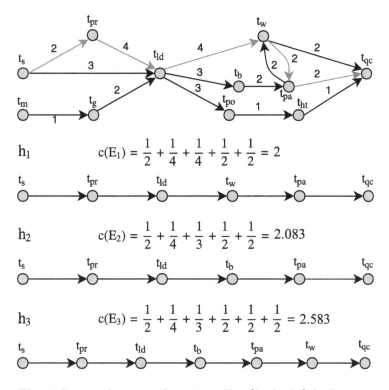

Fig. 6. Generated process alternatives $H = \{h_1, h_2, h_3\}$ for Drainer.

In order to obtain γ_H, we need two auxiliary definitions. First one is the set of all possible resources that can execute the task t:

$$R_t^* = \bigcup_{q=(T,E)\in Q} \bigcup_{t\in T} \gamma(q,t)$$

Then, we define the function $p(r,t)$ for returning the probability of resource r's executing task t:

$$p(r,t) = \frac{\sum_{q\in Q} \begin{cases} 1/|\gamma(q,t)| & \text{if } q \in \gamma(q,t) \\ 0 & \text{otherwise} \end{cases}}{\sum_{q\in Q} \begin{cases} 1 \text{ if } t \in T \text{ where } q = (T,E) \\ 0 \text{ otherwise} \end{cases}}$$

Finally, $\gamma_H(t)$ is defined as:

$$\gamma_H(t) = \bigcup_{r\in R_t^*} (r, p(r,t))$$

In our running example, after generating three alternative processes for the goal product *drainer*, we need to assign resources to tasks in the candidate processes so that they can become executable. For this purpose, we use the function $\gamma_H(t)$ which computes the probability of execution for each resource that has already executed the task t in the past processes. For example, the task t_s has been executed by the resources $\{r_1, r_2, r_4\}$ in q_{sp}, and by $\{r_2, r_3\}$ in q_{fo}. The resources $\{r_1, r_2, r_3, r_4\}$ are therefore feasible resources for executing t_s. However, we also know that r_2 has been assigned for executing t_s in both past processes, which should make r_2 a better option than others. We incorporate this knowledge by deriving probabilities from the assignment frequencies of these resources to tasks. For example, the resource assignment for t_s is $\gamma_H(t_s) = \{(r_1, p(r_1, t_s)), (r_2, p(r_2, t_s)), (r_3, p(r_3, t_s)), (r_4, p(r_4, t_s))\}$ where

$$p(r_1, t_s) = \frac{\frac{1}{3} + 0 + 0}{1 + 0 + 1} = 0.166 \qquad p(r_2, t_s) = \frac{\frac{1}{3} + 0 + \frac{1}{2}}{1 + 0 + 1} = 0.416$$

$$p(r_3, t_s) = \frac{0 + 0 + \frac{1}{2}}{1 + 0 + 1} = 0.25 \qquad p(r_4, t_s) = \frac{\frac{1}{3} + 0 + 0}{1 + 0 + 1} = 0.166$$

Hence, $\gamma_H(t_s) = \{(r_1, 0.166), (r_2, 0.416), (r_3, 0.25), (r_4, 0.166)\}$. This means r_2 is a good candidate for executing t_s, which is followed by r_3, and by equally r_1 and r_4. For assigning resources to the generated processes in Fig. 6, we need to compute $\gamma_H(t_s)$, $\gamma_H(t_{pr})$, $\gamma_H(t_{ld})$, $\gamma_H(t_w)$, $\gamma_H(t_{pa})$, $\gamma_H(t_{qc})$, and $\gamma_H(t_b)$.

4   Validation

We have successfully applied our approach in the turbine manufacturing domain to automatically generate processes for new gas turbine types that have specific modifications upon the existing gas turbine production processes. These modifications in product specifications (e.g. blades and vanes of different sizes) require different sequences of tasks (i.e. assembly and logistics) that are executed by specific resources. In this application, the process base (i.e. existing set of processes) has 173 existing processes, and a typical process has more than 30 tasks. We have generated 10 processes for different turbine configurations which have been reviewed by a domain expert. Afterwards, these generated processes have been used as input in the manufacturing simulation and validation tool Tecnomatix[3]. This simulation tool has also validated the feasibility of the generated processes from the functional, behavioral and organizational perspectives. The domain expert has experienced a considerable time saving in by employing our method in final process design task. Due to company policy, further comments in this section are omitted. Please contact the authors in case of any requests for instance details of the validation.

---

[3] https://www.plm.automation.siemens.com/global/de/products/tecnomatix/logistics-material-flow-simulation.html.

5   Related Work

Several methods have been proposed and developed to automatically explore and evaluate a range of different product designs, their options and parameters [8,9]. These methods ensure feasibility of the resulting products (i.e. their configuration must be consistent and complete regarding the constraints and requirements imposed in the product specifications).

Efforts have been made in modeling decision structures and dependencies to steer the execution of processes [4]. Such efforts are extended towards including product data [3] and for configuring processes [10] to offer flexibility in run-time process execution.

There are general process generation approaches which are taking into account data-flow elements of a process and the part-of relationships of products (i.e. BoMs) by ensuring an executable ordering of tasks that describe how to produce a goal product [1,2].

An alternative approach in automated process design stems from the research in Product Lifecycle Management (PLM). PLM tools support the modeling and representation of product portfolios. A product portfolio (or *product line*) stands for a whole set of product instances. A product instance (i.e. an individualized product) is usually derived from the product portfolio by a *product configuration* step with the help of a configuration tool. Recent research extends this product configuration by encompassing also the needed production processes for manufacturing the individualized product [11–14]. The main differences to the approach proposed in this work are: (i) Production processes are created only for instances of the product portfolio but not for completely new product designs, represented by a new vector of features; and (ii) Production operations for the different product parts are manually modeled and represented in the PLM tool or an extra knowledge base is used in the production workflow generation tool, whereas in our case such knowledge is derived from existing production processes of products.

6   Conclusions

We have introduced a novel method which derives a statistical model from the existing production processes to generate new processes with resource assignments for new products. Our method can be employed as a building block in a smart production ecosystem aiming at flexibility and automation. It has been validated in an industrial gas turbine production setting and proven to facilitate the process design efforts for new gas turbines.

As future work we plan to apply our method in other manufacturing settings (e.g. mobility and medical hardware) and also generalize the method towards the service operations to explore the capabilities of our method in a new domain.

References

1. van der Aalst, W.M.P.: On the automatic generation of workflow processes based on product structures. Comput. Ind. **39**(2), 97–111 (1999)
2. van der Aa, H., Reijers, H.A., Vanderfeesten, I.: Composing workflow activities on the basis of data-flow structures. In: Daniel, F., Wang, J., Weber, B. (eds.) BPM 2013. LNCS, vol. 8094, pp. 275–282. Springer, Heidelberg (2013). https://doi.org/10.1007/978-3-642-40176-3_23
3. Vanderfeesten, I., Reijers, H.A., Van der Aalst, W.M.P.: Product-based workflow support. Inf. Syst. **36**(2), 517–535 (2011)
4. Wu, F., Priscilla, L., Gao, M., Caron, F., De Roover, W., Vanthienen, J.: Modeling decision structures and dependencies. In: Herrero, P., Panetto, H., Meersman, R., Dillon, T. (eds.) OTM 2012. LNCS, vol. 7567, pp. 525–533. Springer, Heidelberg (2012). https://doi.org/10.1007/978-3-642-33618-8_69
5. Kluza, K., Nalepa, G.J.: Automatic generation of business process models based on attribute relationship diagrams. In: Lohmann, N., Song, M., Wohed, P. (eds.) BPM 2013. LNBIP, vol. 171, pp. 185–197. Springer, Cham (2014). https://doi.org/10.1007/978-3-319-06257-0_15
6. Cabanillas, C.: Process- and resource-aware information systems. In: International Conference on Enterprise Distributed Object Computing (EDOC), pp. 1–10 (2016)
7. van der Aalst, W.M.P.: Process Mining: Discovery, Conformance and Enhancement of Business Processes, vol. 2. Springer, Heidelberg (2011). https://doi.org/10.1007/978-3-642-19345-3
8. Paraguai, L., Candello, H., Costa, P.: Collaborative system for generative design: manipulating parameters, generating alternatives. In: Marcus, A., Wang, W. (eds.) DUXU 2017, Part I. LNCS, vol. 10288, pp. 727–739. Springer, Cham (2017). https://doi.org/10.1007/978-3-319-58634-2_52
9. Dhungana, D., Falkner, A., Haselböck, A., Taupe, R.: Enabling integrated product and factory configuration in smart production ecosystems. In: Proceedings of the 43rd Euromicro Conference on Software Engineering and Advanced Applications (SEAA), Vienna, Austria (2017)
10. Parody, L., Gómez-López, M., Varela-Vaca, A., Gasca, R.: Business process configuration according to data dependency specification. Appl. Sci. **8**(10), 2008 (2018)
11. de Silva, L., Felli, P., Chaplin, J.C., Logan, B., Sanderson, D., Ratchev, S.: Realisability of production recipes. In: European Conference on Artificial Intelligence (2016)
12. Dhungana, D., Haselböck, A., Taupe, R.: A marketplace for smart production ecosystems. In: Hankammer, S., Nielsen, K., Piller, F., Schuh, G., Wang, N. (eds.) Customization 4.0. SPBE, pp. 103–123. Springer, Cham (2018). https://doi.org/10.1007/978-3-319-77556-2_7
13. Dhungana, D., Falkner, A., Haselböck, A., Schreiner, H.: Smart factory product lines: a configuration perspective on smart production ecosystems. In: Proceedings of the 19th International Conference on Software Product Line, pp. 201–210. ACM (2015)
14. Campagna, D., Formisano, A.: Product and production process modeling and configuration. Fundam. Inform. **124**(4), 403–425 (2013)

Data-Driven Workflows for Specifying and Executing Agents in an Environment of Reasoning and RESTful Systems

Benjamin Jochum, Leonard Nürnberg, Nico Aßfalg, and Tobias Käfer[✉]

Institute AIFB, Karlsruhe Institute of Technology (KIT), Karlsruhe, Germany
{uzebb,ujeng,uberq}@student.kit.edu, tobias.kaefer@kit.edu

Abstract. We present an approach to specify and execute agents on Read-Write Linked Data that are given as Guard-Stage-Milestone workflows. That is, we work in an environment of semantic knowledge representation, reasoning and RESTful systems. For specifying, we present a Guard-Stage-Milestone workflow and instance ontology. For executing, we present operational semantics for this ontology. We show that despite different assumptions of this environment in contrast to the traditional environment of workflow management systems, the Guard-Stage-Milestone approach can be transferred and successfully applied on the web.

The environment of the web is finally at a stage where hypermedia agents could be applied [5]: We see that dynamic, open, and long-lived systems are commonplace on the web forming a highly distributed system. For instance, microservices [16] build on the web architecture and provide fine-grained read-write access to business functions. Moreover, Internet of Things devices are increasingly equipped with web interfaces, see e.g. the W3C's Web of Things effort[1]. Furthermore, users' awareness for privacy issues leads to the decentralisation of social networks from monolithic silos to community- or user-hosted systems like SoLiD[2], which builds on the web architecture. The web architecture offers REST [8], or its implementation HTTP[3], as uniform way for system interaction, and RDF[4] as uniform way for knowledge representation, where we can employ semantic reasoning to integrate data. To facilitate agents in this environment called Read-Write Linked Data [3], we need to embrace the web architecture and find a suitable way to specify behaviour. As according to REST, the exchange of state information is in the focus on the web, we want to investigate a data-driven approach for specifying behaviour. Moreover, data-driven approaches to workflow modelling can be both intuitive and actionable, and hence are suited to a wide range of audiences with different experience with

---

[1] https://www.w3.org/2016/07/wot-ig-charter.html.
[2] https://solid.mit.edu/.
[3] http://tools.ietf.org/rfc/rfc7230.txt.
[4] http://www.w3.org/TR/rdf11-concepts/.

© Springer Nature Switzerland AG 2019
C. Di Francescomarino et al. (Eds.): BPM 2019 Workshops, LNBIP 362, pp. 93–105, 2019.
https://doi.org/10.1007/978-3-030-37453-2_9

information technologies [13]. Hence, we want to tackle the research question of *how to specify and execute agent behaviour in the environment of Read-Write Linked Data in a data-driven fashion?*

As the environment determines why different workflow approaches are used in different circumstances [7], we need to look at the particularities of Read-Write Linked Data, whose basic assumptions are fundamentally different from traditional environments where workflow technologies are applied, e.g. databases:

The absence of events in HTTP Of the many HTTP methods, there is no method to subscribe to events. Hence, for our Read-Write-Linked-Data native approach, we rely on state data and to resort to polling to get informed about changes in the environment.

Reasoning and querying under the Open-World Assumption While in databases, typically the closed-world assumption is made, i.e. we conclude from the absence of information that it is false, RDF is based on the open-world assumption. Hence, we have to explicitly model all options.

Mitigation strategies would introduce complexity or restrict the generality of our approach: **The absence of events** could, e.g., be addressed by (1) generating events from differences between state snapshots and to process these events, which would add unnecessary complexity if we can do without; (2) assuming server implementations that implement change events using the Web-Socket protocol, which would restrict the generality of our approach and, for uniform processing, would require clear message semantics, which, in contrast to HTTP, event-based systems do not have [8][5]. **The open-world assumption** could, e.g., be addressed by introducing assumptions such as negation-as-failure once a certain completeness class [12] has been reached, which also would add complexity.

In previous work, we defined ASM4LD, a model of computation for the environment of Read-Write Linked Data [14], which allows for rule-based specification of agent behaviour. Based on this model of computation, we provided an approach to specify flow-driven workflows [15]. In contrast, we present a data-driven approach in this paper. Hull et al. present the Guard-Stage-Milestone (GSM) approach [13], which serves as basis for our work. While GSM builds on events sent to a database, which holds the information model consisting in status and data attributes, in our approach, distributed components with web interfaces that supply state information hold the information model. In contrast to Pautasso, who presented an approach to retrofit REST into the BPEL approach to workflow modelling [17], our approach rather retrofits a workflow modelling approach into REST, here GSM.

---

[5] Note that a client's (polling, state-based) application logic can, without changes, benefit from HTTP/2 server push (events): such specific events have been standardised to allow a server to update a client's cached state representations.

Our approach consists in two main parts:

GSM Ontology We present an ontology to specify GSM workflows and instances in the ontology language RDFS. Using this ontology, we can specify, reason over, and query workflow models and instances at run-time.

Operational Semantics We present ASM4LD rules to execute workflow instances specified using our GSM ontology. To this end, we build on a Linked Data Platform container[6], i.e. a writable RESTful RDF data store, to store the status attributes, i.e. workflow instances in our ontology.

The paper is structured as follows: In Sect. 1, we survey related work. Next, in Sect. 2, we give basic definitions on which we build our approach. Subsequently, in Sect. 3, we give an example, which we use in our explanations. In Sect. 4, we present our main contributions. Next, we evaluate our approach Sect. 5 by showing its correctness and performance. Last, in Sect. 6, we conclude.

1   Related Work

Workflow Management Previous Workflow languages and also workflow management systems rely on events using Event-Condition-Action (ECA) rules, e.g. [4,13]. We base our approach on REST so there are no events to be processed. Instead we use state machines as an operational semantics to track the polled application state.

Web Services Web Services based on the WS-* standards, on which approaches such as BPEL are built, allow for arbitrary operations. In contrast, REST constrains this set [19,22]. Thus, extensions for BPEL have been proposed to include RESTful services [17,18]. Our approach however exploits the semantics of the constrained set of operations in REST.

Semantic Web Services OWL-S and WSMO mainly focus on descriptions of services and reasoning to allow agents to compose web services. WS-*-based semantic web services completely rely on events [11] while our approach is based on REST where there are no events.

Scientific Workflows Previous Scientific Workflow approaches often set their focus on the data flow between processes of a workflow [9,21]. Our approach however uses a data-driven approach for control flow.

Ontologies for Workflows In previous works, ontologies for describing workflows and processes have been developed, e.g. in various research projects like Super, ASG, among others. Those ontologies require more expressive reasoning or do not allow for execution.

Workflows in Linked Data In previous work, we developed WiLD to specify and execute flow-based behaviour descriptions in Linked Data [15]. In contrast, in this paper, we investigate data-centric behavior descriptions.

Temporal Description Logics We base our approach on ASM4LD [14], which combines the semantics of RDF and HTTP with Abstract State Machines [10],

---

[6] http://www.w3.org/TR/ldp/.

a model of computation based on evolving first-order structures. That is, to introduce temporal aspects into the Semantic Web, we do not extend description logics with time, for a survey see [1], but cover temporal aspects implicitly by relying on state change.

2   Preliminaries

In this section, we introduce the technologies on which we build our approach.

2.1   The Hypertext Transfer Protocol (HTTP)

HTTP (see Footnote 3) is a stateless application-level protocol, where clients and servers exchange request/response message pairs about resources that are identified using Uniform Resource Identifiers (URIs)[7] on the server. Requests are typed, and the type (i.e. the HTTP method) determines the semantics of both the request and the optional message body. We make extensive use of the GET request to request a representation of a resource, the PUT request to overwrite the representation of a resource, and the POST request to append to an existing collection resource.

2.2   The Resource Description Framework (RDF) and SPARQL

RDF (see Footnote 4) is a graph-based data model for representing and exchanging data based on logical knowledge representation. In RDF, we represent data as triples that follow the form *subject, predicate, object*. Such a triple defines a relation of type *predicate* between graph nodes *subject* and *object*. Multiple triples form an RDF graph. Things in RDF are identified globally using URIs (see Footnote 7), or document-locally using so-called blank nodes. Literals can be used to express values. In this paper, we use the following notation for RDF: As triples encode binary predicates, we write *rdf:type*(*:active, :State*) to mean the following triple in Turtle notation[8]: `:active rdf:type :State`. Moreover, for the special case of class assignments, we use unary predicates: That is, above triple becomes *:State*(*:active*). SPARQL[9] is a query language for RDF data and supports a so-called `ASK` queries that return `true` if a condition holds in an RDF graph.

2.2.1   Abstract State Machines for Linked Data (ASM4LD)

ASM4LD is an Abstract State Machine based operational semantics given to Notation3, a rule language for the semantic web [14]. In ASM4LD, we can encode

---

[7] http://tools.ietf.org/rfc/rfc3986.txt.

[8] Turtle allows for abbreviating URIs, where a colon separates the abbreviating prefix from the local name. The example uses the empty prefix, which denotes http://purl.org/gsm/vocab#. We refer to http://prefix.cc/ for other abbreviations.

[9] http://www.w3.org/TR/sparql11-query/.

Fig. 1. A small example of two stages with associated guard and milestone(s).

two types of rules: Derivation rules (to derive new knowledge) and request rules (which cause HTTP requests). Moreover, ASM4LD supports RDF assertions. In [14], we derived the operational semantics based on the semantics of HTTP requests, first-order logic, and Abstract State Machines. The operational semantics can be summarized in four steps to be executed in a loop, thus implementing polling:

1. Initially, set the working memory be empty.
2. Add the assertions to the working memory.
3. Until the fixpoint, evaluate on the working memory:
   (a) Request rules from which HTTP-GET requests follow. For the rules whose condition holds, make the HTTP requests add the data from the responses to the working memory.
   (b) Derivation rules. Add the thus derived data to the working memory.
   This way, we (a) obtain and (b) reason on data about the world state.
4. Evaluate all other request rules on the working memory, i.e. those rules from which PUT/POST/DELETE requests follow. Make the corresponding HTTP requests.
   This way, we enact changes on the world's state.

2.3 The Guard-Stage-Milestone Approach

The Guard-Stage-Milestone approach is an artifact-centric workflow meta-model, presented by Hull et al. in [13]. The key modelling elements for Guard-Stage-Milestone workflows are the following: The **Information Model** contains all relevant information for a workflow instance: *data attributes* maintain information about the system controlled by the workflow instance, and *status attributes* maintain control information such as how far the execution has already progressed. **Stages** contain a task (i.e. the actual activity, an unit of work to be done by a human or machine) and may be nested. **Guards** control whether a stage gets activated, i.e. the activity may execute, and are formulated sentries. **Sentries** are expressions in a condition language, e.g. Event-Condition-Action rules (**on** <event> **if** <condition> **then** <action>). Here, events may be incoming from the system, or be changes to status attributes. **Milestones** are objectives that can be achieved during execution, and are represented using boolean values. Milestones have achieving and invalidating sentries associated: If an achieving sentry is evaluated to *true*, the milestone is set to *achieved*. An invalidating sentry can set a milestone back to *unachieved*. An example can be found in Fig. 1.

To specify the operational semantics, [13] provides a set of six PAC rules. PAC rules are a variation of Event-Condition-Action rules and are described by an prerequisite, antecedent and consequent, respectively. Both prerequisite and antecedent range over the entire information model, and the consequent is an update to the status attributes. The rules can be subdivided into two categories: **explicit rules**, which accomplish the actual progress in a workflow instance, and **invariant preserving rules**, which perform "housekeeping" by, e.g., deactivating child stages if the parent has been deactivated.

3 Example

Figure 2 shows, using a fire alarm as an example, how a workflow execution proceeds. Every line represents a step. Line 2: when smoke has been detected, the start alarm stage is triggered. Line 3: the "close doors"-guard gets activated, which triggers the closing of the doors. Line 4: the invalidating sentry of m1 deactivates g2. Line 5: after all doors had been closed, somebody re-opens a door. This triggers the invalidation of the corresponding milestone m2b (all doors are closed). As a consequence, the alarm stage is re-triggered.

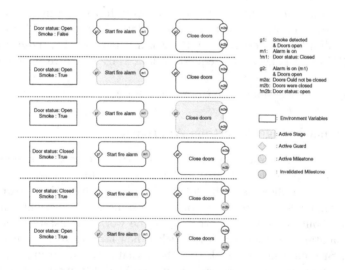

Fig. 2. Example of a workflow execution. Time progresses from top to bottom.

4 Proposed Approach

Our approach consist in an ontology to model GSM workflows and instances and operational semantics in rules with ASM4LD semantics, which interpret those workflows and instances. The rules can be directly deployed on a corresponding interpreter. We first present the ontology and then the operational semantics. We use the syntax described in Sect. 2.

4.1 Ontology for Modelling Entities

We built an ontology[10] to describe the core modelling primitives from [13]. We stay as closely as possible to their definitions and divert only if demanded by the our environment of Read-Write Linked Data: **Tasks** are HTTP requests as atomic activities. **Sentries** contain a SPARQL ASK query in SPIN notation[11]. Correspondingly, we use SPIN's **true** boolean SPARQL query result (we cannot use false, as we excluded negation in the introduction) with sentries and guards. We introduce the class **State** of all states to e.g. model three states of a stage: active, inactive, and done. Stages are set to done after they have been executed once.

In our approach, the information model is not contained in a database, but instead Read-Write Linked Data. Hence, we do not maintain the data attributes: We do not store information about the system we control, but instead, we retrieve the system state live in RDF over HTTP from the system itself. However, we do maintain the status attributes, for uniform access in RDF over HTTP. Thus, from now on we call all information regarding the state of the workflow *status information*. All other information about the system under control, and other relevant information from the environment, e.g. external services, we call *environment information*.

4.2 Operational Semantics

The following operational semantics are based on the PAC-rule-based semantics from [6]. We distinguish between setup and flow conserving (FC) rules. We assume all status information in an collection resource at http://ldpc.example/.

Instance Set-Up Rules. The basic condition for all setup rules is:

$$CS := :StageInstance(i) \land :isInstanceOf(i, s) \land :hasState(i, :uninitialized)$$

Workflow. The setup can be requested by publishing an unitialised instance:

$$CS \longrightarrow \text{PUT}(i, :SuperStageInstance(i) \land \cdots \land :hasState(i, :active))$$

Stage. Then, resources are created for all the model's sub-stages, linked to their counterpart in the model, and with state inactive.

$$CS \land :hasDescendantStage(s, s_{child})$$
$$\longrightarrow \text{POST}(\texttt{http://ldpc.example/}, :isInstanceOf(\texttt{new}, s_{child})$$
$$\land :inSuperStageInstance(\texttt{new}, i) \land :hasState(\texttt{new}, :inactive))$$

---

[10] http://purl.org/gsm/vocab.
[11] http://spinrdf.org/.

Milestone. Also, resources are created for all the model's milestones, linked to their counterpart in the model, and with state unachieved.

$$CS \land :hasDescendantStage(s, s_{child}) \land :hasMilestoneModel(s_{child}, m)$$
$$\longrightarrow \text{POST}(\texttt{http://ldpc.example/}, :isInstanceOf(\textbf{new}, m)$$
$$\land :inSuperStageInstance(\textbf{new}, i) \land :hasState(\textbf{new}, :unachieved))$$

Flow-Conserving Rules. The following condition has to match in every FCR.

$$FC := :SuperStageInstance(S^I) \land :StageModel(s^M) \land :StageInstance(s^I)$$
$$\land :isInstanceOf(s^I, s^M) \land :inSuperStageInstance(s^I, S^I) \land :hasState(S^I, :active)$$

FCR-1. For inactive stages, we check the guards. If a guard holds, the state of the stage is set to active and the task of the stage is executed.

$$FC \land :hasGuard(s^M, g) \land :hasHttpRequest(s^M, req) \land :allAncestorsActive(s^I, \textbf{true})$$
$$\land :hasState(s^I, :inactive) \land :hasCondition(g, c) \land sparql\text{-}results\text{:}boolean(c, \textbf{true})$$
$$\longrightarrow \text{PUT}(s^I, \ldots :hasState(s^I, :active))$$
$$\longrightarrow req(\cdot, \cdot)$$

FCR-2. Set an active stage done and its milestone achieved if the validating sentry holds.

$$FC \land :hasState(s^I, active) \land :hasMilestoneModel(s^M, m^M)$$
$$\land :hasValidatingSentry(m^M, c^a) \land :isInstanceOf(m^I, m^M)$$
$$\land sparql\text{-}results\text{:}boolean(c^a, \textbf{true})$$
$$\longrightarrow \text{PUT}(m^I, \ldots :isAchieved(m^I, \textbf{true}))$$
$$\longrightarrow \text{PUT}(s^I, \ldots :hasState(s^I, :done))$$

FCR-3. When an achieved milestone's invalidating sentry of a completed stage becomes true, three things are done by the following rule: (1) The invalidated milestone is set to unachieved (2) All other milestones of the stage are set to unachieved (3) The milestone's stage is set to inactive.

$$FC \land :hasMilestoneModel(s^M, m^M) \land :isInstanceOf(m^I, m^M) \land :isAchieved(m^I, \textbf{true})$$
$$\land :hasInvalidatingSentry(m^M, c) \land sparql\text{-}results\text{:}boolean(c, \textbf{true})$$
$$\land :hasMilestoneModel(s^M, m_2^M) \land :isInstanceOf(m_2^I, m_2^M) \land :isAchieved(m_2^I, \textbf{true})$$
$$\longrightarrow \text{PUT}(m^I, \ldots :isAchieved(m^I, \textbf{false}))$$
$$\longrightarrow \text{PUT}(m_2^I, \ldots :isAchieved(m_2^I, \textbf{false}))$$
$$\longrightarrow \text{PUT}(s^I, \ldots :hasState(s^I, :inactive))$$

FCR-6. The invalidation of an active stage's milestone sets all descendant stages to inactive and all of the stage's and its descendants' milestones to unachieved.

$$FC \wedge \mathit{:hasMilestoneModel}(s^M, m^M) \wedge \mathit{:isInstanceOf}(m^I, m^M) \wedge \mathit{:isAchieved}(m^I, \mathtt{true})$$

$$\wedge \mathit{:hasInvalidatingSentry}(m^M, c) \wedge \mathit{sparql\text{-}results\text{:}boolean}(c, \mathtt{true})$$

$$\wedge \mathit{:hasMilestoneModel}(s^M, m_2^M) \wedge \mathit{:isInstanceOf}(m_2^M, m_2^M)$$

$$\wedge \mathit{:hasDescendantStage}(s^M, s_d^M) \wedge \mathit{:isInstanceOf}(s_d^I, s_d^M)$$

$$\wedge \mathit{:hasMilestoneModel}(s_d^M, m_d^M) \wedge \mathit{:isInstanceOf}(m_d^I, m_d^M)$$

$$\longrightarrow \mathrm{PUT}(m_2^I, \ldots \mathit{:isAchieved}(m_2^I, \mathtt{false}))$$

$$\longrightarrow \mathrm{PUT}(s_2^I, \ldots \mathit{:hasState}(s_2^I, \mathit{:inactive}))$$

$$\longrightarrow \mathrm{PUT}(m_d^I, \ldots \mathit{:isAchieved}(m_3^I, \mathtt{false}))$$

5 Evaluation

In this section, we show the correctness of the proposed ruleset. We first show that the FCR rules cover the functionality of the PAC rules from [6]. Second we show that our approach does not violate the GSM invariants. This implies the correctness of the rules.

Discussion of Our Rules. PAC-1 activates a stage if its parent and one of its guards are active. FCR-1 completely matches this. PAC-2 sets a milestone to achieved if a milestone's achieving sentry changes to true and the stage the milestone is attached to is active. This also completely matches FCR-2. PAC-3 invalidates all milestones whose invalidating sentries are true. Again, FCR-3 covers this completely. PAC-4 resets milestones of a stage, if the stage's guards are triggered. We omit this rule for the following reasons: (1) the functionality of the rule regarding events is irrelevant for our approach, as we work without events; (2) the functionality of the rule regarding system state can be achieved by adding the guards' sentry as invalidating sentry to all attached milestones. PAC-5 deactivates a stage upon achievement of one of its milestones. We merged this functionality into FCR-2, as otherwise, they would not happen in the same step. PAC-6 determines that all stages nested into another are deactivated, as soon as the latter is deactivated. FCR-6 covers this.

Checking the GSM Invariants. Damaggio et al. [6] provide two invariants to ensure the consistency of their workflow model. To show that our approach does not violate the invariants, we first define the following:

- S is the set of stages
- G is the set of guards, $G_s \subseteq G$ all guards of a stage s
- M is the set of milestones, $M_s \subseteq M$ all milestones of a stage s,
- Φ is the set of Sentries, $\phi : M \rightarrow \{true, false\}$ represents the result of the sentry
- $d(s) = \{s_c \in S | s_c \text{ is child-stage of } s\}$,

- $\Sigma := \{active, inactive, done, achieved, unachieved\}$ is the set of possible states
- $\Sigma^* :=$ The set of all possible Snapshots,
- $f : \Sigma^* \rightarrow \Sigma^*$ representing the result after application of our ruleset,

For reasons of readability, $f(\sigma)$ with $\sigma \in \Sigma^*$ will be abbreviated with σ'. $\sigma(x)$ will be an abbreviation for the state of the instance (of a milestone or stage) x. Note that, as described, there are three possible states for stages: *active*, *inactive* and *done*.

Next, we present the two invariants and give a logical formulation:

GSM-1 "If a stage S owns a milestone m, then it cannot happen that both S is *active* and m has status *true*. In particular, if S becomes active then m must change status to *false*, and if m changes status to *true* then S must become *inactive*" [6].

$$\forall s \in S, m \in M_s : \sigma(s) = active \implies \sigma(m) \neq achieved$$
$$\wedge \sigma(m) = achieved \implies \sigma(s) \neq active$$

GSM-2 "If stage S becomes *inactive*, the executions of all substages of S also become *inactive*" [6].

$$\forall s \in S, s' \in d(s) : \sigma(s) = inactive \implies \sigma(s') = inactive$$

We now show that these invariants are not violated throughout the execution of one of our workflow models. $c : \Sigma^* \rightarrow \{true, false\}$ determines whether a state σ is conform to GSM-1 and 2.

Theorem: GSM-1 and GSM-2 sustain true throughout the workflow execution.

Proof: We apply a set of rules to our distributed information model. We prove the invariants' correctness using mathematical induction. One snapshot σ contains all data concerning the workflow's state, as well as environment values at the beginning of the loop iteration. σ_0 will represent our initial workflow state after the initialization.

Base Case: σ_0: No stage is activated yet $\implies c(\sigma) = true$. ✓

Step Case: $\sigma \rightarrow \sigma'$: In order to get into an inconsistent state one of the invariants must be violated. We will therefore distinct two cases: a violation of GSM-1 and a violation of GSM-2:

GSM-1 To investigate the consistency of the following state we will try to show of the contrary. Assume that there are rules that derive at least one triple so that $c(\sigma') = false$. To achieve that from a consistent state there must be a milestone m, and a stage with either (1) or (2) satisfied:

$$\sigma(m) = unachieved \wedge \sigma'(m) = achieved \wedge \sigma'(s) = active \tag{1}$$

$$(\sigma(s) = done \vee \sigma(s) = inactive) \wedge \sigma'(s) = active \wedge \sigma'(m) = achieved \tag{2}$$

Case (1): The only rule that sets a milestone to *achieved* (*true*) is FCR-2. The rule's condition requires an active stage. If then a corresponding milestone's achieving sentry is *true*, its milestone is set to achieved and FCR-5 is triggered which sets the stage to *done*. This leads to:

$$\phi_\sigma^a(m) = true \land \sigma(s) = active \implies \sigma'(m) = achieved \land \sigma'(s) = done \quad (3)$$

which contradicts to (1). ↯

Case (2): Assuming that (2) holds, there must be a rule that sets a stage's state to *active*, while the state of a milestone attached to it remains *active*. Only FCR-1 is able to change the state of a stage. Its condition requires the state to be inactive prior to being set *active*. If a stage's milestone is *achieved*, the stage is set to *done* instead of *inactive*. Therefore a stage can not be set to *active* while its milestone is *active*. This contradicts (2). ↯

GSM-2 Similar to GSM-1, we will show the contrary by assuming there are rules that derive at least one triple in such a way, that $c(\sigma) = false$. This requires

$$\exists s \in S, s_{child} \in d(s) : \sigma(s) = inactive \land \sigma'(s_{child}) = active \quad (4)$$

Due to FCR-1, if a stage has not been activated in the past, its children can not be active. This combination of states is only possible if a stage becomes inactive, while its children are still active. FCR-6 determines, that for all stages with an attached milestones' invalidating sentry triggered, all descendant stages are set to inactive. This leads to the following:

$$\sigma(s) = inactive \implies \forall s_{child}, s \in d(s) : \sigma'(s_{child}) = inactive \quad (5)$$

Again, we see a clear contradiction to (4). ↯ $\longrightarrow$ **Step case** ✓

These contradictions induce that the rules do not imply a transition from a consistent state σ to an inconsistent state σ', or $f(\sigma)$. □

5.1 Applicability

An implementation of our approach can be found online[12]. We use LDBBC[13] as Linked Data Platform Container implementation, and Linked Data-Fu[14] [20] as N3 rule interpreter with ASM4LD [14] operational semantics. We successfully applied the approach in an use-case with Internet of Things devices having Read-Write Linked Data interfaces [2].

6 Conclusion

We presented an approach to specify and execute agent specifications in the form of data-centric workflows in an environment of semantic knowledge representation and reasoning.

---

[12] http://github.com/nico1509/data-driven-workflows.
[13] http://github.com/kaefer3000/ldbbbc.
[14] http://linked-data-fu.github.io/.

As with all approaches that work on the web architecture, our agent relies on polling to get informed about the world state instead of the environment reporting events. Thus, if the polling rate is not high enough, this sampling approach may miss important system states. For instance, in the case of the Internet of Things, we may miss short button presses. Yet, we showed in [14] that our interpreter can indeed achieve high update rates in the range of milliseconds.

References

1. Artale, A., Franconi, E.: A survey of temporal extensions of description logics. Ann. Math. Artif. Intell. **30**(1–4), 171–210 (2000)
2. Aßfalg, N., Nürnberg, L., Jochum, B., Käfer, T.: Controlling Internet of Things devices with Read-write Linked Data interfaces using data-driven workflows. In: Proceedings of the Posters & Demos at the 15th SEMANTiCS (2019)
3. Berners-Lee, T.: Read-Write Linked Data. Design Issues (2009). http://www.w3.org/DesignIssues/ReadWriteLinkedData.html
4. Casati, F., Ceri, S., Pernici, B., Pozzi, G.: Deriving active rules for workflow enactment. In: Wagner, R.R., Thoma, H. (eds.) DEXA 1996. LNCS, vol. 1134, pp. 94–115. Springer, Heidelberg (1996). https://doi.org/10.1007/BFb0034673
5. Ciortea, A., Mayer, S., Gandon, F., Boissier, O., Ricci, A., Zimmermann, A.: A decade in hindsight: the missing bridge between multi-agent systems and the World Wide Web. In: Proceedings of the 18th AAMAS (2019)
6. Damaggio, E., Hull, R., Vaculín, R.: On the equivalence of incremental and fixpoint semantics for business artifacts with Guard-Stage-Milestone lifecycles. Inf. Syst. **38**(4), 561–584 (2013)
7. Elmroth, E., Hernández-Rodriguez, F., Tordsson, J.: Three fundamental dimensions of scientific workflow interoperability: model of computation, language, and execution environment. Futur. Gener. Comput. Syst. **26**(2), 245–256 (2010)
8. Fielding, R.: Architectural styles and the design of network-based software architectures. Ph.D. thesis, University of California, Irvine, USA (2000)
9. Gil, Y., Ratnakar, V., Deelman, E., Mehta, G., Kim, J.: Wings for pegasus: creating large-scale scientific applications using semantic representations of computational workflows. In: Proceedings of the 19th IAAI/22th AAAI (2007)
10. Gurevich, Y.: Evolving algebras 1993: lipari guide. In: Specification and Validation Methods. Oxford University Press (1995)
11. Haller, A., Cimpian, E., Mocan, A., Oren, E., Bussler, C.: WSMX – a semantic SOA. In: Proceedings of the ICWS (2005)
12. Harth, A., Speiser, S.: On completeness classes for query evaluation on linked data. In: Proceedings of 26th AAAI (2012)
13. Hull, R., et al.: Business artifacts with guard-stage-milestone lifecycles. In: Proceedings of the 5th DEBS (2011)
14. Käfer, T., Harth, A.: Rule-based programming of user agents for linked data. In: Proceedings of the 11th LDOW (2018)
15. Käfer, T., Harth, A.: Specifying, monitoring, and executing workflows in linked data environments. In: Vrandečić, D., et al. (eds.) ISWC 2018. LNCS, vol. 11136, pp. 424–440. Springer, Cham (2018). https://doi.org/10.1007/978-3-030-00671-6_25
16. Newman, S.: Building Microservices-Designing Fine-grained Systems. O'Reilly, Sebastopol (2015)

17. Pautasso, C.: RESTful web service composition with BPEL for REST. Data Knowl. Eng. **68**(9), 851–866 (2009)
18. Pautasso, C., Wilde, E.: Push-enabling RESTful business processes. In: Kappel, G., Maamar, Z., Motahari-Nezhad, H.R. (eds.) ICSOC 2011. LNCS, vol. 7084, pp. 32–46. Springer, Heidelberg (2011). https://doi.org/10.1007/978-3-642-25535-9_3
19. Pautasso, C., Zimmermann, O., Leymann, F.: Restful web services vs "Big" web services: making the right architectural decision. In: Proceedings of the 17th WWW (2008)
20. Stadtmüller, S., Speiser, S., Harth, A., Studer, R.: Data-Fu: a language and an interpreter for interaction with R/W linked data. In: Proceedings of the 22nd WWW (2013)
21. Turi, D., Missier, P., Goble, C., Roure, D.D., Oinn, T.: Taverna workflows: syntax and semantics. In: Proceedings of the 3rd e-Science (2007)
22. Zur Muehlen, M., Nickerson, J.V., Swenson, K.D.: Developing web services choreography standards—the case of REST vs SOAP. Decis. Support. Syst. **40**(1), 9–29 (2005)

The Changing Roles of Humans and Algorithms in (Process) Matching

Roee Shraga and Avigdor Gal[(✉)]

Technion – Israel Institute of Technology, Technion City, Haifa, Israel
shraga89@campus.technion.ac.il, avigal@technion.ac.il

Abstract. Historically, matching problems (including process matching, schema matching, and entity resolution) were considered semiautomated tasks in which correspondences are generated by matching algorithms and subsequently validated by human expert(s). The role of humans as validators is diminishing, in part due to the amount and size of matching tasks. Our vision for the changing role of humans in matching is divided into two main approaches, namely *Humans Out* and *Humans In*. The former questions the inherent need for humans in the matching loop, while the latter focuses on overcoming human cognitive biases via algorithmic assistance. Above all, we observe that matching requires unconventional thinking demonstrated by advanced machine learning methods to complement (and possibly take over) the role of humans in matching.

1 Introduction

The research of data integration spans over multiple decades, holds both theoretical and practical appeal, and enjoys a continued interest by researchers and practitioners. Efforts in the area include, among other things, process matching, schema matching, and entity resolution. Matching problems have been historically defined as a semi-automated task in which correspondences are generated by matching algorithms and subsequently validated by a single human expert. The reason for that is the inherent assumption that humans "do it better."

In recent years, there has been an evolution of data accumulation, management, analytics, and visualization (also known as *big data*). Big data is primarily about collecting volumes of data in an increased velocity from various sources and varying veracity. Big data is characterized by technological advancements such as Internet of things (accumulation), cloud computing (management), and deep learning (analytics). Putting it all together provides us with a new, exciting, and challenging research agenda.

Given the availability of data and the improvement of machine learning techniques, this line of research is devoted to the investigation of respective roles of humans and machines in achieving cognitive tasks in matching, aiming to determine whether traditional roles of humans and machines are subject to change [11]. Such investigation, we believe, will pave a way to better utilize both human and machine resources in new and innovative manners. We consider two

© Springer Nature Switzerland AG 2019
C. Di Francescomarino et al. (Eds.): BPM 2019 Workshops, LNBIP 362, pp. 106–109, 2019.
https://doi.org/10.1007/978-3-030-37453-2_10

possible modes of change, namely *humans out* and *humans in*. *Humans Out* aim at exploring out-of-the-box latent matching reasoning using machine learning algorithms when attempting to overpower human matcher performance. Pursuing out-of-the-box thinking, we investigate the best way to include machine and deep learning in matching. *Humans in* explores how to better involve humans in the matching loop by assigning human matchers with a symmetric role to algorithmic matcher in the matching process.

In following sections we describe each of the two modes of change. Section 2 describes how and where we envision replacing humans in the matching loop. In Sect. 3, we detail our approach to better involve humans in matching by understanding their strengths and weaknesses. Finally, we summarize and discuss future directions in Sect. 4.

2   Humans Out

The *Humans Out* approach seeks matching subtasks, traditionally considered to require cognitive effort, in which humans can be excluded. An initial good place to start is with the basic task of identifying correspondences. We note that many contemporary matching algorithms use heuristics, where each heurisitc associates some semantic cue to justify an alignment between elements. For example, string-based matchers use string similarity as a cue for item alignment. We observe that such heuristics, in essence, encode human intuition about matching. Our earlier work [12] showed that human matching choices can be reasonably predicted by classifying them into types, where a type correspond to an existing heuristic. Moreover, in our experiments, decision making of most human matchers can be predicted well using a combination of two algorithmic matchers. Therefore, we can argue that the cognitive effort of many human matchers can be easily replaced with such heuristics.

Moving to additional cognitive tasks, in [3,5] we suggested a learning algorithm for re-ranking top-K matches so that the best match is ranked at the top. The proposed algorithm has shown good results when tested on real-world as well as synthetic datasets, offering a replacement to humans in selecting the best match, a task traditionally reserved for human verifiers. The novelty of this line of work is in the use of similarity matrices as a basis for learning features, creating feature-rich datasets that fit learning and provide us with a feature aggregation that is needed to enrich algorithmic matching beyond that of human matching. To create a reranking framework, we adopt a learning-to-rank approach [2], utilizing previously suggested matching predictors [4,9] as features. We further show a bound on the size of K, given a desired level of confidence in finding the best match, justified theoretically and validated empirically. This bound is useful for top-K algorithms [7] and, as psychological literature suggests, also applicable when introducing a list of options (as in the traditional top-K setting) to humans [10]. In addition, we provide a set of novel features to complement state-of-the-art and show further improvement in matching and ranking quality.

3 Humans In

The *Humans In* approach aims at investigating whether the current role humans take in the matching process is effective (spoiler: it is not) and whether alternative role can improve overall performance of the matching process (spoiler: it can). A recent study [1], aided by metacognitive models, analyzes the consistency of human matchers. We explore three main consistency dimensions as potential cognitive biases, taking into account the time it takes to reach a matching decision, the extent of agreement among human matchers and the assistance of algorithmic matchers. In particular, we showed that when an algorithmic suggestion is available, humans tend to accept it to be true, in sharp contradiction to the conventional validation role of human matchers.

Interesting enough, all these dimensions were found predictive of both confidence and accuracy of human matchers. This indicates that (1) humans have cognitive biases affecting their ability to provide consistent matching decisions, and (2) that such biases has predictive value in determining to what extent a human matcher's alignment decision is accurate.

4 What's Next?

In this paper we presented our approach for human involvement in the matching loop, introducing tasks where humans can be replaced and emphasizing our envision for understanding human behavior which allows better engagement.

In future work, we intend to show that deep learning can also be applied to "small" matching problems such as process and schema matching, making extensive use of similarity matrices. We believe that applying non-linear transformations over a similarity matrix can reveal hidden relationships between elements that can be used for adjusting a given similarity matrix to better fit a matching problem.

We also aim to predict humans qualification to serve as "experts" for a matching task. We intend to explore predictive behaviors that capture the process of human matching by transforming physical aspects (such as time, screen scrolls, mouse tracking, and eye movement) into features that can be used for examining the role of humans in the matching process. This, in turn, would enable matching systems to carefully select a matching expert that fits the task.

An overarching goal is to propose a common matching framework that would allow treating matching as a unified problem whether we match process activities, schemata attributes, entity's tuples, *etc.* Using machine learning for this purpose immediately raises the issue of shortage of labeled data to offer supervised learning [5,6,8]. Hence, pursuing less-than-supervised (*e.g.*, unsupervised, weakly supervised) methods would be a natural next step to pursue.

Acknowledgments. We would like thank Prof. Rakefet Ackerman, Dr. Haggai Roitman, Dr. Tomer Sagi, and Dr. Ofra Amir, for their involvement in this research.

References

1. Ackerman, R., Gal, A., Sagi, T., Shraga, R.: A cognitive model of human bias in matching. In: Nayak, A.C., Sharma, A. (eds.) PRICAI 2019. LNCS (LNAI), vol. 11670, pp. 632–646. Springer, Cham (2019). https://doi.org/10.1007/978-3-030-29908-8_50
2. Burges, C.J.: From ranknet to lambdarank to lambdamart: an overview. Learning **11**, 23–581 (2010)
3. Gal, A., Roitman, H., Roee, S.: Heterogeneous data integration by learning to rerank schema matches. In: IEEE International Conference on Data Mining, ICDM. IEEE Computer Society (2018)
4. Gal, A., Roitman, H., Sagi, T.: From diversity-based prediction to better ontology & schema matching. In: Proceedings of the 25th International Conference on World Wide Web, pp. 1145–1155. International World Wide Web Conferences Steering Committee (2016)
5. Gal, A., Roitman, H., Shraga, R.: Learning to rerank schema matches. Technical Report IE/IS-2019-01, Technion - Israel Institute of Technology (2019).https:// web.iem.technion.ac.il/images/Technical
6. Jabeen, F., Leopold, H., Reijers, H.A.: How to make process model matching work better? an analysis of current similarity measures. In: Abramowicz, W. (ed.) BIS 2017. LNBIP, vol. 288, pp. 181–193. Springer, Cham (2017). https://doi.org/10.1007/978-3-319-59336-4_13
7. Macdonald, C., Santos, R.L.T., Ounis, I.: The whens and hows of learning to rank for web search. Inf. Retrieval **16**(5), 584–628 (2013)
8. Mudgal, S., et al.: Deep learning for entity matching: a design space exploration. In: Proceedings of the 2018 International Conference on Management of Data, pp. 19–34. ACM (2018)
9. Sagi, T., Gal, A.: Schema matching prediction with applications to data source discovery and dynamic ensembling. The VLDB J. **22**(5), 689–710 (2013)
10. Schwartz, B.: The Paradox of Choice: Why More is Less. Ecco, New York (2004)
11. Shraga, R.: (artificial) mind over matter: integrating humans and algorithms in solving matching problems. In: Proceedings of the 2018 International Conference on Management of Data (SIGMOD). ACM (2018)
12. Shraga, R., Gal, A., Roitman, H.: What type of a matcher are you?: coordination of human and algorithmic matchers. In: Proceedings of the Workshop on Human-In-the-Loop Data Analytics, HILDA@SIGMOD 2018, Houston, TX, USA, 10 June 2018, pp. 12:1–12:7 (2018)

Third International Workshop on Business Processes Meet Internet-of-Things (BP-Meet-IoT)

Third International Workshop on Business Processes Meet Internet-of-Things (BP-Meet-IoT)

The Business Process Management (BPM) discipline, as it is known today, emerged as the result of significant advances experienced since the mid 1990s in business methods, tools, standards, and technology. Since then, this discipline has significantly evolved, but mainly focuses on the business domain with the objective of helping organizations to achieve their goals. However, the arrival of the Internet of Things (IoT) has put into play a huge amount of interconnected and embedded computing devices with sensing and actuating capabilities that are revolutionizing our way of living and doing business. The incorporation of this technology into the BPM field has the potential to create business processes with higher levels of flexibility, efficiency, and responsiveness, providing better support to the evolving business requirements. In addition, the proper combination of these two fields can foster the development of innovative solutions not only in the business domain where the BPM emerged, but also in many different application areas in which IoT is applied (e.g., smart cities, smart agro, or e-health).

While the incorporation of IoT technology into the BPM field has plenty of potentials, it also imposes a set of challenges that need to be addressed. Therefore, the objective of this workshop is to attract novel research at the intersection of these two areas by bringing together practitioners and researchers from both communities that are interested in making IoT-enhanced business processes a reality.

The third edition of this workshop attracted four international submissions, each of them reviewed by three members of the Program Committee. Based on these reviews, the two following papers were accepted to take part in the workshop program: "Enabling the Discovery of Manual Processes using a Multi-modal Activity Recognition Approach" authored by A. Rebmann, A. Emrich, and P. Fettke, and "Integrating IoT with BPM to provide Value to Cattle Farmers in Australia" authored by O. Keates. The first paper presents a multi-modal approach to discover manual business processes in real time from different sources (e.g., wearable sensors and cameras). In addition to this, sources image data captured during process execution and even worker feedback is used to resolve ambiguities for activity recognition. The second paper describes a more practical work developed within the farming domain. In this case, the work describes the steps required to integrate BPM technologies into the increasing IoT implementations across cattle farming operations in Australia, with the target of improving the support to decisions.

In addition to these two papers, the program of the workshop includes two invited talks, one given by Luise Pufahl and a second one by Daniele Mazzei. Finally, the workshop will conclude with a discussion about the challenges of integrating the BP and IoT fields, and a starting initiative on creating a BP-Meet-IoT challenge to be adopted for future research and benchmarking.

Similarly to previous editions of this workshop, the organizers of this event hope that the reader find this selection of papers and talks interesting and useful to get a better insight at the integration of these two fields from a theoretical and practical point of view.

October 2019

Organization

Workshop Chairs

Agnes Koschmider	Christian-Albrechts-Universität zu Kiel, Germany
Massimo Mecella	Sapienza Università di Roma, Italy
Estefanía Serral	KU Leuven, Belgium
Victoria Torres	Universitat Politècnica de València, Spain

Program Committee

Andrea Delgado	INCO, Universidad de la República, Uruguay
Andrea Marrella	Sapienza Università di Roma, Italy
Andreas Oberweis	Karlsruhe Institute of Technology, Germany
Christian Janiesch	University of Wurzburg, Germany
Faruk Hasic	KU Leuven, Belgium
Francesco Leotta	Sapienza Università di Roma, Italy
Felix Mannhardt	SINTEF Digital, Norway
Francisco Ruiz	University of Castilla-La Mancha, Spain
Jianwen Su	University of California at Santa Barbara, USA
Manfred Reichert	University of Ulm, Germany
Matthias Weidlich	Humboldt-Universität zu Berlin, Germany
Mathias Weske	Hasso-Plattner-Institut at the University of Potsdam, Germany
Pnina Soffer	University of Haifa, Israel
Sylvain Cherrier	University Marne-la-Vallée, France
Selmin Nurcan	Universite Paris 1-Pantheon-Sorbonne, France
Udo Kannengießer	Compunity GmbH, Germany
Vicente Pelechano	Universitat Politècnica de València, Spain
Zakaria Maamar	Zayed University, UAE

Keynote Abstracts

BPI Challenge 2019: Discovery of Insightful BPMN Process Models for Procurement

Luise Pufahl (iD)

Hasso Plattner Institute, University of Potsdam, Potsdam, Germany
luise.pufahl@hpi.de

Abstract. The annually organized Business Process Intelligence (BPI) Challenge provides its participants with real-world process event logs and with questions posed by the process owners to analyze the data using different process mining techniques. In the BPI challenge 2019, the event log resulted from the *Purchase-To-Pay* process of a large coats and paints company operating from the Netherlands. In this invited talk, I will describe the main results of the winning report "BPI Challenge 2019: Performance and Compliance Analysis of Procurement Processes Using Process Mining"[1] by Kiarash Diba, Simon Remy and Luise Pufahl. Our main goal of this report was to identify understandable and insightful BPMN diagrams describing the different types of purchase order handling in this log for a subsequent compliance and performance analysis. After filtering and applying different process discovery algorithms, we manually created BPMN process diagrams from the results of automatic discovery using also advanced modeling concepts such as BPMN message events, multi-instance activities, etc. illustrating also the limitations of existing discovery algorithms. In this talk, I will discuss these limitations and the handling of different data quality issues during process discovery. Finally, based on these reflections, I will give prospects for a potential BPM IoT Challenge in the future.

Keywords: Business Process Intelligence Challenge · BPMN · Process discovery

---

[1] https://icpmconference.org/2019/wp-content/uploads/sites/6/2019/07/BPI-Challenge-Submission-6.pdf.

From IOT to Things on Internet, Human-Centered Design in Business Processes Smartification

Daniele Mazzei

Department of Computer Science, University of Pisa, Pisa, Italy
mazzei@di.unipi.it

Abstract. The talk focuses on the Internet of Things concept and tries to explore this "buzz acronym" in the context of industrial data acquisition for Business Process Management applications. Due to the big push of cloud and technology vendors nowadays we are facing with a race towards data acquisition from industrial machines. However, the typical workflow adopted when implementing Industrial Internet of Things to the management of business processes has several limitations. We accepted that things and machines can become smart just because they are connected to the web. But, streaming giga of data into a cloud is not necessarily an added value that turns "stupid" machines into "intelligent" systems. When we talk about smartness we have to start referring to the purpose, to the applications, to the added value that makes this new smart-since-connected device better than its previous version. The talk will introduce a user-centered machine "smartification" paradigm that inverts the typical Industrial Internet of Things workflow by shifting from the extraction of data to the transformation of a machine into a smart appliance. The proposed paradigm starts from the typical DIKW (Data, Information, Knowledge, Wisdom) pyramid adapted to the context of Industrial Internet of Things (I-IOT) used as enabling technology for Business Process Management. The talk will also introduce various use case showing how this knowledge centered approach enabled various business processes restyles aimed at improving the company efficiency.

Integrating IoT with BPM to Provide Value to Cattle Farmers in Australia

Owen Keates[⊠]

Information Systems School, Queensland University of Technology, Brisbane,
Australia
owen.keates@hdr.qut.edu.au

Abstract. This research study covers the journey to date introducing BPM into the increasing IoT implementations across cattle farming operations in Australia, to generate value through improved decision support. The research is supported by Meat and Livestock Australia (MLA) who represent over 50 000 cattle, sheep and goat producers, Hitachi Consulting as well as individual farming businesses from large corporates to family owned farms. The challenges of deploying IoT devices, especially across extensive farms in remote areas, the importance of looking beyond numerous dashboards on various websites and mobile applications to an integrated control centre, with BPM at its core, driving decision support are documented. Capabilities of Business Process Management Systems have been leveraged to drive workflow and process orchestration to extract the right data and run decision models. Learnings from this research study can be applied to any industry or value chain wishing to extract greater value from the opportunities provided by IoT.

Keywords: Internet of things · Business Process Management · Business Process Management Systems · Decision support · Cattle farming

1 Introduction

The scale of cattle farming in Australia is large, the overall herd size in currently 26.2 million with 47 776 businesses having cattle. The size of cattle herds range from the average of 1500 per farm in Northern Australia to 400 per farm in Southern Australia. In Northern Australia there are a small number of very extensive farms, while in Southern Australia there are many relatively small farms. While the average farm size is 4000 hectares, the larger properties exceed one million hectares and can carry up to 85 000 head of cattle. For both the large farming operations, as well as the small, IoT is playing a major role and the opportunity to enhance the value created through integration with Business Process Management Systems (BPMS) is significant. The opportunity extends beyond cattle to the entire meat and livestock industry which comprises the beef, sheep meat and goat meat sectors and has a turnover of $62.3 billion including $14 billion in export revenue, supplying over 100 global markets while contributing 405 000 Australian jobs through direct and indirect employment [1].

C. Di Francescomarino et al. (Eds.): BPM 2019 Workshops, LNBIP 362, pp. 119–129, 2019.
https://doi.org/10.1007/978-3-030-37453-2_11

Meat and Livestock Australia Ltd (MLA)[1] is a producer-owned, not for profit organisation that delivers research, development and market services to the Australia's meat and livestock industry. MLA was founded in 1998 and has over 50 000 cattle, sheep and goat producer members. With funding from transaction levies paid on livestock sales as well as the Australian Government and voluntary contributions from industry, MLA's mission is to deliver value to their stakeholders through initiatives that improve productivity, sustainability and global competitiveness. MLA has a digital value chain strategy which aims to enable the capture, integration and interpretation of data generated within the livestock industry through a range of new technologies [2]. Aligned with this industry wide strategy, several projects were run with the overarching objective to deliver productivity gains through the value chain by the adoption of best practice and appropriate technology.

This study documents the challenges faced by the industry to adopting IoT, with connectivity and cost, major factors. At the core of these projects was the business process management cycle as described by Dumas et al. [3]. Since the BPM cycle was central to each of the meat and livestock projects, with consecutive projects progressing further in the BPM cycle, the stages of the cycle are used as the steps in Sect. 3, which describes the outcomes of this ground breaking IoT project. Section 4 describes the iterations of the BPM cycle as additional projects and implementations spun off from the first and Sect. 5 presents lessons learnt and recommendations.

2   Challenges Faced by Cattle Farmers

Cattle farmers are part of a meat a livestock value chain as shown in Fig. 1, below.

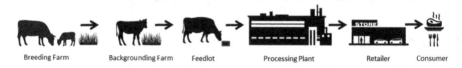

Breeding Farm   Backgrounding Farm   Feedlot   Processing Plant   Retailer   Consumer

Fig. 1. An integrated meat and livestock value chain

Breeding herds, carefully selected according to genetic traits, are kept on breeding farms with management of cow and bull herds. Young heifers and steers are weaned from their mothers and moved to dedicated backgrounding farms where they are managed to optimise growth and weight gain. Once their target weight has been achieved the cattle are either sent straight to the processing plant, to produce grass fed beef products or via a feedlot where they are given increasing quantities of grain and supplements to increase weight gain and improve meat quality, to produce grain fed beef products.

---

[1] https://www.mla.com.au/.

Farmers are paid for their cattle based on the quality of meat which is assessed at the processing plant. The processing plants have a quality – price matrix based on several quality attributes such as marbling, colour and fat content, as well as weight and age of the cattle. Significant discounting of the price paid to farmers occurs if the cattle are sent to processing outside of tight weight specification, which is a difficult parameter to control in large herds of cattle. Individual animals do not gain weight at the same rate and it is not possible to weigh individual animals regularly, especially on the extensive cattle ranches. Similarly, for the eating quality parameters such as marbling and fat content, discounting of price occurs when tight specifications are not achieved, the control factors influencing these specifications are difficult to manage. With this pricing model having a direct impact on profitability of cattle farming businesses, who also must manage the effects of weather patterns that can cause years of drought followed by excessive rains, especially in Australia, the ability to measure and analyse a wide range of environmental factors, as well as individual animal data across the lifecycle of an animal is critical to running a profitable business.

Farmers recognize they require more data to inform their decision making, they are also fully aware of the input costs to raise cattle and the profit per animal which is relatively low. When investing in IoT and solutions to process data, farmers compare the investment and potential return to other investments they can make on their farms, for example, improving infrastructure such as fencing and water supply or increasing the size of their herd. A further significant challenge farmers face is poor connectivity to the internet. At a minimum farmers will have internet connection at the farm house, which doubles as the control centre for the farm. In the remote areas this is achieved via satellite connection, while less remote areas will have 3G or 4G mobile connectivity. Few farms have established connectivity across the farms themselves and are required to establish their own local area networks for on farm connectivity.

Notwithstanding the above challenges, MLA and their stakeholders recognized that effective deployment of IoT in the cattle industry is critical. The approach to identifying the right pathway was key, farmers are very busy and require clear well-planned approaches and actions. To provide such an approach, a decision was made to use the business process management cycle as the core methodology. The BPM cycle developed by Dumas et al. [3] was used and led to the successful implementation of several IoT projects on cattle farms all integrated into a BPMS to provide the required decision support to farmers. The approach and outcomes are described in Sect. 3.

3   Using the BPM Cycle to Integrate IoT with a BPMS

3.1   Developing a Process Reference Model and Process Identification

To identify the appropriate processes which have a high impact on the productivity and profitability of cattle farmers, a process model was developed of the meat and livestock value chain. To capture performance data, processes and practices this a priori model was developed using the Association of Supply Chain Management's SCOR[2] best

---

[2] www.apics.org/apics-for-business/frameworks/scor.

practice reference model as a core framework and documenting the industry specific meat and livestock processes at Levels 4 and 5 under the SCOR hierarchy. To associate practices with processes at Level 4, a literature research was conducted to extract latest cattle management best practices including a review of papers published by MLA. These practices were noted on the respective processes with cross reference to their source. The expanded SCOR model is illustrated in Fig. 2 below. This model was then used to demonstrate the potential opportunities and business value from deploying appropriate IoT and solutions on cattle farms. Using this modelling MLA, in consultation with its stakeholders, selected the Croydon Cattle Station, part of the Australian Country Choice[3] meat and livestock business, which is the most integrated value chain in the country, as the first pilot farm. As discussed in Sect. 4, this adapted best practice reference model as well as process identification step was repeated several times in the program as new project opportunities were created.

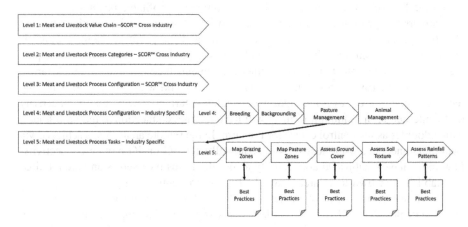

Fig. 2. Expanded process reference model

3.2   Process Discovery

The process discovery stage started with a visit to the Croydon Cattle Station, a medium sized backgrounding farm of 58 419 hectares and a carrying capacity of 16 000 cattle. Farm management took the team on a tour of the farm, explaining processes and important decisions made. These decisions include when to move cattle from one paddock to the next, or onto feedlots and processing, as well as when to buy more cattle. The challenges of estimating weight gain of a herd were explained as well as the financial impact of underestimating the average weight and keeping cattle on pasture longer than required. Emphasis was also placed on the importance of pasture management, ensuring that pastures were not overgrazed and rested (or spelled) to allow grass to regrow and seed effectively. Connected with pasture management is

---

[3] https://www.accbeef.net.au/.

close monitoring of the weather. Good rainfall predictions can influence the decision on when to move cattle. The increased growth of grass following rain can boost cattle weight at a critical time and ultimately improve profitability.

The expanded SCOR model was discussed with farm management who further recommend it's use as a practice knowledge repository to be cross referenced by a decision support solution.

3.3   Process Redesign

The process redesign phase commenced by documenting requirements of farm management at Croydon farm and with reference to the expanded SCOR model. The key requirements were as follows:

- Provide IoT solutions to monitor and manage pastures
- Provide IoT solutions to monitor and manage water sources
- Design a solution capable of capturing RFID (radio frequency identification) data from individual cattle.

To achieve the above the following IoT systems were selected (a) a weather station and soil moisture measurement system (b) water level monitoring of tanks and water troughs. In addition, a solution provider that was able to provide multispectral satellite imagery of farm pastures for determination of available biomass and overall pasture health was found and methods discussed for importing this data into a business process management system. As preparations were underway on the farm to establish infrastructure for the capture of individual animal data at various stock yards, a data model of information that would be available was developed so that the BPMS could be correctly configured to receive this data.

Once key IoT systems were identified, the connectivity challenges had to be addressed. Croydon farm has limited 3G connectivity at the farm house and while it was feasible to establish a local area network to connect the IoT devices to the internet, farm management requested that a satellite solution be deployed as they wished this pilot project to reflect the connectivity situation on their other properties, some of which are much larger (for example Barkly Downs at 942 499 hectares and 85 000 cattle carrying capacity) where local area networks would be more challenging. A satellite solution was developed to connect the IoT devices.

The process redesign, capturing the collection and analysis of data from the IoT devices was documented within a BPM solution using BPMN 2.0 modeling standard with the design principle that parts of the process could later be executed via a BPMS. Capturing the process redesign at this level of detail ensured that MLA and its stakeholders could also deploy similar solutions more broadly within the meat and livestock industry.

3.4   Process Implementation

Process implementation occurred at two levels (a) on farm implementation of IoT devices and connectivity systems (b) configuration of an IoT platform and supporting BPMS to ingest and analyse the data collected. On farm, the weather station was installed measuring barometric pressure, evapotranspiration, humidity, rainfall, rainfall rate, soil moisture, temperature, wind speed and wind direction. Water sensors were installed on a dam and water troughs with solar powered transmitters sending data back to a solar powered satellite hub which transmitted to a cloud based IoT platform.

Configuration of the IoT Platform commenced, key components included a connectivity platform, data ingestion and data processing layer as well as a presentation layer and a dashboard. The IoT Platform also used BPMS components to orchestrate the various processes.

3.5   Process Monitoring and Controlling

The solution was launched at the Croydon Cattle Station. Weather and water information streamed from the sensors via satellite to the IoT platform which then displayed the information in the dashboard format. The cloud based IoT platform was accessed from Croydon Cattle Station via their internet connection and displayed on a large touch screen which has become the Croydon Cattle Station Control Centre. This information is also complemented by multi spectral satellite imagery of the pastures which was analyzed to produce available biomass data. The Croydon Cattle Station Control Centre displays current data as well as historical data enabling the farm team to analyse trends which informs their decision making. The rainfall pattern can be analyzed in conjunction with the biomass data to aid in the decision making of when to move animals to different paddocks or feedlots. The farm team is also able to access the best practice reference model and continue to expand the best practice reference model with their own, detailed farm specific, processes creating a practice knowledge repository which is used to train new farm team members as well as provide a core for the next phase of development, the decision support centre. The Control Centre information can be accessed anywhere which is useful for senior management who often manage many farms.

3.6   Results from the First BPM Cycle

The implementation of the Croydon Cattle Station Control Centre demonstrated the potential of deploying IoT devices using a focused business process management approach. The BPM approach not only provided guidance in the selection of an appropriate farm, and problem type, to commence the digital transformation journey, it also provided a consistent focus that this farm was part of an overall value chain and overall productivity and profitability improvement could only be achieved when information was shared along the value chain.

Locally on the farm the Control Centre provides a central place for farm managers to meet team members, analyse weather and water data and plan the days activities. The automatic water sensing covered two paddocks and demonstrated the business case to expand the monitoring system to all paddocks which will provide significant productivity improvement.

This first project gained global recognition when MLA and Australian Country Choice won the 2017 Hitachi Global Digital Transformation award, it also triggered several further iterations of the BPM cycle as described in Sect. 4.

4 Iterating the BPM Cycle

4.1 Established IoT and BPM Artefacts

The initial project established a best practice reference model, reusable to build business cases for further IoT implementations as well as be extended as a practice knowledge repository for an individual farm. The second artefact was the IoT platform, capable of further expansion as well as multi tenancy to support IoT implementations across the value chain. While the platform uses some BPMS functionality there was an appreciation that this platform could support the deployment of more advanced BPMS capability, ultimately leading to advanced decision support for farmers as well as managers across the end to end meat and livestock value chain.

4.2 Connecting an Entire Farm

Using the artefacts generated in the Croydon project a follow up IoT project was established at the Calliope Cattle Station, a 28300 hectare breeding and backgrounding farm owned by Will Wilson[4]. In this project a local area network was established to connect water level monitors on all water tanks as well as a weather station. Information was relayed to a central point from which connectivity to the IoT platform was established through mobile 3G. A control centre was established at the farm house office with farm management and team members regularly using the control centre for decision making on when to move animals. With all water points automatically monitored, with alerts, there are also productivity benefits as team members no longer travel to each water point for regular 1 inspection of water levels.

The connected farm was taken to a new level on the Carwoola Pastoral Company, a 6000 hectare cattle (1000) and sheep (8000) farm where many IoT sensors were installed, with MLA support, to demonstrate the broader capability of digital solutions applied to a farm[5]. Digital solutions installed include water monitors for troughs, tanks and dams; weather stations and rain gauges; diesel tank monitors; soil probes; sheep and cattle tags; gate and cattle ramp monitors; digital biomass mapping; electric fence monitors; smoke alarms and shed condition monitor devices; vehicle and asset tracking; farm management software; drone technology; silo level monitors and animal handling systems.

While Carwoola Pastoral Company represents an example of what can be achieved with IoT, it also highlighted the challenges of integrating numerous IoT sensors with different connectivity protocols. Some of the IoT sensor providers had not considered information from their sensors would be shared and had not developed application programming interfaces to allow their sensor data to be displayed and analyzed on a common IoT platform. It also reinforced the benefits of using the best practice reference model to evaluate the business case as well as the importance of supporting each IoT

---

[4] www.mla.com.au/news-and-events/industry-news/making-connections/.

[5] www.queenslandcountrylife.com.au/story/5815523/digital-technology-put-to-test/.

implementation with a process redesign, ensuring all process dimensions including change management and technical support are well prepared for process implementation.

4.3 Connecting the Entire Value Chain

Following the successful deployment of IoT technologies on farm, MLA and their stakeholders wish to extend these technologies to connect an entire meat and livestock value chain with the overall objective being to improve productivity, sustainability and global competitiveness. There are many factors that contribute to getting the optimum return on quality and price as illustrated in Fig. 3 below.

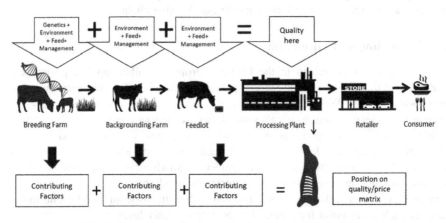

Fig. 3. Factors contributing to Productivity and Profitability

Research is currently underway to develop the appropriate process analytic methodology to integrate IoT with BPMS to achieve process monitoring and control of a meat and livestock value chain. There are several decisions that are required to be made based on Fig. 3 and initial data modeling has shown that there will be gaps in the required data. Collection of individual animal data, for example, is not comprehensive and the lifecycle of an animal can range for two years to ten years, depending on whether it has been retained for breeding purposes. Having established that IoT data can be successfully integrated with a BPMS via an IoT Platform in the connected farm projects the challenge faced is how to manage data gaps across the value chain.

The research team hypothesized that extrapolation and simulation of missing data would be possible and conducted literature research into the emerging methodology of digital twining. According to Tao et al. [4] a digital twin is a real mapping of all components in the product life cycle using physical data, virtual data and interaction data between them, a design solution that can provide designers with information, recommendations and assessment throughout a design process. While this digital twin product framework focuses on the design of a physical product, broader application of the digital twin concept has been used in the design, operation and maintenance of complex systems such as aircraft components and power plants. Here the digital twin has been designed with computer aided design systems, three-dimensional modelling and virtual reality solutions. The research team proposes to follow the same principles using process modelling techniques and process simulation to create a digital twin of a meat and livestock value chain.

To create a digital twin of a meat and livestock value chain as shown in Fig. 1, it will also necessary to generate a digital cow as well, or more specifically a digital cow, digital bull, digital heifer and digital steer, as shown in Fig. 4.

These digital animals will have all the required data attributes, associated with them and will be used to simulate data when real data is not available. Initially these digital twins will be digital avatars, a representation of a connected cow that has all relevant data stored and associated with it since birth. The exact modelling of these avatars is yet to be determined; however, they will have the ability to simulate data for the purposes of developing and testing the methodology when data is not available. The digital twin of the value chain and the digital twin of the above animals will utilize the reference model to ensure they represent industry best practice.

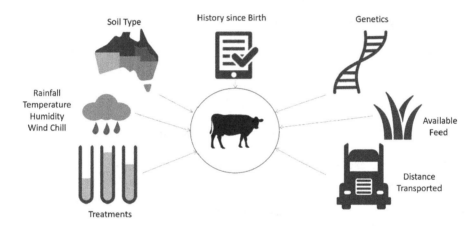

Fig. 4. Digital Cow generated to support the Digital Twin Value Chain

4.4   Bringing It All Together in an Operational System

As the amount of data sensed by IoT devices and harnessed by IoT platforms rapidly increases due to improved connectivity and lower costs of sensors, it becomes important to remain focused on the reason why data is being collected in the first place. In the case of the meat and livestock value chain, data is collected to improve productivity and ultimately profitability of the end to end business. Developing additional dashboards to display more data is not the solution, processing data to provide decision support is. To achieve this goal the researchers have proposed the technical architecture of an enhanced BPMS which includes the following, see Fig. 5:

- An IoT platform which has connectivity, a data ingestion and processing layer as well as a presentation layer including a dashboard.
- Process repository and process modeling solution that documents and retains a best practice reference model which is continuously updated.

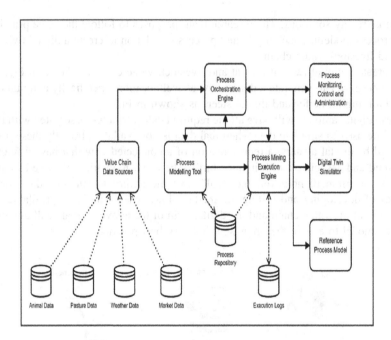

Fig. 5. Technical Architecture of IoT platform integrated with BPMS

- Digital twin simulator which generates and predicts data when key data is not available.
- Process mining execution engine which monitors event logs from the meat and livestock value chain while conformance checking against the best practice reference model.
- Process orchestration engine which manages the above events while also feeding relevant data through decision models to provide decision support and analytics to managers of the meat and livestock value chain.

5 Lessons Learnt and Recommendations

With IoT sensors becoming more pervasive and the amount of available data increasing rapidly it is vital that the focus remains on key questions and decisions managers of meat and livestock value chains are trying to solve. A process centric view, expressed through a value chain reference model, assists in keeping all stakeholders focused on this goal. Following the BPM cycle ensures that process analysis, redesign and implementation is done with the appropriate rigor and a continuous improvement cycle is maintained, all the while building learnings into a best practice reference model.

The researcher presented the following lessons learnt and recommendations to MLA's Red Meat 2018 Conference which showcased the Carwoola Smart Farm[6]:

1. It is not easy - bringing together data from disparate systems and different IoT sensing solutions requires technical skill and resource.
2. IoT solution developers need to build their platforms with the ability to share data, with permission, for the benefit of customers.
3. IoT solution developers are encouraged to build their solutions with open source software, this ensures a greater product lifecycle for their solutions and reduces the maintenance effort required of custom-built platforms.
4. Collaborate or be Disrupted – farmers and value chain managers wish to have a holistic view of the relevant information. The synergy of combining data from various sources to provide decision support is significant, IoT solution providers with "closed" solutions will be disrupted in the future.
5. Never stop listening – the deployment of IoT and BPMS systems across the meat and livestock value chain and agriculture value chains in general, will continue to grow rapidly. Solution providers must never stop listening to the customer, never stop visiting the farms and speaking to the team members who need to make tough decisions every day, they will provide insights that will continuously improve IoT and BPMS.

Acknowledgements. The researcher wishes to acknowledge Meat and Livestock Australia Limited and MLA Donor Company Limited for their funding contributions to these co creation projects, together with Hitachi Consulting Limited. Also, to the meat and livestock businesses, Australian Country Choice, Australian Cattle and Beef Holding and the team members of Croydon Cattle Station who supported and provided deep insights to the projects as well as Will Wilson and team members of the Calliope Cattle Station and Darren Price and team members of Carwoola Pastoral Company. Vodafone Global Enterprises funding and technical support of the Calliope project is also acknowledged. The researcher is currently undertaking a part time professional doctorate degree through the Queensland University of Technology and acknowledges the support of Principal Supervisor, Associate Professor Moe Thandar Wynn and Associate Supervisor: Dr. Wasana Bandara, as well as the researcher's employer Hitachi Consulting Australia Pty Ltd.

References

1. Ernest & Young Report. State of the Industry Report: The Australian Red Meat and Livestock Industry, Meat & Livestock Australia, October 2017. https://www.mla.com.au/globalassets/mla-corporate/research-and-development/documents/industry-issues/state-of-the-industry-v-1.2-final.pdf
2. Meat and Livestock Australia. https://www.mla.com.au/about-mla/
3. Dumas, M., La Rosa, M., Mendling, J., Reijers, H.A.: Fundamentals of Business Process Management, 2nd edn. Springer, Heidelberg (2018)
4. Tao, F., et al.: Digital twin-driven product design framework. Int. J. Prod. Res. **94**(9–12), 3563–3576 (2018)

---

[6] www.mla.com.au/news-and-events/industry-news/watch-all-the-highlights-from-red-meat-2018.

Enabling the Discovery of Manual Processes Using a Multi-modal Activity Recognition Approach

Adrian Rebmann$^{(\boxtimes)}$, Andreas Emrich, and Peter Fettke

Institute for Information Systems (IWi) at the German Center for Artificial
Intelligence (DFKI GmbH) and Saarland University,
Campus D3 2, Saarbrücken, Germany
{adrian.rebmann,andreas.emrich,peter.fettke}@dfki.de

Abstract. The analysis of business processes using process mining requires structured log data. Regarding manual activities, this data can be generated from sensor data acquired from the Internet of Things. The main objective of this paper is the development and evaluation of an approach which recognizes and logs manually performed activities, enabling the application of established process discovery methods. A system was implemented which uses a body area network, image data of the process environment and feedback from the executing workers in case of uncertainties during detection. Both feedback and image data are acquired and processed during process execution. In a case study in a laboratory environment, the system was evaluated using an example process. The implemented approach shows that the inclusion of image data of the environment and user feedback in ambiguous situations during recognition generate log data which well represent actual process behavior.

Keywords: Process discovery · Disambiguation · Human activity recognition

1 Introduction

Process discovery aims at uncovering processes based on event data captured by information systems. Established approaches for process discovery require structured process data recorded in event logs [15]. However, capturing of behavior which is not recorded by existing information systems in an organization remains a challenge. Namely business processes which are executed manually. The increasing digitization of businesses and the advent of the Internet of Things allow for extensive descriptions of business processes in the form of contextual data, as a major objective of the Internet of Things is the capturing of real-world behavior of objects and workflows [9]. Although several approaches focused on using sensor data to uncover workflow models, few have integrated various data sources, aiming at producing an accurate picture of actual process behavior

© Springer Nature Switzerland AG 2019
C. Di Francescomarino et al. (Eds.): BPM 2019 Workshops, LNBIP 362, pp. 130–141, 2019.
https://doi.org/10.1007/978-3-030-37453-2_12

within the context of business process management, specifically in manual business processes.

The goal of this paper is the development and implementation of an **approach for capturing manual processes based on data obtained from the Internet of Things during process execution** to enable the application of process discovery methods. We aim at the **simultaneous incorporation of multiple perspectives** of the executed process in real-time to enable a reliable recognition of performed activities within **ambiguous process contexts**. This leads to the research question: how accurately can manually executed processes be captured using a multi-modal recognition approach?

Thus, this contribution focuses on the challenges *"placing sensors in a process-aware way"* and *monitoring of manual activities*, as well as *"detecting new processes from data"* and *"bridging the gap between event-based and process-based systems"*, as postulated by [6]. Motion data of the workers executing the processes is used to recognize the respective process steps. To improve the recognition of human activities filter mechanisms are used.

The paper follows a design science paradigm. The objective of a design-oriented approach in information systems research is the development of an artifact which addresses a relevant, unresolved problem. The relevance of the artifact developed is derived from the increasing complexity of the business processes of companies and their increasing dynamics, which necessitates an analysis of said business processes and business process models, e.g. using process discovery to observe actual behavior. This might differ from intended behavior and might bear tacit knowledge [11] which could help improve underlying processes. Additionally, the application of established methods both in the development and in the evaluation of the artifact is necessary to ensure the rigor of the research [5].

The remainder of this paper is structured as follows: the application use case is outlined in the next section. The conceptual design is presented in Sect. 3. Section 4 describes the system design of the artifact. The evaluation of the artifact is presented in Sect. 5. Related work is presented in Sect. 6. Section 7 provides a summary of the paper and gives an outlook on future work.

2   Application Use Case

Sensor data sequences, e.g. acquired from wearable sensors capturing the motion of workers, may look very similar. For example, considering a warehouse worker, sensor time series for setting a delivered box on the floor and lifting said box look very similar. Also, the sensor data for tying one's shoes can show a very similar time series. Sensor events which are captured may be directly, indirectly, or not at all be related to actions performed by humans. Hence, there is a need for filtering mechanisms which can exclude activities which are not allowed in certain situations or locations. We created a lab (see Fig. 1) for the execution of use case processes. As stated above, the focus is on resolving ambiguities during recognition. For this purpose, an exemplary process was designed, which deliberately contains ambiguous activities. It consists of gathering assembly parts,

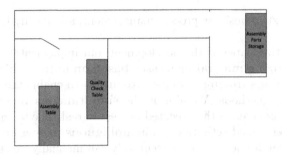

Fig. 1. Floor plan of the laboratory setting. It consists of a storage location, an assembly table and a quality check table.

performing a manual assembly task, a quality check of the assembled product and the packaging and storing of said product. The process model covers all basic control flow patterns [16]: *Sequence, Parallel Split, Synchronization, Exclusive Choice* and *Simple Merge*. The process model is depicted in Fig. 2.

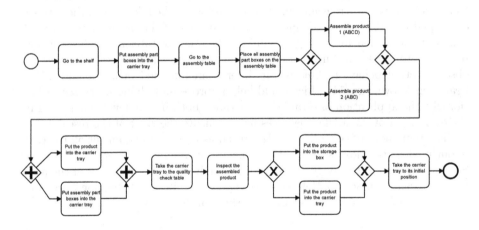

Fig. 2. Process model for the case study.

By simultaneously considering multiple data sources capturing motion and the process execution context we try to resolve the ambiguities present when detecting manually executed business processes. As a final option for disambiguation we request user feedback.

3   Conceptual Design

3.1   Discovery of Manual Business Processes

In order to enable the discovery of manual business processes from IoT data, the following concept is proposed, which can be considered as iterative:

Activity definition: For the scope of this paper the assumption is an existing set of activity types which can occur. This reduces the complexity when recognizing activities during process execution. This step is only executed once before the first iteration.

Configuration: During the first iteration no actual model is available. Thus, the configuration solely consists of the assignment of organizational resources and location context data to activities. In later iterations, once a process model was discovered, there may be the need for further configuration.

IoT data acquisition: Acquiring suitable data for the actual recognition tasks is the most important step as the result of this step determines the quality of the outcome of all subsequent steps. A multi-modal approach is proposed here, which encompasses wearable sensors, cameras as well as user feedback during process execution.

Recognition: During this step an ensemble of recognition procedures is executed to detect the currently running process activity, process instance and process type. The focus here is the activity recognition as the data is highly dependent on the executing actor. This is why several perspectives are combined in this paper, which will be explained in more detail in the next section.

Process logging: The results of the recognition are logged using standard logging formats.

Process mining: In this phase process mining algorithms are applied to the log to generate a process model. During the initial iteration this would be discovery algorithms. In later iterations, the application of conformance and enhancement techniques could be considered as well.

Model visualization: The mined model from the previous step is visualized. The goal is to enable editing and configuring of discovered models, which leads back to the configuration step.

In this paper, the focus is on the IoT data acquisition and the Recognition phases. The following sections outline these phases in more detail.

3.2   Recognizing Manual Activities Using a Multi-modal Approach

Activity Recognition Based on Motion Sensor Data. Motion sensor data is used as an input for machine learning models, which are trained to classify the currently executed activity instance. These models are built using training data, which have to be collected and labeled for existing activity types. For this approach to work, the activity types need to be known a-priori. We used classification methods capable of multi-class-classification for the motion-based activity recognition. The best results were produced by a Random Forest classifier. The input was encoded as numerical feature vectors. These were computed using segments of the time series data produced by the sensors. Each of the sensors produces three-dimensional acceleration data and three-dimensional orientation data. A number of features have proven to result in high accuracy in previous studies. Most features are computed for the time series of individual axes. Besides those, frequency domain features computed on the Fourier transform of the individual sensor segments were used. This approach is frequently applied in human activity recognition (HAR) and is e.g. described in [4,8]. We used a segment size of 32 and 40 features per sensor and an overlap of 50%.

Disambiguation. Motion sensor-based recognition of activities can be fuzzy no matter how good the performance of the machine learning model is. This is because of the fact that some activities have the same time series of sensor measurements and thus these time series are ambiguous. This problem makes it necessary to include other data in the recognition process in order to resolve the ambiguities. Based on image data of the process context taken during process execution, i.e. the scenery in which an activity is executed, we try to resolve ambiguities. The aim here is to filter out impossible activities to improve the accuracy of the activity recognition process and hence the later process discovery. For the scenery classification we retrained a publicly available Convolutional Neural Network [14] on custom categories, which resulted from our laboratory setting (see Fig. 1). If uncertainty remains regarding the currently executed activity, the system actively asks the worker for feedback.

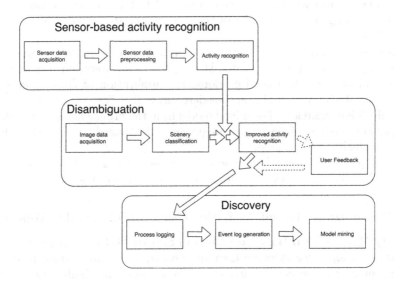

Fig. 3. Data processing and process discovery pipeline.

Recognition Approach. Figure 3 shows the processing approach to recognize manual activities with the goal of creating a process log allowing the application of existing process discovery algorithms. The processing mechanism should not be considered as a linear process. It is rather a decentralized filter mechanism, which continuously tries to minimize the number of activity types which are possible throughout a process instance. Ideally the number of possible instances is equal to one assuming perfect precision.

4   System Design

In order to realize the proposed concept, a system architecture was developed, which is intended to be extendable by arbitrary components. Figure 4 shows the architecture which was implemented in the context of this paper. The architecture corresponds to the procedure introduced in Sect. 3. The *Recognition Server* realizes the motion sensor-based activity recognition and the coordination of the disambiguation components. The scenery classification and the user feedback mechanism are functionally implemented within separate components. The approach presented in this paper does not start from existing process models. Nevertheless, a process engine was adapted, which is controlled by the developed recognition mechanism. The process engine thus controls the execution of process instances while they are detected. Furthermore, the definition of activity types is implemented using an adapted process model repository. This facilitates the definition of activity types and allows for editing and deploying discovered process models. A high-throughput message broker was used for the communication between the software components, so they can operate autonomously. During recognition, the individual components try to classify the current activity based on current motion and image data, by reducing the number of possible activities. The two classification pipelines are independent of each other. In both pipelines data is collected and published to the message broker. In case of the activity classification these data have to be segmented and preprocessed to serve as an input to a trained model. After that, the gathered input data is fed into

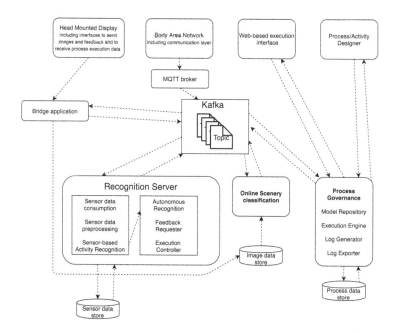

Fig. 4. Proposed system architecture.

the respective classification model, which produces a result. Using activity type information available, possible activity types are derived for each of the classifier results individually. The results of both pipelines are combined in a final step. Here, motion data of the resource responsible for the execution is collected using inertial measurement units (IMU) mounted on their upper and lower arms (see Fig. 5). We combine these data with image data taken from the perspective of the worker, wearing a head-mounted display (HMD), during execution for disambiguation. If no single activity can be found automatically, the HMD is used as feedback channel (see Fig. 5).

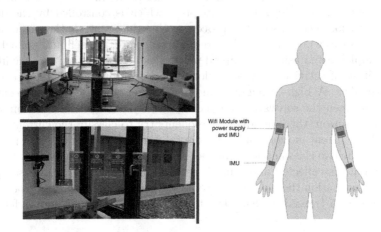

Fig. 5. Laboratory setting (top left), request for user feedback is shown to the worker via holograms over a head-mounted display (bottom left) and Sensor suit consisting of four IMUs, and two Wifi modules transmitting the data (right).

5   Evaluation

This section presents the evaluation of the developed approach. A case study was conducted which yields a proof of concept of the implemented system as well as the underlying concepts.

5.1   Evaluation Approach

The process introduced in Sect. 2 was executed by eight participants in a controlled lab experiment. Each participant executed the process four times, each time a different variant (path through the process model) was executed. We chose a log-centric approach to measure the accuracy of the recognition approach. It is frequently used in conformance checking scenarios. The log of the events which actually occurred, serves as a ground truth. A model generated from this log is used for applying *Alignments*. Given a trace, the goal is to find the closest matching path through the process model for this trace. Hence, an alignment indicates how an event log can be replayed on a process model [1]. To quantify the accuracy, the cost of aligning the traces to the ground truth model is used. Here, a cost of 1 is assigned to all deviations between log and model. Moreover, a comparison of recognition performance was made. This was done by comparing the accuracy of activity recognition using the multi-modal approach with those using only the motion sensor-based recognition. The complete evaluation procedure was as follows: First, motion sensor data and image data were collected and labeled. Then, classifiers were trained on the collected data. After that, the controlled lab experiments were conducted. Based on manually tagged activities which actually took place a ground truth model was mined. Then, the system logs created by the multi-modal recognition approach and the logs of the motion sensor-based approach were aligned to the ground truth.

5.2   Results

Table 1 shows the average trace length, the average cost for aligning a trace to the ground truth process model, and the standard deviation of the alignment cost using the multi-modal recognition approach and using only the motion sensor-based recognition. Considering the per-class accuracy, Fig. 6 shows the results of the recognition using only motion sensor data (left) and the results of the recognition when combining motion sensor-based recognition with scenery classification and user feedback (right).

Table 1. Summary of results using the multi-modal recognition approach (top) and results using only motion sensor-based recognition (bottom).

	Avg. trace length	Avg. cost	Std. dev. of cost
Muti-modal recognition	**9.06**	**4.31**	**2.71**
Motion-based recognition	38.53	33.66	24.84

5.3   Discussion

The results of the case study show that the multi-modal recognition approach produces process instances which can be aligned to the ground truth model at low cost. Especially, when the scenery classification can differentiate the current scenery well, like at the beginning of the use case process, the recognition results are accurate. The activities *Go to shelf*, *Go to the assembly table* and *Put assembly part boxes into the carrier tray* were discovered correctly in almost every case. The wrong classification of *Assemble product 1* in some cases can be traced back to the over-representation in the training data. Since this is the activity which takes the most time to complete. Furthermore, the activity *Put assembled product into the carrier tray* is hardly ever recognized. This can be caused by the duration of this activity. The usefulness of the feedback mechanism has been clearly demonstrated, since in almost every case there was uncertainty

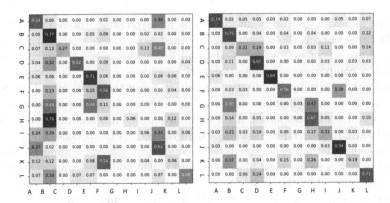

Fig. 6. Normalized confusion matrix for motion sensor-based recognition (left) and for the multi-modal recognition (right). A = Go to the shelf, B = Assemble product 1, C = Take the carrier tray to its initial position, D = Inspect the assembled product, E = Put assembly part boxes into the carrier tray, F = Place all assembly part boxes on the assembly table, G = Put assembled product into the carrier tray, H = Assemble product 2, I = Take the carrier tray to the quality check table, J = Go to the assembly table, K = Put assembly part boxes in the carrier tray, L = Put product into storage box.

about the corresponding product during assembly. This is due to the almost identical sensor sequences and the identical scenery.

The developed approach generates process logs which represent actual behavior. The average alignment cost per trace including the cost for variant mistakes is lower than five. This proves the overall efficacy of the approach. It becomes even clearer when considering the question about the impact of adding scenery classification and user feedback to resolve ambiguous situations during discovery. Results indicate that the multi-modal approach leads to a far more accurate discovery and more stable recognition. The average cost to align a trace to the same ground truth model is significantly higher with the motion sensor-based approach. This outcome is not surprising considering that the activity instances where intended to be ambiguous regarding motion data, but nicely demonstrates the efficacy of combining different data sources in ambiguous use cases. When looking at the confusion matrices shown in Fig. 6 it becomes clear that a **simultaneous consideration of different perspectives of a process** is necessary in such cases. Therefore, regarding the main research question, it can be said that the approach developed generates logs by incorporating different perspectives in real-time. The approach can capture real behavior and the logs can be used for the application of process discovery approaches.

6   Related Work

In this contribution, we propose an approach for the discovery of business processes based on data, which are acquired from the Internet of Things.

The authors of [11] presented a literature review regarding **activity recognition** with the target of using the recognized activity data as input to **process discovery** techniques. They classify each of the analyzed contributions according to a series of criteria, including the type of machine learning approach, which is used, the time the recognition is performed, the granularity of the activities, and the type and placement of the sensors used. They derive a taxonomy for the knowledge extraction from manual industrial processes through activity recognition [11]. [12] perform activity recognition based on data from wearable sensors with the goal of providing workers with ergonomic recommendations during the execution of manual production processes. [7] present the implementation and evaluation of a light-weight system that detects, monitors and logs such worker activities and generates event logs which can be used for process mining. Stationary RGB-cameras are used to detect fine-granular tasks performed by humans.

Comparable approaches aim at discovering process models of **daily routines of humans**. The authors of [10] deal with the identification of research needs in the application of process mining techniques when recognizing activities or habits of people in *Smart Spaces*. The existing problems are illustrated by a case study with Presence Infrared Sensors (PIR). The authors identify the following three challenges: closing the gap between sensor data and actions of humans, improving process mining or the modeling techniques related to often unstructured human habits and the identification of individual process instances. The latter especially regarding several people operating in the same Smart Space. The authors of [13] use an existing sensor log, acquired in a smart space, to mine processes of human habits. PIR sensors where used as a single data source. This complicates the application of the approach in more complex business setting, especially in small environments. Another publication proposed a method using process mining techniques to discover the daily behavior of users in the ambient assisted living domain [2]. The authors of [3] focus on improving the comprehensibility of process models. Fuzzy mining is used to reduce unreadable process models, which result from applying process discovery. The approach was evaluated in a field test by collecting data over four months and comparing the results with users' manually annotated activities.

Summarized, existing work mostly focuses on discovering patterns of human behavior using sensor data. Contributions, considering a complex business context, like a manual assembly process are rare. The approach proposed in this paper combines BPM, IoT and HAR. It aims at resolving ambiguities during process execution to produce an accurate picture of manual business processes.

7 Conclusion and Outlook

In this paper, we proposed an approach to enable the discovery of manual business processes using a multi-modal recognition approach. First, we outlined our application use case before the overall concept was presented. Subsequently, the data processing pipeline was presented. Thereafter, the concrete system architecture and the implementation of its components was outlined. The focus was on the real-time multi-modal recognition of predefined activity types. This included activity type definition, motion sensor-based human activity recognition, scenery classification and the real-time requesting of user feedback during process execution. The overall recognition approach and the individual classification tasks were described. The implementation and evaluation of a case study resulted in a proof of concept of the developed artifact. The results of the case study have shown that the developed multi-modal approach is capable of capturing actual process behavior well in real-time and that there are significant improvements in accuracy and robustness of detection compared to an approach not considering multiple perspectives.

Although the developed approach works well in capturing actual process behavior, there are many possible starting points for future research. Firstly, the underlying classification models can be improved by collecting more training data, especially for underrepresented classes. Moreover, pre-processing approaches to reduce the probability of recognizing multiple sceneries should be researched. Furthermore, scenery classification can be enhanced with object detection, which allows for more detailed differentiations within a scenery and an improved case identification. Moreover, **different sensor setups** can be considered for different use cases which cover scenarios **crossing process instances and multiple workers**. Therefore, even more dimensions need to be considered simultaneously by combining recognition approaches and using data fusion techniques. Process discovery, however, is only one possible application to think of and the concept developed in this paper is the starting point for different applications. For example, the fully automated monitoring of manual process execution can be considered. In such cases process models would already be available. This would enable a pre-filtering of allowed states throughout process execution. In case **compliance problems** are detected, the worker or other process stakeholders can be notified or **appropriate measures can be taken** automatically. Especially the **inclusion of existing model knowledge** of an organization can be used to improve the recognition accuracy. E.g., existing organizational models can be used to filter possible process steps based on the worker who executes it, once the worker is identified. Future work will also examine the acceptance of such a recognition approach among workers. Privacy aspects are particularly important here. In this context, the impairment of the workers by the sensors during the process execution has to be considered and a sensor selection has to be made based on this.

References

1. van der Aalst, W.M.P., Adriansyah, A., van Dongen, B.F.: Replaying history on process models for conformance checking and performance analysis. Wiley Interdiscip. Rev.: Data Min. Knowl. Discov. **2**, 182–192 (2012)
2. Cameranesi, M., Diamantini, C., Potena, D.: Discovering process models of activities of daily living from sensors. In: Teniente, E., Weidlich, M. (eds.) BPM 2017. LNBIP, vol. 308, pp. 285–297. Springer, Cham (2018). https://doi.org/10.1007/978-3-319-74030-0_21

3. Dimaggio, M., Leotta, F., Mecella, M., Sora, D.: Process-based habit mining: experiments and techniques. In: International IEEE Conferences on Ubiquitous Intelligence & Computing, Advanced and Trusted Computing, Scalable Computing and Communications, Cloud and Big Data Computing, Internet of People, and Smart World Congress, pp. 145–152. IEEE (2016)
4. Farkas, I., Doran, E.: Activity recognition from acceleration data collected with a tri-axial accelerometer. Acta Technica Napocensis - Electron. Telecommun. **52**, 38–43 (2011)
5. Hevner, A.R., March, S.T., Park, J., Ram, S.: Design science in information systems research. Des. Sci. IS Res. MIS Q. **28**(1), 75–105 (2004)
6. Janiesch, C., et al.: The Internet-of-Things meets business process management: mutual benefits and challenges (2017)
7. Knoch, S., Ponpathirkoottam, S., Fettke, P., Loos, P.: Technology-enhanced process elicitation of worker activities in manufacturing. In: Teniente, E., Weidlich, M. (eds.) BPM 2017. LNBIP, vol. 308, pp. 273–284. Springer, Cham (2018). https://doi.org/10.1007/978-3-319-74030-0_20
8. Lara, O.D., Labrador, M.A.: A survey on human activity recognition using wearable sensors. IEEE Commun. Surv. Tutor. **15**(3), 1192–1209 (2013)
9. Lasi, H., Fettke, P., Kemper, H.G., Feld, T., Hoffmann, M.: Industry 4.0. Bus. Inf. Syst. Eng. **6**(4), 239–242 (2014)
10. Leotta, F., Mecella, M., Mendling, J.: Applying process mining to smart spaces: perspectives and research challenges. In: Persson, A., Stirna, J. (eds.) CAiSE 2015. LNBIP, vol. 215, pp. 298–304. Springer, Cham (2015). https://doi.org/10.1007/978-3-319-19243-7_28
11. Mannhardt, F., Bovo, R., Oliveira, M.F., Julier, S.: A taxonomy for combining activity recognition and process discovery in industrial environments. In: Yin, H., Camacho, D., Novais, P., Tallón-Ballesteros, A.J. (eds.) IDEAL 2018. LNCS, vol. 11315, pp. 84–93. Springer, Cham (2018). https://doi.org/10.1007/978-3-030-03496-2_10
12. Raso, R., Emrich, A., Burghardt, T., Schlenker, M., Sträter, O., Fettke, P., Loos, P.: Activity monitoring using wearable sensors in manual production processes - an application of CPS for automated ergonomic assessments. In: Drews, P., Funk, B., Niemeyer, P., Xie, L. (eds.) Multikonferenz Wirtschaftsinformatik 2018, Lüneburg, pp. 231–242 (2018)
13. Sora, D., Leotta, F., Mecella, M.: An habit is a process: a BPM-based approach for smart spaces. In: Teniente, E., Weidlich, M. (eds.) BPM 2017. LNBIP, vol. 308, pp. 298–309. Springer, Cham (2018). https://doi.org/10.1007/978-3-319-74030-0_22
14. Szegedy, C., Vanhoucke, V., Ioffe, S., Shlens, J., Wojna, Z.: Rethinking the inception architecture for computer vision. In: 2016 IEEE Conference on Computer Vision and Pattern Recognition (CVPR). IEEE (2016)
15. van der Aalst, W.M.P.: Process Mining-Data Science in Action, vol. 2, 2nd edn. Springer, Heidelberg (2016). https://doi.org/10.1007/978-3-662-49851-4
16. van der Aalst, W.M.P., ter Hofstede, A.H.M., Kiepuszewski, B., Barros, A.P.: Workflow patterns. Distrib. Parallel Databases **14**(1), 5–51 (2003)

15th International Workshop on Business Process Intelligence (BPI)

15th International Workshop on Business Process Intelligence (BPI)

Business Process Intelligence (BPI) is a growing area both in industry and academia. BPI refers to the application of data- and process-mining techniques to the field of Business Process Management. In practice, BPI is embodied in tools for managing process execution by offering several features such as analysis, prediction, monitoring, control, and optimization.

The main goal of this workshop is to promote the use and development of new techniques to support the analysis of business processes based on run-time data about the past executions of such processes. The workshop aims at discussing the current state of research and sharing practical experiences, exchanging ideas, and setting up future research directions that better respond to real needs. We aim to bring together practitioners and researchers from different communities such as business process management, information systems, business administration, software engineering, artificial intelligence, process mining, and data mining who share an interest in the analysis of business processes and process-aware information systems. In a nutshell, it serves as a forum for shaping the BPI area.

The 15th edition of this workshop attracted 21 international submissions. Each paper was reviewed by at least three members of the Program Committee. From these submissions, the top ten were accepted as full papers for presentation at the workshop. The papers presented at the workshop provide a mix of novel research ideas, evaluations of existing process mining techniques, and case studies.

Tsoury, Soffer, and Reinhartz-Berger focus on handling exceptional behavior in the context of conformance checking by first identifying such an exceptional situation and then treating it correspondingly. Boltenhagen, Chatain, and Carmona also look at the problem of conformance checking and at the creation of corresponding artifacts (e.g., alignments, anti-alignments, multi-alignments) with the goal of encoding them as a SAT instance. Li, van Zelst, and van der Aalst propose to compute performance measures for groups of activities in a process in such a way that does not require a priori knowledge and that exploits bipartite graph matching. Klijn and Fahland present conceptual and operational methods to quantify batching behavior in event logs, which is robust to the case of batches overlapping with non-batched instances. Richter, Zellner, Azaiz, Winkel, and Seidl address the problem of comparing processes based on their temporal flow and redefining their problem in terms of Earth mover's distance. Jooken, Creemers, and Jans look at the construction of social networks of collaboration from logs of version control systems by defining an adaptation of the Fuzzy Miner algorithm. Mannel and van der Aalst propose a control-flow discovery technique, which leverages the eST-Miner, for extracting a specific class of Petri nets (i.e., uni-wired nets) out of event logs. Pegoraro, Seran Uysal, and van der Aalst focus on the discovery of process models (i.e., directly-follows graph) where data is recorded together with explicit uncertainty information. Gunnarsson, vanden Broucke, and De Weerdt report on a case study concerned with process monitoring of luggage handling, and corresponding delays in case of connection flights, by presenting a novel stacked prediction model. Lopes and Ferreira present an overview of the BPI Challenges from 2011–2018. Based on this, a comparative analysis is presented reporting about process and data mining techniques, tools, and plug-ins.

As with previous editions of the workshop, we hope that the reader will find this selection of papers useful to keep track of the latest advances in the BPI area. We are looking forward to bringing new advances in future editions of the BPI workshop.

October 2019

Organization

Workshop Chairs

Boudewijn van Dongen	Eindhoven University of Technology, The Netherlands
Jochen De Weerdt	KU Leuven, Belgium
Andrea Burattin	Technical University of Denmark, Denmark
Jan Claes	Ghent University, Belgium

Program Committee

Ahmed Awad	Cairo University, Egypt
Josep Carmona	Universitat Politècnica Catalunya, Spain
Raffaele Conforti	The University of Melbourne, Australia
Johannes De Smedt	The University of Edinburgh, UK
Benoit Depaire	Universiteit Hasselt, Belgium
Claudio Di Ciccio	Vienna University of Economics and Business, Austria
Luciano García-Bañuelos	University of Tartu, Estonia
Gianluigi Greco	University of Calabria, Italy
Gert Janssenswillen	Universiteit Hasselt, Belgium
Anna Kalenkova	Higher School of Economics, Russia
Michael Leyer	University of Rostock, Germany
Fabrizio Maggi	University of Tartu, Estonia
Jorge Munoz-Gama	Pontificia Universidad Católica de Chile, Chile
Pnina Soffer	University of Haifa, Israel
Suriadi Suriadi	Queensland University of Technology, Australia
Seppe vanden Broucke	KU Leuven, Belgium
Eric Verbeek	Eindhoven University of Technology, The Netherlands
Matthias Weidlich	Humboldt-Universität zu Berlin, Germany
Hans Weigand	Tilburg University, The Netherlands
Lieje Wen	Tsinghua University, China
Wil van der Aalst	RWTH Aachen University, Germany

Impact-Aware Conformance Checking

Arava Tsoury[(⊠)], Pnina Soffer[(⊠)], and Iris Reinhartz-Berger[(⊠)]

University of Haifa, Mount Carmel, 3498838 Haifa, Israel
{atsoury, spnina, iris}@is.haifa.ac.il

Abstract. Alignment-based conformance checking techniques detect and quantify deviations of process execution from expected behavior as depicted in process models. However, often when deviations occur, additional actions are needed to remedy and restore the process state. These would seem as further reducing conformance according to existing measures. This paper proposes a conformance checking approach which considers the response to unexpected deviations during process execution, by analyzing the data updates involved and their impact on the expected behavior. We evaluated our approach in an experimental study, whose results show that our approach better captures adapted behavior in response to deviations, as compared to standard fitness measurement.

Keywords: Conformance checking · Alignment · Data impact analysis · Business processes

1 Introduction

Business process models depict the expected and normative course by which processes are expected to be executed. Over the years, process mining techniques [16] were developed for gaining insights into business process behavior from event logs, which capture events that typically correspond to activities performed using an information system. Conformance checking is an area within process mining [6] that analyzes the relations and differences between the expected behavior, specified in a process model, and the actual behavior, reflected in an event log. Most conformance checking techniques (e.g., [1, 15]) focus on control flow aspects, addressing the ordering and flow of activities in the process. Other aspects have been also considered, e.g., [7, 9, 11] and [12] for checking conformance with respect to resource and data-related rules.

Conformance checking is done by comparison of observed behavior with the modeled process. Nevertheless, process models rarely capture the full range of possible behaviors. In particular, exceptions that may occur and possible compensation activities that may be needed due to errors are typically not described in the process model [4]. Hence, comparison may yield deviation from the prescribed process model, as well as involvement of additional data operations that may influence the process state. These may have further consequences in other parts of the process. In other words, when analyzing the event logs using existing conformance checking techniques, process executions can appear to be non-conformant, whereas in fact, they exhibit the expected behavior given the unexpected situation (e.g., of an error and its correction). In contrast, process executions, in which unexpected changes were not fully and appropriately

© Springer Nature Switzerland AG 2019
C. Di Francescomarino et al. (Eds.): BPM 2019 Workshops, LNBIP 362, pp. 147–159, 2019.
https://doi.org/10.1007/978-3-030-37453-2_13

handled, may appear to have better conformance, since the differences between the process model and the event log may be smaller. We claim that the consequences of such a change can be identified by analysing the impact of the changed data values on the process. In this paper, we introduce the concept of *impact-aware conformance checking*, which takes into consideration unexpected events and data changes during process execution for calculating a conformance score. The approach is based on the existence of two main sources: (1) the event log, which may include only basic process control-related information or additional data attributes; (2) the database transaction (redo) log that captures historical data operations performed on the database as a result of business process activities. We have already proposed combining these two sources in [13]. The approach suggested here relies on combining information from event and transaction logs and employing a data-impact analysis technique [14], which propagates an unexpected change in data values along the process model and returns a set of affected process elements. We further introduce a new measure and a new technique for impact-aware conformance checking.

The remainder of the paper is organized as follows. Section 2 presents a running example to illustrate and motivate the need for our approach. Section 3 is devoted to preliminaries required and premises of our approach. Section 4 presents the approach, whose evaluation is reported in Sect. 5. Section 6 discusses related work and, finally, in Sect. 7 we conclude and discuss limitations, raising future research directions.

2   Running Example

To motivate our approach, consider a sales process, in which customers order products that are shipped after payment. Figure 1 depicts a Petri net model of this process. Typically, the data used in the process is stored in a relational database, which includes tables related to customers, orders, shipping companies, employees, and so on.

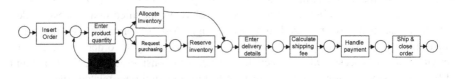

Fig. 1. The sales process model

Let us assume the customer asked to change the quantity of an ordered product *after* payment. This would require repeating the activity *enter product quantity* after *handle payment*. Now consider two possible traces (activities are marked by abbreviations):

(a) *<IO, EPQ, AI, EDD, CSF, HP, EPQ, AI, CSF, HP, SCO>*
(b) *<IO, EPQ, AI, EDD, CSF, HP, EPQ, AI, SCO>*

Clearly, both traces exhibit non-conformant behaviors, addressing unexpected changes after payment. They differ in that trace (a) includes an additional calculation of the shipping fee and additional handling of payment, while trace (b) continues the process normally after performing the change. These differences imply that existing conformance checking methods will find that trace (b) is more conformant than trace (a), since the deviation in (b) is only in the execution of two activities (*enter product quantity* and *allocate inventory*), as opposed to four activities in trace (a) (*enter product quantity, allocate inventory, calculate shipping fee, handle payment*). However, after the unexpected change occurred, trace (a) is more appropriate in business terms than trace (b), since it handles potential consequences of the change in product quantity. Such a change may necessitate additional changes in the shipment arrangements, which may, in turn, lead to changes in the shipping fee. Thus, to properly handle such a change, calculating the shipping fee should be revisited, and, as a result, an additional payment may need to be handled before the process can complete.

To tackle the problems illustrated above, we introduce the notion of impact-aware conformance, for analyzing the differences between the expected behavior and the actual one, in scenarios where unexpected changes may occur. We first elaborate preliminaries and premises and then present the approach.

3 Preliminaries and Premises

As noted, event logs, which are commonly produced by process-aware information systems, are essential for process mining techniques [16]. The content of event logs varies from "thin" logs, which contain only an event label (representing an activity) and a case identifier, to "fat" logs, which may contain data items relevant to the events [10]. Below, we provide some definitions of the basic concepts.

Definition 1 - Event (e), trace (σ): An *event* e is a tuple of the form e = (*caseID, activity, timestamp, d_1, ... d_n*), where *caseID* uniquely identifies a single process execution; *activity* is the event label; *timestamp* holds the time in which the event occurs; $d_1, ...d_n$ are additional (optional) data items. A *trace* σ is a sequence of events referring to the same process execution, $\sigma = <e_1,...,e_m>$.

Besides event logs, conformance checking requires process models against which the event logs are checked. A process model describes the sequences of activities to be performed for reaching a certain business goal [4]. Different representations of process models have been suggested [3, 12]. In this paper we use a Petri net representation.

Definition 2 - Process model (M): A process model (M) is a triplet $M = (P, T, F)$, where P is a set of places; T is a set of transitions; $F \subseteq (P \times T) \cup (T \times P)$ is the flow relation connecting places and transitions.

The state-of-the-art conformance checking approaches construct an alignment between a process model M and a trace σ [2]. An alignment is represented as a two-row matrix, where the first row consists of trace events and the second row includes process model activities.

Definition 3 - Alignment (γ): An alignment γ of a process model $M = (P, T, F)$ and a trace σ is a sequence of pairs of the form (e_i, a_i) where $e_i \in \{e \mid e \text{ is contained in } \sigma\} \cup \{>>\}$, $a_i \in T \cup \{>>\}$, and if $(e_i \neq >>$ and $a_i \neq >>$) then $e_i.$ activity $= a_i$.

Three cases of alignment moves can be defined [6] as follows: (1) *Synchronous move*: involving a matching event and model activity $e_i \neq >>$ and $a_i \neq >>$. (2) *Model move*: a model activity is not matched by an event $e_i = >>$ and $a_i \neq >>$, and (3) *Log move*: an event is not matched by a model activity $e_i \neq >>$ and $a_i = >>$. The last two are generally termed asynchronous moves.

Different alignments may be possible for the same pair of trace and process model. Alignment-based conformance checking commonly constructs "optimal" alignments, with respect to a given cost function that assigns costs to asynchronous moves. The optimal alignment minimizes the total cost of moves. Based on the optimal alignment, conformance can be quantified using the notion of fitness [15].

Definition 4 - Fitness: Given an optimal alignment γ, whose cost is $K(\gamma)$, the *fitness* is calculated as: $Fitness(\gamma) = 1 - \frac{K(\gamma)}{K(\gamma_R)}$, where $K(\gamma_R)$ is the maximal cost of a reference alignment γ_R, which is computed by concatenating moves in log for all events of σ with the alignment of the empty trace.

The basic alignment techniques and fitness measurement relate to a control-flow perspective, correlating a trace with a process model. Recent approaches suggest a multi-perspective data-aware alignment, considering also data and resources in the event log, compared to a model which includes these elements (e.g., [9]). Respective cost functions penalize deviations in data operations as well as in activity executions. Accordingly, optimal alignments are obtained and used in a fitness measure. Data-aware conformance checking is hence more powerful in detecting non-conformant behavior and explaining it. However, its applicability is restricted by the inclusion of data items in the log. Typically, when an event log is created, only a subset of data items used and manipulated by the process is included. In realistic settings, the total number of data items is too large to be included in an event log that should serve for mining and analysis. The selection may reflect expectations of relevance for analysis purposes, but the result is necessarily partial.

To overcome this incompleteness, we suggest using a control-flow perspective as a baseline for alignment, complemented by a *transaction log* for additional explanation and fine-tuning of detected deviations. Transaction logs are typically produced by database management systems and can be provided in diverse formats, depending on the platform. Yet, their structure can be generalized as follows.

Definition 5 - Transaction log: A Transaction log is a set of tuples of the form *(transactionID, beginTime, endTime, operation, caseObject, caseID, attribute, new-Value)*, where *transactionID* uniquely identifies the transaction; *beginTime* and *end-Time* are the timestamps of the transaction's beginning and end, respectively; *operation* $\in$ {insert, update, delete} specifies the change type; *caseObject* and *caseID* identify the changed data item; *attribute* and *newValue* specify the change essence.

The relation between a transaction log and a corresponding event log can be established through two common attributes: (1) the case identifier – which explicitly appears in both event and transaction logs; and (2) the timestamp – the event timestamp should be within the transaction's time frame, assuming that writing to the database is not delayed. Proposals have been made (e.g., [10, 13]) to use a combination of an event log and a transaction log, based on mapping operations between these two sources.

In this paper, the use of a transaction log is to retrieve the data items that have been changed by a specific activity (event). Our study assumes that database operations and activities execution are both recorded. Also, it assumes that data operations are always executed in the context of process activities.

A last element needed for our approach is a mechanism for analyzing data impacts on the process. Data impact analysis addresses dependencies among process elements, stemming from the required data flow. Considering a data item whose value may be changed by a deviating activity, the activities included in the partial trace before the deviation, which were affected by the data item, are its *impact* in the partial trace. The approach suggested in [14], for example, analyzes the effects of a single data item (an attribute or an object) on other process elements, including activities.

Definition 6 - Data Impact (DI): Given a data item[1] d and a trace σ, the set of activities represented by events in σ and affected (either directly or indirectly) by the value of d are termed the *data impact* of d in σ, and marked DI(d, σ).

Returning to our running example, consider the *ordered quantity*, which is determined in the activity *enter product quantity*, and assume its value has been updated unexpectedly after a partial trace σ = <*IO, EPQ, AI, EDD*>. Changing the quantity at this phase may require changes in the inventory allocation and possibly also in the delivery details (a different truck may be needed). Accordingly:
DI(*Ordered quantity, σ) = {AI, EDD}.*

4   The Approach

The main idea of our suggested approach is that once an unexpected change occurs, manifested as a log move, the expected behavior may no longer be the behavior prescribed by the process model. Rather, additional actions may be required as compensation, and these should be augmented as part of the expected behavior, forming a new basis for conformance checking. We now discuss how the expected behavior is recalculated following a deviation from the prescribed behavior.

4.1   Expected Behavior After Initial Deviation

Recall that a *log move* represents an activity that was executed but could not be aligned with the process model, namely, a deviation from the expected behavior. This deviation might occur for many reasons, such as non-compliance of the user, or due to exceptional and unexpected situations that necessitate an ad-hoc change. To understand the essence of the activity classified as a log move, we seek the data operations involved. For an in-depth analysis of the impact of deviating activities, we turn to the full set of data operations (e.g., insert, update, delete) that took place, and are typically available in the transaction log. Consider that a deviating activity a updated the values of a set of data items, denoted by *Aff(a)*. This means that each data item in *Aff(a)* was updated by the event representing a (in the same period of time and by the same case identifier).

---

[1] A data item is an attribute or an object, whose impact is of interest.

Now assume that at least some of these data items have been used earlier in the trace, making impacts on activities and on path selection decisions. This may require revisiting and correcting parts of the process activities. The activities that may potentially be revisited as a response to the deviation are identified by analyzing the data impact of the data items in $Aff(a)$, considering the trace preceding the deviation in a.

Definition 7 - Response set (RS): Given a trace σ, followed by an activity a classified as a log move in an alignment γ, the *response set* to the log move a is $RS(a,\sigma) = \bigcup_{d \in Aff(a)} DI(d,\sigma)$.

Turning to the expected behavior of the process following a deviation (log move), it should include activities from the response set in addition to the ones that correspond to the process model. We cannot, however, anticipate the order in which responses will be attended to. Rather, if the remaining trace includes activities which are not aligned with the process model (additional log moves) but are in the response set, we interpret them as responses and consider them as part of the expected behavior of the process. In other words, assume a baseline expected behavior as all compliant process traces that include the partial trace σ. The expected behavior following an initial deviation a will extend the baseline and also include the set of response activities $RS(a, \sigma)$. Listing 1 provides the algorithm for retrieving the response set, correlating the event log with a transaction log which holds all the data operations performed.

Algorithm 1: Retrieve Response Set (RetrieveRS)

Input: σ is a partial trace, a is a log move occurring immediately after σ

Output: RS is the set of response activities

1. AFF = null
```
> retrieve related data items in the transaction log (based on case
id and timestamp)
```
2. AFF = retrieveDataItems(a.caseId, a.timestamp)
3. **if** *AFF is empty* **then**
4.   **return** `> no data items have changed`
5. **end**
6. RS = $\varnothing$
```
> Iterate over all data items in AFF to retrieve their impact
```
7. **for each** d∈ AFF **do**
8.   RS ← RS ∪ DI(d, σ)
9. **end**
10. **return** RS

Listing 1. Algorithm for retrieving response activities for a specific log move

---

4.2   Impact-Aware Alignment and Fitness

With the extension of the expected behavior, we now turn to adapt the alignment and the fitness measure. Note that calculating a new optimal alignment is not possible since the expected behavior is now extended by an unordered set of response activities. We hence modify the initial given alignment to cater for the extended expected behavior. For this task, the result set and the given alignment are compared. Log moves in the alignment, whose activities are included in the response set, are marked as *response moves* (denoted by ρ). These marks allow for extending the definition of alignment to be impact-aware.

Definition 8 - Impact-Aware Alignment: An *impact-aware alignment* of a process model M = (P, T, F) and a trace σ is an alignment, i.e., a sequence of pairs of the form (e_i, a_i), where a_i can be a response move (i.e., $a_i \in (T \cup \{>>, \rho\})$), and if $(a_i = \rho)$ then there is a preceding log move $a_j =>> $ (j < i) such that e_i.activity $\in RS(a_j, <e_1...e_{j-1}>)$.

While transforming the given alignment to an impact-aware one, we keep track of the activities in RS, but also remove them from RS when encountered along the alignment. Note that this removal takes place whether the activity is part of the original process model (namely, corresponds to a synchronous move in the alignment) or not. The rationale is that an activity is included in RS if its data operations may be required as a response to some deviation. When that activity is performed afterward, additionally to or in-line with the process model, the response is considered done, and the activity can be removed from RS. Having gone through the entire alignment, the activities remaining in RS are *missed response activities*.

To illustrate, consider the following alignment γ of trace (a) in the running example.

IO	EPQ	AL	EDD	CSF	HP	EPQ	AL	CSF	HP	SCO
IO	EPQ	AL	EDD	CSF	HP	>>	>>	>>	>>	SCO

For the first log move, which corresponds to *EPQ*, the calculated response set is RS = {AL, CSF, HP}. Hence, the impact-aware alignment corresponding to γ is:

IO	EPQ	AL	EDD	CSF	HP	EPQ	AL	CSF	HP	SCO
IO	EPQ	AL	EDD	CSF	HP	>>	ρ	ρ	ρ	SCO

In this case the RS remains empty after creating the impact-aware alignment. This means that all expected corrective activities have been performed. This would not be the case with trace (b) in our running example.

As a last note, deviations and unexpected situations may occur more than once when a process is executed. We repeat the analysis for every log move which is not interpreted as a response to a previous one in a trace, recalculating the expected behavior of the process iteratively. Listing 2 provides the algorithm which analyzes an alignment and transforms it to an impact-aware alignment. Note, Algorithm 2 uses Algorithm 1 (Retrieve Response Set – see Listing 1).

We now turn to the calculation of a fitness measure based on the impact-aware alignment. We term this measure *Impact-aware fitness*. For this, the following generic cost function can be used for calculating the cost of an alignment:

$$K(\gamma) = C_M * |RS| + \sum CostOfMove, \text{ Where:}$$

$$CostOfMove = \begin{cases} 0 & Synchronous\ move \\ 1 & Log\ move\ or\ Model\ move \\ C_\rho & Response\ move \end{cases}$$

$C_\rho \in [0, 1]$ and $C_M \in [0, 1]$ are factors indicating the cost associated with a response move and with a "missed" response that remains in RS, respectively; $|RS|$ is the number of elements that remain in the response set after making the alignment impact-aware. With this cost function, the Impact-aware fitness of γ is calculated using the standard fitness definition (see Definition 4). In our running example, if we consider the cost of a response move C_ρ as 0 and the cost of a missed response C_M as 1, the impact-aware fitness of trace (a), where response to a deviation was handled, is 0.94 while its standard fitness is 0.76. The impact-aware fitness of trace (b) (response to deviation not handled) is 0.8 while its standard fitness is 0.87.

Algorithm 2: Construct impact-aware alignment

Input: γ is an alignment

Output: γ' is an impact-aware alignment

---

1. RS = ∅; γ'=γ ▷ Initialize variable
   ▷ Iterate over γ, maintain RS and create γ'
2. **for each** $(e_i. a_i) \in \gamma$, i=1...n, **do**
3.   Response=false   ▷ Initialize a Boolean variable Response
4.   **if** $(e_i.$activity $\in RS)$ **then**
5.   RS←RS - {$e_i.$activity}
6.   Response=true
7.   **end if**
8.   **if** a_i = >> **then**
9.   **if** Response **then** $\gamma'. a_i = \rho$
10.   **else** RS ← RS∪RetrieveRS(<$e_1...e_{i-1}$>, $e_i.$activity)
11.   **end if**
12. **end for**
13. **Return** γ'

Listing 2. An algorithm for transforming an alignment to an impact-aware alignment

5   Evaluation

For feasibility evaluation, we conducted an experimental study using simulated data[2]. We compared our results to those of a standard conformance checking technique [2]. Below, we elaborate on the evaluation design & execution, as well as on the results.

5.1   Evaluation Design

We used the relatively simple sales business process, depicted in Fig. 1. We further defined several scenarios, some conformant and some engaging deviations. When deviations were engaged, we controlled for required compensations (with/without), and for the extent to which these compensations were handled. Last, we generated scenarios where random noise was added. The numbers of simulated cases for these types are

---

[2] The simulated data is available at https://drive.google.com/drive/folders/1RnPxzjO1chO8NEtA3t CCcnKdOuH6_Uhf?usp=sharing

listed in Table 1. For each scenario type, we analyzed the expected differences between standard and impact-aware fitness (column 5 in the table, where standard fitness is marked as F, and impact-aware fitness – as IaF). Particularly, we expect that the impact-aware fitness in *fully handled* scenarios to be higher than the standard fitness, since more compensation activities may imply a greater discrepancy between the model and the event log, thus a lower (standard) fitness. Impact-aware fitness, on the other hand, will count those activities as response moves, whose cost is lower than that of log moves. Following a similar line of thinking, the impact-aware fitness in *not handled* scenarios will be lower than the standard fitness due to missing response activities. For *partially handled* scenarios, the impact-aware fitness depends on the number of compensation activities that have been performed vs. the number of missing responses. For *no required handling* scenarios, we expect both types of fitness to be equal, whereas for *conformant* scenarios, we expect them to be equal to 1. Finally, for *random noise* scenarios, in which activities are performed in a non-conformant manner and without reason, we cannot predict the differences between fitness values.

Following these types of scenarios, we simulated an event log of 70,805 events and a related transaction log[3] with 121,139 records. Overall, we simulated more than 7700 cases, whose distribution among the scenario types is also presented in Table 1.

Table 1. The simulated scenario types

Required compensation	Scenario type	Example in the sales process	Expected fitness differences	# cases
With	Fully handled	• Increase ordered quantity after payment	F < IaF	1141
	Partially handled	• Decrease ordered quantity after payment	Unpredictable	826
	Not handled	• Change shipping type after payment	F > IaF	903
Without	No required handling	• Update shipping address after payment • Update payment type after payment	F = IaF	627
	Conformant		F = IaF	4173
	Random noise	• Activities are added/skipped randomly	Unpredictable	40

5.2   Results

For simplicity we considered the cost of a response move in the impact-aware fitness as 0 and the cost of a missed response as 1.

Figure 2 presents a comparison of the standard and impact-aware fitness values for the different scenario types (shown as different series). Points on the diagonal line are

---

[3] The transaction log was simulated as an audit table, tracking all changes done to the database.

cases where the two types of fitness are equal, points above are where impact-aware fitness is higher than the standard fitness, and points below are where the opposite holds. As can be seen, impact-aware fitness is higher than standard fitness in situations where the deviations were fully or partially handled. In situations where the deviations were not handled at all, the standard fitness is higher than the impact-aware fitness.

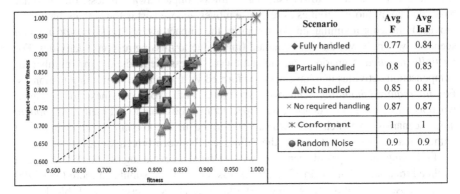

Scenario	Avg F	Avg IaF
◆ Fully handled	0.77	0.84
■ Partially handled	0.8	0.83
▲ Not handled	0.85	0.81
× No required handling	0.87	0.87
✳ Conformant	1	1
● Random Noise	0.9	0.9

Fig. 2. A comparison of the standard and impact aware fitness for the different scenarios

In summary, the obtained results confirm our expectations of how the different scenario types would be assessed in terms of impact-aware fitness. They indicate that impact-aware fitness better captures the extent to which consequences of deviations are handled, as compared to standard fitness.

6 Related Work

Many conformance checking techniques have been proposed over the years, and they mostly focused on control flow aspects. More recently, additional perspectives have been addressed, including data, resources, and time (e.g., [7, 9, 11, 12]). In particular, optimal alignments of a log and a model have been computed using cost functions that consider deviations on these perspectives as well as the control-flow ones. For performing analyses of this kind, an event log with the relevant data attributes is required (i.e., the role of the activity executer or specific values of data attributes).

Our main interest is on studies which take a step toward an explanation of nonconformance. In [8] the authors propose a conformance checking approach which uses declarative models as a basis. Such models specify constraints over process execution and allow flexibility in the expected behavior. Note that the constraints relate to the control flow of the process. They may possibly reflect data-related dependencies among activities, but this is not apparent and thus not considered explicitly. Attempting to provide explanations of non-conformance, the work in [3] focuses on an automated inference of a cost function from an event log as a basis for obtaining an optimal alignment. The data perspective is captured as a process state. Compared to our approach, the main limitation is in the need to rely on the data available in the event log. This limitation does not exist in [4], where data logs (a version of transaction log) and event logs connect the data and process perspectives. The authors propose an algorithm to construct inter-level alignment by linking the data operations recorded in the database log with activities in the control flow alignment. Furthermore, they use a CRUD (create, read, update, delete) matrix, depicting the data operations that can be performed by each activity in the process. As opposed to our approach, they do not consider the impact of deviations on the expected behavior and thus responses to deviations are not addressed. Another recent study which concerns explainability is [5]. In contrast to the common focus of conformance checking approaches (including ours), the focus of that work is on the explainability of conformant traces. The authors suggest metrics that can distinguish between conformant traces and indicate their differences.

To sum up, the literature in the field of conformance checking is rich and diverse. However, the impact of deviations and their data operations on the expected behavior has not been explicitly addressed so far, as is done in this paper. Such an analysis is important when unexpected deviations during process execution occur.

7 Conclusion

This paper has presented a conformance checking approach which considers possible changes in the expected behavior in response to unexpected changes and deviations during process execution. Besides the technical solution we propose, the paper makes two main contributions. First, it highlights the response and compensations that may need to follow unexpected situations, which are manifested as a deviation or a log move. Existing conformance checking techniques do not recognize this need and thus compensation actions, which are normative in business terms, are considered as non-conformant and result in poor conformance assessments (e.g., fitness values).

Second, it proposes a novel use of a combination of an event log and a transaction log. Rather than relying on availability of data in the event log, our approach takes a control flow-based alignment as a baseline, and only upon detection of a deviation, it seeks its data reflection in the full transaction log. By this, two main advantages are achieved: (1) overcoming the need to rely on a partial set of preselected data items that may be available in the event log; (2) avoiding the complexity of dealing with the full set of data throughout the alignment.

The suggested approach uses an impact analysis mechanism. Currently, such mechanisms exist as a stand-alone system, whose availability may form a limitation for the approach. Nevertheless, this can be further integrated into process mining environments and serve for additional analyses. Another limitation is that the approach is based on a given alignment; different alignment algorithms and cost functions may yield different alignments, which may also affect our results. This limitation, however, is inherent in the nature of alignments, which can be optimized but are not considered absolute.

We note that the evaluation reported here used synthetic data, where both the event log and the transaction log were simulated. Additional and extended evaluation of our approach is still needed, using real-life event and transaction logs. In future work, we seek to extend the approach to consider the data-flow along the activities, using a transaction log that stores data values and their changes.

Acknowledgment. This research is supported by the Israel Science Foundation under grant agreements 856/13 and 669/17.

References

1. Adriansyah, A., van Dongen, B.F., van der Aalst, W.M.: Conformance checking using cost-based fitness analysis. In: 2011 IEEE 15th International Enterprise Distributed Object Computing Conference, pp. 55–64. IEEE (2011)
2. Adriansyah, A., van Dongen, B.F., van der Aalst, W.M.: Memory-efficient alignment of observed and modeled behavior. BPM Center Report, 03-03 (2013)
3. Alizadeh, M., De Leoni, M., Zannone, N.: Constructing probable explanations of nonconformity: a data-aware and history-based approach. In: 2015 IEEE Symposium Series on Computational Intelligence, pp. 1358–1365. IEEE (2015)
4. Alizadeh, M., Lu, X., Fahland, D., Zannone, N., van der Aalst, W.: Linking data and process perspectives for conformance analysis. Comput. Secur. **73**, 172–193 (2018)
5. Burattin, A., Guizzardi, G., Maggi, F.M., Montali, M.: Fifty shades of green: how informative is a compliant process trace? In: Proceedings of CAiSE (2019)
6. Carmona, J., van Dongen, B., Solti, A., Weidlich, M.: Conformance Checking: Relating Processes and Models. Springer, Cham (2018). https://doi.org/10.1007/978-3-319-99414-7
7. de Leoni, M., Van Der Aalst, W.M.: Aligning event logs and process models for multi-perspective conformance checking: an approach based on integer linear programming. In: Daniel, F., Wang, J., Weber, B. (eds.) Business Process Management, vol. 8094, pp. 113–129. Springer, Heidelberg (2013). https://doi.org/10.1007/978-3-642-40176-3_10
8. de Leoni, M., Maggi, F.M., van der Aalst, W.M.: An alignment-based framework to check the conformance of declarative process models and to preprocess event-log data. Inf. Syst. **47**, 258–277 (2015)
9. de Leoni, M., Van Der Aalst, W.M., Van Dongen, B.F.: Data-and resource-aware conformance checking of business processes. In: Abramowicz, W., Kriksciuniene, D., Sakalauskas, V. (eds.) Business Information Systems. LNBIP, vol. 117, pp. 48–59. Springer, Heidelberg (2012). https://doi.org/10.1007/978-3-642-30359-3_5
10. de Murillas, E.G.L., Reijers, H.A., van der Aalst, W.M.: Connecting databases with process mining: a meta model and toolset. Softw. Syst. Model. **18**, 1–39 (2018)
11. Ramezani Taghiabadi, E., Fahland, D., van Dongen, B.F., van der Aalst, W.M.P.: Diagnostic information for compliance checking of temporal compliance requirements. In: Salinesi, C., Norrie, M.C., Pastor, Ó. (eds.) CAiSE 2013. LNCS, vol. 7908, pp. 304–320. Springer, Heidelberg (2013). https://doi.org/10.1007/978-3-642-38709-8_20
12. Taghiabadi, E.R., Gromov, V., Fahland, D., van der Aalst, W.M.P.: Compliance checking of data-aware and resource-aware compliance requirements. In: Meersman, R., et al. (eds.) OTM 2014. LNCS, vol. 8841, pp. 237–257. Springer, Heidelberg (2014). https://doi.org/10.1007/978-3-662-45563-0_14
13. Tsoury, A., Soffer, P., Reinhartz-Berger, I.: A conceptual framework for supporting deep exploration of business process behavior. In: Trujillo, J., et al. (eds.) Conceptual Modeling. ER 2018. LNCS, vol. 11157, pp. 58–71. Springer, Cham (2018). https://doi.org/10.1007/978-3-030-00847-5_6
14. Tsoury, A., Soffer, P., Reinhartz-Berger, I.: Towards impact analysis of data in business processes. In: Schmidt, R., Guédria, W., Bider, I., Guerreiro, S. (eds.) BPMDS 2016. LNBIP, vol. 248, pp. 125–140. Springer, Cham (2016). https://doi.org/10.1007/978-3-319-39429-9_9

15. Van der Aalst, W., Adriansyah, A., van Dongen, B.: Replaying history on process models for conformance checking and performance analysis. Wiley Interdiscip. Rev. Data Min. Knowl. Discov. **2**(2), 182–192 (2012)
16. Van Der Aalst, W.: Process Mining: Discovery, Conformance and Enhancement of Business Processes, vol. 2. Springer, Heidelberg (2011). https://doi.org/10.1007/978-3-642-19345-3

Encoding Conformance Checking Artefacts in SAT

Mathilde Boltenhagen[1(✉)], Thomas Chatain[1(✉)], and Josep Carmona[2(✉)]

[1] LSV, CNRS, ENS Paris-Saclay, Inria, Université Paris-Saclay, Cachan, France
{boltenhagen,chatain}@lsv.fr
[2] Universitat Politècnica de Catalunya, Barcelona, Spain
jcarmona@cs.upc.edu

Abstract. Conformance checking strongly relies on the computation of artefacts, which enable reasoning on the relation between observed and modeled behavior. This paper shows how important conformance artefacts like alignments, anti-alignments or even multi-alignments, defined over the edit distance, can be computed by encoding the problem as a SAT instance. From a general perspective, the work advocates for a unified family of techniques that can compute conformance artefacts in the same way. The prototype implementation of the techniques presented in this paper show capabilities for dealing with some of the current benchmarks, and potential for the near future when optimizations similar to the ones in the literature are incorporated.

1 Introduction

On its core, conformance checking relies on the computation of artefacts that link observed and modeled behavior, which are used with different purposes: spotting deviations, evaluating quality metrics of a process model, extending a process model with evidence-based information, among others [11].

Conformance checking is expected to be the fastest growing segment within Process Mining in the years to come.[1] Still, the field is facing several challenges. Among them, we highlight two important ones: techniques for a sound replay of event data on top of process models, and the advent of faithful metrics for evaluating process models with respect to observed behavior.

The former challenge is strongly related to the notion of *alignment* artefact: given a trace and a process model, find a run in the process model that is as close as possible (in edit distance terms) to the observed trace. The seminal work in [1] describes an algorithm for computing alignments based on A^*. Alternatives to this algorithm have appeared recently: in [15], the alignment problem is mapped as an *automated planning* instance. Automata-based techniques were proposed in [19,22]. Finally, the work in [5] proposes using binary decision diagrams to alleviate the computation of alignments. In this paper, we provide an alternative to the aforementioned techniques, that is based on encoding the computation of an alignment as a SAT instance. We also show how

---

[1] https://www.marketsandmarkets.com/Market-Reports/process-analytics-market-254139591.html.

© Springer Nature Switzerland AG 2019
C. Di Francescomarino et al. (Eds.): BPM 2019 Workshops, LNBIP 362, pp. 160–171, 2019.
https://doi.org/10.1007/978-3-030-37453-2_14

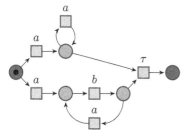

(a) Model with two types of runs :
$\langle a, a, \dots, a \rangle$ and $\langle a, b, a, b, \dots, a, b \rangle$

Log trace : $\langle b, a, b, a, b \rangle$	Hamming distance	Edit distance
run (size = 6) $\langle a, a, a, a, a, a \rangle$	3	6
run (size = 6) $\langle a, b, a, b, a, b \rangle$	6	2

(b) Comparision of Hamming and edit distances

Fig. 1. How Hamming distance penalizes alignment and anti-alignment: Trace $\langle b, a, b, a, b, a \rangle$ is closer to $\langle a, a, a, a, a, a \rangle$ than $\langle a, b, a, b, a, b \rangle$ according to Hamming distance. However there is only a shift of letter a and b between $\langle b, a, b, a, b, a \rangle$ and $\langle a, b, a, b, a, b \rangle$. The model run $\langle a, b, a, b, a, b \rangle$ seems to be a better anti-alignment for trace $\langle b, a, b, a, b, a \rangle$, as the edit distance shows.

to encode also in SAT *multi-alignments* [13], that generalize the notion of alignments so that not one but several traces are considered when computing the closest process model run.

The latter challenge is mainly concerned with the proposal of sound and meaningful metrics for *precision* and *generalization*, nowadays acknowledged as the quality dimensions with less convincing estimations (e.g., [23]). *Anti-alignments*, presented in [12], are an effective conformance artefact to foresee those model runs that deviate most (again, in terms of edit distance) with respect to a trace or a complete log. In [27] it is shown how anti-alignments can be used to provide a more consistent estimation to both precision and generalization. In this paper we provide for the first time an encoding in SAT of anti-alignments that relies on the edit distance between traces, in contrast to the Hamming distance used in [12,27]; to the best of our knowledge this is the first implementation of anti-alignments over the edit distance. The use of Hamming distance for anti-alignment shows important limit highlighted by Fig. 1. In summary, this paper presents SAT as a common means to compute important conformance checking artefacts. The prototype implementation, although not mature and therefore without having any optimization, shows interesting features that can make it a good alternative for the state of the art techniques for the same task in the near future.

The paper is organized as follows: next section provides related techniques for the tasks considered in this paper. Then in Sect. 2 we provide the background for understanding the paper. Section 3 shows how to encode the computation of a run at a given edit distance of a trace. Then Sect. 4 shows how to adapt this encoding to particular conformance checking artefacts. Section 5 reports experiments, whilst Sect. 6 concludes the paper.

Related Work. Levenshtein's edit distance is commonly used in Process Mining to define similarities between traces [8,9,24], and to align log traces to model traces [1,7,12]. Works of the literature on alignments use the definition of edit distance more

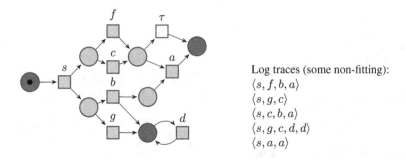

Fig. 2. Example of process model of the behaviors of users rating an app. s represents the start activity, f and c indicate if the user sent a file or wrote a comment. Transitions g and b separate good and bad marks. Bad ratings get apologies noted by activity a. Finally, the d loop is enabled when a user donates to the developer of the app.

or less implicitly [2,7,16]. Alternatively, distance between process models have been characterized in several works [4,17]. The latter uses a dependency edit distance which is computed in a similar way as our approach.

Recent studies focus on SAT implementation of Data Mining algorithm in order to satisfy all the constrains and get optima [14,20]. By introducing a SAT implementation of alignments in this paper, we hope to push a new family of algorithmic methods for conformance checking in the line of [6,12]. However, these works mostly consider Hamming distance between log traces and process models, which is usually considered less appropriate than edit distance (c.f. Fig. 1).

The SAT encoding of the edit distance between words has already been introduced in previous works [3,18]. The work in [3] studies the computation of edit distance for its interest in complexity theory.

2 Preliminaries

We use labeled Petri nets as process models.

Definition 1 (Process Model (Labeled Petri Net)) [21]). *A Process Model defined by a labeled Petri net system (or simply Petri net) is a tuple $N = \langle P, T, F, m_0, m_f, \Sigma, \Lambda \rangle$, where P is the set of places, T is the set of transitions (with $P \cap T = \emptyset$), $F \subseteq (P \times T) \cup (T \times P)$ is the flow relation, m_0 is the initial marking, m_f is the final marking, Σ is an alphabet of actions and $\Lambda : T \rightarrow \Sigma \cup \{\tau\}$ labels every transition by an action or as silent.*

A *marking* is the set of places that contain tokens for a given instant. A transition x can *fire* if all the places before x, noted $^\bullet x \overset{\text{def}}{=} \{y \in P \mid (y, x) \in F\}$, are marked. When a transition fires, all the tokens in $^\bullet x$ are removed and all the places in $x^\bullet \overset{\text{def}}{=} \{y \in P \mid (y, x) \in F\}$ become marked. A marking m' is *reachable* from m if there is a sequence of firings $\langle t_1 \ldots t_n \rangle$ that transforms m into m', denoted by $m[t_1 \ldots t_n \rangle m'$.

Definition 2 (Full runs). *A* full run *of a model N is a firing sequence $\langle t_1 \ldots t_n \rangle$ of transitions that can transform the initial marking m_0 of N to the final marking m_f of N. We note $Runs(N)$ the full runs of N.*

The run $\langle s, f, b, a \rangle$ is a full run of the model of Fig. 2. Now we formalize logs:

Definition 3 (Log). *A* log L *over an alphabet Σ is a finite set of words $\sigma \in \Sigma^*$, called* log traces.

2.1 Alignments

Aligning log traces to model traces has been identified as a central problem in Process Mining [1]. The problem is to find a run in the process model that is as close as possible to the observed trace. This closeness is usually defined in terms of the edit distance:

Definition 4 (Edit distance). *The* edit distance $dist(u,v)$ *between two words u and $v \in \Sigma^*$ is the minimal number of edits needed to transform u to v. In our case, edits can be deletions or additions of a letter in words.*

Example 1. Considering words $u = \langle s, g, c \rangle$ and $v = \langle s, b, c, a \rangle$ the number of edits to transform u to v is indeed 3. The letter g has to be removed and the letters b and a inserted. Then the two words are at distance 3.

Formally, we define alignments in a way that not only achieves the optimal edit distance between the log trace and a model trace, but also makes explicit where they match and mismatch.

Definition 5 (Alignment, optimal alignment). *An alignment of a log trace $s = \langle s_1, \ldots, s_m \rangle \in L$ to a run $u = \langle u_1, \ldots, u_n \rangle$ of process model $N = \langle P, T, F, m_0, m_f, \Sigma, \Lambda \rangle$ is a sequence of moves $\langle (s'_1, u'_1), \ldots, (s'_p, u'_p) \rangle$ with $p \leq m + n$ such that*

– *each move (s'_i, u'_i) is either:*
  - *(e_i, t_i) with $e_i = \Lambda(t_i)$ for a synchronous move*
  - *$(e_i, \gg)$ for a log move ($\gg$ is a skip symbol which indicates the mismatch), i.e. e_i is deleted in the trace*
  - *$(\gg, t_i)$ for a model move, i.e. $\Lambda(t_i)$ is inserted in the trace;*
– *projection of $\langle s'_1, \ldots, s'_p \rangle$ to Σ^* (which drops the occurrences of $\gg$), yields s*
– *projection of $\langle u'_1, \ldots, u'_p \rangle$ to T^* (which drops the occurrences of $\gg$), yields u*

An alignment between $u \in Runs(N)$ and s is optimal *if it minimizes the number of occurrences of $\gg$. This minimal number of mismatches corresponds to the edit distance $dist(u,s)$ between u and s.*

Alignments can be represented in a two-row matrix. Next figure shows an alignment between the second log trace ($\langle s, g, c \rangle$) and the run $\langle s, b, c, a \rangle$ of the model of Fig. 2.

Trace	s	g	≫	c	≫
Run	s	≫	b	c	a

Aligning traces to model often means searching the minimal number of moves, i.e. *the optimal alignment*. Different cost functions are used but the most common one applies the weights 0 for a synchronous move and 1 for a log move or a model move. Next figure shows an alignment for the trace $\langle s, g, c \rangle$ and the run $\langle s, g, c, \tau \rangle$ of the model of Fig. 2. Since τ moves incur into no cost, the alignment has cost 0, and hence it is optimal.

Trace	s	g	c	≫
Run	s	g	c	τ

3   SAT Encoding of the Edit Distance

We first introduce our SAT encoding of the edit distance between two words, which will serve as a building block for alignment, multi-alignment and anti-alignment.

The Boolean satisfiability (or SAT) problem, is the problem of determining, for a given Boolean formula, if there exists a combination of assignments to the variables that satisfies it. In the case of alignment, a SAT formula would encode the following question: *Does it exist an alignment between the trace σ and the model N for a cost d?* In other words, we are looking for a formulas that encodes *the edit distanced between a trace and a run of the model*. Our encoding is based on the same relations that are used by the classical dynamic programming recursive algorithm for computing the edit distance between two words $u = \langle u_1, \ldots, u_n \rangle$ and $v = \langle v_1, \ldots, v_m \rangle$:

$$
\begin{cases}
dist(\langle u_1, \ldots, u_i \rangle, \epsilon) = i \\
dist(\epsilon, \langle v_1, \ldots, v_j \rangle) = j \\
dist(\langle u_1, \ldots, u_{i+1} \rangle, \langle v_1, \ldots, v_{j+1} \rangle) = \\
\quad \begin{cases}
dist(\langle u_1, \ldots, u_i \rangle, \langle v_1, \ldots, v_j \rangle) & \text{if } u_{i+1} = v_{j+1} \\
1 + \min(dist(\langle u_1, \ldots, u_{i+1} \rangle, \langle v_1, \ldots, v_j \rangle), \\
\qquad\qquad dist(\langle u_1, \ldots, u_i \rangle, \langle v_1, \ldots, v_{j+1} \rangle)) & \text{if } u_{i+1} \neq v_{j+1}
\end{cases}
\end{cases}
$$

We encode this computation in a SAT formula ϕ over variables $\delta_{i,j,d}$, for $i = 0, \ldots, n$, $j = 0, \ldots, m$ and $d = 0, \ldots, n + m$. The formula ϕ will have exactly one solution, in which each variable $\delta_{i,j,d}$ is $\texttt{true}$ iff $dist(\langle u_1 \ldots u_i \rangle, \langle v_1 \ldots v_j \rangle) \geq d$.

In order to test equality between the u_i and v_j, we use variables $\lambda_{i,a}$ and $\lambda'_{j,a}$, for $i = 0, \ldots, n$, $j = 0, \ldots, m$ and $a \in \Sigma$, and we set their value such that $\lambda_{i,a}$ is $\texttt{true}$ iff $u_i = a$, and $\lambda'_{j,a}$ is $\texttt{true}$ iff $v_j = a$. Hence, the test $u_{i+1} = v_{j+1}$ becomes in our formulas: $\bigvee_{a \in \Sigma} (\lambda_{i+1,a} \wedge \lambda'_{j+1,a})$. For readability of the formulas, we refer to this coding by $[u_{i+1} = v_{j+1}]$. We also write similarly $[u_{i+1} \neq v_{j+1}]$.

In the following, we describe the different clauses of the formula ϕ of our SAT encoding of the edit distance.

$$\delta_{0,0,0} \quad \wedge \quad \bigwedge_{d>0} \neg \delta_{0,0,d} \tag{1}$$

$$\bigwedge_d \bigwedge_{i=0}^{n} (\delta_{i+1,0,d+1} \Leftrightarrow \delta_{i,0,d}) \tag{2}$$

$$\bigwedge_d \bigwedge_{j=0}^{n} (\delta_{0,j+1,d+1} \Leftrightarrow \delta_{0,j,d}) \tag{3}$$

$$\bigwedge_d \bigwedge_{i=0}^{n} \bigwedge_{j=0}^{n} [u_{i+1} = v_{j+1}] \Rightarrow (\delta_{i+1,j+1,d} \Leftrightarrow \delta_{i,j,d}) \tag{4}$$

$$\bigwedge_d \bigwedge_{i=0}^{n} \bigwedge_{j=0}^{n} [u_{i+1} \neq v_{j+1}] \Rightarrow (\delta_{i+1,j+1,d+1} \Leftrightarrow (\delta_{i+1,j,d} \wedge \delta_{i,j+1,d})) \tag{5}$$

Example 2. At instants $i = 1$ and $j = 1$ of words $u = \langle s, g, c \rangle$ and $v = \langle s, b, c, a \rangle$, the letters are the same, then, by (4), the distance is only higher or equal to 0 : $(u_1 = v_1) \Rightarrow (\delta_{1,1,0} \Leftrightarrow \delta_{0,0,0})$.

However at instants $i = 2$ and $j = 2$, the letters u_2 and v_2 are different. A step before, $\delta_{1,2,1}$ and $\delta_{2,1,1}$ are true because of the length of the subwords. Then, by (5), the distance at instants $i = 2$ and $j = 2$ is higher or equal to 2: $\delta_{2,2,2}$. The result is understandable because the edit distance costs the deletion of g and the addition of b to transform u to v.

4   SAT Encoding of Conformance Checking Artefacts

The distance between log traces and a process model is not only a distance between words, since process model describe a (possibly infinite) language. In this part, we recall the SAT encoding of full runs of Petri nets [12] and combine the implementation with the edit distance.

SAT Implementation of Process Models. For a Petri net $N = \langle P, T, F, m_0, m_f, \Sigma, \Lambda \rangle$ and n the size of the full runs, the variables $m_{i,p}$, with $i \in \{0..n\}$ and $p \in P$, represent the marking at instant i. The variables $\tau_{i,a}$ encode a firing transition $t \in T$ labelled by $a \in \Sigma$ at instant $i \in \{0..n\}$.[2] The following constraints encode the semantics of the Petri net.

- Initial marking:
$$\left(\bigwedge_{p \in m_0} m_{0,p} \right) \wedge \left(\bigwedge_{p \in P \setminus m_0} \neg m_{0,p} \right) \tag{6}$$
- Final marking:
$$\left(\bigwedge_{p \in m_f} M_{n,p} \right) \wedge \left(\bigwedge_{p \in P \setminus m_f} \neg m_{n,p} \right) \tag{7}$$
- One and only one t_i for each i:
$$\bigwedge_{i=1}^{n} \bigvee_{a \in \Sigma} (\tau_{i,a} \wedge \bigwedge_{a' \in \Sigma \setminus t} \neg \tau_{i,a'}) \tag{8}$$

---

[2] For the sake of simplicity of the encoding, we are abusing a bit the notation, i.e., assuming that labels identify transitions. This can be generalized easily for the general case when severals transitions exist for the same label.

- The transitions are enabled when they fire:

$$\bigwedge_{i=1}^{n} \bigwedge_{a\in\Sigma} (\tau_{i,a} \implies \bigwedge_{p\in \bullet t} m_{i-1,p}) \tag{9}$$

- Token game (for safe Petri nets):

$$\bigwedge_{i=1}^{n} \bigwedge_{a\in\Sigma} \bigwedge_{p\in t\bullet, \Lambda(t)=a} (\tau_{i,a} \implies m_{i,p}) \tag{10}$$

$$\bigwedge_{i=1}^{n} \bigwedge_{a\in\Sigma} \bigwedge_{p\in \bullet t\setminus t\bullet, \Lambda(t)=a} (\tau_{i,a} \implies \neg m_{i,p}) \tag{11}$$

$$\bigwedge_{i=1}^{n} \bigwedge_{a\in\Sigma} \bigwedge_{p\in P, p\notin \bullet t, p\notin t\bullet, \Lambda(t)=a} (\tau_{i,a} \implies (m_{i,p} \iff m_{i-1,p})) \tag{12}$$

Process models are now implemented in a CNF formula and can be combined with log traces for alignments.

4.1 SAT Edit Distance for Alignments

Log traces are sequences of activities that can be considered as words and implemented as presented in Sect. 3. SAT encoding of process models has been recalled in previous section. All the above clauses are considered in the SAT implementation of alignments. A last series of constraints is needed to be appended, to relate the fired transitions, represented by the $\tau_{i,a}$, with the actions in the corresponding model trace, represented by variables $\lambda_{i,a}$ from the encoding of Sect. 3:

$$\bigwedge_{i=1}^{n} \bigvee_{a\in\Sigma} (\lambda_{i,a} \iff \tau_{i,a}) \tag{13}$$

Example 3. All the full runs of the process model of Fig. 2 contain a s at the first instant. So the variable $\tau_{1,s}$ is true. If the log trace is $\sigma = \langle s, f, g \rangle$ then, $\lambda_{1,s}$ is true which implies $\delta_{1,1,0}$ by (4).

Minimization of the Edit Distance. The conjunction of the previous clauses for the full runs of the model and the their edit distance to a given log trace σ, gives a formula which has one solution per full run of the model. With each solution, the values of the $\delta_{n,m,d}$ determine the edit distance between the corresponding model trace and σ. Our goal for optimal alignments is to minimize this distance, which corresponds to the number of variables assigned to true among the $\delta_{n,m,d}$. Pseudo-Boolean solvers like MINISAT+ deal with minimization objectives under the form of a weighted sum of variables; in our case: $\sum_{d} 1 \times \delta_{n,m,d}$.

How to Deal with Runs of Different Length. In order to consider different sizes of traces and different sizes of runs, we added a loop on a *wait* activity on the final marking of the model. The SAT encoding of the edit distance is adjusted so that skipping a *wait* activity does not increment the distance between words.

Example 4. Figure 3 shows optimal alignments of every trace of the log of Fig. 2. The deviating trace $\langle s, a, a \rangle$ is then highlighted by the distance to its alignment.

Trace	Alignment	Distance
$\langle s, f, b, a \rangle$	$\langle s, f, b, a \rangle$	0
$\langle s, g, c \rangle$	$\langle s, g, c, \tau \rangle$	0
$\langle s, c, b, a \rangle$	$\langle s, c, b, a \rangle$	0
$\langle s, g, c, d, d \rangle$	$\langle s, g, c, d, \tau, d \rangle$	0
$\langle s, a, a \rangle$	$\langle s, b, c, a \rangle$	3

Fig. 3. Alignment of each trace and the model of Fig. 2

4.2 SAT Implementation for Multi-alignments

Multi-alignments were introduced in [13] as a generalization of alignments. Multi-alignments were used to define a model-based trace clustering method. Instead of aligning a trace to a run of a process model, they align a set of log traces (typically from the same cluster) to a common run of the model.

Definition 6 (Multi-alignment). *Given a finite collection C of log traces and a model N, an (optimal) multi-alignment of C to N is a full run $u \in Runs(N)$ which minimizes the sum $\sum_{\sigma \in C} dist(\sigma, u)$.*

The SAT implementation of multi-alignment requires us to duplicate the variables $\lambda_{i,a}^{\sigma}$ that represent actions in the log traces $\sigma \in L$ and the variables $\delta_{i,j,d}^{\sigma}$ that measure the edit distance to the model trace. Similarly to Sect. 4.1, the optimal multi-alignment is found by minimizing the number of variables assigned to `true` in the following objective: $\sum_d \sum_{\sigma \in L} 1 \times \delta_{n,|\sigma|,d}^{\sigma}$.

Example 5. We computed the multi-alignment of the model and the full log of Fig. 2. The optimal multi-alignment is the full run $\langle s, f, b, a \rangle$ which is at distance $d \leq 3$ to all the log traces.

4.3 SAT Implementation for Anti-alignments

Anti-alignment was introduced in [12]. Contrary to multi-alignments, the aim of anti-alignments is to get, for a given log, the run of a model which differs as much as possible to all the traces in the log. The notion of anti-alignment is used in some quality metrics like precision and generalization [26].

Definition 7 (Anti-alignment). *Given a finite collection L of log traces and a model N, an anti-alignment is a run $u \in Runs(N)$ which maximizes its distance $\sum_{\sigma \in L} dist(\sigma, u)$ to the log.*

The encoding is then very similar to the multi-alignment version. Instead of minimizing the distance to the set of log traces, we maximize it using the opposite minimization objective: $\sum_d \sum_{\sigma \in L} -1 \times \delta_{n,|\sigma|,d}^{\sigma}.$[3]

---

[3] Only the total sum of δ are minimized/maximized in our tool.

Example 6. We computed the anti-alignment of model and the set of log traces of Fig. 2. Limited by a maximum size of run to 8, the optimal anti-alignment found by our tool is $\langle s, b, d, f, d, d, d, \tau \rangle$. The minimal distance between each trace and this full run is 9. We compared our result with the module *Anti-Alignment* of ProM,[4] that computes anti-alignment over Hamming distance. For the same size of run, the algorithm returned the sequence $\langle s, b, d, d, d, c, a, d \rangle$ that is indeed linearly far from the log traces. However as either letters c or a are present in every trace, the run looks more similar to the log than the one found with edit distance.

5 Experiments

The construction of SAT formulas for alignments, multi-alignments and anti-alignments is implemented in Ocaml in our tool DARKSIDER available on github.[5] The software invokes a SAT solver, by default MINISAT+. Examples of the previous section have been fully computed by the SAT formulas to get optimal solutions. Since the approach is heavily influenced by the size of the formulas constructed (which is large even for small models), in this section we focus on heuristics to simplify the formulas, at the expense of sacrifing optimality eventually. Those are preliminary results oriented towards illustrating how to instruct the current tool.

A first heuristic is to remove $\delta_{i,j,d}$ variables, for d large . Limiting the maximal number of edits may not change the result when the traces are close to the model. A second approximation is to get optimal prefix alignment by limiting the size of the run.

Table 1(a)–(c) summarizes the experiments. Notice that while for alignments we show average numbers over one hundred traces, for multi- and anti-alignments (where only a run is computed for the whole log) total numbers are provided. The first two columns (Model and $|L|$) describe the model and the log size, respectively. Column "Size of run" shows the maximal size allowed for the run in the model (which will be an alignment in (a), a multi-alignment in (b), and an anti-alignment in (c)). Sometimes the length is limited when PRE is specified, as explained above. Then the fourth column reports the maximal number of edits allowed, sometimes with a bound as explained above. When LIM is indicated, the distance between the model and the trace is larger than what was tested. The last three columns show the time to construct the formula and the total execution time for our approach, and the time needed in ProM. Notice that since we are providing results for edit-distance based conformance artefacts, only ProM results for alignments are shown.

The goal of this paper was to demonstrate the interest of our approach based on a SAT encoding of Petri net executions and edit distance between traces. For alignments, indeed, the current SAT encoding shows bad execution times compared to the optimized algorithms implemented in ProM. But we obtain, for the first time, an implementation of the anti-alignment and multi-alignment artefacts defined over the edit distance. We believe that there is a lot of space for improvement, for instance by optimizing the encoding or by using heuristics to very efficiently find approximations of the results.

---

[4] Anti-alignment Precision/Generalization of package AntiAlignments of ProM software version 6.8, http://www.promtools.org/.

[5] https://github.com/BoltMaud/darksider.

Table 1. Experimental results for the computation of optimum and approximations of alignments and anti-alignments with our tool DARKSIDER, obtained on a virtual machine with CPU Intel Xeon 2.67 GHz and 50 GB RAM.

Model			$|L|$	Size of run	Maximal number of edits	Formula construction time (sec)	Total execution time (sec)	ProM execution time (sec)
Reference	$|T|$	$|P|$						
Fig. 2	8	7	100	7	5	0.239	0.349	0.002
M8 of [25]	15	17	100	PRE: 20	LIM:10	10.139	15.530	0.001
M1 of [25]	40	39	100	PRE: 7	LIM:10	4.924	7.16	0.005
Loan [10]	15	16	100	PRE: 19	LIM: 10	14.047	20.915	0.002

(a) Alignments (showing averages).

Model				Size of run	Maximal number of edits	Formula construction time	Total execution time	ProM
Fig. 2	8	7	10	8	7	10.101	15.362	/
			100	8	7	99.602	200.569	
M8 of [25]	15	17	10	18	LIM:6	252.471	414.174	/
			100	PRE:15	LIM:6	516.391	741.162	
M1 of [25]	40	39	10	PRE: 13	LIM:10	115.706	172.500	/
			100	PRE: 13	LIM: 5	681.95	1066.94	
Loan [10]	15	16	10	PRE: 19	15	252.572	373.683	/
			100	PRE: 9	LIM:10	359.982	508.542	

(b) Multi-alignments.

Model				Size of run	Maximal number of edits	Formula construction time	Total execution time	ProM
Fig. 2	8	7	10	8	LIM: 10	13.802	21.502	/
			100	8	LIM: 10	137.213	243.842	
M8 of [25]	15	17	10	18	LIM:10	103.812	148.271	/
			100	PRE: 10	LIM: 10	343.529	496.733	
M1 of [25]	40	39	10	39	LIM:10	1337.806	2069.505	/
			100	PRE:13	LIM:5	680.556	995.361	
Loan [10]	15	16	10	PRE: 19	LIM: 10	140.840	203.257	/
			100	PRE:19	LIM: 10	1526.048	2185.785	

(c) Anti-alignments.

6   Conclusion

This paper has shown a unified approach to compute important conformance artefacts over SAT. Thanks to its high versatility, the encoding as SAT formula allows one to compute exact solutions to various problems (here alignments, anti-alignments and multi-alignments) under various optimality criteria. In particular, we show for the first time how anti-alignments can be computed for the edit distance by formulating the problem as a SAT instance. Although technically sound, the encodings proposed in this paper suffer from the explosion of the SAT formulas created. As main research direction, we are working into finding better encodings that can alleviate significantly the size of such formulae, but also incorporating other optimizations and heuristics (possibly inspired from optimization techniques used in automated planning) that can make the approach more efficient in practice.

Acknowledgments. This work has been supported by Farman institute at ENS Paris-Saclay and by MINECO and FEDER funds under grant TIN2017-86727-C2-1-R.

References

1. Adriansyah, A.: Aligning observed and modeled behavior. Ph.D. thesis, Department of Mathematics and Computer Science (2014)
2. Adriansyah, A., Munoz-Gama, J., Carmona, J., van Dongen, B.F., van der Aalst, W.M.P.: Alignment based precision checking. In: La Rosa, M., Soffer, P. (eds.) BPM 2012. LNBIP, vol. 132, pp. 137–149. Springer, Heidelberg (2013). https://doi.org/10.1007/978-3-642-36285-9_15
3. Backurs, A., Indyk, P.: Edit distance cannot be computed in strongly subquadratic time (unless SETH is false). In: Proceedings of the Forty-Seventh Annual ACM Symposium on Theory of Computing, pp. 51–58. ACM (2015)
4. Bae, J., Liu, L., Caverlee, J., Zhang, L.-J., Bae, H.: Development of distance measures for process mining, discovery and integration. Int. J. Web Serv. Res. (IJWSR) **4**(4), 1–17 (2007)
5. Bloemen, V., van de Pol, J., van der Aalst, W.M.P.: Symbolically aligning observed and modelled behaviour. In: 18th International Conference on Application of Concurrency to System Design, ACSD, Bratislava, Slovakia, 25–29 June, pp. 50–59 (2018)
6. Boltenhagen, M., Chatain, T., Carmona, J.: Generalized alignment-based trace clustering of process behavior. In: Donatelli, S., Haar, S. (eds.) PETRI NETS 2019. LNCS, vol. 11522, pp. 237–257. Springer, Cham (2019). https://doi.org/10.1007/978-3-030-21571-2_14
7. Jagadeesh Chandra Bose, R.P., van der Aalst, W.: Trace alignment in process mining: opportunities for process diagnostics. In: Hull, R., Mendling, J., Tai, S. (eds.) BPM 2010. LNCS, vol. 6336, pp. 227–242. Springer, Heidelberg (2010). https://doi.org/10.1007/978-3-642-15618-2_17
8. Jagadeesh Chandra Bose, R.P., van der Aalst, W.M.P.: Abstractions in process mining: a taxonomy of patterns. In: Dayal, U., Eder, J., Koehler, J., Reijers, H.A. (eds.) BPM 2009. LNCS, vol. 5701, pp. 159–175. Springer, Heidelberg (2009). https://doi.org/10.1007/978-3-642-03848-8_12
9. Jagadeesh, R.P., Bose, C., Van der Aalst, W.M.P.: Context aware trace clustering: towards improving process mining results. In: Proceedings of the 2009 SIAM International Conference on Data Mining, pp. 401–412. SIAM (2009)
10. Buijs, J.C.A.M.: Loan application example. 4TU. Centre for Research Data. Dataset (2013). https://doi.org/10.4121/uuid:bd8fcc48-5bf3-480e-8775-d79d6c700e90
11. Carmona, J., van Dongen, B., Solti, A., Weidlich, M.: Conformance Checking - Relating Processes and Models. Springer, Cham (2018). https://doi.org/10.1007/978-3-319-99414-7
12. Chatain, T., Carmona, J.: Anti-alignments in conformance checking – the dark side of process models. In: Kordon, F., Moldt, D. (eds.) PETRI NETS 2016. LNCS, vol. 9698, pp. 240–258. Springer, Cham (2016). https://doi.org/10.1007/978-3-319-39086-4_15
13. Chatain, T., Carmona, J., van Dongen, B.: Alignment-based trace clustering. In: Mayr, H.C., Guizzardi, G., Ma, H., Pastor, O. (eds.) ER 2017. LNCS, vol. 10650, pp. 295–308. Springer, Cham (2017). https://doi.org/10.1007/978-3-319-69904-2_24
14. Davidson, I., Ravi, S.S., Shamis, L.: A SAT-based framework for efficient constrained clustering. In: Proceedings of the 2010 SIAM International Conference on Data Mining, pp. 94–105. SIAM (2010)

15. de Leoni, M., Marrella, A.: Aligning real process executions and prescriptive process models through automated planning. Expert Syst. Appl. **82**, 162–183 (2017)
16. De Leoni, M., van der Aalst, W.M.P.: Data-aware process mining: discovering decisions in processes using alignments. In: Proceedings of the 28th Annual ACM Symposium on Applied Computing, pp. 1454–1461. ACM (2013)
17. Dijkman, R., Dumas, M., Garcia-Banuelos, L., Kaarik, R.: Aligning business process models. In: 2009 IEEE International Enterprise Distributed Object Computing Conference, pp. 45–53. IEEE (2009)
18. Groce, A., Chaki, S., Kroening, D., Strichman, O.: Error explanation with distance metrics. Int. J. Softw. Tools Technol. Transf. **8**(3), 229–247 (2006)
19. Leemans, S.J.J., Fahland, D., van der Aalst, W.M.P.: Scalable process discovery and conformance checking. Softw. Syst. Model. **17**(2), 599–631 (2018)
20. Métivier, J.-P., Boizumault, P., Crémilleux, B., Khiari, M., Loudni, S.: Constrained clustering using SAT. In: Hollmén, J., Klawonn, F., Tucker, A. (eds.) IDA 2012. LNCS, vol. 7619, pp. 207–218. Springer, Heidelberg (2012). https://doi.org/10.1007/978-3-642-34156-4_20
21. Murata, T.: Petri nets: properties, analysis and applications. Proc. IEEE **77**(4), 541–574 (1989)
22. Reißner, D., Conforti, R., Dumas, M., La Rosa, M., Armas-Cervantes, A.: Scalable conformance checking of business processes. In: Panetto, H., et al. (eds.) OTM 2017. LNCS, vol. 10573, pp. 607–627. Springer, Cham (2017). https://doi.org/10.1007/978-3-319-69462-7_38
23. Tax, N., Lu, X., Sidorova, N., Fahland, D., van der Aalst, W.M.P.: The imprecisions of precision measures in process mining. Inf. Process. Lett. **135**, 1–8 (2018)
24. Tax, N., Sidorova, N., Haakma, R., van der Aalst, W.M.P.: Event abstraction for process mining using supervised learning techniques. In: Bi, Y., Kapoor, S., Bhatia, R. (eds.) IntelliSys 2016. LNNS, vol. 15, pp. 251–269. Springer, Cham (2018). https://doi.org/10.1007/978-3-319-56994-9_18
25. Taymouri, F., Carmona, J.: Model and event log reductions to boost the computation of alignments. In: Proceedings of the 6th International Symposium on Data-Driven Process Discovery and Analysis (SIMPDA 2016), Graz, Austria, 15–16 December 2016, pp. 50–62 (2016)
26. van Dongen, B.F., Carmona, J., Chatain, T.: A unified approach for measuring precision and generalization based on anti-alignments. In: La Rosa, M., Loos, P., Pastor, O. (eds.) BPM 2016. LNCS, vol. 9850, pp. 39–56. Springer, Cham (2016). https://doi.org/10.1007/978-3-319-45348-4_3
27. van Dongen, B., Carmona, J., Chatain, T., Taymouri, F.: Aligning modeled and observed behavior: a compromise between computation complexity and quality. In: Dubois, E., Pohl, K. (eds.) CAiSE 2017. LNCS, vol. 10253, pp. 94–109. Springer, Cham (2017). https://doi.org/10.1007/978-3-319-59536-8_7

Performance Mining for Batch Processing Using the Performance Spectrum

Eva L. Klijn and Dirk Fahland(✉) (iD)

Eindhoven University of Technology, Eindhoven, The Netherlands
`e.l.klijn@student.tue.nl`, `d.fahland@tue.nl`

Abstract. Performance analysis from process event logs is a central element of business process management and improvement. Established performance analysis techniques aggregate time-stamped event data to identify bottlenecks or to visualize process performance indicators over time. These aggregation-based techniques are not able to detect and quantify the performance of time-dependent performance patterns such as batches. In this paper, we propose a first technique for *mining* performance features from the recently introduced *performance spectrum*. We present an algorithm to detect batches from event logs even in case of batches overlapping with non-batched cases, and we propose several measures to quantify batching performance. Our analysis of public real-life event logs shows that we can detect batches reliably, batching performance differs significantly across processes, across activities within a process, and our technique even allows to detect effective changes to batching policies regarding consistency of processing.

Keywords: Process mining · Performance mining · Batch processing · Performance spectrum

1 Introduction

Analyzing process performance based on event data is a central activity in business process intelligence and process management. Its aim is to identify performance characteristics and problems such as bottlenecks, unusual variations in performance, or permanent drifts and changes in performance. In the following we focus on time-based performance analysis. Standard techniques to performance analysis are to implement key performance indicators (KPIs) in the running system to measure throughput times, waiting times, etc. [5, Ch.7], or to derive performance information from event logs [5, Ch.10] which is then *aggregated*, e.g., average waiting times of all cases at a task, and then visualized in charts or by annotation of process models [1].

The recently introduced *performance spectrum* [4] showed that all processes exhibit non-stationary performance leading to local performance variations, drifts in performance, and performance patterns. Even the same process step may exhibit two or more different discernible perfomance characteristics. For

© Springer Nature Switzerland AG 2019
C. Di Francescomarino et al. (Eds.): BPM 2019 Workshops, LNBIP 362, pp. 172–185, 2019.
https://doi.org/10.1007/978-3-030-37453-2_15

example, *batch processing* allows grouping several cases arriving at a task into a queue and to process them together in a short period [8,13]. While batching is an effective technique to improve process performance [12] the performance spectrum in Fig. 2 shows that batch performance at one task may vary greatly over time and batching and non-batching behavior may overlap. Aggregation-based performance techniques such as KPIs are not able to discern and quantify local performance characteristics of this kind.

In this paper, we consider the problem of *mining performance characteristics* directly from the performance spectrum. Specifically focusing on batching, our aim is to (1) detect and discern batching from non-batching behavior in any process step, allowing (2) to specifically quantify and compare batching performance and non-batching performance. We contribute an *algorithm to detect batches* from event logs – even when overlapping with non-batching behavior; our only parameter is the minimum batch size. We introduce several *measures to quantify batches* such as size, waiting time in/outside a batch, arrival time in/outside a batch, and batching frequency. We implemented the algorithm and the measures and applied them on several public real-life event logs. We found significant batching in 4 of the event logs; batching is in most cases overlaid with non-batching behavior; the batching characteristics of different process steps are significantly different and subject to large variations over time our technique allows to detect.

Next, we first discuss related work on analyzing batch processing in Sect. 2. We then recall the various types of batch processing and how they are described in the performance spectrum in Sect. 3. We define the batching measures in Sect. 4 and the detection algorithm in Sect. 5 which we evaluate in Sect. 6. We conclude in Sect. 7.

2   Related Work

Batch processing has been studied in several domains. In operations management it is used to solve supply chain scheduling problems [15], order picking problems [6] and machine scheduling problems [2], with the aim of reducing costs in holding, inventory, and machine setup as a trade-off to increased delivery and service times.

In business process management, batch processing received increased attention over the past five years. Pufahl and Weske [13] proposed to explicitly model batch activities, which allows to evaluate batch activity performance [12]. Thereby, batch performance is not just activity-dependent but subject to additional process perspectives such as resource involvement [14], which is confirmed by the performance spectrum [4].

The problem of discovering batch process models from event logs has been explored by Wen et al. [17]. The proposed algorithm requires a prior knowledge of where batching takes place, which has to be provided in a pre-processing step. Nakatumba explores the problem of identifying batch processing from a resource perspective [11] by assuming resources work in batches and discovering batching moments from resource availability; the technique cannot measure detailed characteristics of batching behavior. Martin et al. [10] distinguishes and formalizes three types of batch processing by analyzing a matrix of resource-activity pairs for "footprints" corresponding to the three batching types. By calculating metrics, the influence of the batching behavior on the process performance can be derived. The technique was extended to discover batch activation rules from event logs [8] describing for example the time duration or number of cases required to cause processing of a batch. The work also shows that correct batch activation rules aid in more precise performance prediction (in simulation models).

This research elaborates on [8,10] in several ways. We adopt the three types of batch processing proposed and lift them to the performance spectrum, and we propose several batch measures to quantify batch characteristics more precisely. We differ by detecting batching exclusively from the control-flow perspective (not requiring resource information or any other a priori input) by transforming event log data into the performance spectrum [4]. Further, our method specifically preserves the time-characteristics of all cases in batch detection allowing to quantify the detected batches.

Batching behavior and performance is also relevant in queue mining from event logs [16]. Our analysis of batching behavior between two subsequent tasks can also be seen as analyzing the batching behavior in a queue; however we do not consider other parameters considered in queue mining [16].

3 Batch Processing and Performance

In this section, we introduce the types of batch processing and the corresponding pattern in the performance spectrum. We then combine these concepts into a definition of a batch processing pattern, based on control-flow event logs only.

The processing of cases in a simultaneous manner or in batches, often occurs when they share properties or are processed periodically. These structural and behavioral properties of the process and its instances also influence the way cases are batched. Martin et al. [9] distinguishes three types of batching. *Simultaneous*: Cases are executed by the same resource and have an identical timestamp for the start and completion of the processing step, as illustrated in Fig. 1 (left). *Sequential*: Cases are executed by the same resource and the completion timestamp of each case must (roughly) correspond to the start timestamp of the case that follows, as illustrated in Fig. 1 (middle). *Concurrent*: Cases are executed by the same resource and the start timestamp of a case must be earlier than the complete timestamp of the preceding case, meaning cases should partially overlap in time. This is illustrated in Fig. 1 (right).

The *performance spectrum* [4] visualizes process behavior over time as illustrated in Fig. 2. It describes how cases transition from activity a to activity b over time t, also called the *segment* (a, b). Whenever in a case a is directly followed by b, we observe an *occurrence* of this segment taking place from time t_a (moment of occurrence of a) to time t_b. Each occurrence of a segment is plotted as a line from (t_a, y_a) to (t_b, y_b).

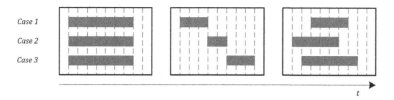

Fig. 1. Schematic representation illustrating three types of batching: simultaneous (left), sequential (middle) and concurrent (right)

Fig. 2. Detailed performance spectrum example of single process segment of Road Traffic Fine Management log

Fig. 3. Schematic representations of batch processing types of Fig. 1 in the performance spectrum: simultaneous (left), sequential (middle) and concurrent (right)

The arrangement of the lines in Fig. 2 forms distinct patterns which can be classified by a taxonomy [4]. This taxonomy defines batching as FIFO behavior where batching occurs at either at the preceding step a, the succeeding step b or both. We will focus on the case where batching occurs at the succeeding step b, called batch(e) [4]. On the performance spectrum, this will show up as lines starting at various points on y_a and converging to a more or less single point in time on y_b, as can also be identified in Fig. 2.

To detect batch(e) from a process segment, the type of batch processing applied in the succeeding processing step must be taken into account. Also, a process segment only holds information regarding the start and end timestamp of cases within that segment, meaning that of each adjacent processing step, only a single timestamp is known. Thus, to distinguish the different types of batch processing for the succeeding processing step, we are limited to the distribution of timestamps at b. Figure 3 illustrates how the different types of batch processing of Fig. 1 and the corresponding distributions of timestamps at b are represented in the performance spectrum.

We see that simultaneous batch processing, having identical start and end timestamps for all cases, clearly distinguishes itself from the other types. Sequential and concurrent batch processing both have distinct start and end timestamps between cases, of which the distribution is dependent on case or processing step characteristics. Being that of each processing step either the start timestamp or the complete timestamp is known within a segment, no distinction can be made between sequential and concurrent batch processing. In the following, we therefore refer to the latter two types as *disjoint* batch processing to distinguish them from *simultaneous* batching processing.

Activity life-cycle information has limited influence on the batch patterns. In most cases, events are recorded using their start timestamp, complete timestamp or both, meaning that a segment will either be between two start events, two complete events or between the complete event of the preceding activity and the start event of the succeeding activity. When considering Fig. 3 once more, we see that for each of the batch processing types, the distribution of start timestamps is similar to the distribution of end timestamps, implying that different procedures of event log recording will not influence the process of detecting batches.

We now project the possible batch(e) patterns w.r.t. the batch processing types on the performance spectrum, illustrated in Fig. 4. Simultaneous batch processing shows non-crossing lines, all ending at the same time-point on y_b, illustrated in Fig. 4 (left). Disjoint batch processing shows lines ending at distinct time-points on y_b, either non-crossing (FIFO-ordered) or crossing (unordered), illustrated in Fig. 4, middle and right, respectively. For this paper, we decide to conform to the FIFO constraint of batch(e) as it is defined in [4] and will therefore discard the latter case.

As a result, we define *batching on end* as follows: *Multiple cases are batched on end when the next processing step handles cases either (1) simultaneously, or (2) in the order they arrived in and in a very short period of time.*

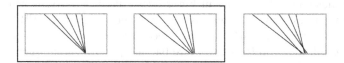

Fig. 4. Schematic representations of batching on end pattern: simultaneous (left), disjoint FIFO ordered (middle) and disjoint unordered (right)

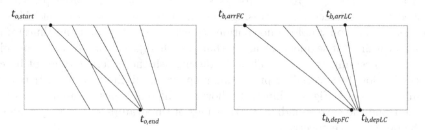

Fig. 5. Performance spectrum visualization of observation (left) and batch (right) time parameters

4 Measuring Batch Performance in the Performance Spectrum

This section discusses several measures for quantifying the performance of batch processing for a given set of batches; Sect. 5 then describes the procedure to identify batches from the performance spectrum. We will first present the three main objects embodied by the algorithm along with their parameters, since the batch processing measures will be derived thereof.

Segment. A segment $s = (x, y)$ is defined by a pair of activities x and y and a start $t_{s,start}$ and end $t_{s,end}$ time for the batching analysis.

Observation. An observation $o = (o_{id}, t_{o,start}, t_{o,end})$ in segment (x, y) describes that the case o_{id} had activities $\ldots, a, b, \ldots$ follow each other with x and y occurring at $t_{o,start}$ and $t_{o,end}$, respectively, as illustrated in Fig. 5 (left).

Batch. A batch $b = o_1, \ldots, o_k$ is a set of observations in (x, y) where the $t_{o_i,end}$ end timestamps of all observations in b are equal (simultaneous batch) or more or less equal (disjoint batch). In the latter case, observations within b must also satisfy the FIFO rule, meaning that if $t_{o_i,end}$ is smaller than $t_{o_j,end}$, $t_{o_i,start}$ must also be smaller than $t_{o_j,start}$. The interest lies with finding all batches in a segment (x, y), i.e., multiple sets $b_1, \ldots, b_m$, of a minimum batch size $k_{min} \leq |b|$.

In the following, we define the measures that are derived of the objects parameters defined in the previous, both for batch- and segment-level perspectives.

One Batch. Each batch has a first case arrival time $t_{b,arrFC}$ and a last case arrival time $t_{b,arrLC}$, derived from $t_{o,start}$ of the first observation and $t_{o,start}$ from the last observation in b. Additionally, each batch also has a first case departure time $t_{b,depFC}$ and a last case departure time $t_{b,depLC}$, derived from $t_{o,end}$ of the first observation and $t_{o,end}$ of the last observation in b. In the case of a simultaneous batch, $t_{b,depFC}$ and $t_{b,depLC}$ are equal. Figure 5 (right) shows an illustration of these time parameters.

Case Waiting Time. We define W_o as the case waiting time or the time a case is pending before the next processing step. On batch-level, we will also define $W_{i,min}$ and $W_{i,max}$, expressing the minimum and maximum W_o of observations within batch i, respectively. This will give insights in the batching time-window used. On segment-level, we additionally define $W_o \in b$ as the waiting time of a batched case and $W_o \notin b$ as the waiting time of a non-batched case. W_o can directly be derived from the performance spectrum, as illustrated in Fig. 6.

Intra-Batch Case Interarrival Time. We define $IBIA$ as the intra-batch case interarrival time, expressing the amount of time that lapses between the arrival of two successive cases within a batch. Each batch consists of k observations, each observation having a time of arrival $t_{o,start}$ within that segment. From these times, $k - 1$ intra-batch interarrival times can be derived, as illustrated in Fig. 7. From a batch-level perspective, solely the $IBIA$ measures of either the green or orange highlighted batch would be taken into account, whereas from a segment level perspective, the $IBIA$ measures of both batches would be taken into account.

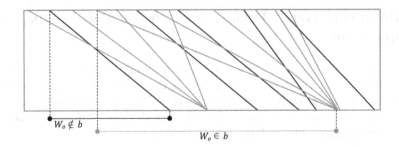

Fig. 6. Illustration of waiting time measures on the performance spectrum

Case Interarrival Time. Where $IBIA$ measures arrival in a batch, the case inter-arrival time IA measures the time between two cases considering *all* segment observations. We additionally define $IA \in b$ and $IA \notin b$, expressing the inter-arrival time of batched cases (independent of which batch it belongs to) and non-batched cases, respectively. Figure 7 illustrates these measures and their difference to $IBIA$.

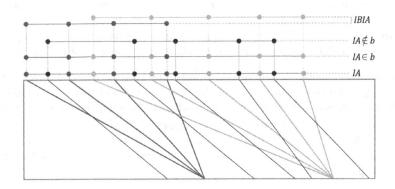

Fig. 7. Illustration of interarrival time measures on the performance spectrum

Batch Interval. We define the batch interval BI as the time lapse between two successive moments or periods of batching, as illustrated in Fig. 8. A moment or a period of batching from a segment perspective means the moment or period in which cases are taken into processing. When considering the batch parameters, this moment or period ranges from $t_{b,depFC}$ to $t_{b,depLC}$, where in the case of a simultaneous batch, the values of these parameters are equal. Intervals between batches are calculated on segment-level and, given there are m batches in a segment and a segment contains at least one batch $m \geq 1$, $m-1$ batch intervals can be derived. The duration of a batch interval between batch i and its successive batch is calculated as follows: $BI_i = t_{b_{i+1},depFC} - t_{b_i,depLC}$.

Batch Size. The batch size k expresses the amount of observations in batch b.

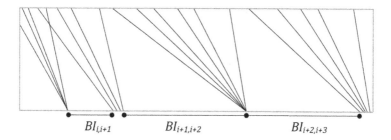

$BI_{i,i+1}$ $\qquad$ $BI_{i+1,i+2}$ $\qquad\qquad$ $BI_{i+2,i+3}$

Fig. 8. Illustration of batch interval BI measure on the performance spectrum

Batching Frequency. We define BF as the batching frequency or percentage, expressing the fraction of observations that are part of a batch out of the total amount of observations within a segment. For a segment containing n observations and m batches, each with k_m observations, the batching frequency is calculated as follows: $BF = \sum_{i=1}^{m} k_i/n$

5 Detecting Batches in the Performance Spectrum

To evaluate the presence of batching and provide performance insights, we propose the following Batch Detection Algorithm.

As input, we use the data from the performance spectrum miner [3], which is represented as a table with columns for s, o_{id}, $t_{o,start}$ and W_o. We first transform this data by (1) calculating $t_{o,end}$, (2) converting data to lists of observations per segment, containing columns for o_{id}, $t_{o,start}$, W_o and $t_{o,end}$ and (3) sorting each list of observations twice, first by $t_{o,start}$ and subsequently by $t_{o,end}$, during which the sorting on $t_{o,start}$ is preserved, resulting in a list L_s per segment s.

The batch detection algorithm for segment s takes as input the list L_s, the minimum batch size k_{min}, and the maximum delay parameter γ defining the maximum amount of time allowed between the two successive cases leaving a batch. For a simultaneous batch, $\gamma = 0$. For a disjoint batch, γ should remain small to adhere to the concept of batch processing and should be chosen based on domain knowledge.

For each list, the algorithm will (4) add the first/next observation to a temporary batch and (5) check for each subsequent pair of observations the following constraint:

$$t_{o_{i-1},end} \leq t_{o_i,end} \leq \gamma + t_{o_{i-1},end} \wedge t_{o_i,start} \geq t_{o_{i-1},start} \qquad (1)$$

If the constraints holds, we (6a.1) add the next observation to the temporary batch. Otherwise we (6b.1) check whether the size of the temporary batch is $\geq k_{min}$ and if so, create a definitive batch b from it and (6b.2) clear the temporary batch and start over. See [7] for pseudocode.

6 Experimental Results

We implemented the batch detection algorithm of Sect. 5 and all measures of Sect. 4 in a Java command-line tool. The tool, available at https://github.com/ multi-dimensional-process-mining/psm-batchmining, takes the PSM data [3] as input and returns for each segment two event logs containing batched and non-batched cases and several statistics tables. We applied our implementation on all BPI Challenge event logs except BPIC'16 and BPIC'18 and the Road Traffic Fine Management (RF) log, available at https://data.4tu.nl/repository/collection: event_logs_real, using $k_{min} = 10$ to evaluate noteworthy presence of batching. In the RF log, we found that 15 out of 70 segments have a batch frequency $BF \geq 20\%$ and 8 have a mean batch size $\mu_k \geq 20$. This amounts to 6 out of 4231 with $BF \geq 20\%$ for the BPIC'11 log, but none with $\mu_k \geq 20$. The BPIC'17 log contains 56 out of 178 segments with $BF \geq 20\%$ and 42 segments with $\mu_k \geq 20$; in BPIC'19, 102 out of 498 segments have $BF \geq 20\%$ and 53 have $\mu_k \geq 20$.

Here, we discuss the RF log and the BPIC'17 log. We show how the algorithm detects batches and provides insights in batching performance from a global (segment) perspective for both logs; for the segment *Create Fine:Send Fine* (RF log) we show how a detailed batch performance analysis gives insights into changes in batching behavior.

We have selected 6 out of 70 segments from the RF log and 1 out of 178 segments from the BPIC'17 log, for which we illustrate the presence of batching and the performance thereof. We have tested various minimum batch sizes to detect clear and distinct batches, resulting in $k_{min} = 20$ for all segments except the third, for which $k_{min} = 30$. Figure 9 shows for each of the chosen segments S1-S7 the performance spectrum of the detected batches (top) and of all non-batched cases (bottom); for S1 we also show the spectrum for the entire log for illustration. Next, we discuss the quality of the batch identification for S1-S7 in Fig. 9 together with the performance measures for the identified batches which are reported in Tables 1 and 2.

We discover unusual batching behavior in **S1**, where batching is seemingly absent at first glance. However, the projection with batched traces shows distinct larger batches, revealing a meaningful presence of batch processing.

In **S2** we are also able to detect the obvious larger batch among the non-batched traces, while also detecting a few that are significantly smaller. This is confirmed by the high standard deviation for k in Table 1. Additionally, Table 1 also reveals one of the highest $\sigma_{W_o \in b}$ relative to $\mu_{W_o \in b}$ compared to other segments, which indicates very distinct performance in terms of waiting time when comparing the large batch to the smaller ones.

In **S3** we are able to detect a fairly large amount of small batches. Though distinct, the behavior seems non-controlled and most probably indicates high workload on those specific days.

In **S4** we discover frequent processing of primarily large batches, which, interestingly, almost all exclusively overlap one another. We also see that non-batched

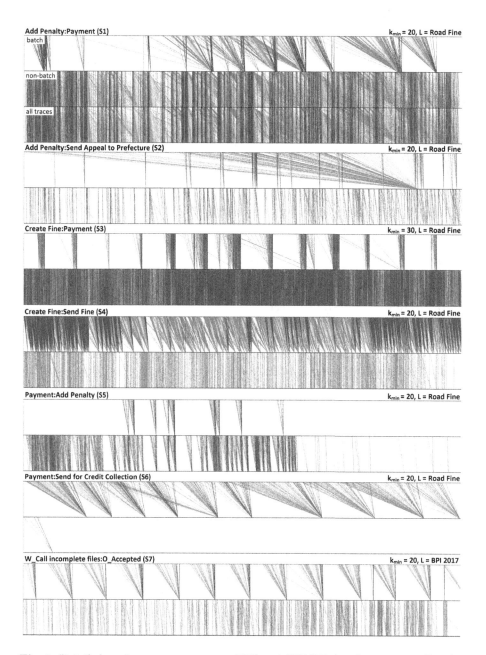

Fig. 9. Detailed performance spectrum of RF and BPIC'17 log for segments S1: *Add Penalty:Payment*, S2: *Add Penalty:Send Appeal to Prefecture*, S3: *Create Fine:Payment*, S4: *Create Fine:Send Fine*, S5: *Payment:Add Penalty*, S6: *Payment:Send for C.C.* and S7: *W_Call incomplete files:O_Accepted*.

Table 1. General performance measurements for S1-S7

Segment	n	BF (%)	m	μ_k	σ_k	μ_{BI}	σ_{BI}	$\mu_{W_o \in b}$	$\sigma_{W_o \in b}$	$\mu_{W_o \notin b}$	$\sigma_{W_o \notin b}$
S1	18621	19.55	104	35.01	25.57	1008.69	2884.14	7047.29	4265.92	3091.11	4070.80
S2	2915	42.54	26	47.69	74.22	4182.72	4233.06	12168.77	20262.79	584.53	910.60
S3	46952	17.67	227	36.55	6.33	472.89	1581.85	241.39	870.23	253.76	481.78
S4	103392	97.15	206	487.61	305.99	569.09	581.72	2161.32	969.51	98.93	589.78
S5	3902	16.27	23	27.61	7.33	1578.50	2807.62	814.83	427.97	792.74	440.05
S6	1538	99.22	10	152.60	41.05	10752.00	3340.28	10697.86	4781.92	6232.00	1340.79
S7	4783	61.61	18	163.72	117.19	518.03	285.79	198.12	140.28	16.61	42.96

traces are processed almost immediately, which is also revealed by the low $\mu_{W_o \notin b}$, meaning that in terms of performance, batched cases have higher waiting times.

While we do detect a few batches in **S5**, Fig. 9 shows more undetected batches, which is probably a result of the FIFO constraint. Additionally, $\mu_{W_o \in b}$ and $\mu_{W_o \notin b}$ and also $\sigma_{W_o \in b}$ and $\sigma_{W_o \in b}$ are nearly equal for this segment, indicating similar performance for batched and non-batched traces.

In **S6** we see clear batches that are processed regularly. The non-batch projection shows one small batch is probably not completely recorded in the log, and therefore too small to be detected.

In **S7** (from BPIC'17), we initially did not detect the apparent batches because the FIFO constraint is violated. By a more detailed analysis of the data, we relaxed the batching constraint 1 in our algorithm from FIFO to a 12-hour time-window: cases handled from 0:00 to 11:59 or from 12:00 to 23:59 are part of the same batch. We now detect clear non-overlapping batches, which seem to be processed each month. Unfortunately, the values for μ_{BI} (≈ 22 days) and σ_{BI} (too high), do not correspond. When evaluating batch-level output, we see that three of the larger batches are split in two, due to the batching-window cut-off at 12:00.

We discover that the three segments with the highest BF (S4, S6 and S7) also have the largest μ_k. While this serves as a good indication for distinct batches, we also found that lower values for these measures do not rule out evident presence of batching. We see that relative to μ_{BI}, σ_{BI} is the lowest for (S6 and S7), which are also the segments for which batching shows up regularly or even periodically on the performance spectrum. For all segments, we find that $\mu_{W_o \in b}$ and $\mu_{W_o \notin b}$ can differ greatly. While this for a part a result of n and BF, we also see that σ relative to μ for these measures varies, indicating that the underlying arrival time distribution differs for batched and non-batched cases. Finally, for segments with batches overlapping (S1-S4 and S6), we see that μ_{IBIA} is always larger than $\mu_{IA \in b}$ and for S5 and S7, this is the other way around. This can be explained by the fact that $IBIA$ is an aggregation of the interarrival time within batches and does not take the arrivals of crossing cases that belong to different batches into account. We think contrasting these measures may help in describing overlapping behavior.

Table 2. Interarrival time measurements for S1–S7

Segment	μ_{IA}	σ_{IA}	$\mu_{IA \in b}$	$\sigma_{IA \in b}$	$\mu_{IA \notin b}$	$\sigma_{IA \notin b}$	μ_{IBIA}	σ_{IBIA}
S1	6.16	31.10	29.68	214.13	7.65	36.78	352.79	861.52
S2	39.12	145.10	89.94	412.52	68.10	198.43	112.69	498.96
S3	2.51	7.99	12.94	203.86	3.05	8.72	48.74	424.87
S4	1.14	5.77	1.17	7.54	39.26	84.84	9.60	104.23
S5	29.62	184.15	57.01	482.74	35.38	200.96	49.61	62.90
S6	71.25	102.70	71.50	103.01	336.00	266.27	115.73	344.86
S7	1.90	7.26	3.06	15.43	4.95	12.91	2.98	13.79

We illustrate a more detailed batch performance analysis by the example of segment *Create Fine:Send Fine* (S4 in Fig. 9). The results are presented in the form of a time series plot in Fig. 10, describing the trend of the measures k, $W_{i,min}$ and $W_{i,max}$. On the left we see that k fluctuates moderately up to nearly the end of 2001, after which it shows excessive increases and decreases until 2005, ranging from a peak of 1806 on 11/9/2001 to a drop of 35 on 12/24/2001. From 2005 onward, the batch size remains predominantly steady and does (for the most part) not exceed 600 cases. On the right we see that $W_{i,max}$ heavily fluctuates between approximately 1000 and 14000 hours ranging the complete time span and a change in trend cannot clearly be observed. However, $W_{i,min}$ does show a small change: it remains close to 0 hours up to 2005, continues to show some higher increases from 2005 to 2010, after which it decreases again. Interestingly, when reviewing the length of the window in which batched cases arrive ($W_{i,max} - W_{i,min}$), we see that due to the trend of $W_{i,min}$, we also observe shorter batching windows between 2005 and 2010, giving insights in batching policy. Detailed batch statistics for this segment are available in [7].

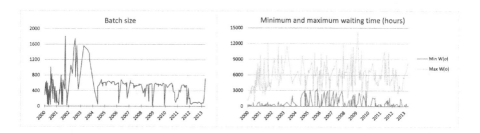

Fig. 10. Time series of k and $W_{i,min}$ and $W_{i,max}$ for segment *Create Fine: Send Fine*

7   Conclusion

We have shown that converting event log data into the performance spectrum allows to mine batches from business process event logs without additional information about resources or prior knowledge. The technique is robust to detect batches even in the presence of overlapping with non-batching behavior. By grouping all observations into batches, detailed performance characteristics can be derived which reveals strong performance variations on real-life data. A current limitation is that the technique does not yet incorporate a clear-cut boundary as to when sets of disjoint cases are considered batches or not, nor is it robust to other types of batching behavior, such as non-FIFO. Potential avenues for future work include incorporating the resource perspective - both resource involvement and efficiency - in the batch detection and evaluation and developing additional detection techniques to cover a greater extent of the performance patterns defined in [4]. Also, detected batches - or even other detected patterns - could be used as features in further performance analysis. Not only could they be used as parameters in queuing networks, complementing [16], they could also prove useful in the area of predictive process monitoring.

References

1. van der Aalst, W.M.P., Adriansyah, A., van Dongen, B.F.: Replaying history on process models for conformance checking and performance analysis. Wiley Interdisciplinary Rev. Data Min. Knowl. Disc. **2**(2), 182–192 (2012)
2. Cachon, G., Terwiesch, C.: Matching Supply with Demand. McGraw-Hill, New York (2013)
3. Denisov, V., Belkina, E., Fahland, D., van der Aalst, W.M.P.: The performance spectrum miner: Visual analytics for fine-grained performance analysis of processes. In: BPM 2018 Demos. CEUR Workshop Proceedings, vol. 2196, pp. 96–100 (2018). CEUR-WS.org
4. Denisov, V., Fahland, D., van der Aalst, W.M.P.: Unbiased, fine-grained description of processes performance from event data. In: Weske, M., Montali, M., Weber, I., vom Brocke, J. (eds.) BPM 2018. LNCS, vol. 11080, pp. 139–157. Springer, Cham (2018). https://doi.org/10.1007/978-3-319-98648-7_9
5. Dumas, M., Rosa, M.L., Mendling, J., Reijers, H.A.: Fundamentals of Business Process Management. Springer, Heidelberg (2013). https://doi.org/10.1007/978-3-642-33143-5
6. Henn, S., Koch, S., Wäscher, G.: Order batching in order picking warehouses: a survey of solution approaches. In: Manzini, R. (ed.) Warehousing in the Global Supply Chain, pp. 105–137. Springer, London (2012). https://doi.org/10.1007/978-1-4471-2274-6_6
7. Klijn, E.L.: Batch pattern detection in the performance spectrum. Capita selecta research project., Eindhoven University of Technology (2019). https://doi.org/10.5281/zenodo.3234102
8. Martin, N., Solti, A., Mendling, J., Depaire, B., Caris, A.: Mining batch activation rules from event logs. IEEE Trans. Serv. Comput. 1 (2019). https://doi.org/10.1109/TSC.2019.2912163

9. Martin, N., Swennen, M., Depaire, B., Jans, M., Caris, A., Vanhoof, K.: Batch processing: definition and event log identification. In: Proceedings of the 5th International Symposium on Data-driven Process Discovery and Analysis (2015)
10. Martin, N., Swennen, M., Depaire, B., Jas, M., Caris, A., Vanhoof, K.: Retrieving batch organisation of work insights from event logs. Decis. Support Syst. **100**, 119–128 (2017)
11. Nakatumba, J.: Resource-aware business process management: analysis and support. Ph.D. thesis, Eindhoven University of Technology (2013)
12. Pufahl, L., Bazhenova, E., Weske, M.: Evaluating the performance of a batch activity in process models. In: Fournier, F., Mendling, J. (eds.) BPM 2014. LNBIP, vol. 202, pp. 277–290. Springer, Cham (2015). https://doi.org/10.1007/978-3-319-15895-2_24
13. Pufahl, L., Weske, M.: Batch Activities in Process Modeling and Execution. In: Basu, S., Pautasso, C., Zhang, L., Fu, X. (eds.) ICSOC 2013. LNCS, vol. 8274, pp. 283–297. Springer, Heidelberg (2013). https://doi.org/10.1007/978-3-642-45005-1_20
14. Pufahl, L., Weske, M.: Requirements framework for batch processing in business processes. In: Reinhartz-Berger, I., Gulden, J., Nurcan, S., Guédria, W., Bera, P. (eds.) BPMDS/EMMSAD -2017. LNBIP, vol. 287, pp. 85–100. Springer, Cham (2017). https://doi.org/10.1007/978-3-319-59466-8_6
15. Selvarajah, E., Steiner, G.: Approximation algorithms for the supplier's supply chain scheduling problem to minimize delivery and inventory holding costs. Oper. Res. **57**(2), 426–438 (2009)
16. Senderovich, A., Weidlich, M., Gal, A., Mandelbaum, A.: Queue mining – predicting delays in service processes. In: Jarke, M., et al. (eds.) CAiSE 2014. LNCS, vol. 8484, pp. 42–57. Springer, Cham (2014). https://doi.org/10.1007/978-3-319-07881-6_4
17. Wen, Y., Liu, J., Chen, J.: Mining batch processing workflow models from event logs. Concurrence Comput. Pract. Experience **25**(13), 1928–1942 (2013)

LIProMa: Label-Independent Process Matching

Florian Richter[✉], Ludwig Zellner, Imen Azaiz, David Winkel,
and Thomas Seidl

Ludwig-Maximilians-Universität München, Munich, Germany
{richter,zellner,winkel,seidl}@dbs.ifi.lmu.de, imen.azaiz@campus.lmu.de

Abstract. The identification of best practices is an important methodology to improve the executions of processes. To determine those best practices process mining techniques analyze process entities and model specific views to highlight points for improvements. A major requirement in most approaches is a common activity space so events can be related directly. However there are instances which do provide multiple activity universes and processes from different sources need to be compared. For example in corporate finance, strategic operations like mergers or acquisitions cause processes with similar workflow but different descriptions to be merged. In this work we develop LIProMa, a method to compare processes based on their temporal flow of action sequences by solving the correlated transportation problem. Activity labels are purposely omitted in the comparison. Hence our novel method provides a similarity measure which is capable of comparing processes with diverging labels often caused by distributed executions and varying operators. Therefore it works orthogonal to conventional methods which rely on similarity between activity labels. Instead LIProMa establishes a correspondence between activities of two processes by focusing on temporal patterns.

Keywords: Process similarity · Transportation problem · Temporal flow · Cross-bin distance

1 Introduction

Managing processes demands improvements considering execution time performances, resource allocations or implementations of regulations. In an optimal scenario we have recourse to previous knowledge, either as experts or as advanced process mining algorithms on information systems. In both cases a rich history of previous process executions is the baseline of the decisions for the next steps. However what can we do without or with insufficient previous process knowledge?

As an example we imagine being a manager with the task to integrate a smaller company that has been taken over. We probably have some basic idea of the foreign process and can mine the event logs of the process to be integrated.

© Springer Nature Switzerland AG 2019
C. Di Francescomarino et al. (Eds.): BPM 2019 Workshops, LNBIP 362, pp. 186–198, 2019.
https://doi.org/10.1007/978-3-030-37453-2_16

Sometimes we can also rely on related process knowledge as we are dealing with similar workflows. Collecting process knowledge from our company is a good starting point and will probably lead to rich data, containing different levels of activity granularity and diverging labels. The next goal is to compare the data of both sources by aggregating them into a common model and benefit from the related knowledge.

This correspondence problem occurs in many other cases if processes are executed independently without direct reconciliation. In decentralized organizations, between competing groups or for different products within the same company, processes are performed differently because there was no initial demand for a regularization. The field of process matching tackles this problem. However the focus lays on matching similar activity labels.

Our contribution here is a method to match activities of different processes by using a label-independent cross-bin distance computation. Our method aggregates for two given processes occurrences of activities in corresponding bins on a timeline according to the average activity timestamp and represents processes based on their logs as characterizing histograms. Solving the transport optimization problem between those histograms provides a flow from activities in the first process to activities in the second one. This flow defines a similarity value between both processes and more importantly directly infers the correspondent activity labels of both processes. This bipartite matching can then be used as a translation system to match the different activity domains.

2   Related Work

The increasing trend within the industry to gather data about processes has led to a rise in both data quality and availability in the field of process comparison. This development as well as the industry's demand for improved methods to analyze and to subsequently utilize the gathered data in a way that allows a more in-depth look into a systems interior (inconsistency detection, process querying etc. [1]) turn process comparison into an increasingly active, yet growing area of research. However, as stated by Syamsiyah et al. [11] "process comparison remains an interesting but insufficiently researched topic".

The related research area can be split into two main categories: log-based and model-based comparison methods. Nevertheless mixed variants are possible [1].

Model-based comparison methods analyze different process models. These process models can be either given or they can be derived from event logs via process mining. Buijs et al. [5] propose an approach that compares two process models representing almost identical processes executed by different organizations. The comparison of the process models is based on their respective alignment to an explicitly defined process model. Cordes et al. [6] introduce a method to compare multidimensional process models. Such are used to analyze processes similar in nature but distinguishable in certain parameters, e.g. the same medical treatment applied to different age groups. Their method enables the visualization of differences between those similar processes.

Log-based methods analyze event-logs to extract information about the process. Bolt et al. [4] use transition systems to detect differences in the variants of the same process. The visualization of these transition systems then allow for detection of significant differences. Van Beest et al. [3] present the *log delta analysis method* that compares two event logs and returns a set of statements in natural language about behavior that is frequent in one log, but infrequent in the other one. Syamsiyah et al. [11] provide a stepwise methodology for applying process comparison in practice. Their methodology guides the user from the initial raw data, which are translated into event logs, to the results of the analysis. Van der Aa et al. [1] proposes multiple similarity metrics which aim at bringing various types of information given by the event log into use. These are: position, occurrence, duration, attribute name, attribute value and prerequisites. These similarity metrics evidentially result in a relation between two processes which is more reasonable. The mentioned duration similarity shows a notion of time by using the average durations of an event class in a log. This notion is taken up again with the difference that occurrence timestamps lead to a more useful application for binning. Dijkman et al. [7] also tries to align two process models by proposing three similarity metrics. Among label-based and structural similarity a causal footprint is used.

Another approach contrasting with the aforementioned methods is the usage of user feedback to support the process matching process. This resembles semi-supervised methods from other fields of research. Klinkmüller et al. [8] introduce an iterative strategy which alternates between correspondence detection and user feedback to ultimately improve the matching procedure. Since this approach faces reality by considering the absence of information required to perform an optimal process matching it heads into a similar direction like this work's method. LIProMa does not rely on event labels but uses a more semantically independent information however. Hence, this paper introduces a method that allows for the comparison of event logs with diverging event descriptors.

The problem is distantly related to conformance checking into which insights can be gained in the literature [2]. Although conformance checking relates single traces to one process and process comparison relates two processes the connection reveals itself in the matching procedure. In the former a matching in the fashion of trace validation on a model which has its origins in the source of the corresponding trace is sought. The latter describes a matching of different but also related processes in ways that validates their correspondence i.e. similarity on activity level.

3   Preliminaries

An event $e = (c, a, t, X)$ is a tuple of features specifying that an activity with label $a \in \mathbb{A}$ has been executed at time $t \in \mathbb{N}$. Multiple but finitely many events that occur in a common context are called case and then share a common case identifier c. Events can carry additional features $x \in X$, however we ignore those for the most part of this work. An event log $\mathcal{L}$ is an aggregation of events

Case ID	Trace
1	$\langle a, b, d, e, h \rangle$
2	$\langle a, d, c, e, g \rangle$
3	$\langle a, c, d, e, f, b, d, e, g \rangle$
4	$\langle a, d, b, e, h \rangle$
5	$\langle a, c, d, e, f, d, c, e, h \rangle$
6	$\langle a, c, d, e, g \rangle$

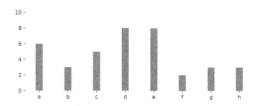

Fig. 1. A process given as a set of traces is transformed into a histogram using the bag-of-activity embedding. In this variant, multiple occurrences per case are not differentiated.

which share the same activity space $\mathbb{A}$. All events contained in an event log are witnesses of a process execution and therefore a descriptive representation of the process. The following simple notation is used for the feature access on events. For an event $e = (c, a, t, X)$ we define $c(e) = c$, $a(e) = a$ and $t(e) = t$. We also need to count occurrences hence $|a|_{\mathcal{L}} = |\{e \in \mathcal{L} \mid a(e) = a\}|$ is the number of occurrences of activity a in log $\mathcal{L}$.

A histogram $h : Y \rightarrow \mathbb{R}^+$ is a mapping from a set of bins Y within a specified domain to positive real values. A simple way to represent a log $\mathcal{L}$ as a histogram is the mapping $h' : \mathbb{A} \rightarrow \mathbb{N}^+$ with $h'(a) = |a|_{\mathcal{L}}$. We give a brief example in Fig. 1 where a trace log is transformed into an activity occurrence histogram. If we define a distance between histogram bins, we call it a ground distance.

As sketched in Fig. 2, a ground distance can be defined bin-wise or crossbin-wise. Due to the potential of work in a process being split into subtasks, we aim for a crossbin-distance. The cross-bin distance we use is the Earth Mover's Distance [10] EMD that is related to the solution of the transportation problem: For given weights $w_1, \ldots, w_m \in \mathbb{R}^+$ on spatial positions $p_1, \ldots, p_m \in P$ and empty bins with capacities $v_1, \ldots, v_n \in \mathbb{R}^+$ on positions $q_1, \ldots, q_n \in V$, find the flow $F = (f_{i,j}) \in \mathbb{R}^{m \times n}$ with minimal costs $c_F = \sum_{i=1}^{m} \sum_{j=1}^{n} f_{i,j} d(p_i, q_j)$ to transport all weights to the bins regarding a ground distance d on the position space under the following constraints:

1. The flow has to be non-negative, so weights are only sent from P to Q and not vice versa: $f_{i,j} \geq 0$, $\forall 1 \leq i \leq m, 1 \leq j \leq n$
2. The sent flow is bounded by the weights in P: $\sum_{j=1}^{n} f_{i,j} \leq w_i$, $\forall 1 \leq i \leq m$
3. The received flow is bounded by the weights in Q: $\sum_{i=1}^{m} f_{i,j} \leq v_j$, $\forall 1 \leq j \leq n$
4. All weights possible have to be sent: $\sum_{i=1}^{m} \sum_{j=1}^{n} f_{i,j} = \min\left(\sum_{i=1}^{m} w_i, \sum_{j=1}^{n} v_j\right)$

The distance is then defined as

$$EMD(P, Q) = \min \frac{\sum_{i=1}^{m} \sum_{j=1}^{n} f_{i,j} d(p_i, q_j)}{\sum_{i=1}^{m} \sum_{j=1}^{n} f_{i,j}}$$

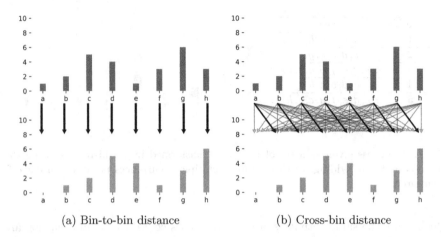

(a) Bin-to-bin distance   (b) Cross-bin distance

Fig. 2. Histograms can be compared using two different paradigms. Bin-to-bin distances compare dimensions individually, which is suitable for bins representing independent features. Cross-bin distances on the other hand use all combinations of bins. This is better suited for correlated features. In this example the green histogram below is translated by one position. Many cross-bin distances detect this shift. However it is only useful for common bin domains and widely used for color histograms in image retrieval for example.

4   Label-Independent Process Matching

Setting two processes into relation requires to define a similarity measure between both objects. Since processes are often given descriptively as event logs, a distance on logs is needed. The most basic similarity measure is the discrete metric $\rho(\mathcal{L}_1, \mathcal{L}_2)$ which is 1 if $\mathcal{L}_1 = \mathcal{L}_2$ and 0 otherwise. This is too strict for many applications. Indeed we aim for a distance function that differentiates more between similar and dissimilar objects. Secondly this distance function has to state the points of dissimilarity, which becomes important for further analysis.

Embedding the log into a vector space by feature extraction provides a suitable step into this direction. A prominent example is the bag-of-activity embedding, which counts activity occurrences and maps the frequencies into a vector space. Usually applied to single traces in trace clustering, we can expand the same idea to logs. All trace-wise activity frequency vectors have to be summed up. Measuring the similarity between two logs is equivalent to computing the distance between the representing activity frequency vectors.

However there are some problems remaining. First of all, the ordering of activities is not embedded in this representation. Due to concurrent executions of certain activities, like the supervision or mentoring in this example, the sequence of activities is not a total order. Although the ordering of activities does not cover the complete structure of a process, it is a meaningful indicator and should therefore be included in the similarity computation. To overcome this issue, the bags-of-activities can be extended to bag-of-n-grams, which store occurrences of trace sub-sequences of up to length n.

The second problem has a greater relevance. When comparing two processes with identical activity spaces, process owners are probably more interested in deviations on the event level like comparing time performances of two activities. The requirement for process matching emerges when we assume similarities at some points but we cannot align activities of both processes easily. The reasons for this mismatch are, for instance, structural changes in the process workflow, activity splits and merges, or language-based differences in the process descriptions. We will now present an approach which is capable of transforming one log into the other by matching the activities independently of the actual labels. LIProMa yields a flow of activity weights so activities with similar occurrence timestamps are matched. Splits and merges of activities are identified as well. Finally the distance is determined by the flow costs, so advanced data mining methods, that require a distance measure like clustering and anomaly detection, can be applied. However, the last applications are future work and not covered here.

4.1 Process Histograms

LIProMa transforms an event log $\mathcal{L}$ into a histogram $h_{\mathcal{L}}$ representing the occurrences of all activities analogously to the bag-of-activity model: $h_{\mathcal{L}}(a^i) = \left|a^i\right|_{\mathcal{L}}$. To decide between the same activity and the ith occurrence in a case, we extend the activity label by an index i to define the histogram, but omit the index for better readability.

For a cross-bin distance we set a ground distance on the bins. Activities of the compared processes are potentially contained in different activity spaces. However, events share a common time space. Hence we use the average temporal position of an activity relative to its framing case. First we shift the temporal starting point of each case to zero first. The remaining activities of each case are shifted accordingly. That way we left-align all timestamps. Next we normalize the case durations by min-max normalization so we can compare processes of different overall runtime. All timestamps are mapped into the interval $[0, 1]$. Finally all cases contain an event with timestamp $t(e) = 0$ and the longest one ends with $t(e) = 1$. For each activity a we compute the average of all timestamps $t_a = \frac{\sum\{t(e)|e\in\mathcal{L}\wedge a(e)=a\}}{|\{t(e)|e\in\mathcal{L}\wedge a(e)=a\}|}$.

The ground distance $d \in \mathbb{R}_{\geq 0}^{\mathbb{A}\times\mathbb{A}}$ between two bins is then defined by the temporal difference $d(a, b) = |t_a - t_b|$ between both average occurrence timestamps.

4.2 Transportation Flow

LIProMa transforms the problem of process comparison to a transportation problem. Each occurrence of an activity can be interpreted as a unit of work. All work units have to be transported from one histogram to the other histogram. Since we normalized the weights so that the weight sums of both histograms are

---

Algorithm 1: Log Transportation Flow

Data: Event Logs $\mathcal{L}_1$ and $\mathcal{L}_2$
Result: Transport Flow F

1 **foreach** $\mathcal{L}_1$, $\mathcal{L}_2$ **do**
2   Initialize mapping $h_{\mathcal{L}_i}$;
     /* Collect timestamps for later distance computations   */
3   Initialize mappings T_i, N_i;
4     $t_1 = 0$;
5   **foreach** *Case* $c \in \mathcal{L}_i$ **do**
6        $t_0 = \min(\{t(e) \mid e \in c\})$;
7        $t_1 = \max(\{t_1\} \cup \{t(e) \mid e \in c\})$;
8   **foreach** *Event* $e \in c$ **do**
9   Increment $h_{\mathcal{L}_i}(a(e))$;
10   Increment N_i;
11   Append $t(e) - t_0$ to $T_i(a(e))$;
12   **end**
13   **end**
     /* normalize histogram weights: the weight sum is 1   */
14     $n = sum(h_{\mathcal{L}_i})$;
15   **foreach** *Value* $v \in h_{\mathcal{L}_i}$ **do**
16        $v = v/n$;
17   **end**
     /* normalize timestamps: case starts are 0, longest ends at 1   */
18   **foreach** *Activity* a **do**
19        $t_a = T_i(a)/t_1$;
20   **end**
21 **end**
22 $w_1 = h_{\mathcal{L}_1}$;
23 $w_2 = h_{\mathcal{L}_2}$;
24 **foreach** *Activity pair* $a, b \in \mathbb{A}(\mathcal{L}_1) \times \mathbb{A}(\mathcal{L}_2)$ **do**
25     $d(a,b) = |t_a - t_b|$;
26 **end**
27 $F = EMD(w_1, w_2, d)$;

---

equal, all weights in the first histogram have to be used and transported to the second histogram's bins. It is explicitly allowed to split one activity as tasks can be modeled as subtasks in another process. Hence we choose EMD as a suitable distance function as weights can be moved to separate bins.

4.3   Limitations

Our novel method works well for processes with point events. Interval events which have a large degree of overlapping are more difficult to align with the current method. If an activity does not follow a near Gaussian distribution or the variance is quite high, the situation is similar to an interval event. In such cases we extend our histogram model to account for other event types.

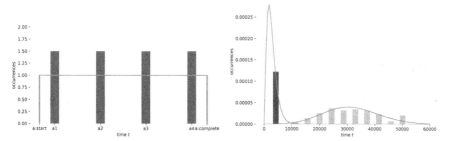

(a) Slicing of interval events so the dura-(b) Slicing of long-term events covers dif-
tion is used in the transport problem.   ferent distributions.

Fig. 3. Long-term activities require adequate modeling instead of using the mean occurrence timestamp only.

In case of interval activities, e.g. activities with life-cycle states, we use these life-cycle events individually and create more bins for one activity. To model activities with large durations a suitable strategy is to apply a more granular binning to this activity. In Fig. 3a the left sketch illustrates this by distributing the long-term activity on four bins. The red line indicates an activity a with life-cycle transitions *start* and *complete*. Slicing this event into several bins gives more flexibility as heterogeneous processes regarding the event durations can be handled. Further we modify the ground distance function so it penalizes if both bins are mapped to different events.

A more complex issue arises for activities with non-consistent occurrence distributions or activities spanning a long duration concurrent to other large process regions. In this very difficult scenario we slice the activity bin into many smaller bins, spreading the occurrences over time. In Fig. 3b the first activity has a very low occurrence variance and thus is well represented with one bin. The second activity has a high variance. Using only one bin causes a bias that impairs the representation of the process log. As before, slicing is again a suitable strategy. However bins do not have equal height here. The solving strategy remains the same and the transportation problem is still well-defined. Due to the increased number of bins, complexity is increased as well. The ground distance function has to account for increased costs if the same long term activity is translated to different target activities.

Another limitation is the handling of events with similar properties. If two activities are concurrent and have the same duration, we can not distinguish between them while ignoring the labels. In some cases this can lead to positive flows from one activity to the other activity if the same process log is compared to itself. Only using timestamps it is not possible to solve this problem. Any suitable strategy has to use additional data like resources for the distance computation. Operators in a process processing the same activities are usually aggregated into roles. By using the role information besides the temporal position, the characterization of activities is enriched and can improve the matching result, which has to be shown in future work as well.

5   Experiments

To the best of our knowledge, there are no other approaches that directly compare process logs without utilizing the activity labels. Existing approaches align process models by matching nodes having the same activity labels or match labels by using textual alignment methods. However, for evaluation purposes here we compare processes with the same activity domain as we are lacking a ground truth otherwise. For the algorithm we do not use an alignment between different processes. This is only used to determine the accuracy of the transportation flow. We perform two experiments. The first one uses cross-validation on 5 domain-similar process logs, which are partitioned again into 5 sublogs each, yielding 25 process logs. Our claim is that partitions of the same log should yield a smaller distance than partitions from different logs. The second experiment on a second dataset shows the matching quality when considering the flow of the solution to the transport problem. Here we use the activity labels as ground truth.

We apply our novel approach on the publicly available datasets of the Business Process Intelligence Challenges BPI2015 [12] and BPI2017 [13]. BPI2015 contains five subdatasets with logs of a building permit application process in five municipalities. Although the processes are theoretically very similar there are differences in the process executions for distinct municipalities. We choose this dataset as this scenario is quite close to our problem statement of having similar processes with distinct logs. However all subprocesses share a common pool of activity labels. The log consists of about 400 activities executed in over 260000 events. We used the BPI2017 dataset for the matching accuracy evaluation. The loan application process contains 1200000 events and 66 activities. Hence it is well suited for performance analysis and the evaluation of our approach regarding different sized activity spaces.

As described before we constructed 25 data logs by partitioning and determined our distance value by cross-validation. The resulting confusion matrix is given in Fig. 4. As one would expect, the diagonal is zero as identical histograms are compared. Also the matrix is symmetrical due to the performed operations. The values range from zero, meaning strong similarity of both logs, to 14 in this case, which refers to higher dissimilarity. Regarding the dataset itself we know that there are five municipalities A, B, C, D, and E. These five municipalities create a grid in the confusion matrix, so a distinction is possible without using any activity label. The distances within each 5×5 block are rather low which is a desired property. The dataset was investigated by various authors and there is a strong similarity between the control flow patterns of log A, C, and E. Our orthogonal approach supports the similarity of A and C, while D is more similar than E from a temporal distribution perspective. The interaction between control flow and temporal flow needs to be investigated further.

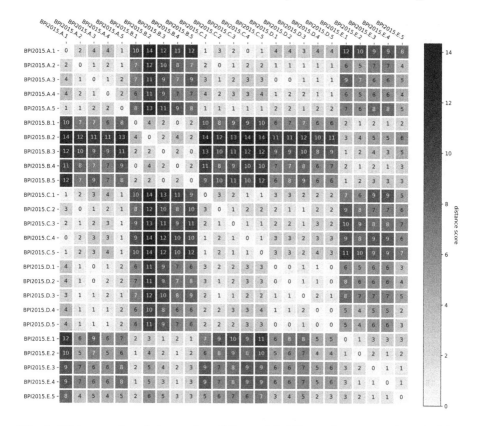

Fig. 4. Distance values between 5-partitions of 5 different logs prove the suitability of LIProMa to differentiate between similar and dissimilar process logs. Further, it highlights that two groups $\{A, C, D\}$ and $\{B, E\}$ have a small intra-distance while both groups have a higher inter-distance.

In a second experiment we evaluate the matching quality and performance of our method. Obviously the complexity increases for a high number of activities. More activities and also repetitions of activities increase the number of bins N in the histograms. The performance of EMD grows with $\mathcal{O}(N^3 \log N)$ and thus suffers from complex activity spaces. Pruning the activity space is a beneficial step to improve time performance. Counter-intuitively accuracy is also raised because fewer labels need to be matched thus fewer matching partners for each activity have to be considered. We apply LIProMa to the BPI2017 dataset and use the previous cross-validation technique again and partition the dataset into 8 sublogs. Next we introduce a pruning parameter a_{min} and omit all activities which occur less than a_{min} in the histogram generation step.

Using low thresholds already omits a great number of activities. Most activities are executed rarely while few activities build the core of the process. We show this in Table 1 and give for various threshold niveaus the average number of activities in all partitions. Without pruning we count 380 activities, which consist of the 66 original activity labels and their repetitions. Using the FastEMD implementation [9] the execution time is already below one second. However after pruning all activities that occur less than 100 times in the log, the number of activities is already one-third of the complete activity space. The execution time is improved by a factor of 36. For further pruning and with decreasing numbers of activities, the execution time is only improved slightly.

Table 1. Quality of activity matchings for BPI2017 is low if all activities are used. Many activities are rare and decrease the matching accuracy like noise.

Threshold	Avg. Nr. of activities	Time	Balanced accuracy	Recall	Precision	F1-score
0	380	646 ms	0.582	0.162	0.165	0.159
100	106	18 ms	0.615	0.237	0.235	0.236
200	88	11 ms	0.624	0.257	0.255	0.256
300	78	8 ms	0.630	0.269	0.267	0.268
400	72	6 ms	0.636	0.283	0.281	0.282
500	63	5 ms	0.642	0.296	0.293	0.294
1000	50	3 ms	0.660	0.333	0.333	0.333
1500	41	2 ms	0.700	0.415	0.411	0.413
2000	29	1 ms	0.766	0.549	0.542	0.546
2500	25	<1 ms	0.840	0.694	0.693	0.693
3000	16	<1 ms	0.917	0.844	0.844	0.844

Considering the quality of the matching, we interpret the matching defined by the flow as a classification problem. We identify for each activity of the first process log the highest flow in the distance computation yielding a 1-to-1 matching. We can compute the F1-Score where the according target indicates the predicted activity label. For $a_{min} = 0$ we have to solve a binary set matching problem with set sizes of 380 activities. Especially the rare activities cause problems here as they pollute the histograms like noise. Having less but more frequent activities improves the prediction accuracy drastically. However, the goal is not to prune all activities as the matching was the focused problem. This evaluation shows that pruning is a suitable tool to improve quality and computation performance and should be applied up to the level where the size of the matching problem is still sufficiently expressive to derive the relation between the two process domains.

6   Conclusion

LIProMa solves the process matching problem for non-comparable activity spaces. A distance metric allows to apply data mining methods like clustering to organize the landscape of processes. Previous process matching focused on aligning activities with the same labels and measuring the activity distance for transforming one process model into the other model. LIProMa provides an orthogonal approach for comparing logs using the temporal perspective only. However, as we described, the accuracy drops depending on the activity sets due to two effects: Rare activities influence the approach like noise while non-punctual and fuzzy temporal occurrences cause inconclusive matchings. In future work we will address the matching quality by extending the histogram models and by adjusting the matching algorithm. Although we sketched some strategies for these limitations, we can not cover those extensions sufficiently here.

References

1. van der Aa, H., Gal, A., Leopold, H., Reijers, H.A., Sagi, T., Shraga, R.: Instance-based process matching using event-log information. In: Dubois, E., Pohl, K. (eds.) CAiSE 2017. LNCS, vol. 10253, pp. 283–297. Springer, Cham (2017). https://doi.org/10.1007/978-3-319-59536-8_18
2. Van der Aalst, W.: Data science in action. Process Mining, pp. 3–23. Springer, Heidelberg (2016). https://doi.org/10.1007/978-3-662-49851-4_1
3. van Beest, N.R.T.P., Dumas, M., García-Bañuelos, L., La Rosa, M.: Log delta analysis: interpretable differencing of business process event logs. In: Motahari-Nezhad, H.R., Recker, J., Weidlich, M. (eds.) BPM 2015. LNCS, vol. 9253, pp. 386–405. Springer, Cham (2015). https://doi.org/10.1007/978-3-319-23063-4_26
4. Bolt, A., de Leoni, M., van der Aalst, W.M.P.: A visual approach to spot statistically-significant differences in event logs based on process metrics. In: Nurcan, S., Soffer, P., Bajec, M., Eder, J. (eds.) CAiSE 2016. LNCS, vol. 9694, pp. 151–166. Springer, Cham (2016). https://doi.org/10.1007/978-3-319-39696-5_10
5. Buijs, J.C.A.M., Reijers, H.A.: Comparing business process variants using models and event logs. In: Bider, I., et al. (eds.) BPMDS/EMMSAD -2014. LNBIP, vol. 175, pp. 154–168. Springer, Heidelberg (2014). https://doi.org/10.1007/978-3-662-43745-2_11
6. Cordes, C., Vogelgesang, T., Appelrath, H.-J.: A generic approach for calculating and visualizing differences between process models in multidimensional process mining. In: Fournier, F., Mendling, J. (eds.) BPM 2014. LNBIP, vol. 202, pp. 383–394. Springer, Cham (2015). https://doi.org/10.1007/978-3-319-15895-2_32
7. Dijkman, R., Dumas, M., Van Dongen, B., Käärik, R., Mendling, J.: Similarity of business process models: metrics and evaluation. Inf. Syst. 36, 498–516 (2011)
8. Klinkmüller, C., Leopold, H., Weber, I., Mendling, J., Ludwig, A.: Listen to me: improving process model matching through user feedback. In: Sadiq, S., Soffer, P., Völzer, H. (eds.) BPM 2014. LNCS, vol. 8659, pp. 84–100. Springer, Cham (2014). https://doi.org/10.1007/978-3-319-10172-9_6
9. Pele, O., Werman, M.: Fast and robust earth mover's distances. In: 2009 IEEE 12th International Conference on Computer Vision, pp. 460–467. IEEE, September 2009
10. Rubner, Y., Tomasi, C., Guibas, L.J.: The earth mover's distance as a metric for image retrieval. Int. J. Comput. Vis. 40(2), 99–121 (2000)
11. Syamsiyah, A., et al.: Business process comparison: a methodology and case study. In: Abramowicz, W. (ed.) BIS 2017. LNBIP, vol. 288, pp. 253–267. Springer, Cham (2017). https://doi.org/10.1007/978-3-319-59336-4_18

12. Van Dongen, B.F. (Boudewijn): BPI challenge 2015 (2015). https://doi.org/10. 4121/UUID:31A308EF-C844-48DA-948C-305D167A0EC1.   https://data.4tu.nl/ repository/uuid:31a308ef-c844-48da-948c-305d167a0ec1
13. Van Dongen, B.F. (Boudewijn): BPI challenge 2017 (2017). https://doi.org/10. 4121/UUID:5F3067DF-F10B-45DA-B98B-86AE4C7A310B.   https://data.4tu.nl/ repository/uuid:5f3067df-f10b-45da-b98b-86ae4c7a310b

A Generic Approach for Process Performance Analysis Using Bipartite Graph Matching

Chiao-Yun Li[1]($\boxtimes$), Sebastiaan J. van Zelst[1,2]($\boxtimes$),
and Wil M. P. van der Aalst[1,2]($\boxtimes$)

[1] Fraunhofer Institute for Applied Information Technology (FIT),
Sankt Augustin, Germany
{chiao-yun.li,sebastiaan.van.zelst,wil.van.der.aalst}@fit.fraunhofer.de
[2] Chair of Process and Data Science, RWTH Aachen University, Aachen, Germany
{s.j.v.zelst,wvdaalst}@pads.rwth-aachen.de

Abstract. The field of process mining focuses on the analysis of event data, generated and captured during the execution of processes within companies. The majority of existing process mining techniques focuses on process discovery, i.e., automated (data-driven) discovery of a descriptive process model of the process, and conformance and/or compliance checking. However, to effectively improve processes, a detailed understanding in differences of the actual performance of a process, as well as the underlying causing factors, is needed. Surprisingly, few research focuses on generic techniques for process-aware data-driven performance measurement, analysis and prediction. Therefore, in this paper, we present a generic approach, which allows us to compute the average performance between arbitrary groups of activities active in a process. In particular, the technique requires no a priori knowledge of the process, and thus does not suffer from representational bias induced by any underlying process representation. Our experiments show that our approach is scalable to large cases and especially robust to recurrent activities in a case.

Keywords: Process mining · Process performance analysis · Bipartite graph matching · Integer Linear Programming

1 Introduction

The field of *process mining* has gained its significance as a technology to objectively obtain insights into business processes by exploiting *event logs*, i.e., records of events executed in the context of a business process. To further understand the execution of business processes, *process performance analysis* aims to measure and analyze the performance of business processes, e.g., throughput time, by extracting information from event logs [19].

The techniques for process performance analysis [2,5,12] can be classified into two approaches: model-based and log-based approaches [19]. The first one typically projects the event log on a predefined or discovered process model

© Springer Nature Switzerland AG 2019
C. Di Francescomarino et al. (Eds.): BPM 2019 Workshops, LNBIP 362, pp. 199–211, 2019.
https://doi.org/10.1007/978-3-030-37453-2_17

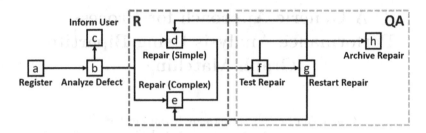

Fig. 1. Process model of repairing telephones in a company [15]. A telephone can be fixed by either the team for simple defects or the other team for complex defects whereas some defects can be handled by either team.

[2,12,13]. The advantage is that the performance analysis results can be interpreted in the context of a process model. However, this assumes that there exists a suitable model with high conformance.

The other approach is purely based on an event log [5,11,19]. The techniques based on such approach are not limited to the constraints of the model-based techniques and more flexible to users' needs. However, current work fails to provide robust performance metrics in the presence of repeated activities in a *case*, i.e., a run of a business process.

In this paper, we present a novel approach for business process performance analysis. Figure 1 serves as a motivating example of a process of repairing telephones in a company [15]. The process starts with a registration of a telephone sent by a customer (a). The telephone is then analyzed (b) and the problem is reported to the customer (c). Afterwards, one of the two teams in the Repair (R) department repairs the defect; one for simple defect (d) and the other for complex defect (e) while some defects can be handled by either team. Then, the Quality Assurance (QA) department tests if the defect is repaired (f) and decides whether to restart another repair (g) or to archive the case (h).

Suppose we are interested in the performance of R and QA department, i.e., the average duration between d or e to g or h. Given a case in which the activities are executed in the order of $\langle a, b, c, d, f, g, e, f, g, e, f, h \rangle$, Fig. 2 shows the events on a timeline for an overview of the duration of interest.

Intuitively, the average duration would be $\frac{1+60+60\times2}{3} = 60.33$ minutes. We propose a novel approach which supports the intuition and the flexibility of analysis while being robust to recurrent activities, e.g., all d, e, g and h are included for analysis. Our approach applies bipartite graph matching to pair the events for computation. Moreover, by allowing multiple selection of activities, our approach can be applied to analysis at a higher abstraction level, e.g., performance of department R and QA, without building another model in advance.

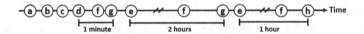

Fig. 2. A case of the process in Fig. 1. The events are plotted on a timeline with the duration of interest shown, i.e., duration between d or e to g or h.

It is common that real processes do not conform with a pre-defined structured process model. Interchangeability and concurrency among activities impose challenges on the existing methods for process performance analysis. These methods either need to exclude the non-conforming cases [2], or explicitly specify the relationships between events [5]. For instance, considering processes in a hospital, suppose we are interested in the duration between an injection and a medical examination for the reaction after the injection. It is impossible to pre-define what and how many injections that a patient would need in advance. By specifying two sets of activities, our approach allows for a certain degree of uncertainty among the activities in the event log, and further provides the flexibility when measuring the performance of complex processes.

To assess our approach, we conduct a quantitative evaluation of its scalability and a qualitative evaluation by means of comparing our approach with analysis obtained from commercial tools. The first experiment shows that our approach is scalable to large cases while the comparative evaluation shows its robustness against recurrent activities in a case.

The remainder of this paper is organized as follows. Section 2 compares our approach with existing work in academia and with the typical functionalities provided by commercial tools. Section 3 introduces the definitions and notations used in the paper. We present our approach in Sect. 4, followed by the evaluation and discussion in Sect. 5. Finally, Sect. 6 concludes the paper.

2   Related Work

Existing work on process performance analysis provides metrics at different levels. Most of them focus on the level of cases, e.g., throughput time, and/or the level of individual activities, e.g., waiting time. Despite various metrics, most techniques can be classified into model-based and log-based approaches [19].

Model-based performance analysis accounts for most techniques proposed [2,3,13,14,16,17]. These techniques attempt to map an event log on a given process model. Regardless of the amount of work [4,6,20], relatively few methods provide solutions on performance analysis for cases deviating from the model. Hornix provides the option to include the deviated cases when computing performance using token-based replay technique [12]. However, the results can be misleading when there are activities with the same label or artificial activities in a model. Adriansyah measures the performance of activities from events with transaction type attributes, e.g., start, complete, and suspend [2]. Nevertheless, there is no generic rule to determine the timestamps of the missing events in the work. Generally speaking, model-based approach confronts the challenges of (1) the reliability of the prescribed model, (2) the need for recalculation if the model changes, (3) complex and flexible business processes which result in low conformance of the model, and (4) the flexibility of the analysis, e.g, the metrics provided are dependent on the model.

Alternatively, one can analyze process performance based on an event log only [5,11,19]. This approach provides the flexibility and static results due to

Table 1. Comparison of works on process performance analysis.

Algorithm/Method		Model independent	Two arbitrary activities	Two sets of activities
[2]	Alignments	−	−	−
[3]	Robust Performance Analysis	−	−	−
[5]	First/Last to First/Last events	+	+	−
[9]	Analysis on Segmented Journey Model	+	+	−
[10]	Disco	−	−	−
[11]	Context-aware KPI	+	−	−
[12]	Log reply on Petri Nets	−	−	−
[13]	Hierarchical Performance Analysis	−	−	−
[14]	Analysis with Advanced Constructs	−	−	−
[16]	Alignments with Stochastic Petri Nets	−	−	−
[17]	Log Replay on Petri Nets	−	+	−
[18]	Queue Enabling CSPNs (QCSPN)	+	−	−
[19]	Dotted Chart	+	−	−
This Paper: Bipartite Graph Matching		+	+	+

the independence of a model. In this paper, we propose a novel performance measurement technique which is scalable to large event logs. We apply bipartite graph matching for pairing as many events for measuring as possible. By the design of the algorithm, users can analyze the performance between two arbitrary groups of activities. Compared to the existing log-based techniques, our approach is more flexible to the need of analysis and, supported by the evaluation, more robust to recurrent activities. Table 1 lists some representative work on process performance analysis. We compare if a method is dependent to a model, i.e., suffers from the reliability and limitations, and flexible to the analysis, i.e., the selection of activities.

3   Preliminaries

In this section, we introduce the related concepts and notations used in this paper, including process mining and bipartite graph matching.

As a preliminary, we first define the required mathematical notations for a *sequence*. We introduce a function: $\sigma : \{1, 2, ...n\} \to X$ for a finite *sequence* of length n over an arbitrary set X. The function assigns every element $x \in X$ to an index $i \in \{1, 2, ..., n\}$. We write $\sigma = \langle x_1, x_2, ..., x_n \rangle \in X^*$, where $x_1 = \sigma(1)$, $x_2 = \sigma(2)$, ..., $x_n = \sigma(n)$ and X^* denotes the set of all possible sequences over X of arbitrary length. Given a sequence $\sigma \in X^*$ and an element $x \in X$, let $|\sigma|$ denotes the length of the sequence, we overload the notation and write $x \in \sigma$ if and only if $\exists 1 \leq i \leq |\sigma| (\sigma(i) = x)$.

3.1   Event Logs

An *event log* is a collection of sequences of *events*, i.e., *traces*. Each event describes the execution of an *activity*, i.e., a well-defined step in a process [1]. During the execution of processes, i.e., *cases*, additional information, i.e., *attributes*, are associated to the events and/or cases. In the following part of this section, we explain the relationships among *event*, *activity*, *trace*, and *event log*. For simplicity, in the remainder of the paper, $\mathscr{E}$ denotes the universe of events and $\mathscr{C}$ denotes the universe of cases.

Definition 1 (Event, Event Attributes). *Let $\mathscr{A}$ be the universe of activities and $\mathscr{T}$ be the universe of timestamps. We define the projection functions:*

- $\pi_{act} : \mathscr{E} \to \mathscr{A}$, *which $\pi_{act}(e)$ represents the activity of the event $e \in \mathscr{E}$,*
- $\pi_{ts} : \mathscr{E} \to \mathscr{T}$, *which $\pi_{ts}(e)$ represents the occurring time of the event $e \in \mathscr{E}$.*

Definition 2 (Trace). *We define a projection function $\pi_{tra} : \mathscr{C} \to \mathscr{E}^*$ for a trace which describes a finite sequence of events in a case such that*

- $\forall 1 \le i < j \le |\sigma| \big(\pi_{ts}(\sigma(i)) \le \pi_{ts}(\sigma(j)) \big)$
- $\forall 1 \le i < j \le |\sigma| \big(\sigma(i) \ne \sigma(j) \big)$

We define an event index as the order of the event in a trace.

Definition 3 (Event Log). *Given a collection of cases $\mathscr{L} \subseteq \mathscr{C}$, an event log L is a collection of traces, i.e., $L = \{ \sigma \in \mathscr{E}^* | \exists c \in \mathscr{L} \big(\sigma = \pi_{tra}(c) \big) \} \subseteq \mathscr{E}^*$ s.t. $\forall \sigma, \sigma' \in L \big(\exists e \in \mathscr{E} (e \in \sigma \wedge e \in \sigma' \Rightarrow \sigma = \sigma') \big)$.*

For the sake of performance analysis, we additionally define the notation of measurements between events in the context of a trace.

Definition 4 (Measurement). *Let M denote all the measurable entities, e.g., duration, cost. Given a measurable entity $m \in M$ and a trace σ, we define the function: $\phi_m : \mathscr{E}^* \times \mathscr{E} \times \mathscr{E} \to \mathbb{R}$ for which $\forall e, e' \in \sigma$.*

In this paper, we assume that any measurable entities for performance analysis are always available given an event log.

3.2   Bipartite Graph Matching

A matching in a bipartite graph is to select the edges, i.e., pairs of nodes, such that no edges share the same nodes. This section formally defines a (weighted) bipartite graph and the corresponding matching problem as follows.

Definition 5 ((Weighted) Bipartite Graph). *Let $G = (V, E, w)$ be a weighted graph where V is a set of nodes, E is a set of edges, and w is a weight function $w : E \to \mathbb{R}$. Given a weighted graph G, it is bipartite if and only if $\exists V_1, V_2 \subseteq V \big(V_1 \cup V_2 = V \wedge V_1 \cap V_2 = \emptyset \big)$ s.t. $\nexists (v, v') \in E \big(v, v' \in V_1 \vee v, v' \in V_2 \big)$. We denote a weighted bipartite graph as $G = (V_1 \cup V_2, E, w)$.*

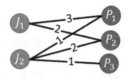

Fig. 3. An example of maximum weighted bipartite matching. The edges (P_1, J_1) and (P_2, J_2) are selected due to maximum weights.

Definition 6 (Maximum Weighted Bipartite Matching). *Given a weighted bipartite graph* $G = (V_1 \cup V_2, E, w)$, *a bipartite graph matching is the selection of* $E' \subseteq E$ *s.t.* $\forall (v, v') \in E' \big(\exists (v, v'') \in E' \Rightarrow v' = v'' \wedge \exists (v'', v') \in E' \Rightarrow v'' = v \big)$.

Let E_M *denote all possible matching. A maximum weighted bipartite matching is a matching* $E' \in E_M$ *such that* $\nexists E'' \in E_M \big(\sum_{e \in E''} w(e) > \sum_{e \in E'} w(e) \big)$.

An example is shown in Fig. 3 as a bipartite graph which J_i is a set of jobs and P_j is the sets of applicants. An edge indicates that an applicant applies for the job with the qualification score implied as its weight. A maximum weighted bipartite matching is the optimal assignments of the applicants to the jobs such that the total qualification score is maximum, i.e., (P_1, J_1) and (P_2, J_2).

In this paper, Integer Linear Programming (ILP) should be applied to find the maximum weighted bipartite matching. ILP is a mathematical optimization method that, given a set of variables, assigns an integral value to every variable to achieve the best result, e.g., minimum cost, maximum benefits, with the *objective* and the *constraints* formulated in linear relationships [7].

4   Generic Process Performance Analysis Using Bipartite Graph Matching

Our approach aims for measuring the performance between two arbitrary groups of activities. By projecting the events of interest into a bipartite graph for each trace, we find the maximum weighted bipartite matching indicating the pairs of events from which the performance metric of a case is derived. Figure 4 depicts a global overview of our approach and we illustrate each step in more detail in the sections specified on the arcs.

4.1   Bipartite Graphs Generation

Given an event log, a user specifies two groups of activities of interest for analyzing the performance from one to the other. We construct a weighted bipartite graph for each case by taking the events of interest as nodes and connecting them according to the direction specified. The weights of the edges are computed using a monotonic function of the event index of the nodes. Figure 5 shows an example of a bipartite graph for the trace which we only show the events of interest. The projection of a trace to a bipartite graph is formalize as follows.

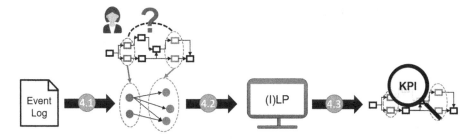

Fig. 4. An overview of the proposed approach. We analyze the performance of a case by finding a maximum weighted bipartite graph matching using ILP.

Definition 7 (Trace Projection). *Let $\mathscr{A}$ be the universe of activities. Given two sets of activities $A_S, A_T \subseteq \mathscr{A}$ and an event log $L \subseteq \mathscr{E}^*$, we define a trace projection function $\rho(\sigma)$ which generates a weighted bipartite graph $G = (S \cup T, E, w)$ from $\sigma \in L$ such that*

- $S = \{e \in \sigma \mid \pi_{act}(e) \in A_S\}$,
- $T = \{e \in \sigma \mid \pi_{act}(e) \in A_T\}$,
- $E = \{(s,t) \in S \times T | \exists 1 \leq i < j \leq |\sigma| (\sigma(i) = s \wedge \sigma(j) = t)\}$, *and*
- *given $(s,t) \in E$ and let $1 \leq i < j \leq |\sigma|$ s.t. $\sigma(i) = s$ and $\sigma(j) = t$, $w(s,t) = \frac{1}{j-i}$.*

Given a graph $\rho(\sigma)$, $\rho(\sigma)_E$ denotes the edges of the graph.

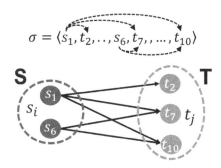

$$\sigma = \langle s_1, t_2, .., s_6, t_7, , ..., t_{10} \rangle$$

Fig. 5. An example of the bipartite graph given a trace

4.2   Maximum Weighted Bipartite Graph Matching Formulation

Given a bipartite graph built as described in Definition 7, we select pairs of events for measurements by finding a maximum weighted bipartite matching using ILP. The ILP model is formulated as below.

Definition 8 (ILP Formulation). *Given a weighted bipartite graph $G = (S \cup T, E, w)$, we assign a variable $x_{(s,t)}$ to every edge $(s,t) \in E$ and limit each variable to 0 or 1, indicating whether the edge is selected provided the integral solutions. We formulate the constraints and the objective as follows to find the maximum weighted bipartite matching (Definition 6).*

$$\text{Maximize } Z = \sum_{(s,t) \in E} w_{(s,t)} x_{(s,t)}$$

$$\text{Subject to } \sum_{s \in S} x_{(s,t)} \le 1, \qquad \forall t \in T$$

$$\sum_{t \in T} x_{(s,t)} \le 1, \qquad \forall s \in S$$

$$x_{(s,t)} \in \{0, 1\}$$

We denote the value for every variable $x_{(s,t)}$ as $x^*_{(s,t)}$ such that Z is maximum.

In principle, the problem above should be solved by ILP. However, we alternatively relax the integral constraints by applying Linear Programming (LP), i.e., $x_{(s,t)} \in [0,1]$, and compare their performance in Sect. 5.1. We found that LP always selects as much fraction of an edge as possible, i.e., 1, due to the constraints on the nodes. Thus, given the integral solutions provided by LP, we use LP in our implementation for better performance.

4.3   Aggregation

For each case, we average the desired measurements, e.g., duration between two events, given the pairs of events selected as described in Sect. 4.2. The performance metrics in terms of case and event log is computed as follows.

Definition 9 (Case Performance). *Let M denote all the measurable entities of a trace. Given a trace $\sigma = \pi_{tra}(c)$ for which $c \in \mathscr{C}$, referring to Definition 8, we compute the case performance in terms of $m \in M$ as:*

$$\delta_m(\sigma) = \frac{\displaystyle\sum_{(s,t) \in \rho(\sigma)_E} \phi_m(\sigma, s, t) * x^*_{(s,t)}}{\displaystyle\sum_{(s,t) \in \rho(\sigma)_E} x^*_{(s,t)}}.$$

Suppose all the cases in an event log are independent to each other, then we compute the log performance as defined below.

Definition 10 (Log Performance). *Let M denote all the measurable entities of a trace. Given an event log L, for every trace $\sigma \in L$, $\delta_m(\sigma)$ denotes the case performance in terms of $m \in M$. We select traces $T = \{\sigma \in L | |\rho(\sigma)_E| \ge 1\}$ and compute the log performance regarding $m \in M$ as:*

$$\Delta_m(L) = \frac{\displaystyle\sum_{\sigma \in T} \delta_m(\sigma)}{|T|}$$

By transforming a trace into a bipartite graph, our approach is capable of measuring the performance between two arbitrary groups of activities. The grouping can also be applied to analyze the performance at a higher level of abstraction without constructing another model at the desired granularity.

5 Evaluation

In this section, we evaluate the scalability using both ILP and LP in Sect. 5.1, followed by a comparative evaluation in Sect. 5.2.

5.1 Scalability with LP and ILP

We evaluate the performance and scalability of our approach by measuring the runtime of solving LP and ILP. Since the complexity of the problem is reflected by the number of variables (edges) and the number of constraints (nodes), we generate a synthetic event log with different number of nodes and edges. The event log simulates the results after filtering out events not of interest. Each event is randomly assigned as either an event in the start or target activity group and an event index between 1 and 1000, indicating the maximum length of a trace before filtering. The length of the trace is limited to 2 to 200 representing the log after the filtering. For each length, 15 traces are generated. Figure 6 shows the number of nodes and the corresponding number of edges.

Figure 7 shows the runtime using LP and ILP in terms of the number of edges and nodes. As expected, the runtime increases with complexity of the graph, i.e., the number of variables (edges) as shown in Fig. 7b and constraints (nodes) as shown in Fig. 7a. However, the edges selected by using LP and ILP are different. Nevertheless, compared to using ILP, the optional solutions from LP provide the same number of edges given integral solutions. Given the better performance and scalability of LP, we apply LP instead of ILP for our approach.

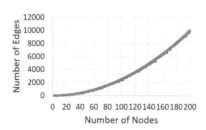

Fig. 6. Number of nodes and edges of the synthetic dataset.

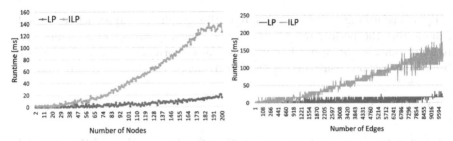

(a) Runtime[ms] versus Number of Nodes (b) Runtime[ms] versus Number of Edges

Fig. 7. Performance of the approach using LP and ILP. The performance using LP is better than ILP and highly scalable.

5.2   Comparative Evaluation

We compare our approach with other techniques with the dataset from the BPI Challenge 2019 [8]. The methods from Celonis [5] and Disco [10] are used for comparison as the representatives of log-based and model-based techniques, respectively. In the experiment, we evaluate the performance between *Vendor creates invoice* and *Record Invoice Receipt*. The two activities are chosen due to the limitation of analyzing with Disco [10], i.e., only selection of two adjacent activities in the model is allowed. To emphasize the difference of analysis result in the presence of recurrent activities, we further select the cases in which each activity is executed at least twice. After the filtering, we obtain 4915 cases with 14498 *Vendor creates invoice* and 17210 *Record Invoice Receipt* in total.

Table 2 shows the performance statistics in terms of the frequency and duration from different approaches. The case frequency shows that the number of cases considered while the absolute frequency shows the number of measurements used for calculation. These two frequencies are the same using Celonis due to only one measurement out of four possible configurations is selected. Figure 8a depicts the four possible configurations: first to first ($c1$), first to last ($c2$), last

(a) Four configurations.   (b) Configuration selected.   (c) Analysis result.

Fig. 8. Performance analysis using Celonis [5]

Table 2. Comparative evaluation results. Our approach computes the performance based on the most measurements, i.e., absolute frequency.

		Celonis				Disco	Our Approach
		$F \to F$	$F \to L$	$L \to F$	$L \to L$		
Case Freq.		4881	4915	2613	4889	2684	4915
Absolute Freq.		4881	4915	2613	4889	3797	14441
$\Delta_{dur}(L)$	Min.	5h	19h	5h	84m	84m	8h
	Max.	70y	70y	159d	447d	15.7w	35y
	Median.	14d	41d	12d	21d	3.7d	18d
	Mean.	61d	105d	17d	33d	10.2d	43.7d

F: Timestamp of the earliest occurrence of *Vendor creates invoice* in a case
L: Timestamp of the latest occurrence of *Record Invoice Receipt* in a case
$\Delta_{dur}(L)$: Duration between two activities of the event log L

to first ($c3$), last to last ($c4$). Each refers to the duration between the first/last event of one activity to the first/last event of the other activity in a trace. For each trace, only one configuration is chosen as the performance metrics. Thus, it produces skewed results depending on the configuration, e.g., the measurement between $c2$ and $c3$ can differ much. Figure 8c shows the analysis from one of the configuration as in Fig. 8b. From the frequency information in Table 2, it shows that some cases are ignored if the events in the cases do not have the configured order, e.g., the last execution of *Vendor creates invoice* is after the first execution of *Record Invoice Receipt*.

Figure 9 shows the mean duration and absolute frequency between activities from Disco [10]. The cases that do not comply to the model are ignored, i.e., only 2684 out of 4915 cases are considered. In addition, the resulting metrics tends to be smaller since it only considers the measurement if *Record Invoice Receipt* directly follows *Vendor creates invoice* without other activities in between.

The results show that our approach is robust for that it covers as many measurements as possible given two activities. Moreover, Table 2 shows that even a simple question as the time between two activities is not as simple as it seems. Besides, our approach allows for measuring the performance between two groups of activities.

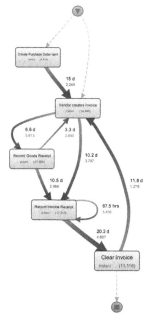

Fig. 9. Analysis using Disco [10]

6   Conclusion

Process performance can be analyzed based on a model or an event log only. Most techniques are model-based and are, thus, limited to the representational bias of the underlying models and not flexible to the need of analysis, e.g., measuring the performance between two milestones. Existing techniques that are based on an event log face the challenge of providing a robust result in the presence of recurrent activities. This paper introduces a novel and generic process-aware log-based technique which is robust to recurrent activities and can be further extended to analysis of two groups of activities. By applying bipartite graph matching, we pair as many events as possible and measure them accordingly. Our experiments show that the proposed approach is scalable to huge event logs and outperforms the existing methods by providing a robust metrics given recurrent activities.

To better locate and diagnose the performance of a process, we aim to extend our approach by incorporating the control-flow constructs into the bipartite graph and clustering the cases based on the measured performance. Furthermore, an investigation on different business contexts or rules can be applied to determine complementary performance indicators and the weights of the bipartite graph. Meanwhile, we also aim to automate the selection of the activities to facilitate the application of our approach.

References

1. Aalst, W.: Data science in action. Process Mining, pp. 3–23. Springer, Heidelberg (2016). https://doi.org/10.1007/978-3-662-49851-4_1
2. Adriansyah, A.: Aligning observed and modeled behavior (2014)
3. Adriansyah, A., van Dongen, B., Piessens, D., Wynn, M., Adams, M.: Robust performance analysis on YAWL process models with advanced constructs. J. Inf. Technol. Theory Appl. (JITTA) **12**(3), 5–26 (2012)
4. Castellanos, M., Casati, F., Shan, M.C., Dayal, U.: iBOM: a platform for intelligent business operation management. In: 21st International Conference on Data Engineering (ICDE 2005), pp. 1084–1095. IEEE (2005)
5. Celonis, S.E.: Academic cloud (2019). https://academiccloud.celonis.com. Accessed 26 Apr 2019
6. Costello, C., Molloy, O.: Building a process performance model for business activity monitoring. In: Wojtkowski, W., Wojtkowski, G., Lang, M., Conboy, K., Barry, C. (eds.) Information Systems Development. Springer, Boston (2009). https://doi.org/10.1007/978-0-387-68772-8_19
7. Dantzig, G.: Linear Programming and Extensions. Princeton University Press, Princeton (2016)
8. van Dongen, B.F.: Dataset BPI Challenge 2019. 4TU Centre for Research Data (2019). https://doi.org/10.4121/uuid:d06aff4b-79f0-45e6-8ec8-e19730c248f1. Accessed 10 Apr 2019
9. Gal, A., Mandelbaum, A., Schnitzler, F., Senderovich, A., Weidlich, M.: Traveling time prediction in scheduled transportation with journey segments. Inf. Syst. **64**, 266–280 (2017)

10. Günther, C.W., Rozinat, A.: Disco: discover your processes. In: BPM (2012)
11. Hompes, B.F., Buijs, J.C., van der Aalst, W.M.P.: A generic framework for context-aware process performance analysis. In: Debruyne, C., et al. (eds.) OTM 2016. LNCS, vol. 10033, pp. 300–317. Springer, Cham (2016). https://doi.org/10.1007/978-3-319-48472-3_17
12. Hornix, P.T.: Performance analysis of business processes through process mining. Master's Thesis, Eindhoven University of Technology (2007)
13. Leemans, M., van der Aalst, W.M.P., Van Den Brand, M.G.: Hierarchical performance analysis for process mining. In: Proceedings of the 2018 International Conference on Software and System Process, pp. 96–105. ACM (2018)
14. Piessens, D., Wynn, M.T., Adams, M.J., van Dongen, B.: Performance analysis of business process models with advanced constructs (2010)
15. ProM: Running example - process to repair telephones in a company (2019). http://www.promtools.org/prom6/downloads/example-logs.zip. Accessed 16 May 2019
16. Rogge-Solti, A., van der Aalst, W.M.P., Weske, M.: Discovering stochastic petri nets with arbitrary delay distributions from event logs. In: Lohmann, N., Song, M., Wohed, P. (eds.) BPM 2013. LNBIP, vol. 171, pp. 15–27. Springer, Cham (2014). https://doi.org/10.1007/978-3-319-06257-0_2
17. Rozinat, A.: Process mining: conformance and extension (2010)
18. Senderovich, A., et al.: Data-driven performance analysis of scheduled processes. In: Motahari-Nezhad, H.R., Recker, J., Weidlich, M. (eds.) BPM 2015. LNCS, vol. 9253, pp. 35–52. Springer, Cham (2015). https://doi.org/10.1007/978-3-319-23063-4_3
19. Song, M., van der Aalst, W.M.P.: Supporting process mining by showing events at a glance. In: Proceedings of the 17th Annual Workshop on Information Technologies and Systems (WITS), pp. 139–145 (2007)
20. Wetzstein, B., Leitner, P., Rosenberg, F., Brandic, I., Dustdar, S., Leymann, F.: Monitoring and analyzing influential factors of business process performance. In: 2009 IEEE International Enterprise Distributed Object Computing Conference, pp. 141–150. IEEE (2009)

Extracting a Collaboration Model from VCS Logs Based on Process Mining Techniques

Leen Jooken$^{(\boxtimes)}$, Mathijs Creemers, and Mieke Jans

Hasselt University, Martelarenlaan 42, 3500 Hasselt, Belgium
leen.jooken@uhasselt.be

Abstract. A precise overview on how software developers collaborate on code could reveal new insights such as indispensable resources, potential risks and better team awareness. Version control system logs keep track of what team members worked on and when exactly this work took place. Since it is possible to derive collaborations from this information, these logs form a valid data source to extract this overview from. This concept shows many similarities with how process mining techniques can extract process models from execution logs. The fuzzy mining algorithm [5] in particular holds many useful ideas and metrics that can also be applied to our problem case. This paper describes the development of a tool that extracts a collaboration graph from a version control system log. It explores to what extend fuzzy mining techniques can be incorporated to construct and simplify the visualization. A demonstration of the tool on a real-life version control system log is given. The paper concludes with a discussion of future work.

Keywords: Process mining · Social network analysis · Data mining · Version control systems

1 Introduction

Companies engaged in software development have to manage a great deal of code on a daily basis. It gets hard to keep track of who has knowledge of certain aspects of the code and which programmers are working together on which parts. This lack of awareness can also cause files that are at risk of becoming unknown to any programmer, to go unnoticed. This is the case when you have a non-static file that only gets changed by a very small amount of people. If this select group of programmers were to quit, the organization can end up with code that nobody really knows the details of. An effective way to improve this awareness is through visualization of these collaboration relationships. This can be achieved through a social network graph that represents programmers as nodes and relationships between them as edges. Since we wish to discover this collaboration from real-life project scenarios, we believe that logs drawn from version control systems serve

© Springer Nature Switzerland AG 2019
C. Di Francescomarino et al. (Eds.): BPM 2019 Workshops, LNBIP 362, pp. 212–223, 2019.
https://doi.org/10.1007/978-3-030-37453-2_18

as an ideal primary data source. This idea of extracting a social network graph from a version control system log strongly resembles the rationale of applying process discovery to process event logs. The two types of logs are similar in the way that every commit message from the former can be seen as an event, and both specify a resource for every event. Aside from these similarities, there are also some aspects in which the two differ. In an event log the events follow a certain process flow, while in the case of a version control system there is no clear process notion involved. The act of developing software might be better described as a project, rather than a process [1]. This lack of a clear process notion is also the reason why we do not build on the implementation by van der Aalst et al. [12].

Since our visualization should make it possible to easily identify insights that are relevant to the business, like the core collaboration teams and isolated nodes, it has to be clear and easy to understand. This shows similarities with another aspect within the process mining domain, namely fuzzy mining [5]. The fuzzy miner's goal is to discover process models from event logs of unstructured processes. This comes down to using event log data to produce a visualization of the links between these events [5]. The goal of our research is using log data to build a visualization of the interaction between people. To achieve this, we will examine to what extend the techniques from fuzzy mining are also applicable to our visualization. We will focus on the metrics used in the weight calculation and simplification approach of the graph.

The main contribution of this paper is the development of a tool that can be used to extract social networks from version control system logs, inspired by the algorithm of the fuzzy miner. The tool has been implemented and demonstrated on a real-life version control system log.

The remainder of this paper is organized as follows. Section 2 discusses the state of the art, while Sect. 3 elaborates the design choices for the new tool. This is followed by a detailed description of the tool's algorithms for weight calculation and simplification in Sects. 4 and 5. Section 6 holds a demonstration of our tool on a real-life example and we conclude this paper with a discussion in Sect. 7.

2 Related Work

Most of the literature covering potential uses of version control system logs focuses on using the files stored in such a system to visualize the source code hierarchy, rather than collaboration aspects [3,4,7–10,14]. In contrast to the numerous amount of tools that deal with code structure and flow, there are almost no tools that make use of data related to social aspects.

A tool that does involve this type of data up to a certain level, is the Manhattan tool, developed by Lanza et al. [6]. This tool produces a 3D visualization of the code with the main goal of supporting team activity by increasing workspace awareness. In order to achieve this, the system visualizes in real-time what each team member is working on and notifies the user about emerging conflicts. Although the tool makes use of collaboration data, it does not explicitly visualize these relationships between programmers. There is no functionality to query the system about how often and how closely members work together.

3   Design of the Collaboration Visualizer

3.1   Solution Requirements and Choice of Tools

In order to gain valuable insights about the social structure of a development team, we set the necessary requirements for our solution. After a comparison between 23 different social network analysis tools, we decided to use a combination of a self-written Python program and Gephi [2] as the tools for our solution. Gephi is an Open Source interactive graph exploration and manipulation software, suited for all kinds of networks [2]. Of all possible network visualization tools, Gephi was chosen because it offers functionalities that best meet the visualization requirements that were set. These requirements will not be discussed here, due to page limitations. With the help of Gephi we can shift our focus to the development of a program that will handle the following remaining requirements:

1. The program is able to read and process the data from a version control system log.
2. It uses the aforementioned data to construct a graph, in which programmers are represented by nodes and the collaboration between them by edges. The edges are labeled according to their collaboration type. There are three possible types an edge can classify as: (1) pair programming, (2) disjunct programming, (3) pair and disjunct programming (when the two programmers worked separately, but also engaged in pair programming).
3. Both the programmers and the edges between them have a weight that reflects their importance.
4. The graph is clear and easy to understand, which implies that some simplification measures have to be taken.

3.2   Detail of the Program Structure

The program will comprise of several algorithms that carry out the steps in the following list in order to generate the resulting graph. Step 2, 4, 5 and 6 will be explained more into detail in Sects. 4.1 to 5 respectively.

1. Parse the log into a usable data structure
2. Calculate the importance of each file
3. Build the base graph by including every programmer that is mentioned in the log as a node and adding an edge if they collaborated at least once. This collaboration can be in the form of pair programming or having an altered file in common
4. Calculate the edge weights
5. Calculate the node weights
6. Simplify the resulting graph
7. Write the simplified graph to CSV format, so Gephi can carry out the visualization

4 Constructing the Base Graph: Weight Calculation Algorithms

4.1 Calculating the File Importance

The program should calculate a value that reflects the importance of a file. This value is necessary to create a well-substantiated weight for both the programmers and their relationships. Rather than to focus on the business importance of a file, we have chosen to let this value represent how important the file is for collaboration. We cannot base this importance of how large a certain file is, since this information is not standard available in a version control system log. So let us consider a file important if it continues to *'grow over time'*. A project is dynamic, it evolves over time to accommodate for new features and new customer requirements. As a result of this, files that make up the core of the application will get altered regularly. These files are good candidates for collaboration since multiple people having knowledge of them secures their further evolution.

Based on these assumptions, we will develop a formula to calculate the file importance. To ensure that files that have been around for a long time are not favoured over relatively new ones, we will work with a ratio that takes the life span of the file into account. This ratio considers the number of months the file in question exists or existed, and calculates in how many of those months the file got changed. We do not factor in the number of commits related to this file that occur within a month, for a programmer is free to choose the size of the chunks he commits. A downside to this approach is that files that were created towards the end of the log will in most cases have a larger importance than files that have been around for a long time. We do not consider this larger importance to be a problem since these files are recent and therefore important at this very moment in the developing stage. This brings us to the following formula that calculates the importance value of a specific file:

$$File\ importance = \frac{the\ number\ of\ months\ in\ which\ the\ file\ got\ altered}{the\ number\ of\ months\ the\ file\ existed} \quad (1)$$

4.2 Calculating the Edge Weights

The edge weight is an indicator of how strong the collaboration between two programmers is. In order to logically substantiate this weight, we examined and selected some metrics used in fuzzy mining [5] that we believe are also applicable to our problem. These metrics are the **binary frequency significance** and the **proximity correlation**. Both of these contribute a value to different aspects that together determine how important the collaboration is. Because of the limited length of this paper, we will only discuss how we adapted these metrics to our problem case. The implementation used by the fuzzy miner can be found in the research paper by Günter et al. [5].

The Binary Frequency Significance. This metric describes the relative importance of the relationship between two nodes. We adapt this idea to our problem context as *'the more files are worked on together, the stronger the relationship'*. However, some constraints have to be added. First of all, we have to take the file importance into account. Omitting this importance can cause two programmers that worked on only a handful of, but very important, files to have a weaker relationship than two that worked on a lot of trivial files together. Secondly, we have to be careful not to favour a programmer with the habit of committing his current work regularly over one that likes to commit very big chunks of code more sparsely. Lastly, we will only consider files from commits that were not pair programming activities between the two programmers under consideration. The pair programming aspect will be included in the edge weight calculation as a separate factor, further explained in the next subsection.

Keeping in mind these three constraints, we suggest the following algorithm to calculate the frequency significance value for a certain edge between two programmers:

1. Collect the commits of the first and second programmer and omit those that concern pair programming
2. For each file in these remaining commits, that both programmers worked on do the following:
   (a) Count the number of times they have worked together on this file, with the constraint that everything that happens within the time span of one week counts as the same collaboration session
   (b) Calculate the frequency significance for this file as:

   $$\textit{the number of sessions calculated in the previous step}$$
   $$\times$$
   $$\textit{the importance of the file under consideration}$$

3. The final frequency significance of this edge is the sum of the frequency significances of all the aforementioned files

The Proximity Correlation. Within the fuzzy mining algorithm, the proximity correlation evaluates event classes which occur shortly after one another, as highly correlated [5]. In terms of the timing aspect, we have already taken the proximity into account during the calculation of the frequency significance by limiting the permitted time span between two consecutive commits. However, there is also the physical proximity in the form of pair programming. This will serve as our proximity correlation, as we have earlier decided to not include this aspect in the frequency significance. Including this as a separate factor allows for pair programming, which is a stronger form of collaboration, to have a greater influence on the final edge weight. Because this metric has the same constraints as the frequency significance, we will again work with the file importance and a predefined time span between commits. This results in the following algorithm:

1. Compose a list of files that appear in the pair programming commits concerning these two programmers
2. For each of these files, do the following:
   (a) Collect the time stamps of all the commits concerning this file and order them chronologically
   (b) Count the number of times the programmers collaborated on this file by going through the chronologically ordered time stamps and tallying whenever there is at least a predefined time span in between two consecutive time stamps. The default value of this predefined time span is one month
   (c) Calculate the proximity correlation for this file as:

the number of times they collaborated, as calculated in step (b)

×

the importance of the file under consideration

3. The final proximity correlation of this edge is the sum of the proximity correlations of all the aforementioned files

The Final Edge Weight. The final weight of an edge is calculated according to the following formula. The proximity correlation has a greater influence on the final weight, because the metric concerns a closer collaboration and there is less uncertainty involved.

$$Final\ edge\ weight = 0.40 \cdot normalized\ frequency\ significance\ +$$
$$0.60 \cdot normalized\ proximity\ correlation \quad (2)$$

4.3   Calculating the Node Weights

The **unary frequency significance** is the only metric stemming from the fuzzy mining algorithm [5] that seemed useful for the calculation of our node weights, that represent the importance of a programmer. The exact implementation of this metric can be found in the research paper by Günter et al. [5]. In order to better substantiate the weight, we looked for additional applicable metrics stemming from traditional graph theory. The three metrics that were eventually selected are the **betweenness centrality**, the **eigenvector centrality** and the **degree centrality**.

The Unary Frequency Significance. We will adopt this metric as '*the more often a programmer appears in the log, the more significant he is*', but we will need to set some constraints. Again, we work with a list of distinct files (so every file is counted only once) to counter the '*free choice of when to commit*' problem. Further, we will also factor in the file importance. This leads us to the following formula that calculates the frequency significance value of a certain programmer:

$$Frequency\ significance\ =\ normalization\ of\ the\ sum\ of\ the\ importance$$
$$value\ of\ each\ file\ this\ programmer\ worked\ on \quad (3)$$

The Betweenness Centrality. We chose this specific metric to handle programmers that are a part of several different teams. This metric considers a node highly important if it forms bridges between many other nodes. The value is computed as the sum of the fraction of all-pairs shortest paths that pass through that node [11]. We normalize the results based on the minimum and maximum centrality value, instead of using the number of nodes. Otherwise the values will be too small to make a real impact on the final weight.

The Eigenvector Centrality. According to this metric, a node is highly important if many other highly important nodes link to it. However, the centrality also depends on the quality of those edges [13]. So this metric reveals teams of important nodes that work closely together.

The Degree Centrality. The degree centrality looks at the number of edges incident upon the node [13]. We will use this metric to identify (nearly) isolated nodes. This, combined with the frequency significance, can uncover programmers that are indispensable sources of knowledge and not easily replaced. So these nodes should have larger weights than other leaf nodes, to prevent them from being pruned during the simplification phase. At first glance, this seems to contradict the eigenvector centrality, described in the previous paragraph. We cannot, however, use a low eigenvector centrality value as an indicator to give extra weight to a nearly isolated node. Imposing a larger weight on a node with a low eigenvector centrality will cause the core team cluster nodes to have a low weight, which is not what we want. For this reason we use the eigenvector centrality to identify the core teams and the degree centrality as a separate metric to identify the (nearly) isolated nodes and give them the weight they deserve.

To calculate our final degree centrality value, we make a distinction between pair programming and non-pair programming edges. Since we want nodes that hardly collaborated with others, and therefore have few incident links, to have a larger weight, we use 1 - the degree centrality for the non-pair programming edges. If a programmer engaged in pair programming his value for this metric should be smaller. So the final value for this metric is

$$1 - the\ degree\ centrality\ for\ the\ non\ pair\ programming\ edges$$
$$- the\ degree\ centrality\ for\ the\ pair\ programming\ edges \quad (4)$$

The Final Node Weight. Using the following formula, the final node weight can be calculated. The eigenvector and the degree centrality have the greatest influence because the main goal of our visualization is identifying the core teams and isolated nodes. We also deemed the betweenness centrality more important than the frequency significance because programmers that are a part of multiple teams form an important insight and this should be emphasized in the visualization.

$Final\ node\ weight =\ 0.15\ \cdot\ normalized\ frequency\ significance\ +\ 0.25\ \cdot$
$normalized\ betweenness\ centrality\ +\ 0.30\ \cdot\ normalized\ eigenvector\ centrality$
$$+\ 0.30\ \cdot\ normalized\ degree\ centrality \qquad (5)$$

5 Graph Simplification Approach

To prevent the graph from being too large and unclear to derive any useful insights from, it needs to be simplified. Our simplification approach is inspired by the fuzzy miner as well. The exact implementation the fuzzy miner uses, will not be explained in this paper, due to page limitations, but can be found in the paper of Günter et al. [5]. We adopt three of the fuzzy miner's graph simplification methods, namely (1) edge filtering, (2) aggregation and (3) abstraction.

5.1 Phase 1: Edge Filtering

In this first phase we filter out edges that are weak and therefore not well established. The fuzzy miner [5] tackles its edge filtering by calculating a utility value for every edge and comparing these values with an edge cutoff parameter. We will adapt this algorithm to better fit our problem case.

Since our graph is undirected, unlike the fuzzy miner one, we perform our filtering algorithm only once on the entire set of edges incident on the node. This creates a new problem: an edge can be unimportant for one node, but important for the target node it is connected to. We solve this problem by only filtering an edge when both its connected nodes agree on it. Another important aspect to mention is that it is impossible to create new isolated nodes, because the edge cutoff parameter is compared to the *normalized* utility values. Since the utility value is meant to be an indication of how important the edge is, we can let the edge weight, as calculated in Sect. 4.2, represent it. Lastly we also have to determine the edge cutoff parameter. This parameter determines how careful the algorithm is with removing edges. The value of it depends on the log under consideration and the purpose of the desired visualization. This brings us to the following algorithm:

1. For very node, compare the normalized weight of its incident edges to the edge cutoff parameter. If the normalized weight < edge cutoff parameter add the edge to the list of possible candidates for filtering
2. If this edge is already present in the list of candidates, it means that the other node it is connected to also suggested it as a candidate for filtering. Move this edge to the list of edges to filter
3. Once all the nodes are handled, remove the edges that appear in the list of edges to filter, from the graph

5.2   Phase 2: Aggregation

In this second phase we form coherent clusters of programmers that are not very important but have very strong relationships between them. Such a group of programmers is represented by a single node in the graph, because their individual importance does not outweigh the added value of simplification.

Our aggregation approach consists of the same two phases the fuzzy miner [5] uses for this purpose: initial cluster building and cluster merging. After carrying out these two phases we also have to redefine the weights of both the cluster and its connected edges and determine the collaboration type of these edges. In the next subsections we will explain each of these phases more into detail, including when and why we deviate from the original fuzzy miner approach. Determining the new collaboration type is omitted due to page limitations.

Initial Cluster Building. We start off by checking for each node whether it is unimportant enough to aggregate. This comes down to comparing the node weight to a node cutoff parameter. By default, this cutoff parameter is set as the average weight of all the nodes in the graph. Once the candidates are selected, each of them has to be evaluated to determine whether they have an edge that is strong enough to carry out the aggregation. We deviate here from the original fuzzy miner [5] approach, which is clustering a candidate regardless of the edge's weight. Because we want a cluster to represent a group of strongly connected programmers, an additional check is added. Only when an edge's aggregation power exceeds an aggregation-correlation parameter, a cluster can be formed. The edge's aggregation power is represented by its *distance significance*, which looks at how much the significance of the edge differs from the significances of its connected nodes [5]. Since we know that the candidate nodes have low weights and we are searching for strong edges, the bigger the difference between the weights of these two, the more the edge qualifies to carry out aggregation. We choose the aggregation-correlation parameter to be the difference between the average node weight of all the nodes and the average edge weight of all the edges in the graph. So for each candidate node we search the edge with the strongest aggregation power that exceeds this aggregation-correlation parameter. If no edge is deemed strong enough, the node remains untouched. If we now resume where we left off while following the fuzzy miner approach, we obtain the following algorithm:

For each node that has a node weight < node cutoff parameter:

1. Calculate the distance significance for each edge incident on this node by using the following formula: | *edge weight - node weight* |
2. Select the edge with the largest distance significance that > aggregation-correlation parameter
3. If the target node this selected edge connects to is a cluster → add the node and its incident edges to this cluster; Else make a new cluster with this node as the first member

Cluster Merging. The fuzzy miner [5] only merges clusters when its pre- or postset contains only cluster nodes. However, it is in our case not necessary that all the neighbouring nodes are cluster nodes. If two nodes have a very strong connection, they will most likely have more or less the same knowledge. Nodes that collaborate with one of these two can theoretically also be seen as working with the other. This implies that in order to merge, not all the neighbouring nodes have to be clusters. Taking this into account, be obtain the following algorithm:

For every cluster, do the following:

1. Calculate the distance significance for every edge that connects this cluster to a neighbouring cluster node, by using the formula: | *edge weight - node weight* |
2. Select from these candidates edges the edge with the largest distance significance value that is > aggregation-correlation parameter to actually carry out the merging
3. Merge both clusters and transitively preserve the edges

Redefining the Cluster's Weights. All edges that connect a node (or cluster) to nodes within the target cluster will be removed and replaced by one edge that is representative of all these connections. This replacing edge needs a new weight. Due to page limitations, only the used formula is given:

Collect the weights of all the edges connecting this node (or cluster) to a node within the target cluster. The weight of the replacing edge =

$$\frac{the\ average\ edge\ weight\ +\ the\ strongest\ edge\ weight}{2} \tag{6}$$

The calculation of the new weight for the cluster node is carried out in exactly the same way, but with the weights of the nodes that make up the cluster.

5.3 Phase 3: Abstraction

We cannot apply the abstraction approach the fuzzy miner [5] uses, which is removing all the isolated and singular clusters, as these can be useful insights. So in our abstraction approach we only remove programmers that are insignificant and not strongly enough connected to other programmers so they can be aggregated. These programmers hardly contributed to the project, so they do not add any useful insights to the visualization. We do this as follows.

For every node that does not belong to a cluster, do the following:

1. Check whether the node's weight < node cutoff parameter
2. If so, check whether all of its incident edges are too weak to carry out aggregation. This is done by confirming that every incident edge has a distance significance value < aggregation-correlation parameter
3. If so, remove the node and all of its incident edges from the graph

6   Tool Demonstration

The tool has been implemented[1] and tested on a real-life log, containing 1486 chronologically ordered commits, dating from June 2009 to November 2017. All the requirements, that were stated in Sect. 3.1, are fulfilled. After the Collaboration Visualizer analyzes this log, we are presented the graph in Fig. 1, in which all contributors have been anonymized. The graph can be used to identify isolated groups of programmers, the core development teams and weak collaboration relationships. By identifying these indispensable resources, the risk of losing valuable sources of knowledge can be mitigated. A full explanation of the insights that can be gained that are relevant to the business, has been omitted due to page limitations.

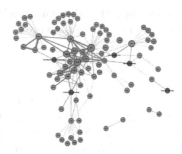

Fig. 1. Collaboration graph that is the result of the tool analyzing a real-life log. The green nodes are clusters of programmers, while a pink one represents a single programmer. The edges are colour-coded to represent the following: the green edges indicate pair programming, the orange ones disjunct programming and the purple ones imply that the programmers engaged in pair as well as disjunct programming. (Color figure online)

7   Discussion and Future Work

In this paper, we described a tool that can extract how programmers collaborated on code from a VCS log and visualize this in the form of a graph. Due to many similarities, we gathered inspiration for the development of the tool from the fuzzy miner algorithm [5]. We combined the metrics from the fuzzy miner that we deemed applicable to our case, with metrics from standard graph theory.

Since this is the very first version of our tool, there are still aspects that can be improved and fine tuned. Further research should be done to provide an improved default parameter setting. Further, the tool currently only works with logs generated from a Tortoise SVN system. Including logs from other version control systems as an input option is a good requirement for the next version of the tool. Lastly, this paper is limited to a demonstration of the applicability of the tool. An extensive interpretation of the insights that can be gained from the graph is omitted due to page limitations, and validation still needs to be added.

---

[1] The source code is available on github.com/LeenJooken/CollaborationVisualizer.git.

References

1. Bala, S., Cabanillas, C., Mendling, J., Rogge-Solti, A., Polleres, A.: Mining project-oriented business processes. In: Motahari-Nezhad, H.R., Recker, J., Weidlich, M. (eds.) BPM 2015. LNCS, vol. 9253, pp. 425–440. Springer, Cham (2015). https://doi.org/10.1007/978-3-319-23063-4_28
2. Bastian, M., Heymann, S.: Gephi: an open source software for exploring and manipulating networks. In: ICWSM (2009). https://doi.org/10.1136/qshc.2004.010033
3. Georget, L., Tronel, F., Tong, V.V.T.: Kayrebt: an activity diagram extraction and visualization toolset designed for the Linux codebase. In: 2015 IEEE 3rd Working Conference on Software Visualization, VISSOFT 2015 - Proceedings (2015). https://doi.org/10.1109/VISSOFT.2015.7332431
4. Gračanin, D., Matković, K., Eltoweissy, M.: Software visualization. Innov. Syst. Softw. Eng. **1**, 221–230 (2005). https://doi.org/10.1007/s11334-005-0019-8
5. Günther, C.W., van der Aalst, W.M.P.: Fuzzy mining – adaptive process simplification based on multi-perspective metrics. In: Alonso, G., Dadam, P., Rosemann, M. (eds.) BPM 2007. LNCS, vol. 4714, pp. 328–343. Springer, Heidelberg (2007). https://doi.org/10.1007/978-3-540-75183-0_24
6. Lanza, M., D'Ambros, M., Bacchelli, A., Hattori, L., Rigotti, F.: Manhattan: supporting real-time visual team activity awareness. In: IEEE International Conference on Program Comprehension (2013). https://doi.org/10.1109/ICPC.2013.6613849
7. Marcus, A., Comorski, D., Sergeyev, A.: Supporting the evolution of a software visualization tool through usability studies. In: Proceedings - IEEE Workshop on Program Comprehension (2005). https://doi.org/10.1109/WPC.2005.34
8. Quante, J.: Do dynamic object process graphs support program understanding? - A controlled experiment. In: IEEE International Conference on Program Comprehension (2008). https://doi.org/10.1109/ICPC.2008.15
9. Sensalire, M., Ogao, P., Telea, A.: Evaluation of software visualization tools: lessons learned, pp. 19–26, October 2009. https://doi.org/10.1109/VISSOF.2009.5336431
10. Storey, M.A., Wong, K., Müller, H.A.: How do program understanding tools affect how programmers understand programs? Sci. Comput. Program. **36**, 183–207 (2000). https://doi.org/10.1016/S0167-6423(99)00036-2
11. Team, N.: NetworkX (2014)
12. Van Der Aalst, W.M., Reijers, H.A., Song, M.: Discovering social networks from event logs. Comput. Support. Coop. Work **14**, 549–593 (2005). https://doi.org/10.1007/s10606-005-9005-9
13. Wasserman, S., Faust, K.: Social Network Analysis: Methods and Applications II (1994). https://doi.org/10.1525/ae.1997.24.1.219
14. Wettel, R., Lanza, M., Robbes, R.: Software systems as cities: a controlled experiment. In: ICSE 2011 Proceedings of the 33rd International Conference on Software Engineering (2011). https://doi.org/10.1145/1985793.1985868

Finding Uniwired Petri Nets Using eST-Miner

Lisa Luise Mannel$^{(\boxtimes)}$ and Wil M. P. van der Aalst

Process and Data Science (PADS), RWTH Aachen University, Aachen, Germany
{mannel,wvdaalst}@pads.rwth-aachen.de

Abstract. In process discovery, the goal is to find, for a given event log, the model describing the underlying process. While process models can be represented in a variety of ways, in this paper we focus on a subclass of Petri nets. In particular, we describe a new class of Petri nets called *Uniwired Petri Nets* and first results on their expressiveness. They provide a balance between simple and readable process models on the one hand, and the ability to model complex dependencies on the other hand. We then present an adaptation of our eST-Miner aiming to find such Petri Nets efficiently. Constraining ourselves to uniwired Petri nets allows for a massive decrease in computation time compared to the original algorithm, while still discovering complex control-flow structures such as long-term-dependencies. Finally, we evaluate and illustrate the performance of our approach by various experiments.

Keywords: Process discovery · Petri nets · Language-based regions

1 Introduction

More and more processes executed in companies are supported by information systems which store each event executed in the context of a so-called *event log*. Each event in such an event log has a name identifying the executed activity (activity name), identification specifying the respective execution instance of the process (case id), a time when the event was observed (time stamp), and often other data related to the activity and/or process instance. In the context of process mining, many algorithms and software tools have been developed to utilize the data contained in event logs: in *conformance checking*, the goal is to determine whether the behaviors given by a process model and event log comply. In *process enhancement*, existing models are improved. Finally, in *process discovery*, a process model is constructed aiming to reflect the behavior defined by the given event log: the observed events are put into relation to each other, preconditions, choices, concurrency, etc. are discovered, and brought together in a process model.

Process discovery is non-trivial for a variety of reasons. The behavior recorded in an event log cannot be assumed to be complete, since behavior allowed by

© Springer Nature Switzerland AG 2019
C. Di Francescomarino et al. (Eds.): BPM 2019 Workshops, LNBIP 362, pp. 224–237, 2019.
https://doi.org/10.1007/978-3-030-37453-2_19

the process specification might simply not have happened yet. Additionally, real-life event logs often contain noise, and finding a balance between filtering this out and at the same time keeping all desired information is often a non-trivial task. Ideally, a discovered model should be able to produce the behavior contained within the event log, not allow for unobserved behavior, represent all dependencies between the events and at the same time be simple enough to be understood by a human interpreter. It is rarely possible to fulfill all these requirements simultaneously. Based on the capabilities and focus of the used algorithm, the discovered models can vary greatly, and different trade-offs are possible.

To decrease computation time and the complexity of the returned process model, many existing discovery algorithms abstract from the full information given in a log or resort to heuristic approaches. Even though the resulting model often cannot fully represent the language defined by the log, they can be very valuable in practical applications. Examples are the Alpha Miner variants ([1]), the Inductive Mining family ([2]), genetic algorithms or Heuristic Miner. On the downside, due to the commonly used abstractions, these miners are not able to (reliably) discover complex model structures, most prominently non-free choice constructs. Algorithms based on region theory ([3–6]) discover models whose behavior is the minimal behavior representing the log. However, they often lead to complex, over-fitting process models that are hard to understand. These approaches are also known to be rather time-consuming and expose severe issues with respect to low-frequent behavior often contained in real-life event logs.

In our previous work [7] we introduce the discovery algorithm eST-Miner, which focuses on mining process models formally represented by Petri nets. This approach aims to combine the capability of finding complex control-flow structures like longterm-dependencies with an inherent ability to handle low-frequent behavior while exploiting the token-game to increase efficiency. Similar to region-based algorithms, the basic idea is to evaluate all possible places to discover a set of fitting ones. Efficiency is significantly increased by skipping uninteresting sections of the search space. This may decrease computation time immensely compared to the brute-force approach evaluating every single candidate place, while still providing guarantees with regard to fitness and precision.

Though inspired by language-based regions, the approach displays a fundamental difference with respect to the evaluation of candidate places: while region-theory traditionally focuses on a globalistic perspective on finding a set of feasible places, our algorithm evaluates each place separately, that is from a local perspective. In contrast to the poor noise-handling abilities of traditional region-based discovery algorithms, this allows us to effectively filter infrequent behavior place-wise. Additionally, we are able to easily enforce all kinds of constraints definable on the place level, e.g. constraints on the number or type of arcs, token throughput or similar.

However, the original eST-Miner has several drawbacks that we aim to tackle in this paper: first, the set of fitting places takes too long to be computed, despite our significant improvement over the brute force approach. Second, the discovered set of places typically contains a huge number of implicit places, that need to be removed in a time-consuming post-processing step. Third, the algorithm finds very complex process structures, which increase precision but at the same time decrease simplicity. In this work, we introduce the new class of *uniwired Petri nets*, which constitutes a well-balanced trade-off between the ability of modeling complex control-flows, such as long-term-dependencies, with an inherent simplicity that allows for human readability and efficient computation. We present a corresponding variant of our eST-Miner, that aims to discover such Petri nets.

In Sect. 2 we provide basic notation and definitions. Afterwards a brief introduction to our original algorithm is given in Sect. 3. We then present the class of *uniwired Petri nets* in Sect. 4, together with first results on their expressiveness and relevance for process mining. In Sect. 5 we describe an adaption of our algorithm that aims to efficiently compute models of this sub-class by massively increasing the amount of skipped candidate places. An extensive evaluation follows in Sect. 6. Finally, we conclude the paper with a summary and suggestion of future work in Sect. 7.

2   Basic Notations, Event Logs, and Process Models

A set, e.g. $\{a, b, c\}$, does not contain any element more than once, while a multiset, e.g. $[a, a, b, a] = [a^3, b]$, may contain multiples of the same element. By $\mathbb{P}(X)$ we refer to the power set of the set X, and $\mathbb{M}(X)$ is the set of all multisets over this set. In contrast to sets and multisets, where the order of elements is irrelevant, in sequences the elements are given in a certain order, e.g. $\langle a, b, a, b \rangle \neq \langle a, a, b, b \rangle$. We refer to the i'th element of a sequence σ by $\sigma(i)$. The size of a set, multiset or sequence X, that is $|X|$, is defined to be the number of elements in X. We define activities, traces, and logs as usual, except that we require each trace to begin with a designated start activity ($\blacktriangleright$) and end with a designated end activity ($\blacksquare$). Note, that this is a reasonable assumption in the context of processes, and that any log can easily be transformed accordingly.

Definition 1 (Activity, Trace, Log). *Let $\mathcal{A}$ be the universe of all possible activities (e.g., actions or operations), let $\blacktriangleright \in \mathcal{A}$ be a designated start activity and let $\blacksquare \in \mathcal{A}$ be a designated end activity. A trace is a sequence containing $\blacktriangleright$ as the first element, $\blacksquare$ as the last element and in-between elements of $\mathcal{A} \setminus \{\blacktriangleright, \blacksquare\}$. Let $\mathcal{T}$ be the set of all such traces. A log $L \subseteq \mathbb{M}(\mathcal{T})$ is a multiset of traces.*

In this paper, we use an alternative definition for Petri nets. We only allow for places connecting activities that are initially empty (without tokens), because we allow only for traces starting with $\blacktriangleright$ and ending with $\blacksquare$. These places are uniquely identified by the set of input activities I and output activities O. Each activity corresponds to exactly one transition, therefore this paper we refer to transitions as activities.

Definition 2 (Petri nets). *A Petri net is a pair $N = (A, \mathcal{P})$, where $A \subseteq \mathcal{A}$ is the set of activities including start and end ($\{\blacktriangleright, \blacksquare\} \subseteq A$) and $\mathcal{P} \subseteq \{(I|O) \mid I \subseteq$*

$A \wedge I \neq \emptyset \wedge O \subseteq A \wedge O \neq \emptyset\}$ *is the set of places. We call I the set of* ingoing *activities of a place and O the set of* outgoing *activities.*

Given an activity $a \in A$, $\bullet a = \{(I|O) \in \mathcal{P} \mid a \in O\}$ and $a\bullet = \{(I|O) \in \mathcal{P} \mid a \in I\}$ denote the sets of input and output places. Given a place $p = (I|O) \in \mathcal{P}$, $\bullet p = I$ and $p\bullet = O$ denote the sets of input and output activities.

Definition 3 (Fitting/Unfitting Places). *Let $N = (A, \mathcal{P})$ be a Petri net, let $p = (I|O) \in \mathcal{P}$ be a place, and let $\sigma \in \mathcal{T}$ be a trace. With respect to the given trace σ, p is called*

- unfitting, *denoted by $\boxtimes_\sigma(p)$, if and only if at least one of the following holds:*
  - $\exists k \in \{1, 2, ..., |\sigma|\}$ *such that*
    $|\{i \mid i \in \{1, 2, ...k - 1\} \wedge \sigma(i) \in I\}| < |\{i \mid i \in \{1, 2, ...k\} \wedge \sigma(i) \in O\}|$
  - $|\{i \mid i \in \{1, 2, ...|\sigma|\} \wedge \sigma(i) \in I\}| > |\{i \mid i \in \{1, 2, ...|\sigma|\} \wedge \sigma(i) \in O\}|,$
- fitting, *denoted by $\square_\sigma(p)$, if and only if not $\boxtimes_\sigma(p)$.*

Definition 4 (Behavior of a Petri net). *We define the* behavior *of the Petri net $(A, \mathcal{P})$ to be the set of all fitting traces, that is $\{\sigma \in \mathcal{T} \mid \forall p \in \mathcal{P}: \square_\sigma(p)\}$.*

Note that we only allow for behaviors of the form $\langle \blacktriangleright, a_1, a_2, \ldots a_n, \blacksquare \rangle$ such that places are empty at the end of the trace and never have a negative number of tokens.

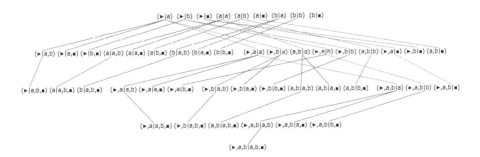

Fig. 1. Example of a tree-structured candidate space for activities $\{\blacktriangleright, a, b, \blacksquare\}$.

3 Introducing the Algorithm

We briefly repeat our discovery approach as presented in [7]. For details we refer the reader to the original paper. As input, the algorithm takes a log L and a parameter $\tau \in [0, 1]$, and returns a Petri net as output. A place is considered fitting, if at the fraction τ of traces in the event log is fitting. A place is perfectly fitting when all traces are fitting. Inspired by language-based regions, the basic strategy of the approach is to begin with a Petri net, whose transitions correspond exactly to the activities used in the given log. From the finite set of unmarked, intermediate places a subset of places is inserted, such that the language defined by the resulting net defines the minimal language containing the input language, while, for human readability, using only a minimal number of places to do so.

The algorithm uses token-based replay to discover all fitting places $\mathcal{P}_{fit}$ out of the set of possible candidate places. To avoid replaying the log on the exponential number of candidates, it organizes the potential places as a set of trees, such that certain properties hold. When traversing the trees using a depth-first-strategy, these properties allow to cut off subtrees, and thus candidates, based on the replay result of their parent. This greatly increases efficiency, while still guaranteeing that all fitting places are found. An example of such a tree-structured candidate space is shown in Fig. 1. Note the incremental structure of the trees, i.e. the increase in distance from the base roots corresponds to the increase of input and output activities. To significantly increase readability, implicit places are removed in a post-processing step.

The running time of the original eST-Miner as summarized in this section, strongly depends on the number of candidate places skipped during the search for fitting places. The approach uses monotonicity results [8] to skip sets of places that are known to be unfitting. For example, if 80% of the cases cannot be replayed because the place is empty and does not enable the next activity in the trace, then at least 80% will not allow for a place with even more output transitions. While this results in a significant decrease of computation time compared to the brute force approach, there are still to many candidates to be evaluated by replaying the log. Moreover, typically, the set of all fitting places contains a great number of implicit places, resulting in very slow post-processing. In the following we explore uniwired Petri nets as a solution to these issues.

4   Introducing Uniwired Petri Nets

In this paper, we aim to discover *uniwired* Petri nets. In uniwired Petri nets all pairs of activities are connected by at most one place. *Biwired* nets are all other Petri nets where at least one pair of activities is connected by at least two places.

Definition 5 (Biwired/Uniwired Petri Nets). *Let* $N = (A, \mathcal{P})$ *be a Petri net.* N *is* biwired *if there is a pair of activities* $a_1, a_2 \in A$, *such that* $|a_1 \bullet \cap \bullet a_2| \geq 2$. N *is* uniwired *if such a pair does not exist.* $wired(\mathcal{P}) = \{(a_1, a_2) \mid \exists_{(I|O) \in \mathcal{P}} \ a_1 \in I \ \wedge \ a_2 \in O\}$ *is the set of wired activities.*

As far as we can tell, the subclass of uniwired Petri nets has not been investigated systematically (like e.g., free-choice nets). However, the class of uniwired nets includes the class of Structured Workflow Nets (SWF-nets) known in the context of process mining. For example, the α-algorithm was shown to be able to rediscover this class of nets [9]. Since we are using an alternative Petri net representation (Definition 2), we define SWF-nets as follows.

Definition 6 (SWF-Nets (Structured Workflow Nets) [9]). *A Petri net* $N = (A, \mathcal{P})$, *is an SWF-net, if the following requirements hold:*

- N has no implicit places (i.e., removing any of the places changes the behavior of the Petri net),
- for any $p \in \mathcal{P}$ and $a \in A$ such that $p \in \bullet a$: (1) $|p\bullet| > 1 \implies |\bullet a| = 1$ (i.e., choice and synchronization are separated), and (2) $|\bullet a| > 1 \implies |\bullet p| = 1$ (i.e., only synchronize a fixed set of activities).

Lemma 1 (Uniwired Petri nets and SWF-nets). *The class of uniwired Petri nets is a strict superset of the class of SWF-nets.*

Proof. Let $N = (A, \mathcal{P})$ be an SWF-net. We show that the assumption that N is biwired leads to a contradiction. If N is biwired, then there are two activities $a_1, a_2 \in A$ and two different places $p_1, p_2 \in \mathcal{P}$ such that $\{p_1, p_2\} \subseteq a_1\bullet$ and $\{p_1, p_2\} \subseteq \bullet a_2$. Since $p_1 \neq p_2$, Definition 6 implies $|\bullet p_1| = |\bullet p_2| = 1$. Hence, $\bullet p_1 = \bullet p_2 = \{a_1\}$. If $a' \in p_1\bullet$ and $a' \neq a_2$, then $|p_1\bullet| > 1$. Hence, Definition 6 implies $|\bullet a_2| = 1$ leading to a contradiction (there are two places). The same applies to p_2. Hence, $p_1\bullet = p_2\bullet = \{a_2\}$. Combining $\bullet p_1 = \bullet p_2 = \{a_1\}$ and $p_1\bullet = p_2\bullet = \{a_2\}$, implies that p_1 and p_2 must have exactly the same connections, making one of these places implicit. Since there are no implicit places in SWF-nets, we conclude that an SWF-net cannot be biwired (i.e., is uniwired).

Figure 2 shows that the class of uniwired Petri nets is a strict superset of the class of SWF-nets. The uniwired Petri net N_1 is not an SWF-net. For example the place $p = (\{e\}|\{f, g\})$ and activity g clearly violate the definition of SWF-nets. □

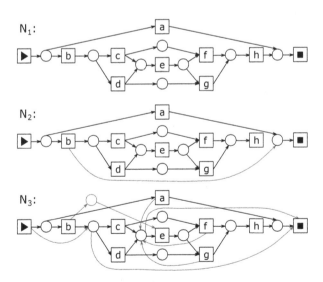

Fig. 2. N_1 shows a uniwired Petri net with long-term dependencies, that is not an SWF-net. The models N_2 and N_3 illustrate the problems resulting from conflicting places of varying (N_3) and similar (N_2) complexity, detailed in Sect. 5, before adding self-loops.

The α-algorithm is able to rediscover SWF-nets, but has well-known limitations with respect to discovering complex control-flow structures, for example long-term dependencies [1]. The class of uniwired Petri nets is clearly more expressive than SWF-nets, as illustrated for example by N_1 in Fig. 2, showing its capability of modeling advanced control-flow structures.

The models included in this class seem to constitute a well-balanced trade-off between human-readable, simple process structures on the one hand, and complex, hard-to-find control-flow structures on the other hand. This makes them a very interesting class of models in the context of process discovery. In Sect. 5, we will show that such models naturally lend themselves to be discovered efficiently by an adaption of our eST-Miner.

5   Discovering Uniwired Petri Nets

The original eST-Miner discovers all non-implicit places that meet a preset quality threshold, and is capable of finding even the most complex control-flow structures. However, the resulting Petri nets can be difficult to interpret by human readers, and might often be more complex then users require. This is indicated for example by the heavy use of Inductive Miner variants in business applications, despite its strong limitations on the discoverable control-flow structures.

In the following, we present an adaption of our eST-Mining algorithm, that aims to discover uniwired Petri nets as introduced in Sect. 4. This approach yields several advantages: the representational bias provided by uniwired Petri nets ensures non-complex and understandable models are discovered, while at the same time allowing for traditionally hard-to-mine control-flow structures such as long-term dependencies. In contrast to many other discovery algorithms, e.g. Inductive Miner, our algorithm is able to discover such structures.

Another advantage of searching for uniwired Petri nets is that incorporating their restrictions in our search for fitting places naturally leads to a massive increase in skipped candidate places and thus efficiency of our discovery algorithm. Additionally, this approach greatly decreases the amount of implicit places found, and thus the complexity of the post-processing step. We will provide details on this variant of our algorithm in the remainder of this section.

Adaption to Uniwired Discovery: The efficiency of our eST-mining algorithm (see Sect. 3) strongly depends on the amount of candidate places skipped by cutting off subtrees in the search space. In the following we optimize this strategy towards the discovery of uniwired Petri nets.

The idea is based on the simple observation, that in a uniwired Petri net there can be only one place connecting two activities. With this in mind, consider our tree-like organized search space. As shown in [7], for every candidate place $p = (I|O)$ it holds that for each of its descendants in the tree $p' = (I'|O')'$ we have that $I \subseteq I'$ and $O \subseteq O'$. In particular, every pair of activities wired by p will also be wired by any descendant p'. Thus, if we include p in our set of fitting places $\mathcal{P}_{fit}$, the whole subtree rooted in p becomes uninteresting and can be skipped.

Additionally, the same two activities will be wired by candidates located in completely independent subtrees. Again, once we have chosen such a place, all these candidates and their children become uninteresting and can be cut off. To keep track of the activities that have already been wired within other subtrees, we globally update the set $wired(\mathcal{P}_{fit})$ (see Definition 5).

Lemma 2 (Bound on Fitting Places). *The set of places discovered by the uniwired variant of our algorithm can contain at most one place for each tree in the search space, that is at most $|A-1|^2$.*

Proof. Our trees are structured in an incremental way: for a root candidate $p_1 = (I_1|O_1)$, for every two descendants $p_2 = (I_2|O_2)$ and $p_3 = (I_3|O_3)$ of p_1 we have that $I_1 \subseteq I_2, O_1 \subseteq O_2$ and $I_1 \subseteq I_3, O_1 \subseteq O_3$. Since we do not allow for candidates with empty activity sets, this implies that $I_2 \cap I_3 \neq \emptyset, O_2 \cap O_3 \neq \emptyset$, and thus a Petri net containing both places would be biwired. We conclude, that for each base root only one descendant (including the root itself) can be part of the discovered set of places. □

While this basic approach is very simple and intuitive, and also results in an enormous decrease in computation time as well as found implicit places, several complications arise. The basic problem is the competition between different fitting candidates, that are wiring the same activities and therefore excluding each other. Our discovery algorithm has to decide which of these conflicting places to include in the final model. Several manifestations of this problem are discussed and addressed in the following, resulting in an incremental improvement of the naive approach.

Included in some of these strategies, we use a heuristic approach to determine how interesting a candidate is. We define a score that is based on the strength of directly-follows-relations expressed by a place $(I|O)$:

$$S((I|O)) = \sum_{x \in I, y \in O} \frac{\#(x,y)}{|I| \cdot |O|},$$

where $\#(x,y)$ denotes the number of times x is directly followed by y in the input log. Applying this heuristic will result in prioritizing places that have a higher token throughput and are therefore expected to better reflect the behavior defined by the log.

Self-looping Places: The first conflict we discuss is related to self-looping places. A place $p = (I|O)$ is *self-looping*, if there is at least one activity a with $a \in I, a \in O$. Self-loops can be important to model process behavior, and should not be ignored. However, our naive approach displays a significant drawback with respect to self-looping places: after discovering this place p, the looping activity a is wired, and thus no other place with this self-loop and additional interesting control-flows can be found, since these places are always contained in the skipped subtrees. For example, a place $p' = (I \cup \{a_1\}|O \cup \{a_2\})$ cannot be discovered, which includes places with more than one self-loop.

Fortunately, we can easily fix these problems related to self-loops. Before starting our search for fitting places, we set $wired(\mathcal{P}_{fit}) = \{(a,a) \mid a \in A\}$. Thus all subtrees containing self-loops are skipped, resulting in a discovered set of fitting places without any self-loops. As an additional post-processing step we replay the log on each fitting place extended by each possible self-loop, resulting in at most $|P_{fit}| \cdot |A|$ additional replays. We then insert a copy of each fitting place with its maximal number of self-loop activities added. Superfluous places created this way are deleted during the final implicit-places-removal. Note, that each activity may be wired at most once, i.e. may be involved in at most one self-loop. This results in conflicts between fitting places that can loop on the same activity. We resolve these conflicts by favoring more interesting, i.e. higher scoring places.

Conflicting Places of Varying Complexity: Consider two fitting places $p_1 = (a|b)$ and $p_2 = (c|d)$, implying the existence of another fitting place $p_3 = (a,c|b,d)$. Obviously, p_1 and p_2 cannot coexist with p_3 in a uniwired Petri net. In this scenario, p_1 and p_2 are also preferable to p_3, since they are much more readable, simpler and constraining. However, the depth-first search of our original eST-Miner is likely to encounter p_3 first, update the set $wired(\mathcal{P}_{fit})$ accordingly, and thus prevent p_1 and p_2 from being found. This can result in overly complex and unintuitive process models, because simple connections end up being modeled by a collection of very complicated places, as illustrated in N_3 of Fig. 2. In comparison to the desired Petri net N_1, this model includes for example the complex places $(a,c,d,f,g|e,\blacksquare)$ and $(\blacktriangleright,b|c,d,\blacksquare)$, as well as an additional place $(b|e)$. Clearly, the simpler places $(c,d|e)$ and $(b|c,d)$ resulting in the place $(b|e)$ being implicit, would be preferable.

This conflict between complex places and more simple places constituting a similar control flow can be avoided by prioritizing simple places in our search. By adopting a traversal pattern more similar to breadth-first-search, we ensure that candidate places with less activities, which are closer to the base roots, are evaluated first. Only after considering all candidates with a certain distance, we proceed further down the trees, thus ensuring that for a pair of activities we always choose the simplest place to wire them. With respect to the introductory example, we would find p_1 and p_2, thus skipping the subtree that contains p_3.

We adapt our original algorithm to this new traversal strategy by transforming the list of base roots to a queue. After removing and evaluating the first candidate in this queue, which in the beginning are the base roots, we add any potentially interesting child candidate to the end of the queue. Rather than going into the depth of the tree, we proceed with evaluating the new first element. We continue to cut off subtrees by simply not adding uninteresting children to the queue. When evaluating a candidate, before replay we check the set $wired(\mathcal{P}_{fit})$, which might have been updated since the candidate was added.

Conflicting Places of Similar Complexity: While the base root level contains only candidate places with non-overlapping pairs of input and output activities, this is not the case for deeper levels. Thus, when evaluating the places of a level it is possible that two places are fitting but cannot coexist in a uniwired Petri net. Simply choosing the first such place can lead to the skipping of very interesting places in favor of places that do not carry as much information. An example is the Petri net N_2 shown in Fig. 2. Here, the place $(a, b|\blacksquare)$ is inserted first, resulting in the much more interesting place $(a, h|\blacksquare)$ of same complexity being skipped.

To circumvent this problem we first find all fitting candidates from the same level (i.e. the same distance from the roots) and then check for conflicts. For each set of conflicting places we chose the one scoring best with respect to the directly-follows-based heuristics described previously. This way we can give preference to candidates with high token-throughput. With respect to the running example in Fig. 2 we rediscover the desired net N_1.

The uniwired variant of eST-Miner described in this section to some degree guarantees the discovery of desired places:

Lemma 3 (Maximal set of discovered places). *Let $\mathcal{P}$ be the set of places discovered by the algorithm. No non-implicit, fitting place can be added to $\mathcal{P}$ without making the Petri net biwired.*

Proof. Assume there would be a non-implicit, fitting place p, that could be added to $\mathcal{P}$ without making the net biwired. There are two possibilities for a fitting candidate not to be discovered: either it is conflicting with another place that was chosen instead or it is located in a subtree that was cut-off.

First assume p was discarded in favor of a conflicting place. This implies that there is a pair of transitions wired by both places. Since the other place was discovered, p cannot be added without making the net biwired.

Now assume p was located in a subtree that was cut-off. Since p is fitting, our original eST-Miner does not cut-off this subtree. Thus the subtree was cut-off because the places it contained were already wired. This implies that p wires a set of transitions that is also wired by another, previously discovered place. Thus adding p would make the net biwired.

We conclude, that no place can be added to $\mathcal{P}$ without making the resulting Petri net biwired.   □

Lemma 3 guarantees that we find a maximal number of fitting places, i. e. no place can be added to the resulting net without making it biwired. However, discovered places might still be extended by adding more ingoing and outgoing activities. The presented optimizations aim to find the most expressive of all possible maximal sets of places.

6 Testing Results and Evaluation

The uniwired Petri nets, that can be discovered using our new variant of eST-Miner, may express advanced behaviors. In particular, we are capable of finding complex structures like long-term dependencies, as illustrated by N_1 in Fig. 2.

A category of conflicting places for which an efficient solution has yet to be found are certain places within the same subtree: assume there is a fitting place $p_1 = (I_1|O_1)$ and a conflicting fitting place $p = (I_1 \cup I_2|O_1 \cup O_2)$ within the subtree rooted in p_1. Then our algorithm will choose the simpler place p_1 and skip p. This is fine in the case described in Sect. 5, with the fitting and non-conflicting place $p_2 = (I_2|O_2)$ being part of a different subtree, but problematic if p_2 is unfitting: the relation between I_2 and O_2 exists, but will not be expressed by the discovered model. This is illustrated in Fig. 3.

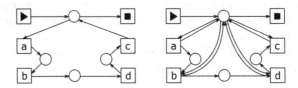

Fig. 3. The left model contains the place $(\blacktriangleright, c|a, \blacksquare)$, that cannot be discovered by our variant. Instead, we discover the model shown on the right, where the missing place is replaced by one having several self-loops.

Table 1. List of logs used for evaluation. The upper part lists real-life logs, the lower part shows artificial logs. Logs are referred to by their abbreviations. The log `HP2018` has not yet been published. The much smaller 2017 version can be found in [10].

Log name	Abbreviation	Activities	Trace variants	Reference
BPI13-incidents	`BPI13`	15	2278	[11]
HelpDesk2018SiavAnon	`HD2018`	12	2179	(see caption)
Sepsis	`Sepsis`	16	846	[12]
Road traffic fine management	`RTFM`	11	231	[13]
Reviewing	`Reviewing`	16	96	[14]
repairexample	`Repair`	12	77	[14]
Teleclaims	`Teleclaims`	11	12	[14]

For the evaluation of efficiency, we use similar logs as in our original paper ([7]) as specified in Table 1. The eST-Miner algorithm increases efficiency by skipping places that are guaranteed to be unfitting. In addition to the places cut off by the original algorithm, the uniwiring variant skips all subtrees that contain places that have been wired already. As illustrated in Fig. 4, this immensely increases the fraction of cut-off candidates: for all our test-logs the percentage of skipped candidates surpasses 99%. The corresponding increase in efficiency in comparison to the original eST-Miner is presented in Fig. 5. For the tested logs, the uniwired variant proves to be up to 60 times faster than the original algorithm and up to 600 times faster than the brute force approach evaluating all candidates [7]. The results of our evaluation show the potential of uniwired Petri nets in combination with our eST-Miner variant: interesting process models that can represent complex structures such as long-term dependencies can be discovered, while immensely increasing efficiency by skipping nearly all of the uninteresting candidate places.

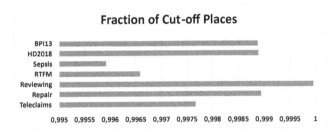

Fig. 4. In comparison to the original algorithm, the uniwired variant is able to increase the fraction of cut-off places dramatically.

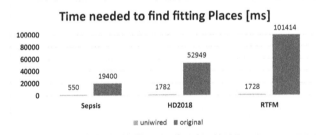

Fig. 5. Comparison of computation time needed by the original eST-Miner and its uniwired variant for computing the set of fitting places. For comparability slightly modified versions of the logs `Sepsis` and `HD2018` were used as detailed in [7].

7   Conclusion

While the original eST-Mining algorithm is capable of discovering traditionally hard-to-mine, complex control-flow structures much faster than the brute force approach, and provides an interesting new approach to process discovery, it still displays some weaknesses. In particular, the computation time for finding the set of fitting places as well as removing implicit places is to high and the resulting models are very precise but hard to read for human users, due to lack of simplicity.

In this paper, we present a new class of Petri nets, the *uniwired Petri nets*. We have shown that they are quite expressive, since they contain the class of SWF-nets, but are strictly larger. In particular, they can model non-free choice constructs. By providing a well-balanced trade-off between simplicity and expressiveness, they seem to introduce a very interesting representational bias to process discovery. We describe a variant of eST-Miner, that aims to discover uniwired Petri nets. While still being able to discover non-free choice constructs, utilizing the uniwiredness-requirement allows us to skip an astonishingly large part of the search space, leading to a massive increase in efficiency when searching for the set of fitting places. At the same time the number of found implicit places is drastically decreased.

For future work, we would like to extend the results on expressiveness and relevance of uniwired Petri nets. Our corresponding discovery algorithm could also be improved in particular with respect to conflicting places within the same subtree, and by refining our scoring system. The influence of the order of candidates within the tree structure should be investigated as well. Additional strategies for skipping even more sets of places, as well as adequate abstractions of the log, can be particularly interesting when analyzing larger logs.

Acknowledgments. We thank the Alexander von Humboldt (AvH) Stiftung for supporting our research.

References

1. Wen, L., van der Aalst, W., Wang, J., Sun, J.: Mining process models with non-free-choice constructs. Data Min. Knowl. Discov. **15**(2), 145–180 (2007)
2. Leemans, S.J.J., Fahland, D., van der Aalst, W.M.P.: Discovering block-structured process models from event logs - a constructive approach. In: Colom, J.-M., Desel, J. (eds.) PETRI NETS 2013. LNCS, vol. 7927, pp. 311–329. Springer, Heidelberg (2013). https://doi.org/10.1007/978-3-642-38697-8_17
3. Badouel, E., Bernardinello, L., Darondeau, P.: Petri Net Synthesis. Text in Theoretical Computer Science, an EATCS Series. Springer, Heidelberg (2015). https://doi.org/10.1007/978-3-662-47967-4
4. Lorenz, R., Mauser, S., Juhás, G.: How to synthesize nets from languages: a survey. In: Proceedings of the 39th Conference on Winter Simulation: 40 Years! The Best is Yet to Come, WSC 2007. IEEE Press (2007)
5. van der Werf, J.M.E.M., van Dongen, B.F., Hurkens, C.A.J., Serebrenik, A.: Process discovery using integer linear programming. In: van Hee, K.M., Valk, R. (eds.) PETRI NETS 2008. LNCS, vol. 5062, pp. 368–387. Springer, Heidelberg (2008). https://doi.org/10.1007/978-3-540-68746-7_24
6. van Zelst, S.J., van Dongen, B.F., van der Aalst, W.M.P.: Avoiding over-fitting in ILP-based process discovery. In: Motahari-Nezhad, H.R., Recker, J., Weidlich, M. (eds.) BPM 2015. LNCS, vol. 9253, pp. 163–171. Springer, Cham (2015). https://doi.org/10.1007/978-3-319-23063-4_10
7. Mannel, L.L., van der Aalst, W.M.P.: Finding complex process-structures by exploiting the token-game. In: Donatelli, S., Haar, S. (eds.) PETRI NETS 2019. LNCS, vol. 11522, pp. 258–278. Springer, Cham (2019). https://doi.org/10.1007/978-3-030-21571-2_15

8. van der Aalst, W.: Discovering the "glue" connecting activities - exploiting mono-tonicity to learn places faster. In: It's All About Coordination - Essays to Celebrate the Lifelong Scientific Achievements of Farhad Arbab (2018)
9. Aalst, W., Weijters, A., Maruster, L.: Workflow mining: discovering process models from event logs. IEEE Trans. Knowl. Data Eng. **16**(9), 1128–1142 (2004)
10. Polato, M.: Dataset belonging to the help desk log of an Italian company (2017)
11. Steeman, W.: BPI challenge 2013, incidents (2013). https://doi.org/10.4121/UUID:500573E6-ACCC-4B0C-9576-AA5468B10CEE
12. Mannhardt, F.: Sepsis cases - event log (2016)
13. De Leoni, M., Mannhardt, F.: Road traffic fine management process (2015)
14. van der Aalst, W.M.P.: Event logs and models used in Process Mining: Data Science in Action (2016). http://www.processmining.org/event_logs_and_models_used_in_book

Discovering Process Models
from Uncertain Event Data

Marco Pegoraro[(⊠)] [iD], Merih Seran Uysal[iD], and Wil M. P. van der Aalst[iD]

Process and Data Science Group (PADS), Department of Computer Science,
RWTH Aachen University, Aachen, Germany
{pegoraro,uysal,wvdaalst}@pads.rwth-aachen.de

Abstract. Modern information systems are able to collect event data in the form of *event logs*. Process mining techniques allow to discover a model from event data, to check the conformance of an event log against a reference model, and to perform further process-centric analyses. In this paper, we consider uncertain event logs, where data is recorded together with explicit uncertainty information. We describe a technique to discover a directly-follows graph from such event data which retains information about the uncertainty in the process. We then present experimental results of performing inductive mining over the directly-follows graph to obtain models representing the certain and uncertain part of the process.

Keywords: Process mining · Process discovery · Uncertain data

1 Introduction

With the advent of digitalization of business processes and related management tools, *Process-Aware Information Systems* (PAISs), ranging from ERP/CRM-systems to BPM/WFM-systems, are widely used to support operational administration of processes. The databases of PAISs containing event data can be queried to obtain *event logs*, collections of recordings of the execution of activities belonging to the process. The discipline of *process mining* aims to synthesize knowledge about *processes* via the extraction and analysis of execution logs.

When applying process mining in real-life settings, the need to address anomalies in data recording when performing analyses is omnipresent. A number of such anomalies can be modeled by using the notion of uncertainty: *uncertain event logs* contain, alongside the event data, some attributes that describe a certain level of uncertainty affecting the data. A typical example is the timestamp information: in many processes, specifically the ones where data is in part manually recorded, the timestamp of events is recorded with low precision (e.g., specifying only the day of occurrence). If multiple events belonging to the same case are recorded within the same time unit, the information regarding the event order is lost. This can be modeled as uncertainty of the timestamp attribute by assigning a time interval to the events. Another example of uncertainty are situations where the activity label is unrecorded or lost, but the events are

© Springer Nature Switzerland AG 2019
C. Di Francescomarino et al. (Eds.): BPM 2019 Workshops, LNBIP 362, pp. 238–249, 2019.
https://doi.org/10.1007/978-3-030-37453-2_20

associated with specific resources that carried out the corresponding activity. In many organizations, each resource is authorized to perform a limited set of activities, depending on her role. In this case, it is possible to model the absence of activity labels associating every event with the set of possible activities which the resource is authorized to perform.

Usually, information about uncertainty is not natively contained into a log: event data is extracted from information systems as activity label, timestamp and case id (and possibly additional attributes), without any sort of meta-information regarding uncertainty. In some cases, a description of the uncertainty in the process can be obtained from background knowledge. Information translatable to uncertainty such as the one given above as example can, for instance, be acquired from an interview with the process owner, and then inserted in the event log with a pre-processing step. Research efforts regarding how to discover uncertainty in a representation of domain knowledge and how to translate it to obtain an uncertain event log are currently ongoing.

Uncertainty can be addressed by filtering out the affected events when it appears sporadically throughout an event log. Conversely, in situations where uncertainty affects a significant fraction of an event log, filtering away uncertain event can lead to information loss such that analysis becomes very difficult. In this circumstance, it is important to deploy process mining techniques that allow to mine information also from the uncertain part of the process.

In this paper, we aim to develop a process discovery approach for uncertain event data. We present a methodology to obtain *Uncertain Directly-Follows Graphs* (UDFGs), models based on directed graphs that synthesize information about the uncertainty contained in the process. We then show how to convert UDFGs in models with execution semantics via filtering on uncertainty information and inductive mining.

The remainder of the paper is structured as follows: in Sect. 2 we present relevant previous work. In Sect. 3, we provide the preliminary information necessary for formulating uncertainty. In Sect. 4, we define the uncertain version of directly-follows graphs. In Sect. 5, we describe some examples of exploiting UDFGs to obtain executable models. Section 6 presents some experiments. Section 7 proposes future work and concludes the paper.

2 Related Work

In a previous work [9], we proposed a taxonomy of possible types of uncertainty in event data. To the best of our knowledge, no previous work addressing explicit uncertainty currently exist in process mining. Since usual event logs do not contain any hint regarding misrecordings of data or other anomalies, the notion of "noise" or "anomaly" normally considered in process discovery refers to outlier behavior. This is often obtained by setting thresholds to filter out the behavior not considered for representation in the resulting process model. A variant of the Inductive Miner by Leemans et al. [6] considers only directly-follows relationships appearing with a certain frequency. In general, a direct way to address infrequent behavior on the event level is to apply on it the concepts of support and confidence, widely used in association rule learning [5]. More sophisticated techniques employ infrequent pattern detection employing a mapping between events [8] or a finite state automaton [4] mined from the most frequent behavior.

Although various interpretations of uncertain information can exist, this paper presents a novel approach that aims to represent uncertainty explicitly, rather than filtering it out. For this reason, existing approaches to identify noise cannot be applied to the problem at hand.

3   Preliminaries

To define uncertain event data, we introduce some basic notations and concepts, partially from [2]:

Definition 1 (Power Set). *The power set of a set A is the set of all possible subsets of A, and is denoted with $\mathcal{P}(A)$. $\mathcal{P}_{NE}(A)$ denotes the set of all the non-empty subsets of A: $\mathcal{P}_{NE}(A) = \mathcal{P}(A) \setminus \{\emptyset\}$.*

Definition 2 (Sequence). *Given a set X, a finite sequence over X of length n is a function $s \in X^* : \{1, \ldots, n\} \to X$, typically written as $s = \langle s_1, s_2, \ldots, s_n \rangle$. For any sequence s we define $|s| = n$, $s[i] = s_i$, $\mathcal{S}_s = \{s_1, s_2, \ldots, s_n\}$ and $x \in s \Leftrightarrow x \in \mathcal{S}_s$. Over the sequences s and s' we define $s \cup s' = \{a \in s\} \cup \{a \in s'\}$.*

Definition 3 (Directed Graph). *A directed graph $G = (V, E)$ is a set of vertices V and a set of directed edges $E \subseteq V \times V$. We denote with $\mathcal{U}_G$ the universe of such directed graphs.*

Definition 4 (Bridge). *An edge $e \in E$ is called a bridge if and only if the graph becomes disconnected if e is removed: there exists a partition of V into V' and V'' such that $E \cap ((V' \times V'') \cup (V'' \times V')) = \{e\}$. We denote with $E_B \subseteq E$ the set of all such bridges over the graph $G = (V, E)$.*

Definition 5 (Path). *A path over a graph $G = (V, E)$ is a sequence of vertices $p = \langle v_1, v_2, \ldots v_n \rangle$ with $v_1, \ldots, v_n \in V$ and $\forall_{1 \leq i \leq n-1}(v_i, v_{i+1}) \in E$. $P_G(v, w)$ denotes the set of all paths connecting v and w in G. A vertex $w \in V$ is reachable from $v \in V$ if there is at least one path connecting them: $|P_G(v, w)| > 0$.*

Definition 6 (Transitive Reduction). *A transitive reduction of a graph $G = (V, E)$ is a graph $\rho(G) = (V, E')$ with the same reachability between vertices and a minimal number of edges. $E' \subseteq E$ is a smallest set of edges such that $|P_{\rho(G)}(v, w)| > 0 \Rightarrow |P_G(v, w)| > 0$ for any $v, w \in V$.*

In this paper, we consider *uncertain event logs*. These event logs contain uncertainty information explicitly associated with event data. A taxonomy of different kinds of uncertainty and uncertain event logs has been presented in [9] which it distinguishes between two main classes of uncertainty. *Weak uncertainty* provides a probability distribution over a set of possible values, while *strong uncertainty* only provides the possible values for the corresponding attribute.

We will use the notion of *simple uncertainty*, which includes strong uncertainty on the control-flow perspective: activities, timestamps, and indeterminate events. An example of a simple uncertain trace is shown in Table 1. Event e_1 has been recorded with two possible activity labels (a or c), an example of strong uncertainty on activities. Some events, e.g. e_2, do not have a precise timestamp but a time interval in which the event could have happened has been recorded: in some cases, this causes the loss of the precise order of events (e.g. e_1 and e_2). These are examples of strong uncertainty on timestamps. As shown by the "?" symbol, e_3 is an indeterminate event: it has been recorded, but it is not guaranteed to have happened.

Table 1. An example of simple uncertain trace.

Case ID	Event ID	Activity	Timestamp	Event type
0	e_1	{a, c}	[2011-12-02T00:00 2011-12-05T00:00]	!
0	e_2	{a, d}	[2011-12-03T00:00 2011-12-05T00:00]	!
0	e_3	{a, b}	2011-12-07T00:00	?
0	e_4	{a, b}	[2011-12-09T00:00 2011-12-15T00:00]	!
0	e_5	{b, c}	[2011-12-11T00:00 2011-12-17T00:00]	!
0	e_6	{b}	2011-12-20T00:00	!

Definition 7 (Universes). *Let $\mathcal{U}_E$ be the set of all the* event identifiers. *Let $\mathcal{U}_C$ be the set of all* case ID identifiers. *Let $\mathcal{U}_A$ be the set of all the* activity identifiers. *Let $\mathcal{U}_T$ be the totally ordered set of all the* timestamp identifiers. *Let $\mathcal{U}_O = \{!, ?\}$, where the "!" symbol denotes* determinate events, *and the "?" symbol denotes* indeterminate events.

Definition 8 (Simple uncertain traces and logs). $\sigma \in \mathcal{P}_{NE}(\mathcal{U}_E \times \mathcal{P}_{NE}(\mathcal{U}_A) \times \mathcal{U}_T \times \mathcal{U}_T \times \mathcal{U}_O)$ *is a* simple uncertain trace *if for any* $(e_i, A, t_{min}, t_{max}, u) \in \sigma$, $t_{min} < t_{max}$ *and all the event identifiers are unique.* $\mathcal{T}_U$ *denotes the universe of simple uncertain traces.* $L \in \mathcal{P}(\mathcal{T}_U)$ *is a simple uncertain log if all the event identifiers in the log are unique. Over the uncertain event $e = (e_i, A, t_{min}, t_{max}, o) \in \sigma$ we define the following projection functions:* $\pi_A(e) = A$, $\pi_{t_{min}}(e) = t_{min}$, $\pi_{t_{max}}(e) = t_{max}$ *and* $\pi_o(e) = o$. *Over $L \in \mathcal{P}(\mathcal{T}_U)$ we define the following projection function:* $\Pi_A(L) = \bigcup_{\sigma \in L} \bigcup_{e \in \sigma} \pi_A(e)$.

The behavior graph is a structure that summarizes information regarding the uncertainty contained in a trace. Namely, two vertices are linked by an edge if their corresponding events may have happened one immediately after the other.

Definition 9 (Behavior Graph). *Let $\sigma \in \mathcal{T}_U$ be a simple uncertain trace. A behavior graph $\beta \colon \mathcal{T}_U \to \mathcal{U}_G$ is the transitive reduction of a directed graph $\rho(G)$, where $G = (V, E) \in \mathcal{U}_G$ is defined as:*

- $V = \{e \in \sigma\}$
- $E = \{(v, w) \mid v, w \in V \wedge \pi_{t_{max}}(v) < \pi_{t_{min}}(w)\}$

Notice that the behavior graph is obtained from the transitive reduction of an acyclic graph, and thus is unique. The behavior graph for the trace in Table 1 is shown in Fig. 1.

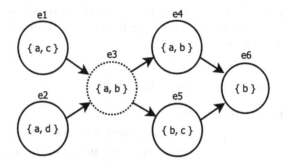

Fig. 1. The behavior graph of the uncertain trace given in Table 1. Each vertex represents an uncertain event and is labeled with the possible activity label of the event. The dotted circle represents an indeterminate event (may or may not have happened).

4 Uncertain DFGs

The definitions shown in Sect. 3 allow us to introduce some fundamental concepts necessary to perform discovery in an uncertain setting. Let us define a measure for the frequencies of single activities. In an event log without uncertainty the frequency of an activity is the number of events that have the corresponding activity label. In the uncertain case, there are events that can have multiple possible activity labels. For a certain activity $a \in \mathcal{U}_A$, the minimum activity frequency of a is the number of events that certainly have A as activity label and certainly happened; the maximum activity frequency is the number of events that may have A as activity label.

Definition 10 (Minimum and maximum activity frequency). *The* minimum *and* maximum *activity frequency* $\#_{min}: \mathcal{T}_U \times \mathcal{U}_A \to \mathbb{N}$ *and* $\#_{max}: \mathcal{T}_U \times \mathcal{U}_A \to \mathbb{N}$ *of an activity* $a \in \mathcal{U}_A$ *in regard of an uncertain trace* $\sigma \in \mathcal{T}_U$ *are defined as:*

- $\#_{min}(\sigma, a) = |\{e \in \sigma \mid \pi_A(e) = \{a\} \wedge \pi_o(v) = !\}|$
- $\#_{max}(\sigma, a) = |\{e \in \sigma \mid a \in \pi_A(e)\}|.$

Many discovery algorithms exploit the concept of *directly-follows relationship* [1,6]. In this paper, we extend this notion to uncertain traces and uncertain event logs. An uncertain trace embeds some behavior which depends on the instantiation of the stochastic variables contained in the event attributes. Some directly-follows relationships exist in part, but not all, the possible behavior of an uncertain trace. As an example, consider events e_3 and e_5 in the uncertain trace shown in Table 1: the relationship "a is directly followed by b" appears once only if e_3 actually happened immediately before e_5 (i.e., e_4 did not happen in-between), and if the activity label of e_3 is a b (as opposed to c, the other possible label). In all the behavior that does not satisfy these conditions, the directly-follows relation does not appear on e_3 and e_5.

Let us define as *realizations* all the possible certain traces that are obtainable by choosing a value among all possible ones for an uncertain attribute of the uncertain trace. For example, some possible realizations of the trace in Table 1 are $\langle a, d, b, a, c, b \rangle$, $\langle a, a, a, a, b, b \rangle$, and $\langle c, a, c, b, b \rangle$. We can express the strength of the directly-follows relationship between two activities in an uncertain trace by counting the minimum and maximum number of times the relationship can appear in one of the possible realizations of that trace. To this goal, we exploit some structural properties of the behavior graph in order to obtain the minimum and maximum frequency of directly-follows relationships in a simpler manner.

A useful property to compute the minimum number of occurrences between two activities exploits the fact that parallel behavior is represented by the branching of arcs in the graph. Two connected determinate events have happened one immediately after the other if the graph does not have any other parallel path: if two determinate events are connected by a bridge, they will certainly happen in succession. This property is used to define a *strong sequential relationship*.

The next property accounts for the fact that, by construction, uncertain events corresponding to nodes in the graph not connected by a path can happen in any order. This follows directly from the definition of the edges in the graph, together with the transitivity of $\mathcal{U}_T$ (which is a totally ordered set). This means that two disconnected nodes v and w may account for one occurrence of the relation "$\pi_A(v)$ is directly followed by $\pi_A(w)$". Conversely, if w is reachable from v, the directly-follows relationship may be observed if all the events separating v from w are indeterminate (i.e., there is a chance that no event will interpose between the ones in v and w). This happens for vertices e_2 and e_4 in the graph in Fig. 1, which are connected by a path and separated only by vertex e_3, which is indeterminate. This property is useful to compute the maximum number of directly-follows relationships between two activities, leading to the notion of *weak sequential relationship*.

Definition 11 (Strong sequential relationship). *Given a behavior graph* $\beta = (V, E)$ *and two vertices* $v, w \in V$, *v is in a* strong sequential relationship *with w (denoted by $v \blacktriangleright_\beta w$) if and only if $\pi_o(v) = !$ and $\pi_o(w) = !$ (v and w are both determinate) and there is a bridge between them: $(v, w) \in E_B$.*

Definition 12 (Weak sequential relationship). *Given a behavior graph* $\beta = (V, E)$ *and two vertices* $v, w \in V$, *v is on a* weak sequential relationship *with w (denoted by $v \triangleright_\beta w$) if and only if $|P_\beta(w, v)| = 0$ (v is unreachable from w) and no node in any possible path between v and w, excluding v and w, is determinate:* $\bigcup_{p \in P_\beta(v,w)} \{e \in p \mid \pi_o(e) = !\} \setminus \{v, w\} = \emptyset$.

Notice that if v and w are mutually unreachable they are also in a mutual weak sequential relationship. Given two activity labels, these properties allow us to extract sets of candidate pairs of vertices of the behavior graph.

Definition 13 (Candidates for minimum and maximum directly-follows frequencies). *Given two activities* $a, b \in \mathcal{U}_A$ *and an uncertain trace* $\sigma \in \mathcal{T}_U$ *and the corresponding behavior graph* $\beta(\sigma) = (V, E)$*, the candidates for minimum and maximum directly-follows frequency* $cand_{min} \colon \mathcal{T}_U \times \mathcal{U}_A \times \mathcal{U}_A \to \mathcal{P}(V \times V)$ *and* $cand_{max} \colon \mathcal{T}_U \times \mathcal{U}_A \times \mathcal{U}_A \to \mathcal{P}(V \times V)$ *are defined as:*

- $cand_{min}(\sigma, a, b) = \{(v, w) \in V \times V \mid v \neq w \wedge \pi_A(v) = \{a\} \wedge \pi_A(w) = \{b\} \wedge v \blacktriangleright_\beta w\}$
- $cand_{max}(\sigma, a, b) = \{(v, w) \in V \times V \mid v \neq w \wedge a \in \pi_A(v) \wedge b \in \pi_A(w) \wedge v \triangleright_\beta w\}$

After obtaining the sets of candidates, it is necessary to select a subset of pair of vertices such that there are no repetitions. In a realization of an uncertain trace, an event e can only have one successor: if multiple vertices of the behavior graph correspond to events that can succeed e, only one can be selected.

Consider the behavior graph in Fig. 1. If we search candidates for "a is directly followed by b", we find $cand_{min}(\sigma, a, b) = \{(e_1, e_3), (e_2, e_3), (e_1, e_5), (e_2, e_4), (e_3, e_4), (e_3, e_5), (e_4, e_6)\}$. However, there are no realizations of the trace represented by the behavior graph that contains all the candidates; this is because some vertices appear in multiple candidates. A possible realization with the highest frequency of $a \to b$ is $\langle d, a, b, c, a, b \rangle$. Conversely, consider "$a$ is directly followed by a". When the same activity appears in both sides of the relationship, an event can be part of two different occurrences, as first member and second member; e. g., in the trace $\langle a, a, a \rangle$, the relationship $a \to a$ occurs two times, and the second event is part of both occurrences. In the behavior graph of Fig. 1, the relation $a \to b$ cannot be supported by candidates (e_1, e_3) and (e_3, e_4) at the same time, because e_3 has either label a or b in a realization. But (e_1, e_3) and (e_3, e_4) can both support the relationship $a \to a$, in realizations where e_1, e_3 and e_4 all have label a.

When counting the frequencies of directly follows relationships between the activities a and b, every node of the behavior graph can appear at most once if $a \neq b$. If $a = b$, every node can appear once on each side of the relationship.

Definition 14 (Minimum directly-follows frequency). *Given* $a, b \in \mathcal{U}_A$ *and* $\sigma \in \mathcal{T}_U$*, let* $R_{min} \subseteq cand_{min}(\sigma, a, b)$ *be a largest set such that for any* $(v, w), (v', w') \in R_{min}$*, it holds:*

$$(v, w) \neq (v', w') \Rightarrow \{v, w\} \cap \{v', w'\} = \emptyset, \qquad \text{if } a \neq b$$

$$(v, w) \neq (v', w') \Rightarrow v \neq v' \wedge w \neq w', \qquad \text{if } a = b$$

The minimum directly-follows frequency $\leadsto_{min} \colon \mathcal{T}_U \times \mathcal{U}_A{}^2 \to \mathbb{N}$ *of two activities* $a, b \in \mathcal{U}_A$ *in regard of an uncertain trace* $\sigma \in \mathcal{T}_U$ *is defined as* $\leadsto_{min}(\sigma, a, b) = |R_{min}|$.

Definition 15 (Maximum directly-follows frequency). *Given* $a, b \in \mathcal{U}_A$ *and* $\sigma \in \mathcal{T}_U$, *let* $R_{max} \subseteq cand_{max}(\sigma, a, b)$ *be a largest set such that for any* $(v, w), (v', w') \in R_{max}$, *it holds:*

$$(v, w) \neq (v', w') \Rightarrow \{v, w\} \cap \{v', w'\} = \emptyset, \qquad \text{if } a \neq b$$

$$(v, w) \neq (v', w') \Rightarrow v \neq v' \wedge w \neq w', \qquad \text{if } a = b$$

The maximum directly-follows frequency $\rightsquigarrow_{max} \colon \mathcal{T}_U \times \mathcal{U}_A{}^2 \to \mathbb{N}$ *of two activities* $a, b \in \mathcal{U}_A$ *in regard of an uncertain trace* $\sigma \in \mathcal{T}_U$ *is defined as* $\rightsquigarrow_{max}(\sigma, a, b) = |R_{max}|$.

For the uncertain trace in Table 1, $\rightsquigarrow_{min}(\sigma, a, b) = 0$, because $R_{min} = \emptyset$; conversely, $\rightsquigarrow_{max}(\sigma, a, b) = 2$, because a maximal set of candidates is $R_{max} = \{(e_1, e_3), (e_4, e_6)\}$. Notice that maximal candidate sets are not necessarily unique: $R_{max} = \{(e_2, e_3), (e_4, e_6)\}$ is also a valid one.

The operator $\rightsquigarrow$ synthesizes information regarding the strength of the directly-follows relation between two activities in an event log where some events are uncertain. The relative difference between the *min* and *max* counts is a measure of how certain the relationship is when it appears in the event log. Notice that, in the case where no uncertainty is contained in the event log, *min* and *max* will coincide, and will both contain a directly-follows count for two activities.

An *Uncertain DFG* (UDFG) is a graph representation of the activity frequencies and the directly-follows frequencies; using the measures we defined, we exclude the activities and the directly-follows relations that never happened.

Definition 16 (Uncertain Directly-Follows Graph (UDFG). *Given an event log* $L \in \mathcal{P}(\mathcal{T}_U)$, *the Uncertain Directly-Follows Graph* $DFG_U(L)$ *is a directed graph* $G = (V, E)$ *where:*

- $V = \{a \in \Pi_A(L) \mid \sum_{\sigma \in L} \#_{max}(\sigma, a) > 0\}$
- $E = \{(a, b) \in V \times V \mid \sum_{\sigma \in L} \rightsquigarrow_{max}(\sigma, a, b) > 0\}$

The UDFG is a low-abstraction model that, together with the data decorating vertices and arcs, gives indications on the overall uncertainty affecting activities and directly-follows relationships. Moreover, the UDFG does not filter out uncertainty: the information about the uncertain portion of a process is summarized by the data labeling vertices and edges. In addition to the elimination of the anomalies in an event log in order to identify the happy path of a process, this allows the process miner to isolate the uncertain part of a process, in order to study its features and analyze its causes. In essence however, this model has the same weak points as the classic DFG: it does not support concurrency, and if many activities happen in different order the DFG creates numerous loops that cause underfitting.

5 Inductive Mining Using Directly-Follows Frequencies

A popular process mining algorithm for discovering executable models from DFGs is the Inductive Miner [6]. A variant presented by Leemans et al. [7], the *Inductive Miner–directly-follows* (IM_D), has the peculiar feature of prepro-cessing an event log to obtain a DFG, and then discover a process tree exclusively from the graph, which can then be converted to a Petri net. This implies a high scalability of the algorithm, which has a linear computational cost over the num-ber of events in the log, but it also makes it suited to the case at hand in this paper. To allow for inductive mining, and subsequent representation of the pro-cess as a Petri net, we introduce a form of filtering called UDFG slicing, based on four filtering parameters: act_{min}, act_{max}, rel_{min} and rel_{max}. The parameters act_{min} and act_{max} allow to filter on nodes of the UDFG, based on how certain the corresponding activity is in the log. Conversely, rel_{min} and rel_{max} allow to filter on edges of the UDFG, based on how certain the corresponding directly-follows relationship is in the log.

Definition 17 (Uncertain DFG slice). *Given an uncertain event log* $L \in \mathcal{P}(\mathcal{T}_U)$, *its uncertain directly-follows graph* $DFG_U(L) = (V', E')$, *and* act_{min}, $act_{max}, rel_{min}, rel_{max} \in [0, 1]$, *an uncertain directly-follows slice is a function* $\overline{DFG_U}: L \to \mathcal{U}_G$ *where* $\overline{DFG_U}(L, act_{min}, act_{max}, rel_{min}, rel_{max}) = (V, E)$ *with:*

$$- V = \{a \in V' \mid act_{min} \leq \frac{\sum_{\sigma \in L} \#_{min}(\sigma, a)}{\sum_{\sigma \in L} \#_{max}(\sigma, a)} \leq act_{max}\}$$

$$- E = \{(a, b) \in E' \mid rel_{min} \leq \frac{\sum_{\sigma \in L} \leadsto_{min}(\sigma, a, b)}{\sum_{\sigma \in L} \leadsto_{max}(\sigma, a, b)} \leq rel_{max}\}$$

A UDFG slice is an unweighted directed graph which represents a filtering performed over vertices and edges of the UDFG. This graph can then be pro-cessed by the IM_D.

Definition 18 (Uncertain Inductive Miner–directly-follows (UIM_D)). *Given an uncertain event log* $L \in \mathcal{P}(\mathcal{T}_U)$ *and* $act_{min}, act_{max}, rel_{min}, rel_{max} \in [0, 1]$, *the* Uncertain Inductive Miner–directly-follows *(UIM_D) returns the pro-cess tree obtained by* IM_D *over an uncertain DFG slice:* $IM_D(\overline{DFG_U}(L, act_{min}, act_{max}, rel_{min}, rel_{max}))$.

The filtering parameters act_{min}, act_{max}, rel_{min}, rel_{max} allow to isolate the desired type of behavior of the process. In fact, $act_{min} = rel_{min} = 0$ and $act_{max} = rel_{max} = 1$ retain all possible behavior of the process, which is then represented in the model: both the behavior deriving from the process itself and the behavior deriving from the uncertain traces. Higher values of act_{min} and rel_{min} allow to filter out uncertain behavior, and to retain only the parts of the process observed in certain events. Vice versa, lowering act_{min} and rel_{min} allows to observe only the uncertain part of an event log.

6   Experiments

The approach described here has been implemented using the Python process mining framework PM4Py [3]. The models obtained through the Uncertain Inductive Miner–directly-follows cannot be evaluated with commonly used met-rics in process mining, since metrics in use are not applicable on uncertain event data; nor other approaches for performing discovery over uncertain data exist. This preliminary evaluation of the algorithm will, therefore, not be based on measurements; it will show the effect of the UIM_D with different settings on an uncertain event log.

Let us introduce a simplified notation for uncertain event logs. In a trace, we represent an uncertain event with multiple possible activity labels by listing the labels between curly braces. When two events have overlapping timestamps, we represent their activity labels between square brackets, and we represent the indeterminate events by overlining them. For example, the trace $\langle \overline{a}, \{b, c\}, [d, e] \rangle$ is a trace containing 4 events, of which the first is an indeterminate event with label a, the second is an uncertain event that can have either b or c as activity label, and the last two events have a range as timestamp (and the two ranges overlap). The simplified representation of the trace in Table 1 is $\langle [\{a, c\}, \{a, d\}], \overline{\{a, b\}}, [\{a, b\}, \{b, c\}], b \rangle$. Let us observe the effect of the UIM_D on the following test log:

$$\langle a, b, e, f, g, h \rangle^{80}, \langle a, [\{b, c\}, e], \overline{f}, g, h, i \rangle^{15}, \langle a, [\{b, c, d\}, e], \overline{f}, g, h, j \rangle^{5}.$$

In Fig. 2, we can see the model obtained without any filtering: it represents all the possible behavior in the uncertain log. The models in Figs. 3 and 4 show the effect on filtering on the minimum number of times an activity appears in the log: in Fig. 3 activities c and d are filtered out, while the model in Fig. 4 only retains the activities which never appear in an uncertain event (i.e., the activities for which $\#_{min}$ is at least 90% of $\#_{max}$).

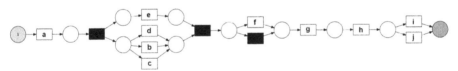

Fig. 2. UIM_D on the test log with $act_{min} = 0$, $act_{max} = 1$, $rel_{min} = 0$, $rel_{max} = 1$.

Fig. 3. UIM_D on the test log with $act_{min} = 0.6$, $act_{max} = 1$, $rel_{min} = 0$, $rel_{max} = 1$.

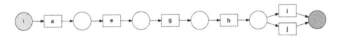

Fig. 4. UIM_D on the test log with $act_{min} = 0.9$, $act_{max} = 1$, $rel_{min} = 0$, $rel_{max} = 1$.

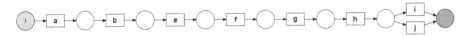

Fig. 5. UIM_D on the test log with $act_{min} = 0$, $act_{max} = 1$, $rel_{min} = 0.7$, $rel_{max} = 1$.

Filtering on rel_{min} has a similar effect, although it retains the most certain relationships, rather than activities, as shown in Fig. 5. An even more aggressive filtering of rel_{min}, as shown in Fig. 6, allows to represent only the parts of the process which are never subjected to uncertainty by being in a directly-follows relationship that has a low $\leadsto_{min}$ value.

The UIM_D allows also to do the opposite: hide certain behavior and highlight the uncertain behavior. Figure 7 shows a model that only displays the behavior which is part of uncertain attributes, while activities h, i and j – which are never part of uncertain behavior – have not been represented. Notice that g is represented even though it always appeared as a certain event; this is due to the fact that the filtering is based on relationships, and g is in a directly-follows relationship with the indeterminate event f.

Fig. 6. UIM_D on the test log with $act_{min} = 0$, $act_{max} = 1$, $rel_{min} = 0.9$, $rel_{max} = 1$.

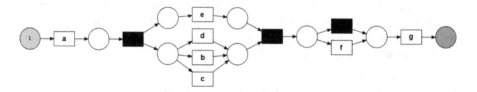

Fig. 7. UIM_D on the test log with $act_{min} = 0$, $act_{max} = 1$, $rel_{min} = 0$, $rel_{max} = 0.8$.

7   Conclusion

In this explorative work, we present the foundations for performing process discovery over uncertain event data. We present a method that is effective in representing a process containing uncertainty by exploiting the information into an uncertain event log to synthesize an uncertain model. The UDFG is a formal description of uncertainty, rather than a method to eliminate uncertainty to observe the underlying process. This allows to study uncertainty in isolation, possibly allowing us to determine which effects it has on the process in terms of behavior, as well as what are the causes of its appearance. We also present a method to filter the UDFG, obtaining a graph that represents a specific perspective of the uncertainty in the process; this can be then transformed in a model that is able to express concurrency using the UIM_D algorithm.

This approach has a number of limitations that will need to be addressed in future work. An important research direction is the formal definition of metrics and measures over uncertain event logs and process models, in order to allow for a quantitative evaluation of the quality of this discovery algorithm, as well as other process mining methods over uncertain logs. Another line of research can be the extension to the weakly uncertain event data (i.e., including probabilities) and the extension to event logs also containing uncertainty related to case IDs.

References

1. Van der Aalst, W., Weijters, T., Maruster, L.: Workflow mining: discovering process models from event logs. IEEE Trans. Knowl. Data Eng. **16**(9), 1128–1142 (2004)
2. Van der Aalst, W.M.: Process Mining: Data Science in Action. Springer, Heidelberg (2016). https://doi.org/10.1007/978-3-662-49851-4
3. Berti, A., van Zelst, S.J., van der Aalst, W.: Process mining for Python (PM4Py): bridging the gap between process- and data science. In: International Conference on Process Mining - Demo Track. IEEE (2019)
4. Conforti, R., La Rosa, M., ter Hofstede, A.H.: Filtering out infrequent behavior from business process event logs. IEEE Trans. Knowl. Data Eng. **29**(2), 300–314 (2017)
5. Hornik, K., Grün, B., Hahsler, M.: arules - a computational environment for mining association rules and frequent item sets. J. Stat. Softw. **14**(15), 1–25 (2005)
6. Leemans, S.J.J., Fahland, D., van der Aalst, W.M.P.: Discovering block-structured process models from event logs - a constructive approach. In: Colom, J.-M., Desel, J. (eds.) PETRI NETS 2013. LNCS, vol. 7927, pp. 311–329. Springer, Heidelberg (2013). https://doi.org/10.1007/978-3-642-38697-8_17
7. Leemans, S.J., Fahland, D., Van der Aalst, W.M.: Scalable process discovery and conformance checking. Softw. Syst. Model. **17**(2), 599–631 (2018)
8. Lu, X., Fahland, D., van den Biggelaar, F.J.H.M., van der Aalst, W.M.P.: Detecting deviating behaviors without models. In: Reichert, M., Reijers, H.A. (eds.) BPM 2015. LNBIP, vol. 256, pp. 126–139. Springer, Cham (2016). https://doi.org/10.1007/978-3-319-42887-1_11
9. Pegoraro, M., van der Aalst, W.M.: Mining uncertain event data in process mining. In: International Conference on Process Mining. IEEE (2019)

Predictive Process Monitoring in Operational Logistics: A Case Study in Aviation

Björn Rafn Gunnarsson$^{(\boxtimes)}$, Seppe K. L. M. vanden Broucke$^{(\boxtimes)}$, and Jochen De Weerdt$^{(\boxtimes)}$

Research Centre for Information Systems Engineering (LIRIS), KU Leuven, Naamsestraat 69, 3000 Leuven, Belgium
{bjornrafn.gunnarsson,seppe.vandenbroucke,jochen.deweerdt}@kuleuven.be

Abstract. The research area of process mining concerns itself with knowledge discovery from event logs, containing recorded traces of executions as stored by process aware information systems. Over the past decade, research in process mining has increasingly focused on predictive process monitoring to provide businesses with valuable information in order to identify violations, deviance and delays within a process execution, enabling them to carry out preventive measures. In this paper, we describe a practical case in which both exploratory and predictive process monitoring techniques were developed to understand and predict completion times of a luggage handling process at an airport. From a scientific perspective, our main contribution relates to combining a random forest regression model and a Long Short-Term Memory (LSTM) model into a novel stacked prediction model, in order to accurately predict completion time of cases.

Keywords: Process mining · Predictive process monitoring · Operations · Long Short-Term Memory (LSTM) · Aviation

1 Introduction

The past decades have seen a considerable rise in the use of process aware information systems within businesses. Today, these systems have become increasingly more intertwined with the operational process they support and often leave a trace of each activity executed by organizations, recorded and stored in the form of event logs. In alignment with the development of these information systems, the field of process mining concerns itself with knowledge discovery from such event logs, aiming to extract insights and information that can be used to understand and improve the underlying processes. Historically, this information has been extracted and analyzed after a business process has been executed. However, more recently, research in process mining has increasingly shifted its focus towards developing and applying process mining techniques in the context

© Springer Nature Switzerland AG 2019
C. Di Francescomarino et al. (Eds.): BPM 2019 Workshops, LNBIP 362, pp. 250–262, 2019.
https://doi.org/10.1007/978-3-030-37453-2_21

of a running process, often with the goal to deliver a prediction or recommendation with regards to the process at hand. These predictive process monitoring methods are constructed by analyzing historical execution traces and are then used to continuously provide users with predictions about the future of a given ongoing process execution. Predictive monitoring can therefore provide businesses with valuable information which enables them to identify violations, deviance and delays within process execution in advance making preventive measures possible [1,7,19].

Predictive process monitoring methods have been successfully applied to a number of different tasks, including cost prediction [25], risk prediction [5,6] and activity sequence predictions [11,22]. Another application that has received increased attention in the last few years is the prediction of the completion time of a process instance [20,21]. Predicting the completion time of a process instance has a number of business applications, including in logistic setting, where such predictions can for instance be used to indicate how long a certain step of a process will take, on the basis of which feedback can be provided to workers and end-users.

This work is inspired by the approaches suggested in [20,21]. More specifically, in a similar manner, we suggest an approach to predict the completion time of a process instance taking as input both the flow of the activities of a running case and other features describing the case. The approach suggested here was developed and applied in the real-life setting of the largest international airport in Belgium, Brussels Airport, and stems from the first phase of an analysis of the airport's baggage system, where the PM^2 process mining methodology [10] (not to be confused with the similarly named project management methodology developed and endorsed by the European Commission) was utilized with the aim of predicting the completion time of transferring bags going through the airport's baggage system. The resulting proposed solution consists of a stacked predictive modeling setup which will be described in depth.

As such, the main research contributions of this paper are as follows: firstly, the usability of the PM^2 process mining framework will be illustrated in the specific setting of airport analytics. Secondly, a novel data-aware modeling approach is proposed for predicting completion time of process instances, which combines two state-of-the-art machine learning models where one model is used to obtain baseline prediction and the other is used to sequentially update the baseline predictions as activities are executed for the process instances.

The remainder of this paper is structured as follows. Section 2 provides a discussion on the literature related to this research. Section 3 details the methodology used. In Sects. 4 and 5, the use of PM^2 framework in the specific setting of airport analytics will be illustrated, the section will provide an explanatory and predictive analysis of the baggage system of Brussels Airport. Finally, Sect. 6 concludes the paper and outlines directions towards future work.

2 Related Work

The first approaches developed to improve business processes and provide support for their execution focused on measuring and monitoring activities. These approaches were attributed to a research area called "Business Activity Monitoring". More recently, the field of process mining has emerged which allowed for the automatic analysis of a running process instance and the prediction of different attributes relating to such instances [14,20].

Since the mid-nineties, various scholars have worked on developing techniques to tackle the problem of accurately predicting the completion time of a process instance [2,21]. One of the first published research articles that focuses on analyzing execution duration times is [23]. However, no detailed prediction algorithm was suggested. Rather, the problem of having cross-trained resources on performance prediction was highlighted. In [8], a non-parametric regression-based approach was suggested that exploits all data available in an event log for predicting the cycle time of a running process instance. Van der Aalst et al. [2] suggested an approach where a finite state machine is constructed, called a transition system, which combines a discovered process model with time information learned from past instances to predict completion times of running instances. In [4] a similar transition system is utilized, though where the authors propose adding nested prediction models to the system which were constructed from event logs. The model can then be used to predict the next activity of a process and the completion time of a running process instance and was found to outperform the approach suggested by [2]. More recently, scholars have started to investigate the applicability of deep learning-based approaches for a wide range of predictive monitoring tasks, stepping away from transition system-based approaches. For instance, [24] investigated the use of LSTM networks for remaining time prediction and predicting the next event of a process and found that they outperformed prior methods for predictive process monitoring.

Whereas most transition system and neural network-based approaches described above are mainly control flow oriented (meaning that the sequence of executed activities is used as the prime information for prediction), more data-driven approaches have also been investigated. In [17], Leitner et al. proposed one of the first data-aware methods for predictive process monitoring, where the authors take advantage of additional data to predict service level agreement violations. Folino et al. [12,13] extended on this research by proposing a method for clustering traces according to corresponding context features and then constructing a predictive model for each cluster. Once trained, the approach can assign new running instances to one of the clusters and then uses the corresponding model constructed for that cluster to make a prediction. In a similar manner [21], suggested a method that uses both the control and data flow perspectives jointly by constructing a process model which is augmented by time and data information in order to enable remaining time prediction.

The research most related to the approach presented in this paper is [20], where the authors propose combining the method of using both control and data flow perspectives jointly as suggested by [21] but extend on this by considering an LSTM network for predictive process monitoring as suggested by [24]. Hence, the method improves on [21] in that it does not require constructing a transition system which can be both time and memory consuming for some real-world data sets and improves on [24] in that it can consider data attributes.

In order to combine information on control flow and data attributes, both [21] and [20] append data features to a vector containing information regarding the control flow of a process instance. The novelty of the approach suggested here is to incorporate data attributes in predictive process monitoring by using a stack of predictive models. That is, a regression-based model is used in a first step to obtain a baseline prediction for the completion time of a process instance at the beginning of its process and then in a second step the baseline predictions are updated to take into account the control flow of the process instance as it moves through the process.

3   Methodology

As discussed above, process mining techniques aim at extracting insights from event data as recorded by an organization's business process information system, which in turn can be used to improve the process performance. However, comparable with the related field of data mining, applying process mining in practice is not a trivial task and oftentimes involves a great deal of data wrangling and experimentation involving different viable approaches and techniques before a satisfactory outcome is reached. Just like with data mining, this has motivated researchers to develop standard methodologies that are tailored towards supporting process mining projects. One such methodology, PM^2, was specifically designed for this purpose and covers a wide range of process mining and related analysis techniques [10]. The six main stages described by PM^2 are shown in Fig. 1.

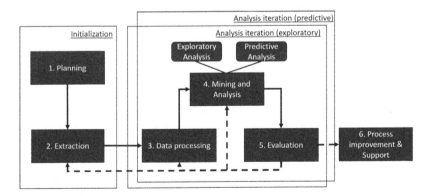

Fig. 1. Overview of the stages described by the PM^2 framework

As shown in the figure the PM^2 methodology can be divided into two main subgroups of stages, *initialization* and *analysis iterations*. During initialization the two first stages of methodology are carried out: planning (1), where initial research questions are defined, and (2) extraction, where the event data needed to answer the defined research questions is extracted from the business information systems. Next, one or more analysis iterations are performed, which focus on answering a specific research question. In general, each analysis iteration executes three stages: preprocessing the extract event data (3), analyzing the data (4), and evaluation of the results (5). If the obtained findings from the analysis iteration are deemed satisfactory, they can be used for improving the business process and supporting its operations (6). In the following two sections the use of the PM^2 methodology will be applied in a real-life setting stemming from an operational logistics context. More specifically, the methodology was used to support the application of process mining activities at Brussels Airport, the largest international airport in Belgium.

4   Initialization

Planning. One of the many services carried out at Brussels Airport is the handling of passenger baggage. One group of bags that are of particular interest in this setting are *transferring* bags, which arrive at the airport with an arrival flight and are then directly rerouted to a departing flight. This group of bags can cause considerable costs if they are—for whatever reason—unable to catch their departing flight. Because of this reality, the primary goal of the research project was to obtain insights with regards to transferring bags. This was achieved by means of a two-fold analysis. First, an exploratory analysis was carried out in order to better understand the variability both in the way these bags are processed in the baggage system and the distributions of duration time. Second, a predictive model was constructed to predict the completion times of transferring bags as they enter and move throughout the luggage system, which could subsequently be used to take preventive measures in order to minimize losses caused by this group of bags.

Extraction. The baggage system at Brussels Airport is almost fully automated, with bags automatically moving through different locations in the system, e.g. through a screening machine or sorting tray. A unique identifier is attached to each bag at the beginning of its journey through the system, which is scanned each time a bag arrives at a specific location in the system with the information being logged into an underlying supporting information system. This results in an event log describing complete and detailed traces for each bag, with the possibility to link this information to additional data, such as flight information, departing and originating airports, gate information for the flights, and airline information. The above information was extracted from the system for all transferring bags that were processed at Brussels Airport throughout 2018. In total, more than 1.5 million bags with their full journeys and related information were extracted from the system and form the basis of the following analysis.

5   Analysis Iterations

5.1   Exploratory Analysis

Preprocessing. In order to carry out an exploratory analysis based on an event log, three main data points need to be present. First, the unique identifier for each bag was used to group events to form a trace of events for each process instance (i.e. a trace describes the steps taken by one transferring bag). Second, as described above, bags are scanned every time they arrive at a location in the baggage system, leading to one event being logged in the underlying information system. To provide names of the activities in our process representation, we hence utilize the names of the locations in the system to construct a set of unique activities that bags can undergo. Third, activities are ordered within each trace by means of an exact recorded time stamp. Note that the time stamps in this setting are atomic, meaning that activities have one single time stamp and effectively have no duration. The duration for a trace as a whole is described by the time between the first time a bag is scanned and the last time, right before it exits the system. Table 1 provides an overview of descriptive statistics for the resulting processed event log.

Table 1. Descriptive statistics for the constructed event log

	Amount			
Number of traces	1548240			
Number of activities	63			
Trace variants	97428			
	Min	Max	Mean	Std.dev.
Date range	2018-01-01	2018-12-31		
Activities per trace	1	384	7.35	7.74
Duration per trace (minutes)	0	1440	178.69	149.96

Exploratory Analysis. The main motivation for carrying out an explanatory analysis was to get a good understanding of the variability of how transferring bags are processed in the baggage system and the distribution of completion times of the bags. Therefore, in a first step, a variant analysis was carried out. A variant describes a group of traces that follow exactly the same end-to-end activity flow. Analyzing their distinct number and frequencies of occurrence provides good initial insights regarding the variability and main flows in the airport's baggage system. Almost 100 thousand different variants were identified in the baggage system during 2018, indicating an extreme variability in how bags are routed through the system. As shown in Fig. 2, most of the process instances are actually captured by a couple of variants (e.g. the single most common variant describes about 300 000 traces), followed by a "long tail" of less frequent behaviour.

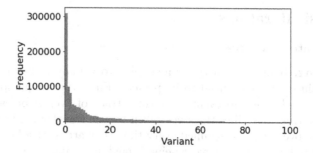

Fig. 2. Frequency overview for the 100 most frequent trace variants

In order to obtain insights into the variability of the completion times for the transferring bags the difference between the last time stamp and first time stamp in the event log for all transferring bags was computed. Figure 3 shows the distribution of the completion time of the transferring bags. As can be seen from the figure, the median completion time is around 143 min and two thirds of transferring bags have a completion time between 10 and 200 min. It can also be observed that some bags have a relatively long completion time, e.g. 5% of bags have a completion time that is longer than 400 min. It should be noted here that this does not necessarily indicate that these bags will miss their departing flight, since it is not infrequent for transferring bags to be stored in the system for relatively long periods if there is a long time difference between the arrival of the flight the bag arrives on and the departure of the flight the bag departs with. The results of both the variant analysis and the analysis on completion times was validated with experts on the baggage system and luggage handling at Brussels Airport. In addition, discovery and visualization of initial process maps helped to increase understanding of the overall process as well as highlight initial bottlenecks. Also, it is important to note that this exploratory process went through a number of iterations—in line with the methodology set forward by the PM^2 framework—based on feedback from domain experts and following the discovery of data quality issues stemming from the underlying information system. The event log being reported on herein is the result of a number of cleaning steps removing spurious and noisy events.

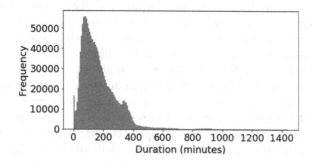

Fig. 3. Histogram of completion times for transferring bags

5.2 Predictive Analysis

Following the exploratory analysis, a number of predictive models were constructed in order to predict the completion times of transferring bags. First, a model was constructed only using the observed control flow for each bag, which will act as a baseline in our comparative analysis. Next, a second model was constructed based only on data elements known about the transferring bags when they arrive at the airport (static features). Lastly, a stacked model was constructed which combines both the static feature-based and control flow-based approaches for predicting the completion time of transferring bags.

Control Flow Model. To construct the control flow-based model, a Long Short-Term Memory (LSTM) neural network was set up to predict the completion time of traces based on last observed activities. LSTMs are a form of recurrent neural networks (RNNs), which are better suited to deal with ordered inputs compared to standard neural network architectures as they can selectively pass information across sequence steps while processing sequential data one element at a time. This enables them to model input data consisting of sequences of elements that are not independent. LSTMs are a special kind of RNNs capable of learning long-term dependencies and have demonstrated ground-breaking performance in a number of fields [15,18], including process mining as previously discussed in Sect. 2. In order to predict the completion time of transferring bags, an LSTM network was constructed using information contained in the trace of each bag. More specifically, the complete trace of each bag was split into multiple sub-sequences of length six, as shown in Fig. 4, left-padded with zeroes where necessary. A window length of six was chosen corresponding with the median number of events per trace. As in [20], the collection of instances was then *one-hot* encoded and used to train a LSTM regression model, using a mean absolute error loss.

Static Features Model. In a second attempt to predict the completion time of transferring bags, a random forest regressor was trained on a feature matrix consisting of data elements known about the bags when they arrive at the airport. Random forests are a well-known ensemble method originally suggested by [3] and have shown great performance in many areas when dealing with traditional, tabular data sets. In short, this method constructs a set (called the

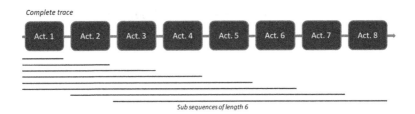

Fig. 4. An example of a complete trace and its sub-sequences

"forest") of decision trees during training and uses that to make predictions. The following features were incorporated in this model to describe bags: (i) time related information (i.e. hour of the day, day of the week, month of the year), (ii) information regarding their arrival flight (e.g. airline, previous airport), and (iii) similar information regarding their departing flight. Remark that no control flow-based features were considered in this model, so that the predictions provided by this model remain unchanged as a bag moves throughout the system and starts to generate events. This is a deliberate choice, as we aim to rely on the LSTM model constructed above to derive control flow-based information, rather than manually performing feature engineering over the traces, since this approach would easily lead to an explosion of features. As such, we opt here for a best of both worlds approach, where a well-understood modeling approach (random forests) is used to provide a robust baseline prediction as an instances arrives and can relatively easily explain its reasoning behind a certain prediction. In this setting for instance, it was observed that the random forest model was able to pick up on hours of the day and certain airlines as being the main important factors influencing completion times. Next, to provide more granular, precise predictions as a bag starts generating activities, the baseline prediction of the random forest has to be updated based on control flow information. To do so, we will utilize the LSTM-based model as described above, which is better suited to work with the more complex sequence data, automatically performing feature engineering as it is being trained. The two models are combined in a "stacked setup" as described below.

Stacked Combined Model. The final model considered for predicting the completion times of transferring bags is constructed using a stacked approach as motivated above. Previous approaches for combining information on control flow and data attributes for predicting completion time of process instances commonly do so by appending data features to a vector containing information on the control flow of each process instance. Here, a stacked model is trained in the following manner: first, the random forest regressor is used to obtain initial baseline estimates of the completion time of transferring bags. Next, a new LSTM model was trained, now with the goal to learn how much the initial baseline predictions deviate from the actual completion times of the transferring bags (i.e. to learn the residuals). On the input side, the LSTM utilizes the same features as the baseline "Control Flow Model" setup described above, with one additional feature per instance being provided with its value set equal to the baseline prediction of the random forest. The final prediction of the stacked model is then obtained by adding the estimated deviation of the prediction random forest, estimated by the LSTM, to the initial baseline prediction of the random forest.

Implementation and Results. To train the three models previously discussed in this section, 75% of transferring bags in 2018 were selected randomly, with the remaining 25% being placed in a hold-out test set to perform a final evaluation comparison. The LSTM models were constructed using two LSTM layers, each having 32 hidden units and trained using the Nadam optimizer [9]. The random forest regressor was constructed using an ensemble of 50 decision trees. Training time encompassed 24 h for the LSTM-based models and 90 min for the random forest. A comparison of the obtained results for each of the three models considered here is shown in Fig. 5. Various novel insights can be obtained from this comparison. First, the control flow-based LSTM network performs drastically worse than both the static feature-based random forest and the stacked model proposed here. As shown in the figure the model has an average mean absolute error (MAE) of around 268 min which is considerably higher than the average MAE of the two other approaches considered here. This suggests that completion times of transferring bags are heavily affected by static features such as time related information regarding when the bag arrives at the airport, information on the airline that arrived and departed with bag and the airport the bag is arriving from and departing to. Research on predictive process monitoring has increasingly focused on data-aware process monitoring where the control flow of process instances is combined with non-process related features in order to improve predictive accuracy. The stacked method suggested in this paper does so by combining a well-understood modelling approach, namely random forests, which provides a robust and explainable baseline prediction based on

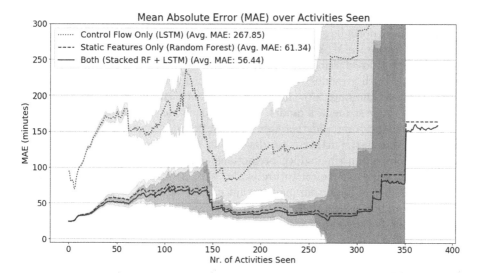

Fig. 5. Comparative overview of the three modeling approaches. The mean absolute error for each approach is shown with 95% confidence intervals over the different number of activities seen so far, together around the mean of absolute errors per different time step. The stacked model is able to improve upon the static modelling approach, especially as traces continue and become lengthier. The difference of means of the stacked versus static MAEs (56.44 versus 61.34) is significant at a 95% confidence level using a correlation-adjusted procedure to compare dependent models [16].

static information, with the deeper learning mechanism provided by an LSTM model to focus on the more complex inputs, namely control flow information. This stacked method is the best performing method of all method considered here and has a average MAE of 56 min which is an improvement of around 5 min compared to the static feature only-based approach.

6 Conclusion and Future Work

For the past decade, research on predictive process monitoring techniques has increasingly focused on developing data-aware methods that combine information on processes with other non-process related information. A novel stacked data-aware predictive process monitoring method was suggested in this paper. The method stacks two predictive models where a robust baseline model, random forest regressor, is used to provide baseline prediction based on static feature information and an LSTM neural network is used to sequential update the initial baseline prediction as the process instance starts generating activities. This approach was developed and applied in a real-life setting where the model was used to predict the completion time of transferring bags at Brussels Airport. The constructed method outperforms both a static feature based approach and a control flow based approach for predicting completion time. The paper also illustrated the usability of the PM^2 process mining framework in the specific setting of airport analytics. This research stems from the first phase of analysis into Brussels Airports baggage system and a number of research avenues can be explored in future work. A key motivation for including a static feature based model in the proposed stacked setup is that its prediction can be relatively easily explained which provides valuable insights into features that influence completion times of the bags. An alternative approach entails including all static features in a single sequential model. This approach has been utilized for predictive process monitoring and therefore a direct comparison of the two methods would be of interest. Furthermore, developing control flow-based methods for predicting completion time which predictions can be easily explained would be another interesting avenue for future work. Also, neural network based models include a number of hyper-parameters that can be tuned. It would therefore be of interest conduct a thorough grid search over a large tune grid of values for these parameters (including the LSTM architecture setup and length of the sub-sequences considered) in an attempt to further improve the performance of the LSTM based networks. Lastly, the research presented here focuses on predicting the duration times of transferring bags. However, it would also be of interest to extend on the model to predict the next likely activities in a running instance.

References

1. Van der Aalst, W.M.P., Weijters, T., Maruster, L.: Workflow mining: discovering process models from event logs. IEEE Trans. Knowl. Data Eng. **16**(9), 1128–1142 (2004)
2. Van der Aalst, W.M.P., Schonenberg, M.H., Song, M.: Time prediction based on process mining. Inf. Syst. **36**(2), 450–475 (2011)
3. Breiman, L.: Random forests. Mach. Learn. **45**(1), 5–32 (2001)
4. Ceci, M., Lanotte, P.F., Fumarola, F., Cavallo, D.P., Malerba, D.: Completion time and next activity prediction of processes using sequential pattern mining. In: Džeroski, S., Panov, P., Kocev, D., Todorovski, L. (eds.) DS 2014. LNCS (LNAI), vol. 8777, pp. 49–61. Springer, Cham (2014). https://doi.org/10.1007/978-3-319-11812-3_5
5. Conforti, R., de Leoni, M., La Rosa, M., van der Aalst, W.M.P.: Supporting risk-informed decisions during business process execution. In: Salinesi, C., Norrie, M.C., Pastor, Ó. (eds.) CAiSE 2013. LNCS, vol. 7908, pp. 116–132. Springer, Heidelberg (2013). https://doi.org/10.1007/978-3-642-38709-8_8
6. Conforti, R., de Leoni, M., La Rosa, M., van der Aalst, W.M.P., ter Hofstede, A.H.: A recommendation system for predicting risks across multiple business process instances. Decis. Support. Syst. **69**, 1–19 (2015)
7. Di Francescomarino, C., Ghidini, C., Maggi, F.M., Milani, F.: Predictive process monitoring methods: which one suits me best? In: Weske, M., Montali, M., Weber, I., vom Brocke, J. (eds.) BPM 2018. LNCS, vol. 11080, pp. 462–479. Springer, Cham (2018). https://doi.org/10.1007/978-3-319-98648-7_27
8. van Dongen, B.F., Crooy, R.A., van der Aalst, W.M.P.: Cycle time prediction: when will this case finally be finished? In: Meersman, R., Tari, Z. (eds.) OTM 2008. LNCS, vol. 5331, pp. 319–336. Springer, Heidelberg (2008). https://doi.org/10.1007/978-3-540-88871-0_22
9. Dozat, T.: Incorporating nesterov momentum into adam. Technical report, Stanford University (2016). http://cs229.stanford.edu/proj2015/054_report.pdf
10. van Eck, M.L., Lu, X., Leemans, S.J.J., van der Aalst, W.M.P.: PM^2: a process mining project methodology. In: Zdravkovic, J., Kirikova, M., Johannesson, P. (eds.) CAiSE 2015. LNCS, vol. 9097, pp. 297–313. Springer, Cham (2015). https://doi.org/10.1007/978-3-319-19069-3_19
11. Evermann, J., Rehse, J.R., Fettke, P.: Predicting process behaviour using deep learning. Decis. Support. Syst. **100**, 129–140 (2017)
12. Folino, F., Guarascio, M., Pontieri, L.: Discovering context-aware models for predicting business process performances. In: Meersman, R., et al. (eds.) OTM 2012. LNCS, vol. 7565, pp. 287–304. Springer, Heidelberg (2012). https://doi.org/10.1007/978-3-642-33606-5_18
13. Folino, F., Guarascio, M., Pontieri, L.: Discovering high-level performance models for ticket resolution processes. In: Meersman, R., et al. (eds.) OTM 2013. LNCS, vol. 8185, pp. 275–282. Springer, Heidelberg (2013). https://doi.org/10.1007/978-3-642-41030-7_18
14. Golfarelli, M., Rizzi, S., Cella, I.: Beyond data warehousing: what's next in business intelligence? In: Proceedings of the 7th ACM International Workshop on Data Warehousing and OLAP, pp. 1–6. ACM (2004)
15. Hochreiter, S., Schmidhuber, J.: Long short-term memory. Neural Comput. **9**(8), 1735–1780 (1997)

16. Jolliffe, I.T.: Uncertainty and inference for verification measures. Weather Forecast. **22**(3), 637–650 (2007)
17. Leitner, P., Wetzstein, B., Rosenberg, F., Michlmayr, A., Dustdar, S., Leymann, F.: Runtime prediction of service level agreement violations for composite services. In: Dan, A., Gittler, F., Toumani, F. (eds.) ICSOC/ServiceWave -2009. LNCS, vol. 6275, pp. 176–186. Springer, Heidelberg (2010). https://doi.org/10.1007/978-3-642-16132-2_17
18. Lipton, Z.C., Berkowitz, J., Elkan, C.: A critical review of recurrent neural networks for sequence learning. arXiv preprint arXiv:1506.00019 (2015)
19. Márquez-Chamorro, A.E., Resinas, M., Ruiz-Cortés, A., Toro, M.: Run-time prediction of business process indicators using evolutionary decision rules. Expert. Syst. Appl. **87**, 1–14 (2017)
20. Navarin, N., Vincenzi, B., Polato, M., Sperduti, A.: LSTM networks for data-aware remaining time prediction of business process instances. In: 2017 IEEE Symposium Series on Computational Intelligence (SSCI), pp. 1–7. IEEE (2017)
21. Polato, M., Sperduti, A., Burattin, A., de Leoni, M.: Data-aware remaining time prediction of business process instances. In: 2014 International Joint Conference on Neural Networks (IJCNN), pp. 816–823. IEEE (2014)
22. Polato, M., Sperduti, A., Burattin, A., de Leoni, M.: Time and activity sequence prediction of business process instances. Computing **100**(9), 1005–1031 (2018)
23. Reijers, H.A.: Case prediction in BPM systems: a research challenge. J. Korean Inst. Ind. Eng. **33**(1), 1–10 (2007)
24. Tax, N., Verenich, I., La Rosa, M., Dumas, M.: Predictive business process monitoring with LSTM neural networks. In: Dubois, E., Pohl, K. (eds.) CAiSE 2017. LNCS, vol. 10253, pp. 477–492. Springer, Cham (2017). https://doi.org/10.1007/978-3-319-59536-8_30
25. Tu, T.B.H., Song, M.: Analysis and prediction cost of manufacturing process based on process mining. In: 2016 International Conference on Industrial Engineering, Management Science and Application (ICIMSA), pp. 1–5. IEEE (2016)

A Survey of Process Mining Competitions: The BPI Challenges 2011–2018

Iezalde F. Lopes and Diogo R. Ferreira[(✉)]

Instituto Superior Técnico (IST), University of Lisbon, Lisbon, Portugal
{iezalde.lopes,diogo.ferreira}@tecnico.ulisboa.pt

Abstract. In recent years, several advances in the field of process mining, and even data science in general, have come from competitions where participants are asked to analyze a given dataset or event log. Besides providing significant insights about a specific business process, these competitions have also served as a valuable opportunity to test a wide range of process mining techniques in a setting that is open to all participants, from academia to industry. In this work, we conduct a survey of process mining competitions, namely the Business Process Intelligence Challenge, from 2011 to 2018. We focus on the methods, tools and techniques that were used by all participants in order to analyze the published event logs. From this survey, we develop a comparative analysis that allows us to identify the most popular tools and techniques, and to realize that data mining and machine learning are playing an increasingly important role in process mining competitions.

Keywords: Process mining · Data mining · Machine learning

1 Introduction

The field of data science is thriving with competitions where participants are asked to perform challenging tasks involving the analysis of real-world data. In the field of process mining, the competition that has brought the community together around real-world event logs is the Business Process Intelligence Challenge (BPI challenge). Since 2011, the BPI challenge has been providing event logs that have served as a testbed for innumerous process mining techniques.

An interesting aspect of the BPI challenge is that it has drawn the attention not only of academic researchers, but also of practitioners working at the intersection between business process management and data analysis. Furthermore, the BPI challenge has served as an introduction to many students entering the field of process mining, and some of its event logs, namely the one from the BPI Challenge 2012, have become a standard of reference for many authors.

In this work, we provide a summary of each BPI challenge, followed by an overview of the tools and techniques that have been most used across all BPI challenges. This paper is intended to share our findings and provide a sense of what sort of techniques and approaches the community is making use of when dealing with real-world event logs.

C. Di Francescomarino et al. (Eds.): BPM 2019 Workshops, LNBIP 362, pp. 263–274, 2019.
https://doi.org/10.1007/978-3-030-37453-2_22

2 BPI Challenges

In this section, we provide a summary of each BPI challenge, including a brief description of the business process domain, the business questions (if applicable), the winning submissions, and an overall impression about the approaches that were used to address the challenge.

2.1 BPI Challenge 2011

In the first edition, the BPI challenge involved an event log from a Dutch Academic Hospital. The event log concerned the diagnosis and treatments performed on patients in a Gynecology department. The names that were given to those treatments do not seem to follow a strict format. This led to a relatively large number of different task names, which means that any direct control-flow analysis of the event log is likely to yield a spaghetti model.

In this first edition, participants could focus on a specific aspect to analyze it in depth, or they could focus on a broader range of aspects without going into much detail. The winning authors were Bose and van der Aalst [1], who used the enhanced fuzzy miner [2] and trace alignment techniques [3] in ProM to group homogenous cases. In the end, they were able to present a compact process model. The trace alignment analysis carried out by the team also yielded common patterns of execution and exceptional/rare behavior.

In general, all authors used ProM combined with different preprocessing techniques to create subsets of cases which could be mined to obtain an understandable process model.

2.2 BPI Challenge 2012

In the second edition, the BPI challenge involved an event log from a Dutch Financial Institute. The event log concerned an application process for personal loans. Here, three distinct subprocesses could be identified by the event prefixes $A\_$ (for application states), $O\_$ (offer states), and $W\_$ (work items).

In this challenge, there were four business questions of interest to the process owner: (1) estimating the total cycle time; (2) determining which resources generate the highest activation rate of applications; (3) discovering the process model; and (4) identifying which decisions have greater influence on the process flow.

The winning authors were Bautista et al. [4] (from a New York-based consulting firm), who used Disco to understand the process, and decision trees to segment loan applications according to their approval result.

In general, authors used both ProM and Disco, and they also focused on several process mining perspectives (control-flow, organizational, performance) using a variety of analysis plug-ins. Preprocessing did not play such a large role as in the first edition, since three different subprocesses were already identified in the event log.

2.3   BPI Challenge 2013

In the third edition, the BPI challenge involved an event log from Volvo IT Belgium. The log contained events from an incident management system called VINST. Each event refers to a change in the status/sub-status of an incident.

In this challenge, there were four business questions of interest to the process owner: (1) whether incidents are pushed too often to second- and third-line support; (2) whether there is ping-pong behavior between teams; (3) whether the wait-user status reveals performance problems; and (4) whether process instances conform across departments.

The winning authors were Kang et al. [5] (a team from a South Korean university), who used a footprint matrix to capture the activity precedence. They found that departments were not conforming to each other, and they analyzed the process at each department with Disco.

In this challenge, there was a noticeable trend towards the use of statistics. In a sense, this was to be expected since the business questions required ranking and comparison among process instances and case attributes. Disco and ProM were the most popular tools. Both the control-flow perspective and the organizational perspective played a key role.

2.4   BPI Challenge 2014

In the fourth edition, the BPI challenge involved an event log from Rabobank Group ICT. The log concerned ITIL processes such as interaction management, incident management and change management.

The main goal was to predict the workload of the Service Desk (SD) and IT Operations (ITO) when a new change is introduced. There were four business questions involving: (1) identification of impact patterns; (2) impact of such patterns on workload; (3) improvement of service level after each change; and (4) creative analysis, where participants could pursue other insights.

The winning authors were Buhler et al. [6] (from the same New York-based firm as in 2012), who used custom metrics to measure performance and improvements after a change. Using decision trees, they classified the impact of those changes on workload. Also, using a multinomial logistic regression model, they were able to determine the probability of a specific change resulting in a given impact.

In this challenge, a new student category was introduced, which targeted BSc, MSc, PhD students. In this category the winning authors were Cacciola et al. [7], who used several custom plots to address the business questions. They also used Disco for performance analysis.

In general, most authors turned to data mining tools and techniques to answer the business questions. It is the first time that we see data mining tools being preferred over process mining tools, although Disco and ProM still played an important role in the analysis.

2.5 BPI Challenge 2015

In the fifth edition, the BPI challenge concerned the application process for construction permits in five Dutch municipalities. There were six business questions involving: (1) roles of people involved in the process; (2) possible improvements to the organizational structures; (3) changes in the process due to relocation of employees; (4) effect of outsourcing in organizational structures; (5) throughput times; and (6) control-flow for each municipality.

The winning author was van der Ham [8] (an independent consultant), who used Disco to analyze the average throughput time per municipality, and decision trees in WEKA to predict resource assignment. Using concept drift analysis [9] in ProM, the author identified the major changes in the process.

In the student category, the winning authors were Teinemaa et al. [10], who used the Kleinberg algorithm [11] to analyze the organization structure and identify key resources. They also analyzed handover of work with Disco and performed concept drift analysis [9] in ProM. Using the heuristics miner [12] and log replay in ProM did not reveal significant differences between the municipalities.

In this challenge, the business questions touched the control-flow, organizational and performance perspectives of process mining. Hence, most authors made use of process mining tools such as Disco, ProM and Minit. The use of data mining techniques (e.g. decision trees) was present in some submissions but had only a secondary role.

2.6 BPI Challenge 2016

In the sixth edition, the BPI challenge involved some very large event logs (~ 1 GB) from the Dutch Employee Insurance Agency (UWV). The logs concerned the customer interaction through different channels (website, messages, and call center) when applying for unemployment benefits. The main goal was to provide insights about the way the website was used.

In this challenge, there were six business questions: (1) identification of usage patterns on the website; (2) change of usage patterns over time; (3) transitions from the website to other channels; (4) change in customer behavior after using other channels; (5) customer behavior leading to complaints; (6) any new insights that could be obtained from the event log.

In this edition, a sponsor provided participants with the opportunity to apply for a free license of a software tool (Minit) for use in the BPI challenge.

The winning author was again van der Ham [13], who used mostly a spreadsheet-based analysis in Excel to collect statistics related to the business questions. The author also used IBM Watson [14] to try to find correlations between complaints and case attributes such as the number of visits to specific webpages.

In the student category, the winning authors were Dadashnia et al. [15], who clustered the traces and then used Disco to analyze the control-flow. In addition, they created a predictive model using a recurrent neural network (RNN) [16] to predict the next user action. They also used sequence clustering in ProM [17] to derive usage patterns.

The event logs for this challenge were much larger than in previous challenges (one of the logs had over 7 million events). Due to the log size and characteristics, it was difficult to apply standard process mining techniques, so most authors focused on simpler techniques based on filtering and clustering. There were some attempts at using machine learning (namely neural networks) but this was not very effective on this dataset.

2.7 BPI Challenge 2017

In the seventh edition, the BPI challenge involved the same Dutch Financial Institute as in the BPI Challenge 2012, and the event log concerned an upgraded version of the loan application process.

There were four business questions: (1) throughput times, in particular the time a customer is waiting for the bank and vice-versa; (2) influence of multiple information requests on offer acceptance; (3) comparison between single-offer and multiple-offer customers; (4) any other interesting trends.

In this edition, there were three categories – student, academic and professional – and there was a record number of submissions (23 in total). The challenge also gave participants the opportunity to use tools provided by the sponsors (Minit and Celonis).

In the student category, the winning authors were Povalyaeva et al. [18], who created a BPMN diagram in Celonis based on their log analysis, and then assessed conformance using the same tool. They used ProM for concept drift analysis, and Disco for performance analysis. The authors also used random forests to analyze the process outcome based on several features.

In the academic category, the winning authors were Rodrigues et al. [19], who used Disco, ProM and Yasper. Disco was used to obtain a process model. ProM was used for conformance checking by calculating the fitness and precision of a Petri net model, and also replaying the log on that model. The authors also modeled parts of the process in BPMN. Yasper was used to assess the performance perspective together with ProM and Disco.

In the professional category, the winning authors were Blevi et al. [20] who combined process mining with KPMG's customer experience methodology. Using Power BI, the authors provided several statistics. They also obtained a process model using fuzzy miner in ProM. Using R and Microsoft Azure Machine Learning Studio, the team created predictive models using logistic regression, random forests, and neural networks.

In general, there was a lot of process mining analysis, combined with data mining and machine learning techniques such as decision trees, random forests, etc. A wide variety of process mining tools were used (Disco, ProM, Celonis, Minit, etc.). Despite the large number of submissions, the analysis was mostly focused on the business questions, with similar results being reported by all authors.

2.8 BPI Challenge 2018

In the eighth edition, the BPI challenge involved an event log from the European Agricultural Guarantee Fund, in Germany. The event log concerns annual payments to farmers. The workflow is based on document types; each document has a state that allows some actions to be performed.

In this challenge, there were four business questions: (1) detection of undesired cases (e.g. late payment); (2) improving the sampling of applications selected for inspection; (3) differences between departments and relation to undesired outcomes; (4) differences across time.

Again, participants were given the opportunity to use tools provided by the sponsors (Minit and Celonis).

In the student category, the winning authors were Brils et al. [21], who obtained a process model with Disco and collected several statistics after filtering the event log with Python. For outcome prediction, they tried several machine learning techniques available in RapidMiner (naive Bayes, logistic regression, neural networks, decision trees, etc.).

In the academic category, the winning authors were Pauwels and Calders [22], who used the competition to test their own method to detect concept drift. Their model is based on Bayesian networks and, after training it, they were able to detect two drift points. They also used attribute-density plots to find differences between departments.

In the professional category, the winning authors were Wangikar et al. [23], who used Celonis to obtain a process model. They used predictive models based on binomial logistic regression to detect undesired outcomes. They also analyzed concept drift in ProM and conformance checking in myInvenio to detect changes in the control-flow.

In this edition, data mining techniques were predominant over process mining. This was due to the nature of the business questions, which involved prediction. These questions received a lot of attention from participants. Process mining was used to get an idea of the control-flow, and to analyze concept drift. Subsequent analysis was performed using machine learning.

3 Comparative Analysis

In this section, we analyze the tools and techniques that were used in the BPI challenges, not only by the winners, but across all submissions. We also analyze the use of specific ProM plug-ins, and of data mining/machine learning techniques.

3.1 Techniques

Typically, process mining focuses on the control-flow, organizational, and performance perspectives. Some BPI challenges involve all these perspectives, while others involve only some of them. In general, however, all BPI challenges go beyond those perspectives to include additional types of analysis, as shown in Table 1.

Table 1. Techniques used in the BPI challenges 2011–2018.

	2011	2012	2013	2014	2015	2016	2017	2018	Total
Control-flow discovery	X	X	X	X	X	X	X	X	**8**
Trace clustering	X	X	X	X		X	X	X	**7**
Social network analysis	X	X	X	X	X	X	X		**7**
Performance perspective	X	X	X	X	X		X	X	**7**
Log statistics		X	X	X	X	X	X	X	**7**
Conformance checking		X	X	X	X		X	X	**6**
Predictive modeling				X	X	X	X	X	**5**
Dotted chart analysis		X	X		X	X	X		**5**
Plotting/visualization			X	X		X	X	X	**5**
Trace alignment	X	X	X				X		**4**
Concept drift analysis					X		X	X	**3**
Spreadsheet-based analysis		X	X			X			**3**

Control-flow discovery techniques were the most used. This is to be expected since no process flow was ever provided in a BPI challenge, and one needs to understand the business process to be analyzed. In general, all authors used at least one control-flow mining technique, either to understand the business process or because a business question required it. Heuristics-based techniques [12] and fuzzy miners [24] were the most used for process discovery.

Trace clustering [25] is also among the most used techniques. Here the main reason is the fact that the derived process models were often too complex to be understood and there was a need to divide the event log into smaller and more homogenous groups of cases in order to mine an understandable process model.

Social network analysis gained an increased popularity, mostly because of the implementation of such techniques in the ProM framework [26], which made them easy to use. The analysis on a resource perspective was also made possible by the data available in the competitions.

Although many authors worked on the performance perspective, most participants only analyzed it thoroughly if a business question required it. Otherwise, basic performance statistics were provided, which were obtained using some tool (more on this below).

Statistics about the event log have been widely collected, and exploratory analysis is the main reason why participants use those statistics. Then techniques such as predictive modeling, data plotting/visualization have also gained special attention. Predictive modeling techniques have been used mainly when a business question required it, but they have also been helpful to study the control-flow perspective.

3.2 Tools

In Table 2, a list of the most popular tools in the BPI challenges is presented. Disco (commercial) and ProM (academic) were definitely the most used. Despite the sponsoring of other tool providers in some editions of the BPI challenge (e.g. Celonis,

Minit), Disco was still the most used. However, the use of competing tools seems to be growing, especially in recent years, possibly due the influence of academic programs that provide students and researchers with the opportunity to use such tools.

Table 2. Popular tools used in the BPI challenges 2011–2018.

	2011	2012	2013	2014	2015	2016	2017	2018	Total
Disco		3	10	9	6	4	18	1	**51**
ProM	3	5	6	3	7	5	17	3	**49**
Excel		1	7	6	4	2	10		**30**
R			2	3	1	2	8		**16**
Celonis						1	13	1	**15**
Python			1				8	1	**10**
WEKA				3	2	1	2		**8**
Oracle		1	1	2	1		2		**7**
RapidMiner			1	1	1		2	1	**6**
Java			1	2	1	1			**5**
SQL Server				2		1	1		**4**
Minit					1	1	1		**3**
C#				1	1		1		**3**

In general, it was observed that the use of these tools follows a cascading pattern: (1) Disco and/or ProM are used to get an overview of the process; (2) additional statistics are collected from the event log using tools such as Excel and R; and (3) other tools are selected depending on the challenge and on the nature of the business questions. In some cases, authors have used database engines (e.g. Oracle, SQL Server) to compute statistics about the event log.

3.3   ProM Plug-ins

Since ProM is a framework that includes a wide variety of plug-ins, it is interesting to check which of those plug-ins have been most used. In Table 3, the focus is on the use of ProM plug-ins only. Here, the heuristics miner [12], the dotted chart analysis [27], and the social network miner [26] take the podium, with a significant lead over the remaining ones.

It is interesting to note that the social network miner is one of the top plug-ins and has been used even in BPI challenges where there were no business questions involving the organizational perspective. It seems that the social network miner is very useful to complement and/or corroborate results obtained in other analysis perspectives.

Following the top three, we find the inductive miner [28] and the fuzzy miner [24], which are two popular control-flow discovery techniques. Both allow some form of abstraction over the control-flow behavior (process trees in the inductive miner, and activity clusters in the fuzzy miner). Finally, it is worth noting that trace alignment [3] and concept drift [9] have been playing an increasingly important role.

Table 3. ProM plug-ins used in the BPI challenges 2011–2018.

	2011	2012	2013	2014	2015	2016	2017	2018	Total
Heuristics miner	2	3	3	1	2		2		**13**
Dotted chart analysis		2	1		4	1	5		**13**
Social network miner	1	1	3		2	1	3		**11**
Inductive miner					1		4		**5**
Fuzzy miner			1	2			2		**5**
Alpha miner		1	1				2		**4**
Sequence clustering			1			2	1		**4**
Trace alignment	1	1	1				1		**4**
Concept drift					1		1	1	**3**
Organizational miner					1		2		**3**
Filter log simple heurist.		1	1		1				**3**
Guide Tree Miner	1	1							**2**
LTL-checker	1	1							**2**
Originator-by-task	1						1		**2**
Pattern abstractions	1	1							**2**
Trace align. w/guide tree	1	1							**2**

3.4 Data Mining/Machine Learning

Since we came across several data mining/machine learning techniques during our survey, it is interesting to analyze their use in the BPI challenges. Table 4 presents the data mining/machine learning techniques used across all BPI challenge submissions.

Although only two BPI challenges (2014 and 2018) included a prediction goal, we can see in Table 4 that this type of analysis was performed in editions where it was not apparently required. Decision trees are by far the most used technique. Participants used tools (for example, RapidMiner) which provide the implementation of the technique, which had only to be parametrized and/or customized to the data. Decision trees were mostly used for classification purposes across the BPI challenges. For example, in the BPI Challenge 2017, decision trees were used to classify the incompleteness of a loan application.

Sometimes, decision trees are used to address specific business questions; other times, they are used to complement the analysis. About half of the times decision trees were used, it was due to the fact that there was a business question requiring to predict a behavior; in the other half, the aim was to provide additional insight to the analysis previously carried out during the process mining phase.

Logistic Regression and Random Forests are also worth mentioning. We have noted that they are usually picked as the second choice after decision trees. They were used for classification tasks as well as for predicting behavior.

In general, data mining and/or machine learning techniques tend to be used to enhance the business process analysis. When there is some aspect that cannot be explained by process mining techniques alone, data mining/machine learning techniques come as an aid to understand those issues.

Table 4. Data mining/machine learning techniques used in the BPI challenges 2011–2018.

	2011	2012	2013	2014	2015	2016	2017	2018	Total
Decision trees		1		4	3		3	1	**12**
Logistic regression				2			2	2	**6**
Random forest				1			3	1	**5**
Neural network						1	2		**3**
Linear regression				2					**2**
Support vector machine				1			1		**2**
Sequential pattern mining				1	1				**2**
Sequence classification				1	1				**2**
Naïve Bayes				1				1	**2**
Ada boost				1					**1**
Apriori algorithm				1					**1**
Association rules				1					**1**
Multilayer perceptron				1					**1**
Binary segmentation					1				**1**
K-means clustering						1			**1**
Bayesian networks								1	**1**
Generalized linear model								1	**1**
Deep learning								1	**1**
Gradient boosted trees								1	**1**

4   Conclusion

The BPI challenges have been not only a testbed for process mining techniques, but have also brought many other approaches into the realm of process mining. Besides the analysis of the control-flow, organizational and performance perspectives, the business questions associated with the BPI challenges often require the use of data mining and machine learning techniques.

Examples are the use of decision trees to find the most important factors that influence the process outcome, and the use of neural networks for next-step prediction. Besides supervised machine learning, unsupervised techniques also play key role, especially with the use of clustering as a means of preprocessing to better understand the process and facilitate the analysis of the event log.

Having observed a growing use of data mining in process mining competitions, we expect that, in the future, the use of data mining techniques will become as important as the use of process mining techniques in the analysis of the event logs.

References

1. Bose, R.P.J.C., van der Aalst, W.M.P.: Analysis of patient treatment procedures. In: Daniel, F., Barkaoui, K., Dustdar, S. (eds.) BPM 2011. LNBIP, vol. 99, pp. 165–166. Springer, Heidelberg (2012). https://doi.org/10.1007/978-3-642-28108-2_17
2. Li, J., Bose, R.P.J.C., van der Aalst, W.M.P.: Mining context-dependent and interactive business process maps using execution patterns. In: zur Muehlen, M., Su, J. (eds.) BPM 2010. LNBIP, vol. 66, pp. 109–121. Springer, Heidelberg (2011). https://doi.org/10.1007/978-3-642-20511-8_10
3. Bose, R.P.J.C., van der Aalst, W.M.P.: Trace alignment in process mining: opportunities for process diagnostics. In: Hull, R., Mendling, J., Tai, S. (eds.) BPM 2010. LNCS, vol. 6336, pp. 227–242. Springer, Heidelberg (2010). https://doi.org/10.1007/978-3-642-15618-2_17
4. Bautista, A.D., Wangikar, L., Akbar, S.M.K.: Process mining-driven optimization of a consumer loan approvals process. In: La Rosa, M., Soffer, P. (eds.) BPM 2012. LNBIP, vol. 132, pp. 219–220. Springer, Heidelberg (2013). https://doi.org/10.1007/978-3-642-36285-9_24
5. Kang, C.J., et al.: Process mining-based understanding and analysis of Volvo IT's incident and problem management processes. In: CEUR Workshop Proceedings, vol. 1052 (2013)
6. Buhler, P., et al.: Service desk and incident impact patterns following ITIL change implementation. In: BPI Challenge 2014 (2014)
7. Cacciola, G., Conforti, R., Nguyen, H.: Rabobank: a process mining case study BPI challenge 2014 report. In: BPI Challenge 2014 (2014)
8. van der Ham, U.: Benchmarking of five dutch municipalities with process mining techniques reveals opportunities for improvement. In: BPI Challenge 2015 (2015)
9. Bose, R.P.J.C., van der Aalst, W.M.P., Žliobaitė, I., Pechenizkiy, M.: Handling concept drift in process mining. In: Mouratidis, H., Rolland, C. (eds.) CAiSE 2011. LNCS, vol. 6741, pp. 391–405. Springer, Heidelberg (2011). https://doi.org/10.1007/978-3-642-21640-4_30
10. Teinemaa, I., Leontjeva, A., Masing, K.-O.: BPIC 2015: diagnostics of building permit application process in dutch municipalities. In: BPI Challenge 2015 (2015)
11. Kleinberg, J.M.: Authoritative sources in a hyperlinked environment. In: Proceedings of the 9th Annual ACM-SIAM Symposium on Discrete Algorithms (1998)
12. Weijters, A.J.M.M., van der Aalst, W.M.P., Alves de Medeiros, A.K.: Process Mining with the HeuristicsMiner Algorithm (2006)
13. van der Ham, U.: Marking up the right tree: understanding the customer process at UWV. In: BPI Challenge 2016 (2016)
14. Chen, Y., Argentinis, E., Weber, G.: IBM Watson: how cognitive computing can be applied to big data challenges in life sciences research. Clin. Ther. **38** (2016). https://doi.org/10.1016/j.clinthera.2015.12.001
15. Dadashnia, S., Niesen, T., Hake, P., Fettke, P., Mehdiyev, N., Evermann, J.: Identification of distinct usage patterns and prediction of customer behavior. In: BPI Challenge 2016 (2016)
16. Evermann, J., Rehse, J.-R., Fettke, P.: A deep learning approach for predicting process behaviour at runtime. In: Dumas, M., Fantinato, M. (eds.) BPM 2016. LNBIP, vol. 281, pp. 327–338. Springer, Cham (2017). https://doi.org/10.1007/978-3-319-58457-7_24
17. Veiga, G.M., Ferreira, D.R.: Understanding spaghetti models with sequence clustering for ProM. In: Rinderle-Ma, S., Sadiq, S., Leymann, F. (eds.) BPM 2009. LNBIP, vol. 43, pp. 92–103. Springer, Heidelberg (2010). https://doi.org/10.1007/978-3-642-12186-9_10
18. Povalyaeva, E., Khamitov, I., Fomenko, A.: BPIC 2017: density analysis of the interaction with clients. In: BPI Challenge 2017 (2017)
19. Rodrigues, A.M.B., et al.: Stairway to value: mining a loan application process. In: BPI Challenge 2017 (2017)

20. Blevi, L., Robbrecht, J., Delporte, L.: Process mining on the loan application process of a Dutch Financial Institute. In: BPI Challenge 2017 (2017)
21. Brils, J.H.H., van den Elsen, N.A.F., de Priester, J., Slooff, T.A.: Business process intelligence challenge 2018: analysis and prediction of undesired outcomes. In: BPI Challenge 2018 (2018)
22. Pauwels, S., Calders, T.: Detecting and explaining drifts in yearly grant applications. In: BPI Challenge 2018 (2018)
23. Wangikar, L., Dhuwalia, S., Yadav, A., Dikshit, B., Yadav, D.: Faster payments to farmers: analysis of the direct payments process of EU's agricultural guarantee fund. In: BPI Challenge 2018 (2018)
24. Günther, C.W., van der Aalst, W.M.P.: Fuzzy mining – adaptive process simplification based on multi-perspective metrics. In: Alonso, G., Dadam, P., Rosemann, M. (eds.) BPM 2007. LNCS, vol. 4714, pp. 328–343. Springer, Heidelberg (2007). https://doi.org/10.1007/978-3-540-75183-0_24
25. Song, M., Günther, Christian W., van der Aalst, W.M.P.: Trace clustering in process mining. In: Ardagna, D., Mecella, M., Yang, J. (eds.) BPM 2008. LNBIP, vol. 17, pp. 109–120. Springer, Heidelberg (2009). https://doi.org/10.1007/978-3-642-00328-8_11
26. Song, M., van der Aalst, W.M.P.: Towards comprehensive support for organizational mining. Decis. Support Syst. **46**, 300–317 (2008)
27. Mans, R.S., Schonenberg, M.H., Song, M., van der Aalst, W.M.P., Bakker, P.J.M.: Application of process mining in healthcare – a case study in a dutch hospital. In: Fred, A., Filipe, J., Gamboa, H. (eds.) BIOSTEC 2008. CCIS, vol. 25, pp. 425–438. Springer, Heidelberg (2008). https://doi.org/10.1007/978-3-540-92219-3_32
28. Leemans, S.J.J., Fahland, D., van der Aalst, W.M.P.: Discovering block-structured process models from event logs - a constructive approach. In: Colom, J.-M., Desel, J. (eds.) PETRI NETS 2013. LNCS, vol. 7927, pp. 311–329. Springer, Heidelberg (2013). https://doi.org/10.1007/978-3-642-38697-8_17

First International Workshop on Business Process Management in the Era of Digital Innovation and Transformation: New Capabilities and Perspectives (BPMinDIT)

First International Workshop on Business Process Management in the Era of Digital Innovation and Transformation: New Capabilities and Perspectives (BPMinDIT)

The fundamental nature of many organizations is being rapidly transformed with the ongoing diffusion of digital technologies. In this era, organizations in many domains are challenged to question their existing business models and to improve or revolutionize them using new technologies. Many IT-based initiatives, such as Uber, Car2Go, DriveNow, Udacity, or Airbnb emerged, and disrupted traditional markets by making use of those digital technologies. To stay ahead of their competitors, even ICT giants, such as Google or Amazon, face the need to constantly evaluate and improve the value they offer their customers.

These developments are also challenging the role of business process management (BPM). Advances in data analytics and AI, uptake of new technologies (such as blockchain, IoT, 3D printing), increased adoption of cloud and mobile technologies, and new business paradigms, such as service-dominant logic, open the path for new business processes and new possibilities – or even necessities – for the application of BPM. We see, for example, how automated BPM can be used to tightly link business analytics and business execution in short process-based iterations to follow quickly changing markets, how real-time data from physical entities ('things' in the IoT sense) is directly injected into decision making in business processes, or how agile, IT-reliant business models are directly mapped to executable business processes.

However, the traditional role of BPM in structuring and optimizing (operational) processes often falls short in making use of these opportunities. This can risk the position of BPM to act as the driving force in digital innovation and transformation initiatives. New BPM capabilities that reflect an explorative-dominant (instead of exploitation-dominant) view may help in addressing the emerging opportunities and challenges of digitalization.

In this workshop, we question and investigate the new role of BPM in the digital era. The goal is to advance our understanding of the BPM capabilities that organizations require to explore emerging opportunities of digital innovation and transformation, and cope with related challenges.

In its first edition, we are excited to receive six submissions. Each paper was reviewed by at least three members of the Program Committee. From these submissions, the top three were accepted for presentation at the workshop. These papers feature highly relevant and novel research ideas.

Laue proposes a method for analyzing the potential for process improvement, where the modeling and analysis initiate at the customer's side in the context of their use of a service or product, instead of organization's current internal processes. Imgrund and Janiesch question the sufficiency of contemporary BPM body of knowledge in equipping organizations with the competitive advantages and operational excellence. Their empirical work indicates that companies have started embracing adaptive and context-sensitive management approaches making use of agile methodologies and modular process improvements. Exler, Mendling, and Taudes studied the concept of distance, and report on a case study performed to explore the different types of distance that are relevant for business process transfer projects. Based on the distance dimensions that they have identified, they propose a model to measure the distance in intra-company transfers.

This first edition of the workshop will also feature a keynote talk by Maximilian Röglinger. His talk will question the traditional BPM capabilities in addressing the new challenges of digitalization, and introduce a new BPM capability framework that reflects the needs of the digital age. We hope that the reader will find the selected papers relevant and interesting.

September 2019   Oktay Turetken
                                                                                   Amy Van Looy
                                                                                   Paul Grefen

Organization

Workshop Chairs

Oktay Turetken	Eindhoven University of Technology, The Netherlands
Amy Van Looy	Ghent University, Belgium
Paul Grefen	Eindhoven University of Technology, The Netherlands

Program Committee

Marco Comuzzi	Ulsan National Institute of Science and Technology, South Korea
Peter Fettke	Saarland University, Germany
Paul Grefen	Eindhoven University of Technology, The Netherlands
Andrej Kovacic	University of Ljubljana, Slovenia
Rob Kusters	Open University, The Netherlands
Peter Loos	Saarland University, Germany
Amy Van Looy	Ghent University, Belgium
Monika Malinova	Vienna University of Economics and Business, Austria
Charles Møller	Aalborg University, Denmark
Baris Ozkan	Eindhoven University of Technology, The Netherlands
Hajo Reijers	Utrecht University, The Netherlands
Maximilian Röglinger	University of Bayreuth, Germany
Estefania Serral	KU Leuven, Belgium
Mojca I. Stemberger	University of Ljubljana, Slovenia
Peter Trkman	University of Ljubljana, Slovenia
Oktay Turetken	Eindhoven University of Technology, The Netherlands
Nils Urbach	University of Bayreuth, Germany

Business Process Management Capabilities for the Digital Age (Keynote Abstract)

Maximilian Röglinger

University of Bayreuth, FIM Research Center
Finance & Information Management, Project Group Business
and Information Systems Engineering of the Fraunhofer FIT,
95444 Bayreuth, Germany
maximilian.roeglinger@fim-rc.de

Process orientation is an accepted paradigm of organizational design that drives corporate success. Hence, business process management (BPM), which deals with the implementation of process orientation, receives constant attention in industry and academia. Today, mature methods and tools support all phases of the BPM lifecycle. Apart from lifecycle models, BPM is commonly structured in terms of BPM capability frameworks that transcend the lifecycle's focus on operational process support by accounting for further core elements such as strategic alignment, governance, people, and culture.

Despite their usefulness, BPM capability frameworks are being challenged by socio-technical changes such as those brought about by digitalization. In line with the uptake of novel technologies, digitalization transforms existing and enables new processes. For example, social collaboration platforms facilitate the assembly of teams independently of time and location. Robotic process automation and cognitive automation enable the automation of unstructured tasks, and the Internet of Things enables decentral production processes as well as an immersion of organizations into their customers' processes.

These examples led us to hypothesize that different capabilities are needed in the future than today for BPM to drive corporate success and that existing BPM capability frameworks need to be updated. Hence, we explored which BPM capabilities will be relevant in the future by conducting a Delphi study with international experts from industry and academia. In my keynote, I present the key results of this Delphi study.

The study resulted in an updated framework of 30 BPM capabilities structured according to the core elements of BPM. We also found six overarching topics (i.e., data, individuals, exploration, networks, context, and change) that will be characteristic of BPM in the future. When analyzing the novelty of the identified BPM capabilities, we found that, on the one hand, there will be a strong link between current and future BPM capabilities and, one the other, BPM requires substantial further development as about half of the identified capabilities are not included in extant frameworks.

The Delphi study on future BPM capabilities was a joint research project with Ulrich König and Georgi Kerpedzhiev from my research group at the FIM Research Center Finance & Information Management, Germany, and Michael Rosemann from

Queensland University of Technology (QUT), Brisbane. For more information, please visit our project website (http://www.digital-bpm.com/) and have a look at the short article published by BPTrends (https://www.bptrends.com/business-process-management-in-the-digital-age/).

The Power of the Ideal Final Result for Identifying Process Optimization Potential

Ralf Laue[1,2(✉)]

[1] University of Applied Sciences of Zwickau, Zwickau, Germany
[2] Department of Computer Science, Dr.-Friedrichs-Ring 2a, 08056 Zwickau, Germany
Ralf.Laue@fh-zwickau.de

Abstract. Modern information and telecommunication technologies provide possibilities not only for improving existing business processes. They can also allow to make processes redundant or to move to new business models. This paper suggests a method for analysing the potential for process improvement and organizational change. Instead of starting with a visualisation of the current process, the starting point is a visualisation of customers' steps in the context of their use of a service or product.

1 Introduction

Successful organizations need a strategy for developing and improving processes to maximise their competitiveness. Exploring better alternatives for providing a service or producing a product can lead to performance and quality leadership. Even more, using modern information and telecommunication technologies can provide opportunities to develop completely new business models.

Against this background, it seems logical that organizations take every effort to organize the innovation process and make use of existing creativity techniques in a systematic way. However, a study by Siemon et al. [13] among German start-up entrepreneurs showed that only a minority of them used creativity techniques for generating ideas related to the business model. Another observation was that a surprisingly small number of people (2.2 on average) was involved in idea generation. The author of the paper at hand is convinced that this means that opportunities will be missed.

In addition to general creativity techniques, there are well-established methods for developing business models such as e^3value [5] or the Business Model Canvas [11]. Both focus on the business perspective. For discussing the process perspective, a common approach is to start an improvement project by analyzing a model of the existing business process. An example of such a method is *ValueApping* by Hoos et al [10]. It is an analysis method to identify value-adding usage scenarios for mobile apps in business processes. A more general method has been suggested by Denner et al. [3]. Their method involves managers, business process experts and end users who evaluate the relative importance of each

© Springer Nature Switzerland AG 2019
C. Di Francescomarino et al. (Eds.): BPM 2019 Workshops, LNBIP 362, pp. 281–287, 2019.
https://doi.org/10.1007/978-3-030-37453-2_23

process step, select digital technologies with the potential to improve the process and value the suitability of the preselected digital technologies. Both mentioned approaches share two disadvantages. First, they involve rather complex calculations which give a pretence of accuracy that is in fact not substantiated. However, a more severe disadvantage is that they start with the current processes with the quest to improve them. It is to be afraid that this will rarely lead to the insight that the current process should be completely replaced or that activities outside the current process can lead to new business chances.

This paper suggests an alternative approach that builds on the concept of the Ideal Process which will be described in the next setion.

2   The Ideal Process

The main idea for the approach was taken from the Theory of Inventive Problem Solving (TRIZ) [1]. TRIZ consists of several methods to support inventors in technical domains. One of these methods is the concept of the *ideal ultimate result*. It describes an ideal situation: A result will be achieved "by itself", i.e. without cost, without energy consumption, etc. While such an ideal result obviously cannot be reached, it is anyway a useful thinking tool for an inventor. For example, an inventor who strives to construct a lawn-mower which does *not make any noise at all* will have a good chance to come up with more creative solutions than an inventor who has the aim to produce a lawn-mower that makes *less* noise. In other words: The aim of thinking about the ideal result is to prevent an inventor from restricting his/her way of thinking too early.

In the area of process improvement, the ideal process is – no process at all (the goal is reached "by itself"). For example, a process for checking data for completeness is redundant if the data is already guaranteed to be always complete. To give another example, if a software does not collect any personal data, no processes for deleting or correcting such data at request of the users are necessary.

In contrast to the ideal process (a goal is reached by itself), existing processes have costs, take time, need resources and have an ecological impact. Traditional process improvement methods aim to improve those process parameters and make a process faster, less costly, etc. A possible outcome of such a process improvement is what Hammer and Champy [8] describe as "paving the cow paths". Computers are used to speed up the processes, but the processes are left intact. Having the ideal process in mind, will automatically lead to the question whether the process is necessary at all. An example for the chances of asking this question is discussed by Hammer in [9]: The Ford Motor Company was in need to cut costs for accounts payable. Instead of organizing the accounts payable clerks' work more efficiently, Ford instituted invoiceless processing. In our terminology, this can be regarded as the ideal process (or at least, it comes close).

From the perspective of an enterprise (the process owner), it could be regarded as an ideal process if the business model "customer self-service" or "automated service" [11] is applied. However, in fact the goal is not achieved "by itself", the burden to achieving it is just transferred to the customer. Badly designed self-service processes can make achieving a goal more difficult for the customer. For this reason, it is imperative that processes are also evaluated from the customers' perspective. The same applies to the employees' perspective. In the short term, it will often be possible to make a process cheaper by forcing the employees to work harder. In the long term, such a policy can destroy the organization. Finally, one more perspective to consider is the ecological perspective which discusses possible negative impacts of a process to the environment.

3   The Method

The method that will be suggested in this paper owns a lot to the idea of the Interaction Room for Digitalization Strategy Development [6]. A key idea of the Interaction Room is to depict the business of an organization in a systematic, but not too restrictive way. The business model is visualized at several canvas using a lightweight notation with visual annotations. This visualization provides a starting point for discussion. One of those canvas is the "touchpoint canvas". It uses customer journey maps which show the order in which a costumer uses different channels in engaging with the organization. Those interactions are usually referred to as customer touch points (or contact points). Customer journeys have been used successfully for understanding the customers' perspective and for improving the service quality [7].

However, analysing existing customer journeys allows only insight into the process "as is". It can be difficult to spot a chance for providing additional services. Therefore, it is useful to embed the observation of touch points into a more general description of the steps that a customer takes in order to reach a certain goal. The activities that can be registered as customer touch points, i.e. those activities where the customer interacts with the process-owner, are usually activities for reaching only a sub-goal. For reaching the main goal, additional activities (which we will call non-touch points) can be necessary.

To give an example, we take the perspective of a hotel owner and discuss the activities of hotel guests. One such guest may need accommodation during a bicycle tour, a selection of his activities is listed in Table 1. For other guests, belonging to different customer groups such as "business traveller", other scenarios will need to be developed. For this reason, *Step 1* in our method is to develop

Table 1. Customer touch points and Non-touch points (Examples)

Touch points	Non-touch points
check-in at the hotel	Use map to find the hotel
ask for a place to lock the bike securely	buy food supply for next day
find room	get massage outside the hotel
open room	bicycle maintenance

Table 2. Visual Annotations for Deviations from the Ideal Process

	error-prone		possible misuse
	expensive *(can be both customer perspective as process owner perspective)*		resource-demanding (non-human resources)
	resource-demanding (human resources)		time-consuming *(can be both customer perspective as process owner perspective)*
	complicated / difficult / many variants		environmental impact
	information disclosure *(customer perspective)*		hard physical work
	dangerous for human workers		tendious / monotonous tasks
	psychological / mental stress		excludes certain groups of people

personas [2] serving as examples for members of common customer groups. *Step 2* is to create what is known as "scenario" in the field of requirements engineering: a description of one out of many possible flows through a process for this person. In *Step 3*, all these activities (contact points as well as non-contact points) will be written on sheets of paper, each representing an activity. Deliberately, we do not use any formal business process modeling language such as BPMN in order to minimize comprehensibility problems.

Step 4, is the core of our method. It should be executed by a group of people, if possible containing members with different experiences and backgrounds. Their task is to visualize *remarkable deviations* of the activities in the scenario under discussion from the ideal process. It is important that all participants understand very well what this means: The objective is not to mark those activities for which one thinks that an improvement is possible or necessary – the objective is to mark activities that show a remarkable deviation from the ideal process (where the goal is reached without costs, without taking time, without using resources, etc.) Visual markers as shown in Table 2 are used for this purpose.

For considering all four perspectives (process owner, customer, employees, environment), two variants are possible. Either, it is the responsibility of selected group members to look at the process from a certain perspective. Alternatively, Step 4 can be executed in four rounds, where in each round all participants take another perspective.

The photo in Fig. 1 shows a selection of activity cards for the hotel example with visual annotations (collected at the International Workshop on Creativity

Fig. 1. Examples of activity cards with visual annotations

in Requirements Engineering 2018). In this example, we see a few noteworthy points: The activity "Ride" does not have any stickers attached to it. While it is obviously time-consuming and effortful, this is not regarded as a problem – the physical exercise is the very reason for undertaking a bike tour. On the other hand "Lock Bike at Safe Place" has a "Time" sticker attached which states that it would be perceived as an improvement if the guest would not have to care about locking the bike at all. "Check-In" has two "Time" stickers, because the check-in activity is regarded as time-consuming both from the perspective of the guest as from the perspective of the employees.

Taking these annotations as input, suggestions for process improvements and additional services can be discussed in *Step 5*: In our example, the activity "ask for a safe place for the bike" is regarded as being time-consuming from the customers' perspective. So maybe "ride in and we will take care of your bike" could be a service for which the customer would be willing to pay. The process of checking-in and entering the room is costly (from the management perspective) and time-consuming (from the customer's perspective). So we can ask why something as check-in is necessary at all? Maybe a guest can use a mobile phone for being registered, being guided to the room, unlock it and pay for the stay in the hotel? Indeed, the use of mobile devices in the pre-trip, on-site and post-trip phase has a huge potential that deserves to be analysed carefully [4]. Of course, not each of such questions will lead to new services or process improvements in reality. However, most likely there will be useful suggestions. Before implementing the improved new process, it can be helpful to discuss the suggested improved process using the proposed method again. This can help to identify potential problems in the improved process. For example, the new process could force the guests to use a smartphone for certain activities. This could result in privacy issues, a possible exclusion of persons who do not own such a device (or simply would have to charge the batteries of their smartphone before) and also in making the task for the guests more difficult.

It has to be noted that the conclusions are valid only for a specific kind of scenario / customer (in our example: the bicycle tourist). Other customers can have different needs which will be addressed by building user stories for different personas. Anyway, if there are considerable many guests who belong to a certain customer group (such as bicycle travellers in our example), it can be a good idea to provide additional services or to change processes according to their needs.

4 Conclusion

Adding visual annotations to depict deficiencies in process steps has been suggested by others as well. For example, Polderdijk et al. [12] provide a BPMN extension to depict human physical risks in manufacturing processes. However, compared to existing work, the paper at hand suggests three new aspects: First, the method forces its users to look at the processes from four perspectives. This tries to avoid solutions that may save time and resources for the enterprise but make life harder for either the customers or the staff. Second, the TRIZ-thinking with respect to the ideal process encourages the participants to put into question whether existing solutions can be replaced by something completely different. And third, the concept of "non-contact points" aims to look for business chances that are currently without the scope of the current organization.

In the future, we are keen to gain experience by using and evaluating the proposed method with scenarios from various domains.

References

1. Altshuller, G.S.: The Innovation Algorithm. Technical Innovation Center, Blaine (1999)
2. Cooper, A., Reimann, R., Dubberly, H.: About Face 2.0: The Essentials of Interaction Design. John Wiley, New York (2003)
3. Denner, M., Püschel, L., Röglinger, M.: How to exploit the digitalization potential of business processes. Bus. Inf. Syst. Eng. **60**(4), 331–349 (2018)
4. Douglas, A., Lubbe, B., van der Merwe, A.: Managing business travellers' use of mobile travel applications. Inf. Commun. Technol. Tourism **2017**, 271–283 (2017)
5. Gordijn, J., Yu, E.S.K., van der Raadt, B.: E-service design using i* and e^{3}value modeling. IEEE Softw. **23**(3), 26–33 (2006)
6. Gruhn, V., Book, M., Striemer, R.: Tamed Agility: Pragmatic Contracting and Collaboration in Agile Software Projects. Springer-Verlag, Switzerland (2016). https://doi.org/10.1007/978-3-319-41478-2
7. Halvorsrud, R., Kvale, K., Følstad, A.: Improving service quality through customer journey analysis. J. Serv. Theory Pract. **26**(6), 840–867 (2016)
8. Hammer, M., Champy, J.: Reengineering the Corporation: A Manifesto for Business Revolution. Harper Business Books, New York (1993)
9. Hammer, M.: Reengineering work: Don't automate, obilerate. Harvard Business Review July/August 1990
10. Hoos, E., Gröger, C., Kramer, S., Mitschang, B.: ValueApping: an analysis method to identify value-adding mobile enterprise apps in business processes. In: Cordeiro, J., Hammoudi, S., Maciaszek, L., Camp, O., Filipe, J. (eds.) ICEIS 2014. LNBIP, vol. 227, pp. 222–243. Springer, Cham (2015). https://doi.org/10.1007/978-3-319-22348-3_13
11. Osterwalder, A., Pigneur, Y.: Business Model Generation: A Handbook for Visionaries, Game Changers, and Challengers. John Wiley, Hoboken (2010)

12. Polderdijk, M., Vanderfeesten, I., Erasmus, J., Traganos, K., Bosch, T., van Rhijn, G., Fahland, D.: A visualization of human physical risks in manufacturing processes using BPMN. In: Teniente, E., Weidlich, M. (eds.) BPM 2017. LNBIP, vol. 308, pp. 732–743. Springer, Cham (2018). https://doi.org/10.1007/978-3-319-74030-0_58
13. Siemon, D., Narani, S.K., Ostermeier, K., Robra-Bissantz, S.: Creativity and entrepreneurship - the role of creativity support systems for start-ups. In: 10th Mediterranean Conference on Information Systems, MCIS 2016 (2016)

Understanding the Need for New Perspectives on BPM in the Digital Age: An Empirical Analysis

Florian Imgrund[1]([⊠]) and Christian Janiesch[2]

[1] University of Würzburg, 97070 Würzburg, Germany
florian.imgrund@uni-wuerzburg.de
[2] TU Dresden, 01187 Dresden, Germany
christian.janiesch@tu-dresden.de

Abstract. The emergence of digital technology is substantially changing the way we communicate and collaborate. In recent years, groundbreaking business model innovations have disrupted industries by the dozen, shifting previously unchallenged global players out of the market within shortest time. Although business process management (BPM) is often identified as a main driver for organizational efficiency in this context, there is little understanding of how its methods and tools can successfully navigate organizations through the uncertainty brought by today's highly dynamic market environments. However, we see more and more contributions emerging that question the timeliness of BPM due to its lack of context sensitivity. In this context, the inflexibility and over-functionalization of hierarchical management structures is often referred to as the primary reason why organizations fail to achieve the flexibility, agility, and responsiveness needed to address today's entrepreneurial challenges. In this research paper, we question whether the contemporary BPM body of knowledge is still sufficient to equip organizations with the competitive advantages and operational excellence that have long yielded sustainable growth and business success. In fact, our empirical observations indicate that the vertical management of functional units inherent to current BPM is increasingly being replaced by adaptive and context-sensitive management approaches drawing on agile methodologies and modular process improvements. From a total of 17 interviews, we derive five criteria that the respondents consider as essential to strengthen the position of BPM in the digital age.

Keywords: Business process management · Status quo report · Digital transformation · Decentralized BPM · Empirical study

1 Introduction

The advent of the digital age, which marks our current historic period built on digital technology, embeds companies in a highly dynamic and fast-paced market environment that is changing the nature of how we conduct business substantially [1]. Following debates in research and practice, both practitioners and researchers are well aware of the groundbreaking characteristics of digital technology [2]. In fact, more and

© Springer Nature Switzerland AG 2019
C. Di Francescomarino et al. (Eds.): BPM 2019 Workshops, LNBIP 362, pp. 288–300, 2019.
https://doi.org/10.1007/978-3-030-37453-2_24

more industries are witnessing digitally enabled business models arising, whose dominance is disrupting them by the dozen. Typically, these companies' superiority fundaments in modern management principles based on agile development cycles [3]. Contrary, however, numerous scholars emphasize that current principles for managing internal structures haven't changed appropriately, lacking the timeliness and responsiveness to remain competitive in today's fast changing market environment [4, 5]. Instead of embracing technology-driven coordination and collaboration to boost operational performance, respective approaches including Business Process Management (BPM) still stick to the vertical management of an organization, relying on a subdivision of power between functional units [6]. To challenge the hypothesis that contemporary BPM becomes increasingly outdated and must reposition itself in order to regain relevance for organizations embarking on their digital transformation journey, we outline the following research questions (RQ).

1. *How do today's organizations deploy BPM from an organizational and technological point of view?*
2. *Which criteria do organizations consider to be essential for the success of BPM in the digital age?*

To answer these RQ, we conducted a qualitative survey involving 17 practitioners from 16 companies. We structure our research as follows. In Sect. 2, we challenge the actuality of BPM by examining its historical evolution and thereby conceptualize its need for new capabilities and perspectives, before we provide insights to the underlying research methodology. In Sect. 3, we give insights on how BPM is currently being deployed in practice, before we derive five criteria that proved to be essential for BPM to remain compatible with the challenges and opportunities of the digital age (Sect. 4). In Sect. 5, we conclude our research with a summary of findings, limitations, and future research opportunities.

2 Background and Research Design

2.1 Historical Evolution of BPM

To assess the timeliness of BPM for the digital age, it is interesting to put current organizational challenges into perspective with information technology developments of recent years. Historically, it was sufficient for organizations to focus on the efficient implementation of physical work based on clear-cut and structured processes. Triggered by the emergence of digital technology and the infusion of the Internet, however, value chain networks increasingly evolved dynamically and organizations started to experience severe limitations due to over-specialization and a lack of overall process control [7]. As a consequence, organizations abandoned from functional approaches toward process-oriented concepts and adapted practices recommended by Total Quality Management, Six Sigma, and Lean Management. Supported by software tools, these practices incorporate today's BPM and provide organizations all over the world with the ability to identify, analyze, monitor, and improve business processes at new levels of efficiency and automation [8].

In recent years, however, an increasing number of practitioners reporting on project failure indicates the approach's deficiency against today's various application contexts and use cases [9, 10]. Due to its formal and hierarchical structures built on control and coercion, the current BPM body of knowledge inherently prioritizes and focuses on processes that maximize their business's profit most, neglecting others whose management does not show a positive proportion of expected surpluses [11].

That is, despite the ongoing diffusion of digital technology that puts organizations in highly dynamic and disruptive market environments, contemporary BPM still seeks to maintain control, predictability, and efficiency of only a limited number of corporate processes by the downward integration of functional managers, process owners, and operational staff [8].

2.2 Decentralization of BPM in Extant Literature

Lacking the responsiveness and flexibility of timely approaches that outsource roles and responsibilities, we argue that hierarchically implemented BPM is no longer synonym to operational efficiency and entrepreneurial advantage [9]. In order to overcome the costly allocation of central resources of contemporary BPM, which is often referred to as its main bottleneck [8, 11], we claim that organizations need to implement an ambidextrous approach to BPM, in which they implement process management projects either centrally or local, depending on where the process's responsibilities and resources are primarily embedded. This also requires BPM to find new ways in challenging and managing organizational complexity by drawing on decentralized resources and leveraging communication, coordination, and co-creation among stakeholders.

Having thoroughly reviewed extant literature to this end, we could identify only three research streams within BPM that are concerned with finding effective ways to re-position its role beyond structuring and optimizing clear-cut and previously prioritized processes. Table 1 lists the identified research streams that we either perceive as fundamental for establishing an explorative view on BPM, or that bear pioneering thoughts, ideas, or concepts to increase the scope of its activities. Though effective, however, these efforts have always been tailored to very specific use cases and implementation scenarios, failing to provide a comprehensive application framework or success factors that put all the various facets of adaptive and context-sensitive BPM into perspective.

2.3 Research Method

We use the grounded theory method (GTM) [17] to evaluate our RQ as its methodology supports the inductive discovery of a yet not sufficiently researched topic that is grounded in data. Further, GTM allows us to focus on data collection and the exploration of new theory instead of validating existing implications [18]. That is, GTM successfully guides us in "accessing other people's interpretations, filtering them through their own conceptual apparatus, and feeding a version of events back to others" [19], p. 77. In the following, we give rationale for *interviewee selection* and describe how we *collected and analyzed data*.

Table 1. Overview of topics and themes for decentralized BPM.

Topic	Description	Example source
Agile BPM	Encompasses all activities that embed modular and asynchronous procedures in process management yielding fast-paced and early project results	e.g. Bruno, et al. [12] and Thiemich and Puhlmann [13]
Collaborative/social BPM	Focuses on implementing social features into BPM and seeks to exploit commons-based peer production	e.g. Brambilla, et al. [14] and Schmidt and Nurcan [15]
Context-sensitive BPM	The deployment of process management in broader business contexts and application scenarios in which organizations cope with unpredictable, knowledge-intense, and often cross-organizational business processes	e.g. vom Brocke, et al. [9] and Rosemann and Recker [16]

Rationale for Interviewee Selection. Consulting the principles of replication logic [20], we selected a total of 12 domain experts from 11 large and medium-sized companies as an initial set of observations. Hereafter, we validated resulting implications by involving a second group of observations consisting of 5 domain experts from 5 different companies. While informants in the first group were employed as BPM experts or an equivalent role in full time, informants in the second group acted as consultants with reasonable expert knowledge. Every organization involved is based in Germany and operates on complex business structures and highly interwoven processes, simultaneously relying on BPM to manage organizational complexity.

Data Collection and Analysis. In line with Yin [21], we started our studies with two precise research questions. Due to the nascent and explorative nature of our research, we used a semi-structured questionnaire as a guideline for our interviews, which allowed us to transport a loose idea of our interest, yet letting emerge and evolve our overall objective throughout the interviews. To analyze audio recordings, we manually categorized the data using open coding preceding to axial coding, enabling us to identify patterns and variances within each interview [22]. Throughout the coding process, we iterated within- and cross-case analysis and constantly validated and cross-checked our finding with extant literature until we reached a state of saturation, i.e. the return of adding data became marginal and neglectable [20].

3 Implementation of Business Process Management in Practice

3.1 Overview of Data Sources and Focus of This Research

In this section, we uncover the current status quo of how organizations deploy BPM in practice and, thereby, answer RQ1. We position our findings along the *organizational*

Table 2. Summary of data sources and company details

Company#/ informant#	Company size	Industry type	Exact job title as defined by their organization	Years of experience
C1/I1	~2,000	Biotechnology	Manager BPM	10+
C2/I2	~6,000	Confectionery	Process Manager	2 to 5
C3/I3	~1,000	Food & non-food retail	Process Manager	5 to 10
C4/I4	~1,000	Chemicals	Process Manager	2 to 5
C5/I5	~16,000	Insurances	Process Consultant	2 to 5
C6/I6	~14,000	Telecommunications	Senior Expert IT Solutions Manager	1 to 2
C6/I7			Customer Experience Manager	10+
C7/I8	~18,000	Retail	Head of Process Management	5 to 10
C8/I9	~2,600	Bank	Manager Process Management	5 to 10
C9/I10	~80,000	Manufacturing	Business Process Manager	1 to 2
C10/I11	~8,000	Telecommunications	Expert BPM	5 to 10
C11/I12	~5,000	Manufacturing	Head of Process Management	10+
C12/I13	~500	Software vendor	Territory Sales Manager	2 to 5
C13/I14	~100	Software vendor	Consultant	10+
C14/I15	~450	Service provider	Managing Director	10+
C15/I16	~2500	Management consulting	Associate Partner	10+
C16/I17	~35	Consulting	Managing Director	10+

and *technological* implementation of BPM. As shown in Table 2, our implications draw on a diversified sample of informants (I) and companies (C), the latter of varying sizes and industries.

3.2   Organizational Implementation of BPM in Practice

Although we did not limit our talks with practitioners to the information given in Table 3, our informants suggested and commonly agreed that these characteristics are crucial for implementing BPM from an organizational perspective. The characteristics emerged from coding the transcribed interviews (cf. Sect. 2.3) and were not predetermined by the interview-guideline or the interviewer. In Tables 3 and 4, the average symbol (Ø) indicates that consultants (C12–C16) reported from aggregated knowledge obtained in multiple projects with different clients.

Management Support and Central BPM. To manage organizational complexity as efficiently as possible, BPM is most often deployed in a central department or authority, often known as BPM "Center of Excellence" (CoE). Due to the central allocation of resources and responsibilities, however, a BPM CoE requires high management attention and the availability of adequate funding. A CoE usually has a clear focus on the management of a limited number of typically high-valued business processes, which are specified by the company's process architecture and whose management is typically accompanied by comprehensive governance standards, roles, and accountabilities (C1, C4, C8, C14). If companies lack management support for BPM (C2, C3, C5, C10), it is hardly possible to establish a CoE. In the case of C5, this was only possible due to the high prestige of the IT-department, which, despite low budgets, operated the CoE on their own expenses and responsibility. C5's informant additionally reported strong conviction and involvement of local stakeholders. Across all companies, informants confirmed that a CoE should have a primary focus on managing processes and responsibilities that are of major strategical importance to the company's success.

Distributed BPM and Direction of Action. Every company included to our sample relied on outsourcing roles and responsibilities to local stakeholders and an increasing number of organizations abandoned from authoritative and hierarchical leadership styles toward the vertical management of corporate processes (C5–C7, C9, C13). In this context, the direction of action describes whether a company deploys BPM top-down ($\downarrow$), bottom-up ($\uparrow$), or in a combination of both approaches ($\leftrightarrow$). According to Table 3, six companies complemented hierarchically implemented BPM with responsibilities at local levels (vertical management), three companies managed processes exclusively in local initiatives without any central supervision, and 7 companies reported of hierarchically implemented BPM. Our informants indicated that the vertical and bottom-up organized management of corporate processes assumes the operation of lean management structures, i.e. the adaptation of Six Sigma (C5, C6, C10) or equivalent principles (C2, C3, C11). Generally, informants stated lean management principles as a very successful means to tackle organizational complexity while not overstraining resources in central authorities. That in mind, all companies embraced the idea of decentralizing BPM activities to local roles and responsibilities. Further, informants stated that local responsibilities should still operate in a more rigid superstructure of common objectives and regulations embodied by a CoE. In this case, numerous informants (C2, C6, C8–C10) strongly recommended the CoE to pursue a servant leadership style and, thus, to guide and not to command its networked structure. Observing a company's individual degree of decentralization, we found that this is considerably affected by the company's strategy toward BPM. Process automatization indicates less decentralization including an organizational setup in which a CoE usually pulls knowledge from local stakeholders without delegating responsibilities. Companies seeking to standardize or document processes still supervise local BPM but tend to delegate full responsibility for local process management to its respective departments. Eventually, deploying BPM to foster collaboration and communication as C10 does, it seems advisable to minimize central responsibilities and supervision where possible to achieve a high degree of self-organization and autonomy among local dependencies.

Central Governance and KPI. In hierarchical BPM, organizations usually limit their process optimization projects to a subset of high-value processes whose management is expected to generate most benefits compared to respective costs. In this case, BPM is subject to a central governance approach including key performance indicators (KPI) that define process-specific goals to evaluate the alignment of strategical objectives and operational business on a regular basis. While hierarchically implemented architectures and governance mechanisms are particularly suitable for organizations that give full control of their corporate processes to a small team of process analysts, our empirical observations suggest that this is no longer viable against the backdrop of today's organization's business structures becoming increasingly complex. Instead, informants reported that their companies either recognized the need to outsource BPM responsibilities to local resources (C1, C4, C12) or are already undergoing transformation initiatives or projects to do so (C8, C13, C15). Further, informants emphasized the necessity for universally valid governance mechanisms for all involved stakeholders, as well as the central availability of outcomes, best practices, and project testimonials.

3.3   Technological Implementation of BPM in Practice

To productize BPM, it is essential to possess technology-enabled tools and services, whose capabilities and benefits are consciously identified, communicated, and made available to relevant stakeholders [8]. Along with our informants, we identified the characteristics presented in Table 4 as crucial parameters to inform about the impact of IT on the renovation and continuous improvement of business processes.

Table 3. Characteristics of organizational implementation of BPM in practice

Company	C1	C2	C3	C4	C5	C6	C7	C8	C9	C10	C11	C12	C13	C14	C15	C16
Management support	high	low	low	high	low	high	high	high	high	low	high	Ø high	Ø high	Ø high	Ø high	Ø high
Central BPM	yes	no	no	yes	yes	yes	yes	yes	yes	no	yes	Ø yes	Ø yes	Ø yes	Ø yes	Ø yes
Distributed BPM	yes	yes	yes	no	yes	yes	yes	yes	yes	yes	yes	Ø yes	Ø yes	Ø yes	Ø yes	Ø no
Direction of action	↓	↑	↑	↓	↔	↔	↔	↓	↔	↑	↔	Ø ↓	Ø ↔	Ø ↓	Ø ↓	Ø ↓
Central governance	yes	no	no	yes	yes	yes	yes	yes	yes	yes	yes	Ø yes	Ø yes	Ø yes	Ø yes	Ø yes
Usage of KPI	yes	no	no	yes	yes	yes	yes	no	yes	no	yes	Ø no	Ø yes	Ø yes	Ø yes	Ø no

Availability of PAIS and Modeling Software. Informants across all companies reported their corporate system infrastructure to include process-aware information systems (PAIS). Although highly appreciating this fact, interviewees suggested that this is mainly due to today's software tools being increasingly interface-based and workflow-oriented by default. Further, virtually all companies (C1–C11, C14, C15) operate process modeling using a Business Process Management Suite (BPMS) possessing domain-specific expert functionalities. It is interesting to mention in this context that none of the companies relies on custom software. Instead, there is a clear tendency toward software as a service (SaaS) of which informants particularly highlighted its simple integration into existing system infrastructures and its high levels of usability and availability as beneficial. There are, admittedly, some companies (C2, C6, C9) that

Table 4. Characteristics of technological implementation of BPM in practice

Company	C1	C2	C3	C4	C5	C6	C7	C8	C9	C10	C11	C12	C13	C14	C15	C16
Availability of PAIS	yes	yes	yes	yes	yes	yes	yes	yes	yes	yes	yes	Ø¹ yes	Ø yes	Ø yes	Ø yes	Ø no
Modeling software	yes	yes	yes	yes	yes	yes	yes	yes	yes	yes	yes	Ø no	Ø no	Ø yes	Ø yes	Ø yes
Global process repository	yes	no	yes	yes	yes	yes	yes	yes	no	yes	yes	Ø no	Ø no	Ø no	Ø yes	Ø no
Integration with extant IT	high	low	low	high	low	high	high	high	high	high	high	Ø low	Ø high	Ø low	Ø high	Ø low
IT-enabled project management	yes	yes	yes	no	no	yes	yes	no	yes	yes	yes	Ø no	Ø yes	Ø no	Ø yes	Ø no

adapted their BPMS to their individual needs. Despite most often included to the BPMS, neither the informants nor their colleagues used any expert functionalities yet. Likewise, although several companies' BPM experts possess theoretical and practical understanding for advanced topics for BPM such as process mining (C3, C6, C7), so far there was neither capacity nor support to implement corresponding projects despite their recognized usefulness.

Global Process Repository and Integration with Related IT Solutions. Most of BPMS offer a global process repository that can be easily connected to a company's present infrastructure. According to informants, the ease of integrating software to the company's existing IT-infrastructure is crucial for top management representatives to award its contract. A *highly* integrated IT infrastructure in this context means that its components are seamlessly integrated and work smoothly with the existing systems in mutual interaction (C1, C4, C6–C9, C10, C13, C15). *Low* integration refers to systems within an IT infrastructure that are incapable to operate reciprocally (C2, C3, C5, C12, C14, C16).

IT-Enabled Project Management. Similar to the functional implementation of BPM, related projects including process discovery have long been subject to hierarchically implemented workshops, interviews, and analysis of work. As a result, outcomes are quite often rather aligned to senior executive's or management's expectations instead of its actual stakeholders' or even the distinct process's needs. The ongoing digitalization, however, embeds organizations in far more complex business structures than ever before. Hence, organizations increasingly adopt IT-enabled project management software whose tools accumulate different stakeholder's inputs, opinions, and knowledge. Informants confirmed that respective tools offer significant value in accessing implicit knowledge quickly and efficiently. However, they also admitted that explicating implicit knowledge is often subject to high effort getting users to participate and, thus, to explicate their process-related knowledge in a structured way.

4   Essential Criteria for the Success of BPM in the Digital Age

Although numerous studies already engaged with the deployment of BPM in practice [23, 24], only little effort is being made to bring these insights into context with the struggles organizations currently face with managing organizational complexity.

Therefore, this study seeks to shed light on the shortcomings of functionally driven management approaches that propagate authority and predictability in hierarchically implemented management structures. At the same time, we aim to conceptualize the need for today's BPM to take on new capabilities and perspectives, especially toward its explorative and context-sensitive usage.

In the following, we introduce to five criteria that emerged as essential for BPM to revitalize its position and to become a driving force for mastering digital innovation and transformation. Each of the 17 practitioners that we interviewed possesses extensive knowledge and experience in the researched field (cf. Table 2). Although their companies might not yet have supported one or more of the distinct criteria to follow, each interviewee confirmed each criteria's full validity for his or her company if implemented appropriately. We iteratively discovered the criteria as they emerged from the interview data and further ensured their viability by deploying within- and cross-case analysis [25]. Ultimately, we ensured the criteria's validity, reliability, and ability to withstand future testing due to cross checking of findings in workshops with practitioners not included to our initial sample. The isolated consideration of the criteria can indeed pick-up common facts and insights already incorporated to contemporary BPM. Yet, it is their collective application that bears the potential to revitalize BPM with new the capabilities and perspectives required to address the emerging opportunities and challenges imposed by the digital age.

Criteria 1: Significantly increase the scope of manageable processes within your organization by attaching local roles and responsibilities to hierarchically implemented BPM. To cope with organizational complexity, organizations seek to expand the scope of manageable processes where possible. In this context, our interviewees strongly suggested to establish local roles and responsibilities as an enhancement to hierarchically implemented BPM, which still operate in a more rigid superstructure of common objectives and regulations embodied by the central BPM authority. This does not imply, however, that organizations can dispense with the hierarchical management of high-value processes and dramatically reduce their resources allocated in a central BPM authority. Instead, they are well advised to ensure a seamless transition and integration of roles and responsibilities between central and local dependencies. Further, interviewees suggested to continuously extend the scope of BPM instead of rushing or enforcing local process improvements due to impatience and zest for action. Hence, organizations can only increase the scope of BPM by gradually deploying and adding key BPM roles and responsibilities in local departments. In this context, a multidirectional communication plan that includes regular meetings and workshops, involves dedicated employees as multipliers, and enables an ongoing bilateral transfer of knowledge and expertise among stakeholders proved to be beneficial.

Criteria 2: Ensure involvement and empowerment of local responsibilities. The implementation of criteria 1 entails the availability of both central and local responsibilities, which means the existence of authoritarian (centrally located) and self-regulative leadership styles (locally) at the same time. Despite their simultaneous existence, we do not postulate these ambidextrous management practices to be beneficial, especially when rivaling each other. Instead, we aspire a seamlessly integrated approach whose stringent implementation and authoritative elements diminish as local roles become increasingly anchored in the organization's structure, finally yielding high degrees of self-regulation, cooperation, and collaboration in local units. Especially when beginning to increase the scope of BPM, however, it will hardly be feasible to fully avoid ambidextrous structures or rivaling leadership styles at any time. Nevertheless, if present, these should be systematically eliminated without haste, until a state of full interconnectedness of stakeholders is reached. In case of the long-term existence of mutually hindering structures, however, this points to missing or non-regulated responsibilities most often located at strategically important areas with high levels of operational responsibility. Any of these, in turn, need to be addressed timely. In this context, it is pivotal to identify committed stakeholders who possess the empowerment and resilience to promote local process optimization, even if the success of their actions is not apparent at any time. To achieve high involvement of stakeholders, informants suggested to foster a can-do attitude that advocates and appreciates personal responsibility, commitment, and autonomy at all levels.

Criteria 3: Follow a consistent governance approach and ensure coordinated knowledge transfers. Our informants strongly suggested that organizations should deploy a corporate governance standard that is authoritative to stakeholders involved to BPM without modifications. This ensures consistent quality standards and interchangeability of results across central and local instantiations of BPM. Nevertheless, local governance should aspire to govern itself through being promoted and implemented by executives in its immediate environment. Local governance adaptations should be strictly avoided, and non-compliant material should not be accessible in organization-wide repositories or platforms. Otherwise, individual process knowledge that is explicitly or implicitly available in local departments might emerge without system and order yielding inconsistencies and inefficiencies instead of being transparent, manageable, and searchable.

Criteria 4: Establish lean management practices to enable adaptive and modular BPM. Assessing the maturity of BPM initiatives from an expert perspective, organizations that pioneer lean thinking and bring agile business practices into focus manage their business remarkably more efficient than organizations that stick to conventional management practices. Consequently, we argue that organizations should attach particular importance to transparency, interconnectedness, and the empowerment of employees to exploit the benefits of communication, collaboration, and peer-production among stakeholders [5]. However, with regard to change management, promoting autonomy and self-responsibility, as well as agile working methodologies should not fully or unexpectedly disrupt an organization's structure. Contrary, over-pacing decentralization can lead to ambidextrous systems or leadership styles as management might obstruct respective activities, feeling to give too much control out of hand overnight, according to our informants.

Criteria 5: Ensure the education and training of stakeholders and pursue change management. Employees involved in an ambidextrous approach to BPM no longer exclusively deal with implementing process management projects. Instead, they are key drivers for enacting organizational change and, thus, are the primary reason whether the overall approach unfolds or suffocates. Thus, besides building basic and advanced knowledge for BPM-related topics, education and training of stakeholders should raise awareness for the cultural transformation that organizations usually undergo following an ambidextrous approach. Therefore, key stakeholders should act as multipliers and take responsibility by disseminating their values and expectations across the organization, simultaneously eradicating political obstacles. Further, according to our informants, this encourages involved employees to perceive ownership of the change process, resulting in higher commitment and increased probability of the project's success. Finally, management must ensure the approach's stakeholders not to fear change but to embrace process optimization as an opportunity to learn, innovate, and strengthen one's own position.

5 Conclusion

Lacking the responsiveness and flexibility of timely approaches that outsource roles and responsibilities, our empirical observations suggest that contemporary BPM is no longer synonym to operational efficiency and entrepreneurial advantage. Instead of expanding its inherent focus on core processes toward a more holistic and universal application including the ability to implement context-specific and modular process optimization, BPM still sticks to the downward integration of functional managers, process owners, and operational staff. Based on qualitative insights of 17 interviews, we sought to conceptualize the new role of BPM in the digital age, which we see as providing the means for digital enterprises to successfully re-group themselves around value-creating processes or services whose highly adaptable and modularizable outputs are created from small and cross-functional teams [26]. As our main result, we discovered five criteria that organizations can follow to move from deploying an inside-out approach to BPM toward facilitating opportunity-driven, proactive, and context-sensitive process improvements. The criteria suggest running BPM on a networked structure of local resources in order to enable modular, adaptive, and asynchronous process improvements. This entails for BPM to embrace agile working methodologies, a culture of involvement, and the establishment of a can-do attitude that appreciates personal responsibility and autonomy.

Eventually, no research is without limitations and nor is the contribution of this paper. On the one hand, deploying BPM as recommended in this paper requires substantial changes of organizational and cultural structures, which need time to manifest and that neither inherently nor instantly amortize the confidence and resources required from the very beginning. Thus, before productizing the decentralization of BPM, we highly recommend to further evaluate the approach toward unforeseen political and cultural challenges. On the part of research, this could be realized by piloting and evolving the criteria in a medium-sized company, leveraging the project's results and learnings as a valuable contribution for future research. On the other hand, we did not reflect the criteria's generalizability beyond European organizations and, thus, cannot judge on their efficiency in varying organizational setups that are subject to different corporate cultures or external market conditions. Besides the need for the criteria's further evaluation, we recommend future contributions to provide precise and accurate design principles and features to properly guide the practical implementation of decentralized BPM. Once this has been achieved, the effectiveness of decentralized BPM has to be evaluated based on long-term perspectives and studies.

References

1. Matt, C., Hess, T., Benlian, A.: Digital transformation strategies. Bus. Inf. Syst. Eng. **57**, 339–343 (2015)
2. Legner, C., et al.: Digitalization: opportunity and challenge for the business and information systems engineering community. Bus. Inf. Syst. Eng. **59**, 301–308 (2017)
3. Rigby, D.K., Sutherland, J., Noble, A.: Agile at scale. Harvard Bus. Rev. **96**, 88–96 (2018)
4. Hamel, G.: The Future of Management. Hum. Res. Manage. Int. Digest **16**, 1–8 (2008)
5. Tapscott, D., Williams, A.D.: Wikinomics: How Mass Collaboration Changes Everything. Penguin, New York (2008)
6. Becker, J., Kugeler, M., Rosemann, M.: Process Management: A Guide for the Design of Business Processes. Springer, Berlin (2013)
7. Bandara, W., Indulska, M., Chong, S., Sadiq, S.: Major issues in business process management: an expert perspective. In: 15th European Conference on Information Systems (ECIS), pp. 1240–1251, St. Gallen, University of St. Gallen (2007)
8. Dumas, M., Rosa, M.L., Mendling, J., Reijers, H.A.: Fundamentals of Business Process Management, 2nd edn. Springer, Berlin (2018)
9. vom Brocke, J., Zelt, S., Schmiedel, T.: On the role of context in business process management. Int. J. Inf. Manage. **36**, 486–495 (2016)
10. Del Giudice, M., Soto-Acosta, P., Carayannis, E., Scuotto, V.: Emerging perspectives on business process management (BPM): IT-based processes and ambidextrous organizations, theory and practice. Bus. Process Manage. J. **24**, 1070–1076 (2018)
11. Imgrund, F., Fischer, M., Janiesch, C., Winkelmann, A.: Managing the long tail of business processes. In: 25th European Conference on Information Systems, pp. 595–610, Guimarães (2017)
12. Bruno, G., et al.: Key challenges for enabling agile BPM with social software. J. Softw. Maintenance Evol.: Res. Pract. **23**, 297–326 (2011)
13. Thiemich, C., Puhlmann, F.: An agile BPM project methodology. In: Daniel, F., Wang, J., Weber, B. (eds.) BPM 2013. LNCS, vol. 8094, pp. 291–306. Springer, Heidelberg (2013). https://doi.org/10.1007/978-3-642-40176-3_25
14. Brambilla, M., Fraternali, P., Ruiz, C.K.V.: Combining social web and BPM for improving enterprise performances: The BPM4People approach to social BPM. In: 21st International Conference on World Wide Web, pp. 223–226, Lyon, ACM (2012)
15. Schmidt, R., Nurcan, S.: BPM and social software. In: Ardagna, D., Mecella, M., Yang, J. (eds.) BPM 2008. LNBIP, vol. 17, pp. 649–658. Springer, Heidelberg (2009). https://doi.org/10.1007/978-3-642-00328-8_65
16. Rosemann, M., Recker, J.C.: Context-aware process design: exploring the extrinsic drivers for process flexibility. In: Proceedings of the 18th International Conference on Advanced Information Systems Engineering. Proceedings of Workshops and Doctoral Consortium, pp. 149–158. Namur University Press (2006)
17. Glaser, B., Strauss, A.: Grounded theory: the discovery of grounded theory. Sociol. J. British Sociol. Assoc. **12**, 27–49 (1967)
18. Seidel, S., Recker, J.C.: Using grounded theory for studying business process management phenomena. In: 17th European Conference on Information Systems, pp. 1–13, Verona, AIS (2009)
19. Walsham, G.: Interpretive case studies in IS research: nature and method. Eur. J. Inf. Syst. **4**, 74–81 (1995)
20. Eisenhardt, K.M.: Building theories from case study research. Acad. Manage. Rev. **14**, 532–550 (1989)

21. Yin, R.K.: Case Study Research: Design and Methods. Sage Publications, Newbury Park (1989)
22. Strauss, A., Corbin, J.M.: Basics of Qualitative Research: Grounded Theory Procedures and Techniques, 2nd edn. Sage Publications Inc, Los Angeles (1998)
23. Anttila, J., Jussila, K.: An advanced insight into managing business processes in practice. Total Qual. Manage. Bus. Excellence **24**, 918–932 (2013)
24. Recker, J., Mendling, J.: The state of the art of business process management research as published in the BPM conference. Bus. Inf. Syst. Eng. **58**, 55–72 (2016)
25. Yin, R.K.: Case Study Research: Design and Methods. SAGE Publications Inc., New Delhi (2013)
26. Christopher, M.: The agile supply chain: competing in volatile markets. Ind. Market. Manage. **29**, 37–44 (2000)

The Use of Distance Metrics in Managing Business Process Transfer - An Exploratory Case Study

Anna-Maria Exler[(⊠)], Jan Mendling[ⓘ], and Alfred Taudes

Vienna University of Economics and Business Administration, Vienna, Austria
{h0854576,jan.mendling,alfred.taudes}@wu.ac.at

Abstract. Business process transfer refers to the innovations of business processes by taking a business process from one organization to another. Such transfer is a prominent technique for improving operations within larger companies that transfer them from one branch or subsidy to another. The success of such transfers can be affected by several factors. A central challenge is the distance between the organizational units, i.e. source organization and target organization. In this context, distance refers to the extent to which the source organization and the target organizations differ, e.g. geographically, culturally or in organizational terms. The factors of distance and their measurement in an intra-organizational context still represent a gap in the literature. For this reason, a case study was conducted in a Central-European financial institution, in which 14 persons from different hierarchical levels, departments and organizational units were interviewed. The study identified eight dimensions of distance, which we integrate into a model for measuring the distance of such intra-company transfers.

Keywords: Business process transfer · Distance · Operationalization

1 Introduction

In order to remain competitive, companies need to continuously innovate their business processes. Triggers for innovations can be internal, for example the desire to improve duration, quality and costs, or externally, for example by new regulatory requirements. A specific mechanism of innovation is the transfer of the new process from the organizational unit that defined the process to the executing units.

So far, the literature has discussed transfer largely from an inter-organizational perspective in connection with outsourcing or Mergers & Acquisitions (M&As). The focus of these works has frequently been placed on decision-making, the influencing factors on the outsourcing process, or the performance after the transfer [1–4]. In comparison, a process transfer carried out within an organization faces presumably different challenges. Also internally, we assume that a specific notion of distance between units might be a potential source of difficulties [5–8]. However, such a distance requires a conceptual and theoretical foundation. With this paper, we address this

© Springer Nature Switzerland AG 2019
C. Di Francescomarino et al. (Eds.): BPM 2019 Workshops, LNBIP 362, pp. 301–312, 2019.
https://doi.org/10.1007/978-3-030-37453-2_25

problem and develop a corresponding distance notion based on an interview-based research design.

This paper presents the findings of a case study of intra-organizational business process transfer. The underlying case study was based on a transfer project of a Central-European company operating in the financial sector, whose head office is located in Austria. The process innovation was initiated by a supervisory authority and comprised both the transfer of a new business process and the roll-out of the related software.

2 Background

Three areas of prior research discuss influencing factors of business process transfers, the distance between the units involved, and respective methods for measurement.

Factors influencing the transfer have been mentioned in the area of business process management (mainly with regard to outsourcing, offshoring and M&As), as well as in the context of routines and knowledge management. A major group of factors comprises cultural differences, e.g. incompatibility of different organizational or national cultures [1, 10–14], the degree of social integration like shared values, religion or behaviors [2, 3] and the so-called cultural fit [17]. Language barriers are also mentioned [5, 18, 19]. Moreover, various organizational issues can affect the transfer of business processes, e.g. the compatibility of the process and its new context [20], the so-called task challenge [3], the team leader's attitude towards the new process, or the interpersonal climate within the teams [21]. Other factors are several personal issues like cognitive capacities and limits [20]. Some framework conditions are stated, like legal or regulatory hurdles [3, 4, 18] or adequate infrastructure and technology [4]. In addition to the factors that influence the business process transfer between two organizational units, it is assumed that these units show a certain degree of distance from each other, which subsequently also affects the business process transfer itself. This is supported by evidence from previous research; Becker [20] mentions that a transfer of routines is more likely if the companies involved share a similar environment.

The distance factor has so far been considered in detail in connection with international business research [22], such as global expansion or the development of new markets or locations abroad [23–28], foreign direct investments [29] or international trade [30], as well as in strategic management [31], marketing [8, 32, 33] or knowledge transfer [19]. In this context, various types of distance were mentioned. The most prominent dimension is the geographical distance, typically between the location of a company in its home country and a foreign place of business [34]. Other attributes are the size of the country, average distances within the country, access to sea routes, topography or transport infrastructure [5]. A very intensively researched area is cultural distance. Hofstede [6] produced the first major work in this field with his framework of four dimensions, on which numerous studies have been based [35, 36]. Closely related to cultural distance is the so-called psychological distance, which encompasses the perceived distance between two countries [7]. Other forms of distance are linguistic [19], economic [5, 38] and the administrative distance [5], as well as institutional distance, which comprises formal rules such as laws as well as informal guidelines such as norms and customs [37]. Although the distance factor has so far been considered extensively in connection with international business research, there is little evidence with regard to intra-company business process transfer.

Regarding the measurement of different types of distance there are numerous approaches and models mentioned, e.g. [7, 33, 38], but these often contain mainly macroeconomic indicators such as Gross Domestic Product, education levels, purchasing power and customer preferences or trade agreements. The methods used to operationalize cultural distance (e.g. [6, 23, 46]) refer to the differences between entire countries, which might be to coarse-grained for individual companies. These methods are therefore not directly applicable to our research question, which focuses on intra-company transfers. Instead, internal factors should be taken into account which can also be actively influenced within the context of the transfer project. For this reason, a suitable measurement model should be designed within the underlying case study.

We structure our further investigation as follows. We consider the case that two units are involved in a business process transfer, i.e. the source organization and the target organization. Due to the large number of influencing factors mentioned in studies on outsourcing, M&As, etc., it is assumed that there are also a number of factors which affect intra-organizational transfers. It is further assumed that the source and target organization exhibit a certain degree of distance, and that this distance is affected by the influencing factors. While the previous literature mainly focused on international business, in this case study the distance factor was examined regarding intra-organizational business process transfer to find out which types of distance have appeared in this context and how they can be converted into measurable items.

3   Methodology

The basic framework of the underlying work was a qualitative research design including an explorative, interpretative single case study. The advantage of case studies is that a certain phenomenon can be intensively investigated in its natural environment over a certain period of time which provides a deep insight into the research object [39–41]. Data collection was conducted by problem-centered expert interviews with narrative sequences, meaning that the interviewees could response in a free manner and to the extent they considered appropriate. In expert interviews, the interviewees can provide the interviewer with an insight into their specific expertise and background knowledge [45]. In preparation of the interviews, a guide consisting of various questions on different aspects of business process transfers was developed. However, the discussion was not strictly aligned to the guide but rather flexibly adjusted to the current course of conversation [9]. The interviews focused on how the business process transfer was carried out, which challenges and problems arose, and what the interviewees would improve if the transfer was carried out again. The factors mentioned in the literature were not directly addressed in the interview guide. Instead, the interviewees were asked to tell about the transfer project as uninfluenced as possible in view of the Grounded Theory analysis. The sample of interviewees consisted of 14 people; to gain a variety of different views on the project, people from different departments and hierarchical levels were interviewed. Furthermore, persons of the head office as well as from different subsidiaries, and from source and target organizations were involved. In line with Grounded Theory, the interviewees were not selected in advance, but rather at short notice in coordination with the results gained so far and depending on the area in which input was still needed. Further people were interviewed until they did not provide new insights anymore. The average duration of the interviews was 40–50 min. The interviews were recorded and afterwards transcribed.

Data analysis was performed following grounding theory methodology according to Strauss and Corbin [42], based on the three-stage procedure open, axial and selective coding. In the first part of analysis, the influencing factors were derived from the data, based on the challenges and problems mentioned by the interviewees. Subsequently, categories were developed by abstracting from the data and grouping concepts. Finally, selective coding was used to identify the core category [33, 42]. According to the basic idea of Grounded Theory [43], this sequence of analysis was performed several times as an iterative process. In this way, the categories were derived solely from the interview data in an explorative way. Only after the complete analysis has been finished they were compared with the literature. It turned out that they are partly congruent with the factors mentioned in research on outsourcing and M&As, but also some new aspects like stakeholder involvement or process blockers could be identified. As already mentioned, it was further assumed that these influencing factors also affect the distance between source and target organization. After the deduction of the influencing factors, they were therefore again subjected to an abstraction process, which finally yielded dimensions of distance. In summary, a total of eight dimensions of distance were identified, and, in a last step, an approach for the measurement of distance was developed by deriving measurable items out of the interview data.

Since qualitative data analyses always represent a subjective interpretation, the concepts and theories developed were shown to interviewees several times, in order to ensure the credibility, consistency and validity of the results. Furthermore, the analysis was made as transparent as possible by underpinning the derived statements with quotations from the interviews in order to gain traceability for the derivation of the results.

4   Findings

4.1   Influencing Factors

The case study identified six categories of influencing factors on business process transfer: communication, acceptance/motivation, organizational structure, process blocker, stakeholder integration and knowledge/comprehension.

The *communication* category includes all interpersonal interaction in conjunction with the business process transfer, e.g. the uncertainty about the person responsible in case of content-related questions during the implementation of the new workflows and the roll-out of the system, as well as available communication channels. Often, the timing of communication was criticized, since some information seemed to be passed on to the specialist departments quite short-term and partly incomplete.

Strongly related to the communication category is the *stakeholder involvement* which stands for the extent of the inclusion of all stakeholders within the head office and at the subsidiaries. The degree of stakeholder involvement decreased during the course of the project, which led to a disregard of the target organizations' needs as well as of the local peculiarities and specifics of the subsidiaries. As a result, partly complex adjustments of their upstream and downstream processes as well as the local systems had to be made, especially at the organization units outside the head office.

The category *organizational structure* includes all factors influencing the business process transfer due to the present organizational conditions or culture within the source and target organization. In the case studied this comprises inter alia the complex, cumbersome decision paths leading to an inefficient decision policy as well as the quite inflexible structures within the source organization. This made an adaptation of the workflows to specific needs of some target organizations nearly impossible. Some employees at the subsidiaries stated the differences between the organizational structure of the head office and those at their own unit, in combination with a mutual lack of knowledge about that issue, as a reason for the partial incompatibility of their own processes and tools with the workflows: "And now you have to merge your own processes with [the head office's] processes somehow, so that you are able to do proper work further on. [...] And this isn't easy if you don't understand [the head office's] side, and even more [the head office] doesn't understand [the subsidiaries'] side."

The category *acceptance/motivation* includes all aspects regarding the target organizations' recognition of the requirements of the new process as well as the personal readiness to deal with the new workflow. A factor negatively influencing the acceptance appeared to be the force to take over the new process. From the subsidiaries' point of view, the lack of perceived added value and the increased effort together with higher costs were negatively influencing the acceptance. Another issue was seen in the increased transparency of the business operations caused by the new workflow.

The category *process blockers* refers to all issues leading to a delay in the implementation and the roll-out. These include the interruption of the testing phase due to the erroneous data migration. Other process blockers resulted from insufficient resources for training, documentation and testing, as well as the lack of coordination of the workflows with the specialist departments, so that the system requirements were partly incomplete and some issues needed for daily business were missing.

As a core category *knowledge/comprehension* was chosen because connections to all other categories could be established and, therefore, it could be regarded as a critical factor or central problem. On the one hand, this category contains all issues connected with the users' content-related, professional understanding of the new workflow, the risk model and the system. This includes factors such as the design of the training or the existence of a common language and terminology for all participants. On the other hand, not only the knowledge of the users turned out to be a crucial factor within the business process transfer, but also the know-how by the project team members during the planning, implementation and roll-out phase.

By using selective coding, links between the categories were derived. Subsequently, theories and hypotheses were formulated for each category, which summarize their influence on the transfer of business processes or on each other.

4.2 Dimensions of Distance

Next, based on the influencing factors, we derived eight dimensions of distance: geographical, organizational/process related, administrative, cultural, personal, content/knowledge based, legal and technical distance. The types of distance mentioned in the literature served as a theoretical basis. A categorization into types of distance reflects Ambos and Ambos [19]: "Units are not equidistant on all dimensions".

Geographical distance includes the physical distance between the source and target organization and it refers to the seating arrangement within the source organization, since the case study showed that common rooms are essential for the exchange of members of the source organization.

All discrepancies regarding the organizational and process structure were summarized under the dimension *of organizational/process related distance*. This type of distance combines influencing factors from the categories organizational structure and stakeholder involvement, such as the difficulties that arose due to the lack of consideration of the subsidiaries' local needs and specifics, the inflexible organizational structure of the head office or the subsidiaries' insufficient knowledge of the head office's processes. The following quotation shows as an example how the dimensions of distance were derived from the interview data: "It was top-down from the beginning, [...] i.e. [the source organization] determines the process and everyone else has to follow. The problem was that it never was clarified: can you follow, and to what extent can you follow." From the quotation it can be deduced that there was a distance between the organizational and process structures. An investigation of the differences was neglected, whereby later partly incompatible or overlapping processes met.

A very similar dimension concerns the so-called *administrative distance*, to which the difficulties due to bureaucratic structures have been attributed. This includes complex communication channels, information at short notice and, above all, the availability of necessary time, personnel and financial resources. The term administrative distance stems from Ghemawat [5], who use it at the national level. Here, it acquired a different meaning by focusing on intra-organizational factors. It includes those factors that can actually be changed by the project owners in contrast to national circumstances.

Regarding the *cultural distance* dimension, too, our study focused on differences between domestic companies and between organizational units within a company. It includes linguistic aspects, such as misunderstandings regarding the terminology used and the language barriers in queries to support, as well as factors like the tone of contact and the subjectively perceived importance of the respective subsidiary, general aversions to change and the different accuracy when introducing the new process.

Based on the items of the category acceptance/motivation, the dimension *personal distance* was derived. It includes factors like the force to take over the new process, the lack of added value and the higher costs, as well as the unexpected outsourcing of activities to certain organizational units. From a cultural point of view, the general openness to change and the personal attitude were identified as important aspects. Furthermore, personal distance was caused by the rejection of the chosen training methods and the complexity of the system's user interface, as well as not least by the considerably increased transparency and the associated traceability of all inputs.

The *content-/knowledge based distance* encompasses all differences with regard to a different knowledge base between source and target organization. Furthermore, the case study also identified a distance between the source organization and the external provider as well as between the teams and persons within the source organization itself. This includes factors like the too theoretical and complex definition of processes, misunderstandings between the external provider and the company, or insufficient training and too complex training material. Here, a parallel can be drawn to the different knowledge base between sender and receiver mentioned in Schiele et al. [44].

The *legal distance* takes into account the extent to which the new process conforms to the (national) laws and guidelines in the countries of the target organizations. The importance of this dimension rises with the number of countries involved.

Finally, the *technical distance* mainly comprises the influencing factors of the process blocker category, like the incompatibility of the systems from which the data originated, the high number of interfaces and data sources, and the differences in data quality depending on the respective subsidiary.

4.3   Operationalization of Distance

Based on the findings of the case study, an approach for the operationalization of distance was developed. Although the literature contains numerous models for measuring different types of distance, these often refer to outsourcing or M&As and contain macro-economic indicators such as gross domestic product, education level, purchasing power and customer preferences or trade agreements. However, the focus of this work was on the intra-organizational level, since the study demonstrated that - apart from the target organizations located abroad - considerable distances already existed between source and target organizations within the head office. As an instrument for the operationalization a questionnaire was chosen which was structured as a five-tier Likert scale from *Completely agree* to *Completely disagree*. The study by Sousa and Bradley [8], in which a five-part scale was similarly used to determine psychological distance, served as a model for this. The individual items were derived from selected phrases of the interviews, which were converted into statements and which are to be evaluated on the basis of the possible answers. For example, the statement of a respondent that "there simply was a lack of awareness or know-how about the needs and processes of the other side. [The head office] saw it from its perspective, and [...] has not considered us. And we [afterwards] had to make a claim about what we actually needed." was converted to the following item in Table 1:

Table 1. Example for operationalization of distance

Item	Completely agree	Mostly agree	Slightly agree	Mostly disagree	Completely disagree	n.a.
The local needs and specifics of the subsidiaries are taken into account when defining the process	□ (5)	□ (4)	□ (3)	□ (2)	□ (1)	□ (0)

In this way, a total of 65 items could be derived. The questionnaire targets project managers in the source organization and should ideally be filled in before the business process transfers. The highest degree of proximity is represented as 5, the lowest as 1. Each of the derived dimensions of distance forms its own cluster, for which the respective mean value is calculated. Finally, the mean values of all clusters can be summed to determine a final value. The higher this value, the greater is the proximity. The results per cluster show in which areas a distance prevails (see Fig. 1).

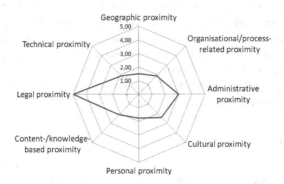

Fig. 1. Network diagram for visualizing the degree of proximity resp. distance

5 Discussion and Implications

The findings of this case study provided a first insight into the various factors influencing the transfer of an intra-organizational business process. There are a number of similarities between our study and prior research. For example, the overload of employees in the core category of knowledge/content-related understanding due to the numerous new contents can be compared with Becker's [20] consideration of cognitive possibilities and limitations; a similar aspect also includes the so-called absorptive capacity [2, 26]. With regard to the choice of the training method, a parallel can be drawn to the theory of the different knowledge base of sender and receiver [44], since the source organization should take the knowledge level of the target organization into account when designing and conducting the training. With regard to the category acceptance/motivation, this factor can also be found in the literature with regard to knowledge transfer [2, 26], and Becker [20] also mentions individual and collective willingness as an important prerequisite. From the identified influencing factors on negative personal attitudes towards the new process and generally towards changes in old habits, one can also draw parallels to the so-called stickiness of routines [47] as well as to the factor of inertia [16] and the breaking through of existing truths [14, 15]. Regarding the organizational structure, the incompatibility of the new process with the local specifics of the subsidiaries with the so-called task challenge [3] and the incompatibility mentioned by Becker [20] can be equated with the new context. Beyond that, we found other aspects including the involvement of stakeholders, organizational requirements in the source organization, or scarce resources.

Regarding the distance factor, eight types were identified in our case study. Interestingly, in terms of geographical distance, all interviewees, whether working in the head office or in more distant target organizations, reported broadly the same problems and challenges. This suggests a contradiction with previous studies on geographic distance, which suggested that the difficulty of transfer increases with geographic distance. A new aspect that could be identified with regard to geographical distance is the physical distance between the employees of the source organization. Common premises appear to support the exchange of project members and thus smoothen distance.

In the concrete context of intra-organizational business process transfers, administrative distance focused on intra-company factors rather than national ones. These include, complex communication channels, information that is too short-term, and, above all, the topic of the necessary time, personnel and financial resources.

Whereas the literature to date has focused primarily on cultural differences in cross-border business activities, the focus of this study has been on cultural differences between domestic companies and between organizational units within a company. Since the organizational units in our case study already belong to the company for a longer time, a certain cultural compatibility could be assumed, since common business practices had already been established, leading to a so-called cultural fit [17]. Still, our results show that even after years of belonging to the group, some cultural differences exist, e.g. misunderstandings regarding the terminology and language barriers, as well as factors such as the tone and subjective importance of the respective subsidiary, a general aversion to changes, and the different accuracy of introducing new process.

Compared to the literature [e.g. 3, 4, 18], the legal distance played a minor role. This may be due to the fact that the companies belonged to the group for a long time.

The dimension referred to as technical distance is mainly based on the influencing factors from the process blocker category. In the literature on distance, this dimension is has not been addressed directly so far, but regarding the influencing factors for example, adequate infrastructure has been mentioned as important prerequisites in connection with M&As [4, 26]. Discrepancies due to a technical incompatibility represent a novel aspect. Especially the support of processes with corresponding tools and systems become more important with increasing automation.

Finally, new dimensions could be identified with regard to organizational/process related and content/knowledge based distance. Also not yet directly mentioned in the literature is the personal distance, which includes the influencing factors of the category acceptance/motivation. Our single case study is a first indication of their relevance. Future research is required to supplement our findings.

The approach of operationalizing distance offers an opportunity to measure differences at the internal level. Since previous research work mainly focuses on the country level [e.g. 7, 38, 33], an approach was developed in this work that focuses on factors that can actually be actively influenced and changed in a company by the project managers. In this way, the distance between source and target organization can be made measurable for future business process transfers. Thus, in those areas where there is a greater distance, appropriate measures can be taken at an early stage. It is assumed that there is a certain set of items that are relevant for all business process transfers. In order to identify them more precisely and to supplement them, further case studies from different companies will be required. With regard to the calculation of the overall distance value and the mean value per category, a finer weighting of the individual items would additionally represent an interesting option for further operationalization of distance.

There are several recommendations that we derive from our work. For instance, attention should be paid to providing a sufficient training to build up knowledge. Furthermore, greater acceptance can be achieved with stronger stakeholder involvement and intensified communication with all units concerned with the new process. Also, the organizational structure should be adapted to simplify the communication as well as the flow of information. The recommended practices can be used as lessons learned for other projects. In order to check which of the procedures are useful for other transfer projects, findings from future case studies are required.

6   Conclusion

Our case study showed that there is a degree of distance between source and target organization which is highly relevant for business process transfer. From our case, six categories of influencing factors and eight dimensions of distance could be identified. Since the literature to date has examined the distance factor primarily with regard to cross-border business activities, an approach for measuring this factor in the context of intra-organizational business process transfers was developed. In total, 65 items were formulated as a questionnaire, which can help to spot possible problems at an early stage. However, since our results are based on a single case study, further research will be necessary to more broadly validate our findings.

References

1. Weber, Y., Tarba, S.Y.: Mergers and acquisitions process: the use of corporate culture analysis. Cross Cultural Manage. Int. J. **19**(3), 288–303 (2012)
2. Björkman, I., Stahl, G.K., Vaara, E.: Cultural differences and capability transfer in cross-border acquisitions: the mediating roles of capability complementarity, absorptive capacity, and social integration. J. Int. Bus. Stud. **38**, 658–672 (2007)
3. Schraeder, M., Self, D.R.: Enhancing the success of mergers and acquisitions: an organizational culture perspective. Manag. Dec. **41**(5), 511–522 (2003)
4. Caiazza, R., Volpe, T.: M&A process: a literature review and research agenda. Bus. Process Manag. J. **21**(1), 205–220 (2015)
5. Ghemawat, P.: Distance still matters: the hard reality of global expansion. Harvard Bus. Rev. **79**(8), 137–147 (2001)
6. Hofstede, G.: Culture and organizations. Int. Stud. Manag. Organ. **10**(4), 15–41 (1980)
7. Hakanson, L., Ambos, B.: The antecedents of psychic distance. J. Int. Manag. **16**(3), 195–210 (2010)
8. Sousa, C.M.P., Bradley, F.: Cultural distance and psychic distance: two peas in a pod? J. Int. Mark. **14**(1), 49–70 (2006)

9. Hopf, C.: Qualitative interviews: an overview. In: Flick, U., von Kardorff, E., Steinke, I. (eds.) A Companion to Qualitative Research, pp. 203–208. SAGE Publications, London (2004)

10. Sarala, R.M., Vaara, E.: Cultural differences, convergence, and crossvergence as explanations of knowledge transfer in international acquisitions. J. Int. Bus. Stud. **41**(8), 1365–1390 (2010)

11. Stahl, G.K., Voigt, A.: Do cultural differences matter in mergers and acquisitions? A tentative model and examination. Organ. Sci. **19**(1), 160–176 (2008)

12. Teerikangas, S., Very, P.: The culture-performance relationship in M&A: From Yes/No to How. Br. J. Manag. **17**, 31–48 (2006)

13. Chakrabarti, R., Gupta-Mukherjee, S., Jayaraman, N.: Mars-venus marriages: culture and cross-border M&A. J. Int. Bus. Stud. **40**(2), 216–236 (2009)

14. Kaplan, S.: Truce breaking and remaking: the CEO's role in changing organizational routines. Adv. Strateg. Manag. **32**, 1–45 (2015)

15. Zbaracki, M.J., Bergen, M.: When truces collapse: a longitudinal study of price-adjustment routines. Organ. Sci. **21**(5), 955–972 (2010)

16. Pentland, B.T., Feldman, M.S.: Organizational routines as a unit of analysis. Ind. Corp. Change **14**, 793–815 (2005)

17. Bauer, F., Matzler, K.: Antecedents of M&A success: the role of strategic complementarity, cultural fit, and degree and speed of integration. Strateg. Manag. J. **35**, 269–291 (2014)

18. Shrestha, N., Sharma, P.: Problems and Prospects of Business Process Outsourcing in Nepal. SAWTEE, Briefing Paper No. 15 (2013)

19. Ambos, T.C., Ambos, B.: The impact of distance on knowledge transfer effectiveness in multinational corporations. J. Int. Manag. **15**(1), 1–14 (2009)

20. Becker, M.C.: Organizational routines: a review of the literature. Ind. Corp. Change **13**(4), 643–677 (2004)

21. Edmondson, A.C., Bohmer, R.M., Pisano, G.P.: Disrupted routines: team learning and new technology implementation in hospitals. Adm. Sci. Q. **46**, 685–716 (2001)

22. Geldes, C., Felzensztein, C., Turkina, E., Durand, A.: How does proximity affect interfirm marketing cooperation? A study of an agribusiness cluster. J. Bus. Res. **68**(2), 263–272 (2015)

23. Shenkar, O.: Cultural distance revisited: towards a more rigorous conceptualization and measurement of cultural differences. J. Int. Bus. Stud. **32**(3), 519–535 (2001)

24. Magnusson, P., Baack, D.W., Zdravkovic, S., Staub, K.M., Amine, L.S.: Meta-analysis of cultural differences: another slice at the apple. Int. Bus. Rev. **17**, 520–532 (2008)

25. López-Duarte, C., Vidal-Suárez, M.M.: External uncertainty and entry mode choice: cultural distance, political risk and language diversity. Int. Bus. Rev. **19**(6), 575–588 (2010)

26. Chang, Y.C., Kao, M.S., Kuo, A., Chiu, C.F.: How cultural distance influences entry mode choice: the contingent role of host country's governance quality. J. Bus. Res. **65**(8), 1160–1170 (2012)

27. Hutzschenreuter, T., Kleindienst, I., Lange, S.: Added psychic distance stimuli and MNE performance: performance effects of added cultural, governance, geographic, and economic distance in MNEs' international expansion. J. Int. Manag. **20**(1), 38–54 (2014)

28. Ambos, B., Hakanson, L.: The concept of distance in international management research. J. Int. Manag. **20**(1), 1–7 (2014)

29. Siegel, J.I., Licht, A.N., Schwartz, S.H.: Egalitarianism, cultural distance, and FDI: a new approach. Organ. Sci. **24**(4), 1174–1194 (2012)

30. Cyrus, T.L.: Cultural distance and bilateral trade. Global Econ. J. **12**(4), 3–28 (2012)

31. Parente, R.C., Baack, D.W., Hahn, E.D.: The effect of supply chain integration, modular production, and cultural distance on new product development: a dynamic capabilities approach. J. Int. Manag. **17**(4), 278–290 (2011)
32. Sousa, C., Bradley, F.: Cultural distance and psychic distance: refinements in conceptualisation and measurement. J. Mark. Manag. **24**(5/6), 467–488 (2008)
33. Brewer, P.A.: Operationalizing psychic distance: a revised approach. J. Int. Mark. **15**(1), 44–66 (2007)
34. Ojala, A.: Geographic, cultural, and psychic distance to foreign markets in the context of small and new ventures. Int. Bus. Rev. **24**, 825–835 (2015)
35. Kogut, B., Singh, H.: The effect of national culture on the choice of entry mode. J. Int. Bus. Stud. **19**(3), 411–432 (1988)
36. Kirkman, B.L., Lowe, K.B., Gibson, C.B.: A quarter century of Culture's Consequences: a review of empirical research in incorporating Hofstede's cultural values framework. J. Int. Bus. Stud. **37**, 285–320 (2006)
37. Li, S., Scullion, H.: Bridging the distance: managing cross-border knowledge holders. Asia Pacific J. Manag. **23**, 71–92 (2006)
38. Tsang, E.W., Yip, P.S.: Economic distance and the survival of foreign direct investments. Acad. Manag. J. **50**(5), 1156–1168 (2007)
39. Recker, J.: Scientific Research in Information Systems: A Beginner's Guide. Springer, Heidelberg (2013)
40. Fernández, W.D.: The grounded theory method and case study data in IS research: issues and design. Inf. Syst. Found. Workshop: Constructing Criticising **1**, 43–59 (2004)
41. Gable, G.G.: Integrating case study and survey research methods: an example in information systems. Eur. J. Inf. Syst. **3**(2), 112–126 (1994)
42. Strauss, A., Corbin, J.: Basics of Qualitative Research: Techniques and Procedures for Developing Grounded Theory, 2nd edn. Sage, Thousand Oaks (1998)
43. Walker, D., Myrick, F.: Grounded theory: an exploration of process and procedure. Qual. Health Res. **16**(4), 547–559 (2006)
44. Schiele, F., Laux, F., Connolly, T.M.: Applying a layered model for knowledge transfer to business process modelling (BPM). Int. J. Adv. Intell. Syst. **7**(1&2), 156–166 (2014)
45. Pfadenhauer, M.: At eye level: the expert interview - a talk between expert and quasi-expert. In: Bogner, A., Littig, B., Menz, W. (eds.) Interviewing Experts. Research Methods Series, pp. 81–97. Palgrave Macmillan, London (2009)
46. Beugelsdijk, S., Nell, P.C., Ambos, B.: When do distance effects become empirically observable? An investigation in the context of headquarters value creation for subsidiaries. J. Int. Manag. **23**, 255–267 (2017)
47. Szulanski, G.: Exploring internal stickiness: Impediments to the transfer of best practice within the firm. Strateg. Manag. J. **17**, 27–43 (1996)

12th International Workshop on Social and Human Aspects of Business Process Management (BPMS2)

12th International Workshop on Social and Human Aspects of Business Process Management (BPMS2)

Social software [1, 2] is a new paradigm that is spreading quickly in society, organizations, and economics. It enables social business that has created a multitude of success stories. More and more enterprises use social software to improve their business processes and create new business models. Social software is used both in internal and external business processes. Using social software, the communication with the customer is increasingly bi-directional. E.g., companies integrate customers into product development to capture ideas for new products and features. Social software also creates new possibilities to enhance internal business processes by improving the exchange of knowledge and information, to speed up decisions, etc. Social software is based on four principles: weak ties, social production, egalitarianism, and mutual service provisioning. Social software is part of social information systems [3] that embrace a variety of software including social networking platforms, collaborative project management tools, or online/content communities. Social information systems differ from traditional information systems by enabling emergent interactions [4].

Up to now, the interaction of social and human aspects with business processes has not been investigated in depth. Therefore, the objective of the workshop is to explore how social software interacts with business process management, how business process management has to change to comply with weak ties, social production, egalitarianism, and mutual service, and how business processes may profit from these principles.

The workshop discussed the three topics below.

1. **Social Business Process Management (SBPM),** i.e., the use of social software to support one or multiple phases of the business process lifecycle
2. **Social Business: Social Software Supporting Business Processes**
3. **Human Aspects of Business Process Management**

Based on the successful BPMS2 series of workshops since 2008, the goal of the 12th BPMS2 workshop is to promote the integration of business process management with social software and to enlarge the community pursuing the theme. Three papers were presented in the workshop.

First, Birger Lantow, Julian Schmitt, and Fabienne Lambusch presented their paper with the title "Mining Personal Service Processes: The Social Perspective." They investigated the use of Process Mining in the domain of Personal Services with a focus on the Social or Organizational Perspective. They presented new objectives and thus approaches for Social Mining in the context of Process Mining.

In their paper, titled "Supporting ED Process Redesign by Investigating Human Behaviors," Alessandro Stefanini, Davide Aloini, Peter Gloor, and Federica Pochiero investigate behavioral factors affecting patient satisfaction in the Emergency Department (ED). The final goal of the research is to support the ED process (re-) design from a holistic perspective.

Iris Beerepoot, Inge van de Weerd, and Hajo Reijers presented their paper with the title "The Potential of Workarounds for Improving Processes." Based on a synthesis of recommendations, they describe five key activities to deal with workarounds that help organizations. In this way they provide concrete recommendations for managing workarounds give a background for future activities on the subject.

We wish to thank all the people who submitted papers to BPMS2 2019 for having shared their work with us, the many participants creating fruitful discussion, as well as the members of the BPMS2 2019 Program Committee, who made a remarkable effort in reviewing the submissions. We also thank the organizers of BPM 2019 for their help with the organization of the event.

October 2019   Selmin Nurcan
                                                                                                         Rainer Schmidt

References

1. Schmidt, R., Nurcan, S.: BPM and Social Software. In: Ardagna, D., Mecella, M., Yang, J., Aalst, W., Mylopoulos, J., Rosemann, M., Shaw, M.J., and Szyperski, C. (eds.) Business Process Management Workshops. pp. 649–658. Springer Berlin Heidelberg (2009).
2. Bruno, G., Dengler, F., Jennings, B., Khalaf, R., Nurcan, S., Prilla, M., Sarini, M., Schmidt, R., Silva, R.: Key challenges for enabling agile BPM with social software. Journal of Software Maintenance and Evolution: Research and Practice. 23, 297–326 (2011). https://doi.org/10.1002/smr.523.
3. Schmidt, R., Alt, R., Nurcan, S.: Social Information Systems. In: Proceedings of the 52nd Hawaii International Conference on System Sciences. Hawaii (2019).
4. Schmidt, R., Kirchner, K., Razmerita, L.: Understanding the Business Value of Social Information Systems – Towards a Research Agenda. In: 2020 53rd Hawaii International Conference on System Sciences (HICSS) (2020).

Organization

Program Committee

Jan Bosch	Chalmers University of Technology, Sweden
Lars Brehm	Munich University of Applied Sciences, Germany
Norbert Gronau	University of Potsdam, Germany
Monique Janneck	Technische Hochschule Lübeck and University of Applied Sciences, Germany
Barbara Keller	Munich University of Applied Sciences, Germany
Ralf Klamma	RWTH Aachen University, Germany
Sai Peck Lee	University of Malaya, Malaysia
Selmin Nurcan	Université Paris 1 Panthéon-Sorbonne, France
Mohammad Rangiha	City, University of London, UK
Gustavo Rossi	LIFIA, UNLP, Argentina
Flavia Santoro	UERJ, Brazil
Rainer Schmidt	Munich University of Applied Sciences, Germany
Miguel-Angel Sicilia	University of Alcalá, Spain
Pnina Soffer	University of Haifa, Israel
Moe Thandar Wynn	Queensland University of Technology, Australia

Mining Personal Service Processes: The Social Perspective

Birger Lantow$^{(\boxtimes)}$, Julian Schmitt, and Fabienne Lambusch

University of Rostock, Albert-Einstein-Str. 22, 18059 Rostock, Germany
{birger.lantow, julian-schmitt,
fabienne.lambusch}@uni-rostock.de

Abstract. The process of digital transformation opens more and more domains to data driven analysis. This also accounts for Process Mining of service processes. This work investigates the use of Process Mining in the domain of Personal Services focusing on the Social or Organizational Perspective respectively. Documenting research in progress, problems of "traditional" organizational mining approaches on knowledge-intensive service processes and possible solutions are shown. Furthermore, new objectives and thus approaches for Social Mining in the context of Process Mining are discussed, addressing the recently increasing focus on workforce well-being.

Keywords: Process Mining · Social Mining · Organizational Mining · Distress · Personal Service · Process Management · Organizational Perspective

1 Introduction

Digitalization is one of the dominant topics of our time. The central task for the future will be to develop concepts for digital change that support not only industries with highly structured processes, but also economic sectors with knowledge-intensive and weakly structured processes [1] and especially for work with a high degree of uncertainty like in Personal Services [2]. From the perspective of Process Mining, this development has some implications. First, there is a new data lake connected with new domains that produce process related data by using information systems. This goes hand in hand with a demand to make these data available for analysis. Second existing Process Mining techniques might not fit to new demands and data. Furthermore, new objectives of Process Mining arise.

This work presents results of an ongoing project on Process Management for family care projects that connects research at Rostock University with several local companies. Process Mining potentially plays a role in all phases of Process Management [3]. Thus it has become a major topic of this project. Although being restricted to a specific domain of Personal Services, there is a big potential of generalization by analyzing local problems and solutions.

In the following we concentrate on the results with regard to the Organizational Perspective for mining Personal Service Processes. It is one out of four perspectives defined by the IEEE Process Mining Task Force [3]. Considering the Organizational Perspective, the focus lies on information about resources within the process data,

© Springer Nature Switzerland AG 2019
C. Di Francescomarino et al. (Eds.): BPM 2019 Workshops, LNBIP 362, pp. 317–325, 2019.
https://doi.org/10.1007/978-3-030-37453-2_26

which actors are involved at all and how they relate to each other. Mining goals are for example finding a corporate structure by identifying employees according to their roles or presenting them in the form of a social network.

Section 2 discusses shortly introduces Personal Services and their characteristics. This is followed by an analysis of the current state of research with regard to objectives and techniques of Organizational or Social Mining in the context of Process Mining in Sect. 3. New objectives and techniques that have been developed with domain experts from family care are discussed in Sect. 4. Section 5 provides a short summary and outlook.

2 Characteristics of Personal Services

The common characteristic of Personal Services is that they are directed at persons instead of things. Already Halmos [4] denoted those services as personal which are concerned with the change of the body or the personality of a client. Bieber and Geiger point out that a great challenge of Personal Services lies in the fact that the recipient of the service is at the same time in the role of a co-producer [2]. They state that Personal Services are characterized particularly by the close, indissoluble connection between persons who have to interact with each other. This interaction as a key element in Personal Services reduces the predictability and increases the uncertainty in the corresponding work processes. Therefore, many of Personal Service processes show characteristics of knowledge-intensive processes [2].

Modeling and analyzing such services requires approaches different from "traditional" Business Process Management. Approaches like Adaptive Case Management (ACM) address the problem of high variability and high autonomy of actors in knowledge-intensive Processes [5]. A structured process model can be provided only at high abstraction levels. Fließ et al. [6] describe three phases: pre-service, service, and post-service. Pre-service describes the preparation of a service, while service describes the actual provision of services in interaction with the customer, and post-service refers to the follow-up. Personal Services are often integrated into a longer-term meta-process, which then also has longer-term objectives. One example would be the restoration of mobility in the context of physiotherapy. This means that direct service encounters may be intertwined with overarching coordination tasks to achieve objectives. We extend this model with two additional phases. From perspective of information processing, the point in time when first information about the client becomes available is important. Therefore, the phase of client intake is added. Moreover, Personal Services are often integrated into a longer-term meta-process, which then also has longer-term objectives. An example would be the restoration of mobility in the context of physiotherapy. This means that phases of direct, short-term service provision may be intertwined with overarching coordination tasks to achieve objectives. Therefore, we add the phase of Coordination. The extended version of the service phases is shown in Fig. 1.

Since the nature of Personal Services is the interaction between people, human interaction and relationships play a strong role for service performance and thus should be considered when analyzing services processes. This long term relationship has great importance for example in therapy and coaching settings. Furthermore, identified

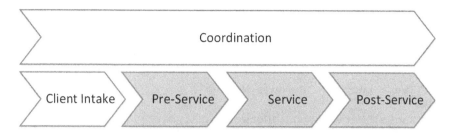

Fig. 1. Extended service phases

activities for process models remain very abstract as stated earlier. For example, the development of reference activities for a group of four German family care companies resulted in a list of twelve activities: Counseling, Escorting, Practical Support, General Activity, Networking, Planning Meeting, Crisis Intervention, Cancelled Contact, Phone Call, Governmental Contact, Documentation, Pre-/Post-processing. Our analysis of 21,823 logged activities of one of the companies resulted in a share of about 70% for Counseling, Networking, Phone Call, and General Activity. These activities may be found in any domain of Personal Services. Furthermore, the application of process discovery tools like DeclareMiner to the sample data resulted in models with little relevancy. The only revealed dependencies were that Escorting is generally connected to a Phone Call activity and that Cancelled Contact activities cause Counseling and Networking activities. Relevant information with regard to the process lies in the communication of the involved actors as well as textual documentation related to clients, describing their condition, behavior, progress with regard to the long-term goals. Generally, such social aspects have a great influence on the performance of Personal Services and should be considered when mining process data. Thus, Social Process Mining is a topic of interest.

3 Social Mining Objectives and Techniques

We performed a Systematic Literature Review (SLR) in order to document the current state of research of Social Mining in the context of Process Mining. The focus was on the objectives and techniques that have been suggested in connection with Social Mining. The main idea of a SLR is to get a representative sample of the scholarly articles in a research field. The following process for a SLR is defined by Ivarsson [7]: (1) Identification of publications, (2) Inclusion/Exclusion based on title and abstract, (3) Data extraction, (4) Data analysis. Steps 1 and 2 are described in Sect. 3.1. Section 3.2 discusses the results with regard to objectives and techniques (steps 3 + 4).

3.1 Identification and Selection of Relevant Publications

For the first step, identification of papers, Scopus (www.scopus.com) was chosen as academic database for the search. Scopus can be considered as a source for publications that meet scientific standards. The following search term has been used:

```
TITLE-ABS-KEY("Organizational Mining"OR"Social Mining" )
OR
TITLE-ABS-KEY(
 ("Enterprise" OR "Business")
 AND
  ("Event-Log" OR "Process Mining")
 AND
  ("Social Network" OR "Network Analysis"))
```

The first part is directly querying for "Organizational Mining" as well as "Social Mining". In order to include publications that do not use these terms directly but fit into the context, the first part has been amended using the "OR" operator. Thus, publications are included in the context of an "Enterprise" or "Business" that either refer to "Process Mining or "Event Logs" and consider "Social Network" or "Network Analysis". The latter addresses the main purpose of Social Mining according to [3]. "Process Mining" or "Event Log" keep the link to Process Mining as an analysis of process related event logs and the first keeps the business context. These narrow context was necessary in order to filter out the vast number of publications with regard to social network analysis in general. Using this search term, 117 potentially relevant, articles were found.

The next step was the inclusion/exclusion of publications based on title and abstract. After a first screening based on the title, 65 potentially relevant publications remained.

The screening of abstracts resulted in a final set of 42 publications for data extraction and analysis.

3.2   Data Extraction and Analysis

According to the found literature, Social Mining in the context of Process Mining has a variety of objectives. The overarching goal is to discover social networks and organizational structures [8]. Table 1 shows the detailed objectives found.

With regard to the techniques or forms on analysis, five can be identified in the literature. All of them result in a graph structure, depicting interactions or relations between the actors.

The first analysis technique is called "Handover of Work". It captures the amount of times when an actor performs a task after another actor within process execution. This results in a directed graph connecting the predecessor with the direct successors [9, 10]. Measures like centrality [11] or betweenness [12] can help to assess the importance of certain actors.

The second analysis technique is called "Working together". It captures the amount of process instances where actors are working together in the same instance [9]. This results in an undirected graph [10]. Statements can be made by looking at the connected persons, meaning they have a stronger relation the more they are working together [9].

Table 1. Social mining objectives and techniques

	Handover of work	Working together	Reassignment	Subcontracting	Similar task
Derive relations of human resources [10]	X	X	X	X	X
Evaluate roles of individuals [13]	X		X		
Track evolution of an organizational structure and relationships [15]	X	X	X	X	X
Find substitutes for human resources [16, 17]					X
Identify teams or organizational units [14, 18]	X	X			X
Strength of connection between actors [18]		X			X
Identification of possible team leaders [19, 20]		X			
Identification of key resources [12]	X				
Find bottlenecks [21]	X				X
Balance workload [13, 21]	X				X
Task assignment [21]	X				
Identification of undesired loops [13]				X	
Security analysis [22]	X			X	

The third analysis technique called "Reassignment metrics" is based on special event types. It explores hierarchical relations by looking at reassignments of tasks by an actor [9]. This results in a hierarchy assuming that those actors that change task assignments have a higher position.

The fourth analysis technique is based on joint activities, called "Subcontracting". Subcontracting occurs, when the same activities are performed in between other activities and the performer of the preceding and succeeding activity is the same actor [9]. The Subcontracting metrics also can be used to detect undesired loops e.g. when a case is routed back to a person with a similar role in the organization/case [13].

The last analysis technique called "Similar Task" is based on the performance of similar activities across processes and process instances. For example, actors who execute similar tasks can substitute each other more easily than others [14].

Table 1 provides a mapping of the found analysis techniques to the respective analysis objectives.

4 Expanding the Limits of Social Mining

In the following we present new approaches to Social Mining in the context of Process Mining that have been developed in workshops with the practitioners from family care. The main driving problems that required new approaches were:

1. Existing analysis objectives of Organizational Mining barely address the nature of organizational resources as human beings that have various intentions, attitudes to each other and states of mind. All these aspects and the general well-being at work have a great influence on the outcome of Personal Services
2. Lack of available data with regard to the aforementioned aspects.
3. High abstraction of activities within the workflow models from the specific competencies and knowledge required in the processes. For example, a "Similar Task" analysis in order to find substitutes will result in the information that all employees of a family care company are able to make phone calls, to provide counseling or to perform networking activities.
4. The role of the client as key resource in the service process is not considered.

New objectives (cf. Sect. 4.1) and data sources as well as mining techniques (Sect. 4.2) have been suggested in order to address these problems.

4.1 New Objectives of Social Mining

New objectives of Social Mining are developed in two dimensions. First, the analysis objects may change. Second, new metrics may be applied. For the first dimension an inclusion of the client in Social Mining activities is straight forward to solve problem number for 4 and to address the importance of the client as a resource in the process. Since Personal Service setting may involve several professionals being internal or external to the service provider. An inclusion of external resources into Social Mining may be worthwhile in general.

Analyzing intentions, attitudes, state of minds as well as the well-being in work life (Problem 1) requires new metrics for Social Mining. For example, changes in personal relationship may cause distress. This also extends to work life. Thus, the change rate in social networks of resources might be evaluated. The achievement of long-term goals in Personal Services (cf. Sect. 2) is bound to a certain quality of the relationship between the client and the service provider. In consequence, these qualities like the level of trust should be tracked. Furthermore, having several actors involved in a process where the actors have a great level of autonomy, their attitudes to each other and their intentions have an influence on the outcome. Diverging intentions should be tracked and could be considered when allocating resources.

4.2 New Data Sources and Techniques for Social Mining

Normally, the data that is required for analysis objectives as suggested in Sect. 4.1 cannot be found in event logs or administrative documentation related to business process execution (Problem 2). Activities of external actors or clients may not be logged in the service provider information systems. Especially in the domain of personal services, the share of information systems use in business activities is rather low. Furthermore, small and medium sized companies dominate the market. Thus, the use of integrated information systems is developing but not common. Last, attitudes and intentions generally aren't tracked by information systems.

We see four sources for the required data: (1) The service provider's employees explicitly collect end enter the data in the information system, (2) Analysis of the communication with the clients, (3) Analysis of internal case documentation related to the client, and (4) The use of hard- and software sensors to collect the data.

The first source will increase effort that is spent on process execution because employees will need time to enter the required data. Sources two and three requires the analysis of textual data. Approaches from Text Mining can be used to analyze the data. For example, Named Entity Recognition (NER) [23] can provide information about actors involved in the process. Opinion Mining [24] can be used to assess attitudes. Communication as well as documentation usually contains time stamps. Thus, gathered data can be mapped to activities or phases in the service process. At last, the use of sensors seems promising. However, when it comes to hardware sensors current applications that track for example emotions based on facial expression require a considerable investment when they should be available company wide and rely on a controlled environment which may not be appropriate for the service provision [26]. Software sensors that analyze for example keystroke features have been proven to be able to measure emotions. However, there was a low accuracy and the specifics of a business context have not been considered [27–29].

At last problem 3, the high abstraction level of activities in workflow models can be addressed by analyzing communication and documentation using Text Mining approaches. Instead of working on activity level, Topic Mining can provide information with regard to the actual content of Counseling sessions for example. Instead of a "Similar Task" analysis (cf. Sect. 3.2) a "Similar Topic" analysis could be used. Furthermore, by tracking the topics over time phases and progress in the long-term service process (cf. Sect. 2) could be identified. First promising results have been achieved by applying LDA [25] on the internal documentation of a family care company. Domain experts agreed, that the found topics and their distribution over time well reflect the progress of selected service process instances. Figure 2 provides an example.

Fig. 2. Distribution of a topic over time for a service process instance based on LDA

5   Summary and Outlook

So far we can only provide ideas how to meet the requirements for Social Mining in the context of Process Mining for the domain of Personal Services. The approach needs further evaluation and validation. Problems like data privacy, acceptance of the approach, and accuracy of the applied methods need to be further investigated.

A deeper look into related work with regard to Social Mining in general and Text Mining might lead to additional suggestions. Finally, the utility of new Process Mining techniques in the domain of Personal Services needs to be proved.

References

1. Kurz, M., Herrmann, C.: Adaptive Case Management – Anwendung des Business Process Management 2.0-Konzepts auf schwach strukturierte Geschäftsprozesse. In: Sinz, E.J., Bartmann, D., Bodendorf, F., Ferstl, O.K. (eds.) Dienstorientierte IT-Systeme für hochflexible Geschäftsprozesse, pp. 241–265. University of Bamberg Press, Bamberg (2011)
2. Bieber, D., Geiger, M.: Personenbezogene Dienstleistungen in komplexen Dienstleistungssystemen – eine erste Annäherung. In: Bieber, D., Geiger, M. (eds.) Personenbezogene Dienstleistungen im Kontext komplexer Wertschöpfung, pp. 9–49. Springer, Wiesbaden (2014). https://doi.org/10.1007/978-3-531-19580-3_1
3. van der Aalst, W., et al.: Process mining manifesto. In: Daniel, F., Barkaoui, K., Dustdar, S. (eds.) BPM 2011. LNBIP, vol. 99, pp. 169–194. Springer, Heidelberg (2012). https://doi.org/10.1007/978-3-642-28108-2_19
4. Halmos, P.: The personal service society. Br. J. Sociol. **18**, 13–28 (1967). https://doi.org/10.2307/588586
5. Motahari-Nezhad, H.R., Swenson, K.D.: Adaptive case management: overview and research challenges. In: Conference on Business Informatics, Wien, pp. 264–269 (2013)
6. Fließ, S., Dyck, S., Schmelter, M., Volkers, M.J.D.: Kundenaktivitäten in Dienstleistungsprozessen – die Sicht der Konsumenten. In: Fließ, S., Haase, M., Jacob, F., Ehret, M. (eds.) Kundenintegration und Leistungslehre, pp. 181–204. Springer, Wiesbaden (2015). https://doi.org/10.1007/978-3-658-07448-7_11
7. Ivarsson, M., et al.: Technology transfer decision support in requirements engineering research: a systematic review of REj. Requirements Eng. **14**(3), 155–175 (2009)
8. van der Aalst, W.M.P.: Exploring the CSCW spectrum using process mining. Adv. Eng. Inform. **21**(2), 191–199 (2006)
9. van der Aalst, W.M.P., et al.: Discovering social networks from event logs. Comput. Support. Coop. Work **14**(6), 549–593 (2005)
10. Ferreira, D.R., Alves, C.: Discovering user communities in large event logs. In: Daniel, F., Barkaoui, K., Dustdar, S. (eds.) BPM 2011. LNBIP, vol. 99, pp. 123–134. Springer, Heidelberg (2012). https://doi.org/10.1007/978-3-642-28108-2_11
11. He, Z., et al.: A process mining approach to improve emergency rescue processes of fatal gas explosion accidents in Chinese coal mines. Saf. Sci. **111**, 154–166 (2019)
12. Kamal, I.M., et al.: Identifying key resources in a social network using f-PageRank. In: IEEE 24th International Conference on Web Services (2017)
13. van der Aalst, W.M.P., et al.: Business process mining: an industrial application. Inf. Syst. **32**(5), 713–732 (2006)
14. Slaninová, K., Vymětal, D., Martinovič, J.: Analysis of event logs: behavioral graphs. In: Benatallah, B., et al. (eds.) WISE 2014. LNCS, vol. 9051, pp. 42–56. Springer, Cham (2015). https://doi.org/10.1007/978-3-319-20370-6_4
15. Appice, A.: Towards mining the organizational structure of a dynamic event scenario. J. Intell. Inf. Syst. **50**(1), 165–193 (2017)

16. Yang, J., Ouyang, C., Pan, M., Yu, Y., ter Hofstede, A.H.M.: Finding the "Liberos": discover organizational models with overlaps. In: Weske, M., Montali, M., Weber, I., vom Brocke, J. (eds.) BPM 2018. LNCS, vol. 11080, pp. 339–355. Springer, Cham (2018). https://doi.org/10.1007/978-3-319-98648-7_20
17. Lee, J., et al.: Dynamic human resource selection for business process exceptions. Knowl. Process Manage. J. Corp. Transform. **26**(1), 23–31 (2018)
18. Matzner, M., et al.: Process mining approaches to detect organizational properties in cyber-physical systems. In: ECIS 2014 Proceedings (2014)
19. Aalst, W.M.P.: Intra- and inter-organizational process mining: discovering processes within and between organizations. In: Johannesson, P., Krogstie, J., Opdahl, A.L. (eds.) PoEM 2011. LNBIP, vol. 92, pp. 1–11. Springer, Heidelberg (2011). https://doi.org/10.1007/978-3-642-24849-8_1
20. Creemers, M., et al.: Social mining as a knowledge management solution. In: CEUR Workshop Proceedings, vol. 1612 (2016)
21. Liu, T., et al.: A closed-loop workflow management technique based on process mining. In: 15th International Conference on Mechatronics and Machine Vision in Practice, Auckland, New Zealand (2008)
22. Accorsi, R., et al.: On the exploitation of process mining for security audits: the process discovery case. In: 28th Annual ACM Symposium on Applied Computing, Coimbra, Portugal (2013)
23. Sanchez-Cisneros, D., Gali, F.A.: UEM-UC3M: an ontology-based named entity recognition system for biomedical texts. In: Second Joint Conference on Lexical and Computational Semantics (*SEM), Volume 2: Seventh International Workshop on Semantic Evaluation (SemEval 2013), 2(SemEval), pp. 622–627 (2013)
24. Liu, B.: Sentiment Analysis and Opinion Mining. Morgan and Claypool Publishers, San Rafael (2012)
25. Blei, D.M., Ng, A.Y., Jordan, M.I.: Latent Dirichlet allocation. J. Mach. Learn. Res. **3**, 993–1022 (2003)
26. Bakhtiyari, K., Taghavi, M., Husain, H.: Implementation of emotional-aware computer systems using typical input devices. In: Nguyen, N.T., Attachoo, B., Trawiński, B., Somboonviwat, K. (eds.) ACIIDS 2014. LNCS (LNAI), vol. 8397, pp. 364–374. Springer, Cham (2014). https://doi.org/10.1007/978-3-319-05476-6_37
27. Epp, C., Lippold, M., Mandryk, R.L.: Identifying emotional states using keystroke dynamics. In: 29th Annual CHI Conference on Human Factors in Computing Systems. Association for Computing Machinery, New York (2011). ISBN 978-1-4503-0267-8
28. Lee, P.M., Tsui, W.H., Hsiao, T.C.: The influence of emotion on keyboard typing: an experimental study using auditory stimuli. PLoS ONE **10**(6) (2015). ISSN 1932-6203
29. Lv, H.R., Lin, Z.L., Yin, W.J., Dong, J.: Emotion recognition based on pressure sensor keyboards. In: IEEE International Conference on Multimedia and Expo, 2008, pp. 1089–1092. IEEE Service Center, Piscataway (2008). ISBN 978-1-4244-2571-6

Supporting ED Process Redesign by Investigating Human Behaviors

Alessandro Stefanini[1]([✉]), Davide Aloini[1], Peter Gloor[2],
and Federica Pochiero[1]

[1] University of Pisa, Lungarno Pacinotti, 43, 56126 Pisa, Italy
a.stefanini@ing.unipi.it
[2] Massachusetts Institute of Technology, Massachusetts Avenue, 77,
Cambridge, MA 02139, USA

Abstract. Human behaviors play a very relevant role in many Knowledge-Intensive Processes (KIPs), as healthcare processes, where the human factors are predominantly involved. Accordingly, health process performances, and particularly patient satisfaction, are highly affected by practitioners' behaviors and interactions.

Leveraging the novelty and potential of wearable sensors, this research aims to investigate behavioral factors affecting patient satisfaction in the Emergency Department (ED), with the final goal of supporting the ED process (re-)design in a holistic perspective. 51 patients and 135 practitioners were systematically monitored in a real emergency department using the Sociometric badges. Preliminary results show that patient satisfaction is greatly influenced by behavioral factors. Specifically, the ED doctors' behaviors seem to assume the most important role in the formation of patient perceptions during ED process. Finally, findings provide to researchers, practitioners, and health managers indications for designing and/or improving the ED service process.

Keywords: Patient satisfaction · Healthcare processes · Human behaviors · Emergency department · Wearable sensors

1 Introduction

The human behaviors, although increasingly important in all contexts (Bendoly et al. 2010; Loch and Wu 2007), play a very relevant role for unstructured processes, also known as Knowledge-Intensive Processes (KIPs), where the human factors are predominantly involved (Di Ciccio et al. 2015; Bendoly et al. 2015). Nowadays, unstructured processes are considerably important as they represent a consistent part of modern business processes, in particular in service field where the delivery process often involves the final customer.

In such contexts, the study of people's behaviors (employees, managers, customers, and, in general, all stakeholders) becomes fundamental to permit better recommendations of how to design/improve processes, management practices, and organizations (Croson et al. 2013; Fitzsimmons et al. 2008; Di Ciccio et al. 2015; Zerbino et al. 2018). Up to date, the BPM community has mostly focused on supporting the process-

© Springer Nature Switzerland AG 2019
C. Di Francescomarino et al. (Eds.): BPM 2019 Workshops, LNBIP 362, pp. 326–337, 2019.
https://doi.org/10.1007/978-3-030-37453-2_27

flow and the coordination structure, and has often neglected collaboration and behavioral aspects, which instead highly influence service process performances.

However, although human behaviors are recognized as central for effective design and management of services (Fitzsimmons et al. 2008), relatively little attention has been paid to quantitatively assess behavioral aspects in healthcare operations (Croson et al. 2013). Other business services are hardly dependent on human behaviors as healthcare, where the relationship practitioners-patient, the medical knowledge, and the team dynamics play a very important role.

Accordingly, individual and team behaviors produce significant "inputs" for many healthcare processes with a relevant impact on patient care, patient satisfaction, and efficiency (Manser 2009; Di Ciccio et al. 2015; Cannavacciuolo et al. 2018). Therefore, the effect of human (individual or group) behaviors on healthcare processes and on their related outcomes emerges as a serious limitation that should be overcome.

Difficulties in quantitatively analyzing behavioral factors in the real operation environments are a probable cause for the poor consideration of such elements in the healthcare context. Indeed, studies of behavioral operations are often conducted through "laboratory experiments" rather than with investigations in the real contexts (Croson et al. 2013; Katok 2011; Fügener et al. 2017). In addition, the data collection, in both simulated and real environments, is usually conducted by interviews, direct observations, questionnaires, and reports rather than by quantitatively measuring the actual behaviors. Consequently, data collection is commonly carried out in a batch way suffering from subjectivity, memory effect, and observer's influence on the system, which implies a lower data quality and trustworthiness (Olguín et al. 2009a; Kim et al. 2012).

Nevertheless, innovative data-driven approaches - e.g. based on wearable sensors or similar tools - may offer a possibility for overcoming these limits. Enabling automatic and objective measurements of human behaviors, these tools do not need the presence of observers and collect data in real time, increasing data richness and reliability and simplifying the analysis of real operation settings (Olguín et al. 2009a; Croson et al. 2013; Kim et al. 2012).

Leveraging the novelty and potential of wearable sensors, this research aims to identify and evaluate the main behavioral factors affecting patient satisfaction in the Emergency Department (ED), with the final goal of supporting the ED service (re-) design in a holistic perspective. Specifically, using the Sociometric Badges, this exploratory study quantitatively investigates the influence of ED medical team behaviors on patient satisfaction. A preliminary case study, within an Italian hospital, systematically and quantitatively monitors 51 patients (episodes) and 135 practitioners in a real emergency department. The obtained findings provide to researchers, practitioners, and health managers new directions for supporting the (re-)design and/or the improvement of ED service process.

This study also provides a relevant contribution from methodological perspective, by proposing a data-driven approach for analytically investigating behavioral factors affecting the service process delivery and, finally, for supporting the service (re-)design in complex socio-technical environments.

2 Theoretical Background

Patient satisfaction can be defined as subjective patient perception deriving from matching the services received and expectations regarding the service process and related outcome (Ross et al. 1987; Jain et al. 2017). Although the improvement of the patient's health is at the core of any healthcare service, the relationship between patient and practitioners also contributes to the service value and might strongly influence the customer perception of the service performance itself (Boquiren et al. 2015). Indeed, not always a good medical result leads to patient satisfaction as well as poor medical outcome might not correspond to a complete patient dissatisfaction. The relationship between patient and practitioners is recognized as an important component of patient satisfaction (Sitzia and Wood 1997; Boquiren et al. 2015), though investigating such connections is still an open issue that has only been addressed by a few researchers quantitatively.

Unquestionably, patient-practitioners relationships are multifaceted, subjective and hard to analyze in a systematic way. Social interactions, leadership, coordination and collaboration attitude, completeness and consistency of information exchanges, courtesy are just some of the many constituent variables which are at the basis of such a construct (LaVela and Gallan 2014; Boquiren et al. 2015). Most of these factors may be identified, at least partially, through unconscious social "honest" signals (Pentland 2008) in verbal and particularly non-verbal communication during teamwork and individual/group interactions with patients, which are typical of healthcare service operations.

Verbal communication in medical care is the fundamental instrument by which the doctor-patient relationship is crafted and by which therapeutic goals are achieved (Roter and Hall 2006). Doctors need information to establish the right diagnosis and treatment plan. Patients need to know and understand what the matter is and to feel known and understood by doctors (Ong et al. 1995). Both the doctors and patients alternate between information-giving and information-seeking phases. The way and the mood in which information is transmitted to the patient, the time devoted to explain and inform him/her, and simplicity in interactions affect the awareness degree of patient about his health condition and the satisfaction for medical care services (Ong et al. 1995; Boquiren et al. 2015). Therefore, communication is recognized as an important element for creating good interpersonal relationship and increasing patient satisfaction (Finney Rutten et al. 2015).

As well as verbal communication, non-verbal behaviors highly affect the relationship between patient and practitioners and, in turn, the patient perception of care services (Bensing 1991; Trout et al. 2000; Robinson 2006). Non-verbal behaviors refer to communicative actions distinct from speech, such as facial expressions, gesturing, body posture, physical distance (proximity), and positioning. Patients are very sensitive to such behaviors that convey the emotional tone of interpersonal interaction (Robinson 2006), in particular in an Emergency Department where patients and relatives feel stronger emotions such as fear, anxiety, and uncertainty (Trout et al. 2000; Chang et al. 2016). Larsen and Smith (1981), for example, investigated the relationship between doctors' non-verbal activities and patient satisfaction discovering that the higher closeness in interactions was, the higher was patient satisfaction. Similarly, Beck et al. (2002) also confirmed that some nonverbal behaviors, like head nodding, leaning forward, direct body orientation, and gaze, positively associate with service perception, while Boissy et al. (2016) proved that training courses on communication skills for physicians can improve patient satisfaction.

Up to date, although many hospitals and policy makers are interested to measure and maximize patient satisfaction (Welch 2010), only a few studies quantitatively analyze the relationship between practitioners-patient behaviors and patient satisfaction in healthcare service and specifically in EDs. Main challenges are still related to systematically measure individual and team behaviors in such dynamic environments (Kim et al. 2012; Rosen et al. 2014). Towards this purpose, this paper proposes a novel approach for evaluating the main behavioral variables (verbal and non-verbal) trying to overcome the limits of past methods in this field.

3   Methodology

For an effective evaluation of influences of ED team behaviors on patient satisfaction, a novel systematic measurement approach powered by the Sociometric Badges is adopted to obtain quantitative and reliable measures of ED team behaviors during the service operations. Indeed, traditional approaches to behavioral studies, like interviews, direct observations, questionnaires and reports, usually suffer of various biases such as subjectivity, memory effects, influence of the observer on the system (Cunningham et al. 2012; Kim et al. 2012; Pronin 2007) that may be overtaken thanks to this innovative approach.

Sociometric Badges are wearable sensors able to automatically and directly measure individual and collective behaviors, exploiting four different sensors: accelerometer, microphones, Bluetooth, and IRDA (Olguín et al. 2009a). In this way, these tools can collect quantitative behavioral measures impossible to gather with surveys/interviews, while still guarantying privacy. Particularly, it is impossible to determine the content of the conversation or to identify the speaker from the sociometric data (Olguín et al. 2009a), an important element in healthcare settings. They are also less intrusive than a human observer limiting any social distortions to the data and potentially enriching the data collected (Olguín et al. 2009a; Rosen et al. 2014). The suitability and usefulness of Sociometric Badges for monitoring behavioral variables is proven by past research (Kim et al. 2012), also in the healthcare field (e.g. Olguín et al. 2009b; Bucuvalas et al. 2014).

Given the lack of research evaluating the relationship between the behaviors of ED teams and their performance and the novelty of the measurement approach, an exploratory case study (Yin 2017) was carried out for this preliminary investigation. The case study was structured in the following three main phases:

- Research setup. The Emergency Department under examination was observed in order to define the research protocol for conducting the study. The preliminary ED investigation outlined all features related to the service, such as the department layout and physical distribution of medical staff in the ED, the tasks assigned to each practitioner and their interactions with the patient, how patients access to service, and finally, patient process paths in the ED. Starting from this information, the research was designed. Specifically, all the relevant aspects for the data collection phase were delineated, as for example the "setting parameters" of Sociometric Badges, the medical staff involved in the evaluation, the sociometric variables to be recorded, the Social Network Analysis metrics applied (e.g. Gloor et al. 2017), the survey measures for the performances to be included in the questionnaires, the control variables, etc. To be noted that the ED is organized for work cells, where each ED team is allocated for a single cell and takes care of 7–8 patients.

- Data collection and preparation. In this phase, all the necessary data for the study were collected and pre-processed with appropriate software for checking correctness and for preparing them for the subsequent analyses. In addition, incorrect or incomplete data were removed.
- Data analysis. Correlation and regression analysis were carried out. The correlation analysis allowed identifying relationships between the behavioral variables, measured with Sociometric Badges, and the outcome in term of patient satisfaction and service performance as perceived by patients. Moreover, this analysis permit to evaluate potential correlations amongst the control variables and the dependent variables. Preliminary regression analysis created models to partially explain the outcome variances with sociometric variables. In this way, it measured the effects of behavioral variables on patient satisfaction. Regression analysis using the control variables also allowed to exclude their significant influence on the outcomes and confirmed the preliminary relationships discovered.

Finally, experts, health managers, and medical staff discussed about the achieved results and the related implications from a managerial viewpoint.

4   Case Study

The research was conducted in an Italian Emergency Department. ED teams composed by a minimum of two practitioners (one doctor and one nurse) to a maximum of four (one doctor, one nurse, one specializing doctor and/or one trainee nurse) were continuously monitored during the service. Data collection involved conscious patients with all emergency severity classification except for red codes (life-threatening, immediate access to care).

For each case investigated, behaviors of the ED team and of the patient were monitored using the Sociometric Badges. Team members and the patient wore the tool for the entire duration of the patient's stay. Exploiting the data recorded by Sociometric Badges, a series of behavioral variables were defined such as body movement, posture activity, speaking activity and network, proximity interaction, and level of audio.

A questionnaire (anonymous) regarding the perception and satisfaction of the service received was submitted to the patient, at the end of his/her stay in the ED. Three "patient perception" variables were extracted from this questionnaire: overall satisfaction, the perceived care effectiveness, and the perceived team responsiveness.

In addition, patient/episode characteristics were collected from the ED information system to achieve the appropriate control variables for excluding potential influences from specific episode characteristics. Specifically, the monitored control variables were: overall length of stay in the emergency department, Patient sex, Patient age, team members' number, and emergency severity (severity color classification).

After discarding incorrect registrations, the final dataset consisted of 51 episodes (patients) with 135 medical staff distinct recordings (total of 293 monitoring hours).

5 Findings

To explore the potential relationships between behavioral variables and patient perceptions, Pearson's correlation analysis was performed. It was useful for getting a first relevant insight of the data and supporting the next phases of regression analysis. The results of correlation analysis are reported in the following three tables (Tables 1, 2, and 3).

In Table 1, the correlations between the Overall Satisfaction, the behavioral variables, and the control variables are shown. For sake of brevity, only the behavioral variables that have a relevant correlation with the Overall Satisfaction are presented.

Two significant correlations, which were deepened through the regression analysis, were discovered for Overall Satisfaction, the *Doctor's Posture Activity* and *Mirroring audio front-Network's deviation*. It is noteworthy to point out that there are no significant correlations between the control variables and Overall Satisfaction.

Table 1. Correlations between Overall Satisfaction, independent and control variables (N = 51)

	Overall Satisfaction	Posture activity - Doctor	Mirroring audio front - Network's deviation	Number of team members	Emergency color classification	Sex	Age	Overall Length Of Stay (LOS)
Overall Satisfaction	1	−.338*	−.452**	−.114	−.065	−.016	−.000	−.005
Doctor's Posture Activity	−.338*	1	.113	.194	.000	.229	.042	.130
Mirroring audio front-Network's deviation	−.452**	.113	1	.228	−.046	.085	−.013	.039
Number of team members	−.114	.194	.228	1	.249	.198	−.058	.136
Emergency color classification	−.065	.000	−.046	.249	1	−.023	−.360**	.075
Sex	−.016	.229	.085	.198	−.023	1	.135	−.024
Age	−.000	.042	−.013	−.058	−.360**	.135	1	.042
Overall Length Of Stay (LOS)	−.005	.130	.039	.136	.075	−.024	.042	1

**Correlation is significant at the 0.01 level (2-tailed).
*Correlation is significant at the 0.05 level (2-tailed).

In Table 2, the correlations between the perceived Care Effectiveness, the behavioral variables, and the control variables are shown. As before, not all the behavioral variables are presented.

Two significant correlations were discovered for Care Effectiveness, the *Doctor's Posture Activity* and *Mirroring audio front-Network's deviation*. Thus, the correlation analysis results for Overall Satisfaction and Care Effectiveness seem to be very similar. It is noteworthy to point out that also in this case there are no significant correlations between the control variables and Care Effectiveness.

Table 2. Correlations between Care Effectiveness, independent and control variables (N = 51)

	Care Effectiveness	Posture activity - Doctor	Mirroring audio front - Network's deviation	Number of team members	Emergency color classification	Sex	Age	Overall Length Of Stay (LOS)
Care Effectiveness	1	−.350*	−.419**	−.128	−.126	−.031	−.038	−.031
Doctor's Posture Activity	−.350*	1	.113	.194	.000	.229	.042	.130
Mirroring audio front-Network's deviation	−.419**	.113	1	.228	−.046	.085	−.013	.039
Number of team members	−.128	.194	.228	1	.249	.198	−.058	.136
Emergency color classification	−.126	.000	−.046	.249	1	−.023	−.360**	.075
Sex	−.031	.229	.085	.198	−.023	1	.135	−.024
Age	−.038	.042	−.013	−.058	−.360**	.135	1	.042
Overall Length Of Stay (LOS)	−.031	.130	.039	.136	.075	−.024	.042	1

**Correlation is significant at the 0.01 level (2-tailed).
*Correlation is significant at the 0.05 level (2-tailed).

In Table 3, the correlations between the perceived Team Responsiveness, the behavioral variables, and the control variables are shown. As before, not all the behavioral variables are presented.

Two significant correlations were discovered for Team Responsiveness, the *Doctor's Posture Activity* and *Doctor's Speech Overlap*. It is noteworthy to point out that also in this case there are no significant correlations between the control variables and Team Responsiveness.

To measure the effects of the behavioral variables on the patient perceptions, a preliminary regression analysis was carried out. Variables recorded by the Sociometric Badges were taken as independent variables, while the Overall Satisfaction, the perceived Care Effectiveness, and the perceived Team Responsiveness were individually introduced in the regression models as dependent variables. Thereby, three significant regression models, one for each dependent variable, were identified in this preliminary phase of regression analysis.

Table 3. Correlations between Team Responsiveness, independent and control variables (N = 51)

	Team Responsiveness	Posture activity - Doctor	Doctor's Speech Overlap	Number of team members	Emergency color classification	Sex	Age	Overall Length Of Stay (LOS)
Team Responsiveness	1	−.333*	−.479**	−.119	−.122	−.042	.088	−.049
Doctor's Posture Activity	−.333*	1	.113	.194	.000	.229	.042	.130
Doctor's Speech Overlap	−.479**	.113	1	.228	−.046	.085	−.013	.039
Number of team members	−.119	.194	.228	1	.249	.198	−.058	.136
Emergency color classification	−.122	.000	−.046	.249	1	−.023	−.360**	.075
Sex	−.042	.229	.085	.198	−.023	1	.135	−.024
Age	.088	.042	−.013	−.058	−.360**	.135	1	.042
Overall Length Of Stay (LOS)	−.049	.130	.039	.136	.075	−.024	.042	1

**Correlation is significant at the 0.01 level (2-tailed).
*Correlation is significant at the 0.05 level (2-tailed).

Here the findings related to the correlation analysis and to the preliminary regression analysis are briefly discussed:

Overall Satisfaction appears positively influenced by a continuous attendance of doctors in the work cell (low *Doctor's Posture Activity* that means low doctor's walking activity) and by the presence of a leader in the communication network (i.e. a dominant figure in the team-speaking network). This evidence confers value to doctor attendance during the ED service delivery and to its role of communication leader within the ED team.

Perceived Care Effectiveness appears positively influenced by two additional factors over the previous ones: patient listening, i.e. time devoted by medical staff listening the patient (high *Patient's Speech*); and patient monitoring, i.e. a frequent check of patient's health conditions by the nurse (high *Nurse's Body Activity*). These two additional factors, not revealed by the correlation analysis, were discovered through the preliminary regression analysis. Consistently with previous evidence about the importance of patient centrality, findings confirm the patient expectations to be actively involved into the communication to express his/her own ideas and doubts to practitioners and to be constantly monitored during his/her stay.

Perceived Team Responsiveness appears influenced positively by continuous attendance of doctors in the work cell (as in the previous models) and negatively by speech overlapping between the doctor and other people (low *Doctor's Speech Overlap*). Again, this evidence enforces the importance of organized communication patterns among team members to achieve completeness and consistency of the information exchange, which is often associated with leadership, coordination, and collaboration attitudes (Boquiren et al. 2015).

To confirm the validity of findings, regressive models using the control variables were tested. The first check was to add the control variables, individually and together, to the regression models obtained. In any case, model performances did not considerably improve and the inserted control variables were non-significant. The second test was to build the model simply using control variables. Also in this case, no model built with the control variables was significant. These tests appear as a confirmation of our findings, excluding any potential effect of the control variables.

In addition, it is interesting to note that, despite expectations, the overall throughput time seems not to influence the patient's judgment in the investigated setting.

6 Conclusions

This paper shed a light on the investigation of behavioral factors, which are often neglected by BPM community and which instead highly influence service process performances. Indeed, healthcare organizations are largely composed of people, thus excluding behaviors from the analysis of this context means to renounce to a holistic, effective, and factual investigation of healthcare processes (Fitzsimmons et al. 2008). By quantitatively exploring team behaviors affecting patient satisfaction in the Emergency Department (ED), this study aims to provide managers with effective service process design and improvement indications.

In line with previous literature (e.g., Sitzia and Wood 1997; Boquiren et al. 2015), results confirm that practitioner (team) behaviors may highly affect patient satisfaction. Specifically, as revealed by statistical analysis, patient evaluations seem to be influenced predominantly by numerous behavioral dynamics which can be associated with service attendance, risk aversion, social interactions, leadership, coordination and collaboration attitude, completeness and consistency of the information exchange in the communication networks (LaVela and Gallan 2014; Boquiren et al. 2015). As shown in the findings section, patient satisfaction is highly influenced by the time devoted to him/her: the presence of the doctor near the patient and the level of interactions (communication) with the staff seem to have a strong impact on the patient perceptions. Moreover, patients also perceive how the team interacts and coordinates itself, favoring teams with a collaborative dialogue. Finally, the patients' judgment seems surprisingly not affected by the overall throughput time.

From a methodological perspective, the proposed data-driven approach may be applied also in other service settings, helping researchers to systematically discover and quantitative evaluate the behavioral elements affecting the service process delivery and finally support the service (re-)design in complex socio-technical contexts. Thus, this research also provides a contribution to the problem of "how" to quantitatively investigate behavioral aspects in the BPM community.

This study also provides a relevant contribution from a managerial perspective. The highlighted findings may offer to hospital managers relevant managerial indications for improving patient satisfaction, based on real data-driven analyses. Specifically, the findings suggest that doctors should remain physically close to the patients (possibly in "eye contact" range) and assume the role of communication leader assuring completeness and consistency of the information exchange within the ED team and between team members and the patients, but they should avoid to speak over other people. Moreover, all the ED team members should pay attention to the patient centrality during the service delivery by frequently monitoring the patients' health conditions and by actively involving the patient into the conversation about his/her illness and, thus, permitting him/her to express his/her own ideas and doubts.

Although it is challenging to control for all the determinants of such behaviors, these indications may support health managers during the service process (re-)design phase and may be useful for training ED staff about leadership, coordination, and collaboration skills (Boquiren et al. 2015). For example, the layout could be redesigned for increasing the proximity of the doctor to the patients, placing the ED doctor desk near to the beds of current ED patients and with the possibility for them to see him continuously, or ED teams configuration may be modified in order to avoid team dispersion during the service delivery.

This research presents also several limitations that point out directions for future research. The first is due to the exploratory nature of the work. Drawing on a single case study, results might be affected by the particular context. This is a common issue for many behavioral studies that limit generalization (Tröster et al. 2014). An extension of the sample in terms of number of investigated patients, teams, time window, and also possible replication to other EDs would be worthwhile.

Moreover, as the number of monitored variables and related indicators is high, the study is clearly not conclusive. Other relevant metrics describing behavioral dynamics of ED teams probably exist and might not be caught by sociometric measures.

Finally, as a suggestion for future development, it would be interesting to investigate behavioral variables more deeply to understand how results are affected by ED team behaviors and properly characterize the relationship behaviors-performances.

References

Beck, R.S., Daughtridge, R., Sloane, P.D.: Physician-patient communication in the primary care office: a systematic review. J. Am. Board Fam. Pract. **15**(1), 25–38 (2002)

Bendoly, E., Croson, R., Goncalves, P., Schultz, K.: Bodies of knowledge for research in behavioral operations. Prod. Oper. Manag. **19**(4), 434–452 (2010)

Bendoly, E., van Wezel, W., Bachrach, D.G.: The Handbook of Behavioral Operations Management: Social and Psychological Dynamics in Production and Service Settings. Oxford University Press, New York (2015)

Bensing, J.: Doctor-patient communication and the quality of care. Soc. Sci. Med. **32**(11), 1301–1310 (1991)

Boissy, A., et al.: Communication skills training for physicians improves patient satisfaction. J. Gen. Intern. Med. **31**(7), 755–761 (2016)

Boquiren, V.M., Hack, T.F., Beaver, K., Williamson, S.: What do measures of patient satisfaction with the doctor tell us? Patient Educ. Couns. **98**(12), 1465–1473 (2015)

Bucuvalas, J., Fronzetti Colladon, A., Gloor, P.A., Grippa, F., Horton, J., Timme, E.: Increasing interactions in healthcare teams through architectural interventions and interpersonal communication analysis. J. Healthc. Inf. Manag. JHIM **28**(4), 58–65 (2014)

Cannavacciuolo, L., Ippolito, A., Ponsiglione, C., Rossi, G., Zollo, G.: How organizational constraints affect nurses' decision in triage assessment performances. Meas. Bus. Excel. **22** (4), 362–374 (2018)

Chang, B.P., Carter, E., Suh, E.H., Kronish, I.M., Edmondson, D.: Patient treatment in emergency department hallways and patient perception of clinician-patient communication. Am. J. Emerg. Med. **34**(6), 1163 (2016)

Croson, R., Schultz, K., Siemsen, E., Yeo, M.L.: Behavioral operations: the state of the field. J. Oper. Manag. **31**(1), 1–5 (2013)

Cunningham, F.C., Ranmuthugala, G., Plumb, J., Georgiou, A., Westbrook, J.I., Braithwaite, J.: Health professional networks as a vector for improving healthcare quality and safety: a systematic review. BMJ Qual. Saf. **21**(3), 239–249 (2012)

Di Ciccio, C., Marrella, A., Russo, A.: Knowledge-intensive processes: characteristics, requirements and analysis of contemporary approaches. J. Data Semant. **4**(1), 29–57 (2015)

Finney Rutten, L.J., et al.: The relation between having a usual source of care and ratings of care quality: does patient-centered communication play a role? J. Health Commun. **20**(7), 759–765 (2015)

Fitzsimmons, J.A., Fitzsimmons, M.J., Bordoloi, S.: Service Management: Operations, Strategy, and Information Technology. McGraw-Hill, New York (2008)

Fügener, A., Schiffels, S., Kolisch, R.: Overutilization and underutilization of operating rooms-insights from behavioral health care operations management. Health Care Manag. Sci. **20**(1), 115–128 (2017)

Gloor, P.A., Fronzetti Colladon, A., Grippa, F., Giacomelli, G.: Forecasting managerial turnover through e-mail based social network analysis. Comput. Hum. Behav. **71**, 343–352 (2017)

Jain, D., et al.: Higher patient expectations predict higher patient-reported out-comes, but not satisfaction, in total knee arthroplasty patients: a prospective multicenter study. J. Arthroplast. **32**, S166–S170 (2017)

Katok, E.: Using laboratory experiments to build better operations management models. Found. Trends Technol. Inf. Oper. Manag. **5**(1), 1–86 (2011)

Kim, T., McFee, E., Olguin, D.O., Waber, B., Pentland, A.: Sociometric badges: using sensor technology to capture new forms of collaboration. J. Organ. Behav. **33**(3), 412–427 (2012)

Larsen, K.M., Smith, C.K.: Assessment of nonverbal communication in the patient-physician interview. J. Fam. Pract. **12**(3), 481–488 (1981)

LaVela, S.L., Gallan, A.: Evaluation and measurement of patient experience. Patient Exp. J. **1**(1), 28–36 (2014)

Loch, C.H., Wu, Y.: Behavioral operations management. Found. Trends® Technol. Inf. Oper. Manag. **1**(3), 121–232 (2007)

Manser, T.: Teamwork and patient safety in dynamic domains of healthcare: a review of the literature. Acta Anaesthesiol. Scand. **53**(2), 143–151 (2009)

Olguin, D.O., Gloor, P.A., Pentland, A.: Wearable sensors for pervasive healthcare management. In: 3rd International Conference on IEEE Pervasive Computing Technologies for Healthcare, Pervasive Health 2009, pp. 1–4 (2009b)

Olguin, D.O., Waber, B.N., Kim, T., Mohan, A., Ara, K., Pentland, A.: Sensible organizations: technology and methodology for automatically measuring organizational behavior. IEEE Trans. Syst. Man Cybern. Part B **39**(1), 43–55 (2009a)

Ong, L.M., De Haes, J.C., Hoos, A.M., Lammes, F.B.: Doctor-patient communication: a review of the literature. Soc. Sci. Med. **40**(7), 903–918 (1995)

Pentland, A.S.: Honest Signals: How They Shape Our World. MIT Press, London (2008)

Pronin, E.: Perception and misperception of bias in human judgment. Trends Cogn. Sci. **11**(1), 37–43 (2007)

Robinson, J.D.: Nonverbal communication and physician–patient interaction. In: The SAGE Handbook of Nonverbal Communication, pp. 437–459 (2006)

Rosen, M.A., Dietz, A.S., Yang, T., Priebe, C.E., Pronovost, P.J.: An integrative framework for sensor-based measurement of teamwork in healthcare. J. Am. Med. Inform. Assoc. **22**, 11–18 (2014)

Ross, C.K., Frommelt, G., Hazelwood, L., Chang, R.W.: The role of expectations in patient satisfaction with medical care. Mark. Health Serv. **7**(4), 16 (1987)

Roter, D., Hall, J.A.: Doctors Talking with Patients/Patients Talking with Doctors: Improving Communication in Medical Visits. Greenwood Publishing Group, Westport (2006)

Sitzia, J., Wood, N.: Patient satisfaction: a review of issues and concepts. Soc. Sci. Med. **45**(12), 1829–1843 (1997)

Tröster, C., Mehra, A., van Knippenberg, D.: Structuring for team success: the interactive effects of network structure and cultural diversity on team potency and performance. Organ. Behav. Hum. Decis. Process. **124**(2), 245–255 (2014)

Trout, A., Magnusson, A.R., Hedges, J.R.: Patient satisfaction investigations and the emergency department: what does the literature say? Acad. Emerg. Med. **7**(6), 695–709 (2000)

Welch, S.J.: Twenty years of patient satisfaction research applied to the emergency department: a qualitative review. Am. J. Med. Qual. **25**(1), 64–72 (2010)

Yin, R.K.: Case Study Research and Applications: Design and Methods. Sage Publications, Thousand Oaks (2017)

Zerbino, P., Aloini, D., Dulmin, R., Mininno, V.: Big data-enabled customer relationship management: a holistic approach. Inf. Process. Manage. **54**(5), 818–846 (2018)

The Potential of Workarounds for Improving Processes

Iris Beerepoot[1,2(✉)], Inge van de Weerd[2], and Hajo A. Reijers[2,3]

[1] ICTZ B.V., Hoorn, The Netherlands
[2] Utrecht University, Utrecht, The Netherlands
{i.m.beerepoot,i.vandeweerd,h.a.reijers}@uu.nl
[3] Eindhoven University of Technology, Eindhoven, The Netherlands

Abstract. Several studies have hinted how the study of workarounds can help organizations to improve business processes. Through a systematic literature review of 70 articles that discuss workarounds by information systems users, we aim to unlock this potential. Based on a synthesis of recommendations mentioned in the reviewed studies, we describe five key activities that help organizations to deal with workarounds. We contribute to the IS literature by (1) providing an overview of concrete recommendations for managing workarounds and (2) offering a background for positioning new research activities on the subject. Organizations can apply these tools directly to turn their knowledge on workarounds into organizational improvement.

Keywords: Workarounds · Information systems · Process improvement

1 Introduction

People often use Information Systems (IS) different from their designed usage. IS users' deviations from designed procedures are also known as *workarounds*, defined by Alter as follows: "A workaround is a goal-driven adaptation, improvisation, or other change to one or more aspects of an existing work system in order to overcome, bypass, or minimize the impact of obstacles, exceptions, anomalies, mishaps, established practices, management expectations, or structural constraints that are perceived as preventing that work system or its participants from achieving a desired level of efficiency, effectiveness, or other organizational or personal goals" [1].

Workarounds are inherently about human agency. No matter how technologies are designed, humans can always choose how they use technologies to perform their work [2–4]. Workarounds are also inherently related to processes. There is always a prescribed process that users deviate from, such as the process of administering medication [5] or accessing patient data [6]. Whereas they have been viewed negatively in the past, current literature calls for a more positive perspective on workarounds [7, 8]. Several studies point out the potential of workarounds for identifying poorly-designed processes [9, 10] and for involving IS users in process improvement efforts [8, 11, 12].

To find out how workarounds can be used for improvement and how IS users can play a role in process improvement efforts, we raised the following research question: *how can organizations unlock the potential of workarounds for improving processes?*

© Springer Nature Switzerland AG 2019
C. Di Francescomarino et al. (Eds.): BPM 2019 Workshops, LNBIP 362, pp. 338–350, 2019.
https://doi.org/10.1007/978-3-030-37453-2_28

Our contribution with this work is twofold. First, by analyzing and synthesizing the literature describing the potential of exploiting workarounds for improving processes, we propose five key activities necessary to unlock this potential, providing organizations with the means to use the workarounds for improvement. Second, we provide a background for positioning new research activities that target workarounds for organizational improvement.

The remainder of this paper is structured as follows. We first describe the methods we used. In the subsequent section, we sketch the preconditions for workarounds, after which our proposed activities for achieving process improvement are discussed. Finally, we present our conclusions and a research outlook.

2 Methods

We performed an in-depth systematic literature review, following the guidelines by [13] and the checklists by [14] and [15]. The aim of this study is to present an integrated and representative overview of existing studies on how organizations can use workarounds for improvement. Figure 1 visualizes the search and selection process.

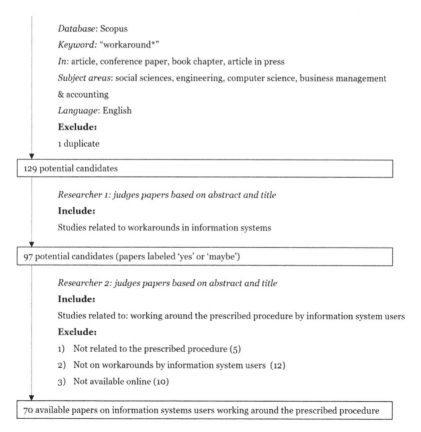

Fig. 1. Search and selection process.

To collect a broad sample of papers, we used the Scopus database to retrieve our candidate papers. The search on Scopus for articles mentioning workarounds resulted in 129 potential candidates. We carried out two screening rounds to narrow our sample. In the first round, the first author judged the papers based on their titles and abstracts. Studies that actually focused on the use of workarounds during the interaction with information systems were selected. Studies that were not included were papers primarily proposing some technical workaround to solve an erroneous software design. Using the workaround definition by Alter mentioned in the introduction, we excluded 32 candidates during the first screening round. These articles were not related to workarounds in information systems. As a result, 97 potential candidates were left for screening in the second round. These articles were labeled either 'yes' or 'maybe' and were further screened by the second author. Studies that were excluded in this round were either not focusing on working around a prescribed procedure, not on workarounds by information system users, or not available via our university's online library. Our final result was a sample of 70 papers on information systems users working around prescribed procedures.

Figure 2 visualizes our analysis and synthesis process. We focused our analysis on the ways in which organizations can exploit workarounds for process improvement. Our aim was to develop a framework that gives insight into both the potential of workarounds for improvement and how this potential can be realized.

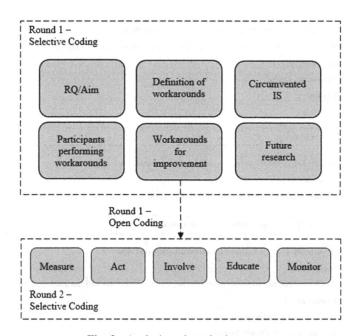

Fig. 2. Analysis and synthesis process.

For our first coding round, we imported all papers into Atlas.ti[1], a software program used to guide qualitative data analysis. The first and second author selectively coded the articles, regularly discussed the codes and adjusted them if necessary.

While selectively coding the literature in the first round, the number of quotations coded 'workarounds for improvement' increased rapidly (529 quotations by the end of the first coding round). Because of this large set of quotations, we decided to use open coding next to selective coding. Doing so, we created sub-codes for the code 'workarounds for improvement'. We found that the studies analyzed include many recommendations for using workarounds for improvement, and that these recommendations could be clustered into five groups. The recommendations related to detecting and gathering information on workarounds (the 'measure'-group), acting on or addressing workarounds (the 'act'-group), involving end users of the information system (the 'involve'-group), training and educating end users (the 'educate'-group) and monitoring workarounds over time (the 'monitor'-group). In the second close-reading iteration, we focused exclusively on the five clusters of recommendations to unlock the potential of workarounds for process improvement. We selectively coded the papers using the five sub-codes.

Before we discuss the five activities in more detail, we give a general introduction to the emergence of workarounds. Based on the literature review, we discuss when they emerge and what their effects are.

3   The Emergence of Workarounds

3.1   Dysfunctionality as a Cause of Workarounds

Several authors believe that the cause of workarounds is a dysfunctional environment [16–18]. There are several reasons why process participants perform workarounds [19]. Our literature review reveals that this is done to overcome 'constraints' [18, 20], 'incompatibilities' [21], 'inadequacies' [22], 'flawed specifications' [23], 'unrealistic processes' [23], 'obstacles' [24, 25], 'mismatches' [26–28] or 'misfits' [29, 30]. In the healthcare setting, for instance, clinicians sometimes feel constraints in achieving their goals: "many workarounds occur because the health IT itself can undermine the central mission of the clinician: serving patients" [31].

Other causes for workarounds are tensions that might exist. An example is the "tension between top-down pressures from the external environment and bottom-up constraints from day-to-day operational work" [11]. Workarounds are used to relieve this tension and balance top-down pressures such as compliance rules and bottom-up time-constraints. Another tension that potentially causes workarounds is the one between standardization and flexibility. Carayon and Gürses [32] found that hospital nurses enact more workarounds when they are coerced into using standardized routines. The same was concluded by Van Beijsterveld and van Groenendaal [28], who established that the inability to customize the system leads participants to engage in

---

[1] https://atlasti.com/ .

workarounds. Without such customizations, they become dissatisfied and start to resist the system [12].

3.2 The Effects of Workarounds

Many workarounds add value [20], save time [24] or improve efficiency [33]. They allow participants to continue work [34–36] by offering a temporary solution to an obstacle [37].

Apart from the positive effects, workarounds can affect an organization negatively in two ways. First, although they can increase efficiency in some situations, they affect efficiency negatively in others [25, 37–39]. When participants feel the need to enact workarounds to achieve their goals, this causes frustration [40], discontent [41] and disengagement [38]. In addition to this, workarounds affect other activities in the process, threatening to decrease the overall outcome of the process [39] and bringing security issues with it. When using the setting of healthcare organizations again, this could mean endangering the safety of patients [18, 42].

The second major negative effect of workarounds is a loss of transparency. Workaround activities are usually hidden [3, 9] and management and IS vendors are often unaware of them [29, 43, 44]. This leads to managers and IS vendors having an inaccurate view of system usage, as workarounds mask "underlying system weakness" [41]. It "creates the illusion that dysfunctional systems are indeed functioning" [45]. Working around bugs in the system, for instance, leaves manufacturers unaware of them [43], which means that nothing is done solve them. Similarly, if management is not made aware of dysfunctionalities, they will not address those either. Alternatively, if they do make decisions on processes, they are "based upon an illusion of actuality and not on the reality of workplace activities" [46]. Managers could be making important decisions based on incomplete information [29], which gives a false sense of compatibility between information systems and work processes.

3.3 Workarounds as Feedback Resources

Organizations can use knowledge of workarounds to improve processes and ISs. The majority of studies on workarounds suggest that workaround activities have the potential to bring about improvement in organizations. They are especially useful for guiding IS redesign since they contain information about necessary customizations of the IS [12]. They "offer a blueprint for identifying the pressing information gaps that need to be resolved when considering improvements in an information flow" [9]. Similarly, workarounds can help improve the design of work processes, because they give insight into the day-to-day activities of participants and their needs to perform these tasks [12]. They may even guide organizations in re-evaluating the entire process environment by challenging "the ability and coherence of processes and systems that no longer serve the organization, its employees, or its customers" [47].

The undertaken improvement efforts, in turn, lead to increased efficiency [25], better communication [47] and improved satisfaction on the part of participants [12, 42, 48]. By approaching workarounds as feedback resources [12], organizations can perform corrective actions and make improvements to processes. In the next section, we derive from literature a set of five activities that help organizations to unlock the potential of workarounds for improvement.

4 Five Activities to Improve Processes

4.1 Measure

Many authors stress the importance of knowing why participants perform workarounds, described as 'motivations' [21, 26], 'reasons' [19, 49], 'obstacles' [24] or antecedents [34]. Others simply call for an understanding of participants' work practices [36, 50, 51] because they consider the way people work and work around prescribed processes imperative for deciding on a strategy. Van Beijsterveld and Van Groenendaal [28], for instance, argue that "actual misfits require a different solution strategy than perceived misfits do". Similarly, Röder et al. [52] debate that whether the intention of the participant is positive or negative should be the basis for deciding on a resolution strategy.

In contrast to focusing on the motivations of workarounds, other authors focus on the consequences instead. Drum et al. [39], for example, state that "the motive underlying the workaround, while interesting, does not afford a satisfactory understanding of workarounds. Rather, we believe it is more beneficial to focus on the outcomes generated by workarounds". Also interesting in terms of consequences of a workaround is its downstream effect [47, 53]. According to Drum et al. [39], "the use of workarounds often constrains or decreases the overall effectiveness of the system, especially for those 'downstream' from the workaround who must deal with its outcomes". Others take both motivations and consequences into account [21, 24, 26]. According to Röder et al. [52], consequences can be further specified into risks and benefits. These risks and benefits can provide a basis for improvement efforts [20, 23, 26, 54, 55].

In terms of the means to measure workarounds, several authors suggest to identify the workarounds in situ, at the practice level [18, 27, 56, 57]. This can be achieved by performing interviews, observations, shadowing and focus groups [31]. Several studies on workarounds, however, pointed out quantitative limitations, for instance not knowing the frequency of workarounds [5] or the expenditure of money, time and effort [28]. A way to overcome this is the use of process mining techniques that "use event data to extract process-related information" [58]. This enables organizations to meet the demand for measuring "the actual value of workaround time and effort compared with the original process" [21]. Outmazgin and Soffer [49] showed that process mining techniques can indeed be used to detect certain types of workaround behaviors, although others were not reflected in the event log. Also, the motivation of participants to perform workarounds and relevant situational factors are difficult to determine using these techniques. Therefore, more traditional techniques such as performing observations remain to have value [31].

In sum, we propose that the first necessary activity to achieve process improvement is to measure workarounds. Specifically, what needs to be measured is the motivation of the participant to perform the workaround and the associated consequences. Our view is that this can best be done in the form of a hybrid approach, by performing qualitative observations of participants and using quantitative process mining techniques.

4.2   Act

According to Drum et al. [59], "workarounds must be addressed". However, as mentioned in the previous section, different types of workarounds must be addressed differently. One rule of thumb that is frequently mentioned in the literature is to manage workarounds by controlling risks and maintaining benefits [20, 21, 53]. Specifically, organizations are advised to facilitate or adopt appropriate workarounds and prevent or block the inappropriate ones [21, 38, 47]. According to Park et al. [33], the evaluation of appropriate and inappropriate workarounds is not an easy task, as "careful internal analysis might be necessary to identify which adaptations [...] should be supported, rather than merely eliminating problematic immediate adaptations". More authors advise organizations against simply eliminating workarounds, as doing so may result in negative outcomes [11, 20, 22]. Eliminating the underlying *reasons* to perform workarounds, however, is recommended and expected to lead to positive results [25].

Acting on workarounds may entail activities such as process redesign, disciplinary actions [49], improvements in the technology or control routines [54]. Usually, these actions fall into two categories: (1) customizations to the information system and (2) changes to the structure of the organization [28]. A concrete example of an organizational action that was suggested in two separate studies is ensuring that participants have physical access to specific process roles. In Halbesleben et al. [5], this entailed relocating a pharmacist to a nursing unit. In Tucker [60], it involved increasing the nurses' access to the process owner. In both cases, this was shown to improve the process: in the first it led to a decreased amount of rework and frequency of workarounds; in the second it caused participants to enact less inappropriate workarounds.

To summarize, we argue that organizations can exploit the measurements of workarounds from the previous section in order to make decisions on how to address them. By evaluating which workarounds are appropriate and which are not, they can facilitate the former and prevent the latter.

4.3   Involve

Various authors comment on the improvement potential of involving participants in designing and diffusing IS. Wheeler et al. [40], for instance, state: "in the case of workarounds, organizations could capitalize on the mindfulness of employees by encouraging employees to share their workarounds in order to improve task design". Insights from users can guide system design [27, 51] and decrease resistance towards the system [42, 48]. Tucker [60] believes that "designing work that considers the natural responses of employees when they encounter operational failures will be helpful in creating improvement programs that are successful over multiple dimensions, such as safety and efficiency". By giving process participants "a way to contribute" [47], allowing them to "reinvent, redefine or modify" [42] and "speak up about operational failures" [60], they participate in forming new work routines that fit their needs. Designers cannot foresee perfectly how their system is used [51], but by involving users in the process, misfits can be resolved. This involvement of participants needs to be facilitated by the organization. Halbesleben et al. [5], however, point out the complexity of gathering different participants with different roles. Safadi and Faraj [12] also indicate that participants often lack the time needed to communicate all the necessary information.

To sum up, we join the view of most authors and propose the involvement of participants in the improvement of processes. They are known to be willing to contribute improvement ideas. We suggest to exploit this willingness and have participants contribute solution strategies, beginning with the participants already known to perform workarounds.

4.4   Educate

What is also stressed in studies on workarounds is the need to set up suitable educational programs [25, 29, 40, 53, 57, 61]. Ongoing training and coaching of participants can enable both the efficient and appropriate way of working [18, 25, 29, 47] and the prevention of workarounds caused by ignorance [28, 57, 60].

One topic that should be addressed in the educational program of participants is the downstream effect of enacted workarounds, which we discussed earlier in the section on measuring workarounds. According to Drum et al. [59], "system users are often unable to fully comprehend their place in the task chain, and thus are unaware of the implications of their actions on information quality". In training and coaching efforts, users need to be explained the broader implications of their actions and how their goals relate to the bigger process [29, 39, 53]. Drum et al. [59] in fact noticed a 'light bulb effect' when participants were made aware of the broader implications of their actions, leading to improved work practices thereafter.

Another topic on the agenda of training programs on workarounds, is the encouragement of users to speak up about obstacles they perceive in their daily work [62]. Only then will their voices reach decision-makers who can then make informed decisions [35]. It also allows the sharing of best practices and the recognition that they are not the only ones struggling [62].

In sum, we propose to focus specifically on educating participants in improvement efforts. Ongoing training and coaching of participants may cause a decrease in resistance and ignorance and eventually in a decrease of workarounds.

4.5   Monitor

In his work on engineering for emergent change, Alter [53] argues an operational work system is dynamic, rather than static and unchanging. A dynamic system that is always in flux, requires a different way of handling than a static system. As such, problems "cannot be easily 'fixed' in a single step (workaround) or using a single, one-time set of measures" [33]. When measures are put in place, additional workarounds may develop [37]. An attempt has to be made in avoiding these additional workarounds [3], although some emerging workarounds simply cannot be avoided [63].

As the development of additional workarounds is unavoidable and their evolution cannot be predicted, the system needs to be monitored over time [20, 53, 64]. Outmazgin [19] suggests monitoring the extent to which participants fail to comply with the prescribed process. Similarly, Alter [53] suggests to track the effectiveness of workarounds and their downstream effects. This could provide decision-makers with the tools to perform corrective measures and notify them whenever workarounds occur [23].

Again, process mining techniques offer a valuable means to accomplish the ongoing monitoring of workarounds [19, 59]. It allows for 'conformance checking', i.e. checking the extent to which participants work around the prescribed process [65]. It would also allow for the tracking of the frequency of workarounds over time, their performance in relation to the prescribed process and its impacts downstream [58]. However, monitoring workarounds using process mining has not been extensively researched yet. This opens up opportunities for future research.

To sum up, we recommend organizations aiming for process improvement to monitor their processes and particularly how participants work around the prescribed process. Using process mining techniques, the evolution of these workarounds can be tracked, together with its frequency, effectiveness and downstream effects. Figure 3 provides the full overview of recommended activities regarding workarounds.

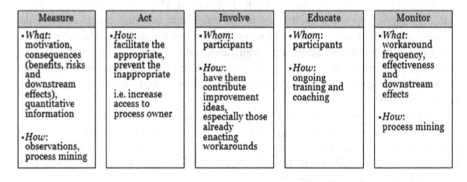

Measure	Act	Involve	Educate	Monitor
•*What:* motivation, consequences (benefits, risks and downstream effects), quantitative information •*How:* observations, process mining	•*How:* facilitate the appropriate, prevent the inappropriate i.e. increase access to process owner	•*Whom:* participants •*How:* have them contribute improvement ideas, especially those already enacting workarounds	•*Whom:* participants •*How:* ongoing training and coaching	•*What:* workaround frequency, effectiveness and downstream effects •*How:* process mining

Fig. 3. Five activities to unlock the potential of workarounds for improving processes.

5   Conclusion and Outlook

Over the years, many studies in IS have discussed the potential of studying workarounds for improving the alignment of IS and work processes. However, they do not provide insight in the necessary activities to achieve this improvement. In order to solve this research gap, we carried out a systematic literature review in which we analyzed existing studies that describe workarounds in organizations. We determined five activities organizations need to perform to unlock the potential of workarounds for improving processes. First, we propose that organizations need to detect these deviations and identify their motivations and consequences by observations and process mining techniques. Second, organizations should use this analysis of motivations and consequences for deciding whether to facilitate or prevent workarounds. Third, organizations can benefit from involving users in the decision-making process by letting them generate improvement ideas. Fourth, we propose to invest in educating and training end users to prevent the deviations in the first place and to make users aware of the broader implications of their actions. Last, monitoring workarounds can lead to continuous improvement in the long run.

References

1. Alter, S.: Theory of workarounds. Commun. Assoc. Inf. Syst. **34**, 1041–1066 (2014)
2. Leonardi, P.M.: When flexible routines meet flexible technologies: affordance, constraint, and the imbrication of human and material agencies. MIS Q. **35**, 147–167 (2011)
3. Azad, B., King, N.: Enacting computer workaround practices within a medication dispensing system. Eur. J. Inf. Syst. **17**, 264–278 (2008)
4. Boudreau, M.-C., Robey, D.: Enacting integrated information technology: a human agency perspective. Organ. Sci. **16**, 3–18 (2005)
5. Halbesleben, J.R.B., Savage, G.T., Wakefield, D.S., Wakefield, B.J.: Rework and workarounds in nurse medication administration process: implications for work processes and patient safety. Health Care Manag. Rev. **35**, 124–133 (2010)
6. Röder, N., Wiesche, M., Schermann, M., Krcmar, H.: Workaround aware business process modeling. In: Wirtschaftsinformatik, pp. 482–496 (2015)
7. Azad, B., King, N.: Rethinking how computer workarounds emerge: taking workarounds seriously but not negatively. In: Academy of Management Proceedings (2017)
8. Cresswell, K.M., Mozaffar, H., Lee, L., Williams, R., Sheikh, A.: Workarounds to hospital electronic prescribing systems: a qualitative study in English hospitals. BMJ Qual Saf. **26**, 542–551 (2017)
9. Petrides, L.A., McClelland, S.I., Nodine, T.R.: Costs and benefits of the workaround: inventive solution or costly alternative. Int. J. Educ. Manag. **18**, 100–108 (2004)
10. Lalley, C., Malloch, K.: Workarounds: The hidden pathway to excellence. Nurse Lead. **8**, 29–32 (2010)
11. Azad, B., King, N.: Institutionalized computer workaround practices in a Mediterranean country: an examination of two organizations. Eur. J. Inf. Syst. **21**, 358–372 (2012)
12. Safadi, H., Faraj, S.: The role of workarounds during an OpenSource Electronic Medical Record System Implementation. In: ICIS, p. 47 (2010)
13. Kitchenham, B., Charters, S.: Guidelines for performing systematic literature reviews in software engineering, vol. 5. Technical report, Ver. 2.3 EBSE Technical Report. EBSE (2007)
14. Webster, J., Watson, R.T.: Analyzing the past to prepare for the future: Writing a literature review. MIS Q. xiii–xxiii (2002)
15. Vom Brocke, J., Simons, A., Riemer, K., Niehaves, B., Plattfaut, R., Cleven, A.: Standing on the Shoulders of Giants. CAIS. **37**, 9 (2015)
16. Dunford, B.B., Perrigino, M.B.: The organization made us do it: demanding formalization and workaround attributions. Academy of Management Proceedings, vol. 2015, pp. 17443–17443 (2015)
17. Morath, J.M., Turnbull, J.E.: To Do No Harm: Ensuring Patient Safety in Health Care Organizations. Wiley, San Francisco (2005)

18. Zhou, X., Ackerman, M., Zheng, K.: CPOE workarounds, boundary objects, and assemblages. In: Proceedings of the SIGCHI Conference on Human Factors Computing System, pp. 3353–3362 (2011)
19. Outmazgin, N.: Exploring workaround situations in business processes. In: La Rosa, M., Soffer, P. (eds.) BPM 2012. LNBIP, vol. 132, pp. 426–437. Springer, Heidelberg (2013). https://doi.org/10.1007/978-3-642-36285-9_45
20. Zimmermann, S., Rentrop, C., Felden, C.: A multiple case study on the nature and management of shadow information technology. J. Inf. Syst. **31**, 79–101 (2017)
21. Nadhrah, N., Michell, V.: A normative method to analyse workarounds in a healthcare environment: motivations, consequences, and constraints (2013)
22. Spierings, A., Kerr, D., Houghton, L.: Issues that support the creation of ICT workarounds: towards a theoretical understanding of feral information systems. Inf. Syst. J. **27**, 775–794 (2017)
23. Alter, S.: Beneficial noncompliance and detrimental compliance: expected paths to unintended consequences. In: AMCIS (2015)
24. Huuskonen, S., Vakkari, P.: I Did It My Way: Social workers as secondary designers of a client information system. Inf. Process. Manag. **49**, 380–391 (2013)
25. Reiz, A., Gewald, H.: Physicians' resistance towards information systems in healthcare: the case of workarounds. In: PACIS, p. 12 (2016)
26. Barata, J., da Cunha, P.R., Abrantes, L.: Dealing with risks and workarounds: a guiding framework. In: Ralyté, J., España, S., Pastor, Ó. (eds.) PoEM 2015. LNBIP, vol. 235, pp. 141–155. Springer, Cham (2015). https://doi.org/10.1007/978-3-319-25897-3_10
27. Blandford, A., Furniss, D., Vincent, C.: Patient safety and interactive medical devices: Realigning work as imagined and work as done. Clin. Risk. **20**, 107–110 (2014)
28. Van Beijsterveld, J.A.A., Van Groenendaal, W.J.H.: Solving misfits in ERP implementations by SMEs. Inf. Syst. J. **26**, 369–393 (2016)
29. Drum, D.M., Pernsteiner, A.J., Revak, A.: Walking a mile in their shoes: user workarounds in a SAP environment. Int. J. Account. Inf. Manag. **24**, 185–204 (2016)
30. Yang, Z., Ng, B.-Y., Kankanhalli, A., Yip, J.W.L.: Workarounds in the use of IS in healthcare. Int. J. Hum Comput Stud. **70**, 43–65 (2012)
31. Koppel, R., Smith, S.W., Blythe, J., Kothari, V.: Workarounds to computer access in healthcare organizations: you want my password or a dead patient? In: ITCH. pp. 215–220 (2015)
32. Carayon, P., Gürses, A.P.: A human factors engineering conceptual framework of nursing workload and patient safety in intensive care units. Intensive Crit. Care Nurs. **21**, 284–301 (2005)
33. Park, S.Y., Chen, Y., Rudkin, S.: Technological and organizational adaptation of EMR implementation in an Emergency Department. ACM Trans. Comput. Interact. **22**, 1 (2015)
34. Ferneley, E.H., Sobreperez, P.: Resist, comply or workaround? An examination of different facets of user engagement with information systems. Eur. J. Inf. Syst. **15**, 345–356 (2006)
35. Kobayashi, M., Fussell, S.R., Xiao, Y., Seagull, F.J.: Work coordination, workflow, and workarounds in a medical context. In: CHI 2005 Extended Abstracts on Human Factors in Computing Systems, pp. 1561–1564 (2005)
36. Saleem, J.J., et al.: Paper persistence and computer-based workarounds with the electronic health record in primary care. In: Proceedings of the Human Factors and Ergonomics Society Annual Meeting, pp. 660–664 (2011)
37. Van Der Sijs, H., Rootjes, I., Aarts, J.: The shift in workarounds upon implementation of computerized physician order entry. Stud. Health Technol. Inform. **169**, 290 (2011)
38. Brooks, J., Ravishankar, M.-N., Oshri, I.: Regulating Vendor-Client Workarounds: An Information Brokering Approach (2015)

39. Drum, D.M., Standifer, R., Bourne, K.: Facing the consequences: examining a workaround outcomes-based model. J. Inf. Syst. **29**, 137–159 (2015)
40. Wheeler, A.R., Halbesleben, J.R.B., Harris, K.J.: How job-level HRM effectiveness influences employee intent to turnover and workarounds in hospitals. J. Bus. Res. **65**, 547–554 (2012)
41. Morrison, B.: The problem with workarounds is that they work: the persistence of resource shortages. J. Oper. Manag. **39**, 79–91 (2015)
42. Barrett, A.K., Stephens, K.K.: Making Electronic Health Records (EHRs) work: informal talk and workarounds in healthcare organizations. Health Commun. **32**, 1004–1013 (2017)
43. Waheed, S.: Identifying medical device workarounds in the hospital and reporting them to the US food and drug administration. J. Clin. Eng. **41**, 30–32 (2016)
44. Woltjer, R.: Workarounds and trade-offs in information security – an exploratory study. Inf. Comput. Secur. **25**, 402–420 (2017)
45. Wears, R.L., Hettinger, A.Z.: The tragedy of adaptability. Ann. Emerg. Med. **63**, 338–339 (2014)
46. Sobreperez, P., Ferneley, E.H., Wilson, F.: Tricks or trompe l'oeil? An examination workplace resistance in an information rich managerial environment. In: ECIS, Regensburg (2005)
47. Alter, S.: A workaround design system for anticipating, designing, and/or preventing workarounds. In: Gaaloul, K., Schmidt, R., Nurcan, S., Guerreiro, S., Ma, Q. (eds.) CAISE 2015. LNBIP, vol. 214, pp. 489–498. Springer, Cham (2015). https://doi.org/10.1007/978-3-319-19237-6_31
48. Malaurent, J., Avison, D.: From an apparent failure to a success story: ERP in China - Post implementation. Int. J. Inf. Manag. **35**, 643–646 (2015)
49. Outmazgin, N., Soffer, P.: Business process workarounds: what can and cannot be detected by process mining. In: Nurcan, S., Proper, Henderik A., Soffer, P., Krogstie, J., Schmidt, R., Halpin, T., Bider, I. (eds.) BPMDS/EMMSAD -2013. LNBIP, vol. 147, pp. 48–62. Springer, Heidelberg (2013). https://doi.org/10.1007/978-3-642-38484-4_5
50. Blaz, J.W., Doig, A.K., Cloyes, K.G., Staggers, N.: The hidden lives of nurses' cognitive artifacts. Appl. Clin. Inform. **7**, 832–849 (2016)
51. Park, S.Y., Chen, Y.: Adaptation as design: Learning from an EMR deployment study. In: Proceedings of the SIGCHI Conference on Human Factors in Computing Systems, pp. 2097–2106. Austin, TX (2012)
52. Röder, N., Wiesche, M., Schermann, M., Krcmar, H.: Toward an ontology of workarounds: a literature review on existing concepts. In: Proceedings of the Annual Hawaii International Conference on System Sciences, pp. 5177–5186. IEEE Computer Society (2016)
53. Alter, S.: Engineering enterprises for emergent change. In: Proceedings - IEEE Conference on Business Informatics, CBI, pp. 113–123. Institute of Electrical and Electronics Engineers Inc. (2014)
54. Gasparas, J., Monteiro, E.: Cross-contextual use of integrated information systems. In: ECIS (2009)
55. Röder, N., Wiesche, M., Schermann, M.: A situational perspective on workarounds in IT-enabled business processes: a multiple case study (2014)
56. Ali, M., Cornford, T., Klecun, E.: Exploring control in health information systems implementation. In: Studies Health Technology Informatics, pp. 681–685. Cape Town (2010)
57. Furniss, D., Masci, P., Curzon, P., Mayer, A., Blandford, A.: 7 Themes for guiding situated ergonomic assessments of medical devices. Appl. Ergon. **45**, 1668–1677 (2014)
58. Van der Aalst, W.M.P.: Process Mining. Springer, Heidelberg (2011). https://doi.org/10.1007/978-3-642-19345-3

59. Drum, D.M., Pernsteiner, A., Revak, A.: Workarounds in an SAP environment: impacts on accounting information quality. J. Account. Organ. Chang. **13**, 44–64 (2017)
60. Tucker, A.L.: The impact of workaround difficulty on frontline employees' response to operational failures: a laboratory experiment on medication administration. Manag. Sci. **62**, 1124–1144 (2016)
61. Hustad, E., Olsen, D.H.: ERP post-implementation issues in small-and-medium-sized enterprises (2011)
62. Campbell, D.: Policy workaround stories are valuable evaluative indicators: but should they be told? Am. J. Eval. **32**, 408–417 (2011)
63. Vieru, D., Arduin, P.-E.: Sharing knowledge in a shared services center context: an explanatory case study of the dialectics of formal and informal practices (2016)
64. Zimmermann, S., Rentrop, C., Felden, C.: Governing IT activities in business workgroups—design principles for a method to control identified shadow IT (2016)
65. Rozinat, A., Van der Aalst, W.M.P.: Conformance checking of processes based on monitoring real behavior. Inf. Syst. **33**, 64–95 (2008)

7th International Workshop on Declarative, Decision and Hybrid Approaches to Processes (DEC2H)

7th International Workshop on Declarative, Decision and Hybrid Approaches to Processes (DEC2H)

Business process models describe the rules, available decisions, premises, and possible outcomes of specific situations. However, those rules and decisions are often implicit in process flows, in process activities, or in the head of employees as tacit knowledge. For computers to analyze or execute business process models, rules and decisions must be made explicit.

The declarative modeling paradigm aims to directly express the business rules or constraints underlying a given process. The paradigm has gained momentum, and in recent years several declarative notations have emerged, e.g., Declare, Dynamic Condition Response (DCR) Graphs, Decision Modelling and Notation (DMN), Case Management Model and Notation (CMMN), Guard-Stage-Milestone (GSM), extended Compliance Rule Graphs (eCRG), and Declarative Process Intermediate Language (DPIL). Recently, there has been a surge of interest in *hybrid* approaches, which combine the strengths of different modeling paradigms.

In this workshop, we are interested in the application and challenges of decision- and rule-based modeling in all phases of the Business Process Management lifecycle (identification, discovery, analysis, redesign, implementation, and monitoring).

DEC2H 2019 is the 7th edition of the workshop formerly known as DeHMiMoP (Workshop on Declarative/Decision/Hybrid Mining and Modelling for Business Processes). It attracted seven high-quality international submissions. Each paper was reviewed by at least three members of the Program Committee. Out of the submitted manuscripts, the top four were accepted for presentation. The presented papers primarily focus on the role of decisions and data/information artefacts both in the modeling and compliance checking of processes.

The invited talk of Marco Montali, with which the workshop began, revolves around the interplay of declarative process specifications with data and decision modeling. The discussion extends across diverse areas of computer science, including conceptual modeling, formal methods, and knowledge representation and reasoning.

Valencia Parra, Parody, Varela Vaca, Caballero, and Gómez López present an approach based on business rules defined in DMN to capture and evaluate data quality. Their approach allows for the definition of data requirements through business rules, as well as the specification of related quality measures and their actual assessment.

Smit and Eshuis report on a case study in which they apply the GSM notation to model rolling stock maintenance logistics processes at Dutch Railways through their business artefacts. The authors highlight that the used declarative notation successfully supports the decision making and re-engineering of the process.

Gutierrez, van Rijswijk, Ruhsam, Kofrak, Kogler, Shadrina, and Zucker present an approach to Adaptive Case Management modeling wherein the goals, a.k.a. "value stream stages", take center stage. The actual process actions follow from the set of actions available to reach those goals, and a set of business rules governing availability of those actions. The work concludes with an application to the construction industry.

Finally, Holfter, Haarmann, Pufahl, and Weske study fragment-based Case Management models. They give a formal semantics in terms of a translation to Petri nets, then use model checking for Petri nets to statically verify compliance properties.

We hope that the reader will benefit from these papers and know more about the latest advances in research about declarative, decision, and hybrid approaches to business process management.

September 2019

Claudio Di Ciccio
Tijs Slaats
Søren Debois
Jan Vanthienen

Organization

Workshop Organizers

Claudio Di Ciccio	WU Vienna, Austria
Tijs Slaats	University of Copenhagen, Denmark
Søren Debois	IT University of Copenhagen, Denmark
Jan Vanthienen	KU Leuven, Belgium

Program Committee

Andrea Burattin	Technical University of Denmark, Denmark
Josep Carmona	Universitat Politècnica de Catalunya, Spain
João Costa Seco	Universidade Nova de Lisboa, Portugal
Massimiliano de Leoni	University of Padua, Italy
Riccardo De Masellis	Stockholm University, Sweden
Johannes De Smedt	University of Edinburgh, UK
Jochen De Weerdt	KU Leuven, Belgium
Rik Eshuis	Eindhoven University of Technology, The Netherlands
Robert Golan	DBmind Technologies Inc., USA
María Teresa Gómez-López	Universidad de Sevilla, Spain
Xunhua Guo	Tsinghua University, China
Thomas Hildebrandt	University of Copenhagen, Denmark
Amin Jalali	Stockholm University, Sweden
Dimitris Karagiannis	University of Vienna, Austria
Krzysztof Kluza	AGH University of Science and Technology, Poland
Fabrizio M. Maggi	University of Tartu, Estonia
Andrea Marrella	Sapienza University of Rome, Italy
Jorge Munoz-Gama	Pontificia Universidad Católica de Chile, Chile
Hajo A. Reijers	Utrecht University, The Netherlands
Stefan Schönig	University of Bayreuth, Germany
Lucinéia H. Thom	Universidade Federal do Rio Grande do Sul, Brazil
Han van der Aa	Humboldt University of Berlin, Germany
Wil M. P. van der Aalst	RWTH Aachen University, Germany
Barbara Weber	Technical University of Denmark, Denmark
Mathias Weske	HPI, University of Potsdam, Germany

Putting Decisions in Perspective

Marco Montali[✉]

Free University of Bozen-Bolzano, Bolzano, Italy
`montali@inf.unibz.it`

Abstract. The advent of the OMG Decision Model and Notation (DMN) standard has revived interest, both from academia and industry, in decision management and its relationship with business process management. Several techniques and tools for the static analysis of decision models have been brought forward, taking advantage of the trade-off between expressiveness and computational tractability offered by the DMN S-FEEL language.

In this short paper, I argue that *decisions have to be put in perspective*, that is, understood and analyzed within their surrounding organizational boundaries. This brings new challenges that, in turn, require novel, advanced analysis techniques. Using a simple but illustrative example, I consider in particular two relevant settings: decisions interpreted the presence of background, structural knowledge of the domain of interest, and (data-aware) business processes routing process instances based on decisions. Notably, the latter setting is of particular interest in the context of multi-perspective process mining. I report on how we successfully tackled key analysis tasks in both settings, through a balanced combination of conceptual modeling, formal methods, and knowledge representation and reasoning.

1 Introduction

Contemporary organizations rely on a variety of management disciplines, with IT as underlying enabling technology, to drive their internal operations and the interactions with customers and other organizations. We consider here three pillars. First, *business process management* focuses on the discovery, modeling, analysis, enactment, and continuous improvement of business processes. Second, *master data management* tackles the business-relevant data produced and/or collected by the organization; structural knowledge is used by the organization to understand which objects and relations are described by such data. Finally, *enterprise decision management* concerns decision making within the organization, in connection with the strategy and processes enacted by the organization. Decision models are typically based on rules that describe how input values correspond to output values.

Interest in enterprise decision management both from academia and industry has been recently revived thanks to the advent of the OMG standardization effort on the Decision Model and Notation (DMN) [11]. This has triggered a

© Springer Nature Switzerland AG 2019
C. Di Francescomarino et al. (Eds.): BPM 2019 Workshops, LNBIP 362, pp. 355–361, 2019.
https://doi.org/10.1007/978-3-030-37453-2_29

number of technical and empirical studies on decision models specified using DMN [7]. From the technical point of view, many analysis tasks to check for correctness, refactor, and optimize decision models are now being attacked, by (re)studying well-established analysis problems [6,12] in the light of the specific modeling choices adopted by the standard [4]. In particular, the S-FEEL DMN language provides an interesting trade-off between expressiveness and computational tractability.

This short paper starts from the obvious but fundamental observation that decisions are never used as such, but are instead interpreted, understood, and applied within organizational boundaries. This requires to *put decisions in perspective*, and in turn to combine enterprise decision management with master data management and business process management. More challenging analysis tasks consequently arise. Specifically, by using a simple but illustrative example, I report on how we successfully tackled key analysis tasks for DMN S-FEEL decisions arising when such decisions are put in two different perspectives: *(i)* background, structural knowledge of the domain of interest, which introduces constraints on the input attributes of a decision; *(ii)* data-aware business processes routing cases based on updatable case data and corresponding decisions. The latter setting is relevant in the context of decision-aware process discovery [10].

2   DMN S-FEEL Decisions

In DMN, decision models are captured using two main artifacts. The fundamental building block is that of a *decision*, declaring input and output data attributes, and one or more rules that map inputs to outputs. Rules may be specified using different languages, the simplest of which is the S-FEEL language, also part of the standard. S-FEEL is quite restrictive, but has two main advantages: it can be graphically visualized in a very intuitive, tabular format, and achieves an interesting trade-off between expressiveness and tractability. The second DMN artifact (which will not be further considered here) is that of *decision requirement graph* (DRG), used to interconnect several decisions.

Figure 1(a) shows a simple DMN S-FEEL decision table used by the courier company TURNAROUND to determine how packages can be transported based on their physical features. For simplicity, we consider only length and weight of packages, abstracting away from other important dimensions (such as girth). Columns with blue and red headers are respectively input and output columns. Length and weight attributes are represented using real, positive numbers, whereas ShipBy is a string that can take two values, car and truck. S-FEEL supports primitive datatypes. The datatype declaration of a column is left implicit in the table, whereas facets (such as "being positive" or "taking only two given values") are declared immediately below the header. The central part of the table (with numbered rows) contains the actual decision rules. Each input cell contains a condition, which amounts to a disjunction of simple tests (such as membership to an interval, or being equal or different from a given value).

Package Declaration			
P	Length (m)	Weight (kg)	ShipBy
	> 0	> 0	car,truck
1	(0.0,1.0]	(0, 5)	car
2	(0.0,0.6]	(5,10]	truck
3	(0.6,1.0]	(4,10]	truck
4	(1.0,1.5]	(0, 3]	car
5	(1.0,2.0]	(3,10]	truck

(a) Decision relating phyisical features of a package to its transportation mode.

A1	There are *only* two types of packages: standard and special.
A2	The minimum weight for a package is 0.5 kg.
A3	A standard package has a length of 0.5 m and bears at most 8 kg.
A4	A special package has a length of 1.2 m and bears at most 9 kg.

(b) A simple ontology of packages

Fig. 1. A DMN S-FEEL decision table, and an ontology constraining the table inputs

In our specific example, the rules are so that the resulting table is *incomplete* and *not unique*. Incompleteness arise because there are combinations of input values that do not match with any rule. Since no default value is defined for the output column, for such non-matching input attributes a null/undefined output value is returned. This is the case, e.g., for a heavy package with a weight greater than 10 kg. Non-uniqueness arises instead from the fact that rules 1 and 3 overlap, that is, there are combinations of input values that match with both rules. This is the case, e.g., for packages with a length of 0.8 m and a weight of 4.5 kg. The presence of overlapping rules call for the definition of a suitable *hit policy*, which either asserts that rules cannot overlap, or that overlaps exist, dictating in the latter case how to unambiguously identify which output is returned in case multiple rules match. The hit policy is shown in the top-left cell of the table; in our case, the hit policy is **P**, which stands for "priority" and indicates that when multiple ruled match, the output `car` has priority over `truck`.

Incompleteness and non-uniqueness can be intuitively visualized from Fig. 2, which provides a geometric interpretation of Fig. 1(a). White regions witness incompleteness, whereas regions covered by multiple rules pinpoint their overlap.

3   Decisions in the Context of Background Knowledge

When a decision is embedded in a concrete organizational context, it is affected by background domain knowledge that explicitly or implicitly constrain the input attributes of the decision. This, in turn, alters the decision logic conveyed by the decision. To show this subtle interplay, let us imagine that the BLACKSHIP company selects TURNAROUND as a partner. BLACKSHIP acts as a mediator between customers and TURNAROUND, on the one hand helping customers in preparing their packages and filling out the necessary forms, on the other hand interacting with TURNAROUND for the actual transportation of packages. Since the rules for transportation are determined by TURNAROUND, BLACKSHIP needs to import Fig. 1(a) in its own organizational context. The decision therein then has to be interpreted considering the types of packages provided by BLACKSHIP. Figure 1(a) illustrates the "package ontology" used by BLACKSHIP.

In [5], we have introduced a logical framework to formalize DMN S-FEEL decisions and DRGs, as well as their integration with background knowledge expressed in a multi-sorted variant of first-order logic. We have then lifted standard analysis tasks previously introduced for single decisions [4] to the more complex case where decisions are connected via DRGs, and interpreted in the presence of background knowledge. Novel analysis tasks also emerge in this sophisticated setting.

Following the approach in [5], Fig. 1(a) and (b) are integrated through a so-called *bridge concept*, that is, an entity type from the ontology that has the input attributes used in the decision. In our case, this is obviously the package entity type. By exploiting the geometric metaphor in Fig. 2, we can see that due to the ontological constraints over the two types of packages used by BLACKSHIP, the **Package Shipment** decision becomes complete and unique: the physical features of standard and regular packages are so that they are fact fully contained in the regions covered by the package shipment rules, and do not intersect regions shared by multiple rules. In this setting, the priority hit policy is then never applied, as output clashes never arise.

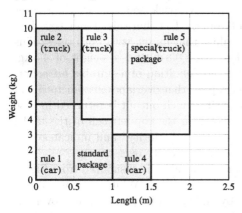

Fig. 2. Geometric interpretation of the **Package Shipment** decision in Fig. 1(a), and of the types in the ontology of Fig. 1(b).

A particularly interesting analysis task in this setting is that of output determinability. This amounts to check whether, in the light of the given background knowledge, it is enough to know some attributes of the bridge concept to be able to fully determine the output configuration induced by a decision. In our running example, knowing the type of package and its weight is enough to determine the output configuration induced by the **Package Shipment** decision table.

While all such analysis tasks cannot be computationally solved in the case of arbitrary first-order logic ontologies, they can all be solved in EXPTIME when the background knowledge is formalized using description logic ALCH($\mathfrak{D}$), which extends the well-known ALCH logic with multiple datatypes. This result is obtained in [5] by showing that DMN DRGs and S-FEEL decisions can be fully encoded in ALCH($\mathfrak{D}$). The same result can be reconstructed when the underlying logic is ALCQI, which is of particular interest since it captures UML class diagrams.

4   S-FEEL Decisions in the Context of Processes

The DMN standard promotes integration with the BPMN via separation of concerns: instead of modeling complex decisions as networks of BPMN gateways, a BPMN rule task should be introduced and linked to a DMN model. This approach leads to the notion of *decision-aware process model* [1]. In my view, a decision-aware process model is also necessarily data-aware: since data are consumed and produced by decisions, the process has to explicitly spell out how data objects are queried and updated.

Data and decisions affect the process execution semantics, and so process models that appear to be sound when analyzed from a pure control-flow perspective could actually become unsound when data, decisions, and data conditions are taken into account. Hence, the well-established notion of process soundness has to be restudied in the light of decision-awareness [1]. Consider the process model in Fig. 4. It describes the shipment preparation process used by the BLACKSHIP company. The process invokes the DMN decision in Fig. 1(a) to determine the transportation mode for a given package, and then invokes the DMN decision in Fig. 3 to determine whether the package must be accompanied by a declaration, and who should sign the declaration. While this process is sound when analyzed from a pure, control-flow perspective, it turns out to be unsound when data and decisions are considered, since the **prepare owner declaration** task can never be executed. This depends on the fact that **owner** is never produced by Fig. 3: rule 1, the only producing **owner** as declaration, is never triggered, since a package weighting 6 kg or more is never transported by car.

Package Declaration			
U	ShipBy	Weight (kg)	Declaration
	car,truck	> 0	none, owner, company
1	car	≥ 6	owner
2	truck	≥ 8	company

Fig. 3. DMN decision table indicating if a package must be accompanied by a declaration, and if so, who has to sign it; none is the default output.

Traditional formal analysis techniques for processes cannot detect this sophisticated types of errors. This is because they solely focus on the process control-flow. In particular, they typically take the input process model, abstract away from data manipulation, reduce decisions to pure nondeterministic choices, and translate the resulting model into a corresponding Petri net. The reachability/coverability graph of the net is then inspected to check whether the process control-flow is sound. Sophisticated approaches for data-aware process analysis have been consequently brought forward to formally analyze integrated models of processes and data [3]. However, such models are much richer than those needed to capture processes like the one in Fig. 4, where only simple variables are used to write and store case-related primitive data items.

This type of "weakly data-aware" processes can be elegantly formalized using Data Petri nets (DPNs) [10]. In [8], we have proposed (and implemented) a constructive technique to check all the decision-aware variants of soundness

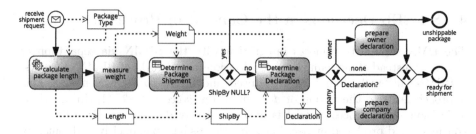

Fig. 4. BPMN diagram of the shipment preparation process of BLACKSHIP; the two decision tasks invoke the corresponding DMN decisions captured in Figs. 1 and 3

introduced in [1] on top of DPNs. DMN S-FEEL decisions are encoded in the DPN by uniqueifying decisions [2], then flattening the so-obtained non-overlapping rules in the DPN as a network of τ-transitions equipped with data conditions. The technique relies on a suitable notion of abstraction to finitely represent the infinite state space induced by the presence of data. Further progress has been made in [9], where conditions going beyond S-FEEL are incorporated. In particular, the analysis technique of [8] is extended to handle variable-to-variable conditions. This paves the way for decision-aware process models where DMN S-FEEL decisions can contain parameters, which are dynamically instantiated at execution time.

5   Conclusion

In this paper, I have argued that decisions have to be put in perspective, and considered the interplay with background knowledge and with business processes. Our research results show how different analysis tasks on these combined models can be successfully tamed when decisions are expressed in the S-FEEL language of the DMN standard, through a mix of conceptual modeling, formal methods, and knowledge representation.

Several challenges are still open. The following three are of particular importance. First of all, it is now time to "push the envelope" and see how we can extend our results to more expressive fragment of the full FEEL language defined in the DMN standard. Second, it is key to move from soundness of decision-aware process models to temporal model checking and synthesis. This will provide the basis to analyze such multi-perspective models in the presence of multiple, possibly non-cooperative actors controlling the different sources of non-determinism (such as choice of next activity and data input). Third, we believe that these two lines of research can be unified into a rich framework accounting at once for decisions, processes, and background knowledge.

Acknowledgments. I am grateful to all co-authors of this research, in particular Diego Calvanese, Massimiliano de Leoni, Marlon Dumas, Paolo Felli, and Fabrizio Maggi.

References

1. Batoulis, K., Haarmann, S., Weske, M.: Various notions of soundness for decision-aware business processes. In: Mayr, H.C., Guizzardi, G., Ma, H., Pastor, O. (eds.) ER 2017. LNCS, vol. 10650, pp. 403–418. Springer, Cham (2017). https://doi.org/10.1007/978-3-319-69904-2_31
2. Batoulis, K., Weske, M.: A tool for the uniqueification of DMN decision tables. In: BPM (Dissertation/Demos/Industry). CEUR Workshop Proceedings, vol. 2196, pp. 116–119. CEUR-WS.org (2018)
3. Calvanese, D., De Giacomo, G., Montali, M.: Foundations of data-aware process analysis: a database theory perspective. In: PODS, pp. 1–12. ACM (2013)
4. Calvanese, D., Dumas, M., Laurson, Ü., Maggi, F.M., Montali, M., Teinemaa, I.: Semantics, analysis and simplification of DMN decision tables. Inf. Syst. **78**, 112–125 (2018)
5. Calvanese, D., Montali, M., Dumas, M., Maggi, F.M.: Semantic DMN: formalizing and reasoning about decisions in the presence of background knowledge. Theory Pract. Log. Program. **19**(4), 536–573 (2019)
6. CODASYL Decision Table Task Group: A Modern appraisal of decision tables: aCODASYL report. ACM (1982)
7. Figl, K., Mendling, J., Tokdemir, G., Vanthienen, J.: What we know and what we do not know about DMN. Enterp. Model. Inf. Syst. Arch. **13**(2), 1–16 (2018)
8. de Leoni, M., Felli, P., Montali, M.: A holistic approach for soundness verification of decision-aware process models. In: Trujillo, J.C., et al. (eds.) ER 2018. LNCS, vol. 11157, pp. 219–235. Springer, Cham (2018). https://doi.org/10.1007/978-3-030-00847-5_17
9. de Leoni, M., Felli, P., Montali, M.: Soundness verification of decision-aware process models with variable-to-variable conditions. In: ER. IEEE Press (2019)
10. Mannhardt, F.: Multi-perspective process mining. In: BPM (Dissertation/Demos/Industry). CEUR Workshop Proceedings, vol. 2196, pp. 41–45. CEUR-WS.org (2018)
11. OMG: Decision Model and Notation (DMN) 1.2 (2019)
12. Vanthienen, J., Dries, E.: Developments in decision tables: evolution, applications and a proposed standard. Research Report 9227, Katholieke Universiteit Leuven (1992)

DMN for Data Quality Measurement and Assessment

Álvaro Valencia-Parra[1]([⊠])(ID), Luisa Parody[2](ID), Ángel Jesús Varela-Vaca[1](ID),
Ismael Caballero[3](ID), and María Teresa Gómez-López[1](ID)

[1] Universidad de Sevilla, Seville, Spain
{avalencia,ajvarela,maytegomez}@us.es
[2] Universidad Loyola Andalucía, Córdoba, Spain
mlparody@uloyola.es
[3] Universidad Castilla-La Mancha, Ciudad Real, Spain
Ismael.Caballero@uclm.es
http://www.idea.us.es/

Abstract. Data Quality assessment is aimed at evaluating the suitability of a dataset for an intended task. The extensive literature on data quality describes the various methodologies for assessing data quality by means of data profiling techniques of the whole datasets. Our investigations are aimed to provide solutions to the need of automatically assessing the level of quality of the records of a dataset, where data profiling tools do not provide an adequate level of information. As most of the times, it is easier to describe when a record has quality enough than calculating a qualitative indicator, we propose a semi-automatically business rule-guided data quality assessment methodology for every record. This involves first listing the business rules that describe the data (data requirements), then those describing how to produce measures (business rules for data quality measurements), and finally, those defining how to assess the level of data quality of a data set (business rules for data quality assessment). The main contribution of this paper is the adoption of the OMG standard DMN (Decision Model and Notation) to support the data quality requirement description and their automatic assessment by using the existing DMN engines.

Keywords: Data quality · Decision Model and Notation · Data quality measurement · Data quality assessment

1 Introduction

Globalization and emerging technologies are bringing new challenges to companies in the contexts in which enterprises should use massive amounts of data, most of them provided by third parties. In order to guarantee the success of the various tasks, and to satisfy the necessities of the business processes that use the data, it is paramount to face up with the quality of these data [12]. Just to mention an example of a critic task, let us think about the integration of data

© Springer Nature Switzerland AG 2019
C. Di Francescomarino et al. (Eds.): BPM 2019 Workshops, LNBIP 362, pp. 362–374, 2019.
https://doi.org/10.1007/978-3-030-37453-2_30

coming from different and heterogeneous sources in complex scenarios [3], and the many problems related to data quality that can arise when it comes to build a new dataset.

To achieve the largest benefits of data, companies will need to find out ways to automatically manage the levels of data quality; this is even more important, in scenarios where it is required high efficiency in terms of computational cost of the operations per second. The assessment of data quality largely depends on the context of the use of data. Despite of the extensive literature around data quality (e.g., definition of methodologies for requirement definitions, selection of criteria to judge data quality- data quality dimensions [1]), assessment and measurement procedures have been typically developed ad-hoc [13]. The data quality context-awareness needs the description of the business rules representing the data quality requirements. In order to describe the data quality requirements, we propose the description of three sets of business rules: those describing the data requirements (BR), those describing how to measure the data quality dimensions (BR.DQM), and those describing how to assess the data quality (BR.DQA) in terms of the measures of the data quality.

The assessment follows a procedure based on the sequencing of the verification in cascade of the three previously-mentioned types of business rules in two phases: first, we conduct the verification of the BRs to estimate a measurement for every data quality dimension according to BR.DQM; and, then, we use the verification of the BR.DQM to produce an estimation of the assessment of the level of data quality, according to the stated BR.DQA. Based on the result of this assessment, the organization should make a decision on the use of the record based on its risk appetite. In this paper, we propose the use of DMN (Decision Model and Notation) [15], a declarative language proposed by OMG to facilitate the description of the business rules, as well as their evaluation [6]. DMN provides human readers with a more understandable and visual representation of business rules [8], being data quality requirements a new scenario where DMN can be used.

The remainder of this paper is organized as follows: Sect. 2 details the foundations involved in this paper; Sect. 3 introduces a case study to make the proposal accessible; Sect. 4 presents the proposal of application of DMN to data quality measurements and assessment respectively; Sect. 5 presents the related work and Sect. 6 concludes and remarks the lessons learned.

2 Foundations

2.1 Data Quality Management: Rules for Measurement and Assessment

The cornerstone of our proposal of data quality management is grounded in the difference between two important concepts typically used as synonymous: "measurement" and "assessment". This differentiation is based on the two definitions of quality: the "meeting requirements" by Crosby (or *"how well data is built"*) and the "fitness for use" by Juran (or *"how usable the data is"*) [19]. The basis

Table 1. Data quality dimensions from Wang [19]

Data quality category	Data quality dimension
Intrinsic	Accuracy, Objectivity, Believability, Reputation
Accessibility	Access, Security
Contextual	Relevancy, Value-Added, Timeliness, Completeness, Amount of data
Representational	Interpretability, Ease of understanding, Concise representation, Consistent representation

of our proposal is to describe by means of business rules when "data is well built" according to several data quality dimensions and when "data is usable" according to the assessment of the quality including all data quality dimensions. Data quality dimensions (criteria used to evaluate the quality of data) are at the core of data quality management [20] because they represent users' data quality requirements. Several authors have proposed their own set of data quality dimensions, both generic ones (like the ones proposed by [19] -see Table 1- or the introduced in ISO 25012 [10]) or for specific context.

When it comes to estimate the amount of data quality that a dataset has, data profiling tools are typically employed to produce some measures that data quality processes use as indicators [9]. But, without knowing how the semantics of data has been implemented in the data model, the results cannot be interpreted, and cannot be used to diagnose the root causes of a low level of data quality. However, it is more than enough to have a qualitative indications of whether data is usable or not. This strategy is specially recommendable when it comes to determine if a record should be used as part of the execution of an instance of a business process. More specifically, when the decision should be made according to the quality of the used data, desirably in an automatic way. At this point, let us recall that every record is a set of attributes a_i; every attribute a_i or every set of attributes $a_i, a_j, a_k, \ldots$ must meet some data requirements specified by means of several Business Rules (BR). Some typical statements of business rules look like:

- BR.01. The attribute a_1 must meet the regular expression RE
- BR.02. The attribute a_2 (datatype numeric) should be lower than the attribute a_3 (datatype numeric)

The measurement implies the verification of the business rules associated to every chosen data quality dimension (e.g. completeness, consistency, ...). To produce a value for measurement, we need a BR.DQM that describes the possible values that every data quality dimension could obtain. Depending on the granularity, a BR.DQM can be defined in terms of one or more attributes and one or more BRs. We propose to use Likert scales to define the possible values for the results of measurement. As an example of possible values of completeness is the set {"Complete Enough", "Not Enough Complete", "Dramatically Non-complete"}. Typically, BR.DQM sentences for completeness can look like:

- *BR.DQM.01. A record can be considered as "Complete Enough" if it meets BR.01 and BR.02*
- *BR.DQM.02. A record can be considered as "Not Enough Complete" if it only meets BR.01*
- *BR.DQM.03. A record can be considered as "Dramatically Non-complete" if it does not meet neither of BR.01 nor BR.02*

Finally, and after defining how to measure the data quality dimensions, it is time for aggregating the measures for the various data quality dimensions to produce an indication of the amount of data quality that a record has. The result of this aggregation will represent the level of usability of a record for a task in terms of the risk appetite of the organization for the underlying task. Once again, a Likert scale (e.g., "Usable", "Potentially usable but risky", "Non-usable") is proposed and the statement of the corresponding business rules for the assessment (BR.DQA) should be done in terms of the chosen data quality dimensions and the corresponding BR.DQM. Typically, BR.DQA can look like:

- *BR.DQA.01. A record can be considered as "Usable" if it meets BR.DQM.01*
- *BR.DQA.02. A record can be considered as "Potentially usable but Risky" if it meets BR.DQM.02*
- *BR.DQA.03. A record can be considered as "Non-usable" if it meets BR.DQA.03*

Please see Sect. 3 for a larger motivating example of the description of an assessment.

2.2   Decision Model and Notation

Decision Model and Notation (DMN) is a modelling language and notation defined to describe business rules [15]. DMN provides a simple way to define the decision logic model understandable by all users, in our case, from the business experts in charge of describing the processes to the data quality experts responsible for defining the quality requirements.

The DMN standard includes two components:

- Decision requirements diagram that defines the decisions to be made, their interrelationships, and their requirements for decision logic.
- Decision logic that defines the required decisions in sufficient details to allow validation and/or automation.

An example of a Decision Requirement Diagram is presented and detailed in Fig. 1. The *decision task* describes a specific task that includes a decision logic, that is, depending on some input values set output values as described in a decision table. The input data, as the name implies, is the necessary data that the decision logic needs to determine the output value. For instance, in the example,

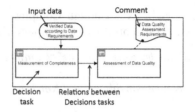

Fig. 1. Decision requirement diagram example

the *Verified Data according to Data Requirements* feeds *Measurement of Completeness* task. On the other side, the arrows that connect two decisions tasks indicates the relationships between these two tasks. In the example presented in Fig. 1, the output of the decision task *Measurement of Completeness* feeds the *Assessment of Data Quality* task, which means that the decision output of the first task is considered as input for the second decision.

Each decision task includes a decision table. The decision table used in this article has an horizontal orientation: the input and outputs are defined in columns and the rules as rows. An example is presented in Fig. 2, which also indicates the different components of the table.

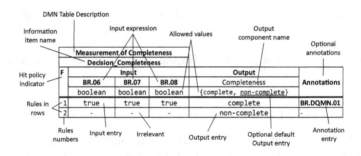

Fig. 2. Decision table example for the *Measurement of Completeness* task (see Fig. 1)

The *information item name* is the name of the variable (i.e. information item) for which the decision table provides the decision logic. The *hit policy indicator* indicates how to handle the multiple matches. In our case, the F means that although multiple rules can match, only the first hit by rule order is returned. There are other possible *hit policy indicators* such as Unique (U), Any (A), and Priority (P) (see [15] for further details). There is also a set of input clauses composed of an *input expression* and optional *allowed values* for the input entries, for instance *BR.06* is an input expression whose possible value is a *boolean* (true and false values). An input entry is contained in a rule: the value *true* corresponds to the input entry for the *BR.06* in the *rule number* 1. The input cell entry '-' means irrelevant, i.e. it can have any of the allowed values. Moreover, a set of output clauses are also included in a decision table. An output clause consists of an *output component name* and its *allowed values*. The allowed value that is underlined corresponds to the default value. Finally, a set of annotations clauses can also be included in the decision table. In our case, rule number 1 is annotated with the entry *DQ.DQMN.01*. The decision table displays the rules in an abbreviated notation organizing the entries in table cells. For example, the rule number 1 can be read as: ***If** (BR.06 and BR.07 and BR.08)* ***then*** *Completeness = 'complete'*.

3 Motivating Example

In order to illustrate how a semi-automatically business rule guided-data quality assessment can be done, we adapted a well-known example of a movie-database introduced in [2] and shown in Table 2. The adaptation just consist of changing some values to have a much better casuistry in the data quality assessment.

Table 2. Example of dataset with data quality problems.

Id	Title	Director	Year	#Remakes	LastRemakeYear
1	Casablanca	Weir	1942	3	1940
2	Dead Poets Society	Weir	1989	0	NULL
3	Rman Holiday	Wylder	1953	0	NULL
4	Sabrin	NULL	1964	0	1985

To illustrate the data quality assessment, we introduce several business rules (*BR*, *BR.DQM*, and *BR.DQA*):

- **Data Requirements**: Associated to the given dataset, some business rules (BR) describing some syntactic and/or semantic data requirements are listed:
  - *BR.01*. The attribute *Title* contains a string no longer than 256 characters
  - *BR.02*. The attribute *Title* must exists in the IMDB database
  - *BR.03*. The attribute *Director* contains an string no longer than 30 characters
  - *BR.04*. The attribute *Director* must appear in the IMDB database associated to the movie having the title specified in the attribute *Title*
  - *BR.05*. The attribute *Year* must be a positive number between 1895 and 2030
  - *BR.06*. The attribute *LastRemakeYear* must be always greater than *Year*
  - *BR.07*, *BR.08* and *BR.09*. The attributes *Title*, *Director*, and *Year* can not be null
  - *BR.10*. If the attribute *#Remakes* is zero, then the attribute *LastRemakeYear* must be null

- **Data Quality Measurement Requirements**: The data quality dimensions along with possible values for expressing the results of measurements are the following:
  - **Completeness**: the possible values are { "Complete", "Non-Complete"}
    * *BR.DQM.01.* A record is *Complete* when meet the business rules *BR.06, BR.07* and *BR.08*
    * *BR.DQM.02.* A record is *Non-complete* when does not meet *BR.DQM.01*
  - **Accuracy**, having the values { "Very Accurate", "Accurate", "Inaccurate"}:
    * *BR.DQM.03.* A record is *Very accurate* when meets *BR.02* and *BR.04*
    * *BR.DQM.04.* A record is *Accurate* when meets *BR.02*
    * *BR.DQM.05.* A record is *Inaccurate* when does not meet neither *BR.DQM.03* nor *BR.DQM.04*
  - **Consistency**, having values from the set { "High Consistency", "Consistency", "Low consistency"}:
    * *BR.DQM.06.* A record is *High Consistency* when meets *BR.04* and *BR.09*
    * *BR.DQM.07.* A record is *Consistency* when meets BR.04
    * *BR.DQM.08.* A record is *Low consistency* when does not meet neither *BR.DQM.06* nor *BR.DQM.07*
- **Data Quality Assessment Requirements**: After measurement, it is the time to aggregate the results previously obtained, in order to generate a judgment about the usability of a record. For this example, we consider the following values: { "suitable quality", "enough-adequate quality", "non-usable"}
  - A record is said to be of *suitable quality* when meet *BR.DQM.01, BR.DQM.02, BR.DQM.04,* and *BR.DQM.05.*
  - *BR.DQA.02.* A record is said to have *adequate quality for use* when meet *BR.DQM.01, BR.DQM.04,* and *BR.DQM.05.*
  - BR.DQA.03. A record is said to be *non-usable* when does not meet any of *BR.DQM.01, BR.DQM.02, BR.DQM.03, BR.DQM.04* or *BR.DQM.05.*

4 Decision Model in DMN for Data Quality

The decision model presented in Fig. 3 establishes two hierarchical levels to reduce the DMN complexity [7]. The bottom level corresponds to the data quality dimensions to be measured. Following the motivating example, these dimensions are *Completeness, Accuracy,* and *Consistency.* The upper level is related to the assessment of the level of data quality.

4.1 Data Quality Measurement

Data quality measurement is at the bottom level in the Decision Model. As aforementioned, it is related to each dimension to be measured in accordance with the Data Quality Measure Requirements presented in Sect. 3.

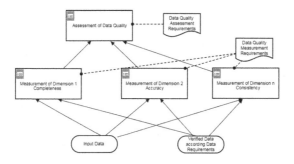

Fig. 3. Decision model diagram for data quality assessment

The first dimension modeled is *Completeness* (see Fig. 2). The input for building the DMN table for *Completeness* are: *BR.06*, *BR.07*, and *BR.08*. Each of these columns might take a *Boolean* value (i.e., true or false) after the verification of the corresponding rule. The different entries (i.e., rows) return an output assessment value for this dimension (cf., Output column). In this example, the completeness dimension has two entries (two rows). The first entry sets *true* to each one of the input value, associating the value *complete* to the output column. It indicates that, in order to consider that the data is *complete*, then it must fulfill the requirements *BR.06*, *BR.07*, and *BR.08*. The second entry basically indicates that if the data does not meet none of the aforementioned requirements, then it is considered *non-complete*.

The same logic applies for building the DMN tables of the other two dimensions. Figure 4 depicts the DMN table for *Accuracy*. It is composed of two input columns, that correspond with *BR.02* and *BR.04*. Three entries establish the data quality measurement requirements for *Accuracy*. The first entry indicates that both requirements must be fulfilled in order to consider the data as *very accurate*. The second entry specifies that only the requirement *BR.02* must be met in order to consider the data as *accurate*. The last entry indicates that the data is *inaccurate* if it does not fulfill none of both requirements.

Measurement of Accuracy				
Decision_Accuracy				
F Input		Output		
BR.02	BR.04	Accuracy		Annotations
boolean	boolean	{very accurate, accurate, inaccurate}		
1 true	true	very accurate		BR.DQMN.02
2 true	-	accurate		BR.DQMN.03
3 -	-	inaccurate		-

Fig. 4. DMN for measurement accuracy.

The last dimension is *Consistency*, shown in Fig. 5. In this case, *BR.04* and *BR.09* are the requirements used to evaluate this dimension. If both requirements are fulfilled, then the data is considered *highly consistent*. If only the requirement *BR.04* is met, then the data is considered *consistent*. In any other case, the consistency of the data is low.

Measurement of Consistency				
Decision_Consistency				
F Input			Output	
BR.04	BR.09		Consistency	Annotations
boolean	boolean		{high consistency, consistency, low consistency}	
1 true	true		high consistency	BR.DQMN.04
2 true	-		consistency	BR.DQMN.05
3 -	-		low consistency	-

Fig. 5. DMN for measurement consistency.

4.2 Data Quality Assessment

Data quality assessment is at the top level in our decision model (cf, Fig. 6). In this level, there is only one DMN table, which takes as input the values returned by the previous DMN tables corresponding to data quality measurement of all dimensions.

Assessment of Data Quality				
Decision_Assessment				
F Input			Output	
Completeness	Accuracy	Consistency	Level of Data Quality	Annotations
{complete, non-complete}	{very accurate, accurate, inaccurate}	{high consistency, consistency,low consistency}	{suitable, enough-adequate, non-usable}	
1 complete	not(inaccurate)	not(low consistency)	suitable	BR.DQA.01
2 complete	-	not(low consistency)	enough-adequate	BR.DQA.02
3 -	-	-	non-usable	BR.DQA.03

Fig. 6. DMN for assessment of data quality

The DMN table consists of three input columns, one per each data quality dimension measured. The value which each column might take depends on the value returned by the corresponding dimension. In this case: *Completeness* could take these values *complete* or *non-complete*; *Accuracy* can be *very accurate*, *accurate* or *inaccurate*; and *Consistency* could take *high consistency*, *consistency* or *low consistency*. The output represents the overall level of data quality which can be one of these values: *adequate*, *suitable*, or *enough*.

The table is composed of three rules. The first rule indicates that the overall level of data quality is *adequate* if *Completeness* is *complete*, *Accuracy* is not *inaccurate*, and *Consistency* is not *low consistency*. It must be highlighted that two of the entries of this rule support more than one valid value. For example, the entry for *Accuracy* indicates that must not be inaccurate, which means the value might be either *very accurate* or *accurate*. The same applies for *Consistency*. Thus, the second rule is similar to the first one, except for the *Accuracy* entry.

In this case, it establishes that this dimension can take any of the possible values. If this rule is met, then the overall level of data quality is *suitable*. The last rule establishes that, for any dimension value, the overall level of data quality is *enough*.

Remark that when the DMN table is analyzed, each decision rule must be unwound, by covering multiples cases for the input values. Thus, it might lead to conflicts between rules. In our example, it does happen since there are four cases covered by the first and the second rules. It means that data fulfilling any of these four conditions might be assessed as either *adequate* or *suitable*. Here is where the hit policy comes to play. In our case study, the policy is *First*, it means that rules might overlap between them and, in the case it occurs, the first hit by rule order is returned. Applied to this scenario, it means that any of the rules exposed would be assessed as *adequate*.

4.3   Results of the Data Quality Assessment

Table 3 shows the results obtained from applying the DMN decision table proposed in Sect. 4 to the tuples of Table 2 (i.e., the motivating example presented in Sect. 3).

Next, we explain how these results have been obtained:

- *Completeness.* The first and fourth tuple are *non-complete*. The first one violates *BR.06* while the fourth one violates *BR.08*.
- *Accuracy.* The first tuple is *accurate* due to the violation of *BR.04*, while the second tuple is *very accurate* because it fulfill all rules. The third and fourth tuples violate *BR.02*, causing them to be valuated as *Inaccurate*.
- *Consistency.* All tuples except the second one are labeled as *low consistency* because they violate *BR.04*.
- *Data Quality.* The first, third and fourth tuples are labeled as *non-usable* because their consistency is low. However, the second tuple is labeled as *Suitable* since its *Completeness* is *complete*, its *Accuracy* is *inaccurate*, and its *Consistency* is *low consistency*.

Table 3. Results of data quality assessment.

Id	Completeness	Accuracy	Consistency	Data quality assessment
1	non-Complete	accurate	low consistency	non-usable
2	complete	very accurate	high consistency	suitable
3	complete	inaccurate	low consistency	non-usable
4	non-Complete	inaccurate	low consistency	non-usable

5   Related Work

Data quality has been considered a key topic in many contexts, what makes that many researchers and practitioners have developed their own data quality models and the underlying assessment methods [1]. Although the idea of data quality dimensions have been widely studied and proposed through literature, only few authors as [9] or [18] have published specific implementable measurement methods for the various data quality management initiative. Many practitioners claims guides for a sound interpretation of the results of data profiling tools to better identify root cause of the problems. In fact, as [21] stated, data profiling are not explicitly presented associated to the idea of data quality dimensions but to the idea of data quality errors.

On the other hand, it is necessary to understand the necessity of having available mechanisms to determine almost in execution-time of the instance of a business process, if a given record has quality enough to be usable for the task at hand. So profiling the whole dataset as a way to assess the level of data quality is not useful for us. Therefore, we need to integrate the data quality assessment into the running instance of the business processes, and desirably, enable this assessment to be done automatically. We observed that many data quality analysts knew how to describe by means of business rules whether a record of data was usable or not.

However, to the best of our knowledge, there are not published similar approaches to the one we presented in this paper. Our proposal establishes a semi-automatically rules-guided waterfall cycle for assessing the usability of individual records in a datasets during the execution of the instance of business processes. The very nature of this approach suggest the use of DMN to implement the measurement methods to make a decision on the use of data. DMN has been used to represent business rules facilitating the understanding and description of business rules [4,5]. DMN is frequently used into BPMN [14] models by means of decision tasks that incorporate the set of rules that must be evaluated during instantiation time [11]. Moreover, some DMN extensions let the integration of the decision making process non-only incorporating dataflow [17]. The necessity to incorporate data quality measurement in business processes was early identified in [16]. However, to the best of our knowledge, there is no solution that use DMN as a mechanism to model and evaluate data quality assessment and measurement requirements and it is still a challenge how bring the gap between the human description of data quality requirement description, with and automatic data quality assessment and measurement.

6   Conclusions

The inclusion of data quality assessment requirements in business process helps organizations to make more reliable decisions on the use of data. We have introduced in the paper the application of DMN (Decision Model and Notation) standard with the aim of facilitating the assessment of data quality requirements.

Thanks to the use of DMN, the automation of the evaluation of the data quality level ceases to be a theoretical contribution and becomes a reality. On the one hand, business experts can easily include their knowledge in DMN since is a common notation readily understandable. On the other hand, DMN facilitates the inclusion of the business rules for data quality measurement and assessment as part of the decision process and feeds the process with the information related to data quality.

Acknowledge. This work has been partially funded by the Ministry of Science and Technology of Spain ECLIPSE (RTI2018-094283-B-C33) and (RTI2018-094283-B-C31) projects, the Junta de Andalucía via the PIRAMIDE and METAMORFOSIS projects, the European Regional Development Fund (ERDF/FEDER), GEMA: Generation and Evaluation of Models for dAta Quality (Ref.: SBPLY/17/180501/000293), and the Cátedra de Telefónica "Inteligencia en la Red" of the Universidad de Sevilla.

References

1. Batini, C., Cappiello, C., Francalanci, C., Maurino, A.: Methodologies for data quality assessment and improvement. ACM Comput. Surv. (CSUR) **41**(3), 16 (2009)
2. Batini, C., Scannapieco, M.: Data Quality: Concepts, Methodologies and Tea niques. Data-Centric Systems and Applications. Springer, New York (2006). https://doi.org/10.1007/3-540-33173-5
3. Ceravolo, P., et al.: Big data semantics. J. Data Semant. **7**(2), 65–85 (2018)
4. Dangarska, Z., Figl, K., Mendling, J.: An explorative analysis of the notational characteristics of the decision model and notation (DMN). In: 20th IEEE International Enterprise Distributed Object Computing Workshop, EDOC Workshops 2016, Vienna, Austria, 5–9 September 2016, pp. 1–9 (2016)
5. Dasseville, I., Janssens, L., Janssens, G., Vanthienen, J., Denecker, M.: Combining DMN and the knowledge base paradigm for flexible decision enactment. In: Supplementary Proceedings of the RuleML 2016 Challenge, 10th International Web Rule Symposium, RuleML 2016, New York, USA, 6–9 July 2016 (2016). http://ceur-ws.org/Vol-1620/paper3.pdf
6. Figl, K., Mendling, J., Tokdemir, G., Vanthienen, J.: What we know and what we do not know about DMN. Enterp. Model. Inf. Syst. Arch. **13**(2), 1–16 (2018)
7. Hasic, F., Craemer, A.D., Hegge, T., Magala, G., Vanthienen, J.: Measuring the complexity of DMN decision models. In: Business Process Management Workshops - BPM 2018 International Workshops, Sydney, NSW, Australia, 9–14 September 2018, Revised Papers, pp. 514–526 (2018)
8. Hasić, F., De Smedt, J., Vanthienen, J.: Towards assessing the theoretical complexity of the decision model and notation (DMN). In: CEUR Workshop Proceedings, vol. 1859, pp. 64–71. CEUR Workshop Proceedings (2017). https://lirias.kuleuven.be/1548429?limo=0
9. Heinrich, B., Klier, M.: Metric-based data quality assessment - developing and evaluating a probability-based currency metric. Decis. Support. Syst. **72**, 82–96 (2015)
10. ISO-25012: Iso/IEC 25012: Software engineering-software product quality requirements and evaluation (square)-data quality model (2008)

11. Janssens, L., Bazhenova, E., Smedt, J.D., Vanthienen, J., Denecker, M.: Consistent integration of decision (DMN) and process (BPMN) models. In: Proceedings of the CAiSE 2016 Forum, at the 28th International Conference on Advanced Information Systems Engineering (CAiSE 2016), Ljubljana, Slovenia, 13–17 June 2016, pp. 121–128 (2016). http://ceur-ws.org/Vol-1612/paper16.pdf
12. Lee, S., Ludäscher, B., Glavic, B.: PUG: a framework and practical implementation for why and why-not provenance. VLDB J. **28**(1), 47–71 (2019)
13. Loshin, D.: The Practitioner's Guide to Data Quality Improvement. Elsevier, Amsterdam (2010)
14. OMG: Business process model and notation (2017). http://www.omg.org/spec/BPMN/2.0
15. OMG: Decision Model and Notation (DMN), Version 1.2, January 2019. https://www.omg.org/spec/DMN
16. Parody, L., Gómez-López, M.T., Bermejo, I., Caballero, I., Gasca, R.M., Piattini, M.: PAIS-DQ: extending process-aware information systems to support data quality in PAIS life-cycle. In: Tenth IEEE International Conference on Research Challenges in Information Science, RCIS 2016, Grenoble, France, 1–3 June 2016, pp. 1–12 (2016)
17. Pérez-Álvarez, J.M., Gómez-López, M.T., Parody, L., Gasca, R.M.: Process instance query language to include process performance indicators in DMN. In: 20th IEEE International Enterprise Distributed Object Computing Workshop, EDOC Workshops 2016, Vienna, Austria, 5–9 September 2016, pp. 1–8 (2016)
18. Sebastian-Coleman, L.: Measuring data quality for ongoing improvement: a dataquality assessment framework. Newnes (2012)
19. Wang, R.Y.: A product perspective on total data quality management. Commun. ACM **41**(2), 58–65 (1998)
20. Wang, R.Y., Reddy, M.P., Kon, H.B.: Toward quality data: an attribute-based approach. Decis. Support. Syst. **13**(3–4), 349–372 (1995)
21. Woodall, P., Oberhofer, M., Borek, A.: A classification of data quality assessment and improvement methods. Int. J. Inf. Qual. **3**(4), 298–321 (2014)

Modeling Rolling Stock Maintenance Logistics at Dutch Railways with Declarative Business Artifacts

Erik Smit and Rik Eshuis[✉]

School of Industrial Engineering, Eindhoven University of Technology,
Eindhoven, The Netherlands
esmit@live.nl, h.eshuis@tue.nl

Abstract. A case study is presented in which the maintenance logistics process of rolling stock (train units) at Dutch Railways is modeled. This decision-intensive process has a high complexity due to its many steps and the different actors that are involved. Declarative Business Artifacts are used as modeling technique to master this complexity. We discuss the key elements of the artifact model and focus on encountered issues and lessons learned. The artifact model is compared with an earlier developed agent-based model of the maintenance logistics process. The case study shows that declarative business artifacts are a promising technique to structure complex decision-intensive processes.

Keywords: Declarative notations · Knowledge-intensive processes · Case study

1 Introduction

Nowadays, a key challenge in engineering decision-intensive processes (DiPs) is to use data generated in DiPs to improve the decision-making, in particular for environments that are already complex in itself. An example DiP is rolling stock maintenance logistics [2], which is the process that ensures that rolling stock (train units) that is marked for maintenance, is at the expected maintenance location at the expected time. This DiP has a high complexity due to the many different actors that are involved and the volatility of the environment, for instance disruptions in the railway network.

This paper presents a case study on mapping the current maintenance logistics process at Dutch Railways, the main public train operator in the Netherlands. Key aim is identifying decision-making points to improve the overall performance. Since the maintenance logistics process is decision-intensive and requires flexibility, we have chosen declarative business artifacts to model the process [8], in particular Guard-Stage-Milestone (GSM) schemas [11], which have been applied before to model DiPs [18]. The paper presents the key elements of

ⓒ Springer Nature Switzerland AG 2019
C. Di Francescomarino et al. (Eds.): BPM 2019 Workshops, LNBIP 362, pp. 375–387, 2019.
https://doi.org/10.1007/978-3-030-37453-2_31

the artifact model of the maintenance logistics process at Dutch Railways, focusing on encountered issues and lessons learned. Additional background and full details are in a separate report [17].

A few other papers describe experiences in applying business artifacts to real-world business processes [3,6]. The maintenance process at Dutch Railways has been modeled before using agents [13]. Section 5 provides an extensive comparison of this agent-based model with the artifact model developed in this paper.

The remainder of this paper is structured as follows. Section 2 briefly explains how maintenance logistics is performed at Dutch Railways. Section 3 introduces the used artifact modeling technique, GSM schemas. Section 4 elaborates the design, focusing on the key aspect and encountered issues. Section 5 evaluates the design by an extensive comparison with an alternative agent-based design [13]. Section 6 ends the paper with conclusions.

2   Maintenance Logistics at Dutch Railways

Rolling stock such as train units require maintenance, in order to keep performing. This paper focuses on short-term maintenance, which concerns maintaining brake linings, wheel axles, filters, etc. of train units [1]. These maintenance tasks are performed at specific maintenance depots (Onderhoudsbedrijf in Dutch, abbreviated OB). Each depot has expertise to maintain certain train unit types.

Due to the high use of the Dutch railway network, it takes much effort to direct a train unit to a maintenance depot for an (un)planned service. Moreover, the Dutch railway system is complex, because train units do not operate between fixed points but follow a complex path through the rail network [5]. Movements of the regular train plan are used to let train units arrive at maintenance locations. Train movements are influenced by unexpected events like accidents or big disruptions that require rescheduling [15], which also may affect train units marked for maintenance [4].

Dutch Railways uses the concept of *train paths* to manage the allocation of train units to train movements or activities. A train path consists of multiple subsequent movements and has a different daily start and final location. In many cases multiple allocation mutations are needed to get a train unit in the right path that ends up in a maintenance location; a path can take a couple of days.

Trains units can switch best from one path to another at night, which means in practice starting in another path the next day. However, switching is sometimes required during the day for example if a train unit must arrive at the maintenance location in the evening. Mutations that take place during the day require perfect alignment or some buffer time at a station.

3   Business Artifacts

GSM schemas are a declarative approach to represent business artifacts [11]. An alternative, related approach is the Case Management Model and Notation (CMMN) [16] standard, which however has a higher modeling complexity as it

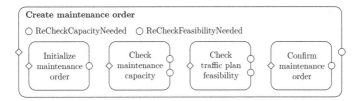

Fig. 1. Example GSM notation

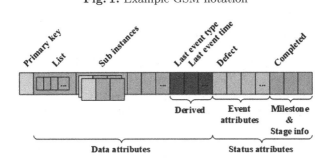

Fig. 2. Structure GSM information model

contains a lot more constructs than GSM. A GSM schema of a business artifact consists of both an information model and a lifecycle model. An example of a GSM lifecycle is given in Fig. 1. Guards, depicted with diamonds, are attached to stages. A guard is a rule that controls the opening of the corresponding stage. A stage, represented as a rounded rectangle, is an amount of work. Stages can be hierarchical. Milestones are visualized with circles and correspond to 'key business-relevant operational objectives that an instance might achieve' [11]. The status (activation) of stages and milestones is controlled by rules, which are called sentries. Each guard is a sentry, but also deactivation of a stage and achievement/invalidation of a milestone is controlled by sentries. Each sentry has the structure "*on* event *if* condition", where both parts are optional.

The GSM lifecycle specifies how data of the corresponding information model is changed. The information model contains for instance status attributes which represents the current state of that particular artifact instance. Execution rules specify the transitions from one state to another state. The information model consists of one record for each artifact (e.g. a maintenance order) and the fields correspond to the data of the GSM lifecycle. These data includes data attributes, event attributes and milestone and stage information [11].

Figure 2 shows the structure of an information model. Each line with a primary key represents one instance. This instance has a set of data attributes and status attributes. The data attributes are written by the business activities in the stages or are derived (so immutable) from other sources [11]. Examples of derived data attributes are the most recent event time or attributes that are managed by other artifacts. The status attributes describe the status of each stage and milestone. If an external event such as completion of a stage occurs, sentries evaluate to true and the status attributes they belong to change value. GSM schemas have a precise execution semantics [7], but this is out of scope for this paper.

4 Design

We next present the GSM design, focusing on the aspects that are non-standard.

4.1 Identifying Business Artifacts

Bhattachary et al. [3] propose an artifact-design method in which first the key artifacts are discovered, e.g., based on scenarios. Kumaran et al. [14] propose a method to derive key artifacts from an activity-centric process model, in which artifacts are input and output of activities. Eshuis and Van Gorp [9] define a method to derive a GSM schema from an activity-centric model.

Inspired by these approaches, we identified business artifacts in the case study in a stepwise fashion. First, a communication model was designed in which the actors and departments at Dutch Railways were modeled and their interactions (Fig. 3), including the applications. Each background box represents a department; the bottom one is the national organization that maintains the rail network (not part of Dutch Railways). Each solid rectangle is a human actor that makes decisions. Each dashed rectangle is an application used by an actor. Lines indicate communication channels; the exchanged messages are annotated on the lines. A communication model resembles a BPMN conversation diagram, but in addition uses message ordering. This approach of starting with actors is known as subject-oriented BPM [10]. Reason for choosing this modeling technique is that the process involves many different departments and roles. A communication model provided a useful starting point for a subsequent detailed analysis.

Next, a class diagram was modeled to specify the content of the messages of the communication model. It identifies the entities, a subset of which are elaborated into business artifacts. Due to space limitations the model is not shown here, but it can be found elsewhere [17].

Next, an activity-centric process model was developed, for each activity describing the input and output information entities. An activity can create, read or update an information entity [9]. Based on this activity-centric model, we use the notion of dominance introduced by Kumaran et al. [14] to create an entity domination graph (Fig. 4). Each oval in this figure represents an information entity and each arrow points to a dominated entity. Entities with only outgoing arrows are dominant information entities, underlined in Fig. 4, and are the key business artifacts [14]. These entities get their own lifecycle in the GSM schema.

We also experimented with two alternative approaches for deriving dominating entities: using multiplicity constraints and domain knowledge. For example, a WorkPackage is connected to exactly one MaintenanceTask, while a MaintenanceTask is connected to zero or one WorkPackage; this implies that a

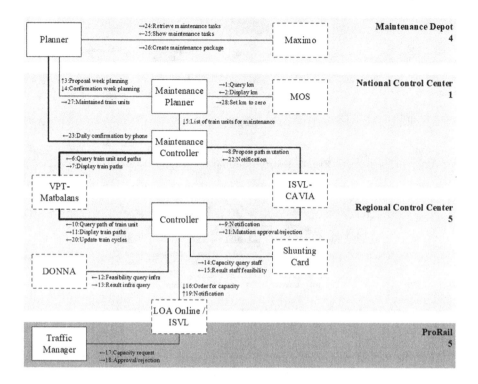

Fig. 3. Communication model

MaintenanceTask provides the context for a WorkPackage, so MaintenanceTask dominates WorkPackage. However, the multiplicity relations do not provide sufficient information to decide domination. For instance, there is a many-to-many relation between Technician and MaintenanceTask, while MaintenanceTask dominates Technician in Fig. 4.

The example in the previous paragraph about the Maintenance task and the Work package is also solvable with domain knowledge. The OB Planner first selects maintenance tasks before creating a work package. Therefore, Maintenance task dominates Work package. Since this approach looks at the practical meaning of entities, it can define domination where multiplicity constraints are not helpful. Another advantage of this approach is the possibility to validate 'domination' choices by the other approaches. The risk of using domain knowledge is that experts may use slightly different dominance definitions. For example, it is possible to define dominance based on causality (entity e1 can create e2) or based on subordination (e1 consist of e2), which are two different definitions.

4.2 Used GSM Variant

We annotated stages in the GSM schemas with the responsible actors [11]. Next, we introduced additional status attributes for the guards (see Fig. 5) compared to the original model (Fig. 2). Guards are only used in the original approach

Fig. 4. Entity domination graph

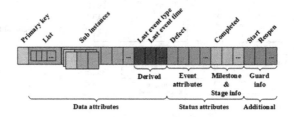

Fig. 5. Information model with additional status attributes

of [7] for the visual representation of sentries, which means that stages open immediately after the guard becomes true. Therefore the original information model only includes status attributes of milestones and stages. However, the opening of a guard in the present approach *enables* the stage for opening by the (human) owner. Figure 6 represents graphically how this works in the GSM notation and gives the default execution rules, where +A indicates the decision of an actor to open the stage.

Our proposal, which is allowed by the stage lifecycle of CMMN [16], is motivated by the many human-driven stages in the GSM schema. It gives actors fine-grained control over the execution of stages and enables the measurement of waiting time between the enabling and actual start of a stage.

4.3   GSM Design

The GSM schema of the maintenance logistics process at Dutch Railways consists of the four artifacts identified using the entity domination approach, marked gray in Fig. 4. Together, these GSM artifacts give an integrated overview of the actual process in a compact way.

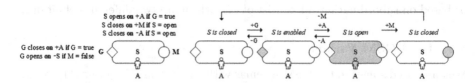

Fig. 6. GSM default rules

Given the limited space, we only present one business artifact, Maintenance Order (MO), in detail; the others are detailed elsewhere [17]. The MO artifact handles requests to the railway network to supply a train unit and activates the maintenance logistics process. Figure 7 presents the lifecycle of the MO artifact; Grey dashed lines visualize execution rules; a fragment of the rules are shown in Table 1; all rules can be found elsewhere [17].

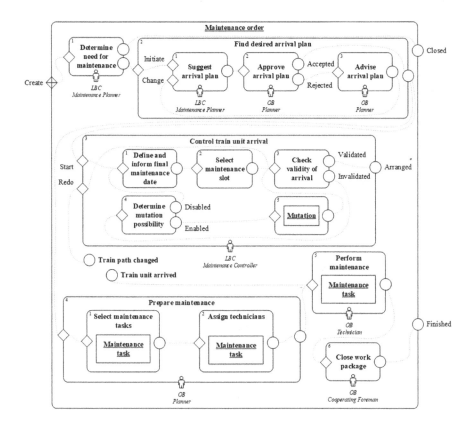

Fig. 7. GSM maintenance order

The plus diamond (⊕) creates every week an MO instance for each train unit, since maintenance is time-driven. This instance only progresses if the LBC Maintenance Planner determines that the train unit needs maintenance again, after which the following main tasks are performed.

Find Desired Arrival Plan. Finding a rough plan for maintenance is an interaction between the LBC Maintenance Planner and the OB (Maintenance) Planner. The OB Planner accepts or rejects suggested visits and gives advice for rescheduling if necessary.

Table 1. Fragment of the rules for Maintenance Order (MO) lifecycle in Fig. 7

Name	Guard/ milestone	Rule
MO_2 (Find desired maintenance plan)	◇ Start	open **on** $+MO_1$(Initialized)
	○ Found	achieve **on** $+MO_{2.2}$(Accepted)
$MO_{2.1}$(Suggest arrival plan)	◇ Initiate	open **on** $+MO_2$(Start)
	◇ Change	open **on** $MO_{2.3}$(Advised)
	○ Suggested	achieve **if** $MO_{2.1}$ is completed invalidate **on** $+MO_{2.3}$(Advised)
$MO_{2.2}$ (Approve arrival plan)	◇ Start	open **on** $+MO_{2.1}$(Suggested)
	○ Accepted	achieve **if** plan meets local requirements
	○ Rejected	achieve **if** plan does not meet local requirements
$MO_{2.3}$ (Advise arrival plan)	◇ Start	open **on** $+MO_{2.2}$(Rejected)
	○ Advised	achieve **if** $MO_{2.3}$ is completed

Control Train Unit Arrival. The second main task, managed by the LBC Maintenance Controller, is getting the train unit on the desired date at the right maintenance location. At the beginning of a day, the OB (Maintenance Depot) is informed about which train units are planned for arrival at the end of the day. Subsequently, the train unit is allocated to a maintenance slot that is connected with a particular train path. A train unit connected to this train path automatically arrives at the right time, while additional (empty) movements may be needed for train units not connected to such a train path.

The LBC Maintenance Controller checks the validity of arrival; in case of invalidity, he checks whether mutations are possible to ensure a proper arrival. Invalidity occurs to train reschedulings in case of incidents.

Prepare and Perform Maintenance. The preparation of the maintenance visit by the OB Planner starts in parallel with controlling the train unit arrival, while the actual execution of maintenance only starts after the arrival event. This preparation consists of selecting maintenance tasks and assigning technicians [17]. Technicians perform these maintenance tasks after the train unit arrival, after which the Cooperating Foreman can close the work package and thereby the entire MO instance.

4.4   Performance Evaluation

One particular use of Business artifacts is that they provide context for the performance analysis of DiPs. To illustrate this, we present the initial part of the performance analysis of the response time on mutation requests. This relates to a GSM stage of the Mutation artifact. Figure 8 gives an indication of the relation between the MO and the Mutation artifact based on the data of 2017 from information systems Maximo and ISVL (Informatie System Verkeersleiding) [17]. The corresponding train units need on average ±3.7 mutations before the arrival to the maintenance location. This number becomes a problem in combination with the duration of processing as explained elsewhere [17].

Ideally, data would be available for all stages of the GSM schema to create a complete heat map. Unfortunately, that is not the case and therefore the performance analysis only gives a first impression of some bottlenecks. It appears that the mutation process is time-consuming for example as a result of the large response times for requests to the RBCs. Still, the performance analysis gives useful pointers for redesigning the maintenance logistics process [17].

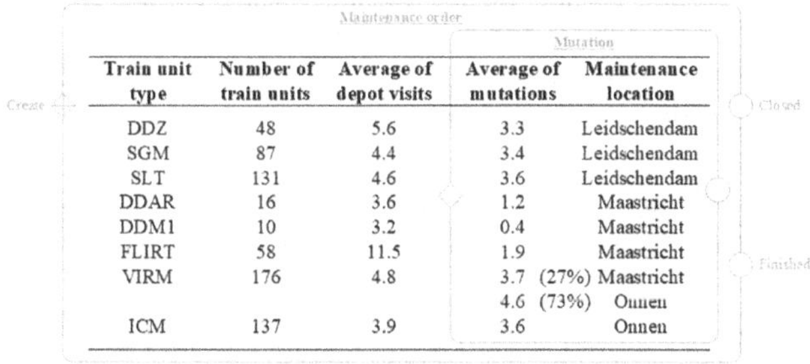

Train unit type	Number of train units	Average of depot visits	Average of mutations	Maintenance location
DDZ	48	5.6	3.3	Leidschendam
SGM	87	4.4	3.4	Leidschendam
SLT	131	4.6	3.6	Leidschendam
DDAR	16	3.6	1.2	Maastricht
DDM1	10	3.2	0.4	Maastricht
FLIRT	58	11.5	1.9	Maastricht
VIRM	176	4.8	3.7 (27%)	Maastricht
			4.6 (73%)	Onnen
ICM	137	3.9	3.6	Onnen

Fig. 8. Overview of maintenance visits and mutations per train unit type

Table 2. Acronyms OperA+ model and their GSM counterparts

Acronym	Full name	GSM Actor
MOP	Maintenance Order Provider	National Control Center
TSP	Transfer Service Provider	National Control Center
MSP	Maintenance Service Provider	OB (Maintenance Depot)
MOPN	Maintenance Order Planner	LBC Maintenance Planner
ODA	Operational Data Administrator	MOS Information System
MRP	Maintenance Routing Planner	LBC Maintenance Controller
MJP	Maintenance Job Planner	OB Planner
MMT	Maintenance Mechanics Team	OB Technician
MPM	Maintenance Production Manager	OB Cooperating Foreman

5   Evaluation: Agent-Based Model Comparison

We evaluate the GSM design by comparing it with an earlier developed design of the maintenance process [13], which is based on the agent modeling approach OperA+ [12]. Agent models are used to describe, analyze and simulate socio-technical systems, and to investigate complex relationships between entities, in order to improve knowledge sharing and communication between different stakeholders [13] The agent-based design is compared with the communication model of Fig. 3 and the GSM schema in Sect. 4. This comparison consists of two parts: (i) the roles from the OperA+ model are linked to the corresponding roles in the communication model and (ii) the mapped process of the OperA+ model is compared with the GSM schema.

Roles. Each role in the OperA+ model corresponds to an agent, department or application in the communication or GSM schema. Table 2 gives an overview of this comparison. The National Control Center appears twice in Table 2 while this is represented by two different roles in the OperA+ model. In reality, each role is played by a different subdepartment of the National Control Center. Besides that, it is interesting to see that three roles (ODA, TDA and MJA) correspond to three information systems in the communication model. The notation of the OperA+ model does not distinguish human and artificial agents. In the GSM schema, we only included the human actors.

Process. The comparison between the processes presented in the OperA+ model and the GSM schema is based on the relationships between the three main roles – the MOP, TSP and MSP – in the OperA+ model, as shown in the left part of Fig. 9. A unique color is used for each of these three roles to show the relation with the GSM schema. The GSM lifecycle of the maintenance order artifact (Fig. 7) is the natural counterpart for the OperA+ model as this artifact directs the complete process from the trigger for maintenance until the actual maintenance.

We illustrate the relation between the OperA+ model and the GSM schema by explaining for one of the three main relations, MOP ↔ MSP, how it is represented in the GSM schema. The maintenance process starts after passing the threshold value for mileage and time. In the OperA+ model, the *Maintenance Order Provider* (MOP) informs the *Maintenance Service Provider* (MSP) that a particular maintenance job must be scheduled for execution before a certain due date. The MSP has to decide whether this maintenance job fits into the maintenance program of the depot. Therefore this relationship is reciprocal, as visible in the OperA+ model. This reciprocal relationship is also visible in stage MO_2 of the GSM schema. The LBC Maintenance Planner makes an arrival plan in stage $MO_{2.1}$ which is checked by the OB Planner in stage $MO_{2.2}$. The OB Planner can either accept this plan or send another plan to the LBC Maintenance Planner.

There is also another relationship between the MOP and the MSP. If the job is accepted by the MSP, the *Maintenance Job Planner* (MJP) has to make a feasible production plan. This relation is also visible in the GSM schema: if the milestone of stage MO_2 gets achieved (which happens if a 'rough' arrival plan is found), then the guard of stage MO_4 becomes true. In practice this means that the OB Planner can start the preparation of the actual maintenance visit.

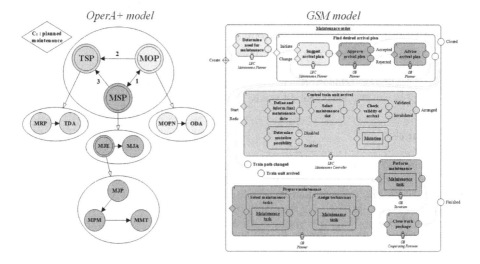

Fig. 9. Comparison between GSM and OperA+ model

Conclusion. This comparison shows that the OperA+ model is less detailed than the business artifact model. Moreover, the OperA+ model focuses on the decision makers while the GSM schema details the decision making process. Although both models represents the maintenance logistics process, the GSM schema is more useful to reengineer a business process. The OperA+ model focuses on the organization of roles, which are less amenable to change than processes and information that are central to the GSM schema.

6   Conclusion

This paper reports on a case study in which a complex DiP, the rolling stock maintenance logistics process at Dutch Railways, was modeled with declarative business artifacts. Though the maintenance logistics process has a high complexity, involving a lot of steps and different actors, declarative business artifacts proved to be a suitable technique to model the process and to structure the information landscape for decision making. The use of auxiliary models such as a communication model proved to be vital to design the artifact model for such a complex DiP. In comparison with an agent-based design, business artifacts focus on the decision-making rather than the decision-maker aspect, which seems more valuable for reengineering the process.

The case study shows that declarative business artifacts are a promising technique to structure complex DiPs, as claimed before [18]. However, additional case studies are needed to analyze this claim in more depth.

Dutch Railways is currently facing the challenge to organize maintenance of their rolling stock assets based on their condition, instead of mileage. Using data to properly perform the maintenance logistics process in a dynamic way then becomes vital. Declarative business artifacts can play a key role to support these dynamic, data-driven processes in an effective way.

Acknowledgments. This research was sponsored by TKI Logistics Netherlands under the program Big Data: Real-time ICT for Logistics.

References

1. Apallius de Vos, J.I., Van Dongen, L.A.: Performance centered maintenance as a core policy in strategic maintenance control. Procedia CIRP **38**, 255–258 (2015)
2. Arts, J., Basten, R., van Houtum, G.: Maintenance service logistics. In: Zijm, H., Klumpp, M., Regattieri, A., Heragu, S. (eds.) Operations Logistics Supply Chain Management, pp. 493–517. Springer, Germany (2019)
3. Bhattacharya, K., Caswell, N.S., Kumaran, S., Nigam, A., Wu, F.Y.: Artifact-centered operational modeling: lessons from customer engagements. IBM Syst. J. **46**(4), 703–721 (2007)
4. Bin Osman, M.H., Kaewunruen, S., An, M., Dindar, S., Bin Osman, M.H.: Disruption: a new component in the track inspection schedule. In: Proceedings 2016 IEEE ICIRT, pp. 249–253. IEEE (2016)
5. Busstra, M., Van Dongen, L.: Creating value by integrating logistic trains services and maintenance activities. Procedia CIRP **38**, 250–254 (2015)
6. Chao, T., et al.: Artifact-based transformation of IBM global financing. In: Dayal, U., Eder, J., Koehler, J., Reijers, H.A. (eds.) BPM 2009. LNCS, vol. 5701, pp. 261–277. Springer, Heidelberg (2009). https://doi.org/10.1007/978-3-642-03848-8_18
7. Damaggio, E., Hull, R., Vaculín, R.: On the equivalence of incremental and fixpoint semantics for business artifacts with guard-stage-milestone lifecycles. Inf. Syst. **38**(4), 561–584 (2013)
8. Eshuis, R.: Modeling decision-intensive processes with declarative business artifacts. In: Lam, H.-P., Mistry, S. (eds.) ASSRI -2018. LNBIP, vol. 367, pp. 3–12. Springer, Cham (2019). https://doi.org/10.1007/978-3-030-32242-7_1
9. Eshuis, R., Van Gorp, P.: Synthesizing data-centric models from business process models. Computing **98**(4), 345–373 (2016)
10. Fleischmann, A., Schmidt, W., Stary, C., Obermeier, S., Börger, E.: Subject-Oriented Business Process Management. Springer, Heidelberg (2012). https://doi.org/10.1007/978-3-642-32392-8
11. Hull, R., et al.: Introducing the guard-stage-milestone approach for specifying business entity lifecycles. In: Bravetti, M., Bultan, T. (eds.) WS-FM 2010. LNCS, vol. 6551, pp. 1–24. Springer, Heidelberg (2011). https://doi.org/10.1007/978-3-642-19589-1_1
12. Jiang, J., Dignum, V., Tan, Y.: An agent based inter-organizational collaboration framework: OperA+. In: Proceedings WI-IAT 2011, pp. 21–24. IEEE Computer Society (2011)

13. Jiang, J., Huisman, B., Dignum, V.: Agent-based multi-organizational interaction design: a case study of the Dutch railway system. In: Proceedings IAT 2012, pp. 196–203. IEEE Computer Society (2012)
14. Kumaran, S., Liu, R., Wu, F.Y.: On the duality of information-centric and activity-centric models of business processes. In: Bellahsène, Z., Léonard, M. (eds.) CAiSE 2008. LNCS, vol. 5074, pp. 32–47. Springer, Heidelberg (2008). https://doi.org/10.1007/978-3-540-69534-9_3
15. Mannhardt, F., Arnesen, P., Landmark, A.D.: Estimating the impact of incidents on process delay. In: Proceedings ICPM 2019. IEEE Computer Society (2019)
16. Marin, M., Hull, R., Vaculín, R.: Data centric BPM and the emerging case management standard: a short survey. In: La Rosa, M., Soffer, P. (eds.) BPM 2012. LNBIP, vol. 132, pp. 24–30. Springer, Heidelberg (2013). https://doi.org/10.1007/978-3-642-36285-9_4
17. Smit, E.: Modeling rolling stock maintenance logistics with business artifacts: a case study at Dutch railways. Master's thesis, Eindhoven University of Technology (2018). https://research.tue.nl/files/108943973/Master_Thesis_Erik_Smit.pdf
18. Vaculin, R., Hull, R., Heath, T., Cochran, C., Nigam, A., Sukaviriya, P.: Declarative business artifact centric modeling of decision and knowledge intensive business processes. Proc. EDOC **2011**, 151–160 (2011)

Applying Business Architecture Principles with Domain-Specific Ontology for ACM Modelling: A Building Construction Project Example

Antonio Manuel Gutiérrez Fernández[1(✉)], Freddie Van Rijswijk[1],
Christoph Ruhsam[1], Ivan Krofak[2], Klaus Kogler[2], Anna Shadrina[3],
and Gerhard Zucker[3]

[1] ISIS-Papyrus, Brunn am Gebirge, Austria
{antonio.gutierrez,freddie.van.rijswijk,
christoph.ruhsam}@isis-papyrus.com
[2] CES Clean Energy Solutions, Vienna, Austria
{i.krofak,k.kogler}@ic-ces.at
[3] AIT Austrian Institute of Technology, Vienna, Austria
{anna.shadrina,gerhard.zucker}@ait.ac.at

Abstract. Adaptive Case Management (ACM) empowers knowledge workers of any industry by utilizing their experience in the execution of business cases. It is designed to handle non-repetitive processes, where the course of actions has to be decided for each case individually, based on the assessment of the current situation. ACM uses domain-specific data and the analysis of rules to provide the relevant information at the right time to the business users. This supports their actions and decision making but does not dictate predefined workflows that are difficult or even impossible to define for knowledge-intensive work and prevent users from efficiently achieving their goals. Solutions for ACM implementations need to focus on business requirements to support business users rather than limit them by extending existing tools such as content management systems, BPM-based software suites, or collaboration tools with ad hoc development that all lack the ability to adapt flexibly to business level changes without requiring significant IT development efforts. To tackle these challenges, this paper proposes a novel approach by applying business architecture concepts for the definition of ACM applications in combination with domain-specific ontologies and business rules. Business domain analysts describe with the domain-specific ontology the complete value stream with its goals, activities and rules which use a formalized natural business language. The resulting business information model is mapped to an underlying data model and instantly enacted during case execution so that adaptations get immediately available to the business users. The behavioral business rules guide business users to guarantee case compliance which facilitates an easy and rapid adaptation of business applications. We apply this approach to a real life example from the construction industry where several parties from different trades have to collaborate in heterogeneous environments.

© Springer Nature Switzerland AG 2019
C. Di Francescomarino et al. (Eds.): BPM 2019 Workshops, LNBIP 362, pp. 388–399, 2019.
https://doi.org/10.1007/978-3-030-37453-2_32

Keywords: Adaptive Case Management · Data based execution · Business Process Management · Behavioral Business Rules · Ontology

1   Introduction

One of the challenges in the execution of business applications is the short term adaption to changing business conditions and the handling of exceptions to cover unforeseen situations. The complexity of formalizing and automating business use cases with variants of process designs makes them difficult to manage and maintain and call for complex integration efforts of different IT systems.

Adaptive Case Management (ACM) provides flexibility to knowledge-based processes where the task flow is not strictly defined and the execution of cases depends on the analysis of the data for each case and on the decisions taken by the business users (knowledge workers) according to their expertise in the business domain and the assessment of the current situation. ACM bases the business case execution on behavioral rules constraining the business user actions instead of predefining tasks in a flow [24]. Existing ACM proposals use the analysis of compliance rules during business case execution [22]. However, care has to be taken that these approaches do not require the interaction between business analysts and IT developers to formalize, manage and implement business application definitions through data, actions and rules. Therefore, the supporting systems usually depend on IT experts, such as database modelling, workflow and process definition systems, rule engines and content management systems that cannot be easily adapted by business users to business changes. Further, such implementations are not transparent to the business users as hiding the definitions in the IT layer of such applications.

In order to overcome such limitations we propose a novel approach by modelling cases in ACM (i) with business architecture methodology as described by the Open Group [17] in combination (ii) with a business domain-specific ontology which we introduced earlier [21] to describe the complete business case with terms and relations as known by business users from their daily work and (iii) the definition of business rules with natural business vocabulary as proposed by Ross [19]. The definitions look at value streams (VSs), a business architecture entity, which are composed from atomic business goals (value stream stages, VSSs) that provide some benefit or product to the customer served by the company. The accomplishment of each goal of a VS is achieved by the execution of actions. And the evaluation of behavioral business rules guide the business users to complete actions in compliance with the regulatory framework for a specific business domain. The actions and rules are formalized using the same underlying business ontology. This formalization together with the action implementation and the evaluation of rules enable the instant enactment of the VSs providing an agile business application environment, as business entities can be directly maintained by business users. The resulting ACM application focuses on business goals over a rigid process flow or other aspects such as IT considerations and guarantees compliance during case execution by guiding the business user with transparently formulated behavioral business rules using business domain-specific terminology. These rules are defined by business domain analysts and avoid any technical notations or other IT dependencies.

This methodology is implemented by the ACM platform Papyrus Converse [18]. The combination of (i) providing the complete information for the current business case with (ii) business rules supported by (iii) our earlier proposed User Trained Agent (UTA) which uses the experience gained from the previous executions, leads to an effective knowledge sharing between all users working on the same system [11]. They receive suggestions of "best next actions" stemming from similar situations which the UTA has learned earlier to complete their goals.

In order to proof the potential of this approach, we apply it to scenarios from the construction industry which lacks a standard for the definition of building processes because of the heterogeneity of this domain involving different parties from different trades and the challenges coming from the unique characteristics of each building. Furthermore, the processes are even susceptible to changes in each building phase (design, implementation, operation) which can be addressed by applying ACM for such applications.

The rest of the paper is organized as follows. The challenges that arise with existing ACM solutions are described in Sect. 2 using scenarios from the construction industry. A discussion about existing approaches in ACM together with the business architecture context are presented in Sect. 3. Our proposal to address these challenges with value streams is described in Sect. 4 and it is applied to the project management of the building construction industry in Sect. 5. Finally, some conclusions about our proposal together with future work are discussed in Sect. 6.

2 Challenges in Adaptive Case Management

In the ACM paradigm [16, 20], the flow of performed activities for each business case depends more on the case information and the business experience and knowledge of users than on conducting strict workflows. This is the case in multiple scenarios such as outsourcing of maintenance services or the management of building construction projects, where many heterogeneous parties are involved with their own procedures, resources, etc. The effort to model these cases with a stable process model is expensive as these models need to include multiple decision points which are difficult to maintain at the pace of daily business changes and can therefore easily become obsolete.

Previous approaches to automate ACM are based on the definition of business rules to analyze the compliance of business tasks with business requirements and support the decision making for each case [12, 21, 24]. However, supporting the user by integrating domain knowledge has not yet been sufficiently considered, which is an obstacle for business user' acceptance. A more general approach is the Semantics of Business Vocabulary and Business Rules (SBVR) standard. It is an adopted standard of the Object Management Group (OMG) which enables the use of domain knowledge for the definition of rules. While this standard enables the definition of a vast amount of different rules, the automated verification of them, especially in a highly flexible context, remains a challenge [6]. It is important that the enactment of the ontology and business rules must not be hindered because their management requires the support of

IT experts following an iterative standard software development lifecycle (design, develop, test, release). A business domain can have hundreds of compliance rules and the dependency from IT makes them expensive to maintain. Furthermore, the case implementation shall enable a direct view of the business domain and not a reflection of the underlying IT system. That is, when the IT model requires changes because of performance or scalability, the business process implementation can be impacted. Therefore, it is important to describe the actions executed by the business users with the language used in their daily business to ensure that there is a direct alignment between business description and its enactment.

In addition to that, the accomplishment of business rules should be observed during the whole case [19] and to make the business constraints transparent to business workers, it is important to describe which actions depend on which conditions by means of behavioral business rules which identify how the case should behave. The behavioral business rules also support the definition of business goals (e.g. KPIs) with quantitative and qualitative targets which is out of the scope of this work.

Therefore, in this work we propose applying a novel methodology for ACM to address these research challenges:

- R1: Minimize the technical gap between the business knowledge and its formalization and enactment in an ACM system by applying business architecture principles based on the definition of an ontology which enables the definition of action and rules with business concepts.
- R2: Trace the relationship between the action execution and business rules to define behavioral business rules.

3 ACM Enactment and Business Architecture Concepts

There are a number of papers dealing with the support of ACM and compliance rules. In previous work [22–24], Tran et al. propose driving the execution of cases based on the evaluation of business rules instead of a predefined business model. Gómez-López et al. [9] monitor the violation of business rules in process aware information systems to support proactive decision taking in next actions. Additionally, in the transport and logistics domain, the monitoring of the compliance of each transport case with agreed KPIs is evaluated to guide the action flow during the case execution [10, 14]. Such approaches usually address only the support of action sequences based on information models but do not propose how to close the gap between the business language and the cases enactment which requires support from IT teams.

In computational domains, Brandic et al. in [5, 8] propose mapping from business KPIs to low level metrics in order to support the monitoring and evaluation of these KPIs. However, these works are specific to computational services. And based on natural language process description, López et al. [13], propose a tool to formalize the natural language concepts as tasks with associated user roles and maintain their relationship through the process lifecycle. However, this proposal does not address the automation aspects, which still require IT efforts.

Therefore, as far as we know, there is no work related to support the complete lifecycle of business cases (i.e. modelling, execution, refining, etc.) avoiding the semantic gap between business and IT domains. In this work, we address this gap by taking advantage of existing business architecture concepts in order to align business concepts with the underlying formal model and its execution. Specifically, this work is based on the approach of VSs proposed by Open Group [17], derived from the proposal by Martin [15].

In business architecture, one of the main challenges is to match business capabilities with business goals in order to ensure business success [3, 4]. In this regard, the identification of strengths and weaknesses supports the strategies for company resource investments and recruitment. Different approaches to implement straight forward business cases have been proposed. In the ACORD framework [1], a proposal to define UML Classes for business concepts and tasks is described, which has been used to define reference system models in domains such as Insurance [2]. Although this framework can be used to implement the data model, it does not support the business actions or process automation, as the business rules have still to be defined and implemented by a separate system.

Open Group [17] proposes to focus on the provided value to customers or stakeholders, namely value streams, as a mechanism to organize the actions to take. We use the involved terminology for our proposal. In this proposal, the main concepts are defined in the following way and as depicted by Fig. 1.

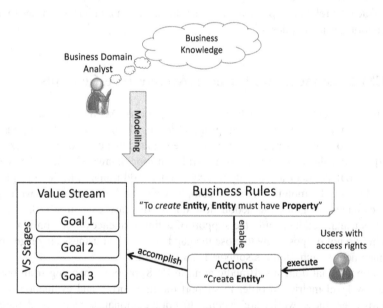

Fig. 1. Modelling business architecture

Definition 1 (Value Streams). Value streams (VSs) are defined as an end-to-end value for the customer, i.e. each single case of the value stream provides (or expects to provide) a benefit when finished. For example, in a building project, a value stream is the creation of a new architecture model which is ready to be delivered to external parties such as the plumbing company.

Definition 2 (Value Stream Stages). A value stream is composed of goals or value stream stages (VSSs). These goals can be in principle independent but it is possible that behavioral rules establish dependencies between goals. For example, different checks can be evaluated without a specific order over a building model but the calculation of thermal properties only makes sense after confirming that there are no gaps in the building envelope. Therefore, a dependency between the goal "creating an envelope without gaps" and the goal "creating a thermal model of the building" exists.

Definition 3 (Rules). The expected behavior to accomplish goals is expressed by means of behavioral business rules which constrain actions to be executed in the course of the work. These rules are evaluated to enable an action (e.g.: if a building model has not been considered finished, no checks can be performed) or to complete an action (e.g. if the building envelope has gaps, thermal calculations can be performed but these calculations should not be used for any purpose).

The accomplishment of goals is achieved with the execution of business tasks. We model these tasks with *actions*, which are either of atomic nature, being finished in the same moment, or actions starting a task and, when completed, finishing the task with an end action, to support decision making, handling of business entities, etc. by the business user. Our work use these definitions to propose a methodology to enact cases, based on the ACM paradigm and using business language for the definition of value streams and their stages together with rules.

4 Modelling of Business Value Streams

Our proposal addresses the alignment of business goals and ACM implementations using the value stream approach and behavioral business rules as shown in Fig. 2.

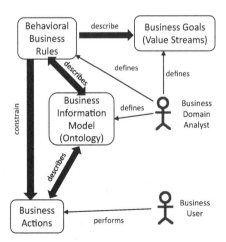

Fig. 2. Business architecture abstraction

Business domain analysts describe VSs with their expected outcome and a trigger action which enacts the case. Both use the domain-specific business ontology in order to ensure uniqueness of terms and their alignment between all involved parties. The accomplishment of goals is performed by actions, which represent the business tasks supported by the ACM platform. These actions can be linked with organizational units, such as stakeholders, customers or domain-specific roles. The actions related to a value stream are constrained by behavioral business rules which are defined by the domain analysts which will commonly manage entities required by that value stream. Actions are executed by business users until all rules are satisfied and the goals of the value stream are reached.

4.1 Modelling the Ontology

Considering the terminology defined in the previous section, our methodology begins with the definition of value streams, actions and rules in natural business language to create the domain-specific ontology. With this set of entities, the business domain analyst describes a full business application which is depicted in the Fig. 3. The ontology is used together with a grammar to formalize the values stream. The business rules which constrain the accomplishment of a value stream stage, enable the execution of actions as declared by the rules. The implementation of these actions is based on the definition of action templates in the ACM platform.

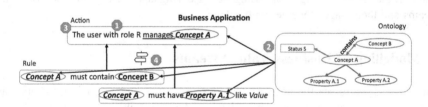

Fig. 3. Formalization of business case

This formalization of the business application is systematically described with the following steps, as showed in Fig. 3:

1. A goal is accomplished as a result of the execution of actions. An action is described with a simple sentence which contain a *subject* and a *predicate*. The subject identifies the business user or assignment role responsible for the action (item (1) in Fig. 3).
2. The predicates are formed by a *verb* and an *object*. This *object* has to be mapped to a concept or property in the ontology. Likewise, the rules relate concepts to properties or to specific properties values. Either the action or rules objects are mapped to concepts, properties and status (item (2) in Fig. 3) in the business ontology. Status are defined with a property identifying the concept lifecycle.
3. The *verb* in the predicate describes the action that the platform executes. These actions can include basic CRUD actions (Create, Retrieve, Update or Delete) for concepts (e.g.: *"Create" a Concept*) or domain-specific actions (e.g. *"Delivery" a*

Purchase). The business actions are directly executed as actions in the case (item (3) in Fig. 3).

4. The enactment of the actions described for the value stream stages are driven by the list of behavioral business rules. Therefore, we use the dependencies between rules and actions to pinpoint when these rules must be evaluated. This relationship is displayed in item (4) in Fig. 3.

4.2 Formalizing Actions and Rules

To formalize the action and rules, the domain ontology obtained (see item (2) in Fig. 3) is extended with a natural language grammar for the definition of rules and their relationship to actions. This grammar includes expressions such as mandatory terms (i.e.: must/exists), conjunctions and disjunctions (and/or) and navigation between concepts and properties (e.g.: Concept C has Property P) which enable to express a wide range of rules and actions using business language.

With the described methodology, business cases can be freely defined by business domain analysts in a formalized way. The implementation and automation of the business entities, i.e. (i) management of ontology, (ii) enactment of actions and (iii) evaluation of rules, can be achieved with common IT resources, whose detailed description is out of the scope of this paper. The management of the domain-specific ontology by business domain analysts can be addressed with ontology editors together with data mapping definitions to the underlying persistent data layer of information systems. After an initial business data definition they are considered as rather stable during the business lifecycle which allows business domain analysts to adapt the ontology to new business needs with no or drastically reduced support from IT developers. The rule evaluation can be performed with the monitoring of business events on system level, as have been addressed by ACM proposals stated in previous sections [22, 23].

We have implemented this methodology using the ISIS Papyrus ACM platform Papyrus Converse [18] to support business domain analysts with the definition of value streams, actions and rules including the maintenance of the domain-specific ontology (Papyrus Converse Composer). Business users manage business cases and their execution with the Papyrus Converse Player, a conversational user interface for intuitive, user guided business collaboration. In next section, we address the management of a building construction project as a running example of this methodology.

5 Applying ACM Enhanced with Business Architecture Concepts to the Construction Industry Information Modelling

Construction industry projects involve a number of different companies and trades acting in several roles, such as architects, building project managers or plumbing workers. Building Information Modeling (BIM) [7] proposes an open, vendor-independent environment to bring together architects, planners, BIM managers and many other disciplines, allowing them to work on a common data model. However, standardization in terms of interfaces and interoperability is weak and incomplete, leaving gaps and ambiguities when passing data from one knowledge worker to the next, leading to efficiency problems, constructions delays and budget problems.

To achieve higher efficiency and quality, the workflow of knowledge workers in the field of architecture, planning and building services shall be optimized by making the information, which is transferred between the domains accessible in a dynamic, case-based process management with ACM. This will improve the overall interoperability and reduce the information loss at the interfaces. ACM empowers the design process to detect model problems and give feedback to knowledge workers from a previous stage; support by machine learning methods helps guiding the knowledge workers from previously observed situations, but also supports fixing model problems and assists the knowledge workers with feasible recommendations for improvements. In this evaluation, we focus on the business cases involved in the calculation of energy consumption during all building phases. These cases include the reviewing of the architecture model quality (to avoid problems in the building envelope) or processing the building physics (material thermal transmittance) in order to obtain the heating load for the building.

5.1 Building Construction Project Modelling

Based on the methodology described, we describe one of the use cases of BIM projects, "Approve an architectural model". Architectural models with a minimum quality are required for later stages, such as defining thermal properties or calculating energy consumption. Following the procedure described in previous section, the business modelling starts with the definition of the goals and related actions for this VS with natural language. This VS handles the iterative refining of an architecture model and includes actions such as the committing a new architecture models to the platform or the reviewing of the model quality by the project manager (BIM manager). Figure 4 depicts a simplified description of all the entities involved. First, this VS is composed of three goals, Commit Model, Check Model and Approve Model (item (1) in Fig. 4). These goals are accomplished through related actions (item (3) in Fig. 4). The execution of these actions depends on business rules, such as "To check model, the model must be committed" (item (4) in Fig. 4). And the formalization of such actions and rules requires the definition on a business ontology including all the concepts (names) and relationships (verbs) addressed in natural language (e.g.: models, check results, etc. as depicted in item (2) in Fig. 4).

5.2 BIM Scenario Architecture

The resulting architecture of the ACM platform for the BIM domain is depicted in the Fig. 5. It is composed by (a) a component to enact business actions based on the monitoring and evaluation of rules and (b) the BIM ontology which is mapped to an underlying data model. New actions specific for the BIM domain are implemented in order to communicate with external tools, such as building model checkers or calculators for thermal properties. The supported data includes also file types, such as BIM standard formats, Industry Foundation Class (IFC) and BIM Collaboration Format (BCF).

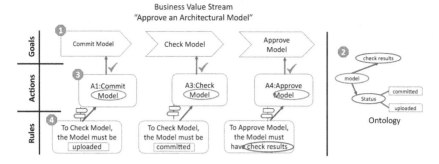

Fig. 4. Definition of value stream "Approve an architectural model"

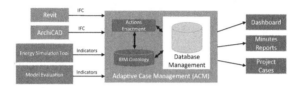

Fig. 5. ACM platform architecture

6 Conclusions

Applying the presented methodology to the building construction domain using Papyrus Converse allowed our project partners from the construction industry to describe the business cases with a language that is natural to them (architects, building physicists, etc.). The initial definition of the domain-specific ontology turned out to require quite some coordination efforts between all stakeholders to describe "how we work", in order to reflect a unique description of used terms and how they are related. Despite these efforts, the definition of values streams, ontology, actions and rules to enact the business cases proofed to empower business domain analysts without IT involvement. Only the mapping of ontology entities to an underlying data model and the definition of actions calling external BIM tools like ArchiCAD, Revit, energy simulation, etc. requires a certain support from IT which is covered by the initial system setup and does not call for continuous adaptations when business needs change. Thus, it facilitates a work environment with considerably reduced IT efforts as business users are able to adapt the system according to changes resulting from daily business needs without the hurdles imposed by rigid business processes definitions or new IT developments. As the case definition is aligned to the business objectives, actions can be clearly related to relevant KPIs, focused on business capabilities and linked with organizational units. The uncoupling of the domain-specific business ontology (business information model) from IT implementation avoids dependencies from IT requirements such as performance or scalability.

As the presented methodology is completely agnostic to any business domain it can be easily applied to new scenarios such as insurance applications or maintenance services. Further work is planned for these domains, together with the enrichment of the Papyrus Converse user interfaces for a seamless and even more natural communication with users supported by techniques such as speech recognition and language machine learning to take advantage of the defined business language.

Acknowledgment. This work is funded by the Austrian Research Promotion Agency (FFG) within project BIMSavesEnergy (grant agreement number 861710) where ISIS Papyrus delivers the business architecture enhanced ACM methodology and CES and AIT the construction industry and energy domain-specific aspects, respectively.

References

1. ACORD reference architecture, November 2018. https://www.acord.org/standards-architecture/reference-architecture
2. Aggarwal, A.: Industry reference blueprint for insurance, January 2015. http://www.referenceblueprint.com/
3. Business Architecture Guild: A Guide to the Business Architecture Body of Knowledge (2014)
4. Business Architecture Guild: The Business Architecture Quick Guide. Meghan-Kiffer Press, The Netherlands (2018)
5. Brandic, I., Music, D., Leitner, P., Dustdar, S.: VieSLAF framework: enabling adaptive and versatile SLA-management. In: Altmann, J., Buyya, R., Rana, O.F. (eds.) Grid Economics and Business Models, pp. 60–73. Springer, Heidelberg (2009). https://doi.org/10.1007/978-3-642-03864-8_5
6. Czepa, C., Tran, H., Zdun, U., Tran, T.T.K., Weiss, E., Ruhsam, C.: Ontology-based behavioral constraint authoring. In: 2016 IEEE 20th International Enterprise Distributed Object Computing Workshop (EDOCW). http://eprints.cs.univie.ac.at/4754/
7. Eastman, C., Teicholz, P., Sacks, R., Liston, K.: BIM Handbook: A Guide to Building Information Modeling for Owners, Managers, Designers, Engineers and Contractors. Wiley, Hoboken (2008)
8. Emeakaroha, V.C., Brandic, I., Maurer, M., Dustdar, S.: Low level metrics to high level SLAs - LoM2HiS framework: bridging the gap between monitored metrics and SLA parameters in cloud environments. In: 2010 International Conference on High Performance Computing Simulation, pp. 48–54, June 2010. https://doi.org/10.1109/HPCS.2010.5547150
9. Gómez-López, M.T., Parody, L., Gasca, R.M., Rinderle-Ma, S., et al.: Prognosing the compliance of declarative business processes using event trace robustness. In: Meersman, R. (ed.) On the Move to Meaningful Internet Systems: OTM 2014 Conferences, pp. 327–344. Springer, Heidelberg (2014). https://doi.org/10.1007/978-3-662-45563-0_19
10. Gutiérrez, A.M., Cassales Marquezan, C., Resinas, M., Metzger, A., Ruiz-Cortés, A., Pohl, K.: Extending WS-agreement to support automated conformity check on transport and logistics service agreements. In: Basu, S., Pautasso, C., Zhang, L., Fu, X. (eds.) ICSOC 2013. LNCS, vol. 8274, pp. 567–574. Springer, Heidelberg (2013). https://doi.org/10.1007/978-3-642-45005-1_47
11. Kim, T.T.T., Ruhsam, C., Pucher, M.J., Kobler, M., Mendling, J.: Towards a pattern recognition approach for transferring knowledge in ACM. In: Proceedings of the 2014 IEEE 18th International Enterprise Distributed Object Computing Conference Workshops and Demonstrations, EDOCW 2014, Washington, DC, USA, pp. 134–138. IEEE Computer Society (2014). http://dx.doi.org/10.1109/EDOCW.2014.28

12. Kim, T.T.T., Weiss, E., Ruhsam, C., Czepa, C., Tran, H., Zdun, U.: Enabling flexibility of business processes using compliance rules: the case of mobiliar. In: vom Brocke, J., Mendling, J. (eds.) Business Process Management Cases. Management for Professionals, pp. 91–109. Springer, Cham (2018). https://doi.org/10.1007/978-3-319-58307-5_6
13. López, H.A., Debois, S., Hildebrandt, T.T., Marquard, M.: The process highlighter: from texts to declarative processes and back. In: Proceedings of the Dissertation Award, Demonstration, and Industrial Track at BPM 2018, Sydney, Australia, 9–14 September 2018, pp. 66–70 (2018)
14. Marquezan, C.C., Metzger, A., Franklin, R., Pohl, K.: Runtime management of multi-level SLAs for transport and logistics services. In: Franch, X., Ghose, A.K., Lewis, G.A., Bhiri, S. (eds.) Service-Oriented Computing, pp. 560–574. Springer, Heidelberg (2014). https://doi.org/10.1007/978-3-662-45391-9_49
15. Martin, J.: The Great Transition: Using the Seven Disciplines of Enterprise Engineering to Align People, Technology, and Strategy. Oxford Early Christian Studies, AMACOM (1995). https://books.google.at/books?id=fFJ4QgAACAAJ
16. Motahari-Nezhad, H.R., Swenson, K.D.: Adaptive case management: overview and research challenges. In: 2013 IEEE 15th Conference on Business Informatics, pp. 264–269, July 2013. https://doi.org/10.1109/CBI.2013.44
17. Open Group: Value Streams. Technical report, The Open Group Architecture Forum Business Architecture Work Stream, January 2017. https://publications.opengroup.org/g170
18. Papyrus Converse ACM Platform. https://www.isis-papyrus.com/e15/pages/business-apps/papyrus-converse.html. Accessed 21 May 2019
19. Ross Jr., R.G.: Business Rule Concepts. Business Rule Solutions, Incorporated (1998)
20. Swenson, K.D.: Mastering the Unpredictable: How Adaptive Case Management Will Revolutionize the Way that Knowledge Workers Get Things Done. Meghan-Kiffer Press, Tampa (2010). http://www.worldcat.org/search?qt=worldcatorgall&q=9780929652122
21. Tran, T., et al.: An ontology-based approach for defining compliance rules by knowledge workers in adaptive case management. In: 5th International Workshop on Adaptive Case Management and Other Non-workflow Approaches to BPM (AdaptiveCM 2016), 20th IEEE International Enterprise Computing Workshops (EDOCW 2016), September 2016. http://eprints.cs.univie.ac.at/4753/
22. Tran, T., Weiss, E., Ruhsam, C., Czepa, C., Tran, H., Zdun, U.: Embracing process compliance and flexibility through behavioral consistency checking in ACM: a repair service management case. In: 4th International Workshop on Adaptive Case Management and Other Non-workflow Approaches to BPM (AdaptiveCM 2015), Business Process Management Workshops 2015, August 2015. http://eprints.cs.univie.ac.at/4409/
23. Tran, T., Weiss, E., Ruhsam, C., Czepa, C., Tran, H., Zdun, U.: Enabling flexibility of business processes by compliance rules: a case study from the insurance industry. In: 13th International Conference on Business Process Management 2015, Industry Track, August 2015. http://eprints.cs.univie.ac.at/4399/
24. Tran, T.T.K., Pucher, M.J., Mendling, J., Ruhsam, C.: Setup and maintenance factors of ACM systems. In: Demey, Y.T., Panetto, H. (eds.) On the Move to Meaningful Internet Systems: OTM 2013 Workshops, pp. 172–177. Springer, Heidelberg (2013). https://doi.org/10.1007/978-3-642-41033-8_24

Checking Compliance in Data-Driven Case Management

Adrian Holfter$^{(\boxtimes)}$, Stephan Haarmann, Luise Pufahl, and Mathias Weske

Hasso Plattner Institute, University of Potsdam, Prof.-Dr.-Helmert-Str. 2-3,
14482 Potsdam, Germany
`adrian.holfter@student.hpi.uni-potsdam.de`
`{stephan.haarmann,luise.pufahl,mathias.weske}@hpi.uni-potsdam.de`

Abstract. Case management approaches address the special requirements of knowledge workers. In fragment-based case management (fCM), small structured parts are modelled and loosely coupled through data dependencies, which can be freely combined at run-time. When executing business processes, organizations must adhere to regulations, to laws, to company guidelines etc. Business process compliance comprises methods to verify designed and executed business processes against certain rules. While design-time compliance checking works well for structured process models, flexible knowledge-intensive processes have been rarely considered despite increasing interest in academia and industry.

In this paper, we present (i) formal execution semantics of fCM models using Petri nets. We also cover concurrently running fragment instances and case termination. We (ii) apply model checking to investigate the compliance with temporal logic rules; finally, we (iii) provide an implementation based on the open-source case modeler Gryphon and the free model checker LoLA.

Keywords: Business process management · Business process compliance · Case management

1 Introduction

Business process management supports organizations with the design, analysis and execution of their processes. One objective is asserting compliance with regulations such as laws, internal guidelines, and standards. Business process compliance comprises methods to check that models adhere to rules at design-time, to assert that process instances behave accordingly at run-time, and to detect and handle violations at run-time or a posteriori [9]. Checking compliance at design-time is usually achieved by formalizing both – the process models and the constraints – to apply model checking techniques afterwards [13].

Compliance checking for traditional business processes has been investigated in depth [9]. However, in today's business environment, the work of employees becomes more and more knowledge-driven. Standard process modeling languages

© Springer Nature Switzerland AG 2019
C. Di Francescomarino et al. (Eds.): BPM 2019 Workshops, LNBIP 362, pp. 400–411, 2019.
https://doi.org/10.1007/978-3-030-37453-2_33

such as Business Process Model and Notation (BPMN) [15] are not well suited to capture these kinds of flexible business processes [1]. In case management, knowledge workers drive the execution of processes based on the case characteristics and their experiences. Generally, two types are distinguished [14]: while adaptive case management (ACM) focuses on unstructured processes; production case management (PCM) is based on identifying common structured patterns at design time and allows re-combining and changing them at run time. Fragment-based case management (fCM) is an implementation framework embracing the hybrid nature of PCM; process *fragments* represent structured segments that can be dynamically combined at run-time within the constraints defined by data.

Case management is less restrictive than traditional business processes and puts more responsibility on the knowledge worker driving the case. The flexibility of fCM comes with a drawback with regards to a formal analysis: at design-time, it is not directly visible how the process will unfold at run-time, which can lead to unexpected and unwanted behavior. At the same time, compliance rules must not be violated despite the freedom granted to the knowledge worker. Here, (semi-)automated design-time compliance checking can help verifying that certain properties hold. This poses some challenges: the flexibility of fCM models must be modeled formally and may lead to infinite state spaces.

In this paper, we present an automated approach to design-time compliance checking of data-driven case management models. To this end, we provide a formal Petri net mapping for fragment-based case management as an exemplary framework for PCM. Subsequently, we show that while compliance of fCM is undecidable because of multiple sources of unboundedness and the resulting infinite state space during case execution; it is possible to stay within the bounds of decidability by restricting the use of certain modeling elements. For evaluation, we showcase an open-source implementation.

The remainder of this paper is structured as follows. Section 2 provides the foundations and definitions of the structures used. In Sect. 3, we present related work. The formal behavioral model is introduced in Sect. 4. We evaluate the approach using a prototypical implementation in Sect. 5. Finally, Sect. 6 concludes the paper and discusses future work.

2   Fragment-Based Case Management

As stated in the introduction, Production Case Management (PCM) can be realized with fragment-based Case Management (fCM) [8,10], in which a *case model* comprises small process *fragments* (comparable to small BPMN process diagrams) that share a set of common *data objects*. These can be initiated and combined by a knowledge worker as long as the data conditions are not violated. This section will introduce fCM concepts based on an example given in Fig. 1 capturing the patient's treatment at a hospital's emergency ward.

Each **fragment** uses a subset of the BPMN notation including activities, gateways, events connected by sequence flow. Each fragment has a start and an end event. The topmost fragment (*Diagnosis*) describes the procedure for

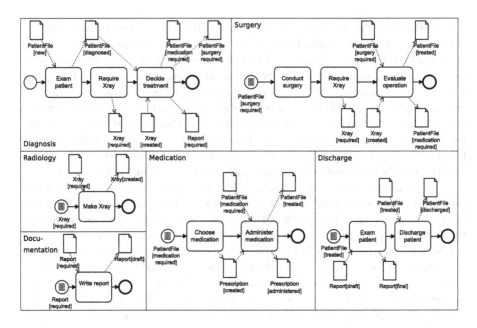

Fig. 1. Example case model of an emergency handling process.

admitting and examining the patient. This fragment starts with a blank start event, such that it can be enabled and used as soon as the case model is initiated. Its first task "Exam patient" reads the *PatientFile* in state *new*. This is the initial state of the data class *PatientFile*, and as soon as the case model is initiated, a data object of this type in state *new* will be available. The complete **object life cycle (OLC)** of the data class *PatientFile* with all possible states is shown in Fig. 2. It describes the allowed states and their order, in which a *PatientFile* can be during the emergency handling. A case model can act on different data classes which are needed for its execution. Thus, each case model describes in a **data model** the set of data classes, which are essential for the execution of its activities and fragments. Each of those data classes has its own OLC.

Coming back to the *Diagnosis* fragment, the next activity "Require Xray" produces a new data object, *Xray[required]*. The upcoming activity is only enabled, if the *Xray* is in state *created*. This is provided by another fragment – the *Radiology* fragment – starting with a

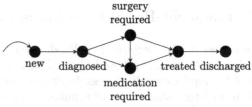

Fig. 2. OLC for data class *PatientFile*

conditional start event, which uses an atomic **data condition**. Thus, this fragment is activated and can be started as long as the *Xray* is in state *[required]*. Multiple atomic conditions can be also conjunctively combined to form a **condition set**.

When the *Xray[created]* is available, "Decide on treatment" can be executed and it produces either *PatientFile* in state *[medication required]* or *[surgery required]* based on the decision made by the medical expert, and additionally the data object *Report[required]*. In case of data objects referring to the same class but in different states, it is assumed that only one of them is produced as data output, or required as data input. Depending on the result of the *Diagnosis* fragment, either the *Medication* or the *Surgery* fragment is enabled by the data entry condition and can be executed. When executing the *Surgery* fragment, the *Radiology* fragment is needed again, and after "Evaluate operation", the *Medication* fragment can be enabled if the *PatientFile[medication required]* is the final output. Thus, fragments can be enabled and also executed multiple times during a case. As soon as a fragment is enabled, it can be also initiated concurrently if allowed by its data entry conditions. This means several instances of one fragment can exist at runtime.

In parallel to the just described fragments, the *Documentation* fragment is enabled as well. It can be executed at any time until discharge. Its final data output is needed for starting "Exam patient" in the *Discharge* fragment. When finally the *PatientFile[discharged]* is available, then the case model has reached its **termination condition**. The termination condition of each case model consists of a set of data conditions which can be connected by logical con-/disjunctions to describe when a case worker can terminate a case. Summarized, each fCM case model consists of a set of fragments, a set of data classes – each of it referencing to an OLC – and a termination condition.

After introducing related work, we describe how to map a fCM case model to a Petri net to describe the behavior precisely. Additionally, rules must be expressed clearly. While visual notations such as BPMN-Q [6] and extended compliance rule graphs [12] exist, we rely on temporal logic. Both temporal logic and the model's formal execution semantics are provided to a modeler checker (e.g., LoLA [17]). The model checker provides feedback—the temporal logic property is either satisfied, or a counterexample is provided.

3 Related Work

Business process compliance aims for conformance of organizations' processes with jurisdictional, internal, and other rules. In this paper, we focus on design-time compliance checking, which validates process models against compliance rules. Numerous works within the BPM community have investigated parts of this question. Hashmi et al. provide an overview of business process compliance research [9]. Awad et al. introduce BPMN-Q—a query language for BPMN process models [2] capable of expressing compliance rules—as well as a model checking-based approach for evaluating them [4]. In subsequent works, Awad and his co-authors extend their approach toward rudimentary data support [5]. In a similar fashion, we apply model checking to fCM case models, which consists of process fragments that are connected by shared data objects. However, we express compliance rules as temporal logic expressions, which may or may not be derived from a visual compliance language.

Fragment-based case management addresses the special requirements of knowledge/intensive processes: while those processes are partly structured, they are mostly driven by the decisions of knowledge workers. In such settings, methods asserting compliance become important due to the processes' complexity and variability. An alternative to compliance checking is the explicit incorporation of compliance rules into the process models. Declarative process languages such as DECLARE [16] and Dynamic Condition Response graphs (DCR graphs) [11] follow this approach. Contrarily, modelling highly structured parts and data dependencies with these approaches is arguably less intuitive than with fCM. In this paper, we address fCM's lack of explicit support for (declarative) compliance rules. While DECLARE and DCR graphs follow a different paradigm (adaptive case management) than fCM (production case management), our approach supports modelers in establishing a comparable degree of compliance.

In this paper, we employ model checking. Model checking requires a formal model of the process. Therefore, we assign formal execution semantics to fCM models using Petri nets. Dijkman et al. map BPMN models to Petri nets [7]; their work is the base for our mapping. A first extension to flexible fCM models supporting the composition of fragments via data objects has been provided by Sporleder [18]. We further cover repeated and concurrent instantiation of fragments as well as data objects and the termination of process instances.

4   Formal Behavioral Model for fCM

In this section, we define the behavior of fCM models formally. For this purpose, we provide a direct translation of fCM to Petri nets. We additionally discuss the unboundedness of the derived Petri nets and provide methods to establish boundedness in most cases.

4.1   Translation to Petri Nets

As introduced in Sect. 2, a case model consists of a **data model**, **fragments** with **activitites**, and a **termination condition**. Combined, these elements specify the allowed behavior. Thus, the formalization of the case model comprises Petri net translations for all these elements. The case model translation is the sum of its parts, as there are no special top-level translation elements. Consequently, the translation does not depend on an order.

Data Model. Data objects are central to fCM as they are used for enablement and synchronization of fragments. Figure 3 shows the translation of the *Patient-File* data class. Since we omit data attributes for the purpose of this work, each state of each data class is translated to a place. A token on such a place represents exactly one data object in the corresponding state. Data object identity is neglected here since tokens are indistinguishable. When a new case is started, a token is produced in the place representing the initial state of the case class (i.e., *PatientFile[new]*). State changes only occur by executing activities.

Fig. 3. Translation of the *PatientFile* data class with a token representing an object in the state *new*.

Fragments. An activity is always part of a fragment. On a fragment level, initialization, checking of the pre-condition, termination, and re-initialization need to be considered. Figure 4 shows the Petri net translation for the *Radiology* fragment. When a new case instance starts, all fragments are instantiated and initialized. This is done by placing one token in the initial place of each fragment.

Some fragments (e.g., Radiology) have a pre-condition. The fragment's pre-condition must be evaluated and satisfied before an initialized fragment instance can begin execution (i.e. before the start event occurs). For each condition set of a fragment's pre-condition, a transition representing this condition set is introduced, the fragment's initial place is added to the transition's pre-set and the place before the start event transition is added to its post-set. As the *Radiology* fragment's pre-condition contains one condition set, one transition is created. Then, for each atomic condition of a condition set (i.e., object *Xray* in state *required*), the place representing the data object in the state corresponding to the atomic condition is added both to the transition's pre- and post-sets. This way, the transition can only fire if the corresponding condition set is satisfied. Note how tokens are consumed and immediately produced again, therefore effectively not modifying any data object's state.

Fragments can be executed multiple times. This is done by re-initializing fragment instances. The Petri net translation in Fig. 4 re-initializes the *Radiology* fragment after its completion. To facilitate fragment re-initialization, a place is introduced after the end event transition (*re-init-ready* in Fig. 4). A token in this place represents a terminated fragment instance. In its post-set, a new transition is added that produces a token in the initial place of a fragment, thereby creating a loop from the end back to the start. This way, a fragment can be executed multiple times sequentially. Later on, we present a translation that allows concurrent instantiation.

Activities. In previous works (e.g. [3,7]), activities are usually represented by single transitions. As concurrency plays a big role in fCM's characteristics, representing ongoing activity execution is of interest since data objects are bound to activities during execution and therefore influence enablement of other activities and fragments. The translation proposed in [3] is suitable for the representation of state changes in data objects. To represent an activity, a transition with pre- and post- condition set pairs is created.

We propose a translation that models the *running* state of an activity and correctly represents bound data objects. For example, the *Decide treatment* activity in the *Diagnosis* fragment is translated as shown in Fig. 5. This activ-

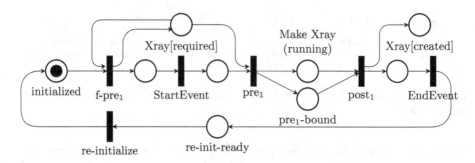

Fig. 4. Translation of the *Radiology* fragment. The token in the *initialized* place indicates that one fragment instance is initialized, but not started yet, as no Xray is required.

ity has a pre-condition consisting of the condition *PatientFile[diagnosed]* and *Xray[created]* and a post-condition with two condition sets. Both condition set contain *Xray[created]* and *Report[required]*, but one set contains *Patient-File[surgery required]* while the other contains *PatientFile[medication required]*.

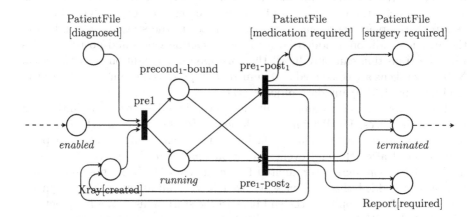

Fig. 5. Translation of the *Decide treatment* activity (in the *Diagnosis* fragment).

An activity's pre-condition is translated similarly to one of a fragment. However, the consumed tokens of places representing data objects are not immediately written back, but need to be bound to the activity during execution. All of the condition set transitions have therefore two places in their output set: First, a shared place that models ongoing activity execution (i.e., *running*), and a second individual place per condition set that represents the bound data objects (i.e., *precond$_1$-bound*). Note that there are no individual places for each atomic condition (i.e. explicitly binding data objects in a specific state), but one per condition set, representing all data objects in their respective states that were consumed to satisfy the corresponding condition set.

Handling post-conditions is done similarly, only that tokens are created in the respective places rather than consumed. We create one transition for each valid combination of input set and output set (i.e., pre_1-$post_1$ and pre_1-$post_2$). A combination is considered valid, if each element in the output set can be derived from an element in the input set through a legal state transition (one that is modeled in the concerned data objects' OLC). Such a transition's pre-set consists of a place representing the bound data objects and the place *running*. Additionally, the transition produces tokens for one output set by putting tokens on the respective places. If a data object is part of the input set but not of the output set, the object's state remains the same. The respective place is therefore also in the transition's post-set.

Besides activities, each fragment has a start event and an end event. End events and start events are represented by a single transition. Although processes can be split into fragments to represent decisions and concurrency, fCM supports exclusive (XOR) gateways and parallel (AND) gateways. Gateways are translated as described in [7]. This means that non-determinism is used to model possible choices by the knowledge worker during case execution.

Termination Condition. While a case unfolds, multiple fragments with possibly multiple instances of each fragment including their activities can be executed. The knowledge worker can terminate the case as soon as and as long as the termination condition holds. The termination condition shown in Fig. 6 is translated similarly to a minimal fragment with the termination condition as pre-condition, but without end

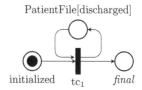

PatientFile[discharged]

initialized tc_1 *final*

Fig. 6. The translated termination condition.

event, re-initialization or actual content. The place in the termination condition transition's post-set is considered to be the *final* place of the translated case model, i.e. placing a token in this place signals successful termination of the case.

4.2   Concurrent Fragment Instances and Unboundedness

So far, fragment re-initialization was handled sequentially, i.e. a new instance of a fragment is created and initialized only when another instance of that fragment terminates. In [10] however, re-initialization happens already when another initialized fragment instance starts execution.

This behavior is translated as depicted in Fig. 7(a) (for comparison, the previously presented sequential re-initialization is shown in Fig. 7(c)). While this guarantees that fragment instances are always available when needed, it can lead to an unbounded number of *concurrent* fragment instances. Therefore, the state space of the translated model becomes infinite and checking compliance becomes undecidable.

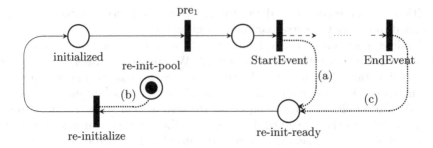

Fig. 7. Fragment re-initialization either after fragment start (a) or after fragment termination (c) and optionally with instance limitation pool (b).

We identified two ways of approaching this problem: One option is to allow concurrent execution of multiple instances of the same fragment, but introduce an upper bound of fragment instances per fragment. This is done by adding a place to the re-initialization transition's pre-set and supplying it initially with a fixed amount of tokens that correspond to the maximum number of fragment instances. This is depicted in Fig. 7(b).

By limiting the number of fragment instances (and therefore also to the number of data object instances, as fragments cannot contain loops), a source of unboundedness is contained. While this restriction theoretically limits the expressiveness of the model, in practice a (comparatively low) upper bound of number of fragment instances can be found that still allows successful execution of all relevant cases. After all, case management is goal-oriented.

However, as fragment instances can exist and execute concurrently, but model checkers usually assume serialized execution when extracting the state space, having concurrent fragment instances causes state space explosion. In most situations, concurrent execution of fragment instances (inside one case instance) is not needed. Therefore, it is usually possible to perform fragment re-initialization not when beginning execution of one fragment instance, but when terminating it. Thus, only one concurrent instance per fragment can exist. This is also the behavior that was described earlier and is depicted in Fig. 7(c).

Still, the Petri net modelling sequential fragment instances can remain unbounded: if an activity creates a new instance of a data object and is executed repeatedly, it produces multiple instances of that data object. If the enclosing fragment does not have a precondition that prevents infinitely many executions, some of the places representing the respective data object are unbounded since infinitely many objects can be created. However, this is usually caused by a fundamental modelling error, which should be detected on a structural level. In most cases, fragments overwrite existing data objects rather than creating new ones. Therefore, we do not consider this case further.

In practice, how to handle fragment re-initialization must be decided case-by-case. Sequential re-initialization works for many use cases and causes a great reduction in state space complexity. If concurrent instance execution of the same fragment is required, introducing an upper limit provides a viable solution.

5 Implementation

The proposed formal behavior model for fCM is based on generating a Petri net for a given case model. This Petri net can then be used to check compliance of the originating model. We applied this approach by integrating a prototypical compliance checking component into an existing system for modeling and executing fCM models. The project page[1] contains more information and a demonstrating screencast. The setup uses *Gryphon*[2] as a modeler, *Chimera*[3] as execution engine and *Low Level Analyzer (LoLA)* [17] as model checker for Petri nets. The system architecture is displayed in Fig. 8.

We added a user interface component to Gryphon that affords checking LTL and CTL formula (formalized compliance rules) against a designed case model. However, Gryphon does not provide any convenient method to access the case model elements individually. Chimera, on the other hand, offers easy access to the model in an object-oriented representation, such that the translation logic was added to Chimera.

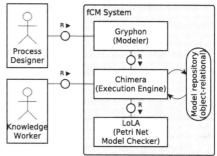

Fig. 8. The *fCM system* comprised of a modeler, an execution engine and a model checker.

In the LTL/CTL query it is possible to refer to the place representing a data object in a certain state by using a term like {DataClass[State]} and to the place representing a running activity by using a term like {ActivityName}. Therefore, exemplary constraints such as "Every patient will eventually be discharged" and "Per patient, a maximum of one Xray is created" can be formalized to the queries F ({Discharge patient} = 1) and G ({Xray[created]} <= 1) respectively. After translating the case model, the compliance query is then checked by LoLA on the generated Petri net. The result is forwarded back to Gryphon, where it is displayed to the user.

When checking the example constraints, Gryphon reports that the first formula is satisfied, while the second is not. For the second query, a witness path is shown that visits the corresponding transition of the *Make Xray* activity twice, during the *Diagnosis* and *Surgery* fragments. More specifically, also the selected pre- and post-condition sets are listed with the corresponding consumed and

---

[1] https://bptlab.github.io/bpm2019ws-fcm-compliance/.

[2] https://github.com/bptlab/gryphon.

[3] https://github.com/bptlab/chimera.

produced data objects in their respective states. Therefore, the two occurrences of *Xray[created]* are easy to spot in the witness path.

6   Discussion and Conclusion

Fragment-based case management (fCM) is a data-centric case management approach. It supports knowledge workers with managing their flexible processes. In this paper, we presented a formal execution semantics of fCM models using Petri nets to allow automated compliance checking at design time. This is realized by checking formalized fCM models against compliance rules given as temporal logic formula. Since compliance-checking of fCM models is generally undecidable, we propose explicit upper bounds of fragment instances or prohibiting multiple concurrent instances of fragments.

We implemented our approach in an existing system-environment for fCM modeling and execution in which we integrated *LoLA* as a Petri net model checker. Our experience shows that for reasonably sized fCM models, design-time compliance checking can be done in under a second. While the prototypical implementation shows the feasibility of checking compliance of data-driven case management models, there are some limitations to our work: First, the provided translation does not consider data attributes nor data object identity, as the model's behavior does not depend on them. In practice however, the valuation of data attributes might influence the decisions of a knowledge worker. Second, to ensure that compliance is decidable, sources of unboundedness in the model that lead to infinite state spaces need to be avoided. These are mainly unlimited concurrent execution of fragment instances. Thus, our approach does not allow compliance checking for all possible case models using fCM's original semantics.

fCM allows to alter the model at run-time. Furthermore, it is assumed that the case is driven by a knowledge worker, who may avoid compliance violations even if the models allow them. In the future, we will investigate compliance checking for fCM at runtime to give recommendations, such as "Starting in the current state, which steps would lead to compliance violations and need to be avoided?".

References

1. van der Aalst, W.M.P.: Process-aware information systems: design, enactment, and analysis. In: Wiley Encyclopedia of Computer Science and Engineering. Wiley (2008). https://doi.org/10.1002/9780470050118.ecse577
2. Awad, A.: BPMN-Q: a language to query business processes. In: EMISA, vol. 119, pp. 115–128 (2007)
3. Awad, A., Decker, G., Lohmann, N.: Diagnosing and repairing data anomalies in process models. In: Rinderle-Ma, S., Sadiq, S., Leymann, F. (eds.) BPM 2009. LNBIP, vol. 43, pp. 5–16. Springer, Heidelberg (2010). https://doi.org/10.1007/978-3-642-12186-9_2

4. Awad, A., Decker, G., Weske, M.: Efficient compliance checking using BPMN-Q and temporal logic. In: Dumas, M., Reichert, M., Shan, M.-C. (eds.) BPM 2008. LNCS, vol. 5240, pp. 326–341. Springer, Heidelberg (2008). https://doi.org/10.1007/978-3-540-85758-7_24

5. Awad, A., Weidlich, M., Weske, M.: Specification, verification and explanation of violation for data aware compliance rules. In: Baresi, L., Chi, C.-H., Suzuki, J. (eds.) ICSOC/ServiceWave -2009. LNCS, vol. 5900, pp. 500–515. Springer, Heidelberg (2009). https://doi.org/10.1007/978-3-642-10383-4_37

6. Awad, A.M.H.A.: A compliance management framework for business process models. Ph.D. thesis, Hasso Plattner Institute, University of Potsdam, Germany (2010)

7. Dijkman, R.M., Dumas, M., Ouyang, C.: Semantics and analysis of business process models in BPMN. Inf. Softw. Technol. **50**(12), 1281–1294 (2008)

8. Gonzalez-Lopez, F., Pufahl, L.: A landscape for case models. In: Reinhartz-Berger, I., Zdravkovic, J., Gulden, J., Schmidt, R. (eds.) BPMDS/EMMSAD -2019. LNBIP, vol. 352, pp. 87–102. Springer, Cham (2019). https://doi.org/10.1007/978-3-030-20618-5_6

9. Hashmi, M., Governatori, G., Lam, H.P., Wynn, M.T.: Are we done with business process compliance: state of the art and challenges ahead. Knowl. Inf. Syst. **57**, 1–55 (2018)

10. Hewelt, M., Weske, M.: A hybrid approach for flexible case modeling and execution. In: La Rosa, M., Loos, P., Pastor, O. (eds.) BPM 2016. LNBIP, vol. 260, pp. 38–54. Springer, Cham (2016). https://doi.org/10.1007/978-3-319-45468-9_3

11. Hildebrandt, T.T., Mukkamala, R.R.: Declarative event-based workflow as distributed dynamic condition response graphs. In: Proceedings Third Workshop on Programming Language Approaches to Concurrency and communication-cEntric Software, PLACES 2010, Paphos, Cyprus, March 21 2010, pp. 59–73 (2010)

12. Knuplesch, D., Reichert, M., Ly, L.T., Kumar, A., Rinderle-Ma, S.: Visual modeling of business process compliance rules with the support of multiple perspectives. In: Ng, W., Storey, V.C., Trujillo, J.C. (eds.) ER 2013. LNCS, vol. 8217, pp. 106–120. Springer, Heidelberg (2013). https://doi.org/10.1007/978-3-642-41924-9_10

13. Kunze, M., Weske, M.: Behavioural Models: From Modelling Finite Automata to Analysing Business Processes. Springer, Cham (2016). https://doi.org/10.1007/978-3-319-44960-9

14. Motahari-Nezhad, H.R., Swenson, K.D.: Adaptive case management: overview and research challenges. In: 2013 IEEE 15th Conference on Business Informatics (CBI), pp. 264–269. IEEE (2013)

15. Object Management Group (OMG): Business Process Model and Notation (BPMN). OMG Document Number formal/13-12-09 (2014). Version 2.0.2

16. Pesic, M., Schonenberg, H., van der Aalst, W.M.P.: DECLARE: full support for loosely-structured processes. In: 11th IEEE International Enterprise Distributed Object Computing Conference (EDOC 2007), Annapolis, Maryland, USA, October 15–19 2007, pp. 287–300 (2007). https://doi.org/10.1109/EDOC.2007.14

17. Schmidt, K.: LoLA a low level analyser. In: Nielsen, M., Simpson, D. (eds.) ICATPN 2000. LNCS, vol. 1825, pp. 465–474. Springer, Heidelberg (2000). https://doi.org/10.1007/3-540-44988-4_27

18. Sporleder, T.: Fragment-based case management: specification and translational semantics. Master's thesis, Hasso Plattner Institute, University of Potsdam, Germany (2016)

Second International Workshop on Methods for Interpretation of Industrial Event Logs (MIEL)

Second International Workshop on Methods for Interpretation of Industrial Event Logs (MIEL)

The objective of the workshop is to discuss novel intelligent data analysis methods suitable for the analysis and interpretation of industrial event logs. We focus on the inclusion and possibly combination of two distinct perspectives. The first one uses methods from data mining and computational intelligence. The second one uses background domain knowledge for conceptual analysis. Hence, we would like to focus on (but not limit) the applications of the mentioned methods according to those two perspectives to Industry 4.0. In this setting, the source of the event logs would be industrial machinery, but also possibly personnel or activity monitoring devices.

The Second edition of this workshop attracted seven international submissions. Each paper was reviewed by at least two members of the Program Committee. From these submissions, we accepted two as full papers for presentation at the workshop. The papers presented at the workshop provide a mix of novel research ideas, evaluations of existing mining, and analysis techniques for interpreting industrial event logs, as well as new methods in that context. The submission of Stephan Sigg, Sameera Palipana, Stefano Savazzi, and Sanaz Kianoush presents an approach for capturing human-machine interaction events from radio sensors in Industry 4.0 environments. In their paper, they describe their current efforts towards recognizing cases in a multi-subject scenario conducting several simultaneous activities. Furthermore, the submission of Stefan Bloemheuvel, Benjamin Kloepper, and Martin Atzmueller targets methods for graph summarization of complex industrial event logs, in order to allow computational sensemaking on such complex data. For that, suitable modeling and mining methods are combined into a novel approach, and demonstrated in a real-world case study.

As with the previous edition of the workshop, we envision that the reader will find this selection of papers useful to keep track of the latest advances in the area of interpreting industrial event logs. We look forward to further new advances in future editions of the MIEL workshop.

October 2019

Grzegorz J. Nalepa
David Camacho
Edyta Brzychczy
Roberto Confalonieri
Martin Atzmueller
Marco Montali

Organization

Program Committee

Andrea Burattin	Technical University of Denmark, Denmark
Szymon Bobek	AGH University of Science and Technology, Poland
Josep Carmona	Universitat Politecnica de Catalunya, Spain
Diego Calvanese	Free Univeristy of Bozen-Bolzano, Italy
Dirk Fahland	Technical University in Eindhoven, The Netherlands
Felix Mannhardt	SINTEF, Norway
Chiara Ghidini	FBK Trento, Italy
Rushed Kanawati	Université Sorbonne-Paris-Cité, France
Benjamin Kloepper	ABB, Germany
Victor Rodriguez	Universidad Autonoma de Madrid, Spain
Dominik Slezak	University of Warsaw, Poland
Jose Tomas Palma Mendez	Universidad de Murcia, Spain
Christopher Turner	University of Surrey, UK
Marcin Szpyrka	AGH University of Science and Technology, Poland

Graph Summarization for Computational Sensemaking on Complex Industrial Event Logs

Stefan Bloemheuvel[1,3]($\boxtimes$), Benjamin Kloepper[2], and Martin Atzmueller[1,3]

[1] Department of Cognitive Science and Artificial Intelligence, Tilburg University,
Warandelaan 2, 5037 AB Tilburg, The Netherlands
{s.d.bloemheuvel,m.atzmuller}@uvt.nl
[2] ABB AG, Corporate Research Center, Wallstadter Str. 59,
68526 Ladenburg, Germany
benjamin.kloepper@de.abb.com
[3] Jheronimus Academy of Data Science (JADS), Sint Janssingel 92,
5211 DA 's-Hertogenbosch, The Netherlands

Abstract. Complex event logs in industrial applications can often be represented as graphs in order to conveniently model their multi-relational complex characteristics. Then, appropriate methods for analysis and mining are required, in order to provide insights that cover the relevant analytical questions and are understandable to humans. This paper presents a framework for such computational sensemaking on industrial event logs utilizing graph summarization techniques. We demonstrate the efficacy of the proposed approach on a real-world industrial dataset.

1 Introduction

In the context of industry 4.0, event logs typically capture complex multi-relational data and information. Here, a novel approach to log analysis is graph summarization, focussing on interconnected data [1]. When analysing event logs, graphs can be used to model the events in the log as nodes and the transitions between these events as the edges. The summarization in the context of an event log is filtering the graph to find the most interesting information. Then, such methods can facilitate computational sensemaking [2] on event logs: given complex data, the aim is to apply appropriate analysis and mining methods to provide important insights that both support and cover the relevant analytical questions, are understandable to humans, and aid decision support.

This paper provides such a computational sensemaking approach on complex event log data, aiming to provide a framework based on graph summarization techniques. Our contributions are summarized as follows: (1) We propose a framework for computational sensemaking using graph summarization on complex event log data, covering multiple/heterogeneous sources, and discuss the methods applied in an analytical process in detail. (2) We present the results of

© Springer Nature Switzerland AG 2019
C. Di Francescomarino et al. (Eds.): BPM 2019 Workshops, LNBIP 362, pp. 417–429, 2019.
https://doi.org/10.1007/978-3-030-37453-2_34

an application to real-world industrial event log data, demonstrating the efficacy of the proposed analytics framework.

The rest of the paper is structured as follows: Sect. 2 discusses related work. After that, Sect. 3 presents the proposed framework for computational sensemaking using graph summarization on complex event log data. Next, Sect. 4 discusses our results. Finally, Sect. 5 concludes with a summary and interesting directions for future work.

2   Related Work

Below, we first describe related work on the analysis and mining of event logs, before we outline graph summarization techniques.

2.1   Mining Event Logs

Process mining tries to discover the process model of log data [3], i. e., it aims at the discovery of business process related events in a sequential event log. The assumption is that event logs contain fingerprints of business processes, which can be identified by sequence analysis. One task of process mining is conformance checking [4] which has been introduced to check the matching of an existing business process model with a segmentation of the log entries. Based on that, also fault detection and anomaly detection can be implemented using log analysis [5]. Fault detection consists of creating a database of fault message patterns. If an event (or a sequence of events) matches a pattern, the log system can take action. Anomaly detection consists of building a model of normal log behaviour to detect unexpected behaviour [5,6]. In addition, data clustering is a machine learning and data mining technique that groups entities into clusters. In log analysis, clustering is an useful first step to lower the number of events to deal with. Further analysis is then computationally feasible due to the reduction in input size [5]. For clustering, there are very many similarity measures that can be applied, cf. [7], including the Jaccard distance, Levenshtein distance and Hamming distance. In addition, a variety of algorithms are available, including the HDBScan algorithm [8], which performs hierarchical density-based clustering.

In contrast to the approaches discussed above, we do not primarily focus on business process mining, fault detection, or anomaly detection. Instead, our goal is to provide a summary on the event log data, in order to enhance its structuring and make relationships understandable – i. e., on providing the data on different levels of abstraction or summary levels. Then, business process analysis, fault detection, and anomaly detection can be considered as a secondary criterion, in the scope of the further computational sensemaking process. In the following, we describe the specific summarization method: Due to the sequential nature of directed graphs and the possibility to cluster nodes together, graph summarization is then a novel approach to summarize event logs, which we will discuss in the following section.

2.2   Graph Summarization

Graph summarization speeds up the analysis of a graph by creating a lossy but concise representation of the graph [9]. Such a representation is smaller in size, making it easier to interpret for operators. A wide variety of techniques have been explored in the literature, each with a different approach. Examples of techniques that are used by researchers are generating grouping-based methods, simplification-based methods, compression-based methods and influence based methods [1]. One of the most popular grouping-based methods is generating a supergraph. A supergraph is a graph where nodes are recursively aggregated into supernodes and edges are aggregated into compressor or virtual nodes [10]. Simplification-based graph summarization consists of removing less important node or edges, which results in a sparsified graph. In comparison with super-graphs, a summary now consists of a subset of the original set of nodes and edges, e. g., [11]. However, there is still an open debate in the literature about what a graph summary should look like. It is always application dependent and can serve countless different goals: preserving the answers to graph queries, finding different graphs structures, merging nodes into supernodes and merging edges into superedges. However, the challenges that each creator of a graph summary faces are shared by each domain [1]:

- It gets increasingly harder to split or merge nodes and edges, if a graph gets more attributes. This can cause problems, since real networks tend to have heterogeneous nodes in increasing manner [1].
- The main goal of graph summarization is to reduce the input graph so further analysis becomes less complex. However, techniques to do so also suffer from the challenge of processing large volumes of data.
- The main end result of a graph summarization process is to return the most interesting observations that appear in a graph. But, what is labelled as interesting is often a subjective task, depending on specific analysis goals.
- Even when we successfully determine how to define what is of interest during graph summarization, another problem that will surface is determining when we have "sufficiently many results". A bit compression technique, for example, will look at how much the size of the graph was reduced in terms of raw bits. As another example, a simplification technique will dive into the difference in the total number of nodes and edges before and after summarization. In general, each technique has an unique set of rules to determine when a success has been achieved. In addition, each technique suffers from the fact that a ground-truth solution is often not available.

In contrast to existing approaches, this paper combines different approaches, i. e., sequential patterns mining, clustering and graph summarization methods into a novel framework – to the best of the authors' knowledge. A paper that comes close to the goals of this study is the end-to-end event log analysis platform called FLAP [12]. However, their approach is more focused on creating visualization on the data than involving either sequential pattern mining or clustering for abstraction and summarization. Instead, we focus on these to ultimately enable computational sensemaking on the complex log.

3   Method

A schematic view on the framework is given in Fig. 1. First, the raw log data was filtered to make the strings suitable for the C-SPADE algorithm [13]. The sequences are then matched against the log data and each occurrence that matches one of the sequences is merged. At the same time, each sequence is clustered and labelled by the domain expert. The leftover log file is then relabelled according to these cluster names. In the following, we will describe the individual steps of the framework in detail.

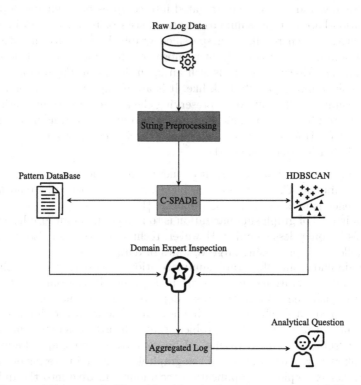

Fig. 1. Proposed Framework: First, identifiers/strings are preprocessed. Next, sequential pattern mining (with C-Spade) is used to produce a pattern database. The patterns are clustered (with HDBSCAN) and both the clusters and pattern database are examined by the domain expert. Finally, the log data is aggregated on the sequential patterns and the sequential patterns are renamed to the cluster labels, resulting in a reduced log.

3.1   Sequence Mining and Clustering

The first step of the framework consists of string filtering, to construct an unified representation of the events. For easier human interpretation, the title of the message (e.g., "Motors ON State") is concatenated to the message number (e.g., "10011") and spaces are removed. The end result of combining and filtering the message number "10011" and title "Motors ON State" will be "10011MotorsONState". Then sequence mining is applied. For that, we applied the C-Spade algorithm [13]. The C-Spade algorithm is suitable for our domain, since we allow variation in the sequences that are found (no closed or generator patterns are mined). For clustering, the HDBSCAN method was used to find clusters in the event sequences generated by C-Spade [8]. HDBSCAN is a clustering algorithm that builds upon DBSCAN extending it to a hierarchical clustering algorithm. Afterwards, a flat clustering was extracted based on the stability of the clusters. Advantages of HDBSCAN are that it is deterministic, each run will result in the same results. In addition, HDBSCAN can handle noise better than other methods such as k-means, which have to label every observation to exactly one cluster.

3.2   Graph Summarization

We consider a graph $G = (V, E)$, where V is the set of nodes and E is the set of edges that connect the nodes. An undirected graph is a graph where the edges between the nodes have no connection. A directed graph is a graph where the edges between the nodes do have a connection. The edges can have properties, e. g., weights and timestamp information. This paper focuses on such timestamped directed and weighted graphs.

The main method to summarize a graph in this paper is graph contraction. Graph contraction itself consists of two main disciplines: node contraction and edge contraction. The node contraction of a pair of vertices V_i and V_j of an event log graph produces a new graph in which the two nodes V_i and V_j are replaced by a new node V if they appear in the sequential pattern database. Another method that was used to generate supernodes was community detection. In community detection, the general problem states that given a network $G(V, E)$, find the optimal communities (clusters of nodes grouped together) that represent more nodes and edges that are inside the communities than nodes and edges that link to nodes in the rest of the graph. In the scope of this paper, we apply the Walktrap algorithm [14], which is based on a random walks technique. The concept is that when performing random walks on a graph, the walks will stay within the same community with a high probability since only a few paths will lead outside a community (bridges appear less often than edges). The algorithm runs short random walks of 3 to 5 steps and uses the result to merge communities from bottom-up. Modularity measures the strength of a network by analysing clusters. Networks that have a high modularity score therefore have dense connections between nodes that are in a community. After calculating the communities, we assign the node with the highest degree as the name for each cluster. Then, the names of all the nodes are changed to the name of the cluster they belong to. Each node gets a weight of 1 and are contracted by their membership with their weights summed up. Lastly, self loops and multiple edges are removed.

4   Results

Below, we first discuss the applied dataset, before we summarize the results of applying the methods in the workflow described in Sect. 3.

4.1   Dataset

The applied dataset was provided by ABB in an anonymized version, providing a real-world event log of an industrial process for analysis. For purpose of understanding, the first few rows of the dataset are shown in Table 1. The attributes that describe an event are the time the event happened, the message number that gives insight into what is done, the message category and severity, the title of an event that give some more detailed information about what event was performed, the description that provides even more information, and the location of the event (performed by a robot, which is in a cell or robots and that lies in a line of cells). The total number of events that were available in the dataset was about 4 million. The events were a snapshot of around 1.5 years of activity monitoring of the factory.

Table 1. The first few events of the log dataset used in this paper.

event_timestamp	Message title	Line	Cell	Robot
2014-6-3 12:29:32	10011MotorsONState	13	2	48
2014-6-3 12:29:33	10012SafetyGuardStopState	13	2	48
2014-6-3 12:29:34	10010MotorsOFFState	13	2	7
2014-6-3 12:29:35	20205AutoStopOpen	13	2	48

A total of about 200 robots were monitored during this study. To generate nodes from the event log, the message numbers were used as nodes and the transitions were considered as edges. For example, if a graph would be generated from the events in Table 1, the nodes will be (10011MotorsONState, 10012SafetyGuardStopState, 10010MotorsOFFState, 20205AutoStopOpen). The edges will be (10011MotorsONState → 10012SafetyGuardStopState), (10012SafetyGuardStopState → 10010MotorsOFFState) and (10010MotorsOFFState → 20205AutoStopOpen). The general events in the dataset consist of operations being performed on AUTO mode or Manual mode. Therefore, common events are those where the motors are turned ON/OFF.

4.2   Implementation: Event Log Data Processing and Analysis

Applying the proposed framework, we analyzed the data using sequential pattern mining, clustering and graph summarization. Supernodes were created with help of the C-SPADE and HDBSCAN algorithms. The results of each incremental step are visible in Table 2.

Table 2. Results in each stage of the graph summarization process. The *Original* column holds the metrics of the raw input graph. The *Title Merge* column consists of the information after merging the event names with the event titles. The *C-SPADE* column are the intermediate results that show the information of the graph after merging events based on the sequential rules. The *HDBSCAN* column consists of the metrics of the graph where event names in the same cluster got named after the cluster label that was provided by the domain expert. The *simplified* graph consists of an edge contracted version of the HDBSCAN graph. Lastly, the *WalkTrap* column consists of information about the graph after nodes that were considered to be in the same community got merged.

Metric	Query				Non-Query	
	Original	Title Merge	C-SPADE	HDBSCAN	Simplified	WalkTrap
Nodes	234	486	880	499	499	145
Edges	3,975,765	1,609,363	633,054	627,127	8876	500
Diameter	7	5	6	6	6	6
Radius	3	2	2	2	2	2
APL	2.57	2.35	2.16	2.33	2.33	2.34
Density	72.92	6.83	0.82	2.52	0.04	0.02
Transitivity	0.67	0.86	0.72	0.84	0.04	0.76

To start, the original graph (the raw event log-based graph) had 234 nodes (event types) and about 4 million edges (events). The diameter of the graph was 7, meaning that the maximum distance a node had to travel to the other side of the graph was 7. The eccentricity (3,60), radius (3) and average shortest path length (2.57) are at their highest in the coming 6 stages of the summarization. Most important, the density of the graph in the original form is a staggering 72.92 (due to the high count of multi-edges, e.g., the edge $10010 \rightarrow 10011$ appears over 50.000 times). In combination with a moderate transitivity score, the original graph shows to be an incredibly dense network where almost all the nodes are connected to each other.

The nodes in the log data increased from 234 to 486 when the title (small description, see Table 1) of the event was concatenated to the message number, in the data preprocessing phase. The reason for this merging was that event number 80002 depended on the title information to distinguish them from each other.

After merging the message numbers and titles, the next step was merging the rows that appear as rules in Table 3, which is a small subset of the total rules that were mined by the C-SPADE algorithm. In total, a number of 396 rules got mined ranging from length 2–10. The rules were filtered on a minimum support value of 0.90. The number of nodes increased from 486 to 880 nodes. The number of edges more than halved, decreasing from 1,609,363 to 633,054 edges. The diameter however, increased from 5 to 6 which is inline with the increase in nodes. The merging of the nodes did decrease the density of the graph by a magnitude again.

Table 3. Example of rules that are mined by the C-SPADE algorithm.

ID	Event 1	Event 2	Event 3	Support
1	10052Regainstart	10053Regainready		0.99
2	10125Programstopped	10052Regainstart	10053Regainready	0.99
3	10125Programstopped	10052Regainstart		0.99
4	10125Programstopped	10012Safetyguardstopstate		0.99
5	10053Regainready	10156Programrestarted		0.98

The sequences were clustered with help of HDBSCAN and TF-IDF in a total of 15 clusters. 65 sequences were estimated as noise points and manually added to the most similar cluster. The average number of nodes in a cluster was 22 (SD = 13.47). The most sequences were clustered in a cluster which was labelled as 'Restart and Operator Interaction' (a total of 50 sequences). The clusters were labelled with help of the domain expert. The nodes were renamed to the cluster they belonged to, which reduced the number of nodes to 499. This version of the graph is the end result and is visible in Fig. 2. Nodes with a high degree have bigger node sizes to visualize their influence in the network. The nodes in the center of the graph are the clustered collection of nodes that got assigned a label by the domain expert. The number of edges stayed roughly equal, decreasing from 633054 to 627127. The diameter stayed equal and the eccentricity, radius and

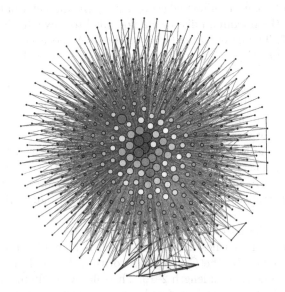

Fig. 2. Plot of the graph after the nodes got relabelled according to the HDBSCAN clustering. Nodes with a high degree have bigger node sizes to visualize their importance in the network. The nodes in the middle of the graph generally are the clustered collection of nodes that were labelled by the domain expert.

average shortest paths length were not altered significantly by cluster merging. However, the transitivity increased from 0.72 to 0.84.

It is important to note that each of the applied alterations of the graph retain the ability to perform queries on the log data. It is therefore still possible to ask for a specific slice of the dataset (since timestamp information of events is still available) to investigate the behaviour of the factory in a given time period. Graph simplification and community detection, however, are techniques that lose timestamp details. However, graph simplification and community detection are useful to decrease the size of the graph. The simplified version of the graph resulted in a huge decrease in graph size. In this step, edges are merged if they occur multiple times, assigning an according weight to the edges. The number of edges decreased from 627127 to 8876. All other metrics stayed the same, which makes sense since the structure of the graph was not altered. The graph simplification procedure does affect the density of the graph, since that is heavily influenced by the number of edges in a graph in ratio to the nodes. The community detection summarization resulted in a drastic reduction in graph size, as well. The nodes were reduced to 145 and the total number of edges from 8876 to 500.

To conclude, a total of five alterations were made to the input graph. First, the title of an event was added to the event number to make it unique. Afterwards, C-SPADE was used to find frequent sequential patterns in the log data [13]. These patterns were clustered, reducing the raw input graph from over 3.8 million edges to 8876 edges without losing query support. We reduced the graph even further to 1329 edges.

4.3   Example Application

An example of an usecase is examining a small part of the event log. When examining the last two days in the event log, a total of 5067 events were executed. We can examine the distribution of events in Fig. 3 and Table in Fig. 3. Table in Fig. 3 shows the Top 10 most occurring events in the last day of the factory. Each event in Table in Fig. 3 that has no numerical value in the event string is a cluster of summed-up sequential patterns that were found by HDBSCAN. The most occurring event is a 'Restart And Operator Interaction', followed by 'Program Restart' and 'Motor & Program Restart'. Such behaviour is expected, since the events that got merged to each of these clusters have been summed-up, resulting in a much higher frequency rate due to being a collection of other sequential patterns.

In order to examine the behaviour of the factory during 2 days even further, it is possible to dive into the timeline of the events. Figure 4 shows the number of events per minute during our subset of the log, which are reduced for increased sensemaking by the graph summarization procedure. There seems to be a spike around 00:00 and a peak between 12:00 and 18:00 during the last day. The same pattern can be found when examining the total time during events in the same period (see Fig. 5). The drop in events during 00:00 is visible in the peak of the difference in time per minute between the events. Another interesting event is

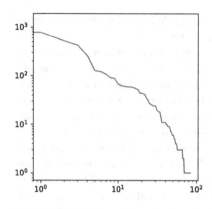

Event	Count
Restart And Operator Interaction	1327
Program Restart	777
Motor & Program Restart	520
Repeated Start & Stop	422
Restart	258
10125 Program Stopped	128
40538 Max Time Expired	122
Program Stop And Operator Interaction	107
Backup	91
10123 Program Stopped	89

Fig. 3. (A): LogLog plot of the event occurrence (Y-axis) and number of events (X-axis) in the last two days of the factory. A Power-law is visible in the distribution of the events occurrences, some events occur often and most events occur infrequently. (B): Top 10 occurring events in the last day of the factory. The most occurring events are clearly events related to restarting the factory. In addition, most of the nodes that are visible in the top 10 are the labelled cluster-nodes by the domain expert.

visible in Fig. 4 between 12:00–18:00. The total of 40 events during this period was the detection of a collision by one robot in the factory (due to maintenance or reprogramming of the robots). Lastly, the big spike at the beginning of the timeline in Fig. 5 is a 'Motor and Program Restart' that took 47 min and 19 s to complete.

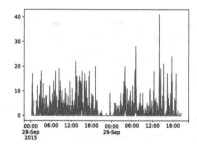

Fig. 4. Per minute events during the last day of the event log. There seems to be a spike around 12:00 on the second day and a drop during the night hours between the two day period.

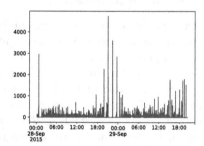

Fig. 5. Per minute difference in of time during the events (y-axis in seconds). The plot fits the lack of events during the night period in Fig. 4.

It could be interesting for the system users to know what caused these downtimes in the factory. Since we have summarized the log data as a graph, it is possible to calculate all the paths to such events. For example, if we would want

to know how many times (and why) event '20205 Auto Stop Open' happened, the predecessors are can be calculated (which are visible in Table 4). The 'Restart And Operator Interaction' → 'Restart And Operator Interaction' seems to be the major reason a '20205 Auto Stop Open' occurred. Second is a 'Repeated Start and Stop' before a 'Restart and Operator Interaction' with 6 occurrences.

Table 4. Output of inputting '20205 Auto Stop Open' in a predecessors analysis.

'20205 Auto Stop Open' Count = 43			Predecessor frequency
Restart And Operator Interaction	→	Restart And Operator Interaction	9
Repeated Start And Stop	→	Restart And Operator Interaction	6
Restart And Operator Interaction	→	Program Restart	4
Program Restart	→	Program Restart	2
⋮	⋮		
10010 Motors OFF State	→	Restart And Operator Interaction	1
Safety Stop	→	Auto Manual	1

The results of Table 4 are visible in Fig. 6. The center node is most logically '20205 Auto Stop Open' since we are focusing on all the paths to this event.

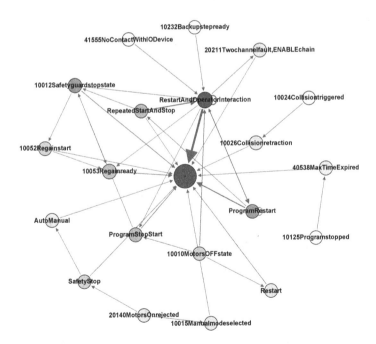

Fig. 6. Flow plot of the last-two-days focused graph with all paths leading to the event '20205 Auto Stop Open' in the middle (the name of the event is removed due to readability issues). A node is colored dark green if it has a high degree. An edge has a larger width if it occurs more often, cf. Table 4. (Color figure online)

The label is only left out due to visibility for the arrows around the node. A node in this graph is larger if it has a high degree. The edges appear larger (bigger arrow head and line width) if they appear often, which corresponds to the reason frequency in Table 4. The high scores in Table 4 are evidently visible when examining 'Restart and Operator Interaction', 'Repeated Start and Stop' and 'Program Restart'. Each of these three event occur very frequently as reasons for a '20205 Auto Stop Open' event.

5   Conclusions

In this paper, we presented a framework for computational sensemaking on industrial event logs utilizing graph summarization techniques. We demonstrated the efficacy of the proposed approach on a real-world industrial dataset. Our results indicate, that graph summarization can be applied to effectively reduce the data complexity and has the potential to answer specific queries. However, it remains a human-centered process requiring background knowledge and iterations with a human-in-the loop.

For future work, we aim to investigate closed sequential pattern mining that is able to process very long sequence data. Another interesting direction considers the inclusion of background knowledge, e. g., contained in knowledge graphs [15–17] or using further declarative specifications, e. g., [18] to be utilized in the human-centered computational sensemaking process. Then, also explanation-aware approaches [19–22] can be further supported by explicative data analysis methods, e. g., [2,23].

Acknowledgements. This work has been supported by Interreg NWE, project Di-Plast - Digital Circular Economy for the Plastics Industry.

References

1. Liu, Y., Safavi, T., Dighe, A., Koutra, D.: Graph summarization methods and applications: a survey. ACM Comput. Surv. (CSUR) **51**(3), 62 (2018)
2. Atzmueller, M.: Declarative aspects in explicative data mining for computational sensemaking. In: Seipel, D., Hanus, M., Abreu, S. (eds.) WFLP/WLP/INAP 2017. LNCS (LNAI), vol. 10997, pp. 97–114. Springer, Cham (2018). https://doi.org/10.1007/978-3-030-00801-7_7
3. Van Der Aalst, W.: Process Mining: Discovery, Conformance and Enhancement of Business Processes, vol. 2. Springer, Heidelberg (2011)
4. Munoz-Gama, J., Carmona, J., van der Aalst, W.M.P.: Single-entry single-exit decomposed conformance checking. Inf. Syst. **46**, 102–122 (2014)
5. Vaarandi, R.: A data clustering algorithm for mining patterns from event Lyuogs. In: Proceedings of the IEEE Workshop on IP Operations & Management, pp. 119–126. IEEE (2003)
6. Burns, L., Hellerstein, J., Ma, S., Perng, C., Rabenhorst, D., Taylor, D.: A systematic approach to discovering correlation rules for event management. In: Proceedings of the IFIP/IEEE IM, pp. 345–359 (2001)

7. Deza, M.M., Deza, E.: Encyclopedia of Distances, pp. 1–583. Springer, Heidelberg (2009). https://doi.org/10.1007/978-3-662-44342-2
8. McInnes, L., Healy, J., Astels, S.: HDBSCAN: hierarchical density based clustering. J. Open Source Softw. **2**(11) (2017)
9. Riondato, M., García-Soriano, D., Bonchi, F.: Graph summarization with quality guarantees. Data Min. Knowl. Discov. **31**(2), 314–349 (2017)
10. LeFevre, K., Terzi, E.: Grass: graph structure summarization. In: Proceedings of SDM, pp. 454–465 (2010)
11. Shen, Z., Ma, K.L., Eliassi-Rad, T.: Visual analysis of large heterogeneous social networks by semantic and structural abstraction. IEEE TVCG **12**(6), 1427–1439 (2006)
12. Li, T., et al.: Flap: an end-to-end event log analysis platform for system management. In: Proceedings of SIGKDD, pp. 1547–1556. ACM (2017)
13. Zaki, M.J.: SPADE: an efficient algorithm for mining frequent sequences. Mach. Learn. **42**(1–2), 31–60 (2001)
14. Pons, P., Latapy, M.: Computing communities in large networks using random walks. In: Yolum, I., Güngör, T., Gürgen, F., Özturan, C. (eds.) ISCIS 2005. LNCS, vol. 3733, pp. 284–293. Springer, Heidelberg (2005). https://doi.org/10.1007/11569596_31
15. Atzmueller, M., et al.: Big data analytics for proactive industrial decision support. atp edition **58**(9) (2016)
16. Wilcke, X., Bloem, P., de Boer, V.: The knowledge graph as the default data model for learning on heterogeneous knowledge. Data Sci. **1**, 1–19 (2017)
17. Sternberg, E., Atzmueller, M.: Knowledge-based mining of exceptional patterns in logistics data: approaches and experiences in an Industry 4.0 context. In: Ceci, M., Japkowicz, N., Liu, J., Papadopoulos, G.A., Raś, Z.W. (eds.) ISMIS 2018. LNCS (LNAI), vol. 11177, pp. 67–77. Springer, Cham (2018). https://doi.org/10.1007/978-3-030-01851-1_7
18. Atzmueller, M., Güven, C., Seipel, D.: Towards Generating Explanations for ASP-Based Link Analysis using Declarative Program Transformations, University of Cottbus, Germany
19. Wick, M.R., Thompson, W.B.: Reconstructive expert system explanation. Artif. Intell. **54**(1–2), 33–70 (1992)
20. Roth-Berghofer, T.R., Richter, M.M.: On explanation. Künstl. Intell. **22**(2), 5–7 (2008)
21. Atzmueller, M., Roth-Berghofer, T.: The mining and analysis continuum of explaining uncovered. In: Proceedings of SGAI International Conference on Artificial Intelligence (AI 2010), Cambridge, UK, pp. 273–278 (2010)
22. Biran, O., Cotton, C.: Explanation and justification in machine learning: a survey. In: IJCAI 2017 Workshop on Explainable AI, pp. 8–13 (2017)
23. Atzmueller, M.: Onto explicative data mining: exploratory, interpretable and explainable analysis. In: Proceedings of Dutch-Belgian Database Day, TU Eindhoven, Netherlands (2017)

Capturing Human-Machine Interaction Events from Radio Sensors in Industry 4.0 Environments

Stephan Sigg[1], Sameera Palipana[1(✉)], Stefano Savazzi[1,2],
and Sanaz Kianoush[1,2]

[1] Aalto University, Espoo, Finland
sameera.palipana@aalto.fi
[2] Consiglio Nazionale delle Ricerche (CNR), IEIIT institute,
Piazza Leonardo da Vinci 32, 20133 Milano, Italy

Abstract. In manufacturing environments, human workers interact with increasingly autonomous machinery. To ensure workspace safety and production efficiency during human-robot cooperation, continuous and accurate tracking and perception of workers' activities is required. The RadioSense project intends to move forward the state-of-the-art in advanced sensing and perception for next generation manufacturing workspace. In this paper, we describe our ongoing efforts towards multi-subject recognition cases with multiple persons conducting several simultaneous activities. Perturbations induced by moving bodies/objects on the electromagnetic wavefield can be processed for environmental perception by leveraging next generation (5G) New Radio (NR) technologies, including MIMO systems, high performance edge-cloud computing and novel (or custom designed) deep learning tools.

Keywords: 5G · Industry 4.0 · Radio sensing · Collaborative Robotics

1 Introduction

We report from ongoing studies in the RadioSense[1] project (Fig. 1). The RadioSense project explores passive radio sensing, or vision technologies which aim to track, recognize and analyse human-robot interactions continuously without requiring workers to wear any devices, and without the need for privacy-intrusive video, while ensuring workers' safety and privacy in industrial environments. RadioSense technology leverages real-time collection and processing of ambient (or stray) radio signal streams (e.g., those found in 5G and WiFi connections) and the Channel State Information (CSI) that form a specific type of "big data".

In this paper, we describe our ongoing efforts towards multi-subject recognition cases with multiple persons conducting several simultaneous activities. Perturbations induced by moving bodies/objects on the electromagnetic field can be

---

[1] http://ambientintelligence.aalto.fi/radiosense/.

C. Di Francescomarino et al. (Eds.): BPM 2019 Workshops, LNBIP 362, pp. 430–435, 2019.
https://doi.org/10.1007/978-3-030-37453-2_35

Fig. 1. Industry 4.0 seamless human-robot interaction targeted in RadioSense

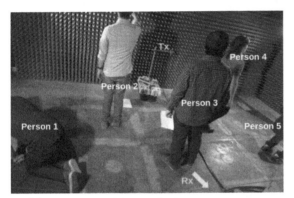 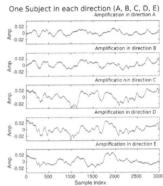

(a) Subjects performing during measurements   (b) Measured RF signals

Fig. 2. Case study conducted with multiple subjects simultaneously performing distinct activities, gestures and movement.

processed for environmental perception. In particular, we will adopt next generation (5G) high-frequency technologies as well as distributed massive MIMO systems.

2 Multi-subject Recognition

To support human-machine interaction in Industry 4.0 scenarios, we propose to exploit Radio sensing technology [6]. In particular, in the RadioSense project, we investigate installations with multi-antenna receive devices.

A common challenge in traditional radio sensing is to distinguish simultaneous movements from multiple subjects. Essentially, perturbations in the signal strength are typically analyzed at a receiver and interpreted as activities, gestures or other relevant motions. The perturbations are caused by reflection, frequency or phase shift during subject movements. In the case of multiple subjects, reflected signals are superimposed (constructively and destructively combined) and it is not trivial to distinguish individual movements apart. We propose to process the received signal such that each antenna element in the receiver antenna array has a unique phase shift so that signals coming from a given direction is amplified.

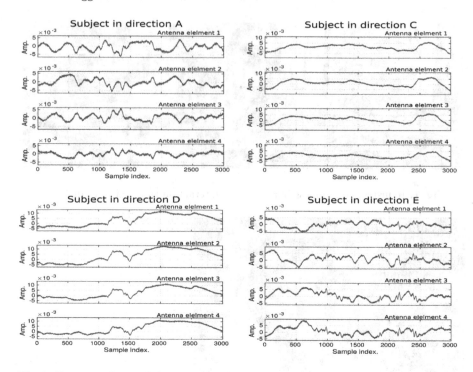

Fig. 3. Raw signal recorded at the four antenna elements while a single subject performs movement in various directions relative to the receive antennas.

Case Study. We conducted a case study (Fig. 2a) in an anechoic chamber of size 4 m × 5.2 m where a signal continuously emitted by a single transmitter was captured by a receiver with 4 phase-synchronized linear antenna elements having a spacing of $\frac{\lambda}{2}$ where λ is the wavelength at a central frequency of 3.42 GHz. In the room, 1 to 5 human subjects performed distinct movements simultaneously while being placed at 5 different locations. We have, in addition, computed 5 sets of phase multipliers for all 4 antennas during the initialization stage to amplify the signals coming from the five directions where the humans are located. In the following, we will refer to these directions as A, B, C, D and E.

Results. For a single subject located in directions A, C, D, or E, Fig. 3 depicts the signals received at the four receive antennas (for location B, the measurements taken at one of the antennas were omitted as corrupted). We can observe that the signal on all antennas is correlated as all antennas receive the correlated signal perturbations from all directions equally.

While it is possible to distinguish the movement conducted by the subject as demonstrated in the literature (e.g. [5]), subject direction or location is not visible from this time-domain data only[2]. When applying data processing to amplify signal perturbations from respective directions (A, B, C, D, E), we

---

[2] We remark that localization might still be possible from phase information.

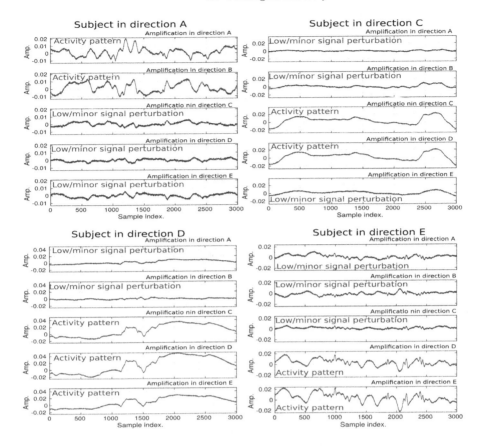

Fig. 4. Directional amplification of signals while a single subject performs movement in various directions relative to the receive antennas.

observe that, for the same data, a rough localization of the subject is possible based on direction of amplified signal (cf. Fig. 4).

While the accuracy of the directional signal amplification is not sufficient to precisely distinguish between the 5 directions, we observe that only 2–3 adjacent directions are excited with signal perturbations while other locations experience only minor signal perturbation. On top of this, activity recognition might be applied as well as phase-based localization to further improve the accuracy.

Similarly, for multiple (here: two) subjects, Fig. 5 indicates that distinct patterns are captured from the two subjects from the amplified directions. In contrast, when observing the raw data only (Fig. 6), no such distinction is possible. Consequently, by amplifying the signals from distinct direction, it shall be possible to recognize distinct workers in Industry 4.0 settings, performing simultaneous activity. We will, in further investigations also study settings with 8, 16 or 32 phase-synchronized antennas in order to increase the directional perception accuracy and the count of people that can be recognized simultaneously.

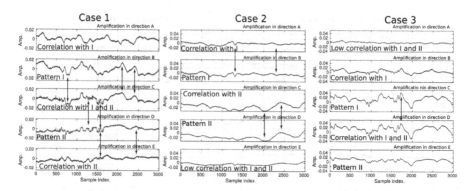

Fig. 5. Two subjects performing movement in different direction (amplification).

In particular, as indicated above already, the perception is limited in the number of individual simultaneous movement that can be recognized by the granularity and accuracy of the environmental perception. In particular, since neighbouring areas are excited too, it was in our setting, for instance, not possible to extract 5 uncorrelated patterns from 5 independently moving subjects at locations A, B, C, D, and E. This is depicted in Fig. 2b.

3   Related Work

Research has demonstrated the use of various radio signal measurements for inference of human motion including time delay [1], phase [4], and signal strength [3]; and these have been used for various purposes such as vital sign monitoring [9], activity and gesture recognition [8], localization [7] and fall detection [2].

4   Discussion and Conclusion

We have reported from ongoing studies in the RadioSense project regarding the rough relative localization and activity recognition of multiple subjects in

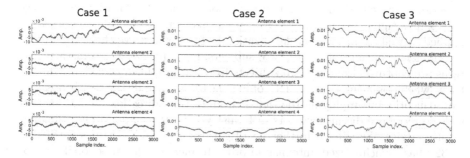

Fig. 6. Two subjects performing movement in different direction (raw).

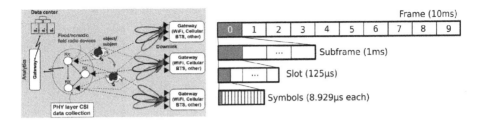

Fig. 7. Integration of Rf-sensing as a service into existing communication systems

Industry 4.0 scenarios. For seamless integration into existing environments, we propose the integration of RF-sensing as a service into beyond 5G cellular systems. A rough calculation (Fig. 7) reveals that the reservation of a single symbol within each subframe of the communication system would translate to a 1kHz sampling frequency. This is easily sufficient for common activity recognition tasks. At the same time, such allocation would deprive only 1% of the overall capacity of such communication system from communication to sensing.

Overall, we could show that with appropriate pre-processing of signal phases at distinct, synchronized receive antenna elements, environmental perception in indoor spaces can be improved with respect to relative direction of movement and perception of simultaneously conducted movement.

References

1. Adib, F., et al.: Smart homes that monitor breathing and heart rate. In: Proceedings of CHI (2015)
2. Palipana, S., et al.: FallDeFi: ubiquitous fall detection using commodity Wi-Fi devices. In: Proceedings ACM Interactive, Mobile, Wearable Ubiquitous Technologies, vol. 1, no. 4, pp. 1–25 (2018)
3. Patwari, N., Agrawal, P.: Effects of correlated shadowing: connectivity, localization, and RF tomography. In: Proceedings of IPSN (2008)
4. Pu, Q., et al.: Whole-home gesture recognition using wireless signals. In: Proceedings of Mobicom (2013)
5. Savazzi, S., Sigg, S., Vicentini, F., Kianoush, S., Findling, R.: On the use of stray wireless signals for sensing: a look beyond 5G for the next generation of industry. Computer **52**(7), 25–36 (2019)
6. Savazzi, S., et al.: Device-free radio vision for assisted living: leveraging wireless channel quality information for human sensing. IEEE Signal Process. Mag. **33**(2), 45–58 (2016)
7. Shi, S., et al.: Accurate location tracking from CSI-based passive device-free probabilistic fingerprinting. IEEE Trans. Veh. Technol. **67**(6), 5217–5230 (2018)
8. Sigg, S., et al.: The telepathic phone: frictionless activity recognition from WiFi-RSSI. In: Proceedings of PerCom, pp. 148–155 (2014)
9. Wang, H., et al.: Human respiration detection with commodity WiFi devices: do user location and body orientation matter? In: Proceedings of Ubicomp (2016)

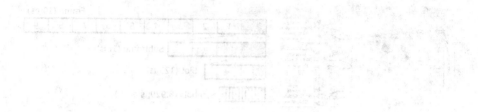

First International Workshop on Process Management in Digital Production (PMDiPro)

First International Workshop on Process Management in Digital Production (PMDiPro)

PMDiPro was the first workshop of the series. It took place on September 2, 2019, in Vienna, Austria, as part of the 19th Business Process Management Conference (BPM 2019). The workshop consisted of one keynote and one paper, selected based on a thorough review process (each paper was reviewed by three members of the Program Committee and subject to subsequent discussion among the reviewers).

The keynote was a joint effort between industry and academia and was held by Florian Pauker from EVVA Sicherheitstechnik GmbH, Thomas Grausgruber by Digitalization Factory GmbH, and Juergen Mangler from the University of Vienna, Workflow Systems and Technology Group.

The goal of PM-DiPro was to establish a forum for researchers and professionals interested in understanding, envisioning, and discussing the challenges and opportunities of utilizing Process Management Systems and Data Analytics in industrial settings. Currently on shop-floors data is mostly collected and stored by focusing on individual machines. The increasing digitization of industrial processes also across organizations sometimes labeled industry 4.0 or smart manufacturing, however, requires a more holistic and connected perspective. The introduction of process management technology to orchestrate business processes from ERPs all the way down to machine control seems an appropriate answer and produces much more interconnected data. Subsequently this opens up the potential for gaining much deeper insight into industrial processes through data analytics than ever before. Furthermore, insights from data analytics can be used to improve business process at design time but also to improve the execution of business processes during run time.

We want to thank all the reviewers and organizers who made the workshop possible, and the workshop participants for their insightful discussion contributions.

The workshop has been partially supported and funded by the Austrian Research Promotion Agency (FFG) via the "Austrian Competence Center for Digital Production" (CDP) under the contract number 854187.

October 2019   Juergen Mangler

Organization

Workshop Chairs

Juergen Mangler	University of Vienna, Austria
Stefanie Rinderle-Ma	University of Vienna, Austria
Christian Huemer	TU Wien, Austria
Gerti Kappel	TU Wien, Austria
Stefan Thalmann	TU Graz, Austria
Stefanie Lindstaedt	Know-Center Graz, Austria
Georg Grossmann	University of South Australia, Australia

Program Committee

Marc Streit Johannes	Kepler University Linz, Austria
René Peinl	Hochschule Hof, Germany
Heimo Gursch	Know-Center Graz, Austria
Karolin Winter	University of Vienna, Austria
Florian Pauker	EVVA Sicherheitstechnik GmbH, Germany
Maria Teresa Gomez Lopez	University of Seville, Spain
Belgin Mutlu	Know-Center Graz, Austria
Tobias Schreck	TU Graz, Austria
Florian Stertz	University of Vienna, Austria
Tatjana Zolotareva	Austrian Center for Digital Production, Austria
Ralph Vigne	Austrian Center for Digital Production, Austria
Sebastien Bougain	Austrian Center for Digital Production, Austria
Ralf Gitzel	ABB, Germany
Wided Guédria	Luxembourg Institute of Science and Technology, Luxembourg
Thomas Setzer	Technical University Munich, Germany
Felix Mannhard	SINTEF Digital, Norway
Daniel Ritter	SAP, Germany

BPMN and DMN for Easy Customizing of Manufacturing Execution Systems

René Peinl$^{(\boxtimes)}$ and Ornella Perak$^{(\boxtimes)}$

Institute of Information Systems, Hof University of Applied Sciences,
Alfons-Goppel-Platz 1, 95028 Hof, Germany
rene.peinl@hof-university.de, ornella.perak@iisys.de

Abstract. Manufacturing execution systems (MES) are the central integration point for the shop floor. They collect information about customer orders from ERP systems (enterprise resource planning), calculate the best plans for production orders and their assignment to machines and monitor the execution of these plans. It is therefore also the central data hub on the shop floor and communicates progress back to the ERP system. However, each company and production facility is a bit different and it is a hard decision to find a good compromise between standard products with a long customization period, industry-specific solutions with less customization need and company-specific solutions with long development times. This paper proposes the use of business process management (BPM) as a means for easy graphical customization of production processes that lead to immediately executable workflows (zero code development) or need very few code additions to get executable (low code development).

Keywords: Manufacturing execution system · Customization · Business process management · Decision table · Low code development

1 Introduction

Within office work contexts, it turned out that well-structured workflows can be used only for small parts of the daily work since most processes are weakly structured and have many variations and exceptions, especially for knowledge intensive work [1]. Looking at industrial production processes, you would expect to have mainly well-structured processes instead. However, on the journey to lot-size one, processes are getting more and more exceptions and variations there as well. To cope with this complexity not only for a single company, but also as a vendor of standard software for many companies, you have to come up with new solutions to the customization problem. Putting 20 additional fields in each database table and calling them custom01 … custom20 together with offering a plug-in architecture or scripting language are not enough. Especially when looking at interfaces to other enterprise systems like enterprise resource planning (ERP) on the one hand and supervisory control and data acquisition (SCADA) or machines directly on the other hand, you really wish for semantic understanding of what data you are dealing with, instead of having a lot of custom fields with unknown meaning that need to be looked up in documentation

© Springer Nature Switzerland AG 2019
C. Di Francescomarino et al. (Eds.): BPM 2019 Workshops, LNBIP 362, pp. 441–452, 2019.
https://doi.org/10.1007/978-3-030-37453-2_36

which is usually missing, vague or incomplete. We therefore propose to use business process management (BPM), especially the Object Management Groups family of standards called BPMN (business process model and notation), CMMN (case management model and notation) and DMN (decision model and notation) for customizing the dynamic part of a manufacturing execution system (MES) as well as a data mapper with semantic capabilities to deal with the static part (data). With this approach, both developers of the MES and administrative personnel in the end user companies can work in a semantic way and do not have to cope with generic fields and code.

The remainder of the paper is structured as follows. First, shortcomings of existing customization approaches for enterprise information systems (EIS) in general and MES in particular are discussed. Then the role of process-aware information systems in customization is reviewed, before our own solution is presented. Finally, requirements for customization of an MES are collected based on five case studies. These are used for evaluating the suitability of our approach before discussing the results and giving an outlook for future development.

2   Customization of Enterprise Information Systems

MES tends to be highly industry and process-specific, which means highly customized for a specific process running at specific plants [2]. Literature on the customization of information systems (IS) is however somewhat limited and mainly deals with customization of ERP systems [3]. Generally speaking, if a company needs to implement an enterprise information system (EIS) to support its business processes, it has the choice between using a standard software or developing an individual solution, or a combination of both [4]. The latter means customization of the standard software. Rothenberger and Srite define customization as "building custom features by using standard programming languages or the ERP system's language, changing the ERP system code, and/or including third-party packages that require some degree of programming to implement" [5]. However, the term is not used consistently in literature and there are both different interpretations for the term customization (e.g. adaptation to a single user's needs [6]) as well as other terms like "configuration" or "modification" with synonymous or at least similar meaning [3]. Most companies do at least a moderate degree of adjustments [4], although the commonly accepted best practice for large, enterprise-wide system implementation has been to limit customization, due to high costs and problems with later system upgrades [3]. Provided a certain level of fit between the standard software and the company's processes, the effect of software modifications becomes negative and related cost like support effort for the IS are rising [4].

However, customization has mainly negative impact if it changes the system in a way that was not foreseen by developers and therefore the changes will hinder future upgrades. Therefore, EIS should be as customizable as possible. To allow for that, Tsoi et al. propose a knowledge-based system that aids in selecting the components of a highly modular EIS [7]. Their approach utilizes an expert system together with case-based reasoning. They also provide a kind of GUI designer for mobile applications, but it is unclear, how the underlying system has to be designed, so that the components can be put together in arbitrary sequence.

Software customization has been argued to be beneficial for both the productivity of software engineers and end users [8]. To achieve that, the usual development methods must be applicable to highly customizable software as well. Furthermore, the customization methods should not rely on specialists to manually tweak individual applications for a specific user group [8], but allow technically savvy end-users to perform the customizations. Using graphical user interfaces to allow customization of business rules is a way to achieve that [9]. Rubart also stresses the use of semantic technologies in this context, especially to map between terms that are used as synonyms, but could confuse the business person during customization.

Customization is especially challenging in multi-tenant environments. Jansen et al. list four ways to customize web applications in such an environment, namely user interface and branding, workflow and business rules, data model extensions, and access control [6]. They also discuss integration with other IS which is often neglected in literature. Our solution, presented later on includes all these aspects except access control.

Parthasarathy and Sharma make a different distinction and differentiate between module, database and source code modifications [10], where module customization means more or less choosing certain options the module provides (configuration). They found that both database and especially source code modifications have a negative impact on quality of the customized system. Song et al. propose to use intrusive microservices to allow tenants to run their own code in a multi-tenant environment [11]. However, this is not much different from using a component-based customization as discussed in other papers. The difference is more or less in the packaging (container vs. component) and the interface (http vs. normal procedure call).

Customization is becoming even more important recently, since it is not only required once during the introduction of a new EIS, which is than run for 5–10 years without modification. Instead, the requirements are changing more frequently now and therefore adaptation of the EIS in general and MES in particular during its normal lifetime is getting crucial. Helo et al. state that current EIS lack the flexibility to adapt to changing needs and often lack a modular system architecture [2]. They also constitute a current lack of semantic interoperability (ibid). For MES, this is especially relevant for the interface to the ERP system [2], but also for the interface to manufacturing equipment. More than 10 years ago, holonic MES were proposed as solutions. They have a structure organized as a loosely coupled network of communicating agents to answer in a flexible way to changing requirements [12]. However, they were not adopted in practice. Similar to that, the vision of adaptability based on OPC-UA and CAEX (Computer Aided Engineering Exchange) as described by Sauer and Jasperneite has not become reality up to now [13]. The most promising available solution is centurio.work [14], a modular orchestration solution that uses BPMN for modeling the whole process from the ERP system to the MES and even to trigger SCADA functionality. However, it is unclear how the interplay between workflow engine and UI is working.

Summing up, MES heavily rely on customization, but despite a number of proposals in scientific literature, the state of practice is still a lack of customization possibilities. Requirements for a highly customizable MES are therefore (1) the ability to model the process steps for a single production step, (2) a way to model the sequence of production steps for different types of products, (3) a means to map data from the ERP system, machines and the UI to the internal data schema and vice vesa and finally (4) an approach to extend the data schema. In the next section the possible contributions of business process management systems as part of process-aware IS will be reviewed.

3   Process-Aware Information Systems

Van der Aalst [15] introduced the term "process-aware information system" more than a decade ago and defined it as "a software system that manages and executes operational processes involving people, applications, and/or information sources on the basis of process models". He used the term to demarcate modern process-centric IS from traditional data-centric IS. Advantages of such systems that represent processes in (graphical) models instead of code include greater flexibility in terms of dealing with change and easier process improvement [15]. Whereas traditionally business process management (BPM) was mainly used for analysis and improvement of business processes before the introduction of a new IS like an ERP [16], nowadays the focus shifts towards later phases that van der Aalst calls system configuration (customizing) and process enactment which means the operational phase of an IS. However, although modern IS like the SAP ERP system "have a workflow engine, most processes are hard-coded into the system and can only be changed by programming or changing configuration parameters" [15]. That may be due to the fact that popular process modelling languages like event-driven process chains (EPC) and business process model and notation (BPMN) are not directly executable [17, 18].

However, this problem is solved in popular open source BPM systems like Camunda, which is seen as having the best support for BPMN constructs [19] and allows for linking BPMN service tasks to Java classes as well as providing a task handling application with HTML forms for humans to interact with the process engine. On the other hand, this leads to platform-specific models as Santos et al. [20] call them. That means they are not portable across tools, although they adhere to the official BPMN specification. The executable parts are not standardized and are represented as extension elements with a tool-specific namespace. The advantage of this approach is that there is no transformation process necessary and other tools can still use the official BPMN elements in the process. Therefore, the same model can be seen both as platform-independent (general process specification) and platform-specific (the executable part). The challenge is, to provide the user of the BPMN modeler with mechanisms to choose the available service task implementations or directly link the decision tables or sub-processes available in the workflow engine (see Fig. 1). To do this, we extended the Web-based Camunda modeler[1] with own code to implement drop-down boxes instead of free-text fields. This leverages the existing mechanism

---

[1] https://bpmn.io.

from Camunda called "element templates", but goes one step further than that. It is similar to the developments of [21], that provide a plug-in architecture for service tasks.

Another aspect is the integration of a user interface (UI) for user tasks. This is an often neglected part of BPEL execution engines since they do not focus user tasks [22]. Even in BPMN, where process engines usually have the capability to automatically create simple task forms or show special tailored HTML forms provided by the developer, there are still deficiencies esp. in dialog modeling and handling the UI for multiple parallel instances of a workflow [22]. In the last ten years, the need for a better mechanism for UI modeling was expressed by multiple researchers [21, 23, 24]. Whereas Guerrero et al. showcase the use of the UI description language with generating a Java Swing UI, Bouchelligua et al. make use of the Eclipse Modeling Framework together with the Graphical Editing Framework. Kammerer et al. delegate the definition of forms to Adobe Acrobat and use the resulting PDF forms in their own process engine. Our own approach combines automatic generated HTML task forms for non-critical tasks with pre-defined HTML UI masks that can be tailored to companies' needs during the customization phase of the MES introduction with a self-developed UI editor.

For customizing the MES system, an important step is to easily model the decision on which products need to flow through which production steps on which machines. [25] show that declarative process languages are not capable of representing such decisions. They propose to use DMN, the decision model and notation for this purpose. Similar to the editor proposed in [26], the Camunda BPM engine comes with a graphical editor for decision models that consists of a part to aggregate sub-decisions to a larger decision using boxes and arrows, as well as a spreadsheet-like editor for decision tables that use process variables as input and output. Rules can be expressed with the friendly-enough expression language (FEEL, [26]). The combination of a spreadsheet-like UI with a process engine has already proven successful in the finance industry [27] and is expected to work as well for the target group in manufacturing companies.

[28] demand the integration of a technical systems model, the production process model and an MES functional view and propose to use BPMN for the latter two. Similar to our approach, they define which BPMN elements should be used for what purpose in the MES functional view. We use e.g. swimlanes for representing machines and sub-processes to group the phases of a single production step. However, Witsch and Vogel-Heuser only use the models for communication during the analysis phase.

Finally, the data used within the processes has to be handled as well. [21] demand that "process models need to be enriched with a sophisticated, yet intuitive data flow perspective to support process execution". In an MES environment, important data that can be integrated from other IS or the machines themselves are often hard to interpret, "because data points are often neither labeled, described nor self-explanatory" [28]. This is still true if data exchange standards like OPC-UA are used [29], since they mainly standardize the transport of data, but only optionally demand certain data structures and semantics (information models in so called companion standards) [14]. "The required transparency for the integration of a new MES into an existing IT-infrastructure can hardly be achieved without an appropriate graphical model" [28]. Therefore, we provide a Web-based graphical mapper to translate between incoming data from an ERP system (Web services) or machine (OPC, OPC-UA) into our internal data model. The data objects required in a production process are passed to the process engine as process variables in an object-oriented way, so that both standard attributes modeled by the developer of the MES as well as custom attributes introduced during the system configuration phase are available and can be both read and written to from the process. [30] call such an approach object-aware process management.

4 Proposed Solution

Our solution is called HiCuMES (highly customizable MES). It is built in a modular way using a microservice architecture with a Java Enterprise backend, an Angular frontend, a relational data store and OpenID connect as SSO protocol. Each production step is modelled as consisting of up to four phases: (1) choosing the next production order to process, (2) equipping the machine with the appropriate tool and setting the machine parameters accordingly (mounting), (3) doing occasional quality checks during production and (4) reporting back manufactured products and wastage. All of these phases have counterparts in the Web UI (user interface) of HiCuMES and can be done either manually by the machine operator or automatically by the machine, a SCADA software or HiCuMES (e.g. setting parameters via OPC-UA or getting reports about produced units). HiCuMES is developed in an agile way. After the first iterations, it was able to support a single production step with all four phases and only few customizable options. It is used in production by company A and is available for download under Apache open source license v2[2]. It supports OPC-UA for reading machine data. The goal for the next iteration is the connection between the chosen BPM solution and HiCuMES. The Camunda BPM suite is used due to its open source code base, support for the OMGs BPM standards and its good integration into Java-based applications. It is run in a standalone way as a separate microservice [11] next to the order, machine and production microservices of HiCuMES. Data is exchanged using REST calls from HiCuMES to Camunda and by triggering Java delegate classes from within Camunda processes that send asynchronous messages to the other microservices.

Figure 1 shows a part of the BPMN model used for company B. Each machine is modeled as a swim lane in BPMN. One machine can perform a single or several production steps. Multiple machines working closely together on a production step can be aggregated in a pool. One workflow instance represents a single production order on specific machine instances. If there are multiple instances of the same machine in the factory, one workflow instance is started for each of them. If one process step creates semi-finished products for multiple orders, it has to be treated separately and gets its own production order. This usually goes along with own product types. Therefore, machine A produces semi-finished products of type A1 as a basis for products B and C that may be sold to the same or different customers. This results in one production order (PO) for A1 and separate ones for B and C. The PO for A1 is represented by a workflow instance of type A. POs for B and C have a workflow instance each. Both workflow instances

---

[2] https://github.com/iisys-hof/dfap.

have the same type. The distinction between different product types requiring different process steps and maybe sequence are modelled with DMN (see below).

A production step is represented by a sub-process in BPMN. Each phase in the step is represented by a task, which can be a manual (see Fig. 1, the tasks with a human icon, called a user task) or an automatic task (see the gears in Fig. 1, called a service task). The flexibility of BPMN can be used to accommodate for different requirements. Whereas company A had a strictly sequential process with obligatory quality checks for every lot produced, company B has a time based model for quality checks, so that short POs can be processed without a mandatory check (see Fig. 1). It would have been quite hard to code that manually in HiCuMES and it is easy to model with BPMN.

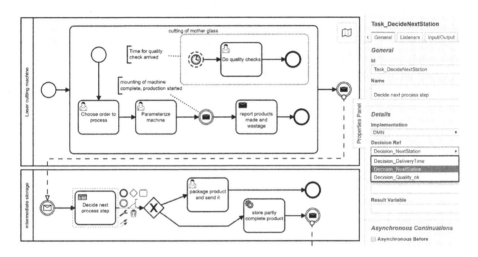

Fig. 1. Part of a BPMN process depicting the production of electronically dimmable glass

The workflow status is synchronized with the rest of the application, so that the appropriate UI form is presented to the machine operator at any time. Vice versa, user interaction with the Web UI update process variables in the workflow. Camunda uses these process variables instead of BPMN data objects to pass data between BPMN elements. Due to the tight integration between Camunda and HiCuMES, the process analysis features of Camunda can be fully utilized to track production progress, determine bottlenecks and do audits.

After completion of a process step, the digital twin of the PO goes to a virtual intermediate storage (see Fig. 1). This can be the representation of a real warehouse, or a conveyor belt that automatically brings products to the next machine and process step. In the latter case, choosing the PO on the next machine has to be modeled as a service task. We have extended the Camunda Web-based modeler to choose implementations of service tasks as well as references to decisions, cases and other BPMN processes directly based on the available ones in the Camunda engine (see "Decision Ref" in Fig. 1) from a drop-down list instead of a text field for free-text input. The DMN model "Decision_NextStation" chosen there is the reference to the DMN model shown in Fig. 2) The modeled processes can also be directly deployed to the Camunda engine. This allows for citizen developers [31] to use it.

In the intermediate storage, a decision task (DMN) is triggered. This task contains the business logic about the decision upon the next process step for the current PO. In that way, the logic does not need to be hard coded and can be easily customized by a team consisting of an IT expert and a production expert. Figure 2 shows an example of such a decision table. In this example, the product type together with the previous step and the information about whether the process is a research process or not are used to decide upon the next process step. You can use any additional process variables as a decision base. Process variables are set from the HiCuMES application when initializing the process and contain the production order with the products to be manufactured, their amount and additional data included by the manufacturing company during customization. A drawback of using the product type for deciding upon the next production step is that companies with large amounts of product types and production steps would have to fill in many lines in the decision table. Therefore, it is advisable to look for common properties of those product types that share the same sequence of production steps in order to limit the number of lines to a reasonable amount.

Decision_NextStation						
U	Input ⊞					Output +
	Produkttyp		letzter Fertigungsschritt	Forschungsprozess		nextProcessStep
	prodType		lastProcessStep	isResearch		
	string		integer	boolean		string
1	Scheibenmodell A	1		false	2	
2	Scheibenmodell A	2		false	3	
3	Scheibenmodell A	3		false	4	
4	Scheibenmodell A	4		false	x	
5	Scheibenmodell B	1		true	2	
6	Scheibenmodell B	2		true	4	
7	Scheibenmodell B	4		true	x	

Fig. 2. Example of a DMN decision table for choosing the next production step

Optional tasks within a production step can be modelled with CMMN that can be included into the BPMN process with a call activity. Alternatively, BPMN 2.0 does also provide so called ad-hoc sub-processes for such use cases. If you put tasks into such an ad-hoc sub-process, it means they are either optional, can be executed multiple times or have no specific order of execution. This might be easier to read for people developing and maintaining the process definitions, since they do not have to change between BPMN and CMMN with their different model element names and symbols.

5 Case Studies and Evaluation

In order to collect requirements regarding the needed flexibility in terms of processes and data, five producing companies in Bavaria, Thuringia and Saxony (all in Germany) have been analyzed. They are volunteers and therefore not representative, but still cover a broad range of different production processes reaching from highly customer-specific processes with a lot of production steps and variations (nearly single-part production) up to highly standardized production of mass goods of a million parts per week (resistors). Implementing those five production processes with the HiCuMES demonstrator will prove its applicability and can be seen as a proof-of-concept.

Company A is producing window profiles in a plastics extrusion process. It has only two production steps and the variability comes mainly from using one of a few hundred tools in step one, combined with a few options in step two. The lot size is usually medium, so that step one takes at least 4–8 h to produce all goods of an order, but can also run for several days on multiple machines. Company A has about 40 different quality measures, but only a few of them (less than a dozen usually) is applicable for a certain product type.

Company B is producing electronically dimmable window glass. It has 13 production steps, which are not all required for every product. Step one is cutting the original glass into the shapes and sizes required for the customer (see Fig. 1). Minimizing the amount of wasted glass is the main goal here, which means orders of multiple customers can be mixed to achieve the best results. Therefore, a separate production order is generated in HiCuMES for this first production step, which results in semi-finished products that serve as input for the next production steps. This means also a new workflow instance for the next steps.

Company C is producing leather for industrial goods like car and plane seats. A specialty of their production process is the long duration of single steps and restrictions on the maximal storage time between process steps. This can be modeled with timer events in BPMN. Additionally, they use the same trick as company B to have intermediate products. These are even sold directly in some cases, but can also serve as input for further production steps with a different production order. Finally, ownership of semi-finished goods can change during the processing or storage time. However, this is not really affecting the MES and more a commercial issue handled in the ERP system.

Company D is producing resistors of varying sizes from a few millimeters to several centimeters. The main challenge for HiCuMES with their requirements is that manufactured goods belonging to one production order can simultaneously be processed on multiple machines in subsequent production steps. In the most challenging cases, some products may already be finished with all production steps, before other ones of the same production order are even through the first step. They also have a very sophisticated self-developed production planning module in place that will not be easy to replace by HiCuMES. However, HiCuMES is built for extensibility and integration with the existing software infrastructure of the company (microservice architecture, [11]). It could therefore replace other modules of the current solution while reusing the production planning module. However, this will likely need code adaptions.

Company E is producing needles and yarn for medical usage under high cleanness requirements that also affects packaging of the produced goods. They have few production steps with low variability. The main challenge is the high degree of variability in the material quality that cannot be influenced by HiCuMES.

As a preliminary evaluation of the system, the production processes of the companies were modeled with BPMN and bound to Java classes in Camunda, so that the workflow can be executed. This was done by the developers themselves and not by employees of the companies. It is planned as the next step to do workshops with the companies and let them fill out questionnaires to find out their satisfaction with the system.

(1) The process steps can be modeled with BPMN service tasks and human tasks. Service tasks can be bound to Java delegates (classes), so that they are executable. (2) The sequence of process steps can be modeled with DMN decision tables, based on process variables that allow access of product meta data. (3) Data can be mapped with a schema mapper from Webservices (ERP system) and OPC-UA (machines) to the internal data schema and with a UI editor to the UI. (4) The internal schema can be edited with a graphical schema mapper that produces Java classes with JPA annotations, so that the respective database schema can be generated by JPA.

Summed up, HiCuMES is able to fulfill the requirements of the five companies despite their heterogeneity. The proposed method can be applied and is suited for the task. It is possible to graphically adapt the system to the companies processes and generate directly executable processes based on BPMN, DMN and a process runtime that allows linking to Java classes. An unsolved challenge is to model processes that are both visible in their whole, but still are able to run sub-processes independently from each other, as required by company D. The idea for this case is to use boundary non-interrupting conditional events thrown by the "report products" task. The start of the next process step would be a catch event instead of being triggered by the completion of the previous step. By doing that, more than one sub-process could run at the same time. However, this part has not been tested yet, so it is not sure this will work out as planned.

6 Discussion and Outlook

This paper presented the use of BPMN and DMN in an industrial production process for customizing a manufacturing execution system. It showed the advantages of graphical modeling for customizing a MES for both the manufacturing company as well as the software developer. The graphical model can be used for better understanding the manufacturing processes and to do reengineering, as well as directly executing the model when running in an appropriate engine like Camunda and links to appropriate implementing code like in the HiCuMES application. Using process models for customization has a long history in ERP systems, but to the best of the authors knowledge was not yet used for MES. Unlike the proposal in [32], HiCuMES uses BPMN not only for requirements analysis, but also for implementing the customization. Furthermore, HiCuMES provides direct executability of the process models, which is not possible with other solutions. In combination with the data modeler, GUI editor and automatic generation of Java sub-classes based on the additional attributes of the manufacturing company, the solution allows for a fully customizable, process-aware enterprise information system without the need of writing code. Of course, this has limitations and in practice, there may still be reasons to extend the code base, which can easily be done due to its open source license and domain-driven development that emphasizes domain terminology in the code base and pushes back boilerplate and technical code. Future development is planned for the integration of a sophisticated production-planning component that is able to consider the different production requirements and usual production times of a product type throughout the whole production process. It will calculate a machine layout plan based on the ability of the machine to produce the product, how many times the product was already produced on the machine, the utilization of the machine, its cost rate and the duration of the production process in relation to the desired delivery date. During customization, the weights for the different planning criteria can be set and additional criteria can be added as a DMN decision table.

References

1. Peinl, R.: IT support for knowledge processes in digital social collaboration. In: North, K., Maier, R., Haas, O. (eds.) Knowledge Management in Digital Change. PI, pp. 113–127. Springer, Cham (2018). https://doi.org/10.1007/978-3-319-73546-7_7
2. Helo, P., Suorsa, M., Hao, Y., Anussornnitisarn, P.: Toward a cloud-based manufacturing execution system for distributed manufacturing. Comput. Ind. **65**, 646–656 (2014)
3. Cox, S.R., Rutner, P.S., Dick, G.: Information technology customization: how is it defined and how are customization decisions made? In: Proceedings of the 15th Southern Association for Information Systems Conference (2012)
4. Koch, S., Mitteregger, K.: Linking customisation of ERP systems to support effort: an empirical study. Enterp. Inf. Syst. **10**, 81–107 (2016)
5. Rothenberger, M.A., Srite, M.: An investigation of customization in ERP system implementations. IEEE Trans. Eng. Manag. **56**, 663–676 (2009)
6. Jansen, S., Houben, G.-J., Brinkkemper, S.: Customization realization in multi-tenant web applications: case studies from the library sector. In: Benatallah, B., Casati, F., Kappel, G., Rossi, G. (eds.) ICWE 2010. LNCS, vol. 6189, pp. 445–459. Springer, Heidelberg (2010). https://doi.org/10.1007/978-3-642-13911-6_30
7. Tsoi, S.K., Cheung, C.F., Lee, W.B.: Knowledge-based customization of enterprise applications. Expert Syst. Appl. **25**, 123–132 (2003)
8. Hui, B., Liaskos, S., Mylopoulos, J.: Requirements analysis for customizable software: a goals-skills-preferences framework. In: Proceedings 11th IEEE International Requirements Engineering Conference, pp. 117–126. IEEE (2003)
9. Rubart, J.: Semantic adaptation of business information systems using human-centered business rule engines. In: 2016 IEEE Tenth International Conference on Semantic Computing (ICSC), pp. 187–193. IEEE (2016)
10. Parthasarathy, S., Sharma, S.: Impact of customization over software quality in ERP projects: an empirical study. Softw. Qual. J. **25**, 581–598 (2017)
11. Song, H., Chauvel, F., Solberg, A.: Deep customization of multi-tenant SaaS using intrusive microservices. In: 2018 IEEE/ACM 40th International Conference on Software Engineering: New Ideas and Emerging Technologies, pp. 97–100. IEEE (2018)
12. Blanc, P., Demongodin, I., Castagna, P.: A holonic approach for manufacturing execution system design: an industrial application. Eng. Appl. Artif. Intell. **21**, 315–330 (2008)
13. Sauer, O., Jasperneite, J.: Adaptive information technology in manufacturing. In: CIRP Conference on Manufacturing Systems (2011)
14. Pauker, F., Mangler, J., Rinderle-Ma, S., Pollak, C.: centurio.work - Modular secure manufacturing orchestration. In: 16th International Conference on BPM, pp. 164–171 (2018)
15. van der Aalst, W.M.P.: Process-aware information systems: lessons to be learned from process mining. In: Jensen, K., van der Aalst, W.M.P. (eds.) Transactions on Petri Nets and Other Models of Concurrency II. LNCS, vol. 5460, pp. 1–26. Springer, Heidelberg (2009). https://doi.org/10.1007/978-3-642-00899-3_1

16. Liu, H., Nan, L.: Process improvement model and it's application for manufacturing industry based on the BPM-ERP integrated framework. In: Dai, M. (ed.) ICCIC 2011. CCIS, vol. 231, pp. 533–542. Springer, Heidelberg (2011). https://doi.org/10.1007/978-3-642-23993-9_77

17. Asztalos, M., Mészáros, T., Lengyel, L.: Generating Executable BPEL Code from BPMN Models. In: GraBaTs 2009 Tool Contest (2009)

18. De Giacomo, G., Oriol, X., Estañol, M., Teniente, E.: Linking data and BPMN processes to achieve executable models. In: Dubois, E., Pohl, K. (eds.) CAiSE 2017. LNCS, vol. 10253, pp. 612–628. Springer, Cham (2017). https://doi.org/10.1007/978-3-319-59536-8_38

19. Geiger, M., Harrer, S., Lenhard, J., Wirtz, G.: BPMN 2.0: the state of support and implementation. Future Gener. Comput. Syst. **80**, 250–262 (2018)

20. Santos, N., Duarte, F.J., Machado, R.J., Fernandes, J.M.: A transformation of business process models into software-executable models using MDA. In: Winkler, D., Biffl, S., Bergsmann, J. (eds.) SWQD 2013. LNBIP, vol. 133, pp. 147–167. Springer, Heidelberg (2013). https://doi.org/10.1007/978-3-642-35702-2_10

21. Kammerer, K., Kolb, J., Andrews, K., Bueringer, S., Meyer, B., Reichert, M.: User-centric process modeling and enactment: the Clavii BPM platform (2015)

22. Trætteberg, H.: UI design without a task modeling language – using BPMN and DiaMODL for task modeling and dialog design. In: Forbrig, P., Paternò, F. (eds.) HCSE/TAMODIA - 2008. LNCS, vol. 5247, pp. 110–117. Springer, Heidelberg (2008). https://doi.org/10.1007/978-3-540-85992-5_9

23. Guerrero, J., Vanderdonckt, J., Gonzalez, J.M., Winckler, M.: Modeling user interfaces to workflow information systems. In: 4th International Conference on Autonomic and Autonomous Systems (ICAS 2008), pp. 55–60. IEEE (2008)

24. Bouchelligua, W., Mahfoudhi, A., Mezhoudi, N., Daassi, O., Abed, M.: User interfaces modelling of workflow information systems. In: Barjis, J. (ed.) EOMAS 2010. LNBIP, vol. 63, pp. 143–163. Springer, Heidelberg (2010). https://doi.org/10.1007/978-3-642-15723-3_10

25. Mertens, S., Gailly, F., Poels, G.: Enhancing declarative process models with DMN decision logic. In: Gaaloul, K., Schmidt, R., Nurcan, S., Guerreiro, S., Ma, Q. (eds.) CAISE 2015. LNBIP, vol. 214, pp. 151–165. Springer, Cham (2015). https://doi.org/10.1007/978-3-319-19237-6_10

26. Biard, T., Le Mauff, A., Bigand, M., Bourey, J.-P.: Separation of decision modeling from business process modeling using new "Decision Model and Notation" (DMN) for automating operational decision-making. In: Camarinha-Matos, Luis M., Bénaben, F., Picard, W. (eds.) PRO-VE 2015. IAICT, vol. 463, pp. 489–496. Springer, Cham (2015). https://doi.org/10.1007/978-3-319-24141-8_45

27. Stach, M., Pryss, R., Schnitzlein, M., Mohring, T., Jurisch, M., Reichert, M.: Lightweight process support with spreadsheet-driven processes: a case study in the finance domain. In: Teniente, E., Weidlich, M. (eds.) BPM 2017. LNBIP, vol. 308, pp. 323–334. Springer, Cham (2018). https://doi.org/10.1007/978-3-319-74030-0_24

28. Witsch, M., Vogel-Heuser, B.: Formal MES modeling framework-integration of different views. IFAC Proc. **44**, 14109–14114 (2011)

29. Pauker, F., Frühwirth, T., Kittl, B., Kastner, W.: A systematic approach to OPC UA information model design. Procedia CIRP **57**, 321–326 (2016)

30. Steinau, S., Andrews, K., Reichert, M.: Flexible data acquisition in object-aware process management. In: 7th International Symposium on Data-Driven Process Discovery and Analysis (SIMPDA 2017), Neuchâtel, Switzerland (2017)

31. Gartner: Citizen developer. http://www.gartner.com/it-glossary/citizen-developer/

32. Michalik, P., Štofa, J., Zolotová, I.: The use of BPMN for modelling the MES level in information and control systems. Qual. Innov. Prosper. **17**, 39–47 (2013)

Second International Workshop on Process-Oriented Data Science for Healthcare (PODS4H)

Second International Workshop on Process-Oriented Data Science for Healthcare (PODS4H)

The world's most valuable resource is no longer oil, but data. The ultimate goal of data science techniques is not to collect more data, but to extract knowledge and insights from existing data in various forms. For analyzing and improving processes, event data is the main source of information. In recent years, a new discipline has emerged combining traditional process analysis and data-centric analysis: Process-Oriented Data Science (PODS). The interdisciplinary nature of this new research area has resulted in its application for analyzing processes in different domains such as education, finance, and especially healthcare.

The International Workshop on Process-Oriented Data Science for Healthcare 2019 (PODS4H 2019) aimed at providing a high-quality forum for interdisciplinary researchers and practitioners (both data/process analysts and a medical audience) to exchange research findings and ideas on healthcare process analysis techniques and practices. PODS4H research includes a wide range of topics from process mining techniques adapted for healthcare processes, to practical issues on implementing PODS methodologies in healthcare centers' analysis units. For more information visit pods4h.com.

The second edition of the workshop attracted 29 regular paper proposals, a remarkably number of high quality submissions, from which 13 Regular Papers (45% acceptance rate) were selected for presentation. The papers included a wide range of topics: privacy, interactive healthcare process discovery, process-oriented medical education and surgical training, sepsis and breast cancer, outpatient appointments, performance and queues in emergency rooms, returning patient costs, clinical guidelines, simulation, change detection, and standard definitions and codes, among others. The conference also included a number of success cases and a discussion panel.

This edition of the workshop included two awards, the Best Paper Award and the Best Student Paper Award. The PODS4H 2019 Best Paper Award was given to "Towards Privacy-Preserving Process Mining in Healthcare" by Anastasiia Pika, Moe Wynn, Stephanus Budiono, Arthur Ter Hofstede, Wil van der Aalst, and Hajo A. Reijers. The PODS4H 2019 Best Student Paper Award was given to "Understanding Undesired Procedural Behavior in Surgical Training: the Instructor Perspective" by Victor Galvez, Cesar Meneses, Gonzalo Fagalde, Jorge Munoz-Gama, Marcos Sepúlveda, Ricardo Fuentes, and Rene de la Fuente. The price included a voucher for a professional Process Scientist Training, provided by the Celonis Academic Alliance.

The workshop was an initiative of the Process-Oriented Data Science for Healthcare Alliance (PODS4H Alliance). The goal of this international alliance is to promote the research, development, education, and understanding of process-oriented data science in healthcare. For more information about the activities and its members visit pods4h.com/alliance.

The organizers would like to thank all the Program Committee members for their valuable work in reviewing the papers, and the BPM 2019 Organizing Committee for supporting this successful event.

October 2019

Organization

Workshop Chairs

Jorge Munoz-Gama	Pontificia Universidad Católica de Chile, Chile
Carlos Fernandez-Llatas	Universitat Politècnica de València, Spain
Niels Martin	Hasselt University, Belgium
Owen Johnson	University of Leeds, UK

Program Committee

Robert Andrews	Queensland University of Technology, Australia
Joos Buijs	Eindhoven University of Technology, The Netherlands
Andrea Burattin	Technical University of Denmark, Denmark
Daniel Capurro	Pontificia Universidad Católica de Chile, Chile
Josep Carmona	Universitat Politècnica de Catalunya, Spain
Claudio di Ciccio	Vienna University of Economics and Business, Austria
Marco Comuzzi	Ulsan National Institute of Science and Technology, South Korea
Benjamin Dalmas	École des Mines de Saint-Étienne, France
Carlos Fernandez-Llatas	Universitat Politècnica de Valencia, Spain
René de la Fuente	Pontificia Universidad Católica de Chile, Chile
Roberto Gatta	Università Cattolica del Sacro Cuore, Italy
Emmanuel Helm	University of Applied Sciences Upper Austria, Austria
Zhengxing Huang	Zhejiang University, China
Owen Johnson	University of Leeds, UK
Felix Mannhardt	SINTEF, Norway
Ronny Mans	VitalHealth Software, The Netherlands
Niels Martin	Hasselt University, Belgium
Renata Medeiros de Carvalho	Eindhoven University of Technology, The Netherlands
Jorge Munoz-Gama	Pontificia Universidad Católica de Chile, Chile
Ricardo Quintano	Philips Research
David Riaño	Universitat Rovira i Virgili, Italy
Stefanie Rinderle-Ma	University of Vienna, Austria
Eric Rojas	Pontificia Universidad Católica de Chile, Chile
Lucia Sacchi	University of Pavia, Italy
Fernando Seoane	Karolinska Institutet, Sweden
Marcos Sepúlveda	Pontificia Universidad Católica de Chile, Chile
Minseok Song	Pohang University of Science and Technology, South Korea

Emilio Sulis	Università di Torino, Italy
Pieter Toussaint	Norwegian University of Science and Technology, Norway
Vicente Traver	Universitat Politècnica de Valencia, Spain
Wil van der Aalst	RWTH Aachen University, Germany
Rob Vanwersch	Maastricht University Medical Center, The Netherlands
Moe Wynn	Queensland University of Technology, Australia

Analysis and Optimization of a Sepsis Clinical Pathway Using Process Mining

Ricardo Alfredo Quintano Neira[1,2(✉)], Bart Franciscus Antonius Hompes[2,3],
J. Gert-Jan de Vries[2], Bruno F. Mazza[4], Samantha L. Simões de Almeida[4],
Erin Stretton[2], Joos C. A. M. Buijs[3], and Silvio Hamacher[1]

[1] Industrial Engineering Department,
Pontifícia Universidade Católica do Rio de Janeiro,
Rio de Janeiro, Brazil
[2] Philips Research, Eindhoven, The Netherlands
ricardo.quintano@philips.com
[3] Department of Mathematics and Computer Science,
Eindhoven University of Technology, Eindhoven, The Netherlands
[4] Hospital Samaritano de São Paulo, São Paulo, Brazil

Abstract. In this work, we propose and apply a methodology for the management and optimization of clinical pathways using process mining. We adapt the Clinical Pathway Analysis Method (CPAM) by taking into consideration healthcare providers' needs. We successfully applied the methodology in the sepsis treatment of a major Brazilian hospital. Using data extracted from the hospital information system, a total of 5,184 deviations in the execution of the sepsis clinical pathway were discovered and categorized in 43 different types. We identified the process as it was actually executed, two bottlenecks, and significant differences in performance in cases that deviated from the prescribed clinical pathway. Furthermore, factors such as patient age, gender, and type of infection were shown to affect performance. The analysis results were validated by an expert panel of clinical professionals and verified to provide valuable, actionable insights. Based on these insights, we were able to suggest optimization points in the sepsis clinical pathway.

Keywords: Clinical Pathways · Sepsis · Process analytics

1 Introduction

Clinical Pathways (CPs) consist of well-defined care plans which describe the ordering and timing of interventions and their expected outcomes. Healthcare providers typically model their CPs after evidence-based medical guidelines supplied by communities such as the Society of Critical Care Medicine, promoting the improvement and standardization in patient care while reducing expenses and consumption of resources [4,13,25,39]. The correct implementation of CPs is known to promote positive outcomes such as early disease recognition, correct treatment, reduction of mortality and length of stay, and hospital costs [24].

© Springer Nature Switzerland AG 2019
C. Di Francescomarino et al. (Eds.): BPM 2019 Workshops, LNBIP 362, pp. 459–470, 2019.
https://doi.org/10.1007/978-3-030-37453-2_37

However, the management of CPs has proven to be a challenging task. Time constraints, lack of qualified resources, administrative burdens, and absence of tools for evaluating and reporting are common contributing factors [8,19]. The analysis of data extracted from hospital information systems provides an interesting direction for the optimization efforts of healthcare professionals.

Sepsis is defined as a "life-threatening organ dysfunction caused by a dysregulated host response to infection" [32], and is considered a worldwide health issue and economic burden. The mortality rates from different countries varied in a wide range, but can go as high as 64.5% [6,9,12,21,22,27,30,34]. Chalupka and Talmor found that the mean costs per case ranged from USD 4,888 (Argentina) to USD 103,529 (United States) [7]. Improving performance and compliance to sepsis CPs can simultaneously save lives and reduce financial impact.

Process mining is a research field concerned with the analysis of business processes using data extracted from information systems [1,23,38]. It combines model-based process analysis with data-driven analysis techniques. It enables automatic extraction of process models (process discovery), recognition of compliance to protocol (conformance checking), performance analysis, and many other applications. Process mining has been successfully applied to the healthcare domain in several studies [23,31,37]. For example, Augusto et al. presented an automated method to simulate and evaluate CPs in a severe heart failure case study [3]. Fernandez-Llatas et al. demonstrated the use of process mining techniques to support the analysis of stroke cases in the emergency room [11].

In [5], Caron et al. presented the Clinical Pathway Analysis Method (CPAM). This method enables the acquisition of insights regarding CPs using event data and process mining techniques. It is a flexible method supporting a plethora of analyses such as workflow discovery, social network analysis, task allocation, and bottleneck analysis. In this paper, we adapt CPAM based on the needs and challenges of the healthcare professionals, and apply it to the sepsis CP of Hospital Samaritano, a large private Brazilian hospital located in São Paulo. The proposed methodology – entitled Simplified Clinical Pathway Analysis Method (SCPAM) – was intentionally kept simple to facilitate its adoption by healthcare organizations and helps answer the following set of questions: *How does the actual executed process look like? What are the deviations from protocol in the execution of the CP? What is the performance of the CP? What are the performance bottlenecks in the CP? Which factors affect the performance?*, and *How to optimize the CP?*

The remainder is organized as follows. We describe the analyzed sepsis CP in Sect. 2. In Sect. 3 we describe the analysis performed according to the SCPAM methodology and the obtained results. In Sect. 4 we discuss the validation of results by medical experts. In Sect. 5 we conclude and discuss future perspectives.

2   Adult Emergency Department Sepsis Clinical Pathway

Hospital Samaritano de São Paulo is a large private hospital in Brazil with over three hundred beds. It stands out for excellence in healthcare [18,29]. One of the

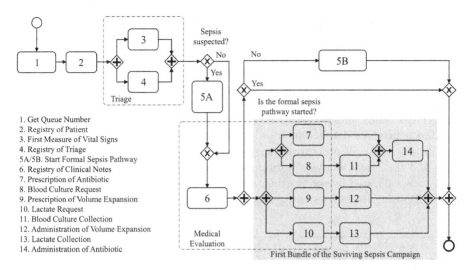

1. Get Queue Number
2. Registry of Patient
3. First Measure of Vital Signs
4. Registry of Triage
5A/5B. Start Formal Sepsis Pathway
6. Registry of Clinical Notes
7. Prescription of Antibiotic
8. Blood Culture Request
9. Prescription of Volume Expansion
10. Lactate Request
11. Blood Culture Collection
12. Administration of Volume Expansion
13. Lactate Collection
14. Administration of Antibiotic

Fig. 1. Prescriptive sepsis treatment process model of the emergency department, modeled using Business Process Model and Notation (BPMN).

main CPs of the hospital is the sepsis pathway. Figure 1 depicts the sepsis CP as it should be executed in the adult emergency department. The process was mapped and validated by interviewing the hospital staff during two visits.

The process starts when a patient arrives in the emergency department and receives a queue number (activity 1). The patient waits until a receptionist calls them. The receptionist registers the patient's admission in the hospital information system (2). The patient then waits at the reception until a nurse calls them to start the triage. During the triage, the nurse measures the patient's vital signs (temperature, blood pressure, heart rate and respiratory rate) (3), asks specific questions to the patient, prioritizes the patient according to their severity level, and registers all the triage information in the Electronic Health Record (EHR) (4). If the nurse suspects that the patient has sepsis, they must start the formal sepsis pathway (5A). The nurse must fill out a specific sepsis CP paper-based form, must register a clinical note in the EHR that the sepsis CP was started and must immediately communicate with the first available physician that the patient is sepsis suspicious. The physician evaluates the patient (anamnesis and physical examination) and records all information in the clinical notes (6). If the physician defines a sepsis diagnostic hypothesis, then antibiotics and volume expansion must be prescribed (7 and 9) and blood culture and lactate exams must be requested (8 and 10). If the sepsis CP was not formally started previously, then the physician must do so (5B). Finally, the patient goes to the medication room and a nurse technician collects their blood for the blood culture and lactate exams (11 and 13) and administers the antibiotics (14) and the volume expansion (12). The designed CP represents all sepsis treatment steps until the administration of antibiotics as described in the first bundle of the Surviving Sepsis Campaign [33].

3   Clinical Pathway Analysis and Optimization

In order to support the management and evaluation of the sepsis CP, we developed and applied the SCPAM methodology (see Fig. 2), which is a simplification of CPAM and aims to facilitate adoption by healthcare organizations. The main differences between the two methodologies are that SCPAM predefines a set of research questions, process perspectives, PM analyses and their execution order and that it presents detailed instructions on how to perform each step and analysis. In this work, all process mining analyses were conducted using ProM and Disco and statistical analyses were performed using R[1]. The results of this case study were validated by six healthcare professionals who all actively work in the sepsis CP, using interviews and structured questionnaires.

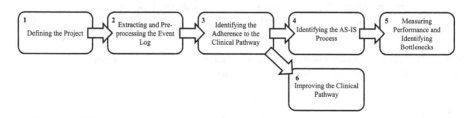

Fig. 2. Overview of the SCPAM methodology followed to analyze the sepsis CP.

3.1   Defining the Project

In the first step we define the scope of the project (i.e. the CP). A prescriptive CP model is created, mapping the selection and order of activities to be executed for the treatment of a given disease. This model guides the data extraction and is used to identify CP compliance and to create the AS-IS model.

In this study, we focus on the sepsis CP described in Sect. 2. We are interested in the analysis of the compliance and performance of the process. For example, the administration of antibiotics must happen after the collection of the blood culture and, as prescribed by the hospital, this activity must happen within one hour after the sepsis presentation. The faster the patient receives the antibiotics, the greater their chance of survival [20].

3.2   Extracting and Pre-processing the Event Log

The aim of the second step is to extract all necessary data from support systems and prepare the event log for analysis. This step is composed of the following activities. First, for each activity in the prescriptive CP model, the attributes that are relevant for the analysis are identified (e.g. date and time of execution of the activity, the age of patient). Next, the selected attributes are mapped to

---

[1] See https://promtools.org, https://fluxicon.com, and https://r-project.org.

the available systems databases (e.g. hospital information system or EHR) and the relevant data is extracted. For most process mining analyses, the minimal set of attributes per recorded event consists of the case identifier (e.g. the admission or patient), the executed activity (e.g. triage, administration of medicine), and the order in which the activities were executed (date and time). The event log is converted to the XES format supported by process mining tools [2,36]. Tools such as Disco and ProM provide this conversion. Ideally, the event log should be as simple as possible. Duplicated or otherwise unnecessary activities for the intended analysis should be removed.

We extracted two years of hospitalizations of patients either diagnosed with sepsis or who died of sepsis, identified through the list of International Statistical Classification of Diseases and Related Health Problems tenth revision (ICD-10) sepsis codes provided by the Canadian Institute for Health Information [9]. Cases in which the sepsis prescription template was used were added as well. All selected patients are over eighteen years of age, have two or more Systemic Inflammatory Response Syndrome (SIRS) signals, and are identified as sepsis suspicious by a physician. We enriched the event log with the following attributes: patient age, divided into 8 groups ([18–26], [27–35], [36–44], [45–53], [54–62], [63–71], [72–80] and [≥81]), patient gender, patient priority (following the Manchester triage system), type of infection (either community or hospital-acquired), presence of SIRS signal for each of the six clinical parameters (temperature, chills and shivering, respiratory rate, heart rate, leukocytes, and rod cells percentage), and total number of SIRS signals present. This information was translated to case attributes. The extracted data was converted to an event log that contains all complete sepsis cases that started and ended in the emergency department, as well as cases that started in the emergency department and then were transferred to the wards. Regarding the event log creation, our main challenges were: there was no registry of volume expansion in the EHR (activities 9 and 12) - we created these events taking in consideration the amount of saline solution prescribed to each patient; event times were derived from the time that the action was entered in the system rather than the time that the action was actually performed (the latter one was rarely available in the EHR) - in our visits to the hospital we observed that the staff registers the information in a very close time of executing the action; registry of the collection of blood and the administration of antibiotics were in most cases performed in the same time - we added one minute to all events of activity 14. The resulting event log contains 1,710 cases, 20,605 events, 14 activities (those described in Fig. 1) and 292 unique trace variants (activity orderings). The granularity of time is in minutes. The extraction phase is further detailed in previous work [28].

3.3 Identifying the Adherence to the Clinical Pathway

The aim of the third step is to identify how well the process was actually executed, i.e. the identification of potential deviations in the execution with regards to the prescribed CP. The deviations can be identified using conformance checking tools (e.g. the "Replay a Log on Petri Net for Conformance Analysis" plugin in ProM), by discovering the AS-IS model and comparing it manually to the prescribed CP, or by looking for events performed by professionals with different roles than expected (e.g. a receptionist registering the triage instead of a nurse). Conformance checking results are typically expressed using measures such as fitness and precision. Trace fitness for example is represented by a number between zero and one that indicates how well a case can be replayed by the model. The higher the fitness, the more compliant is the executed process.

Confronting the event log with the sepsis CP model defined in Sect. 2, we identified an average trace fitness of 0.85, indicating that the process as performed in the hospital is reasonably close to the one defined in its CP. In our study, deviations were associated to sub-sequences executed in a different order than expected, activities not executed, activities executed by different roles, and the non-compliance to the target of one hour to give the antibiotics. As such, a single case may exhibit multiple deviations. We identified 43 different types of deviations in the execution of the CP, constituting 5,184 deviation instances. The non-prescription and non-administration of volume expansion were among the most frequent type of deviations (89% of cases), followed by the start of the formal CP during triage (45%).

3.4   Identifying the AS-IS Process

The aim of the fourth step is to identify the process as it is actually executed. There are several process discovery algorithms that can be used to automatically discover a process model from the event log. However, due to the complexity of healthcare processes and the variation typically found in the data extracted from hospital information systems, these algorithms typically generate models that are not easily understandable by clinicians. Furthermore, such discovered models tend to differ substantially from the prescriptive CPs, complicating further analysis. As such, we recommend manually and iteratively updating the prescribed CP by incorporating the most frequent deviations identified in the previous step and calculating the trace fitness with respect to the updated model. The incorporation of deviations into the CP should be done in such a manner to bring the trace fitness as close to one as possible, without sacrificing other quality metrics, most notably simplicity. When only few cases deviate substantially from the process model, it is better to sacrifice fitness in favor of a more legible process model with less activities or infrequent paths.

To obtain a process model describing how the sepsis process was really executed, we manually adjusted the CP model by incorporating the most frequent types of deviations and observed behaviour from the event log. The adjusted model has an improved average trace fitness of 0.97, and can be seen in Fig. 3. The differences between the prescribed CP and the AS-IS model are as follows. The first measure of vital signs (activity 3) usually happens before the registration of triage (4), rather than in parallel with it. Furthermore, the formal start of the sepsis pathway (5A) usually happens during triage. Lastly, the prescription and administration of volume expansion (9 and 12) are not always executed.

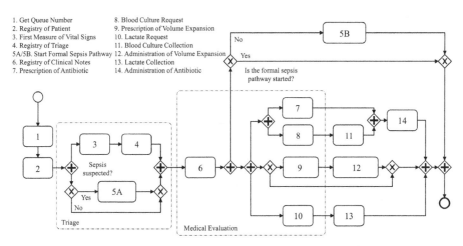

1. Get Queue Number
2. Registry of Patient
3. First Measure of Vital Signs
4. Registry of Triage
5A/5B. Start Formal Sepsis Pathway
6. Registry of Clinical Notes
7. Prescription of Antibiotic
8. Blood Culture Request
9. Prescription of Volume Expansion
10. Lactate Request
11. Blood Culture Collection
12. Administration of Volume Expansion
13. Lactate Collection
14. Administration of Antibiotic

Fig. 3. AS-IS sepsis treatment process model of the emergency department, modeled using Business Process Model and Notation (BPMN).

3.5 Measuring Performance and Identifying Bottlenecks

In the fifth step, we measure the performance of the process and identify which factors contribute most to any identified bottlenecks. The step is composed of the following activities. First, the relevant performance measures and metrics are defined. Again, several process mining techniques exist that measure the performance using an event log and a process model (e.g. the "Replay a Log on Petri Net for Performance/Conformance Analysis" plugin in ProM, and the lifecycle-based technique presented in [14], which allows to define custom performance metrics). Next, each metric should be carefully analyzed by answering questions such as *Does this waiting time make sense for this specific activity?*, *Is the duration for this selected set of patients significantly different?*. A technique to interactively analyze healthcare process performance using visual analytics is presented and applied in [10,17]. High waiting times often indicate potential bottlenecks. For example, while an average waiting time of three hours for the administration of medicines is not out of the ordinary, one hour is a long waiting time for a medical encounter with a physician in the emergency department, and as such might indicate a bottleneck. Contextual factors that affect performance should be identified as well. Process performance is generally affected by a plethora of factors, such as the amount of scheduled staff, the number of patients under care, or the clinical parameters of a patient [15]. The identification of these factors can support clinicians in tailoring their CPs. The ProM plugin "Find context-based differences in process performance" supports this type of analysis [15,17]. It tries to automatically discover statistically significant differences in performance by looking at the context. Similarly, causal relations between factors can be identified [16]. We further recommend removing outlying cases as they may affect performance statistics.

To analyse the performance of the sepsis process, we looked at the duration of activities, the time between activities, and the total case duration. We removed cases with outlier performance using Tukey's method [35] and did not consider those pairs of activities with a case frequency less than 5% (for example, activity "13. Lactate Collection" was only followed by "7. Prescription of Antibiotic" in 0.8% of cases). We identified two potential bottlenecks. Firstly, patients waiting in the reception before triage wait an average of 18 min between getting a queue number (activity 1) and getting their first measure of vital signs (3). Next, between the registry of clinical notes (6) and treatment prescription (7, 8, and 10), physicians spent an average of 5 min. These two bottlenecks may seem to represent a small amount of time. However, compared to the target time to administrate antibiotics – which is one hour – they represent a significant amount of time (30% and 8% respectively).

In order to identify contextual factors correlated with performance, we divided the process in six phases: waiting in reception, triage, waiting for medical evaluation, medical evaluation, treatment, and observation. Again, Tukey's method was used to remove outliers for each phase. We discovered that the triage phase takes significantly more time for cases that started during a period of medium or high demand compared to when the emergency department was not busy. Female patients and patients in the age groups [72–80] and [≥81] spent more time in the observation phase compared to male and younger patients, respectively. Patients that showed four SIRS signals and patients with a SIRS signal relating to respiratory rate or that had an infection acquired in the hospital also exhibited a significantly longer observation time.

3.6   Improving the Clinical Pathway

The aim of this step is to update the prescriptive CP to optimize its expected performance and outcome in terms of mortality, length of stay, costs, etc. Using the deviations and the performance results identified in the previous steps, those deviations that can improve the CP are identified. Per deviation, we apply the appropriate statistical tools. In case the identified deviation correlates with improved performance or favorable outcomes, the healthcare provider may update the prescribed CP, incorporating those positive deviations. Alternative available approaches can be applied to improve the CP, such as the Multi-Criteria Analysis Technique [26] that identifies activities and sub-sequences that contribute in a positive or negative way to the outcomes. This technique can be very useful when multiple simultaneous criteria need to be considered in the optimization (e.g. minimize the length of stay and mortality rate).

In our study, the analysis of the impact on performance outcomes of the identified deviations indicated that sepsis cases in which the registration of clinical notes occurs after the prescription of medicine and request of exam, spent on average 3.5 min less time waiting for the administration of antibiotics, with a median of 3 min[2].

---

[2] Two-tailed Mann-Whitney U test: $U = 261,472.5$; $n1 = 366$; $n2 = 1,199$; $p < 0.001$.

4   Validating Results with Hospital Staff

We performed the validation of the results with 3 physicians, 2 nurses, and 1 quality analyst who all actively work in the sepsis CP. The validation happened in 2017 by means of a group interview and structured questionnaires. We validated the AS-IS process, four deviations, two bottlenecks, five of the context factors correlated to performance, and the deviation that could potentially optimize the execution of the CP. The panel considered three deviations as real, actual deviations (registry of clinical notes after prescription of treatment, cases without prescription of volume expansion, and antibiotic not being administrated until one hour after identification). The panel considered blood culture being requested by a nurse technician as invalid since in practice, this should never happen. The panel validated that the AS-IS model in Fig. 3 indeed represents the process as it is actually executed in the emergency department. Concerning the bottlenecks, the panel considered both as real. Furthermore, they did not provide any additional bottleneck from their own experience. Regarding contextual factors, the expert panel agreed with three results, was neutral regarding the finding that the case duration is higher for patients with a heart rate $<90\,\mathrm{BPM}$, and disagreed with the finding that female patients tend to stay in observation longer. The panel agreed that prescribing the treatment before registering the clinical notes clearly reduces the administration target time since the delivery of medication process starts early. However, in general, they believe that the CP should not be updated since this deviation is associated with severe patients. All professionals considered it important to have access to the outcomes presented in this research as they help hospital staff identify problems in the execution of the sepsis CP. For them, this is a key feature in managing their CP. Regarding the bottleneck of patients waiting in the reception, the panel suggested a simplification in the registration of triage information. In addition, the process will be updated such that the triage is performed before the registration of the patient. This will help the hospital to quickly identify sepsis suspicious patients. With respect to the bottleneck of the prescription of the treatment, the panel suggested a simplification in the registration of the prescription.

5   Conclusions and Further Considerations

The employment of Clinical Pathways (CP) is a viable direction towards improving outcomes of medical processes. Yet, the management and evaluation of the execution of CPs presents multiple challenges. In this work, we adapted a data-based method employing process mining techniques in order to support the management and optimization of CPs and applied it to a sepsis process of a large Brazilian hospital. We consider the results of this study to be quite promising. Firstly, a simplified methodology can aid healthcare providers in the management and optimization of their CPs as was confirmed by structured interviews and questionnaires with a panel of medical experts. Secondly, we were able to successfully apply the methodology to real-life data from a sepsis process and discover deviations from protocol, performance bottlenecks and contextual factors that influence performance, leading to improvement suggestions.

As future steps, we suggest the application of SCPAM method in different healthcare facilities and for different diseases to verify its generalization and scalability. We believe SCPAM is particularly promising for the evaluation of CPs in the ED such as stroke and acute chest pain. SCPAM could be used by communities (e.g. the Latin American Sepsis Institute) to create an overview of the treatment of the diseases as executed in different healthcare facilities, supporting the update, management and deployment of clinical guidelines and clinical pathways. A prerequisite for the replication of this research study in different contexts is that the registration of healthcare information (e.g. triage, medical evaluation, prescription) is done using integrated systems, allowing the creation and extraction of the event data.

Acknowledgments. The authors thank A. Medeiros, D. Brizida Dreux, M. Santos, R. da Silva Santos, S. Barbosa, W.M.P. van der Aalst and all professionals involved from Hospital Samaritano and Philips for their help in the development of this research. This study was partly financed by the Coordenação de Aperfeiçoamento de Pessoal de Nível Superior - Brasil (CAPES) - Finance Code 001, Conselho Nacional de Desenvolvimento Científico e Tecnológico - CNPq [grant numbers 140511/2018-0, 306802/2015-5, 403863/2016-3], and Philips Research.

References

1. van der Aalst, W.M.P.: Process Mining - Data Science in Action, 2nd edn. Springer, Heidelberg (2016). https://doi.org/10.1007/978-3-662-49851-4
2. Acampora, G., Vitiello, A., Di Stefano, B., van der Aalst, W.M.P., Günther, C.W., Verbeek, E.: IEEE 1849TM: The XES Standard. IEEE Comput. Intell. Mag., 4–8 (2017)
3. Augusto, V., Xie, X., Prodel, M., Jouaneton, B., Lamarsalle, L.: Evaluation of discovered clinical pathways using process mining and joint agent-based discrete-event simulation. In: Winter Simulation Conference, WSC 2016, Washington, DC, USA, pp. 2135–2146. IEEE (2016)
4. Baker, K., et al.: Process mining routinely collected electronic health records to define real-life clinical pathways during chemotherapy. I. J. Med. Inform. **103**, 32–41 (2017)
5. Caron, F., Vanthienen, J., Vanhaecht, K., Van Limbergen, E., Deweerdt, J., Baesens, B.: A process mining-based investigation of adverse events in care processes. Health Inf. Manag. J. **43**(1), 16–25 (2014)
6. Carrillo-Esper, R., Carrillo-Cordova, J.R., Carrillo-Cordova, L.D.: Epidemiological study of sepsis in Mexican intensive care units. Cir. Cir. **77**(4), 301–308 (2009)
7. Chalupka, A.N., Talmor, D.: The economics of sepsis. Criti. Care Clin. **28**(1), 57–76 (2012)
8. Chawla, A., Westrich, K., Matter, S., Kaltenboeck, A., Dubois, R.: Care pathways in US healthcare settings: current successes and limitations, and future challenges. Am. J. Manag. Care **22**(1), 53–62 (2016)

9. CIHI: Canadian Institute for Health Information: In focus: A national look at sepsis (2009). https://secure.cihi.ca/estore/productFamily.htm?&pf=PFC1564
10. Dixit, P.M., Garcia Caballero, H.S., Corvò, A., Hompes, B.F.A., Buijs, J.C.A.M., van der Aalst, W.M.P.: Enabling interactive process analysis with process mining and visual analytics. In: BIOSTEC: HEALTHINF, Porto, Portugal, vol. 5, pp. 573–584. SciTePress, February 2017
11. Fernandez-Llatas, C., et al.: Analyzing medical emergency processes with process mining: the stroke case. In: Daniel, F., Sheng, Q.Z., Motahari, H. (eds.) BPM 2018. LNBIP, vol. 342, pp. 214–225. Springer, Cham (2019). https://doi.org/10.1007/978-3-030-11641-5_17
12. Finfer, S., Bellomo, R., Lipman, J., French, C., Dobb, G., Myburgh, J.: Adult-population incidence of severe sepsis in Australian and New Zealand intensive care units. Intensive Care Med. **30**(4), 589–596 (2004)
13. Fujino, Y., et al.: Impact of regional clinical pathways on the length of stay in hospital among stroke patients in Japan. Med. Care **52**(7), 634–640 (2014)
14. Hompes, B.F.A., van der Aalst, W.M.P.: Lifecycle-based process performance analysis. In: Panetto, H., Debruyne, C., Proper, H., Ardagna, C., Roman, D., Meersman, R. (eds.) OTM 2018 Conferences. LNCS, vol. 11229, pp. 336–353. Springer, Cham (2018). https://doi.org/10.1007/978-3-030-02610-3_19
15. Hompes, B.F.A., Buijs, J.C.A.M., van der Aalst, W.M.P.: A generic framework for context-aware process performance analysis. In: Debruyne, C., et al. (eds.) OTM 2016 Conferences. LNCS, vol. 10033, pp. 300–317. Springer, Cham (2016). https://doi.org/10.1007/978-3-319-48472-3_17
16. Hompes, B.F.A., Maaradji, A., La Rosa, M., Dumas, M., Buijs, J.C.A.M., van der Aalst, W.M.P.: Discovering causal factors explaining business process performance variation. In: Dubois, E., Pohl, K. (eds.) CAiSE 2017. LNCS, vol. 10253, pp. 177–192. Springer, Cham (2017). https://doi.org/10.1007/978-3-319-59536-8_12
17. Hompes, B., Dixit, P., Buijs, J.: Using process analytics to improve healthcare processes. Data Science for Healthcare: Methodologies and Applications, pp. 305–325. Springer, Cham (2019). https://doi.org/10.1007/978-3-030-05249-2_12
18. Hospital Samaritano, Com 125 anos de atividades, o Hospital Samaritano Higienópolis, destaca-se pela excelência e humanização no atendimento à saúde (2019). http://samaritano.com.br/institucional/institucional/
19. Khalifa, M., Alswailem, O.: Clinical pathways: identifying development, implementation and evaluation challenges. In: ICIMTH, pp. 131–134. IOS Press (2015)
20. Kumar, A., et al.: Duration of hypotension before initiation of effective antimicrobial therapy is the critical determinant of survival in human septic shock. Crit. Care Med. **34**(6), 1589–1596 (2006)
21. Liao, X., Du, B., Lu, M., Wu, M., Kang, Y.: Current epidemiology of sepsis in mainland China. Ann. Transl. Med. **4**(17), 324 (2016)
22. Machado, F.R., et al.: The epidemiology of sepsis in Brazilian intensive care units (the Sepsis PREvalence Assessment Database, SPREAD): an observational study. Lancet Infect. Dis. **17**(11), 1180–1189 (2017)
23. Mans, R.S., van der Aalst, W.M.P., Vanwersch, R.J.B.: Process Mining in Healthcare - Evaluating and Exploiting Operational Healthcare Processes. SBPM. Springer, Cham (2015). https://doi.org/10.1007/978-3-319-16071-9
24. Palleschi, M.T., Sirianni, S., O'Connor, N., Dunn, D., Hasenau, S.M.: An interprofessional process to improve early identification and treatment for sepsis. J. Healthc. Qual. **36**(4), 23–31 (2014)
25. Panella, M., Marchisio, S., Di Stanislao, F.: Reducing clinical variations with clinical pathways: do pathways work? Int. J. Qual. Health Care **15**(6), 509–521 (2003)

26. Quintano Neira, R.A.: A multi-criteria process mining optimization tool and its application in a sepsis clinical pathway. Ph.D. thesis, Pontifícia Universidade Católica do Rio de Janeiro, Rio de Janeiro (2018)
27. Quintano Neira, R.A., Hamacher, S., Japiassú, A.M.: Epidemiology of sepsis in Brazil: incidence, lethality, costs, and other indicators for Brazilian Unified Health System hospitalizations from 2006 to 2015. PLoS ONE **13**(4), e0195873 (2018)
28. Quintano Neira, R.A., de Vries, G., Caffarel, J., Stretton, E.: Extraction of data from a hospital information system to perform process mining. Stud. Health Technol. Inform. **245**, 554–558 (2017)
29. Revista exame: Os hospitais brasileiros de excelência em 2014. https://exame.abril.com.br/seu-dinheiro/os-hospitais-brasileiros-de-excelencia-em-2014
30. Rodríguez, F., et al.: The epidemiology of sepsis in Colombia: a prospective multi-center cohort study in ten university hospitals. Crit. Care Med. **39**(7), 1675–1682 (2011)
31. Rojas, E., Munoz-Gama, J., Sepúlveda, M., Capurro, D.: Process mining in healthcare: a literature review. J. Biomed. Inform. **61**, 224–236 (2016)
32. Singer, M., et al.: The third international consensus definitions for sepsis and septic shock (sepsis-3). JAMA **315**(8), 801–810 (2016)
33. Society of Critical Care Medicine: Surviving sepsis campaign. http://www.survivingsepsis.org
34. The EPISEPSIS Study Group: EPISEPSIS: a reappraisal of the epidemiology and outcome of severe sepsis in French intensive care units. Intensive Care Med. **30**(4), 580–588 (2004)
35. Tukey, J.: Exploratory Data Analysis. Addison-Wesley, Reading (1977)
36. XES Working Group: IEEE Standard for eXtensible Event Stream (XES) for achieving interoperability in event logs and event streams. IEEE Std 1849-2016, pp. 1–50 (2016)
37. Yang, W., Su, Q.: Process mining for clinical pathway: literature review and future directions. In: 2014 11th International Conference on Service Systems and Service Management (ICSSSM), pp. 1–5 (2014)
38. Yoo, S., et al.: Assessment of hospital processes using a process mining technique: outpatient process analysis at a tertiary hospital. I. J. Med. Inform. **88**, 34–43 (2016)
39. Zhang, Y., Padman, R., Patel, N.: Paving the COWpath: learning and visualizing clinical pathways from electronic health record data. J. Biomed. Inform. **58**, 186–197 (2015)

Understanding Undesired Procedural Behavior in Surgical Training: The Instructor Perspective

Victor Galvez[1($\boxtimes$)], Cesar Meneses[1], Gonzalo Fagalde[1,3], Jorge Munoz-Gama[1],
Marcos Sepúlveda[1], Ricardo Fuentes[2], and Rene de la Fuente[2]

[1] Department of Computer Science, School of Engineering,
Pontificia Universidad Católica de Chile, Santiago, Chile
{vagalvez,cnmeneses,gfagalde,jmun}@uc.cl, marcos@ing.puc.cl
[2] Department of Anesthesiology, School of Medicine,
Pontificia Universidad Católica de Chile, Santiago, Chile
{rfuente,rdelafue}@med.puc.cl
[3] Department of Engineering Sciences, Universidad Andres Bello, Santiago, Chile
g.fagaldeseguel@uandresbello.edu

Abstract. In recent years, a new approach to incorporate the process perspective in the surgical procedural training through Process Mining has been proposed. In this approach, training executions are recorded, to later generate end-to-end process models for the students, describing their execution. Although those end-to-end models are useful for the students, they do not fully capture the needs of the instructors of the training programs. This article proposes a taxonomy of activities for surgical process models, analyzes the specific questions instructors have about the student execution and their undesired procedural behavior, and proposes the *Procedural Behavior Instrument*, an instrument to answer them in an easy-to-interpret way. A real case was used to test the approach, and a preliminary validity was developed by a medical expert.

Keywords: Undesired procedural behavior · Process perspective ·
Healthcare · Medical training · Surgical procedures · Surgical process models

1 Introduction

In medical education, a common task is to teach surgical procedures. To analyze how students perform it, videos of their executions are recorded. Then, the instructor watches them to evaluate their students and provide feedback to them.

Recently, Process Mining has been used to incorporate the process perspective into the medical training of surgical procedures [9]. Process Mining is a novel technique that allows to discover the different variants in which a procedure is executed, compare the executions versus a reference model of the procedure, and also identify opportunities to redesign the procedure [1]. In order to do that, the

© Springer Nature Switzerland AG 2019
C. Di Francescomarino et al. (Eds.): BPM 2019 Workshops, LNBIP 362, pp. 471–482, 2019.
https://doi.org/10.1007/978-3-030-37453-2_38

aforementioned videos are tagged by a human observer, who then uses a software to transform the tagged videos into an event log [9]. Process Mining allows a procedural analysis on the execution performed by each student, so as to provide feedback about their performance. The end-to-end process models allow to provide feedback to them so as they focus their efforts to avoid the same mistakes in the future [9]. Although these end-to-end models are useful for the students, they do not fully capture the needs of the instructors of the training programs, because they do not provide global information about the whole class. It also does not allow instructors to observe stages of the procedure that need more supervision during training, or to easily visualize undesired execution patterns.

To address this problem, in this article we propose a two-part approach. The first part is a taxonomy of activities that allows categorizing the activities of a process model of the procedure considering its semantics, and a set of relevant procedural questions based on such taxonomy. The second part is the design of the *Procedural Behavior Instrument*, an instrument that shows undesired patterns in a twofold way, i.e., for a specific student and for the whole class.

The structure of the article is as follows. First, the real case used as a running example to show the proposed approach is explained. Then, the taxonomy of activities for surgical procedures is proposed. Afterward, the questions of interest for instructors are presented. Later, components of the *Procedural Behavior Instrument* are shown, how to compute them and how they are represented. Then, results obtained from the preliminary validation carried out by an instructor (medical expert) are presented. Finally, the main conclusions of this article are highlighted.

2   The Running Case: Central Venous Catheter

To illustrate the proposed approach, the installation of a Central Venous Catheter (CVC), hereinafter the *CVC case*, is used. This procedure has an initial stage that is the preparation of implements and the patient. Then, a trocar (instrument used to access or drain body fluids) is installed in the vein with help of ultrasound. Next, a guidewire is passed through the trocar, and after checking that it was installed correctly, the trocar is removed. After, the catheter is installed with help of the guidewire. Finally, the guidewire is removed and the catheter is installed [10].

A BPMN model representing the procedure was developed establishing the consensus of physicians using the Delphi method [5]. The model can be seen in the following section (Fig. 1). Twenty medical residents of the specialty of anesthesiology at the Pontificia Universidad Católica de Chile participated in a training program in which process-oriented feedback was provided to the students in order to enhance their learning [2,9].

3   Surgical Procedure Activities and Undesired Behavior

The first part of the approach proposes a taxonomy of surgical procedure activities, because it has been seen that enriching the formalization of surgical process models benefit the teaching of these procedures [8]. Then, based on this taxonomy, the approach defines three questions of interest for instructors related to procedural undesired behavior.

3.1   Surgical Procedure Activities

If we talk about processes in general and abstractly, a process is composed of a series of steps (activities) that must be performed in a certain order, e.g., first execute a, then b or c, and finally d. This definition also applies to surgical procedures, but with one difference: surgical procedures are accompanied by a semantic and a domain where they are applied, which allows to establish differentiating characteristics among the activities. For example, in the *CVC case*, to puncture the patient to place the trocar is not the same as to check that the trocar is correctly placed. The first is a step with an active role in the process and has effects on the patient's body, while the second has a controlling role on the correct performance of the first step.

Our approach proposes the distinction of activities in four groups:

Action *Activities that progress the status of the surgical procedure.*
These activities are associated with invasive activities for the patient or that may cause some level of pain, such as punctures, inserting or installing different elements, among others. In the *CVC case*, "Puncture", "Guidewire install" and "Advance catheter" are examples of action activities.

Identification *Activities that identify an element before an action activity.*
These activities correspond to the detection of some element that will be intervened in an action activity, such as the identification of a vein or some organ. In the *CVC case*, "Anatomic identification", "Doppler identification" and "Compression identification" are used to identify the vein.

Control *Activities that control the progress after performing an action activity.*
Surgical procedures can include the realization of a control activity after an action activity (or a sequence of action activities) to verify that it was executed correctly (for example, that something has been installed correctly). In the *CVC case*, "Blood return" controls that it has been punctured the vein in the right place.

Preparation *Activities that define the steps to prepare the execution of the surgical procedure.*
These activities are carried out before the execution. They are any activity prior to the execution of the procedure, such as preparing the implements, tools, supplies and the patient for the procedure. For example, in the *CVC case*, "Hand washing", "Ultrasound configuration" and "Put sterile gel" are examples of preparation activities.

Our approach defines the four groups of activities as disjoint, i.e., an activity cannot belong to two groups at once. This is sufficient for most processes, but other more complex processes may require expert domain knowledge. In turn, an activity difficult to classify can be transformed into two or more new activities, which can be categorized in one of the four proposed groups.

Figure 1 shows the application of this approach to the *CVC case*. The BPMN model [5] has been enriched according to the defined taxonomy.

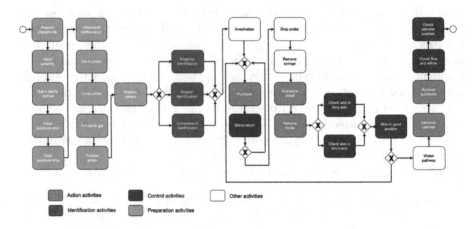

Fig. 1. BPMN model with the taxonomy of activities.

3.2 Undesired Behavior

We define undesired behavior as any pattern that indicates little fluidity through-out the execution of the procedure, e.g., perform again an activity that is specific to the beginning of the procedure, repeat a part of the process, or do not check an action activity. Detection of undesired behavior is relevant in the instruction of the procedure, so as to correct errors promptly. Therefore, instructors need to identify them quickly and easily.

Based on expert knowledge in the teaching of surgical procedures, our approach proposes the following three questions of interest to discover undesired behavior [11]:

Q1 *What patterns show an undesired sequence of action activities during the execution of the procedure performed by a student?*
  The answer to this question allows to discover in which action activity students usually go backward in the procedure, if they skip any of them, and if they repeat them many times.

Q2 *What patterns show an undesired use of identification and control activities during the execution of the procedure performed by a student?*
  This allows the instructor to know if students perform the necessary checks after an action activity, and where in the process they perform control

and identification activities. It also allows to know if a student execute the procedure with confidence.

Q3 *What patterns show an undesired use of preparation activities during the execution of the procedure performed by a student?*
Preparation activities are carried out before the procedure begins. Their execution during the procedure is an undesired pattern, since it shows that the student did not plan the procedure in advance.

There are other types of undesired behavior when performing the steps of the procedure, e.g., the amount of anesthesia injected is not correct, or releasing the guidewire while installing. However, they are not related to the execution of the procedure from a process perspective. These behaviors can be analyzed with other instruments, such as checklists or global scales [6, 7].

4   Procedural Behavior Instrument

The second part of the approach considers the design of an instrument that allows answering the three questions of the previous section in an easy and direct way for an instructor, which is not necessarily expert in process analysis. The instrument, called *Procedural Behavior Instrument*, is composed of 4 components: rework, checking, verification and preparation. Each component is twofold, i.e., it provides a view at the student level and a view at the level of the whole class. The rework and preparation components have been designed to answer questions Q1 and Q3, respectively, while Q2 is answered with the checking and verification components.

This article defines, for each component, how to compute and represent the different procedural behaviors using the taxonomy of activities defined in the previous section. Calculation and representation is done for a particular student and for the whole class, so the instructor can observe the undesired patterns at both levels. To understand the components, lets define $\sigma_k = \langle e_1, \ldots, e_m \rangle$ as the sequence of steps (i.e., activities) executed by the student k, and e_i as the event performed in the position i of the sequence σ_k.

4.1   Rework Component

This component allows the instructor to know in which activity of the procedure students go backward, if they skip any activity, and if they repeat some part of the process many times. It shows the number of times a certain pair of action activities is executed.

For this component, the events of σ_k that are action activities are used, because they are the ones that determine the progress of the procedure execution.

In the case of student k, rework is the number of times each pair of action activities are followed directly in σ_k, i.e. $\langle e_i, e_{i+1} \rangle$. Figure 2 shows that student k executed twice the pair "Puncture" → "Guidewire install", and once the pair

"Remove guidewire" → "Puncture". This means that student k executed the action activities until the end, but then went backward and executed all the action activities again.

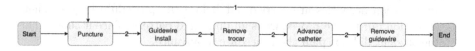

Fig. 2. Rework component for student k.

For the whole class view, the number of students performing each pair of action activities at least once is calculated. Figure 3 shows that three students consecutively repeat "Puncture", one student did not perform "Advance catheter", and three students after performing "Remove Guidewire" returned to "Puncture", so they probably had to execute all action activities again.

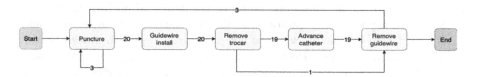

Fig. 3. Rework component for the whole class.

4.2   Checking Component

This component shows the number of identification and control activities performed during the procedure. This allows the instructor to determine if the identifications and controls established by the model are being performed.

For this component, σ_k events that are action, identification and preparation activities are used.

For student k, the number of identification and control activities that were performed before and after each action activity are calculated. Figure 4 shows student k performed three identification activities before "Puncture" and two control activities after "Puncture". The number to the left and to the right of each action activity indicate the number of identification and control activities that were performed, respectively. If there is no number means that, according to the model, it is not necessary to carry out those activities.

For the whole class view, the number of students who performed identification and control activities before and after each action activity at least once are calculated. Figure 5 shows that the twenty students of the class performed the identification of the vein before "Puncture", and that only eighteen of them controlled that the guidewire was installed correctly. The number to the left

Fig. 4. Checking component for student k.

and to the right of each action activity indicate the number of students who performed the activities. If there is no number means that, according to the model, these activities are not required.

Fig. 5. Checking component for the whole class.

4.3 Verification Component

This component shows whether the identification and control activities are being performed each time an action activity is performed. Compared to the checking component, it considers not how many identification/control activities are performed but whether they are performed or not each time the action activity is performed.

For the calculation of this component, events of σ_k that are action, identification and control activities are used.

To determine if student k performed the identification activities required to perform an action activity, the following should be considered:

– If it is one identification activity, it must be performed at least once just before the action activity.
– If it is more than one identification activity, they must be performed in the order defined in the model.

To determine if the student k controlled that he/she performed the control activities to verify if the action activity was executed correctly, the following should be considered. Let's suppose we have a set of control activities to control the correct execution of an specific action activity:

– If the set has only one control activity, it must be executed at least once just after the action activity.
– If there are two or more control activities in the set:
  • Control activities should be executed in the order defined by the process model of the procedure, just after the action activity.

- If the student executed at least one of the control activities of the set just after the action activity, and then execute a prior activity (according to the order defined by the process model of the procedure), we consider that student controlled the action activity. This is because if the student performed at least one of the control activities of the set and then executed a prior activity, we assume the performed control activity was enough for the student to realize that he/she was not performing the procedure correctly.

In the representation for student k, it is marked with 'X' if the student did not perform either an identification or a control activity each time the action activity was executed, and with '✓' if it was performed. At the center is the action activity; on the left, the percentage of times the student correctly performed identification activities; and on the right, the percentage of times the student correctly controlled the right execution of the action activity. If there is nothing, it means that, according to the model, it is not necessary to perform either the identification or the control activities.

Figure 6 shows the verification component for the student k. The student performed "Puncture" twice, the first time the identification was performed while in the second time it was not. It also shows that the two times "Puncture" was performed, the control activities were performed correctly. In the case of "Advance catheter", the single time the student placed the catheter, he/she did not control whether it was installed correctly.

Fig. 6. Verification component for student k.

In the representation for the whole class, at the center the action activity is shown; on the left, the percentage of cases in which the identification was performed every time the action activity was executed; and on the right, the percentage of cases in which the students controlled that they had performed correctly the action activity. If there is nothing, it means that, according to the model, it is not necessary to perform either the identification or the control activities.

Figure 7 shows the verification component for the whole class. 90% of the students performed the identification of the vein before puncturing, and 80% of the students controlled that they had punctured correctly.

Fig. 7. Verification component for the whole class.

4.4 Preparation Component

This component allows the instructor to know if students successfully completed the preparation of the procedure. An unsuccessful preparation would be detected if there are preparation activities performed during the execution of the procedure.

For this component, the σ_k events that are action and preparation activities are used.

For student k, the number of preparation activities performed before an action activity (excluding the first action activity, since prior to it all the preparation activities should have been completed) is calculated. Figure 8 shows that student k executed the procedure in an undesired way, since a preparation activity was performed before "Guidewire install"; and another one, before "Remove trocar".

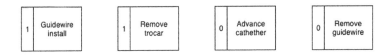

Fig. 8. Preparation component for student k.

For the whole class view, the number of students performing preparation activities at least once before each action activity (excluding the first one) is calculated. Figure 9 shows an undesired pattern in the executions of the training program, since four students performed preparation activities before "Guidewire install" and three students before "Remove trocar".

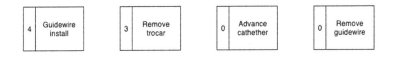

Fig. 9. Preparation component for the whole class.

5   Preliminary Evaluation

The simplest way to validate an artifact is by expert opinion, in order to evaluate if the artifact would perform correctly in the contexts imagined by the expert [12]. With this in mind, a preliminary assessment of the validity of the proposed approach was performed by one of the instructors (a medical expert) of the training program of the *CVC case*.

The instructor considered the proposed taxonomy of activities to be significant to understand the procedure, as well as to analyze it and refine it. In addition, the instructor considered that the visualization of the BPMN model (see Fig. 1) allows to comprehend the *Procedural Behavior Instrument* more easily. The instructor highlighted the use of different colors for each type of activity in the BPMN model.

The *Procedural Behavior Instrument* was preliminarily validated by the instructor through the adaptation of the questionnaire PUEU (Perceived Usefulness and Ease of Use) [3], widely used to evaluate the perception of usefulness and ease of use of IT systems.

Regarding to the perceived usefulness, the overall assessment of the instructor was positive. The instructor emphasized that the *Procedural Behavior Instrument* would allow greater effectiveness in the identification of undesired patterns in the executions carried out by students. Other positive aspects are: the *Procedural Behavior Instrument* allows to identify the undesired patterns more quickly and easily. Regarding the perception of ease of use, the instructor's overall opinion was also positive, highlighting the ease of interacting with the *Procedural Behavior Instrument*. However, the instructor considers that it is not easy to learn to interpret the components; in this sense, a prior introduction would be required before using it.

Additionally, the instructor was asked open questions regarding the *Procedural Behavior Instrument*, considering both positive aspects and aspects to improve.

Among the general positive aspects, the instructor highlighted that the *Procedural Behavior Instrument* allows a panoramic view of the execution of the procedure, allows to compare a particular student versus the whole class, and the distinction of different kind of activities in the execution of the procedure. Regarding the specific components, the instructor commented that they were all useful, highlighting the possibility of identifying when students go backward in the process (rework in Fig. 2), and to identify when checks were done more times than expected (see Fig. 6).

Some aspects to improve are design details typical of a prototype version, such as the incorporation of context information (e.g., the number of students in the class, how many students were able to complete the procedure, and providing the model of the process in advance). However, the overall evaluation is positive in terms of its potential to support the instructor's work, and its learning for future executions.

6 Conclusions and Future Work

As [4] mentions, Interactive Pattern Recognition can support physicians in their daily practice. This article shows the use of a process perspective to recognize patterns on students' performance in medical training, and afterwards it analyzes how instructors interact with the patterns discovered.

In this paper, we propose a taxonomy of activities for surgical procedures and the development of the *Procedural Behavior Instrument*, an instrument that allows the instructor of a training program (in a surgical training context) to know what are the undesired patterns performed by their students. The taxonomy contributes to the formalization of surgical process models, helps to establish a hierarchy of activities to evaluate students, and facilitates the analysis of surgical process models. In turn, the *Procedural Behavior Instrument* allows instructors to easily discover undesired patterns in the execution of each student's procedure and in the executions of the whole class. With the *Procedural Behavior Instrument*, the instructor can easily observe patterns related with action activities (Q1) using the rework component, patterns related with identification and control activities (Q2) using the checking and verification component, and finally, patterns related with preparation activities (Q3) using the preparation component.

The approach was applied to a real case, the *CVC case*, and was preliminarily validated by an instructor. A limitation of this approach is that the *Procedural Behavior Instrument* is a prototype, so there are details that need to be tuned. The preliminary validation was very important to know what things to improve of the *Procedural Behavior Instrument*.

In the future, we expect to generate a *Procedural Behavior Dashboard* designed for instructors, which is going to allow them to know the undesired behaviors in a more friendly way. In addition, we plan to test the proposed approach with a larger group of instructors.

Acknowledgments. This work is partially supported by CONICYT FONDECYT 11170092, CONICYT REDI 170136, VRI-UC INTERDISCIPLINA II170003, and FOND-DCC 2017-0001. CONICYT-PFCHA/Doctorado Nacional/2019-21190116.

References

1. van der Aalst, W.M.P.: Process Mining - Data Science in Action. Springer, Heidelberg (2016). https://doi.org/10.1007/978-3-662-49851-4
2. Corvetto, M.A., Pedemonte, J.C., Varas, D., Fuentes, C., Altermatt, F.R.: Simulation-based training program with deliberate practice for ultrasound-guided jugular central venous catheter placement. Acta Anaesthesiol. Scand. **61**, 1184–1191 (2017)
3. Davis, F.D.: Perceived usefulness, perceived ease of use, and user acceptance of information technology. MIS Quart. 319–340 (1989)
4. Fernández-Llatas, C., Meneu, T., Traver, V., Benedi, J.: Applying evidence-based medicine in telehealth: an interactive pattern recognition approximation. Med. J. Islamic Repub. Iran (MJIRI) **10**, 32–45 (2018)

5. de la Fuente, R.: Entrenamiento en Destrezas Procedurales: Oportunidades desde el análisis de Minería de Procesos. Master's thesis, School of Engineering, Pontificia Universidad Católica de Chile (2018)
6. Ilgen, J.S., Ma, I.W., Hatala, R., Cook, D.A.: A systematic review of validity evidence for checklists versus global rating scales in simulation-based assessment. Med. Educ. **49**(2), 161–173 (2015)
7. Khanghahi, M.E., Azar, F.E.F.: Direct observation of procedural skills (dops) evaluation method: Systematic review of evidence. Med. J. Islamic Repub. Iran (MJIRI) 32–45 (2018)
8. Lalys, F., Jannin, P.: Surgical process modelling: a review. Int. J. Comput. Assist. Radiol. Surg. **9**(3), 495–511 (2014)
9. Lira, R., et al.: Process-oriented feedback through process mining for surgical procedures in medical training: The ultrasound-guided central venous catheter placement case. Int. J. Environ. Res. Publ. Health **16**(11) (2019). https://doi.org/10.3390/ijerph16111877, https://www.mdpi.com/1660-4601/16/11/1877
10. Munoz-Gama, J., de la Fuente, R., Sepúlveda, M., Fuentes, R.: Conformance checking challenge 2019 (2019). https://doi.org/10.4121/uuid:c923af09-ce93-44c3-ace0-c5508cf103ad
11. Rojas, E., Sepúlveda, M., Munoz-Gama, J., Capurro, D., Traver, V., Fernandez-Llatas, C.: Question-driven methodology for analyzing emergency room processes using process mining. Appl. Sci. **7**(3), 302 (2017)
12. Wieringa, R.J.: Design Science Methodology for Information Systems and Software Engineering. Springer, Heidelberg (2014). https://doi.org/10.1007/978-3-662-43839-8

Towards Privacy-Preserving Process Mining in Healthcare

Anastasiia Pika[1(✉)], Moe T. Wynn[1], Stephanus Budiono[1],
Arthur H. M. ter Hofstede[1], Wil M. P. van der Aalst[2,1], and Hajo A. Reijers[3,1]

[1] Queensland University of Technology, Brisbane, Australia
{a.pika,m.wynn,sn.budiono,a.terhofstede}@qut.edu.au
[2] RWTH Aachen University, Aachen, Germany
wvdaalst@pads.rwth-aachen.de
[3] Utrecht University, Utrecht, Netherlands
h.a.reijers@uu.nl

Abstract. Process mining has been successfully applied in the healthcare domain and helped to uncover various insights for improving healthcare processes. While benefits of process mining are widely acknowledged, many people rightfully have concerns about irresponsible use of personal data. Healthcare information systems contain highly sensitive information and healthcare regulations often require protection of privacy of such data. The need to comply with strict privacy requirements may result in a decreased data utility for analysis. Although, until recently, data privacy issues did not get much attention in the process mining community, several privacy-preserving data transformation techniques have been proposed in the data mining community. Many similarities between data mining and process mining exist, but there are key differences that make privacy-preserving data mining techniques unsuitable to anonymise process data. In this article, we analyse data privacy and utility requirements for healthcare process data and assess the suitability of privacy-preserving data transformation methods to anonymise healthcare data. We also propose a framework for privacy-preserving process mining that can support healthcare process mining analyses.

Keywords: Process mining · Healthcare process data · Data privacy

1 Introduction

Technological advances in the fields of business intelligence and data science empower organisations to become "data-driven" by applying new techniques to analyse large amounts of data. Process mining is a specialised form of data-driven analytics where process data, collated from different IT systems typically available in organisations, are analysed to uncover the real behaviour and performance of business operations [1]. Process mining was successfully applied in the healthcare domain and helped to uncover insights for improving operational efficiency of healthcare processes and evidence-informed decision making [4,6,11,12,14]. A recent literature review [6] discovered 172 articles

© Springer Nature Switzerland AG 2019
C. Di Francescomarino et al. (Eds.): BPM 2019 Workshops, LNBIP 362, pp. 483–495, 2019.
https://doi.org/10.1007/978-3-030-37453-2_39

which report applications of various process mining techniques in the healthcare domain.

While the potential benefits of data analytics are widely acknowledged, many people have grave concerns about irresponsible use of their data. An increased concern of society with protecting the privacy of personal data is reflected in the growing number of privacy regulations that have been recently introduced or updated by governments around the world. Healthcare data can include highly sensitive attributes (e.g., patient health outcomes/diagnoses, the type of treatments being undertaken), and hence privacy of such data needs to be protected.

The need to consider data privacy in process mining and develop privacy-aware tools was raised at an early stage in the Process Mining Manifesto [3]. However, the process mining community has, until recently, largely overlooked the problem. A few recent articles highlight "a clear gap in the research on privacy in the field of process mining" [10] and make first attempts to address some privacy-related challenges [5,7,9,10,13] yet, significant challenges remain.

Privacy considerations are quite well-known in the field of data mining and a number of privacy-preserving data transformation techniques have been proposed [2,17] (e.g., data swapping, generalisation or noise addition). *Although there are many similarities between data mining and process mining, some key differences exist that make some of the well-known privacy-preserving data mining techniques unsuitable to transform process data.* For example, the addition of noise to a data set may have an unpredictable impact on the accuracy of all kinds of process mining analyses.

In this article, we present related work (Sect. 2), analyse data privacy and utility requirements for process data typically recorded in the healthcare domain (Sect. 3) and then assess the suitability of privacy-preserving data transformation methods proposed in the data mining and process mining fields to anonymise healthcare process data (Sect. 4). We show that the problem of privacy protection for healthcare data while preserving data utility for process mining analyses is challenging and we propose a privacy-preserving process mining framework as a possible solution to address this problem in Sect. 5. Section 6 concludes the paper.

2   Related Work

Privacy-Preserving Data Mining. Privacy, security, and access control considerations are quite well-known in the general field of data mining. A number of data transformation techniques, access control mechanisms and frameworks to preserve data privacy have been proposed [2,8,17]. In order to preserve data privacy, privacy-preserving methods usually reduce the representation accuracy of the data [2]. Such data modifications can affect the quality of analyses results. The effectiveness of the transformed data for analyses is often quantified explicitly as its *utility* and the goal of privacy-preserving methods is to "*maximize utility at a fixed level of privacy*" [2]. For example, privacy guarantees can be specified in terms of k-*anonymity*: each record in a data set is indistinguishable from at least k-1 other records.

Privacy-preserving data mining techniques can be generic or specific [17]. *Generic* approaches modify data in such a way that "the transformed data can be used as input to perform any data mining task" [17]. These approaches can provide anonymisation[1] by modifying records without introducing new values (e.g., data swapping) or they can modify original values (e.g., by adding noise). In *specific* approaches privacy preservation is embedded in specific data mining algorithms (e.g., privacy-preserving decision tree classification) [17]. Furthermore, *outputs* of some data mining algorithms can also be sensitive and methods that anonymise such outputs have been proposed (e.g., association rule hiding) [2]. Finally, *distributed* privacy-preserving methods are proposed for scenarios in which multiple data owners wish to derive insights from combined data without compromising privacy of their portions of the data [2]. Such methods often use cryptographic protocols for secure multi-party computations (SMC) [2].

Below, we describe traditional generic privacy-preserving data transformation approaches, such as data swapping, suppression, generalisation and noise addition [2]. *Data swapping* involves enacting privacy to a dataset by the existence of uncertainty. Uncertainty is introduced into individual records by swapping the true values of sensitive attributes between subsets of records [8]. *Suppression* anonymises data by omission. Values can be removed under three types of data suppression [2]. The most common type is column suppression which targets the presence of highly sensitive attributes whose values directly identify an individual (e.g., patient names). Alternatively, row suppression is used when outlier records are infrequent and difficult to anonymise. Value suppression omits selected sensitive attribute values. *Generalisation* methods define values approximately making it difficult for adversaries to identify records with full confidence [2]. The process of generalising usually includes the construction of a generalisation hierarchy, which is a predefined classification of values at decreasing levels of granularity. For numeric data, values are sorted into numerical ranges. For categorical data, a domain expert creates semantically meaningful generalisations using a tree structure. *Noise addition* can be used for both numerical and categorical data [17]. Numerical values are often anonymised by factoring randomly and independently generated "white noise" into the original data [2]. White noise is generated using a random distribution, often either uniform or Gaussian. Adding noise to categorical values is more complex, and can be achieved, for example, using clustering-based techniques [17].

Privacy-Preserving Process Mining. A few recent articles made first attempts to address some privacy-related process mining challenges [5,7,9,10,13,15,16]. Mannhardt et al. [10] analysed privacy challenges in human-centered industrial environments and provided some generic guidelines for privacy in process mining. Liu et al. [9] presented a privacy-preserving cross-organisation process discovery framework based on access control. Tillem et al. [15,16] presented interactive two-party protocols for discovery of process models from encrypted data, which are based on multiple communication rounds (and have high computation costs). The first privacy-preserving data transformation approach presented in the

---

[1] In this article, *anonymisation* refers to any method that can protect data privacy.

process mining community [5] proposes to use deterministic encryption methods for anonymisation of event log attribute values. (Such methods are also a part of the confidentiality framework proposed by Rafiei et al. [13]). Timestamps are treated as numeric values and are encrypted in a way that preserves the order of events. Deterministic encryption methods produce "the same ciphertext for a given plaintext" and preserve differences between values, which is important for process mining [13]. Encryption only provides weak data privacy protection and "could be prone to advanced de-anonymization techniques" [5]. More advanced privacy-preserving process mining approaches proposed by Rafiei et al. [13] and Fahrenkrog-Peterse et al. [7] will be discussed in detail in Sect. 4.

In this article, we focus on protecting privacy of process data in a healthcare organisation. Distributed privacy scenarios are not considered in this work.

3   Data Privacy and Utility Requirements: Healthcare

In order to realise our objective of privacy-preserving process mining for the healthcare domain, we first analyse privacy requirements for process data typically recorded in the healthcare domain, which is then followed by a discussion of data requirements of process mining approaches to analyse healthcare processes.

Healthcare Process Data. Process mining uses process data in the form of an event log, which represents collated and aggregated data from IT systems available in organisations. An event log contains events where each event refers to a case, an activity, a point in time, transaction type (e.g., *start* or *complete*) and (optionally) a resource and data attributes. An event log can be seen as a collection of cases and a case can be seen as a sequence of events.

Cases in healthcare processes typically refer to patients receiving treatments in a healthcare setting (e.g., a patient's pathway) and resources refer to medical personnel involved in the process. Figure 1 depicts an example event log which contains six events (represented by rows) related to two cases (*1* and *2*) where patient identifiers are already hidden. For example, we can see that case *1* refers to a patient whose age is *56*, who speaks English and was diagnosed with pancreatitis; activity *Register* is completed in this case; activity *Blood test* was started on *13/01/2019* at *17:01* by *Robert*; and treatment code *3456* is associated with activity *Triage* in case *1*. Data attributes can refer to cases (e.g., age, language and diagnosis) or to events (e.g., treatment codes are recorded for events associated with activity *Triage*). Data attributes used in this example are recorded in two publicly available healthcare logs. The healthcare MIMIC data set[2] contains information about language and diagnosis (as well as ethnicity, religion, marital status and insurance). The Dutch academic hospital event log[3] contains information about age, diagnosis and treatment codes.

Legislative Requirements. An increased concern of people with protecting the privacy of their data is reflected in the growing number of privacy regulations that have been recently introduced (e.g., the EU General Data Protection

---

[2] https://mimic.physionet.org/mimicdata/.

[3] https://data.4tu.nl/repository/uuid:d9769f3d-0ab0-4fb8-803b-0d1120ffcf54.

Case ID	Activity	Type	Time	Resource	Age	Language	Diagnosis	Treatment Code
1	Register	complete	12/01/2019 11:03	Ann	56	EN	Pancreatitis	-
1	Triage	start	12/01/2019 14:55	Michael	56	EN	Pancreatitis	3456
1	Blood test	start	13/01/2019 17:01	Robert	56	EN	Pancreatitis	-
2	Register	complete	14/01/2019 9:30	Ann	44	IT	Pneumonia	-
2	X-ray	complete	14/01/2019 11:00	Mary	44	IT	Pneumonia	-
2	Triage	start	14/01/2019 11:37	Michael	44	IT	Pneumonia	6543

Fig. 1. Example of an event log with typical healthcare data attributes.

Regulation (GDPR) 2018, the California Consumer Privacy Act of 2018) or updated by governments around the world (e.g., Australian Privacy Regulation 2013 under the Privacy Act 1988). In addition, data privacy requirements are often included in legislation governing specific sectors, e.g., Australian Healthcare Identifiers Act 2010.

Guidance for de-identification of protected health information in the US is provided in the Health Insurance Portability and Accountability Act (HIPAA) Privacy Rule. For example, the "safe harbor" de-identification method of the HIPPA Privacy Rule prescribes removal of all elements of dates (except year) related to an individual (e.g., admission or discharge dates)[4]. In Australia, the Office of Australian Information Commissioner provides guidelines for the use of health information for research. The guidelines prescribe de-identification of personal information by "removing personal identifiers, such as name, address, d.o.b. or other identifying information" and "removing or altering other information that may allow an individual to be identified, for example, because of a rare characteristic of the individual, or a combination of unique or remarkable characteristics"[5]. Furthermore, the recently introduced My Health Records Amendment (Strengthening Privacy) Bill 2018 allows Australians to opt out of having an electronic health record and allows the deletion of their records permanently at any time. Whilst providing strong privacy protections for Australians; for analysis purposes, they also introduce data quality issues such as missing and incomplete data; thus reducing the utility of data and the accuracy of results.

Privacy of public healthcare data is typically protected by replacing sensitive attribute values with anonymised values (e.g., treatment codes are used in a publicly available Dutch academic hospital event log and subject IDs are used in the healthcare MIMIC data set) or by removing sensitive attributes from data (e.g., employee information is removed from both Dutch hospital and MIMIC data sets). All timestamps in the MIMIC data set were shifted to protect privacy: dates are randomly distributed, but consistent for each patient. The former method only provides weak privacy protection while the latter methods can significantly decrease data utility.

---

[4] https://www.hhs.gov/hipaa/for-professionals/privacy/special-topics/de-identification/index.html\#protected.

[5] https://www.oaic.gov.au/engage-with-us/consultations/health-privacy-guidance/business-resource-collecting-using-and-disclosing-health-information-for-research.

Privacy Requirements for Healthcare Process Data. Healthcare process data can contain sensitive information such as patient or employee names or identifiers. Other attributes in the event log can also reveal patient or employee identities when combined with background knowledge about the process. For example, accident or admission time, a rare diagnosis or treatment, or a combination of age and language could potentially identify a patient. An employee could be identified by a combination of an activity name and execution time (e.g., when a blood test is always performed by the same employee during a shift). Hence, typical event log attributes such as *case ID, activity, time, resource* and many data attributes (e.g., a patient's personal and treatment information) can contribute to identity disclosure.

Furthermore, relations between events in a log can contribute to identity disclosure and this is especially pertinent for a healthcare event log due to the high variability of process paths typical for the sector [4]. Consider, for example, the Dutch hospital event log where 82% of cases follow unique process paths. Hence, someone with knowledge of the process could link these cases to individual patients. Moreover, cases which follow the same process path can include other atypical behaviors. In the Dutch hospital log, the fifth most frequent process variant is followed by 8 cases: 7 cases are related to only one organisational group ("Obstetrics and Gynecology clinic") and only one case is also related to the "Radiotherapy" group. Although the case does not follow a unique process path, the relation to the "Radiotherapy" group is unique and could be used by someone with knowledge of the process to identify the patient. Other examples of *atypical process behaviour* which could contribute to a patient's identity disclosure include abnormally short or long execution times of activities or cases, or an abnormally low or high number of resources involved in a case.

Data Requirements for Process Mining Approaches. All process mining algorithms require case IDs and activities to be recorded accurately in the log and most algorithms also require (accurate) timestamps. A recent literature review [6] discovered that the following types of process mining analyses were frequently used in healthcare: discovery techniques (which include process discovery as well as organisational mining approaches such as social network mining), conformance checking, process variant analysis and performance analysis.

- Process discovery techniques usually take as input a multi-set of traces (i.e., ordered sequences of activity labels) and do not require timestamps; however, timestamps are typically used to order events.
- Most academic process conformance and performance analysis techniques (e.g., alignment-based approaches) use formal models and require that complete traces are recorded in the log. Most commercial process mining tools (as well as some ProM plugins) convert the log to Directly Follows Graphs (DFG) annotated with frequencies and times, which show how frequently different activities follow each other and average times between them. DFG-based tools do not require complete traces and only require that "directly-follows" relations between activities are preserved in the log.

- Organisational mining techniques require resource information to be recorded in the log (in addition to case IDs, activities and timestamps). Moreover, resource and data attributes can also be required by conformance checking approaches that consider different process perspectives.
- Process variant analysis, which is concerned with comparing process behaviour and performance of different cohorts, often uses case data attributes to distinguish between cohorts.

In order to comply with strict privacy requirements for healthcare data, one would need to consider *anonymising (1) event log attribute values and (2) atypical process behaviour.* However, many process mining techniques require that healthcare process data is accurate and representative. That is: *(1) all events belong to a particular case*; *(2) attributes that represent case identifiers and activity labels are accurate; and (3) timestamps are reliable and accurate.* Thus, the need to balance the privacy requirements of healthcare data and the utility requirements of process mining techniques is paramount. In the following section, we assess whether existing privacy-preserving data transformation approaches can preserve the attribute values and relations between events discussed above.

4 Anonymising Healthcare Process Data

4.1 Anonymising Sensitive Attribute Values

As discussed in Sect. 3, typical event log attributes such as *case, activity, time, resource* and many data attributes could contribute to identity disclosure. Below, we discuss how these attributes could be anonymised using generic data transformation approaches described in Sect. 2. We evaluate the suitability of deterministic encryption (referred to here as encryption), which was used to anonymise event log data [5,13], and other traditional data transformation approaches proposed in the data mining community such as data swapping, value suppression, generalisation and noise addition (which, to the best of our knowledge, have not been applied to event logs). Figure 2 depicts how some of these techniques can be applied to the event log in Fig. 1.

Case identifiers can be encrypted (as well as other event log attributes); however, encryption does not provide strong data privacy protection (and may not be suitable to protect sensitive healthcare data). An underlying assumption of all process mining algorithms is that case identifiers are unique, which makes the application of value suppression and generalisation not suitable (these methods are used to hide infrequent attribute values). Adding noise to case identifiers can yield values that are no longer unique, which can decrease the accuracy of all process mining algorithms. Data swapping can be applied to case IDs without impact on process mining results.

Activity labels can be encrypted; however, encrypted labels can be identified by someone with knowledge of the process (e.g., most or least frequent activities [13]). Moreover, encryption makes it difficult to interpret analysis results. In addition, one must also encrypt process model labels when applying process

Case ID	Activity	Type	Time	Resource	Age	Language	Diagnosis	Treatment Code
2	Register	complete	12/01/2019 11:20	Team A	56	EN	-	-
2	Triage	start	12/01/2019 15:00	Team B	56	EN	-	3456
2	Blood test	start	13/01/2019 17:03	Team C	56	EN	-	-
1	Register	complete	14/01/2019 9:35	Team A	44	IT	-	-
1	X-ray	complete	14/01/2019 11:10	Team C	44	IT	-	-
1	Triage	start	14/01/2019 11:48	Team B	44	IT	-	6543

Fig. 2. Application of data transformation techniques to the event log in Fig. 1: Case ID: swapping; Time: noise addition; Resource: generalisation; Diagnosis: suppression.

mining algorithms that use process models as input (e.g., many process performance and conformance analysis approaches). Application of value suppression and generalisation to activity labels may affect the accuracy of process mining results where the utility loss depends on the process mining algorithm used. For example, removing infrequent activity labels may not have a significant effect on process discovery results (as process models often capture mainstream process behavior); however, process conformance analysis results may become invalid. One can use generalisation to hide some sensitive activities (e.g., replace activities "HIV test" and "Hepatitis C test" with activity "Blood test"). The result of process discovery performed on such logs will be correct; however, the discovered process model will be on a higher level of granularity. Noise addition and swapping activity labels will invalidate the results of all process mining algorithms. For example, if activity labels in a log are swapped, the resulting traces will consist of random activity sequences; hence, discovered process models will be incorrect, as well as other process mining results.

Timestamps can be treated as numerical values and encrypted using methods which preserve the order of events. Such encryption will not affect the results of process mining algorithms that work with ordered events and do not require timestamps (such as many process discovery algorithms). On the other hand, an event log with encrypted timestamps will not be suitable for performance analysis. Value suppression and generalisation can be used to anonymise sensitive timestamps (e.g., as discussed in Sect. 3, according to the HIPAA Privacy Rule admission and discharge times must be anonymised). This will affect the accuracy of most process mining algorithms. For example, if value suppression is applied to admission times, the discovered process model will not include activity "Admission". On the other hand, if generalisation is applied to admission times (by only leaving year as prescribed by the HIPAA Privacy Rule), process discovery may not be affected; however, process performance analysis results may become invalid (as time between admission and other activities in the process will no longer be correct). Adding noise to timestamps or swapping their values will yield incorrect process mining results (as the order of events in the transformed log is no longer preserved).

Resource information can be encrypted without impacting organisational mining results, while noise addition and swapping will invalidate such results (as resources will no longer be related to correct events and cases). One can apply generalisation to resource information (e.g., by replacing individual identifiers with team identifiers), which will yield the analysis on a team level. Value suppression can affect the accuracy of organisational mining techniques (e.g., a discovered social network may have fewer nodes).

Data attributes can be encrypted, though encryption of numerical values can make it difficult to conduct some analyses. For example, if *age* is encrypted, one can no longer compare process variants for different age cohorts. Value suppression can decrease the accuracy of process mining algorithms that use data (e.g., when infrequent age values are removed, the corresponding cases will not be included in process variant analysis). Using generalisation may decrease the accuracy of conformance analysis that consider data; however, it may not have any impact on variant analysis (e.g., when comparing different age groups). Noise addition and data swapping will nullify results of the methods that use data.

Table 1 summarises the suitability of different data transformation approaches to anonymising event log attribute values. Encryption has a minimal effect on data utility for most process mining algorithms; however, it may not provide a required level of privacy protection. Data swapping can be used to anonymise case IDs; however, application of this method to other event log attributes will invalidate process mining results. Noise addition will nullify all process mining results. Value suppression and generalisation are not suitable for case IDs (as they have unique values), these methods can be applied to other attributes; however, the accuracy of process mining results may be affected.

Table 1. Suitability of privacy-preserving data transformation approaches to anonymising event log attributes: NA: not applicable; '+': does not affect process mining results; '−': can be used to anonymise an attribute, however invalidates process mining results; '+/−': can decrease the accuracy of some process mining methods.

	Case ID	Activity	Time	Resource	Data
Encryption (deterministic)	+	+	+/−	+	+/−
Swapping	+	−	−	−	−
Noise addition	−	−	−	−	−
Value suppression	NA	+/−	+/−	+/−	+/−
Generalisation	NA	+/−	+/−	+/−	+/−

4.2 Anonymising Atypical Process Behaviour

As discussed in Sect. 3, relations between events in the log (such as event order or grouping of events by case identifiers) can be used to identify atypical process behaviour (which could be linked to individuals). There could be many different types of atypical process behaviour (e.g., infrequent activity sequences, abnormal number of resources or atypical durations). Below, we evaluate two approaches which target anonymisation of atypical process behaviour: a confidentiality framework [13] and PRETSA [7].

The *confidentiality framework* for process mining [13] combines a few data transformation techniques. The first step of the framework is filtering out all cases "that do not reach the minimal frequencies" [13]. The framework changes the structure of an event log: a new attribute "previous activity" is added (which specifies for each event the preceding activity in a case) and case IDs are removed. Since events in the transformed log are no longer related to cases, it is impossible to identify traces (and atypical process behaviour). However, the transformed log can no longer be used by process mining algorithms that require complete traces; it is only suitable for DFG-based tools (e.g., commercial process mining tools). Moreover, as discussed in Sect. 3, healthcare processes are often highly variable and in some processes all traces in the log may be unique. The confidentiality framework (which proposes to filter out traces with infrequent process behaviour) may not be suitable to anonymise event log data from such healthcare processes.

PRETSA [7] is a log sanitisation algorithm, which represents a log as a prefix tree and then transforms the tree until given privacy guarantees are met while striving to preserve directly follows relations. The approach allows to anonymise two types of atypical process behaviour: infrequent traces and atypical activity execution times. The article [7] evaluates the impact of the log transformation on the results of process discovery and performance analysis algorithms using three real-life logs. It also compares the performance of PRETSA with a "baseline" approach which filters out infrequent traces. The evaluation showed that PRETSA outperforms the baseline approach on all logs and data utility losses are minimal for event logs which do not have many unique traces. However, for a log in which most traces are unique the utility of the transformed log is significantly decreased, even more so for stricter privacy requirements (which means that the algorithm may not be suitable for healthcare process data).

5   Privacy-Preserving Process Mining Framework

On the one hand, the healthcare sector needs to comply with strict data privacy requirements; on the other hand, healthcare process data often contain many sensitive attributes and highly variable process behaviour that presents additional threats to privacy. Ensuring high levels of privacy protection for such data while also preserving data utility for process mining purposes remains an open challenge for the healthcare domain.

The analysis of the suitability of existing data transformation approaches to anonymise healthcare process data (presented in Sect. 4) highlighted the trade-off between data privacy and utility. The methods that preserve higher data utility for process mining purposes (e.g., encryption) do not provide strong privacy protection. On the other hand, the methods that can satisfy stricter privacy requirements (e.g., value suppression and generalisation) can decrease the accuracy of results. The magnitude of the data utility loss depends on characteristics of a particular log and varies for different process mining algorithms. Furthermore, performing analyses on anonymised process data without understanding how the data was transformed can yield unpredictable results.

We propose a privacy-preserving process mining framework (Fig. 3) which uses a history of privacy-preserving data transformations to quantify their impact and improve the accuracy of process mining results. The proposed framework can be applied to the healthcare domain as well as other domains with high privacy needs. The first two steps of the framework (i.e., data anonymisation and creation of privacy metadata) are performed by the data owner or a trusted representative. The third step (i.e., conducting privacy-preserving process mining analysis) can be performed by (not trusted) third parties.

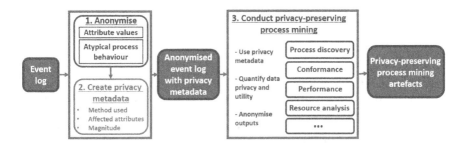

Fig. 3. Privacy-preserving process mining framework.

The first step of the framework is *anonymising* sensitive information such as sensitive attribute values and atypical process behavior. Anonymisation of sensitive attribute values could be achieved using data transformation approaches discussed in Sect. 4.1. Some atypical process behaviours can be anonymised using approaches discussed in Sect. 4.2; however, methods which could anonymise different types of atypical process behaviour in highly variable processes while preserving data utility for different algorithms are yet to be developed.

The second step of the framework is *creating privacy metadata*, which maintains the history of privacy-preserving data transformations in a standardised and machine readable way. Such metadata can be stored in a privacy extension to the IEEE XES log format used for process mining. This privacy metadata will also assist in formally capturing the log characteristics that influence the anonymisation efforts for various forms of process mining.

The third step of the framework is *conducting privacy-preserving process mining* analysis of the anonymised event log with privacy metadata. The privacy metadata can be exploited by new "privacy-aware" process mining techniques to improve mining results. Privacy-aware process mining methods could also quantify data privacy and utility (e.g., by providing confidence measures). Finally, results of process mining techniques could also threaten privacy (by identifying patterns which are linked to individuals). To the best of our knowledge, anonymisation methods for process mining outputs are yet to be developed.

6 Conclusion

Keeping healthcare process data private while preserving data utility for process mining presents a challenge for the healthcare domain. In this article, we analysed data privacy and utility requirements for healthcare process data, assessed the suitability of existing privacy-preserving data transformation approaches and proposed a privacy-preserving process mining framework that can support process mining analyses of healthcare processes. A few directions for future work include: an empirical evaluation of the effects of privacy-preserving data transformation methods on healthcare logs, the development of privacy extensions for logs and the development of privacy-aware process mining algorithms.

References

1. van der Aalst, W.: Process Mining: Data Science in Action. Springer-Verlag, Heidelberg (2016). https://doi.org/10.1007/978-3-662-49851-4. http://www.springer.com/978-3-662-49850-7
2. Aggarwal, C.C.: Data Mining: The Textbook. Springer, Cham (2015). https://doi.org/10.1007/978-3-319-14142-8
3. van der Aalst, W., et al.: Process mining manifesto. In: Daniel, F., Barkaoui, K., Dustdar, S. (eds.) BPM 2011. LNBIP, vol. 99, pp. 169–194. Springer, Heidelberg (2012). https://doi.org/10.1007/978-3-642-28108-2_19
4. Andrews, R., Suriadi, S., Wynn, M., ter Hofstede, A.: Healthcare process analysis. Process Modelling and Management for HealthCare. CRC Press, USA (2017)
5. Burattin, A., Conti, M., Turato, D.: Toward an anonymous process mining. In: FiCloud 2015, pp. 58–63. IEEE (2015)
6. Erdogan, T.G., Tarhan, A.: Systematic mapping of process mining studies in healthcare. IEEE Access 6, 24543–24567 (2018)
7. Fahrenkrog-Petersen, S.A., van der Aa, H., Weidlich, M.: PRETSA: event log sanitization for privacy-aware process discovery. ICPM (accepted) (2019)
8. Fienberg, S.E., McIntyre, J.: Data swapping: variations on a theme by Dalenius and Reiss. In: Domingo-Ferrer, J., Torra, V. (eds.) PSD 2004. LNCS, vol. 3050, pp. 14–29. Springer, Heidelberg (2004). https://doi.org/10.1007/978-3-540-25955-8_2
9. Liu, C., Duan, H., Qingtian, Z., Zhou, M., Lu, F., Cheng, J.: Towards comprehensive support for privacy preservation cross-organization business process mining. IEEE Trans. Serv. Comput. 12(4), 639–653 (2016)
10. Mannhardt, F., Petersen, S.A., Oliveira, M.F.: Privacy challenges for process mining in human-centered industrial environments. In: IE 2018, pp. 64–71. IEEE (2018)
11. Mans, R.S., van der Aalst, W.M.P., Vanwersch, R.J.B.: Process Mining in Healthcare: Evaluating and Exploiting Operational Healthcare Processes. SBPM. Springer, Cham (2015). https://doi.org/10.1007/978-3-319-16071-9
12. Partington, A., et al.: Process mining for clinical processes: a comparative analysis of four Australian hospitals. ACM (TMIS) 5(4), 19 (2015)
13. Rafiei, M., von Waldthausen, L., van der Aalst, W.: Ensuring confidentiality in process mining. In: SIMPDA 2018 (2018)
14. Rojas, E., Sepúlveda, M., Munoz-Gama, J., Capurro, D., Traver, V., Fernandez-Llatas, C.: Question-driven methodology for analyzing emergency room processes using process mining. Appl. Sci. 7(3), 302 (2017)

15. Tillem, G., Erkin, Z., Lagendijk, R.L.: Privacy-preserving alpha algorithm for software analysis. In: SITB 2016 (2016)
16. Tillem, G., Erkin, Z., Lagendijk, R.L.: Mining sequential patterns from outsourced data via encryption switching. In: PST 2018, pp. 1–10. IEEE (2018)
17. Toshniwal, D.: Privacy preserving data mining techniques for hiding sensitive data: a step towards open data. In: Munshi, U.M., Verma, N. (eds.) Data Science Landscape. SBD, vol. 38, pp. 205–212. Springer, Singapore (2018). https://doi.org/10.1007/978-981-10-7515-5_15

Comparing Process Models for Patient Populations: Application in Breast Cancer Care

Francesca Marazza[1](✉), Faiza Allah Bukhsh[2](✉) (iD), Onno Vijlbrief[3](✉),
Jeroen Geerdink[3](✉) (iD), Shreyasi Pathak[2](✉) (iD), Maurice van Keulen[2](✉) (iD),
and Christin Seifert[2](✉) (iD)

[1] Università di Genova, Genoa, Italy
francesca.marazza93@gmail.com
[2] University of Twente, Enschede, Netherlands
{f.a.bukhsh,s.pathak,m.vankeulen,c.seifert}@utwente.nl
[3] Hospital Group Twente (ZGT), Hengelo, Netherlands
{o.vijlbrief,j.geerdink}@zgt.nl

Abstract. Processes in organisations such as hospitals, may deviate from intended standard processes, due to unforeseeable events and the complexity of the organisation. For hospitals, the knowledge of actual patient streams for patient populations (e.g., severe or non-severe cases) is important for quality control and improvement. Process discovery from event data in electronic health records can shed light on the patient flows, but their comparison for different populations is cumbersome and time-consuming. In this paper, we present an approach for the automatic comparison of process models extracted from events in electronic health records. Concretely, we propose to compare processes for different patient populations by cross-log conformance checking, and standard graph similarity measures obtained from the directed graph underlying the process model. Results from a case study on breast cancer care show that average fitness and precision of cross-log conformance checks provide good indications of process similarity and therefore can guide the direction of further investigation for process improvement.

Keywords: Process mining · Process comparison · Quality control · Breast cancer care

1 Introduction

Quality of health care can be assessed by observing the structure, processes and the outcomes of healthcare [7]. Processes in organisations, such as hospitals, may deviate from intended standard processes, due to unforeseeable events and the complexity of the organisation. For hospitals, the knowledge of actual patient streams for different patient populations (e.g., severe or non-severe cases) is important for quality control and improvement. Electronic health records (EHR)

© Springer Nature Switzerland AG 2019
C. Di Francescomarino et al. (Eds.): BPM 2019 Workshops, LNBIP 362, pp. 496–507, 2019.
https://doi.org/10.1007/978-3-030-37453-2_40

contain a wealth of information about patients, including timestamps of diagnoses and treatments. Thus, EHR can serve as input data to discover the as-is processes [2] in healthcare. To investigate patient processes for patient populations of interest (e.g., severe or non-severe cases) their process models have to be constructed and compared. Manual comparison of process models requires expertise in understanding process models and is time-consuming, which makes it unfeasible for many populations of interest. In this paper, we present an approach to automatically compare process models obtained from EHR event logs in order to have an initial quantification of the degree of similarity between different patient subgroups. More specifically, we compare three different approaches for obtaining a similarity between process models: (i) visual inspection, i.e., human judgment, (ii) cross-log conformance checking, and (iii) similarity measures on directed graphs extracted from the process model. For cross-log conformance checking, we apply replay technique using the event log of one population against the process model discovered for a second population (and vice versa). Then, we evaluate which of the methods for measuring process similarities best approximate human judgment. The contribution of this paper is the following:

- We present new methods for quantitative process model comparison based on conformance checking and graph similarities
- We evaluate the methods in the domain of breast cancer care to compare different patient populations.

In the following, we first introduce the application domain breast cancer care (cf. Sect. 2), review related work (cf. Sect. 3). We then present the approach in detail (cf. Sect. 4), and its evaluation for breast cancer populations (cf. Sect. 5).

2   Application Background

Electronic health records (EHRs) can be a solution for improving the quality of medical care. EHRs represent digitally collected longitudinal data, like reports, images, sensitive data and clinical information about patients and their provided treatments [13]. The considered EHR of breast cancer patients contains 4 general types of reports. Radiology reports communicate the findings of imaging procedure by describing the radiology images (e.g., X-rays). In case of a patient with a suspicion of breast cancer, it also contains a BI-RAD score. Pathology reports are free medical texts where a diagnosis based on the pathologists examination of a sample of the suspicious tissue is given. The narrative operative surgery reports document breast cancer surgery. The multidisciplinary reports (MDO) are free-texts written during a multidisciplinary expert team meeting.

One of the most important data included in the radiology report is the BI-RADS category. Breast Imaging-Reporting and Data System (BI-RADS) [18] is a classification system proposed by the American College of Radiology (ACR) to represent the malignancy risk of breast cancer of a patient in a standardized manner. A BI-RADS category can range from 0 to 6, with 0 being benign and 6 being the most malignant. Patients with different BI-RADS follow different processes e.g. patients with BI-RADS category 0 need additional imaging evaluation, BI-RADS category 3 needs initial short-interval follow-up and BI-RADS category 4 may be recommended for biopsy. In Netherlands, women between the ages of 50 and 75 are solicited for screening once every 2 years. The purpose of the screening program is to detect the breast cancer at an early stage, before symptoms appear.

3   Related Work

Quality in health care and the corresponding reporting and evaluation is an issue of national and international importance. Donabedian [7] proposed that the quality of health care can be assessed by observing the structure, processes and the outcomes of healthcare. The Institute of Medicine (IOM) defines health care quality as "the degree to which health services for individuals and populations increase the likelihood of desired health outcomes and are consistent with current professional knowledge" [10]. Process-based quality measures are more suited to explain how and what is required to improve health care processes, as compared to outcome-based measures [14, 17]. Measuring the quality of various processes can also answer questions like accuracy of diagnosis, disease monitoring and therapy and percentage of patients, who received care as recommended [14]. Better process quality can also lead to better patient satisfaction with the series of transactions occurring during their hospital visit [12]. To summarize, the previous works state that quality of health care can be improved by measuring the health care processes, which can further lead to better health outcome.

The goal of **process mining** is to extract process models from event logs [6], also known as transactional logs or audit trails [1]. An event corresponds to an activity (i.e., a well-defined step in the process) affiliated with a particular case (i.e., process instance) [3] and particularly consist of a time stamp and optional information such as resources or costs. Process mining as a discipline consists of three dimensions, process discovery, conformance checking and process enhancement [19]. *Process discovery* refers to the construction of a comprehensive process model, e.g., Petri-Nets or State-charts, to reproduce the behavior seen in the log file [3]. We use the Inductive Visual Miner (IvM) [8] plug-in of ProM[1] since provides an user-friendly visualization, with an opportunity to investigate deviations. The deviations represent cases which do not follow the most common behaviours and thus correspond to event log traces that the process does not explain. *Conformance checking* is applied to compare process models and event logs in order to find commonalities and discrepancies between the modeled behavior and the observed behavior [16]. Our goal is to compare two processes, therefore we use the (ProM) plug-in "Replay a Log on Petri Net for Conformance Analysis" to play the event log of one patient population to the discovered process model of another population and use the obtained fitness, precision and generalization measures for our similarity analysis. A case study explored the applicability of **process mining in health care** and raised the concern that

---

[1] promtools.org last accessed 2019-05-06.

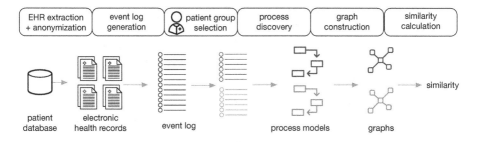

Fig. 1. Overview of the approach.

traditional process mining techniques have problems with unstructured processes as they can be found in hospitals [11]. In this paper, we will focus on a very small sub-domain, breast cancer care, to reduce the complexity of the EHR extraction and resulting process models.

When **comparing two graphs** G_1 and G_2 one is either interested in exact matches of full or sub-graphs (graph homomorphisms) [5] or a measure of structural similarity (e.g. [15]), among which the graph edit distance (GED) [5] is most widely adopted. The GED is defined as the minimum number of operations (add/remove/substitute nodes and edges) needed to transform G_1 to G_2. The problem of calculating the GED is NP-hard in general, making it unfeasible to solve for larger graphs and giving rise to heuristic approximation approaches [20]. More recently, supervised machine learning approaches have been suggested. For instance, Li et al. use a combination of two neural networks to learn a similarity score for G_1 and G_2 [9]. As the authors demonstrate, supervised machine learning approaches can generate highly accurate similarity scores, but they require ground-truth graph-similarity data for training the models. In this work, a graph-similarity ground-truth is not available, therefore we use approximations of GED and an unsupervised machine learning approach based on hand-crafted features for G_1 and G_2 and standard distance metrics on these features.

4 Approach

In this work, we are interested in care processes followed by different patient populations and how these processes compare to each other. In this section we describe our approach for obtaining process models from EHR.

Figure 1 provides an overview of the approach. The available data are EHR, extracted and anonymized from a patient database. Then, event logs are constructed for populations of interest, the corresponding process models are constructed (cf. Sect. 4.1) and transformed to weighted directed graphs Sect. 4.2. Two process models are then compared by (i) visual inspection using obtaining a similarity measure based on human judgment, (ii) cross-log conformance checking and (iii) graph comparison methods (cf. Sect. 4.3).

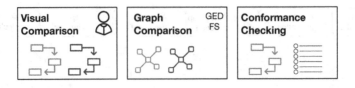

Fig. 2. Overview of methods for pairwise comparison of process models.

4.1 Process Discovery on EHR

In a first step, the EHR of patients subgroups have to be extracted from the hospital data base. From those EHRs, only information related to events of interest have to be extracted. Depending on the hospital data base, this process involves a combination of hand-crafted filtering and extraction rules. To preserve the privacy of hospital patients, personal data has to be anonymized.

In order to apply process mining techniques for analyzing models, data has to be transformed in event logs. EHR, a collection of reports, are our event logs. Each event is associated with a case, a patient, and they are chronologically ordered. Each case can have multiple events. Based on the question that we want to answer, data can be divided in different ways, creating various populations of interest. Consequently, event logs and process models are generated and they should reflect the logs. The algorithm used for producing process models is Inductive Visual Miner, that take care about the deviations. The deviations are patients whose behaviour is different from the most common path.

4.2 Graph Construction

In order to use graph comparison methods, we converted the process models to weighted directed graphs. We constructed $G(V1, V2, E)$ as follows: nodes in $V1$ represent activities and nodes in $V2$ correspond to logical operators in the process model. A directed edge is inserted between two nodes if it in between activities or activities and operators in the underlying process model. We used Boolean function operators to capture the semantic meaning of the process model in the directed graph. Operators can either be AND or XOR, the addition of "-split" or "-join" indicates the start and end of the respective paths. Thus, an "AND-split" means that patients have to follow both paths that the operator outlines, without particular order. The "AND-join" indicates the end of the paths that must be executed in parallel. The XOR operators work similarly, but states that the patient takes either one path, but not both. "Loop" indicates a cycle in the process. Each edge has an associated weight that is set to the frequencies of the connection in the process model.

4.3 Process Model Comparison

Figure 2 illustrates the approaches for process model comparison.

For **visual similarity assessment** of pairs of process models were judged by humans for their similarity. We used a 5-point Likert scale (0: Identical, 1: Slightly Different, 2: Somewhat Different, 3: Very Different, 4: Extremely Different). 3 of the co-authors (no medical background) of this paper judged pairs of process models independently, and were given the instruction to focus on the structure of the process and not on the numbers of the edges when assessing the similarity.

We applied **cross-log conformance checking** as follows: we used the (ProM) plug-in "Replay a Log on Petri Net for Conformance Analysis" to play the event log of one subpopulation with the process model generated by the second subgroup (and vice versa) and recorded standard conformance checking metrics fitness, precision and generalization.

For the **graph-based comparison** of process models, we used the networkx graph library for python for calculation graph metrics and similarities[2]. On the weighted directed graphs constructed from the process models (see Sect. 4.2), we calculate the graph edit distance (GED) using an approximation algorithm [4]. For feature-based comparison of two graphs, we generated a feature vector for each graph with the following graph metrics as features: number of nodes, number of edges, average degree, average weighted degree, average clustering coefficient, average shortest path, average closeness and average betweenness centrality. We then obtained the similarity score for two graphs by first normalizing the feature vectors to unit length and then calculating the Euclidean distance of the two normalized vectors. The similarity score is then 1 minus the obtained distance.

5   Experiments

In our experiments, we collected EHRs from the hospital database (cf. Sect. 5.1, created event logs for populations of interest (cf. Sect. 5.2), and obtained process models for these populations (cf. Sect. 5.3). We then compared these process models pairwise, by (i) visual inspection, (ii) cross-log conformance checking and (iii) graph-based similarity measures and investigate how the obtained similarity measures reflect the similarity obtained by human judgment (cf. Sect. 5.4).

5.1   EHRs Extraction and Event Log Preparation

Free-text reports on breast cancer patients from 2012–2018 were collected from the hospital database. The following rules defined whether a patient was included in the analysis: with the purpose of identifying the complete health path of the patients inside the hospital, patient has to be a "new patient": in the range of time considered, the patient must have the first visit, that can not be described in the referring report by key words like "MRI", "punctie" (biopsy), "mamma-tumor" (tumor in the breast). Also, it can not be represented by a MDO or

---

[2] https://networkx.github.io/, last accessed 2019-05-20.

Table 1. Overview of the selected sub-populations.

#Reports in EHR						
Population	#Events	#Cases	Radiology	Pathology	MDO	Surgery
SVOB	17,677	5,793	10,429	6,987	199	62
NoSVOB	26,542	6,427	15,254	10,208	784	296
Age ≥ 50	31,157	7,740	18,132	12,894	819	312
Age < 50	12,062	4,480	7,551	4,301	164	46
Birad12	23,393	8,612	15,356	7,874	131	32
Birad3-6	20,019	3,365	9,805	9,041	849	324

Surgery report, because it is impossible that the first visit of a patient is one of that. So, the start event is the first report for each patient. On the other hand, the criteria to understand if a patient has finished the treatments in the hospital was not found due to the complexity of the problem. Therefore, there is no end condition point in health path analysis. The gathered patients are 12220, for a total of 44219 reports.

5.2 Patient Populations

We considered the following patient populations:

SVOB: Patients coming from a national breast cancer screening program,
NoSVOB: Patients sent to the hospital by the general practitioner,
Birad12: Patients with a BIRAD score 1 or 2 (0% likelihood of cancer),
Birad3-6 Patients with a BIRAD score of 3 (probably benign), 4 (suspicious), 5 (highly suggestive of malignancy) or 6 (known biopsy-proven).
Age ≥ 50: Patients of age 50 or older,
Age < 50: Patients younger than 50.

The age boundary is set to 50 due to an empirically increased risk for breast cancer development at this age.

Table 1 provides an overview for the selected populations. It shows the number of events with a breakdown on event type, i.e. the specific type of report and the number of cases, i.e., the number of patients. For each population, the event log contains at least 12,000 events, with radiology reports being the most frequent and surgery being the least frequent event type in all populations. Note that radiology report is always present for each case. This result will be confirmed graphically by process figures where an AND condition is always existing for radiology events. Surgery events are generally occurring less frequent than the others event types. In particular, the percentage of surgeries for Birad12 and Birad3-6 populations are significantly different.

Table 2. Quantitative population comparison. Reporting Jaccard similarity of sets of patient ids in the two groups.

Group 1	Group 2	Jaccard similarity
SVOB	NoSVOB	0.00
Age ≥ 50	Age < 50	0.00
Birad12	Birad3-6	0.00
NoSVOB	Age < 50	0.30
SVOB	Birad12	0.42

We compared process models of populations that are of clinical interest, namely screening vs. non-screening patients (SVOB/NoSVOB), low vs. high probability of malignancy (Birad12 vs. Birad3-6) and age groups before and after screening age (Age ≥ 50 vs. Age < 50). We also included a comparison of before screening age patients and non-screening patients (NoSVOB vs. Age < 50) and screening patients that were transferred to the hospital, but had a low probability of malignancy (SVOB vs. Birad12). Table 2 gives an overview of the compared groups and shows the Jaccard similarity for different populations pairs. To investigate the patient overlap between these populations, we calculated the Jaccard coefficient as follows: for each population, we created a set with anonymized patient ids. The Jaccard coefficient is the ratio of the number of patients two populations have in common (set intersection) and the total number of unique patient ids (set union). As can be seen, there is a 30% overlap between NoSVOB patients and patients younger than 50 as well as between SVOB patients and patients with low likelihood of cancer (Birad12).

5.3   Process Discovery and Graph Construction

Process models obtained by the IvM plug-in of ProM (noise filtering set to 90%) for three populations are shown in Figs. 3 and 4 (top)[3]. Blue rectangles represent events, the intensity of colour is associated with the number of events. It can be seen that the process models for SVOB and Birad12 have the same structure (but different frequencies) although their patient populations are not the same (Jaccard similarity of 0.42, cf. Table 2). The process model for Birad3-6 patients has a complicated structure, indicating that patients with non-zero probability of malignancy follow complex paths in the hospital. Many deviations, represented by red dashed lines, exist in all processes.

Figure 4 shows the process model for NoSVOB patients and the correspondent translated graph. Circles denote the logic operators while rectangles represent events. The un-normalized feature vector f obtained from the graph for NoSVOB patients is $f = (18, 28, 2.89, 3.09, 0.19, 0.81, 0.13, 0.03)$, the features are in the order as described in Sect. 4.3.

---

[3] Process models for the other populations were omitted due to space constraints.

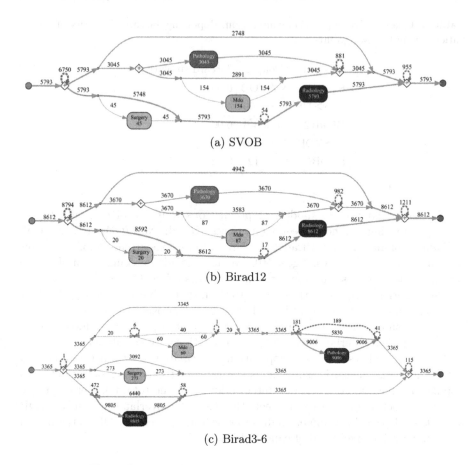

(a) SVOB

(b) Birad12

(c) Birad3-6

Fig. 3. Discovered process models for 3 sub-populations.

Table 3. Process comparison. Comparing visual impression (Visual), Process conformance checking metrics (F - Fitness, P - Precision) when group 1 event log is checked against the process model (PM) constructed for group 2 and vice versa. Graph similarities (GED - graph edit distance, FS - feature-based similarity)

Populations			Conformance checking						Graph sim.	
			Log1-PM2		Log2-PM1		Average			
Group 1	Group 2	Visual	F_{12}	P_{12}	F_{21}	P_{21}	$\overline{F}$	$\overline{P}$	GED	FS
SVOB	NoSVOB	3 ± 0	1.00	0.47	0.71	0.79	0.86	0.63	16	0.97
Age ≥ 50	Age < 50	4 ± 0	0.69	0.94	1.00	0.43	0.85	0.69	30	0.96
Birad12	Birad3-6	3 ± 1	0.81	0.82	0.96	0.52	0.89	0.67	24	0.90
NoSVOB	Age < 50	3 ± 1	0.65	0.94	1.00	0.47	0.83	0.71	24	0.95
SVOB	Birad12	0 ± 0	0.80	0.82	0.78	0.79	0.79	0.81	3	0.98

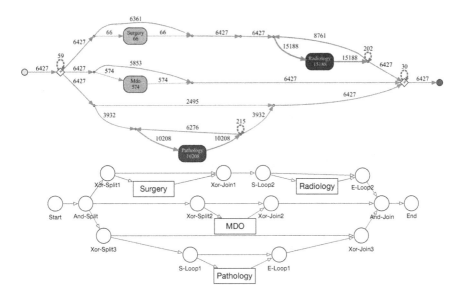

Fig. 4. Process model for NoSVOB population (top) and corresponding directed graph (bottom). Edge weights and loops omitted in the graph for readability.

5.4   Process Model Comparison

Table 3 summarizes the comparison of process models. For visual assessment, we reported the average similarity score and standard deviation obtained by human judgment on a 5-point Likert scale (0 - very similar). Further we reported the fitness and precision values obtained by cross-log conformance checking and their averages. We also reported the GED and the similarity obtained by feature-based graph comparison (FS). As FS values are generally all above 0.9 and similar, they do not seem useful for comparing the underlying process models. Fitness and precision varies more and interestingly differs depending on whether the log of group 1 is played on the process model of group 2 or vice versa.

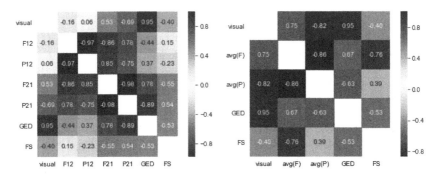

Fig. 5. Correlation between similarity measures. Based on the measures for the variables reported in Table 3.

Figure 5 shows the Pearson correlations between the similarity measures for both, the original conformance-checking values and their averages. The GED shows the highest positive correlation with the visual judgment ($\rho = 0.95$) and the average fitness is also strongly correlated ($\rho = 0.67$). We observed a strong negative correlation between average precision and human judgment ($\rho = 0.82$). From this, we can conclude that at least for the tested event logs and process models, cross-log conformance checking measures provide an indication of process similarity, the higher the average fitness, the more similar the processes and the lower the average precision, the more dissimilar processes are. While the GED provides the highest correlation, the results need to be judged with care. We used an iterative algorithm for calculating the GED, and stopped the approximation after 1 h of calculation if the values did not change anymore. This means, (i) reported values might not be the true GED (and are for certain not in the case of SVOB - Birad12 comparison, where GED should be 0) and (ii) this approach is not practical due to its computational inefficiency.

6   Summary

In this paper, we raised the problem of automatic quantitative comparison of process models generated from event logs with the same types of events. We proposed comparisons based on cross-log conformance checking and standard graph similarity measures obtained from the directed graph underlying the process model. We applied these methods on process models from different subgroups of breast cancer care patients. Results show that average fitness and precision of cross-log conformance checking provide good indications of process similarity and therefore can guide the direction of further investigation for process improvement. In our application scenario, the compared process models were rather small; they contained maximally 4 different event types. At a higher level of event granularity (e.g. when differentiating between different types of radiology reports, such as Diagnostic, Biopsy) process models easily get more complex and thus automatic comparison becomes even more desirable. While the results indicate that cross-conformance checking metrics are indicative for process model similarity, further research in different domains and/or different event types in the same domain remains to be done.

Acknowledgement. This work was supported by the Hospital Group Twente (ZGT) by providing data, secure server infrastructure and domain advice.

References

1. Van der Aalst, W.M., van Dongen, B.F., Herbst, J., Maruster, L., Schimm, G., Weijters, A.J.: Workflow mining: a survey of issues and approaches. Data Knowl. Eng. **47**(2), 237–267 (2003)
2. Van der Aalst, W.M., van Hee, K.M., van Werf, J.M., Verdonk, M.: Auditing 2.0: using process mining to support tomorrow's auditor. Computer **43**(3), 90–93 (2010)

3. Van der Aalst, W.M., Weijters, A.: Process mining: a research agenda (2004)
4. Abu-Aisheh, Z., Raveaux, R., Ramel, J.Y., Martineau, P.: An exact graph edit distance algorithm for solving pattern recognition problems. In: 4th International Conference on Pattern Recognition Applications and Methods 2015, Lisbon, Portugal, January 2015. https://doi.org/10.5220/0005209202710278
5. Berretti, S., Del Bimbo, A., Vicario, E.: Efficient matching and indexing of graph models in content-based retrieval. IEEE Trans. Pattern Anal. Mach. Intell. 23(10), 1089–1105 (2001). https://doi.org/10.1109/34.954600
6. Bogarín, A., Cerezo, R., Romero, C.: A survey on educational process mining. Wiley Interdiscip. Rev.: Data Min. Knowl. Discov. 8 (2017). https://doi.org/10.1002/widm.1230
7. Donabedian, A.: Evaluating the quality of medical care. Milbank Mem. Fund Q. 44(3), 166–206 (1966)
8. Leemans, S.J.J., Fahland, D., van der Aalst, W.M.P.: Discovering block-structured process models from event logs - a constructive approach. In: Colom, J.-M., Desel, J. (eds.) PETRI NETS 2013. LNCS, vol. 7927, pp. 311–329. Springer, Heidelberg (2013). https://doi.org/10.1007/978-3-642-38697-8_17
9. Li, Y., Gu, C., Dullien, T., Vinyals, O., Kohli, P.: Graph matching networks for learning the similarity of graph structured objects. In: Proceedings of International Conference on Machine Learning (2019)
10. Lohr, K.N., Schroeder, S.A.: A strategy for quality assurance in medicare. N. Engl. J. Med. 322(10), 707–712 (1990)
11. Mans, R.S., Schonenberg, M.H., Song, M., van der Aalst, W.M.P., Bakker, P.J.M.: Application of process mining in healthcare – a case study in a Dutch hospital. In: Fred, A., Filipe, J., Gamboa, H. (eds.) BIOSTEC 2008. CCIS, vol. 25, pp. 425–438. Springer, Heidelberg (2008). https://doi.org/10.1007/978-3-540-92219-3_32
12. Marley, K.A., Collier, D.A., Meyer Goldstein, S.: The role of clinical and process quality in achieving patient satisfaction in hospitals. Decis. Sci. 35(3), 349–369 (2004)
13. Noumeir, R., Pambrun, J.F.: Images within the electronic health record, pp. 1761–1764 (2009). https://doi.org/10.1109/ICIP.2009.5414545
14. Palmer, R.H.: Process-based measures of quality: the need for detailed clinical data in large health care databases. Ann. Intern. Med. 127(8_Part_2), 733–738 (1997)
15. Raymond, J.W., Gardiner, E.J., Willett, P.: RASCAL: calculation of graph similarity using maximum common edge subgraphs. Comput. J. 45(6), 631–644 (2002). https://doi.org/10.1093/comjnl/45.6.631
16. Rozinat, A., van der Aalst, W.M.P.: Conformance checking of processes based on monitoring real behavior. Inf. Syst. 33(1), 64–95 (2008)
17. Rubin, H.R., Pronovost, P., Diette, G.B.: The advantages and disadvantages of process-based measures of health care quality. Int. J. Qual. Health Care 13(6), 469–474 (2001)
18. Sickles, E.A., D'Orsi, C.J., Bassett, L.W., et al.: ACR BI-RADS Atlas. American College of Radiology, Reston (2013)
19. Van Der Aalst, W.: Process Mining: Discovery, Conformance and Enhancement of Business Processes, vol. 2. Springer, Heidelberg (2011). https://doi.org/10.1007/978-3-642-19345-3
20. Zeng, Z., Tung, A.K.H., Wang, J., Feng, J., Zhou, L.: Comparing stars: on approximating graph edit distance. Proc. VLDB Endow. 2(1), 25–36 (2009). https://doi.org/10.14778/1687627.1687631

Evaluating the Effectiveness of Interactive Process Discovery in Healthcare: A Case Study

Elisabetta Benevento[1]([✉]), Prabhakar M. Dixit[2],
M. F. Sani[4], Davide Aloini[3], and Wil M. P. van der Aalst[4]

[1] University of Rome Tor Vergata, Rome, Italy
elisabetta.benevento@students.uniroma2.eu
[2] Eindhoven University of Technology, Eindhoven, The Netherlands
[3] University of Pisa, Pisa, Italy
[4] Rheinisch-Westfälische Technische Hochschule (RWTH), Aachen, Germany

Abstract. This work aims at investigating the effectiveness and suitability of Interactive Process Discovery, an innovative Process Mining technique, to model healthcare processes in a data-driven manner. Interactive Process Discovery allows the analyst to interactively discover the process model, exploiting his domain knowledge along with the event log. In so doing, a comparative evaluation against the traditional automated discovery techniques is carried out to assess the potential benefits that domain knowledge brings in improving both the quality and the understandability of the process model. The comparison is performed by using a real dataset from an Italian Hospital, in collaboration with the medical staff. Preliminary results show that Interactive Process Discovery allows to obtain an accurate and fully compliant with clinical guidelines process model with respect to the automated discovery techniques. Discovering an accurate and comprehensible process model is an important starting point for subsequent process analysis and improvement steps, especially in complex environments, such as healthcare.

Keywords: Interactive Process Discovery · Business process modeling · Healthcare · Process Mining

1 Introduction and Background

Thanks to the pervasive adoption of Information Systems within healthcare organizations and the raising amount of patient and process-data, recent research has started focusing on data-driven approaches for investigating patient-flows through automatic or semi-automatic ways. Particularly, Process Mining (PM) has emerged as a suitable approach to analyze, discover, improve and manage real processes, by extracting knowledge from event logs [1]. Among the different PM perspectives, Process Discovery (PD) focuses on automatically discovering process models based on the event log, without using any apriori knowledge [1, 2]. Of course, to gain significant outcomes, the event log should contain all the necessary information. A considerable number of PD techniques has been proposed by researchers for automatically discovering process models [1]. The most promising techniques for healthcare processes

C. Di Francescomarino et al. (Eds.): BPM 2019 Workshops, LNBIP 362, pp. 508–519, 2019.
https://doi.org/10.1007/978-3-030-37453-2_41

are Heuristic Miner [3], Fuzzy Miner [4], Split Miner [5], and Inductive Miner [6], as they can handle noisy and incomplete event log [7, 8]. Most of them produce formal models (Petri nets, transition systems, process trees, etc.), having clear semantics. In addition, there are available also several commercial tools (Disco, Celonis, QPR, ProcessGold, etc.) to support PD. They return process models that either have no formal semantics or correspond to so-called Directly-Follows Graphs (DFGs) that cannot express concurrency. These models provide valuable insights but cannot be used to capture the casual relationships of the activities in the process and draw reliable considerations [9].

PD is particularly critical for healthcare processes due to their intrinsic complexity, high variability and continuous evolution over time [2, 10]. Specifically, case hetero-geneity typically leads to extract extremely complex, and often incomprehensible, process models, i.e. the so-called "spaghetti-like models" [1, 11]. Besides, healthcare processes are highly dependent on clinicians' experience and expertise, i.e., they are knowledge-intensive [12], involving semi-structured and unstructured decision making. Such deep knowledge is not recorded in the event log, and, thus, it results difficult to elicit [13]. As a result, the mined models do not provide a meaningful representation of the reality, leading to a significant interpretation challenge for healthcare manager. To improve model quality, domain experts and analysts heuristically perform a refinement at the end of the discovery phase. Such a refinement is based on their knowledge, and it has turned out to be a time-consuming as well as iterative task [14].

Recently, new interactive PD approaches have been emerging, that allow to incorporate domain knowledge into the discovery of process models [13, 15, 16]. Combining domain knowledge and process-data may improve process modeling and lead to better results [17, 18]. Interactive approaches are particularly useful in healthcare context, where physicians typically have a deep domain knowledge, whose integration within the process discovery phase can provide critical advances with respect to traditional automated discovery techniques [13, 14].

This work aims at demonstrating the effectiveness and suitability of *Interactive Process Discovery* (IPD), an innovative interactive technique developed by Dixit [19], to model healthcare processes. IPD allows the user (i.e., the analyst or the expert) to interactively discover the process model, exploiting the domain knowledge along with the event log (Fig. 1). In so doing, a comparative evaluation against the existing state-of-the-art process discovery techniques is carried out, in order to assess the potential benefits that domain knowledge brings in improving the quality and understandability of process models. The comparison is performed by using a real dataset from an Italian Hospital, in collaboration with medical staff.

The results confirm that IPD can outperform the existing process discovery techniques, providing a more accurate, comprehensible, and guideline compliant process model. Appropriate modeling of patient-flows may support healthcare managers in taking decisions related to capacity planning, resource allocation, and for making necessary changes in the process of care.

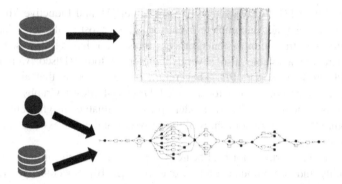

Fig. 1. Traditional automated Process Discovery (top) and Interactive Process Discovery (bottom) (process models are not intended to be readable).

2 Interactive Process Discovery

Interactive Process Discovery (IPD), developed by Dixit [19], is an innovative interactive technique for modeling knowledge-intensive processes based on the domain knowledge along with the event log. In IPD, the user (i.e., the analyst/expert) has total control over the discovery approach, and can model the process incrementally, at the desired complexity level, exploiting his deep knowledge. Information extracted from the event log is used to guide the user in making decisions about where to place a specific activity within the model. To enable the interactive discovery of a process model, the IPD miner uses the synthesis rules [19, 20], which allow expanding a minimal synthesized net[1] by adding one transition and/or one place at a time. A brief description of IPD approach is reported here following (for more details see [19]). During the modeling phase, the user interacts with the synthesized net by applying arbitrarily three synthesis rules: (a) the abstraction rule, (b) the place rule and (c) the transition rule [20]. All possible applications of these rules are projected on the synthesized net, based both on the user interaction and on the information from the activity log[2].

More in detail, the user selects the activity to be added into the net from the activity log. Depending on the selected one, the status of the current synthesized net is updated. Specifically, IPD indicates to the user if the selected activity occurs before or after the other activities within the synthesized net. Alternatively, it highlights that the selected activity and the others in the network never take place at the same time. In so doing, IPD suggests where to place the activity, depending on the insights gained from the activity log. The user can decide to take assistance from the data or ignore the suggestion. The projected information can be based either on the eventually follows

---

[1] A synthesized net is a free-choice workflow net containing a source place, a sink place, a start transition, and an end transition. For more details see [19, 20].

[2] An activity log is a multi-set (or bag) of sequences of activities. Every sequence of activities in the activity log is called an activity trace [19].

(precedes) relation or on the directly follows (precedes) relation, as desired by the user. The user labels the newly-added transitions in the synthesized net with an activity from the activity log. If the transition does not represent an activity, it is depicted as a silent transition. The activity label of the new transition is pre-selected by the user, after which the rule is applied.

3 Case Study: Objective and Methodology

In this work, IPD was applied to a real case of an Italian Hospital to show both the effectiveness of the approach and its suitability in a complex and knowledge-intensive environment. More in detail, we carried out a comparative evaluation against automated discovery techniques, to assess the potential benefit that domain knowledge brings in improving the quality of process models. The evaluation was performed in terms of accuracy and compliance with clinical guidelines.

The approach followed for the evaluation goes through three main steps: (a) data collection and preparation, (b) model building, and (c) model comparison (as depicted in Fig. 2).

3.1 Data Collection and Preparation

We collected and pre-processed data of all lung cancer patients treated by the hospital during the years 2014 and 2015. The management of lung cancer is complex and requires the integration of decisions made by practitioners from different disciplines. Decisions are mainly based on the practitioner's deep knowledge and expertise.

Data were mostly gathered from the Hospital Information Systems. The initial database consisted of 995 cases, 820 types of activities and more than 90,000 events. Before modeling, we decided to refine the raw event log, in order to guarantee its quality. As a matter of fact, it is directly related to the quality and the applicability of results. Data cleaning and preparation included: (a) outliers and incomplete cases removal, (b) low level activities aggregation, (c) less significant activities abstraction, (d) activity redundancy detection. As an example, we kept only the 21 most frequent activities to simplify the event log, since it contained a huge amount of different and fine-grained activities. In so doing, we aimed at building models with a comparable yet meaningful number of activities.

In the end, the refined event log consisted of 990 patient cases, 21 activities, and more than 14,000 events.

3.2 Model Building

Firstly, we applied IPD, as implemented in ProM 6.8, to extract the process model for lung cancer patients, with the collaboration of medical staff. To obtain the resulting model, on several occasions, we took assistance from insights of the event log gained via IPD (e.g., for positioning "radiotherapy" and "nuclear medicine" within the model). On some other occasions, we chose to ignore the information from the data, deeming it inadequate (e.g., for placing the "x-ray" within the model).

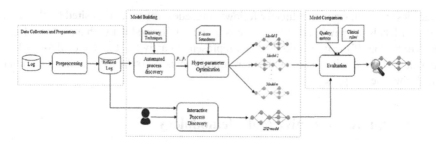

Fig. 2. Comparative evaluation approach.

Following, among the state-of-the-art automated discovery techniques, we chose and applied the Inductive Miner (IM) [6], as implemented in ProM 6.8, and the Split Miner (SM) [5], as implemented in Apromore. As the 30 commercials tools produce Directly-Follows Graphs (DFGs) [9], we also applied the Directly-Follows Graphs Miner, as implemented in PM4Py [21].

As each discovery technique came with several parameters to be tuned, we optimized it by testing different parameter values, to find the best results. The optimization was carried by using a Rapid Miner extension, called RapidProM, and was based on the F-score metric [22] to find the solution with the optimal balance between fitness and precision. The F-score was computed on Petri nets since the measuring tools work only on Petri nets. Conversions of the process model in Petri nets were done using ProM's package.

3.3   Model Comparison

We measured and compared the quality of the model produced by IPD and of the best configurations of the automated discovery techniques in terms of accuracy and compliance with clinical guidelines.

To evaluate the accuracy of the process model, we experimented with two well-known metrics: fitness and precision [1], which both range between 0 and 1. The higher the fitness value, the more the model can replay the log. Conversely, the higher the precision value, the fewer behaviors (i.e., traces) are probable not to appear in the event log [1, 9]. To compute fitness and precision, we resorted to the state-of-the-art alignment-based approaches described in [23, 24]. Due to the trade-off between fitness and precision [22, 25], we used the F-score as an evaluation metric, to take into account the ability of the model to equally fulfill and balance fitness and precision goals [21].

To assess the compliance of the model with the AIOM (Italian Association of Medical Oncology) guideline [26], we carried out a qualitative analysis with the collaboration of medical staff. Specifically, each process model was investigated from a "semantic point of view" and was evaluated on the capability to respect a set of medical rules. Specifically, the evaluation was based on the number of rules that were met by each process model. These medical rules were defined starting from the AIOM clinical guidelines and formalized by using a subset of (Declare) templates. Table 1 shows an overview and an interpretation of the templates that we considered. Each template provides a way to specify a dependency between two different classes of activities (e.g., a precedence constraint between the activities involved in the classes "surgery" and "medical examination") [27].

Table 1. Templates interpretation.

Type	Template interpretation based on the domain
Chain response (A, B)	**R1:** X-ray (B) should occur immediately after the Surgery (A)
Precedence (A, B) & Not Succession (B, A)	**R2:** Invasive diagnostic examination (B) should be preceded by radiological examinations (A) & (B) should not be followed by (A) **R3:** Surgery (B) should be preceded by invasive diagnostic examinations (A) & (B) should not be followed by (A) **R4:** Surgery (B) should be preceded by medical examinations (A) & (B) should not be followed by (A)
Precedence (A, B)	**R5:** If the removal of therapeutic aid (B) occurs, it should be preceded by the x-ray (A)
Init(A)	**R6:** The process should start with a general physical examination (A)
Existence (2, A)	**R7:** Lab test (A) should occur at least 2 times inside the process

4   Results

Table 2 reports the results related to the quantitative evaluation, i.e. the accuracy values obtained by IPD process model and by the best configurations for the automated discovery techniques. More in detail, the table summarizes, for each discovered model: (a) fitness, precision and F-score values; (b) the best configuration parameters provided as input (only for the automated discovery techniques).

Table 2. Quantitative evaluation of IPD process model and the best configurations for three representative automated discovery techniques.

PM technique		Best configuration parameters	Accuracy		
			Fitness	Precision	F-score
Interactive process discovery		–	0.70	0.64	0.67
Automated discovery techniques	Inductive miner	1.0	0.59	0.71	0.64
	Split miner	0.9 & 0.0	0.81	0.61	0.69
	DFG miner	0.2	1	0.21	0.36

As shown in Table 2, IPD, IM, and SM miners provide similar results in terms of F-score, unlike the DFG miner that is unable to balance fitness and precision values. Note that we used the DFG model as a proxy for the models generated by commercial tools like Celonis, Disco, etc. As regards fitness and precision, all the techniques achieve different performance. Specifically, the DFG Miner strikes the best fitness with a value of 1, followed by the SM. However, the DFG Miner is less precise than the others, allowing behaviors not recorded in the event log. On the other hand, IM obtains a model that is slightly less able to reproduce the different behaviors in the log but more precise (with a value of 0.71). With IPD, experts could obtain a model with a quite high value of fitness, without penalization in precision. This is, definitely, a promising result in a knowledge-intensive domain such as the medical one.

Table 3 reports the scores obtained by the discovered process models in terms of satisfied rules.

Table 3. Number of rules satisfied by the models generated by the IPD, IM, SM, and DFG miners.

Rules	Interactive process discovery	Inductive miner	Split miner	DFG miner
R1	1	0	1	0
R2	1	0	0	0
R3	1	0	0	0
R4	1	0	0	0
R5	1	1	1	0
R6	1	0	0	0
R7	1	0	1	1
Total value	7/7	1/7	3/7	1/7

Despite similar performance in terms of model accuracy, less than half of the rules were respected by the models generated by the IM, SM, and DFG miners, unlike IPD model. This is due to the fact that IM, SM, and DFG miners do not take the organizational information and the domain knowledge on the treatment process into account; as a result, their models fail to properly keep the structure of the process in line with clinical guidelines.

To better clarify this statement, let us drill down the behavior of each model with respect to rules R2 and R6. Figures 3, 4, 5 and 6 show the process models produced by IM, DFG miner, IPD, and SM respectively. In a healthcare context, some activities must follow a specific order of execution (see R1-R5 in Table 3). For example, clinical guidelines suggest that invasive diagnostic procedures (e.g., bronchoscopy) must be executed immediately after radiological exams (x-ray and CT scan) and not vice-versa (R2), to confirm the diagnosis and evaluate the extent of the disease. Yet, IM, SM, and DFG miner models seem not to be able to capture this restrictive relationship, allowing also the inverse behavior for some process instances. Indeed, they use parallelism or exclusive choice with loops to represent the activities within the model. In such cases, the activities can take place in a different order from case to case, not respecting the restrictive condition (Figs. 3, 4 and 6). On the other hand, in the process model produced by IPD, invasive diagnostic procedures are directly preceded by radiological exams (Fig. 5). Furthermore, the lung cancer process should start with a general physical examination executed by the specialist (R6). Unlike IPD, the IM, SM, and DFG miner seem to violate this rule. Specifically, in both models produced by IM and DFG miner, the process may start with an activity other than the general physical examination (Figs. 3 and 4). Similarly, in the SM model, the starting activity is the lab test, rather than the general physical examination (Fig. 6). On the contrary, the correct order is properly captured in the IPD model (Fig. 5).

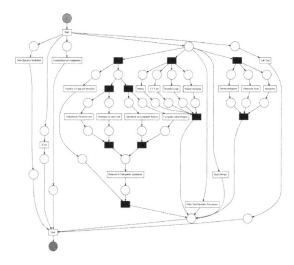

Fig. 3. Petri net for lung cancer patients generated by the Inductive Miner.

Fig. 4. Petri net for lung cancer patients converted from the DFG Miner (Note that the DFG Miner can be seen as a representative example of the discovery technique used by commercial systems).

These preliminary results confirm that leveraging domain knowledge and information recorded in the event log helps obtaining a process model that fully complies with the guidelines and is comparable in accuracy to the models produced by the automated techniques.

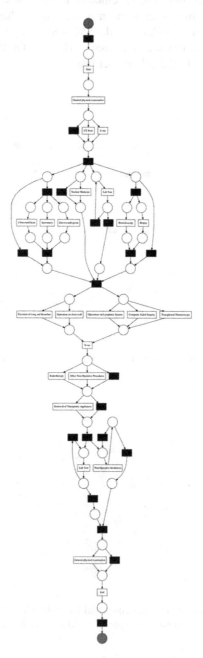

Fig. 5. Petri net for lung cancer patients generated by Interactive Process Discovery.

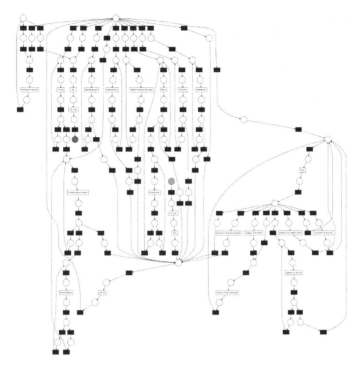

Fig. 6. Petri net for lung cancer patients converted from the Split Miner.

5 Discussion and Conclusions

In this study, we demonstrated the effectiveness and suitability of IPD [19] to model healthcare processes. IPD provides the analyst with a flexible way to interact with model construction, directly exploiting the domain knowledge along with the event log. Prior knowledge from domain experts represents a valuable resource in the discovery of process models, providing critical advances with respects to automated discovery techniques [13, 19]. This is especially true in healthcare, where physicians typically have deep domain knowledge, not recorded in the event log, and, thus, difficult to elicit [14]. Therefore, both the automated discovery techniques fail in producing meaningful and comprehensible process models, resulting in a significant interpretation challenge for healthcare manager [19]. In contrast, IPD technique tries to structure the process data by using domain knowledge.

Our evaluation demonstrates that IPD can be used to obtain a guideline compliant process model, without penalizing its accuracy. Specifically, IPD achieves satisfactory results in terms of model accuracy, comparable to those of IM and SM. It also outperforms the DFG Miner. In addition, since IM, SM, and DFG Miner do not take the domain knowledge on the treatment process into account, their models fail to properly keep the structure of the process in line with clinical guidelines. On the contrary, IPD provides the ability to discover patient pathways that cover the most frequent medical behaviors which are regularly encountered in clinical practice, according to medical staff.

From a managerial viewpoint, discovering an accurate and comprehensible process model is an important starting point for subsequent process analysis and improvement steps, especially in complex environments, such as healthcare. Specifically, appropriate modeling of patient-flows may help healthcare managers to identify process-related issues (e.g., bottlenecks, process deviations, etc.) and main decisions.

While the initial experimental evaluation has provided satisfactorily results, in the future we aim at conducting a more extensive evaluation, replicating the study in different healthcare contexts to test the applicability and generalizability of IPD.

References

1. van der Aalst, W.M.P.: Process Mining: Data Science in Action. Springer, Heidelberg (2016). https://doi.org/10.1007/978-3-662-49851-4
2. Hompes, B., Dixit, P., Buijs, J.: Using process analytics to improve healthcare processes. In: Consoli, S., Reforgiato Recupero, D., Petković, M. (eds.) Data Science for Healthcare, pp. 305–325. Springer, Cham (2019). https://doi.org/10.1007/978-3-030-05249-2_12
3. Weijters, A.J.M.M., Ribeiro, J.T.S.: Flexible Heuristics Miner (FHM). In: IEEE SSCI 201: Symposium Series on Computational Intelligence-CIDM 2011: 2011 IEEE Symposium on Computational Intelligence and Data Mining, pp. 310–317 (2011)
4. Günther, C.W., van der Aalst, W.M.P.: Fuzzy mining – adaptive process simplification based on multi-perspective metrics. In: Alonso, G., Dadam, P., Rosemann, M. (eds.) BPM 2007. LNCS, vol. 4714, pp. 328–343. Springer, Heidelberg (2007). https://doi.org/10.1007/978-3-540-75183-0_24
5. Augusto, A., Conforti, R., Dumas, M., La Rosa, M., Polyvyanyy, A.: Split miner: automated discovery of accurate and simple business process models from event logs. Knowl. Inf. Syst. **59**, 1–34 (2018)
6. Leemans, S.J.J., Fahland, D., van der Aalst, W.M.P.: Discovering block-structured process models from event logs containing infrequent behaviour. In: Lohmann, N., Song, M., Wohed, P. (eds.) BPM 2013. LNBIP, vol. 171, pp. 66–78. Springer, Cham (2014). https://doi.org/10.1007/978-3-319-06257-0_6
7. Mans, R.S., Schonenberg, M.H., Song, M., van der Aalst, W.M.P., Bakker, P.J.M.: Application of process mining in healthcare – a case study in a dutch hospital. In: Fred, A., Filipe, J., Gamboa, H. (eds.) BIOSTEC 2008. CCIS, vol. 25, pp. 425–438. Springer, Heidelberg (2008). https://doi.org/10.1007/978-3-540-92219-3_32
8. Rojas, E., Munoz-Gama, J., Sepúlveda, M., Capurro, D.: Process mining in healthcare: a literature review. J. Biomed. Inform. **61**, 224–236 (2016)
9. van der Aalst, W.M.P., De Masellis, R., Di Francescomarino, C., Ghidini, C.: Learning hybrid process models from events. In: Carmona, J., Engels, G., Kumar, A. (eds.) BPM 2017. LNCS, vol. 10445, pp. 59–76. Springer, Cham (2017). https://doi.org/10.1007/978-3-319-65000-5_4
10. Rebuge, Á., Ferreira, D.R.: Business process analysis in healthcare environments: a methodology based on process mining. Inf. Syst. **37**(2), 99–116 (2012)
11. Diamantini, C., Genga, L., Potena, D.: Behavioral process mining for unstructured processes. J. Intell. Inf. Syst. **47**(1), 5–32 (2016)
12. Di Ciccio, C., Marrella, A., Russo, A.: Knowledge-intensive processes: characteristics, requirements and analysis of contemporary approaches. J. Data Semant. **4**(1), 29–57 (2015)

13. Bottrighi, A., Canensi, L., Leonardi, G., Montani, S., Terenziani, P.: Interactive mining and retrieval from process traces. Expert Syst. Appl. **110**, 62–79 (2018)
14. Canensi, L., Leonardi, G., Montani, S., Terenziani, P.: Multi-level interactive medical process mining. In: ten Teije, A., Popow, C., Holmes, J.H., Sacchi, L. (eds.) AIME 2017. LNCS (LNAI), vol. 10259, pp. 256–260. Springer, Cham (2017). https://doi.org/10.1007/ 978-3-319-59758-4_28
15. Rembert, A.J., Omokpo, A., Mazzoleni, P., Goodwin, R.T.: Process discovery using prior knowledge. In: Basu, S., Pautasso, C., Zhang, L., Fu, X. (eds.) ICSOC 2013. LNCS, vol. 8274, pp. 328–342. Springer, Heidelberg (2013). https://doi.org/10.1007/978-3-642-45005-1_23
16. Xu, X., Jin, T., Wei, Z., Wang, J.: Incorporating domain knowledge into clinical goal discovering for clinical pathway mining. In: 2017 IEEE EMBS International Conference on Biomedical & Health Informatics (BHI), pp. 261–264 (2017)
17. Dixit, P.M., Buijs, J.C.A.M., van der Aalst, W.M.P., Hompes, B.F.A., Buurman, J.: Using domain knowledge to enhance process mining results. In: Ceravolo, P., Rinderle-Ma, S. (eds.) SIMPDA 2015. LNBIP, vol. 244, pp. 76–104. Springer, Cham (2017). https://doi.org/ 10.1007/978-3-319-53435-0_4
18. Mannhardt, F., de Leoni, M., Reijers, H.A., van der Aalst, W.M.P., Toussaint, P.J.: Guided process discovery–a pattern-based approach. Inf. Syst. **76**, 1–18 (2018)
19. Dixit, P.M., Verbeek, H.M.W., Buijs, J.C.A.M., van der Aalst, W.M.P.: Interactive data-driven process model construction. In: Trujillo, J.C., Davis, K.C., Du, X., Li, Z., Ling, T.W., Li, G., Lee, M.L. (eds.) ER 2018. LNCS, vol. 11157, pp. 251–265. Springer, Cham (2018). https://doi.org/10.1007/978-3-030-00847-5_19
20. Desel, J., Esparza, J.: Free Choice Petri Nets, vol. 40. Cambridge University Press, Cambridge (2005)
21. Berti, A., van Zelst, S.J., van der Aalst, W.M.P.: Process Mining for Python (PM4Py): Bridging the Gap Between Process-and Data Science. arXiv preprint arXiv:1905.06169 (2019)
22. De Weerdt, J., De Backer, M., Vanthienen, J., Baesens, B.: A robust F-measure for evaluating discovered process models. In: IEEE Symposium on Computational Intelligence and Data Mining (CIDM 2011), pp. 148–155 (2011)
23. Adriansyah, A., Munoz-Gama, J., Carmona, J., van Dongen, B.F., van der Aalst, W.M.P.: Measuring precision of modeled behavior. Inf. Syst. e-bus. Manag. **13**(1), 37–67 (2015)
24. Adriansyah, A., van Dongen, B.F., van der Aalst, W.M.P.: Conformance checking using cost-based fitness analysis. In: 2011 IEEE 15th International Enterprise Distributed Object Computing Conference, pp. 55–64. IEEE (2011)
25. Sani, M.F., van Zelst, S.J., van der Aalst, W.M.P.: Improving process discovery results by filtering outliers using conditional behavioural probabilities. In: Teniente, E., Weidlich, M. (eds.) BPM 2017. LNBIP, vol. 308, pp. 216–229. Springer, Cham (2018). https://doi.org/10. 1007/978-3-319-74030-0_16
26. Linee guida AIOM 2018. https://www.aiom.it/wp-content/uploads/2018/11/2018_LG_ AIOM_Polmone.pdf. Accessed 15 Apr 2019
27. Maggi, F.M., Bose, R.P.J.C., van der Aalst, W.M.P.: A knowledge-based integrated approach for discovering and repairing declare maps. In: Salinesi, C., Norrie, M.C., Pastor, Ó. (eds.) CAiSE 2013. LNCS, vol. 7908, pp. 433–448. Springer, Heidelberg (2013). https:// doi.org/10.1007/978-3-642-38709-8_28

Developing Process Performance Indicators for Emergency Room Processes

Minsu Cho[1], Minseok Song[1]($\boxtimes$), Seok-Ran Yeom[2], Il-Jae Wang[2],
and Byung-Kwan Choi[2]

[1] Pohang University of Science and Technology, Pohang, Korea
{mcho,mssong}@postech.ac.kr
[2] Pusan National University Hospital, Busan, Korea
{seokrany,jrmr9933,spine}@pusan.ac.kr

Abstract. In a healthcare environment, it is essential to manage the emergency room process since its connectivity to the quality of care. In managing clinical operations, quantitative process performance analysis is typically performed with process mining, and there have been several approaches to utilize process mining in emergency room process analysis. These research provide a comprehensive methodology to analyze the emergency room processes using process mining; however, performance indicators for directly assessing the emergency room processes are lacking. To overcome the limitation, this paper proposes a framework of process performance indicators utilized in emergency rooms. The proposed framework starts with the devil's quadrangle, i.e., time, cost, quality, and flexibility. Based on four perspectives, we suggest specific process performance indicators with a formal explanation. To validate the applicability of this research, we present a case study result with the real-life clinical data collected from a tertiary hospital in Korea.

Keywords: Process performance indicators · Process mining ·
Emergency Room · Healthcare · Performance measurements

1 Introduction

In a healthcare environment, systematic management of clinical processes is inevitable since it is closely connected to medical service quality [13]. In particular, the Emergency Room (ER) process is one of the main processes in the healthcare system [18,20]. The ER process is as highly complicated as other major healthcare processes, including outpatients and inpatients [1,15]. Besides, it is often overcrowded and out of control since it is exposed to the risk of unexpected factors [6]. For these reasons, comprehensive process management for

This work was supported by clinical research grant from Pusan National University Hospital and the MSIT (Ministry of Science and ICT), Korea, under the ITRC (Information Technology Research Center) support program (IITP-2018-0-01441) supervised by the IITP (Institute for Information & communications Technology Promotion).

C. Di Francescomarino et al. (Eds.): BPM 2019 Workshops, LNBIP 362, pp. 520–531, 2019.
https://doi.org/10.1007/978-3-030-37453-2_42

efficient operation and care quality management is considered to be essential in the ER context.

Performance measurements are of paramount importance in managing processes [7]. Above all, the quantitative performance analysis is getting a keen interest due to the abundance of data and advances of data-driven methods [2,9]. Process mining, i.e., a relatively young discipline focused on deriving knowledgeable process-related insights from event logs, has enabled to perform data-driven process analyses [14]. In the ER environment, there has been numerous research utilized in process mining. For example, Rojas et al. [20] proposed the comprehensive question-driven methodology for analyzing the ER processes using process mining with the four schemes, including process discovery, conformance analysis, performance analysis, and organization analysis. Despite its novelty and applicability in practice, process performance indicators to directly assess the ER process are lacking. In other words, it has revealed what to analyze using process mining, but it is insufficient to provide details on what aspects of the emergency room will be analyzed for a specific purpose. Therefore, it is necessary to develop indicators and analytical methods that can measure ER performances (e.g., time, cost, quality, and flexibility) considering the purpose of analysis.

Using *the devil's quadrangle* [7], i.e., time, cost, quality, and flexibility, this paper proposes a framework of process performance indicators (PPIs) utilized in ER processes. Our previous work suggested PPIs that can measure the effects of business process redesigns using process mining functionalities [3]. Based on the approach, this study develops performance indicators for evaluating ER process management. To this end, we first prepare PPIs that can be analyzed using clinical event logs and verify them with a thorough discussion with clinical experts in the emergency department. To validate the applicability of our framework, we performed a case study with the real-life clinical data collected from a tertiary hospital in Korea.

The remainder of this paper is organized as follows. Section 2 summarized related works. Then, in Sect. 3, we describe our framework and process performance indicators in detail. Section 4 shows the application of the proposed framework in the case study. Finally, Sect. 5 concludes the paper and describes future directions.

2   Related Work

There have been numerous research efforts to apply process mining into a healthcare setting, e.g., outpatients, inpatients, and emergency room processes [8,14,15,17–20]. Especially, regarding the ER process, existing research have presented several approaches to analyzing the care flows and the insights by applying them. In [20], the authors proposed a comprehensive question-driven methodology, which provides a data reference model, frequently-posed questions, and the detailed stages to solve the questions using process mining. Also, there was an approach to evaluate the capabilities of process mining to the ER process on the stroke case [8]. [18] proposed the six-phase method for performance analysis of emergency room episodes. However, they do not focus on detailing process performance indicators to evaluate the ER process.

Regarding the process performance indicators, some related works exist in the context of the quantitative process analysis in business process management. [11] proposed how to determine the performance indicators and suggested six requirements on indicators: quantifiability, sensitivity, linearity, reliability, efficiency, and improvement-oriented. [5] proposed *PPINOT*, i.e., a metamodel to define PPIs comprehensively, which includes how to connect elements in business processes and PPIs and provides an implementation of the metamodel using description logics. [3] presented a framework of process performance indicators to assess the effects of business process redesigns in four perspectives: time, cost, quality, and flexibility.

The studies reviewed above do not sufficiently suggest process performance indicators to be utilized immediately to manage the ER processes using process mining techniques.

3   Method

In this section, we explain a framework for defining *emergency room process performance indicators* (i.e., ERPPIs) for managing emergency department processes. Figure 1 presents an overview of the proposed framework. In developing EDPPIs, this research employed four perspectives mainly utilized in the quantitative business process performance analysis, i.e., devil's quadrangle: time, cost, quality, and flexibility [7]. Starting from the four perspectives, we suggested 9 performance indicators as presented in Fig. 1. To define these indicators, we prepared process performance indicators that can be derived from clinical event logs collected in electronic health records and by process mining techniques. After that, they were verified by clinical experts in the emergency department, whether they are meaningful and applicable in practice. The remaining section will describe the clinical event logs as preliminaries and the detailed explanation about the proposed ERPPIs for each perspective.

3.1   Preliminaries

Prior to detailing on PPIs, we explain clinical event logs, i.e., the inputs of process mining in healthcare, utilized for a formal definition of them. Definition 1 presents a formal explanation of event logs, variants, emergency values, and activity relations.

Definition 1 (Event, Case, Event Log). *Let A, O, T, ET be a finite set of activities, originators, timestamps, and event types, respectively. $E = A \times O \times T \times ET$ is the set of events, i.e., combinations of an activity, an originator, a timestamp, and an event type (e.g. $e_i = \{a_i, o_i, t_i, et_i\}$). Let L be an event log which has a multiset of traces, and $C = \{c_1, c_2, c_3, ..., c_k\}$ be the set of cases. A trace $\sigma_k = \{e_{k,1}, e_{k,2}, e_{k,3}, ..., e_{k,n}\}$ is mapped into a case c_k, where $e_{k,n}$ denotes*

Process mining-supported ERPPIs

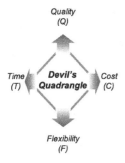

Time	ERPPIT1	Length of stay for patients
	ERPPIT2	Length of stay for patients of a variant
	ERPPIT3	Time of a clinical activity in ER
Cost	ERPPIC1	Number of resources
Quality	ERPPIQ1	Workload of resources
	ERPPIQ2	Variation of length of stay for patients
	ERPPIQ3	Triage-based patient response rate
Flexibility	ERPPIF1	Number of variants
	ERPPIF2	Relations in the process model

Fig. 1. Overview of the proposed framework

n-th event of the k-th case. Let $V = \{v_1, v_2, v_3, ..., v_o\}$ be a finite set of variants where v_i is a nonempty subset of all possible combinations of activities. var is a function mapping each case to a variant (e.g. $var(c_k)$ is the variant of k-th case). Let $EM = \{1, 2, 3, 4, 5\}$ be a finite set of emergency values. emer is a function mapping each case to a emergency value (e.g. $emer(c_k)$ is the emergency value of k-th case). Activity Relation $(AR) \subseteq A \times A$ is a set of activity relations where two events have causal relations (e.g. $ar_{k,ij} = \{(a_{k,i}, a_{k,j}) | a_{k,i}, a_{k,j} \in A\}$ where $e_{k,i}$ is the predecessor of $e_{k,j}$ (i.e. $e_{k,i} > e_{k,j}$)).

3.2 Time-Related ERPPIs

Most organizations aim at managing their business processes by improving time-related indicators, such as decreasing the processing time and waiting time. In the clinical processes for the emergency room, time-related performances are essential since they are highly relevant to the clinical results. In this research, we suggest three indicators in the time perspective: length of stay for patients ($ERPPIT1$), length of stay for patients of a variant ($ERPPIT2$), and cycle time of a clinical activity ($ERPPIT3$). Followings are the definitions of time-related process performance indicators (Definitions 2, 3 and 4).

Definition 2 (ERPPIT1: Length of stay for patients). *Let LOS be the length of stay for patients in an event log L.*

- $LOS = \{t_{k,n} - t_{k,1} | \forall_{0 < k \leq |c|} \forall_{0 < i \leq n} c_k \in L \wedge e_i \in c_k\}$

Definition 3 (ERPPIT2: Length of stay for patients of a variant). *Let $LOS(v_1)$ be the length of stay for patients of a variant v_1 in an event log L.*

- $LOS(v_1) = \{t_{k,n} - t_{k,1} | \forall_{0 < k \leq |c|} \forall_{0 < i \leq n} c_k \in L \wedge e_i \in c_k \wedge var(c_k) = v_1\}$

Definition 4 (ERPPIT3: Cycle time of a clinical activity in a event log). *Let $CT(a_1)$ be cycle time of a clinical activity a_1 in an event log L.*

- $CT(a_1) = \{t_{k,i+1} - t_{k,i} | \forall_{0 < k \leq |c|} \forall_{0 < i \leq n} c_k \in L \wedge e_i \in c_k \wedge a_{k,i} = a_1\}$

The first two time-related indicators, i.e., *ERPPIT1* and *ERPPIT2*, are the performance of passing the entire clinical processes from start to end. Thus, it is necessary to manage them since they are highly related to the congestion of the emergency department. *ERPPIT3* measures the performance of leading clinical activities, including treatments, medical tests, and consultations, which are of great importance to prohibit bottlenecks in the process. All time-related indicators are combined with aggregation functions, e.g., average (f^{AVG}), median (f^{MED}), minimum (f^{MIN}), and maximum (f^{MAX}).

3.3   Cost-Related ERPPIs

As far as the cost-related PPIs are concerned, clinical event logs have to include cost information to identify how expensive the emergency clinical process is for operating it. However, it is often unfeasible to obtain clinical data enhancing the cost information. Therefore, based on the assumption that all resources are full-time equivalents, we develop an alternative indirect cost-related PPI, i.e., the total number of originators in the log (*ERPPIC1*); thus, it can be calculated from the commonly available clinical event logs. Definition 5 gives the formal definitions of the cost-related PPI.

Definition 5 (ERPPIC1: Total number of originators). *Let N_o be the total number of originators in an event log L.*

$$- N_o = \sum_{q=1}^{m} \begin{cases} 1 & \text{if } O_q \in \{\sum_{0<k\leq|c|} \sum_{0<i\leq n} \pi_o(e_{k,i})\} \\ 0 & \text{otherwise} \end{cases}$$

With the assumption that wages are similar among full-time employees, *ERPIC1* becomes the significant cost indicator since labor cost is usually essential.

3.4   Quality-Related ERPPIs

In business process management, quality-related performance analysis can be differentiated as external and internal aspects. More in detail, the external quality focuses on client's angles (e.g., patients), while the process participant's viewpoint is relevant to the internal quality. In an emergency room environment, external quality can include patient's satisfaction and clinical results (e.g., mortality rate or re-visit rate). These indicators, however, cannot be easily derived since clinical event logs generally do not hold the relevant information. Therefore, this research proposes three indicators in the internal quality perspective that can be measured from clinical data.

First, the workload of resource (*ERPPIQ1*) indicates how much works an originator gets. In an emergency room environment, workload management is essential since it is highly crowded in general. Definition 6 gives a detailed explanation of how to measure the workload of resources. As described in the Definition, it requires two different values within a specific time period (tp_j): frequency of events started $(o_q, start, tp_j)$ and terminated $(o_q, complete, tp_j)$ by a specific originator (o_q). In the initial stage (tp_1), the workload is computed by checking the difference between the number of started and completed events. From the second time period $(j > 1)$, the workload of the immediately preceding stage is also considered.

Definition 6 (ERPPIQ1: Workload of Resources). *Let* $C = O \times T \times TP \rightarrow \mathbb{R}$ *be a function that computes the number of events from a log* L *for a given resource* $(o_q \in O)$, *a type* $(\{start, complete \in T\})$, *and a time period* $(tp_j \in TP)$. $C(o_q, start, tp_j)$ *denotes the number of events started by the resource* o_q *within the time period* tp_j. *Here, if the event type only holds complete, the complete time of the immediately preceding event in the same case becomes the start time.* $C(o_q, complete, tp_j)$ *denotes the number of events completed by the resource* o_q *within the time period* tp_j. *The workload for the resource* o_q *within the time period* tp_j *is defined as follows.*

- $Workload_{o_q,tp_j} =$

$$\sum_{0<q\le|m|} \sum_{0<j\le|p|} \begin{cases} C(o_q, start, tp_j) - C(o_q, complete, tp_j) & \text{if } j = 1 \\ Workload_{o_q,tp_{j-1}} \\ \quad + C(o_q, start, tp_j) - C(o_q, complete, tp_j) & \text{otherwise} \end{cases}$$

The second quality-related indicator, i.e., *ERPPIQ2*, shows how diverse are the variations of the cycle time in the emergency clinical process. It implies to identify whether the process is stable and standardized. Definition 7 presents the formal explanation of *ERPPIQ2*.

Definition 7 (ERPPIQ2: Variation of length of stay for patients). *Let* $\sigma(LOS)$ *be the standard deviation of the length of stay for patients in an event log* L.

- $\sigma(LOS) = f^{STD}(\{t_{k,n} - t_{k,1} | \forall_{0<k\le|c|} \forall_{0<i\le n} c_k \in L \wedge e_i \in c_k\})$

The last indicator of the quality perspective, i.e., *ERPPIQ3*, is closely linked to the emergency department. In the initial step of the emergency room process, the patient's emergency degree is measured by the *triage* activity, e.g., Korean Triage and Acuity Scale (KTAS) [12], and hospitals provide the clinical services based on it. That is, it is a fundamental policy that higher emergency of the patient acquires a higher priority. In this paper, we suggest the triage-based patient response rate by considering two policies: (1) first come, first served (FCFS) and (2) more urgent people are allowed to be treated first even if they are late. Definition 8 presents the formal explanation of this indicator.

Definition 8 (ERPPIQ3: Triage-based patient response rate). *Let* $emer(c_k)$ *denotes the emergency degree of a case* $c_k \in C$. *Assume that all cases* $c_k \in C$ *are sorted by the completed time of their first event* $t_{k,1}$. *Then, triage-based patient response rate* TPR_{tp_j} *for a specific time period* $tp_j \in TP$ *is formally defined as follows.*

$$- TPR_{tp_j} = 1 - \frac{\sum_{0<k\leq|c|} \begin{cases} 1 & \text{if } t_{k,2} > t_{k+1,2} \\ & \wedge \; emer(c_k) > emer(c_{k+1}) \\ & \wedge \; min(tp_j) \leq t_{k,2}, t_{k+1,2} \leq max(tp_j) \\ 0 & otherwise \end{cases}}{|c|-1}$$

3.5   Flexibility-Related ERPPIs

To assess the flexibility of emergency room clinical processes, we introduce two indicators, i.e., *ERPPIF1* and *ERPPIF2*, which evaluate whether a process can react to changes. Definitions 9 and 10 provide the formal explanation of each flexibility-related indicator.

Definition 9 (ERPPIF1: The total number of variants in a log). *Let N_v be the total number of variants in the log.*

$$- N_v = \sum_{r=1}^{o} \begin{cases} 1 & \text{if } V_r \in \{\sum_{0<k<|c|} \pi_{var}(c_k)\} \\ 0 & otherwise \end{cases}$$

Definition 10 (ERPPIF2: The total number of relations in a process model). *Let N_{ar} be the total number of relations in the process model.*

$$- N_{ar} = \sum_{0<k\leq|c|} \sum_{0<i<j\leq n} \begin{cases} 1 & \text{if } c_k \in L \wedge e_{k,i}, e_{k,j} \in c_k \wedge a_l, a_m \in A \\ & \wedge e_{k,i} > e_{k,j} \wedge \pi_a(e_{k,i}) = a_l \\ & \wedge \pi_a(e_{k,j}) = a_m \\ 0 & otherwise \end{cases}$$

These two indicators signify to identify whether the process model has an ability to handle a higher variety of cases with different control-flows.

4   Case Study

To demonstrate the applicability of the proposed process performance indicators, we performed a case study with the real-life clinical data collected from the electronic health records (EHR) system in a tertiary hospital in Korea.

4.1   Context

In the case study, we collected the clinical event log of the ER patients during 2018. The event log contained 15 medical tasks: entry, basic treatment, first aid treatment, other treatment, diagnostic test, visual test, consultation, cooperation request, cooperation arrival, the decision on hospitalization or discharge, prescription request, prescription receiving, certificate issuing, discharge, and hospitalization. Also, in the log, around 460,000 events were included for about 30,000 patients who visited the emergency room. To measure the PPIs, we used the Fluxicon Disco [10] and ProDiscovery [16].

4.2   Results

Time Perspective. For the time perspective, we first measured the length of stays for ER patients, i.e., *ERPPIT1*. As a result, we identified that patients stayed 8.35 h in the ER on average; while the median is 4.8 h. Figure 2 presents the distribution of the length of stays of patients. In the figure, we can identify that most of the patients, i.e., 97%, stayed just within a single day; while only 3% of patients stayed over a day. Here, interestingly, we were able to find out a couple of patients who remain for around 40 days.

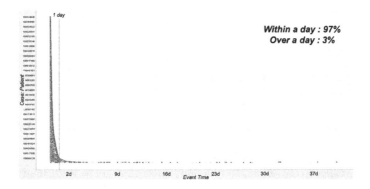

Fig. 2. The length of stay analysis result using the dotted chart

Concerning the second time-related indicator, there was a significant difference of variants according to its characteristic on whether it connects to hospitalization or discharge. More in detail, inpatients tended to stay longer than discharged patients (hospitalized: 11.52 h and discharged: 5.42 h on average).

Also, we measured the cycle time for clinical activities in the ER process, i.e., *ERPPIT3*, and several points required relatively long cycle time: decision on hospitalization and discharge, hospitalization, and prescription receiving. To identify the preceding activities that cause problematic points, we performed the model-based performance analysis as depicted in Fig. 3. As a result, we identified that it takes a long time from the diagnostic test to hospitalization; besides, it was followed by the decision on hospitalization and discharge to discharge, visual test to prescription request, and diagnostic test to the decision on hospitalization and discharge.

Cost Perspective. Regarding the cost perspective, we measured the number of doctors. From the log, we identified that there were 6.8 doctors on average in a single day. More in detail, we identified that it has a time pattern of the value; thus, there was a variation according to time. Figure 4 depicts the resource-related analysis result. In the figure, blue dots and grey lines signify the number of doctors working in the emergency room by each hour and patients staying

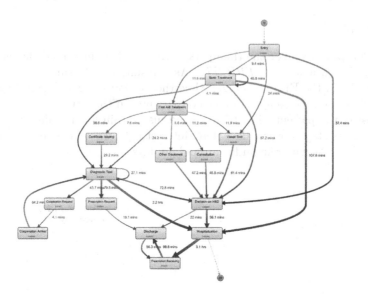

Fig. 3. The time-related analysis result with the process model

in the ER, respectively. Thus, it was confirmed that there were many medical personnel as the number of patients increased during the daytime.

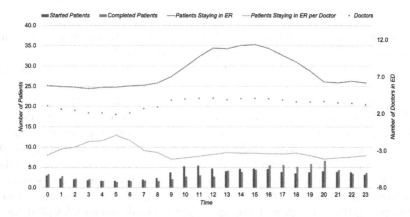

Fig. 4. The resource-related analysis result for the cost and quality perspectives (Color figure online)

Quality Perspective. With regard to the quality perspective, we calculated the workloads for doctors in the ER, i.e., *ERPPIQ1*. Due to the limitation of the collected data, the number of patients in the emergency room, rather than the number of medical activities performed by the physician, was assumed as a workload. As a result, we identified that each person averagely managed 11.9

patients in a single day. Also, there was a variation of workload by each hour as depicted in Fig. 4. More in detail, there was a trend that clinicians were busier at dawn despite the small number of patients, as shown in the yellow line.

Then, the variation of the length of stay for ER patients was computed, i.e., *ERPPIQ2*, and the standard deviation was 8.61 h. More in detail, we analyzed that the value of hospitalized patients (9.99 h) was more diverse than discharged patients (5.44 h).

For *ERPPIQ3*, the triage-based patient response rate was measured, and the average value was 0.74. To this end, we performed further analysis with the dotted chart, as depicted in Fig. 5. The figure shows the response pattern of the well-managed day with the high value of 0.91. As presented in the blue box in the figure, we observed that a more urgent person, i.e., the red dots, was treated faster than a less urgent person, i.e., the yellow dots.

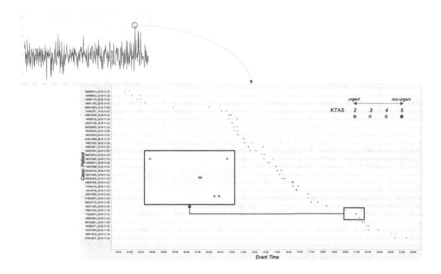

Fig. 5. The triage-based patient response analysis result (Color figure online)

Flexibility Perspective. Lastly, in the flexibility perspective, we identified that there were 25,004 variants in the log (i.e., *ERPPIF1*). Thus, considering the total number of patients, i.e., 30,000 patients, it was confirmed that most people have different variants from each other. Then, we produced the ER process model with a discovery technique, i.e., frequency mining [4], and measured the number of relations in the model, i.e., *ERPPIF2*. As a result, there were 188 activity relations in the discovered model.

Discussion with ER Experts. These case study results were discussed with ER experts to assess the overall state of the ER and identify the rationales of them. First, the hospital has well managed the time-related indicators. Notably, it was confirmed that a couple of points that take a long time are the results of not determining hospitalization at night for the convenience of other inpatients. Regarding the cost perspective, the number of clinicians was maintained at the appropriate level. In terms of the quality, the variation of the length of stay and the response to the emergency patients were satisfactory, but the proper allocation of the work considering the number of patients was necessary. Finally, we confirmed that the emergency room was operated with sufficient flexibility to care for various patients.

5   Discussion and Conclusion

In this paper, we suggested ERPPIs (Emergency Room Process Performance Indicators) for the emergency room to assess emergency room processes based on four perspectives: time, cost, quality, and flexibility. Also, we validated the proposed indicators by applying to the real-life clinical event log collected from a tertiary hospital in Korea.

Our work has an outstanding implication of presenting a method to directly measure the performance of the ER processes with the four perspectives, unlike other existing general frameworks, which consist of data preparation, preprocessing, analysis, and evaluation. As such, we believe that our framework has sufficient applicability in practice, and its results can become a solid basis for decision making in the ER.

Our work also has several limitations. First, it is necessary to advance ERPPIs by diversifying and elaborating them. The indicators presented in this study are just the most fundamental metrics. Therefore, it should be expanded in consideration of various aspects of the context in the emergency room, as well as aspects that can take advantage of the advanced techniques of process mining. Besides, despite that the criteria for determining the analysis results are needed to evaluate the emergency room process, this study did not present them but merely chose to discuss them with the medical experts. Therefore, we will work on developing a method to derive the appropriate criteria, based on the results of analysis by period (e.g., weekly or monthly).

Furthermore, for future works, we will implement a tool to support the application of the proposed framework. Also, we will make a connection of this research to the process redesign that presents improvement plans according to the results of the performance analysis. Lastly, more case studies will be conducted to validate our framework.

References

1. Baek, H., Cho, M., Kim, S., Hwang, H., Song, M., Yoo, S.: Analysis of length of hospital stay using electronic health records: a statistical and data mining approach. PLoS ONE **13**(4), e0195901 (2018)
2. Berner, E.S.: Clinical Decision Support Systems, vol. 233. Springer, Cham (2007). https://doi.org/10.1007/978-3-319-31913-1

3. Cho, M., Song, M., Comuzzi, M., Yoo, S.: Evaluating the effect of best practices for business process redesign: an evidence-based approach based on process mining techniques. Decis. Support Syst. **104**, 92–103 (2017)
4. Cho, M., Song, M., Yoo, S., Reijers, H.A.: An evidence-based decision support framework for clinician medical scheduling. IEEE Access **7**, 15239–15249 (2019)
5. Del-Río-Ortega, A., Resinas, M., Cabanillas, C., Ruiz-Cortés, A.: On the definition and design-time analysis of process performance indicators. Inf. Syst. **38**(4), 470–490 (2013)
6. Derlet, R.W., Richards, J.R.: Overcrowding in the nation's emergency departments: complex causes and disturbing effects. Ann. Emerg. Med. **35**(1), 63–68 (2000)
7. Dumas, M., La Rosa, M., Mendling, J., Reijers, H.A., et al.: Fundamentals of Business Process Management, vol. 1. Springer, Heidelberg (2013). https://doi.org/10.1007/978-3-642-33143-5
8. Fernandez-Llatas, C., et al.: Analyzing medical emergency processes with process mining: the stroke case. In: Daniel, F., Sheng, Q.Z., Motahari, H. (eds.) BPM 2018. LNBIP, vol. 342, pp. 214–225. Springer, Cham (2019). https://doi.org/10.1007/978-3-030-11641-5_17
9. Gray, J.A.M.: Evidence-Based Healthcare and Public Health: How to Make Decisions About Health Services and Public Health. Elsevier Health Sciences (2009)
10. Günther, C.W., Rozinat, A.: Disco: discover your processes. BPM (Demos) **940**, 40–44 (2012)
11. Kueng, P.: Process performance measurement system: a tool to support process-based organizations. Total Qual. Manag. **11**(1), 67–85 (2000)
12. Kwon, H., et al.: The Korean triage and acuity scale: associations with admission, disposition, mortality and length of stay in the emergency department. Int. J. Qual. Health Care **31**(6), 449–455 (2019)
13. Leu, J.D., Huang, Y.T.: An application of business process method to the clinical efficiency of hospital. J. Med. Syst. **35**(3), 409–421 (2011)
14. Mans, R.S., van der Aalst, W.M., Vanwersch, R.J.: Process Mining in Healthcare: Evaluating and Exploiting Operational Healthcare Processes. Springer, Heidelberg (2015). https://doi.org/10.1007/978-3-319-16071-9
15. Mans, R.S., Schonenberg, M.H., Song, M., van der Aalst, W.M.P., Bakker, P.J.M.: Application of process mining in healthcare – a case study in a Dutch hospital. In: Fred, A., Filipe, J., Gamboa, H. (eds.) BIOSTEC 2008. CCIS, vol. 25, pp. 425–438. Springer, Heidelberg (2008). https://doi.org/10.1007/978-3-540-92219-3_32
16. PuzzleData: Prodiscovery. http://demo.prodiscovery.co.kr. Accessed 17 May 2019
17. Rebuge, Á., Ferreira, D.R.: Business process analysis in healthcare environments: a methodology based on process mining. Inf. Syst. **37**(2), 99–116 (2012)
18. Rojas, E., Cifuentes, A., Burattin, A., Munoz-Gama, J., Sepúlveda, M., Capurro, D.: Performance analysis of emergency room episodes through process mining. Int. J. Environ. Res. Public Health **16**(7), 1274 (2019)
19. Rojas, E., Munoz-Gama, J., Sepúlveda, M., Capurro, D.: Process mining in healthcare: a literature review. J. Biomed. Inform. **61**, 224–236 (2016)
20. Rojas, E., Sepúlveda, M., Munoz-Gama, J., Capurro, D., Traver, V., Fernandez-Llatas, C.: Question-driven methodology for analyzing emergency room processes using process mining. Appl. Sci. **7**(3), 302 (2017)

Interactive Data Cleaning for Process Mining: A Case Study of an Outpatient Clinic's Appointment System

Niels Martin[1]([✉]), Antonio Martinez-Millana[2], Bernardo Valdivieso[3],
and Carlos Fernández-Llatas[2]

[1] Hasselt University, Agoralaan Building D, 3590 Diepenbeek, Belgium
`niels.martin@uhasselt.be`
[2] Universitat Politècnica de València, Camí de Vera, s/n, 46022 València, Spain
{`anmarmil,cfllatas`}`@itaca.upv.es`
[3] Hospital Universitario y Politécnico de La Fe,
Avinguda de Fernando Abril Martorell, 106, 46026 València, Spain
`valdivieso_ber@gva.es`

Abstract. Hospitals are becoming increasingly aware of the need to improve their processes and data-driven approaches, such as process mining, are gaining attention. When applying process mining techniques in reality, it is widely recognized that real-life data tends to suffer from data quality problems. Consequently, thorough data quality assessment and data cleaning is required. This paper proposes an interactive data cleaning approach for process mining. It encompasses both data-based and discovery-based data quality assessment, showing that both are complementary. To illustrate some key elements of the proposed approach, a case study of an outpatient clinic's appointment system is considered.

Keywords: Process mining · Data quality · Interactive data cleaning · Process discovery · Outpatient clinic

1 Introduction

Hospitals are confronted with a multitude of challenges including reduced budgets contrasted to augmenting care needs [15]. To cope with these challenges, hospitals are becoming increasingly aware of the need to improve their processes [16]. This awareness, combined with the increased availability of data about these processes, led to an increased attention for process mining in the healthcare domain. Process mining refers to the extraction of knowledge from an event log containing process execution information originating from a process-aware information system such as a hospital information system (HIS).

Process mining research has a predominant focus on the development of new techniques or the innovative application of existing techniques [6]. However, when applying these techniques in reality, it has been widely recognized

© Springer Nature Switzerland AG 2019
C. Di Francescomarino et al. (Eds.): BPM 2019 Workshops, LNBIP 362, pp. 532–544, 2019.
https://doi.org/10.1007/978-3-030-37453-2_43

that real-life data tends to suffer from a multitude of data quality problems [2,6,21]. This especially holds in a flexible and dynamic environment as healthcare, where commonly observed data quality issues include missing events and incorrect timestamps [11,15,16]. Given the biases that data quality issues can introduce in the results, research attention on data quality assessment is increasing in the process mining field in recent years [3,13,21–23]. Data quality assessment research tends to focus on the identification of problematic patterns in the data, which are candidate for mitigation before proceeding to process discovery. However, process discovery can also be leveraged to detect data quality issues that might otherwise remain hidden from the analyst. Despite its potential, the use of process discovery within the context of data cleaning has not been given explicit research attention.

This paper proposes an interactive data cleaning approach consisting of three key components: data-based data quality assessment, discovery-based data quality assessment, and data cleaning heuristics. To illustrate these key components, a case study of an outpatient clinic's appointment system is used. The available dataset suffers from several timestamp-related data quality issues, requiring data cleaning to make it usable for process mining purposes. The proposed data cleaning approach and the case study position process discovery as an integral part of data cleaning, which is a new angle in literature. Moreover, the interactive character of the approach supports stepwise data quality improvement in close collaboration with domain experts.

The remainder of this paper is structured as follows. Section 2 highlights some key prior research on data quality in the process mining field. Section 3 presents the interactive data cleaning approach. In Sect. 4 the case study is outlined. The paper ends with a discussion and conclusion in Sect. 5.

2 Related Work

Data quality has been widely studied in several domains such as statistics and data mining [4]. This section will focus on related research within the process mining field, which can be subdivided in: (i) the identification of data quality issues, and (ii) the mitigation of data quality issues.

2.1 Identification of Data Quality Issues

With respect to the identification of data quality issues, Verhulst [23] reviews existing literature to develop a data quality framework for event logs consisting of 12 high-level dimensions such as completeness and consistency. Bose et al. [6] distinguish 27 more specific event log data quality problems in four categories: missing data, incorrect data, imprecise data, and irrelevant data. Examples of issues are missing events, incorrect timestamps, and imprecise resource information. Using the framework of Bose et al. [6], Mans et al. [15] evaluate the data quality in the HIS-database of the Maastricht University Medical Centre.

Following an interview-based approach, they conclude that the three most frequently occurring issues are missing events, imprecise timestamps, and imprecise resource information. Similar to Mans et al. [15], Kurniati et al. [14] assess the data quality of the publicly available MIMIC-III database for process mining purposes. This involves, amongst others, determining whether case identifiers, activity labels and timestamps are available, and whether duplicated data is present.

While Bose et al. [6] outline event log quality issues at a generic level, Suriadi et al. [21] define 11 event log imperfection patterns in a more detailed way. This involves, amongst others, a description of how an issue manifests itself in the log and its side effects [21]. Similarly, Vanbrabant et al. [22] present a set of specific data quality assessment techniques which can be used to identify data flaws prior to its use for process analysis purposes.

At a methodological level, Andrews et al. [3] describe a cyclical methodology aiming to support the initial stages of a process mining project while taking data quality explicitly into account. Another approach, presented in Fox et al. [13], stresses, amongst others, the importance of keeping a structured data quality register.

2.2   Mitigation of Data Quality Issues

Several authors propose heuristics for data cleaning in an effort to improve the quality of an event log. These approaches range from conceptual recommendations to formal methods. For an extended reference list on event log cleaning methods, the reader is referred to the recent review by Solti [20].

An example of a conceptual approach are the recommendations of Suriadi et al. [21] to tackle each of the defined event log imperfection patterns. When, for instance, multiple events share the same timestamp because they are recorded by saving a single form, they propose to merge these events in a single [21]. Similarly, Martin [16] conceptually argues the potential of data integration to tackle quality problems in HIS-data.

More formal methods are developed to impute missing data in an event log [5,8] or to correct wrong data [9,18]. While doing so, these methods typically require domain knowledge, often in the form of a process model. This holds, e.g., for Rogge-Solti et al. [18], where a process model is used to repair timestamp errors [18]. Similarly, Di Francescomarino et al. [8] assume that a correct and complete process model is available in an effort to complete event log traces using action languages. Besides a process model, Bayomie et al. [5] also requires activity duration information as an input to detect missing case identifiers by means of decision trees. For the purpose of repairing event ordering issues in an event log, Dixit et al. [9] do not require a full process model. In their approach, users are required to establish ordered relationships between activities, which are matched with the event log using alignment techniques [9].

While the aforementioned approaches require domain knowledge to operate, e.g. in the form of a process model, Nguyen et al. [17] recently developed approaches to detect anomalies in an event log and to add missing values without a need for prior knowledge. To this end, autoencoders, which is a specific type of neural networks, are used. Even though preliminary results on structured artificial data are promising, their approaches still experience difficulties to manage the variability and complexity of real-life data [17].

3   Interactive Data Cleaning

This paper proposes an interactive data cleaning approach, which is visualized in the boxed area in Fig. 1. The approach centers around (i) data quality assessment to identify data quality problems and (ii) data cleaning heuristics to mitigate these problems. Taking a raw log as an input, a user will perform both data-based and discovery-based data quality assessment. Existing assessment approaches in the process mining field have a strong data-based focus [21,22], i.e. they only concentrate on retrieving problematic patterns in the dataset. However, discovery-based assessment, aiming to identify data quality issues by discovering process models, can enable the identification of data inconsistencies which might remain hidden during data-based assessment. The discovered process models can relate to the control-flow, but also to other process mining perspectives such as the organizational perspective. The potential of discovery-based assessment will be illustrated in the case study in Sect. 4.

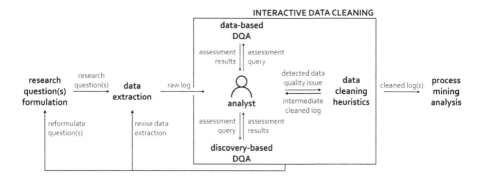

Fig. 1. Interactive data cleaning approach

Based on the assessment results, the analyst can specify appropriate data cleaning heuristics to rectify the detected issues. This generates an intermediate cleaned log, which can, once again, be subject to data quality assessment. After several iterations between assessment and cleaning, a cleaned event log is obtained, enabling the analyst to proceed with the process mining analysis.

Figure 1 positions the interactive data cleaning approach within the broader context of a question-driven process mining project [1] because data cleaning is likely to depend on the question(s) under consideration. When questions relate solely to the process control-flow, the order of activities is essential and the exact timestamp values are of secondary importance. If, in contrast, a process performance analysis is required for a particular question, the analyst should be more reluctant towards changing timestamp values during data cleaning. This shows that, when research questions cover several types of process mining analyses, it might be necessary to generate several cleaned event logs. Data quality assessment can also instigate a reformulation of the research questions as a question might no longer be answerable due to data quality issues. Alternatively, data quality assessment might indicate that the data extraction process should be revised to, e.g., add additional data to the raw log.

The approach in Fig. 1 is complementary to existing process mining methodologies. It presents an extension of the log inspection stage of the Process Diagnostics Method, which is the stage that aims to familiarize the analyst with the event log [7]. When considering the L* life-cycle model, interactive data cleaning can be positioned between data extraction (stage 1) and the creation of a control-flow model (stage 2) [1]. Within the PM^2 methodology, it can strengthen the data processing stage as no explicit attention is attributed to data cleaning [10]. Figure 1 is also consistent with the quality-driven process mining preparation methodology of Andrews et al. [3]. In particular, it proposes an interactive approach to operationalize the event quality, pre-study process mining, and evaluation and feedback steps [3].

4   Case Study: Appointment Data of an Outpatient Clinic

This section illustrates some key elements of the interactive data cleaning approach using a case study with appointment data of an outpatient clinic. First, the case study is described (Sect. 4.1). Afterwards, data-based data quality assessment (Sect. 4.2) and discovery-based data quality assessment (Sect. 4.3) are illustrated. Pointers to data cleaning heuristics are added to these last two sections. Due to space limitations, the case study will focus on the components included in the boxed area in Fig. 1 and will center around the control-flow. A more extensive case study, encompassing the entire approach in Fig. 1, will be conducted in future work.

4.1   Case Study Description

The dataset for the case study was provided by University Hospital La Fe (Valencia, Spain), which provides healthcare services to more than 300,000 people and accounts for over 1,100 doctors. The hospital has an area devoted to outpatient services of a wide variety of clinical specialties, which is the specific focus of the case study.

The outpatient clinic is equipped with an automatic registration system using magnetic cards. This magnetic card is used to identify a patient during his/her visit to the hospital. More specifically, four reference times are recorded:

- **Arrival time (At):** time at which the patient arrives at the waiting room and he/she registers using the magnetic card
- **Call time (Ct):** time at which the doctor calls the patient to the consultation
- **Entry time (Et):** time at which the patient enters the room for the consultation
- **Departure time (Dt):** time at which the patient leaves from the consultation

The studied dataset contains anonymized information on 1.6 million appointments that took place at the hospital for 262,061 unique patients. Even though the data shows that a patient can have several appointments during a single visit, the majority of the visits consist of only one or two consultations (87.70% and 10.80% of all visits, respectively).

4.2   Data-Based Data Quality Assessment and Data Cleaning Heuristics

The available dataset suffers from several data quality issues. This subsection highlights the key timestamp-related data quality issues that were detected during data-based data quality assessment.

Missing Timestamps. Missing timestamps imply that one or more of the reference timestamps are absent for an appointment. This quality issue is explicitly recognized in the framework of Bose et al. [6], and belongs to the missing values category defined by Vanbrabant et al. [22].

The call time is absent the most often (for 24.44% of the appointments), followed by the arrival time (15.43%), entry time (4.16%) and departure time (4.15%). Table 1 provides richer insights on this matter as it studies the combinations of missing timestamps. From this table, it follows that all timestamps

Table 1. Missing timestamps

Missing timestamps	n	%
None	1,222,289	74.64
At, Ct	169,534	10.35
Ct	162,589	9.93
At, Ct, Et, Dt	67,766	4.14
At	15,042	0.92
At, Ct, Et	261	0.02
At, Et, Dt	88	<0.01
At, Ct, Dt	13	<0.01
Dt	3	<0.01
Ct, Et, Dt	1	<0.01
Ct, Et	1	<0.01

are recorded for 74.64% of the appointments. For 10.35% of the appointments both the arrival time and the call time are missing, while for 0.92% only the former is missing and for 9.93% only the latter is absent. Note that for 4.4% of the consultations, none of the reference times are recorded. Other missing timestamp combinations occur less frequently.

Overlapping Timestamps. Another data quality problem are overlapping timestamps, which means that several of the reference times share the same timestamp. This can, at least partly, be attributed to the fact that timestamps are recorded at the granularity level of minutes. When timestamps are expected to be close to each other, e.g. for the call time and the entry time, these times can be correct. In other situations, e.g. between the entry and the departure time, overlapping timestamps are more problematic. In relation to the data quality framework of Bose et al. [6], this problem can belong to either the imprecise or the incorrect timestamp category (depending on whether the issue is problematic). In Vanbrabant et al. [22], it is positioned within the inexactness of timestamps group.

Table 2 provides an overview of the occurrence of overlapping timestamps. When assuming that absent timestamps do not overlap, all reference times differ in 33.46% of the appointments. The most frequently occurring overlap is situated between the call time and the entry time (in 61.89% of the appointments this is the only overlap). While the former could be perceived as normal behavior, this does not hold for an overlap between the entry and departure time (the only overlap in 3.74% of the appointments). As a consultation is unlikely to end in the same minute it started, an overlap indicates that one of these timestamps are incorrect.

Table 2. Overlapping timestamps

Overlapping timestamps	n	%
Ct, Et	1,013,461	61.89
None	547,943	33.46
Et, Dt	61,259	3.74
Ct, Et, Dt	7,691	0.47
At, Ct, Et	2,660	0.16
At, Et	1,972	0.12
At, Ct	1,347	0.08
{At, Ct} and {Et, Dt}	618	0.04
At, Ct, Et, Dt	315	0.02
Ct, Dt	150	0.01
At, Et, Dt	78	<0.01
At, Dt	73	<0.01
{At, Dt} and {Ct, Et}	17	<0.01
At, Ct, Dt	3	<0.01

Time Ordering Violation. The correct time ordering for an appointment is known, i.e. arrival time → call time → entry time → departure time. Hence, appointments for which this order is not respected are likely to contain incorrect timestamps. This quality issue belongs to the incorrect timestamp category according to Bose et al. [6] and constitutes a violation of logical order in the framework of Vanbrabant et al. [22].

Time order violations occurring at least 250 times are included in Table 3. Note that appointments which violate the correct time ordering due to missing timestamps are not taken into consideration. From Table 3, it can be observed that the most frequent violation constitutes the recording of the arrival time between the start and the completion of a consultation.

Table 3. Time order violations

Time order violation	n	%
Et → At → Dt	12,267	0.75
At → Ct → Dt → Et	1,630	0.10
Et → Dt → At	1,531	0.09
Dt → At → Ct → Et	892	0.05
Ct → At → Et → Ct	463	0.03

Appointment Overlap in the Same Room. Besides the aforementioned issues, several other data quality issues were discovered. An example involves the observation of appointment overlaps in the same room, implying that the next appointment in a particular room appears to have started before the current one has ended. For instance: the appointment of patient A in a specific room is recorded from 10:58 until 11:06, while the consultation of patient B in that same room takes place from 11:00 until 11:15. Even though this is not sensible in practice, this pattern has been identified in the dataset for 605,188 appointments.

Data Cleaning Heuristics. Based on the results of the data-based data quality assessment, several data cleaning heuristics are defined taking into account expected behavior and knowledge about the process. Considering the quality problems outlined above, some exemplary cleaning heuristics are:

- **At and Ct missing:** Issue present in 10.35% of the appointments. Under the assumption that these timestamps occur right before Et, both timestamps can be imputed right before Et.
- **Ct missing:** Issue present for 9.93% of the appointments. Under the assumption that Ct happens immediately before the consultation starts, Ct can be imputed right before Et.
- **Ct and Et overlapping:** Issue occurs for 61.21% of the appointments. Under the assumption that both timestamps reflect distinct states, timestamps can be corrected such that entry happens after the call.

– **Appointment overlap in same room:** Issue occurs for 605,188 appointments. When making the assumption that a room can only host one appointment at the same time, timestamps can be corrected such that the departure time of the previous appointment in a particular should precede the entry time of the current appointment in that room.

4.3   Discovery-Based Data Quality Assessment and Data Cleaning Heuristics

To illustrate discovery-based data quality assessment, two distinct process models will be considered. The first one, shown in Fig. 2, will act as a reference model as it is retrieved from the full dataset without applying data cleaning heuristics. The second model, shown in Fig. 3, is mined from the full dataset after applying the data cleaning heuristics mentioned at the end of Sect. 4.2.

Both models are discovered using PALIA [12]. Arrow colours indicate the number of times that a path is followed (with green referring to less frequent paths and red to more frequent paths). The colors of the circles reflect the average time that elapses between that event and the next event. When mining a process model, we enforced PALIA with two basic conditions to produce a model: (i) reliable time and date formats, and (ii) reject parallel branches as a patient cannot be in several states at exactly the same time. Consequently, Fig. 2 is based on 1,090,808 appointments (66.61% of the dataset) and Fig. 3 on 1,406,326 appointments (85.88% of the dataset).

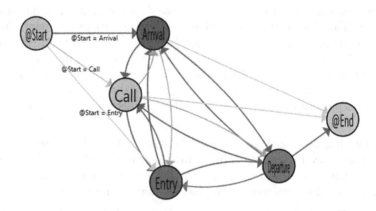

Fig. 2. Discovered process model before data cleaning heuristics (Color figure online)

When comparing Figs. 2 and 3, it becomes clear that the data cleaning heuristics based on data-based data quality assessment succeed in removing some anomalous behavior caused by data quality problems. Compared to Fig. 2, where almost every connection is marked as frequent, the majority of the patients follow the expected trajectory (Arrival → Call → Entry → Departure).

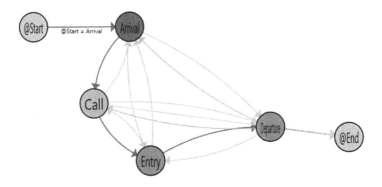

Fig. 3. Discovered process model after data cleaning heuristics (Color figure online)

Despite the effectiveness of the data cleaning heuristics, Fig. 3 shows patterns that require further investigation to determine whether they originate from further data quality issues or just represent unexpected process characteristics. Examples of such patterns include:

– **At after Ct:** These patients may have been called by the doctor before arriving to the appointment, or they forgot to record their arrival with their magnetic card and only do this once they have been called.
– **At after Et:** This pattern suggests that the arrival takes place after the consultation has started, or that the appointment has ended without the departure being recorded.
– **Ct after Et:** Some appointments show this anomalous behavior as a patient who has entered should no longer be called for that appointment. Further analysis should determine whether the call is related to another appointment of this patient, or whether the doctor performs the calling action once the patient has entered the cabinet.
– **Departure connected to Arrival, Call and Entry:** This exemplifies the process variability. The model shows that some patients with multiple appointments during one visit only seem to arrive after their first appointment has finished. Moreover, some patients are immediately called, without an arrival preceding it. Other patients immediately go to the doctor's cabinet for their next appointment (Departure → Entry).

The potential anomalies that appear from analyzing the mined control-flow model need to be sorted out before a process mining analysis can start. The identified issues exemplify the added-value of discovery-based data quality assessment compared to only using data-based data quality assessment. The need for domain expertise to investigate the aforementioned issues supports the need for an interactive approach, as visualized in Fig. 1 by the central position of the analyst. When these issues can be attributed to a data quality issue, suitable data cleaning heuristics need to be specified.

5   Discussion and Conclusion

This paper proposed an interactive data cleaning approach consisting of three key components: data-based data quality assessment, discovery-based data quality assessment, and data cleaning heuristics. Existing research on data quality assessment in process mining focuses on the identification of problematic patterns in the data, i.e. data-based data quality assessment. The proposed approach argues that process discovery should also be considered during assessment as this can unveil data quality issues that might otherwise have remained hidden. This was illustrated in a case with a fairly simple process structure. The more complex the process structure becomes, the likelier it becomes that some data problems are overlooked during data-based quality assessment. In such contexts, the added-value of discovery-based insights will become even more significant.

The interactive character of the proposed approach requires an analyst with sufficient domain expertise in control of interactive data cleaning. When applying the approach, an inherent risk for confirmation bias exists. Confirmation bias refers to a person's tendency to search for information that confirms his/her beliefs and to avoid contradicting information [19]. Within the context of data cleaning, this would imply that an analyst might attribute patterns that conflict his/her process beliefs to data quality issues and take according data cleaning measures. Consequently, a prudent approach towards data cleaning is required. Future research could develop guidelines to support analysts and researchers on this matter. A potential guideline could relate to maintaining detailed data cleaning records to ensure that each cleaning step is fully traceable.

Besides guidelines for data cleaning, several other research challenges can be distinguished. Foremost, formal techniques for discovery-based data quality assessment should be developed, similar to existing techniques for data-based assessment. This would enable the development of an integrated toolkit to support interactive data cleaning. Besides enabling the application of assessment techniques and cleaning heuristics, this toolkit should also make the effect of the heuristics explicit. Related to the latter, another promising direction for future work is a benchmark study on the effect of various data cleaning methods on process mining outcomes. Besides Dixit et al. [9] as a notable exception, limited research attention is attributed to this topic.

References

1. van der Aalst, W.M.P.: Process Mining: Data Science in Action. Springer, Heidelberg (2016). https://doi.org/10.1007/978-3-662-49851-4
2. van der Aalst, W.M.P., et al.: Process mining manifesto. In: Daniel, F., Barkaoui, K., Dustdar, S. (eds.) BPM 2011. LNBIP, vol. 99, pp. 169–194. Springer, Heidelberg (2012). https://doi.org/10.1007/978-3-642-28108-2_19
3. Andrews, R., et al.: Leveraging data quality to better prepare for process mining: an approach illustrated through analysing road trauma pre-hospital retrieval and transport processes in Queensland. Int. J. Environ. Res. Public Health 16(7), 1138 (2019)

4. Batini, C., Scannapieco, M.: Data Quality: Concepts, Methodologies and Techniques. Springer, Heidelberg (2006). https://doi.org/10.1007/3-540-33173-5
5. Bayomie, D., Helal, I.M.A., Awad, A., Ezat, E., ElBastawissi, A.: Deducing case IDs for unlabeled event logs. In: Reichert, M., Reijers, H.A. (eds.) BPM 2015. LNBIP, vol. 256, pp. 242–254. Springer, Cham (2016). https://doi.org/10.1007/978-3-319-42887-1_20
6. Bose, R.J.C.P., Mans, R.S., van der Aalst, W.M.P.: Wanna improve process mining results? It's high time we consider data quality issues seriously. Technical report, BPM Center Report BPM, 13 February 2013
7. Bozkaya, M., Gabriels, J., van der Werf, J.M.: Process diagnostics: a method based on process mining. In: 2009 International Conference on Information, Process, and Knowledge Management, pp. 22–27. IEEE (2009)
8. Di Francescomarino, C., Ghidini, C., Tessaris, S., Sandoval, I.V.: Completing workflow traces using action languages. In: Zdravkovic, J., Kirikova, M., Johannesson, P. (eds.) CAiSE 2015. LNCS, vol. 9097, pp. 314–330. Springer, Cham (2015). https://doi.org/10.1007/978-3-319-19069-3_20
9. Dixit, P.M., et al.: Detection and interactive repair of event ordering imperfection in process logs. In: Krogstie, J., Reijers, H.A. (eds.) CAiSE 2018. LNCS, vol. 10816, pp. 274–290. Springer, Cham (2018). https://doi.org/10.1007/978-3-319-91563-0_17
10. van Eck, M.L., Lu, X., Leemans, S.J.J., van der Aalst, W.M.P.: PM^2: a process mining project methodology. In: Zdravkovic, J., Kirikova, M., Johannesson, P. (eds.) CAiSE 2015. LNCS, vol. 9097, pp. 297–313. Springer, Cham (2015). https://doi.org/10.1007/978-3-319-19069-3_19
11. Fernández-Llatas, C., Lizondo, A., Monton, E., Benedi, J.M., Traver, V.: Process mining methodology for health process tracking using real-time indoor location systems. Sensors 15(12), 29821–29840 (2015)
12. Fernandez-Llatas, C., Valdivieso, B., Traver, V., Benedi, J.M.: Using process mining for automatic support of clinical pathways design. In: Fernández-Llatas, C., García-Gómez, J.M. (eds.) Data Mining in Clinical Medicine. Methods in Molecular Biology, vol. 1246, pp. 79–88. Springer, New York (2015). https://doi.org/10.1007/978-1-4939-1985-7_5
13. Fox, F., Aggarwal, V.R., Whelton, H., Johnson, O.: A data quality framework for process mining of electronic health record data. In: 2018 IEEE International Conference on Healthcare Informatics, pp. 12–21 (2018)
14. Kurniati, A.P., Rojas, E., Hogg, D., Hall, G., Johnson, O.A.: The assessment of data quality issues for process mining in healthcare using Medical Information Mart for Intensive Care III, a freely available e-health record database. Health Inform. J. 25, 1878–1893 (2018)
15. Mans, R.S., van der Aalst, W.M.P., Vanwersch, R.J.B.: Process Mining in Healthcare: Evaluating and Exploiting Operational Healthcare Processes. SpringerBriefs in Business Process Management. Springer, Cham (2015). https://doi.org/10.1007/978-3-319-16071-9
16. Martin, N.: Using indoor location system data to enhance the quality of healthcare event logs: opportunities and challenges. In: Daniel, F., Sheng, Q.Z., Motahari, H. (eds.) BPM 2018. LNBIP, vol. 342, pp. 226–238. Springer, Cham (2019). https://doi.org/10.1007/978-3-030-11641-5_18
17. Nguyen, H.T.C., Lee, S., Kim, J., Ko, J., Comuzzi, M.: Autoencoders for improving quality of process event logs. Expert Syst. Appl. 131, 132–147 (2019)

18. Rogge-Solti, A., Mans, R.S., van der Aalst, W.M.P., Weske, M.: Repairing event logs using timed process models. In: Demey, Y.T., Panetto, H. (eds.) OTM 2013. LNCS, vol. 8186, pp. 705–708. Springer, Heidelberg (2013). https://doi.org/10.1007/978-3-642-41033-8_89
19. Sanderson, C.A.: Social Psychology. Wiley, Hoboken (2010)
20. Solti, A.: Event log cleaning for business process analytics. In: Sakr, S., Zomaya, A. (eds.) Encyclopedia of Big Data Technologies. Springer, Heidelberg (2018). https://doi.org/10.1007/978-3-319-63962-8
21. Suriadi, S., Andrews, R., ter Hofstede, A.H., Wynn, M.T.: Event log imperfection patterns for process mining: towards a systematic approach to cleaning event logs. Inf. Syst. **64**, 132–150 (2017)
22. Vanbrabant, L., Martin, N., Ramaekers, K., Braekers, K.: Quality of input data in emergency department simulations: framework and assessment techniques. Simul. Model. Pract. Theory **91**, 83–101 (2019)
23. Verhulst, R.: Evaluating quality of event data within event logs: an extensible framework. Master's thesis, Eindhoven University of Technology (2016)

Clinical Guidelines: A Crossroad of Many Research Areas. Challenges and Opportunities in Process Mining for Healthcare

Roberto Gatta[1]([✉])[ID], Mauro Vallati[2][ID], Carlos Fernandez-Llatas[3][ID],
Antonio Martinez-Millana[3], Stefania Orini[8], Lucia Sacchi[10],
Jacopo Lenkowicz[4], Mar Marcos[11], Jorge Munoz-Gama[9], Michel Cuendet[5],
Berardino de Bari[12], Luis Marco-Ruiz[6], Alessandro Stefanini[7],
and Maurizio Castellano[1]

[1] Dipartimento di Scienze Cliniche e Sperimentali dell'Universitá
degli Studi di Brescia, Brescia, Italy
roberto.gatta.bs@gmail.com
[2] School of Computing and Engineering, University of Huddersfield, Huddersfield, UK
[3] ITACA, Universitat Politécnica de Valéncia, Valéncia, Spain
[4] Fondazione Policlinico Universitario A. Gemelli IRCCS, Rome, Italy
[5] Department of Oncology, University Hospital of Lausanne, Lausanne, Switzerland
[6] Norwegian Centre for E-health Research, University Hospital of North Norway,
Tromsø, Norway
[7] Dipartimento di Ingegneria dell'energia dei sistemi del territorio e delle costruzioni,
Universita degli Studi di Pisa, Pisa, Italy
[8] Alzheimer Operative Unit, IRCCS Istituto Centro San
Giovanni di Dio Fatebenefratelli, Brescia, Italy
[9] Pontificia Universidad Católica de Chile, Santiago de Chile, Chile
[10] Universitá di Pavia, Pavia, Italy
[11] Department of Computer Engineering and Science, Universitat Jaume I,
Castelló de la Plana, Spain
[12] Radiation Oncology Department, Centre Hospitalier Régional Universitaire
Jean Minjoz, Besançon, France

Abstract. Clinical Guidelines, medical protocols, and other healthcare indications, cover a significant slice of physicians daily routine, as they are used to support clinical choices also with relevant legal implications. On the one hand, informatics have proved to be a valuable mean for providing formalisms, methods, and approaches to extend clinical guidelines for better supporting the work performed in the healthcare domain. On the other hand, due to the different perspectives that can be considered for addressing similar problems, it lead to an undeniable fragmentation of the field. It may be argued that such fragmentation did not help to propose a practical, accepted, and extensively adopted solutions to assist physicians. As in Process Mining as a general field, Process Mining for Healthcare inherits the requirement of Conformance Checking. Conformance Checking aims to measure the adherence of a particular (discovered or known) process with a given set of data, or vice-versa.

© Springer Nature Switzerland AG 2019
C. Di Francescomarino et al. (Eds.): BPM 2019 Workshops, LNBIP 362, pp. 545–556, 2019.
https://doi.org/10.1007/978-3-030-37453-2_44

Due to the intuitive similarities in terms of challenges and problems to be faced between conformance checking and clinical guidelines, one may be tempted to expect that the fragmentation issue will naturally arise also in the conformance checking field. This position paper is a first step on the direction to embrace experience, lessons learnt, paradigms, and formalisms globally derived from the clinical guidelines challenge. We argue that such new focus, joint with the even growing notoriety and interest in PM4HC, might allow more physicians to make the big jump from user to protagonist becoming more motivated and proactive in building a strong multidisciplinary community.

Keywords: Conformance checking · Clinical guidelines · Computer interpretable clinical guidelines

1   Introduction

Process Mining (PM) is the discipline focusing on techniques, tools and methods to discover, monitor and improve real processes by extracting knowledge from event logs commonly available in today's information systems [1]. This sort of analysis of operational data can extract knowledge from the underlying sequences of activities and model the actual organizational workflow; for this reason, PM is sometimes described as a bridge between data mining and Business Process Management (BPM). There are three main areas subsumed by Process Mining: Process Discovery, Conformance Checking (CC), and enhancement. Buijs et al. [8] explain how automatic Process Discovery allows process models to be extracted from an event log; how CC allows monitoring deviations by comparing a given model with the event log; and how enhancement allows extending or improving an existing process model using information about the actual process recorded in the event log.

One of the most prominent domains of application of PM is healthcare, as suggested by a recent review of the area [48], based on 1,278 articles. Additionally, it is to be acknowledged that Process Mining in healthcare poses unique, non-trivial challenges because *hospital is not a factory and patients cannot be cured using a conveyor belt system*, as correctly reported by [15,28]. Indeed, the care pathway of a patient is often a long and demanding journey, whose complexity is tightly linked to the high number of professional figures, diagnostic opportunities and therapeutic strategies that are available for each particular clinical need. Multidisciplinary teams are often involved in the care process, and choices have to be made among several treatment options and based on a variety of evidence such as laboratory tests, imaging data, medical visits. The patients can usually play a role according to their values, beliefs or expectations. In coping with this complexity, even if the domain has well-established strategies, each single task – treatment or diagnostic procedure – can be seen as a chess move, where the physician and the patient wait to see the results, before deciding the next move. In a nutshell, to appropriately apply Process Mining for Healthcare

(PM4HC), there is the need to reckon *medical treatment processes are, in fact, highly dynamic, highly complex, increasingly multidisciplinary, and often ad hoc* [44]. Therefore, the dedicated field of PM4HC has been identified to mediate general PM with the needs of the clinical application domain.

According to our experience, what physicians are mainly and foremost keen on, when exposed to this kind of analysis techniques, is monitoring how patients flow through Clinical Guidelines (CG), in order to check not only the conformance agreement, but also to spot a light on those groups of patients that did not follow it, in the quest of understanding why that was the case and what are the implications [30]. This kind of needs has been also identified, in the last decades, by other research areas of Computer Science applications, such as BPM, Computer Interpretable Clinical Guidelines, Clinical Decision Support System (CDSS), Case Based Reasoning (CBR), among other disciplines. On the one hand, this fragmentation of perspectives and disciplines created a rich set of initiatives. On the other hand, the fragmentation may have reduced the concrete real-wold impact that a unified and compelling vision could have allowed to deliver. From this point of view, physicians need to be assisted in coping with CG, and their growing interest in PM4HC can really be a fertile ground to capture physicians engagement and make them take the leap to be proactive in leading effective application in the daily clinical practice.

In this paper, 12 research centers (out of which 6 are hospitals) propose their ideas on Conformance Checking in PM4HC, meant to provide a unified view on the future of the discipline in coping with the challenges of CG. The paper is organised as follows. We describe the notion of CG. Then, we introduce one of the most active areas of research in the field of CG, namely Computer Interpretable Clinical Guidelines (CICG). We then show the overlap between CICG and BPM, and how CG and PM have co-evolved in the last decade. Finally, discussions on opportunities and challenges posed by CG to the PM4HC discipline are given.

2 (Not Only) Clinical Guidelines

CG are defined as *statements that include recommendations intended to optimize patient care that are informed by a systematic review of evidence and an assessment of the benefits and harms of alternative care options* [23]. CG are growing in importance due to their potentially positive effect on the quality of care, efficiency in the use of resources, and in their ability to precisely define the legal duties and responsibility of providers and institutions. In the last years numerous evidence-based guidelines have been developed by a wide range of organizations and bodies. Such heterogeneity, often caused an enrichment of the general definition, according to specific needs. For example, the World Health Organization (WHO) defined three different types of guidelines: Rapid advice guideline, Standard guideline, Full guidelines. Other International organizations, individual state health policy departments, medical specialty organizations, and profit and no-profit entities can adopt personalized enrichment of the definition.

In addition, due to the need to put emphasis on specific aspects of the definition, a quite rich set of different satellite concepts has been defined over time (e.g., Indications, Recommendations, Standards, Consensus Statements, Expert Advice, etc.). Even if such items can be significantly different in terms of aims, they are often quite similar in terms of tools and methods for their representation (e.g. workflows, rules, decision trees, etc.).

Some of the commonly accepted issues concerning clinical guidelines are:

- they must be relevant to the care setting, clear, easy to access and apply, and auditable for feedback and reporting [58];
- they should be based upon the best available research evidence and practice experience, developed using clear, explicit processes to minimize bias and optimize transparency. Possibly, the quality of guidelines should be measures with one of the existing system (e.g. GRADE, AGREE II, etc.);
- they tend to address the common or average patient, and do not evaluate the impact of multiple chronic conditions, socio-personal context, etc.;
- they need to be updated and re-evaluated over time to be re-validated when new clinical evidence is available;
- they need to be adapted from international to local context. Adaptation is *the systematic approach to the modification of a guideline produced in one cultural and organisational setting for application in a different context* [59];
- their purpose is to support and inform, not to dictate. There may be the temptation to use them as legally-binding documents, but clinicians are the only Decision Makers, and have the responsibility of decisions;
- as guidelines have been used for decisions about insurance coverage and standards for measuring quality of care, they have become increasingly influential, and conflict of interest (COI) in developing guidelines has become an important potential source of bias in the development of CGs [53].

In dealing with other workflow-based similar concepts, the heterogeneity of definition can also be more dramatic: Care Pathway, for example, is defined by the European Pathway Association as *a complex intervention for the mutual decision making and organisation of care processes for a well-defined group of patients during a well-defined period*. In 264 articles, the concept was referred to with 84 different definitions [11], with differences mainly based on three features: nouns, characteristics and aims and outcomes.

Summarizing, even if CG (and related concepts) have relatively well established definitions, practical needs and goals can induce in some *ad hoc* re-definition or interpretation: this should generally be discouraged by a stronger consensus, aiming at a reduction of the fragmentation of terms and ambiguity.

3   Computer Interpretable Clinical Guidelines

Given the rising attention to evidence-based medicine, and the wide-spread adoption of electronic health records, the development of clinical decision support systems (CDSSs), in general, and Computer Interpretable Guidelines (CIGs), in particular, have emerged as relevant fields of research [38]. Starting from the late '90s, several research groups have devoted their attention to the development of languages for the representation and of tools for the execution of CIGs. Well-known examples of such languages include Asbru [51], GLIF [7], GLARE [55], PROforma [54], EON [37], and GUIDE [10]. These languages, which can be considered as task-network models, allow formally representing CGs, and executing them through an execution engine, which delivers recommendations by coupling the represented guideline knowledge to patient's data. The systematic comparison of CIG models carried out by Peleg et al. [40] highlighted that, although using different computational models, all these methodologies allow formalizing guidelines through a set of actions (the so-called plan) that are executed over time. The control-flow is defined by organizing plan components according to different routing schemes (e.g. sequence, parallel, etc), and all the models support nesting of processes and the explicit management of temporal constraints. A common challenge for those formalisms are the complexities derived from their local adaption [38]. Although formalisms such as GLIF were designed for reusing procedural knowledge across organizations, the complexity related to their adaption to local contexts and their connection with the Electronic Health Record (EHR) has limited their broad adoption [27,38]. This is actually a challenge shared by all computerized CDSS [27,57]. With the raise of interest into the use of standards for EHR, some researchers focused on the need for a proper connection between the CDSS/CIG and the EHR. Various standards have been proposed for defining summarized EHR data views that the CIG accesses (a.k.a. virtual medical record). Clinical information standards allow the procedural component (i.e. the decision algorithm) of the CDSS to reference standard data schemas (defined with HL7 CDA, openEHR, etc.) rather than proprietary data schemas. One of the original works to overcome the challenges for connecting the CIG and the EHR data schemas in a flexible manner was the GELLO language [52]. GELLO allowed for defining restrictions over object oriented models to allow the Arden syntax rules accessing data that could be represented in HL7 v3 [52]. In the United States, the adoption of HL7 CDA as a part of the meaningful use initiative has significantly contributed to boost the adoption of standard-based CDSS [12,16].

Many CDSS use clinical information standards for defining their virtual medical records, however CDSS often use a summary of the information contained in the EHR, for this reason CDSS-specific standards have been defined. Kawamoto et al. led an international collaboration that elicited a standard specifically designed for the definition of virtual medical records (HL7 vMR) [26]. In European nations such as Sweden, Denmark, or the UK, openEHR-based CIGs have also defined mechanisms for better scaling and decoupling procedural knowledge from EHR data schemas.

The Guideline Definition Language (GDL), introduced by Chen et al. is a rule-based language that allows to directly reference openEHR archetypes [9]. This allows the seamless integration of CDSS modules with openEHR-based EHRs. GDL has been used at large scale for classifying population according to their risk of suffering a particular disease [2]. Recently, the openEHR community published the specification for Task Planning that complements the GDL language by enabling the definition of workflows and actions as archetypes [4]. Both GDL and the openEHR Task Planning models are designed to run over openEHR compliant repositories, thus posing a requirement on the data format to be supplied to the CIG decision algorithm. When the clinical information is in a format (proprietary or standard) different from openEHR, a pre-processing stage can be performed for making it openEHR compliant [32, 33].

Latest developments not only take care of the information format, but also define a common service interface for exposing CIGs functionality. Examples of this are OPEN CDS, SMART ON FHIR, CDS HOOKS, openEHR REST specification, among others.

Another important challenge that is currently being addressed by the research community is related to the management of patients with multiple health conditions [5]. From a technical perspective, using guideline-based CDSSs to handle comorbidities requires the integration of multiple disease-specific CIGs, by preserving patient safety and maximizing efficiency during execution. The approaches proposed in this area include on the one hand the manual definition of a comprehensive guideline starting from separate CGs, and on the other the automatic integration of multiple CIGs [13, 45, 60], considering the temporal and runtime aspects as well [3, 24].

4   CIGs and BPM

Languages from the CIG field provide a wide range of constructs to accommodate the rich variety of CG knowledge. Peleg et al. distinguish two main dimensions, namely knowledge about structuring of CG procedures in plans of decisions and actions, and about linking to patient data and medical concepts [40]. The parallels with the BPM and workflow fields as regards the former dimension have been recognized and exploited for some time in several works. A seminal work is the analysis of CIG languages based on the implementability of workflow control-flow patterns, by Mulyar et al. [35]. In the same line, another work provides a formal method to determine the implementability of patterns in a CIG language, with illustrations in the PROforma language [19]. To take another example, BPM notations have been advocated as a tool to facilitate the acquisition of CG procedural knowledge [34], motivated by the fact that CIG languages are not always comprehensible for clinicians.

In addition, the growing interest into the application of business process modeling and workflow management systems (WfMs) to the representation of clinical workflows [18], brought to the integration of CIGS into WfMS for the definition of the so-called careflows [43, 50] or care pathways [17, 49], which constitute the implementation of CGs or protocols in specific healthcare environments, consider the resource and the organizational settings.

To sum up, although there has been some exchange of ideas between the fields of CIGs and BPM, the benefits from an actual cross-fertilisation have not yet been achieved. Several authors argue that both CIG and workflow systems fail to address important aspects of healthcare processes when used individually (see e.g. [39]). The fact that CIG languages stand out for their expressive power in some regards, e.g. to represent the logic of decisions, may explain why this field has not embraced to a greater extent the methods and tools of the BPM one. In doing so, however, the GCs do not allow their conjoint application with BPM tools and, thus, do not provide an effective support to the hospitals from a managerial point of view (e.g. resource allocation, performance analysis, etc.).

In a recent review about clinical decision-support models and frameworks, Greenes et al. question whether it is possible at all (and even desirable) to develop an over-arching framework integrating all related aspects (design, modeling, formalization, integration into workflow, deployment, etc.) [20]. Then, a possible path is the co-existence and coordination of different frameworks for each one of these CIG aspects.

5 Conformance Checking and (Not Only) Clinical Guidelines

The Process Mining field was established to bridge the gap between the process-oriented nature of BPM and the need of a more data-driven approach to build processes. Originally, the prominent role was played by Process Discovery, while Conformance Checking was relegated to represent an ancillary activity on automatically mined processes [36].

PM4HC inherited from PM the culture of being Process Discovery oriented: the most recent and extended meta-review [21] shows that only the 20–30% of the papers on PM4HC deals with conformance checking. However, while PM is data-driven and domain-agnostic, PM4HC is data and domain-driven and it has to face domain specific needs, issues and culture. It is quite common, for physicians, to see in PM4HC an opportunity to deal with the problem of CG, protocols, workflows, pathways, all concepts quite invasive in their daily routine and seems to have a solution in the languages we use to deal with processes. This is increasingly evident, but was also clear at the dawn of PM in healthcare, when Mans and van der Aalst [31], in defining four typical questions to be answered by medical process specialists, asked *Do we comply with internal and external guidelines?*. Due to this kind of need, in 2015 the same authors [47] represented a Clinical Guideline with DECLARE [41] and performed Conformance Checking with ProM [56]. A 2016 review [46] reveals that Conformance Checking to a pre-determined model (not automatically mined), has been applied in 14 of the 71 reviewed studies. Another review [29], more specific for oncology, counts 7/37 papers where PM4HC was adopted in measuring the distance between expectation and data evidences on CG compliance. More recently, [25] used Conformance Checking on CG for alcoholism, [30] and [6] for the treatment of rectal and skin (melanoma) cancer respectively.

Generally speaking, CG contributes in coping with the measure of conformance on a known clinical process and, as mentioned, can be found in PM4HC, BPM, CICGs, DSS, WfMs but also in other research areas, such as Case Based Reasoning (e.g. [42]).

Summarising, the overall picture reveals that (i) the clinical activity can benefit by PM4HC when dealing with known clinical processes; (ii) previous attempts to propose models and tools can be found in many areas of Computer Science and this fragmentation led to a plethora of solutions and point of views. A significant part of the PM4HC community intercepted (ii) and saw a possible solution in what they name Conformance Checking, even if this means that PM4HC needs to be enriched of formalisms, models, etc., to tackle with this specific issue.

This represents a new challenge for PM4HC specialists because requires to collect the previous experiences from many research areas and re-shape such knowledge in the perspective of their discipline. However, this awareness represent an opportunity to avoid to *re-invent the wheel*, wherever this is possible and reasonable.

6 Conclusion and Future Trends

Clinical decision making is complex and of high responsibility, and a professional (or a team) must balance the benefit chances against the chances of harm. Diagnostic, especially therapeutic prescriptions, entails risks for patient's health: therefore, decisions must be supported by strong evidence, sufficient information and professional expertise.

CIGs can support the clinical decision making when some of the mentioned factors is missing or lacks reliability. However, even though they are built on the basis of consensus, evidence, and wide agreements, CIGs application can be compromised when the clinical context outstrips their constraints and assumptions. BPM provides tools to overcome these limitations by inferring and describing longitudinal data at different levels of granularity and multiple perspectives.

CIG was the ICT solution to traditional Evidence Based Medicine thesis in order to support medical care. Process Mining provides a great opportunity to fill the traditional gap between engineers and Health Professionals in this field. However, we should leverage on this opportunity not only to close this gap, but also to evolve CIGs to a new concept that provides a solution to new trends and paradigms in health care beyond Evidence Based Medicine. Combining CIGs and CC, we can push to create Value-Based Health Care solutions that provides not only better guidelines, but also being more personalized, providing Better Care, Better Health and Lower Cost [22]. This is far to be a trivial question. To engage health professionals in the ICT world it is mandatory to provide real solutions for real scenarios. Taking into account that there not exists one-fit-all solutions working in healthcare. For that, the models should be adapted to the final scenario in an iterative and interactive way [14]. That means the we need to provide models formal but understandable; complete, but usable; standard, but adaptable; specific, but flexible; general, but personalized... Otherwise, we will fail in the application of ICT to the healthcare domain.

The community of PM4HC is young and dynamic, close to the real world problem. It has the potential to give an important contribution in dealing with

CG, also thanks to the attitude to be real-data oriented and the extensive use of Machine Learning. The exploitation of Machine Learning, for example, can lead to a vision of a CIG describing tools and methods to analyze clinical data and suggesting possible decision scenarios based on confidence indicators, instead of depicting decision work flows with concrete thresholds.

Future work will focus on fostering *consensus* about the role of PM4HC in dealing with CG (and related fields), by developing initiatives aimed at sharing experience and results, and the inclusion of other centers.

References

1. van der Aalst, W.M.P.: Process Mining: Discovery, Conformance and Enhancement of Business Processes, 1st edn. Springer, Heidelberg (2011). https://doi.org/10. 1007/978-3-642-19345-3
2. Anani, N., et al.: Applying openEHR's Guideline Definition Language to the SITS international stroke treatment registry: a European retrospective observational study. BMC Med. Inform. Decis. Mak. **17**(1), 7 (2017)
3. Anselma, L., Piovesan, L., Terenziani, P.: Temporal detection and analysis of guideline interactions. Artif. Intell. Med. **76**, 40–62 (2017)
4. Beale, T., et al.: openEHR Task Planning Specification (2017). https:// specifications.openehr.org/releases/PROC/latest/task_planning.html
5. Bilici, E., Despotou, G., Arvanitis, T.N.: The use of computer-interpretable clinical guidelines to manage care complexities of patients with multimorbid conditions: a review. Digit Health **4** (2018)
6. Binder, M., et al.: On analyzing process compliance in skin cancer treatment: an experience report from the evidence-based medical compliance cluster (EBMC^2). In: Ralyté, J., Franch, X., Brinkkemper, S., Wrycza, S. (eds.) CAiSE 2012. LNCS, vol. 7328, pp. 398–413. Springer, Heidelberg (2012). https://doi.org/10.1007/978-3-642-31095-9_26
7. Boxwala, A.A., et al.: GLIF3: a representation format for sharable computer-interpretable clinical practice guidelines. J. Biomed. Inform. **37**(3), 147–161 (2004)
8. Buijs, J.C.A.M., van Dongen, B.F., van der Aalst, W.M.P.: On the role of fitness, precision, generalization and simplicity in process discovery. In: Meersman, R., et al. (eds.) OTM 2012. LNCS, vol. 7565, pp. 305–322. Springer, Heidelberg (2012). https://doi.org/10.1007/978-3-642-33606-5_19
9. Chen, C., Chen, K., Hsu, C.Y., Chiu, W.T., Li, Y.C.J.: A guideline-based decision support for pharmacological treatment can improve the quality of hyperlipidemia management. Comput. Methods Progr. Biomed. **97**(3), 280–285 (2010)
10. Ciccarese, P., Caffi, E., Quaglini, S., Stefanelli, M.: Architectures and tools for innovative Health Information Systems: the Guide Project. Int. J. Med. Inform. **74**(7–8), 553–562 (2005)
11. De Bleser, L., Depreitere, R., De Waele, K., Vanhaecht, K., Vlayen, J., Sermeus, W.: Defining pathways. J. Nurs. Manag. **14**(7), 553–563 (2006)
12. Dixon, B.E., et al.: A pilot study of distributed knowledge management and clinical decision support in the cloud. Artif. Intell. Med. **59**(1), 45–53 (2013)
13. Fdez-Olivares, J., Onaindia, E., Castillo, L., Jordan, J., Cozar, J.: Personalized conciliation of clinical guidelines for comorbid patients through multi-agent planning. Artif. Intell. Med. **96**, 167–186 (2018)

14. Fernández-Llatas, C., Meneu, T., Traver, V., Benedi, J.M.: Applying evidence-based medicine in telehealth: an interactive pattern recognition approximation. Int. J. Environ. Res. Public Health **10**(11), 5671–5682 (2013)
15. Ghasemi, M., Amyot, D.: Process mining in healthcare: a systematised literature review. Int. J. Electron. Healthc. **9**, 60 (2016)
16. Goldberg, H.S., et al.: A highly scalable, interoperable clinical decision support service. J. Am. Med. Inform. Assoc. (JAMIA) **21**(e1), e55–e62 (2014)
17. Gonzalez-Ferrer, A., ten Teije, A., Fdez-Olivares, J., Milian, K.: Automated generation of patient-tailored electronic care pathways by translating computer-interpretable guidelines into hierarchical task networks. Artif. Intell. Med. **57**(2), 91–109 (2013)
18. Gooch, P., Roudsari, A.: Computerization of workflows, guidelines, and care pathways: a review of implementation challenges for process-oriented health information systems. J. Am. Med. Inform. Assoc. **18**(6), 738–748 (2011)
19. Grando, M.A., Glasspool, D., Fox, J.: A formal approach to the analysis of clinical computer-interpretable guideline modeling languages. Artif. Intell. Med. **54**(1), 1–13 (2012)
20. Greenes, R.A., Bates, D.W., Kawamoto, K., Middleton, B., Osheroff, J., Shahar, Y.: Clinical decision support models and frameworks: seeking to address research issues underlying implementation successes and failures. J. Biomed. Inform. **78**, 134–143 (2018)
21. Gurgen Erdogan, T., Tarhan, A.: Systematic mapping of process mining studies in healthcare. IEEE Access **6**, 1 (2018)
22. Ibanez-Sanchez, G., et al.: Toward value-based healthcare through interactive process mining in emergency rooms: the stroke case. Int. J. Environ. Res. Public Health **16**(10), 1783 (2019)
23. Institute of Medicine: Clinical Practice Guidelines We Can Trust. The National Academies Press, Washington, DC (2011)
24. Jafarpour, B., Abidi, S.R., Woensel, W.V., Abidi, S.S.R.: Execution-time integration of clinical practice guidelines to provide decision support for comorbid conditions. Artif. Intell. Med. **94**, 117–137 (2019)
25. Johnson, O.A., Ba Dhafari, T., Kurniati, A., Fox, F., Rojas, E.: The ClearPath method for care pathway process mining and simulation. In: Daniel, F., Sheng, Q.Z., Motahari, H. (eds.) BPM 2018. LNBIP, vol. 342, pp. 239–250. Springer, Cham (2019). https://doi.org/10.1007/978-3-030-11641-5_19
26. Kawamoto, K., et al.: Multi-national, multi-institutional analysis of clinical decision support data needs to inform development of the HL7 virtual medical record standard. AMIA Annu. Symp. Proc. **2010**, 377–381 (2010)
27. Kawamoto, K., Greenes, R.A.: The role of standards: what we can expect and when. In: Greenes, R.A. (ed.) Clinical Decision Support, chap. 21, 2nd edn, pp. 599–615. Academic Press, Oxford (2014)
28. Kaymak, U., Mans, R., van de Steeg, T., Dierks, M.: On process mining in health care. In: 2012 IEEE International Conference on Systems, Man, and Cybernetics (SMC), pp. 1859–1864 (2012)
29. Kurniati, A.P., Johnson, O., Hogg, D., Hall, G.: Process mining in oncology: a literature review. In: 2016 6th International Conference on Information Communication and Management (ICICM), pp. 291–297 (2016)
30. Lenkowicz, J., et al.: Assessing the conformity to clinical guidelines in oncology: an example for the multidisciplinary management of locally advanced colorectal cancer treatment. Manag. Decis. **56**(10), 2172–2186 (2018)

31. Mans, R.S., van der Aalst, W.M.P., Vanwersch, R.J.B., Moleman, A.J.: Process mining in healthcare: data challenges when answering frequently posed questions. In: Lenz, R., Miksch, S., Peleg, M., Reichert, M., Riaño, D., ten Teije, A. (eds.) KR4HC/ProHealth - 2012. LNCS (LNAI), vol. 7738, pp. 140–153. Springer, Heidelberg (2013). https://doi.org/10.1007/978-3-642-36438-9_10
32. Marco-Ruiz, L., Moner, D., Maldonado, J.A., Kolstrup, N., Bellika, J.G.: Archetype-based data warehouse environment to enable the reuse of electronic health record data. Int. J. Med. Inform. **84**(9), 702–714 (2015)
33. Marcos, M., Maldonado, J.A., Martínez-Salvador, B., Boscá, D., Robles, M.: Interoperability of clinical decision-support systems and electronic health records using archetypes: a case study in clinical trial eligibility. J. Biomed. Inform. **46**(4), 676–689 (2013)
34. Martínez-Salvador, B., Marcos, M.: Supporting the refinement of clinical process models to computer-interpretable guideline models. Bus. Inf. Syst. Eng. **58**(5), 355–366 (2016)
35. Mulyar, N., van der Aalst, W.M., Peleg, M.: A pattern-based analysis of clinical computer-interpretable guideline modeling languages. J. Am. Med. Inform. Assoc. **14**(6), 781–787 (2007)
36. Munoz-Gama, J.: Conformance Checking and Diagnosis in Process Mining - Comparing Observed and Modeled Processes, vol. 270. Springer, Heidelberg (2016). https://doi.org/10.1007/978-3-319-49451-7
37. Musen, M.A., Tu, S.W., Das, A.K., Shahar, Y.: EON: a component-based approach to automation of protocol-directed therapy. J. Am. Med. Inform. Assoc. **3**(6), 367–388 (1996)
38. Peleg, M.: Computer-interpretable clinical guidelines: a methodological review. J. Biomed. Inform. **46**, 744–763 (2013)
39. Peleg, M., González-Ferrer, A.: Guidelines and Workflow Models, chap. 16, 2nd edn, pp. 435–464. Academic Press, Oxford (2014)
40. Peleg, M., et al.: Comparing computer-interpretable guideline models: a case-study approach. J. Am. Med. Inform. Assoc. **10**, 52–68 (2003)
41. Pesic, M., van der Aalst, W.M.P.: A declarative approach for flexible business processes management. In: Eder, J., Dustdar, S. (eds.) BPM 2006. LNCS, vol. 4103, pp. 169–180. Springer, Heidelberg (2006). https://doi.org/10.1007/11837862_18
42. Qu, G., Liu, Z., Cui, S., Tang, J.: Study on self-adaptive clinical pathway decision support system based on case-based reasoning. In: Li, S., Jin, Q., Jiang, X., Park, J.J.J.H. (eds.) Frontier and Future Development of Information Technology in Medicine and Education. LNEE, vol. 269, pp. 969–978. Springer, Dordrecht (2014). https://doi.org/10.1007/978-94-007-7618-0_95
43. Quaglini, S., Stefanelli, M., Cavallini, A., Micieli, G., Fassino, C., Mossa, C.: Guideline-based careflow systems. Artif. Intell. Med. **20**(1), 5–22 (2000)
44. Rebuge, Á., Ferreira, D.R.: Business process analysis in healthcare environments: a methodology based on process mining. Inf. Syst. **37**, 99–116 (2012)
45. Riaño, D., Collado, A.: Model-based combination of treatments for the management of chronic comorbid patients. In: Peek, N., Marín Morales, R., Peleg, M. (eds.) AIME 2013. LNCS (LNAI), vol. 7885, pp. 11–16. Springer, Heidelberg (2013). https://doi.org/10.1007/978-3-642-38326-7_2
46. Rojas, E., Munoz-Gama, J., Sepulveda, M., Capurro, D.: Process mining in healthcare: a literature review. J. Biomed. Inform. **61**, 224–236 (2016)
47. Rovani, M., Maggi, F.M., de Leoni, M., van der Aalst, W.M.: Declarative process mining in healthcare. Expert Syst. Appl. **42**(23), 9236–9251 (2015)

48. dos Santos Garcia, C., et al.: Process mining techniques and applications - a systematic mapping study. Expert Syst. Appl. **133**, 260–295 (2019)
49. Schadow, G., Russler, D.C., McDonald, C.J.: Conceptual alignment of electronic health record data with guideline and workflow knowledge. Int. J. Med. Inform. **64**(2–3), 259–274 (2001)
50. Shabo, A., Peleg, M., Parimbelli, E., Quaglini, S., Napolitano, C.: Interplay between clinical guidelines and organizational workflow systems. Experience from the MobiGuide project. Methods Inf. Med. **55**(6), 488–494 (2016)
51. Shahar, Y., Miksch, S., Johnson, P.: The Asgaard project: a task-specific framework for the application and critiquing of time-oriented clinical guidelines. Artif. Intell. Med. **14**(1–2), 29–51 (1998)
52. Sordo, M., Ogunyemi, O., Boxwala, A.A., Greenes, R.A.: GELLO: an object-oriented query and expression language for clinical decision support. In: AMIA Annual Symposium Proceedings, p. 1012 (2003)
53. Sox, H.C.: Conflict of interest in practice guidelines panels. JAMA **317**(17), 1739–1740 (2017)
54. Sutton, D.R., Fox, J.: The syntax and semantics of the PROforma guideline modeling language. J. Am. Med. Inform. Assoc. **10**(5), 433–443 (2003)
55. Terenziani, P., Molino, G., Torchio, M.: A modular approach for representing and executing clinical guidelines. Artif. Intell. Med. **23**(3), 249–276 (2001)
56. van Dongen, B.F., de Medeiros, A.K.A., Verbeek, H.M.W., Weijters, A.J.M.M., van der Aalst, W.M.P.: The ProM framework: a new era in process mining tool support. In: Ciardo, G., Darondeau, P. (eds.) ICATPN 2005. LNCS, vol. 3536, pp. 444–454. Springer, Heidelberg (2005). https://doi.org/10.1007/11494744_25
57. Van de Velde, S., et al.: A systematic review of trials evaluating success factors of interventions with computerised clinical decision support. Implement. Sci. **13**(1), 114 (2018)
58. Wall, E.: Clinical practice guidelines–is "regulation" the answer? J. Am. Board Family Med. **29**(6), 642–643 (2016)
59. Wang, Z., Norris, S.L., Bero, L.: The advantages and limitations of guideline adaptation frameworks. Implement. Sci. **13**(1), 72 (2018)
60. Wilk, S., Michalowski, W., Michalowski, M., Farion, K., Hing, M.M., Mohapatra, S.: Mitigation of adverse interactions in pairs of clinical practice guidelines using constraint logic programming. J. Biomed. Inform. **46**(2), 341–353 (2013)

Predicting Outpatient Process Flows to Minimise the Cost of Handling Returning Patients: A Case Study

Marco Comuzzi$^{(\boxtimes)}$, Jonghyeon Ko, and Suhwan Lee

School of Management Engineering,
Ulsan National Institute of Science and Technology (UNIST),
Ulsan, Republic of Korea
{mcomuzzi,whd1gus2,ghksdl6025}@unist.ac.kr

Abstract. We describe an application of process predictive monitoring at an outpatient clinic in a large hospital. A model is created to predict which patients will wrongly refer to the outpatient clinic, instead of directly to other departments, when returning to get treatment after an initial visit. Four variables are identified to minimise the cost of handling these patients: the cost of giving appropriate guidance to them, the cost of handling patients taking a non-compliant flow by wrongly referring to the outpatient clinic, and the false positive/negative rates of the predictive model adopted. The latter determine the situations in which patients have not received guidance when they should have had or have been guided even though not necessary, respectively. Using these variables, a cost model is built to identify which combinations of process intervention/redesign options and predictive models are likely to minimise the cost overhead of handling the returning patients.

Keywords: Predictive monitoring · Business process · Case study · Outpatient clinic

1 Introduction

Predictive monitoring of business processes aims at developing techniques to predict aspects of interests of a process using data stored in so-called process event logs. The aspects predicted range from the next activity in a running case, timestamps, that is, the instant of occurrence of future events in running cases, or process outcomes, such as the satisfaction of service level objectives or constraints predicated over the order and occurrence of activities in running cases. Most frequently, predictive monitoring models are obtained by training machine learning models using data in event logs, appropriately pre-processed, as training and test sets. By empowering decision makers with insights about the possible occurrence of specific situations in the future, predictive monitoring supports proactive decision making, such as developing early warning systems or supporting resource reallocation and rescheduling.

© Springer Nature Switzerland AG 2019
C. Di Francescomarino et al. (Eds.): BPM 2019 Workshops, LNBIP 362, pp. 557–569, 2019.
https://doi.org/10.1007/978-3-030-37453-2_45

A large number of techniques focused on different predictive monitoring tasks have been developed. These often focus on generating new features from event logs that may improve the accuracy of existing methods [6]. From an evaluation standpoint, the developed techniques are tested on a variety of publicly available event logs commonly used by the research community for benchmarking. Generally, the application of these techniques to specific contexts for the solution of real world problems remains limited. Commercial process mining tools do not normally incorporate advanced predictive monitoring into their functionality, and there are only very few examples of academic publications that have applied predictive monitoring in real world settings, particularly in the health care.

In this paper, we discuss an application of process predictive monitoring in a real world scenario, i.e., predicting outpatient flows at a large hospital in South Korea. While this outpatient clinic strives to complete patient treatments in one single day, inevitably some patients, after an initial visit, are required to come back on a different day to receive additional treatment. However, when returning, instead of referring directly to the department handling the required additional treatment, many of these patients refer again to the reception desk of the outpatient clinic. This creates unnecessary load for the reception desk and, at the same time, it deteriorates patient satisfaction and level of service. The reception desk, in fact, often struggles to quickly redirect patients to the appropriate department from which they should receive treatment.

We build a simple model for minimising the overhead cost of handling patients wrongly returning to the outpatient clinic reception desk. The model considers, as cost-related variables, the cost of giving appropriate guidance to returning patients (in order for them to refer directly to the correct department when returning) and the cost of handling patients not referring directly to the treatment department. Additionally, we also devise two different classes of predictive models to predict whether patients will wrongly refer to the reception desk of the outpatient clinic. Models in the first class are trained using the event log as extracted from logging information of systems at the hospital. These models show high accuracy, but also a large rate of false positives. Being the returning patients a minority of cases, the second class of models is trained using oversampling of this minority class in the original event log. Models in this class achieve lower accuracy, but also lower false positive rates. Finally, considering the performance of the predictive models and different redesign options, which result in different relative values of the cost variables, the proposed model can be used to identify the combinations of redesign options and predictive monitoring models that are likely to minimise the cost overhead of handling returning patients for the hospital.

The paper is organised as follows. Section 2 introduces the case study. Section 3 describes the methods and experimental settings used in our predictive monitoring analysis, while Sect. 4 presents the model of overhead costs of returning patients. The experimental results are discussed in Sect. 5. Related work is briefly outlined in Sect. 6 and conclusions are drawn in Sect. 7.

2   Case Description

We consider the treatment process of outpatients at the Pusan National University Hospital (PNUH), which is one of the largest tertiary hospitals in Korea. The hospital hosts 42 medical departments and 341 rooms with 1351 beds. With a 25% increase of total outpatient visits in the last 5 years, quality of outpatient service has been issued as a critical agenda item at PNUH. The Korea National Patient Survey Program, firstly implemented in 2018, indicates that outpatients at PNUH experience low quality of clinical services, especially as far as the treatment process is concerned. As a consequence, PNUH has recently started a collaboration with academic partners to analyse and improve outpatient operations. The work presented in this paper is part of this initiative, and it relies on the analysis of an event log of outpatient visits recorded in a period from December 2017 to September 2018.

Table 1. Attributes of event log

Variable	Definition	Data type
caseID	Combined patient id and visit number	Categorical
patientID	ID of a patient	Categorical
Time stamp	Complete timestamp of event	Date
Activity	Name of executed activity	Categorical
Resource	ID of resource executing activity	Categorical
Resource_department	Department of the resource	Categorical
Clinic_department	Department providing treatment	Categorical
Disease_code	Type of diagnosed disease	Categorical
DoctorID	ID of doctor giving treatment	Categorical
Clinic room	Room number for consultation with a doctor	Categorical
Remaining time by next visit	Time remaining before the next visit	Continuous

Starting from logs of systems at the outpatient clinic and other departments, an event log has been firstly extracted with records of 145,638 outpatient visits at PNUH between December 2017 and September 2018 in 10 departments. As shown later, the departments chosen for this case study are the 10 with the highest share of patients following a non-compliant flow when returning for a visit. The event log considers standard attributes, such as a case id, patient id, activity label, and timestamp, and other domain specific categorical and numerical attributes, as shown in Table 1.

As mentioned in the Introduction, the work presented in this paper is part of a larger initiative to identify and fix issues with outpatient operations. In particular, PNUH requires solutions to handle increasing number of patients, which cause low quality of service. The objective, in this case, is to improve the clinical processes under the constraint that the number and type of resources, e.g., nurses, doctors and administrative staff, remains unchanged.

The first step, in our case, has been process discovery and, in particular, reaching consensus on a process model of outpatient visits with all stakeholders. Even though the outpatient clinic process at PNUH has slightly different flows for each department, activities in the process are recorded using common labels across departments. This has allowed us to integrate logs from 10 different departments to discover an overall common process. After a preliminary process discovery phase (using the process mining tool Disco), the standard process equally followed by all 10 departments has been identified in several meetings with stakeholders at PNUH, including nurses, IT staff, other administrative personnel, and doctors. Consensus was reached on the process shown in Fig. 1. The process starts with a patient referring to the reception desk of the outpatient clinic after having reserved an appointment online or by phone. During patient reception, administrative staff schedule an appointment with the department where patients need treatment, and instructs them on where and when to wait for such an appointment. Then, the patient goes to a waiting room in the specific department. Then, administration staff call the patient for consulting with a doctor. During the consultation, the doctor may decide that the patient needs to revisit for an additional medical examination on a different day, if needed. After finishing the consultation, the patient pays for the visit and then may print a certificate of the treatment or get a prescription before returning home. For patients who have to visit again the hospital to get treatment, the patient should directly refer to the department providing the required treatment (the revisit part of the process is not depicted in Fig. 1).

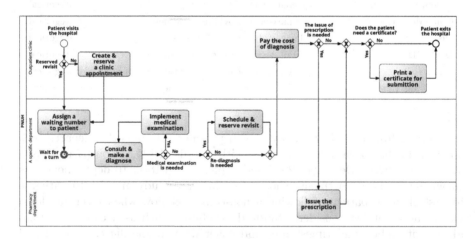

Fig. 1. BPMN 2.0 model for outpatient clinic process at PNUH hospital

In this paper, we investigate a particular problem of outpatient operations, i.e., non-compliant process cases in which patients, when returning to receive additional treatment after a visit, wrongly refer to the reception at the outpatient clinic, rather than directly at the reception of the department where they should receive treatment (see Fig. 2, which models compliant and non-compliant process flows from a patient point of view). This results in disruption for the department giving treatment and longer waiting times for patients, since it is problematic for the outpatient clinic to quickly refer them to the correct department. A solution to this problem is to implement a predictive monitoring system to identify, at the moment a next visit is scheduled, and based on historic data, which patients are likely to wrongly refer to the outpatient clinic after the initial visit, i.e., following the non-compliant process flow. If an accurate model could be developed, then an appropriate guidance system could be put in place (e.g., sending reminders and/or instructing nurses/doctors to clarifying the correct referral procedure) to substantially reduce the likelihood of patients wrongly referring to the outpatient reception desk at the next visit.

To achieve this goal, the second step in the case study has been to quantify the number of patients following the non-compliant flow in the process. Figure 2 shows that, across all departments, 17% of patients referred to the outpatient clinic reception desk rather than the specialist department when returning for a visit. Using information in the attribute *Remaining time by next visit* in the event log, we calculated that patients in this group spend on average 69 min more waiting to receive treatment compared to 83% of patients referring directly to the correct department when returning.

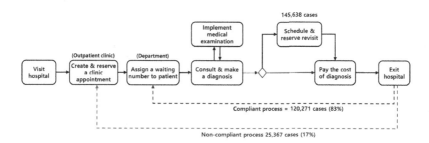

Fig. 2. Compliant and non-compliant process of returning outpatients

Table 2 shows the ratio of compliant and non-compliant cases for the 10 departments that we consider in this case study. In Sect. 5, results will be discussed for Department A and B, i.e., the two departments with highest number of non-compliant cases and, therefore, more critical from a performance analysis/customer satisfaction point of view for PNUH.

Note that, because of a non disclosure agreement with PNUH, the event log used in this paper cannot be made public and we are not allowed to reveal the actual names of the departments involved in our analysis.

Table 2. Non-compliant flows in analyzed departments

Department (Anonymised)	Proportion of non-compliant cases
Department A	32.5%
Department B	27.4%
Department C	19.9%
Department D	19.5%
Department E	18.9%
Department F	18.3%
Department G	16.1%
Department H	14.8%
Department I	14.0%
Department J	14.0%

3 Building Predictive Monitoring Models

In order to create a predictive model for the problem that we have identified, we pre-processed the event log to obtain a feature-label dataset to feed into a classification algorithm. The predictive model considered in this paper adopts a simple feature vector using, for each case, the attributes of the activity *Reserve next visit*. After extensive test, among all the activities in the process, this is the one whose attributes better predict the behaviour of returning patients. The label, in this case, is binary, capturing whether the patient has followed the compliant process flow (referring to the correct department) or the non-compliant one (wrongly referring to the outpatient clinic reception). We leave the adoption of more *process-oriented* predictive models, such as predicting compliant/non-compliant returns by using next activity prediction [8] or as a process outcome [9], to future work.

Using these feature vector and label, we build predictive monitoring models with two classification techniques, i.e., decision tree (DT) and random forest (RF). Note that tree-based classifiers are known to be high performing techniques for binary classification problems and have also been shown to be effective for classification problems that use data in event logs [8]. We develop decision tree classifiers with parameters $max\_depth = 30$ and $min\_split = 2$ using the C4.5 algorithm (R package *rpart*), which involves tree pruning that reduces misclassification errors due to noise, and can be applied with both continuous and categorical attributes [2]. For random forest, we consider the implementation in the R package *randomForest* with parameters of $num\_tree = 1000$ and $mtry(split) = 3$, which is one of the most efficient implementations from the perspective of prediction accuracy and runtime performance [3]. To validate whether the developed models are trained without bias, the two R packages (*rpart*, *randomForest*) provide built-in cross validation algorithms, named *x-error* and *out-of-bag error*, respectively. These cross validation algorithms, which are designed for reasonable approximation of test error, stabilise the error rate efficiently during tuning. The dataset is divided into train set and test set in proportion of 75:25 and the train set is used for both fitting and validating the two models with the built-in cross validation algorithms.

Since the labels of the classifier input are imbalanced, we also consider oversampling techniques to decrease the imbalance rate. In many practical situations, in fact, classifiers tend to perform better when there is a balance between samples across the classes that have to predicted. This because, if one label is much more frequent than others, then the classifier may be able to generalise only the samples in the majority class. In this work, we adopt random oversampling of the minority class [1].

The performance of classifiers is evaluated using standard measures. Given the number of true positives $\#TP$, true negatives $\#TN$, false positives $\#FP$, and false negatives $\#FN$ obtained on the test set, we consider the ratio of true positives $TP = \frac{\#TP}{\#TP+\#FN}$, ratio of true negatives $TN = \frac{\#TN}{\#TN+\#FP}$, ratio of false positives $FP = \frac{\#FP}{\#TN+\#FP}$, ratio of false negatives $FN = \frac{\#FN}{\#TP+\#FN}$; and then the accuracy $A = \frac{\#TP+\#TN}{\#TP+\#TN+\#FP+\#FN}$ and F-1 score $F = \frac{2\#TP}{2\#TP+\#FP+\#FN}$.

4   An Overhead Cost Model of Handling Returning Patients

Before presenting and discussing the experimental results, in this section we devise a simple optimisation model, which in the next section is used to identify the combinations of process intervention options and predictive monitoring models that are likely minimise the overhead cost of handling returning patients at the wrong location.

This model is characterised by 4 variables:

– CG is the cost of giving appropriate guidance to one patient who has to return for treatment on a different day, in order for this patient to refer to the appropriate department, rather than the outpatient clinic. The magnitude of this cost depends on the way in which this guidance is implemented. Options are discussed later in this section;
– CR is the cost of handling one patient referring to the outpatient clinic reception desk, instead of directly returning to the department providing the treatment required. This cost is implicitly determined by the fact that patients returning at the wrong location create disruption at both the outpatient clinic and the referral department. In fact, these patients will be late for their appointments at the referral department, which creates obvious issues with appointment scheduling and execution;
– FP is the false positive rate of the predictive model;
– FN is the false negative rate of the predictive model.

False positives, in this model, are those patients that are predicted not to return at the wrong location – and, therefore, who have not been given appropriate guidance – but that, actually, return for treatment on a different day at the wrong location, i.e., referring to the outpatient clinic reception desk. Because these patients have to be handled by the reception desk, for each false positive patient, the hospital sustains a cost equal to CR;

False negatives are patients that are predicted to return for treatment at a later date at the wrong location, but that, in the end, do not make this mistake, and refer correctly to the department providing the treatment that they need. From a practical standpoint, the hospital has to give guidance to these patients, as instructed by the predictive model. The cost of giving this guidance, however, can be seen as an unnecessary cost for PNUH, since guidance is not necessary in the end. Therefore, for each false negative patient, the hospital sustains an opportunity cost equal to CG.

The objective of the hospital is to minimise the following overhead cost function, $OVH(CG, CR, FN, FP)$, which represents the cost overhead of handling improperly returning patients at the outpatient clinic:

$$OVH = CR \cdot FP + CG \cdot FN$$

Note that the variables CG and CR are cost variables and, as such, depend on process intervention, i.e., on how guidance and handling of returning patients are implemented in the process at PNUH, respectively. The variables FN and FP depend on the performance of the predictive model that is adopted for the implementation of the predictive monitoring system.

Different types of process interventions for both (i) the guidance given to patients predicted to take the non-compliant flow and (ii) the handling of patients actually taking the non-compliant flow, i.e., wrongly referring at the outpatient clinic reception when returning, can be considered.

Regarding the guidance given to non-compliant returning patients, the following options, classified in order of cost and expected effectiveness (from low to high) can be identified:

- Improved guidance from doctor performing the initial visit: if signalled from the predictive monitoring system that a patient is likely to take the non-compliant flow, doctors can be instructed in real time to take particular care in explaining the correct procedure when returning to patients;
- Phone call reminders: every day, the reception desk may call the patients expected to return on that day to remind them of the correct location where to return;
- Social media: the correct flow can be reminded through automated messages on the social network services preferred by returning patients;
- Monitors at reception desk: monitors with clear instructions for patients returning on the day can be installed at the outpatient clinic to ensure that, even if entering the outpatient clinic, returning patients may read from the monitor to which department they should actually refer.

Note that these options are not mutually exclusive, but they can be combined. Obviously, the more options implemented, the higher the cost CG.

The options available for handling of patients wrongly referring at the outpatient clinic reception desk when returning are more limited and expensive. Specifically, it may be possible to integrate to a different degree the reception desk information system with the scheduling system in each department. The integration can range from simple notifications to each department of patients referring to the reception desk at the outpatient clinic, to full scale integration where personnel at the reception desk could re-schedule treatments at different departments on the same day for this type of patients.

5   Experimental Results and Discussion

We first presents the experimental results of the predictive models. Then, we discuss a simulation of the cost model introduced in Sect. 4.

Tables 3 and 4 show the performance of the predictive model developed for department A and B, respectively. Results are reported for the original event log (default) and for 4 different oversampling configurations, with the frequency of the minority class increasing from the actual value in the default event log to 60%. The performance in terms of accuracy and F1-score is remarkable, even on the default imbalanced event logs. While DT and RF show similar performance on the original logs, accuracy and F1-score in the oversampled configurations tend to be higher for RF models.

Table 3. Result of classification in department A

Event log	Decision tree			Randomforest		
	(TP, TN, FP, FN) rate	ACC	F1-score	(TP, TN, FP, FN) rate	Accuracy	F1-score
Default (68:32)	(0.84, 0.42, 0.58, 0.16)	0.757	0.798	(0.90, 0.33, 0.67, 0.10)	0.741	0.813
OS1 (60:40)	(0.64, 0.69, 0.31, 0.36)	0.860	0.818	(0.72, 0.64, 0.36, 0.29)	0.911	0.840
OS2 (55:45)	(0.60, 0.73, 0.27, 0.40)	0.867	0.806	(0.69, 0.68, 0.32, 0.31)	0.921	0.837
OS3 (50:50)	(0.57, 0.75, 0.25, 0.43)	0.868	0.759	(0.65, 0.73, 0.27, 0.35)	0.922	0.831
OS4 (45:55)	(0.57, 0.74, 0.26, 0.43)	0.872	0.759	(0.63, 0.75, 0.25, 0.37)	0.927	0.829
OS5 (40:60)	(0.56, 0.76, 0.24, 0.44)	0.876	0.750	(0.59, 0.78, 0.22, 0.41)	0.935	0.824

Table 4. Result of classification in department B

Event log	Decision tree			Randomforest		
	(TP, TN, FP, FN) rate	ACC	F1-score	(TP, TN, FP, FN) rate	Accuracy	F1-score
Default (73:27)	(0.84, 0.63, 0.37, 0.16)	0.857	0.850	(0.85, 0.66, 0.34, 0.15)	0.867	0.861
OS1 (60:40)	(0.78, 0.67, 0.33, 0.22)	0.816	0.717	(0.77, 0.80, 0.20, 0.23)	0.809	0.759
OS2 (55:45)	(0.75, 0.70, 0.30, 0.25)	0.826	0.698	(0.77, 0.83, 0.17, 0.23)	0.823	0.749
OS3 (50:50)	(0.67, 0.73, 0.27, 0.33)	0.831	0.674	(0.76, 0.83, 0.17, 0.24)	0.838	0.732
OS4 (45:55)	(0.67, 0.74, 0.26, 0.33)	0.822	0.672	(0.75, 0.85, 0.15, 0.25)	0.845	0.719
OS5 (40:60)	(0.66, 0.76, 0.24, 0.34)	0.834	0.688	(0.74, 0.87, 0.13, 0.26)	0.854	0.698

Regarding the variables FP and FN of the cost model, it is worth noticing that oversampling of the minority class (i.e., wrongly returning patients) of the default event log clearly helps to reduce the false positive rate. For instance, in the case of department A (Table 3), the evenly oversampled configuration (50:50) more than halves the ratio of false positive when compared with the default log. At the same time, oversampling of the minority class also increases the false negative rate FN, even though to a lower degree than the decrease of FP.

Table 5. Overhead cost of patients wrong returns, department A (minimum values in bold)

Method	Event log	Rate		Cost: $OVH = CR \cdot FP + 0.5 \cdot FP$				
		FP	FN	$CR = 0$	$CR = 0.25$	$CR = 0.5$	$CR = 0.75$	$CR = 1$
Decision tree	Default (68:32)	0.58	0.16	**0.079**	**0.223**	0.368	0.512	0.657
	OS (60:40)	0.31	0.36	0.180	0.257	0.334	0.411	0.489
	OS (55:45)	0.27	0.40	0.198	0.266	**0.334**	0.402	0.470
	OS (50:50)	0.25	0.43	0.217	0.278	0.340	0.401	0.463
	OS (45:55)	0.26	0.43	0.216	0.282	0.348	0.414	0.479
	OS (40:60)	0.24	0.44	0.221	0.281	0.340	**0.400**	**0.459**
Random forest	Default (68:32)	0.67	0.10	**0.050**	**0.218**	0.387	0.556	0.724
	OS (60:40)	0.36	0.29	0.143	0.233	0.323	0.413	0.504
	OS (55:45)	0.32	0.31	0.157	0.236	0.315	0.394	0.473
	OS (50:50)	0.27	0.35	0.175	0.242	**0.310**	0.377	0.444
	OS (45:55)	0.25	0.37	0.187	0.249	0.310	0.372	0.433
	OS (40:60)	0.22	0.41	0.205	0.259	0.313	**0.367**	**0.421**

Table 6. Overhead cost of patients wrong returns, department B (minimum values in bold)

Method	Event log	Rate		Cost: $OVH = CR \cdot FP + 0.5 \cdot FP$				
		FP	FN	$CR = 0$	$CR = 0.25$	$CR = 0.5$	$CR = 0.75$	$CR = 1$
Decision tree	Default (73:27)	0.37	0.16	**0.079**	**0.170**	**0.262**	0.354	0.445
	OS (60:40)	0.33	0.22	0.110	0.192	0.275	0.358	0.440
	OS (55:45)	0.30	0.25	0.123	0.199	0.274	**0.349**	0.424
	OS (50:50)	0.27	0.33	0.163	0.230	0.296	0.362	0.429
	OS (45:55)	0.26	0.33	0.164	0.228	0.292	0.356	0.420
	OS (40:60)	0.24	0.34	0.172	0.233	0.293	0.354	**0.414**
Random forest	Default (73:27)	0.34	0.15	**0.073**	**0.158**	0.243	0.328	0.413
	OS (60:40)	0.20	0.23	0.115	0.165	0.214	0.264	0.313
	OS (55:45)	0.17	0.23	0.117	0.160	0.203	0.246	0.289
	OS (50:50)	0.17	0.24	0.122	0.164	0.205	0.247	0.289
	OS (45:55)	0.15	0.25	0.125	0.164	0.202	0.240	0.279
	OS (40:60)	0.13	0.26	0.132	0.165	**0.198**	**0.231**	**0.264**

Tables 5 and 6 show the results of a simulation of the cost of handling patients returning to the wrong location for Department A and B, respectively. In particular, we consider the rate FP and FN from the predictive models presented above. As far as the costs CR and CG are concerned, we fix the value of CG (arbitrarily chosen to be 0.5) and we calculate OVH for different values of CF ($a = 0, 0.5, 1, 1.5, 2$).

We can notice that as the cost CR increases in respect of CG, the event log oversampling the minority class becomes the one associated with lower overhead costs for handling patients at the wrong location. In practical terms, this means that, as the cost CR associated with patients following the non-compliant flow increases in respect of the cost CG of the process intervention chosen for giving guidance to patients predicted to take the non-compliant flow, PNUH will be better off by adopting a predictive model that largely oversamples non-compliant cases.

The choices of process intervention options depends naturally on a number of other factors not considered by this model, such as available budget, target customer satisfaction, or the actual effectiveness of the patient guidance intervention (in this work, in fact, we implicitly assumed that any intervention in this regard will work 100% effectively, so that all patients receiving guidance will refer to the correct department). However, it is worth noticing that, given a specific configuration of process interventions, the work presented in this paper can be used to suggest a predictive model (i.e., default or oversampling the minority class to a certain degree) that helps to minimise the overhead overhead cost of the hospital.

6   Related Work

Within the process mining literature, process predictive monitoring has emerged recently as a new set of techniques that aim at adapting traditional data mining and machine learning techniques to predict aspects of interest in a process [6]. Process data in event logs are seen as a knowledge base, which is used to train a machine learning model to predict at runtime some aspects of interest of currently executing process cases. There are 3 classes of aspects of interests that can be predicted using predictive process monitoring: (i) the activities that will occur in a running process case in the future, (ii) the time at which activities in running process cases will occur and (iii) outcomes of processes. Predicting next activities in running process cases enables the design of warning systems, in which stakeholders can be warned should a critical or risky activity be predicted to happen with high likelihood in the near future. Classification techniques are adopted to solve this problem [9].

Predicting times of next activities can also be used to design warning systems, i.e., to warn stakeholders that delays in process execution can be expected, and to support pro-active decision making by having reliable estimates of when processes may terminate. Regression techniques are used to solve this problem [9].

Finally, outcome-based predictive monitoring deals with predicting outcome of running case. There are different examples of outcomes that can be predicted, such as whether given performance or service level objectives will be eventually met, or constraints predicted on the order and occurrence of activities in a running process case. Since different outcomes are usually categorical variables, classification techniques are used to solve this problem [10].

Business processes in the health care have often been target of process mining initiatives in the literature. This because they are usually very challenging, being characterised by high variability and unpredictability, while at the same time being sufficiently digitised to make available data to compile high quality event logs [7]. As far as predictive analytics is concerned, simulation emerges as a techniques which is often used [4,5]. However, it is generally adopted as a complementary method to process mining for analysing and improving offline the care pathway, rather than for improving the online reaction to unexpected situations, which is the more natural application of predictive monitoring.

7   Conclusions

This paper has presented a case study about the application of predictive monitoring techniques for improving the handling of returning patients at an outpatient clinic in a large hospital in Korea. This work is currently being improved in several ways. First, we are extending the cost model, by considering, for instance, that different guidance interventions may be effective to a different degree in changing patient behaviour. Then, we are experimenting with more accurate prediction models, by applying different process-oriented paradigms, such as next activity and process outcome prediction. Finally, we are currently working with PNUH to translate the results of the proposed model into actionable process redesign options.

References

1. Ali, A., Shamsuddin, S.M., Ralescu, A.L.: Classification with class imbalance problem: a review. Int. J. Adv. Soft Compuy. Appl. **7**(3), 176–204 (2015)
2. Anyanwu, M.N., Shiva, S.G.: Comparative analysis of serial decision tree classification algorithms. Int. J. Comput. Sci. Secur. **3**(3), 230–240 (2009)
3. Crisci, C., Ghattas, B., Perera, G.: A review of supervised machine learning algorithms and their applications to ecological data. Ecol. Model. **240**, 113–122 (2012)
4. Johnson, O.A., Ba Dhafari, T., Kurniati, A., Fox, F., Rojas, E.: The clearpath method for care pathway process mining and simulation. In: Daniel, F., Sheng, Q.Z., Motahari, H. (eds.) BPM 2018. LNBIP, vol. 342, pp. 239–250. Springer, Cham (2019). https://doi.org/10.1007/978-3-030-11641-5_19
5. Konrad, R., et al.: Modeling the impact of changing patient flow processes in an emergency department: insights from a computer simulation study. Oper. Res. Health Care **2**(4), 66–74 (2013)
6. Marquez-Chamorro, A.E., Resinas, M., Ruiz-Corts, A.: Predictive monitoring of business processes: a survey. IEEE Trans. Serv. Comput. **11**, 962–977 (2017)
7. Rojas, E., Munoz-Gama, J., Sepúlveda, M., Capurro, D.: Process mining in healthcare: a literature review. J. Biomed. Inform. **61**, 224–236 (2016)

8. Tama, B.A., Comuzzi, M.: An empirical comparison of classification techniques for next event prediction using business process event logs. Expert Syst. Appl. **129**, 233–245 (2019)
9. Teinemaa, I., Dumas, M., Rosa, M.L., Maggi, F.M.: Outcome-oriented predictive process monitoring: review and benchmark. ACM Trans. Knowl. Discov. Data (TKDD) **13**(2), 17 (2019)
10. Verenich, I., Dumas, M., La Rosa, M., Maggi, F., Teinemaa, I.: Survey and cross-benchmark comparison of remaining time prediction methods in business process monitoring. arXiv preprint arXiv:1805.02896 (2018)

A Data Driven Agent Elicitation Pipeline for Prediction Models

John Bruntse Larsen[1]([⊠]) [ID], Andrea Burattin[1] [ID], Christopher John Davis[1,2] [ID],
Rasmus Hjardem-Hansen[3], and Jørgen Villadsen[1] [ID]

[1] DTU Compute, Technical University of Denmark, 2800 Kongens Lyngby, Denmark
{jobla,andbur,jovi}@dtu.dk
[2] University of South Florida, Saint Petersburg, FL 33701, USA
davisc@mail.usf.edu
[3] Greve Muncipality, Greve, Denmark

Abstract. Agent-based simulation is a method for simulating complex systems by breaking them down into autonomous interacting agents. However, to create an agent-based simulation for a real-world environment it is necessary to carefully design the agents. In this paper we demonstrate the elicitation of simulation agents from real-world event logs using process mining methods. Collection and processing of event data from a hospital emergency room setting enabled real-world event logs to be synthesized from observational and digital data and used to identify and delineate simulation agents.

Keywords: Agent-based simulation · Process mining · Emergency Rooms

1 Introduction

Emergency medicine is characterized by time and life critical episodes: the range of cases from sprained ankles to trauma patients with life threatening injuries and illnesses create a complex and dynamic working environment. This entry point to the hospital is characterized by unique and substantial challenges for resource planning and scheduling. Changes in workload can happen very quickly and bottlenecks can emerge in a matter of minutes. Our work is motivated by the aspiration of clinicians and administrators to be able to more precisely predict what the situation will look like minutes or hours from now and, ideally, understand the impact of moving staff or changing clinical priorities.

The setting for the research reported in this paper is a Joint Emergency Room (JER) in Denmark (Fælles Akutmodtagelse). In addition to major trauma and minor injuries, the JER is the entry point for patients admitted by their General Practitioner and brought by ambulance: this further increases the diversity of the symptoms and injuries presented. Nursing skills in the JER are transferable, allowing them to be readily deployed to alleviate a bottleneck. Doctors and other staff are less easily redeployed, primarily tending to patients within their specialty. Some staff are always on duty, others on call.

Short term planning in the JER is therefore highly complex. Prior attempts to predict patient flow using simulation [3,6] highlight the difficulty of customizing

© Springer Nature Switzerland AG 2019
C. Di Francescomarino et al. (Eds.): BPM 2019 Workshops, LNBIP 362, pp. 570–582, 2019.
https://doi.org/10.1007/978-3-030-37453-2_46

models for a given department. The JER presents a limiting case, since the flow of patients here involves unpredictable and complex interactions between clinical specialties. We use agent-based simulation [12,16] as a means to reveal and articulate these complex interactions.

Our goal is to provide explanatory insight into JER roles—doctors, nurses, administrators etc. – as they follow triage and other clinical protocols to assign patients to 'tracks' and schedule procedures. By focusing on the clinical agents themselves, we ground our data collection and model development directly into the JER context. This approach minimizes the risk that the data and model are abstracted from the clinical environment, increasing the tractability of model genesis and the credibility of the simulation offered by the model. Clinical event data were gathered in the JER using a custom-designed research instrument. These were combined with data extracted from the Hospital Information System (HIS) (following Mans et al. [13]) to develop a multi-phase protocol for real-world, time-specific clinical event data acquisition and logging.

Section 2 relates our research approach to prior work; in Sect. 3 we outline our method for eliciting simulation agents; in Sect. 4 we set out our data collection methods; in Sect. 5 we present our pipeline of methods and tools used to process the various data types into event logs; in Sect. 6 we demonstrate use of the pipeline in an emergency medicine setting and assess the validity and cohesion of the event log and the elicited agents. The concluding Sect. 7 summarizes our contributions, and explores limitations and future work.

2 Related Work

In this work, we combine agent-based simulation with process mining to elicit a set of agents that supports simulation. This approach is similar to that of Ito et al. [8] who used process mining to analyze data generated by a multi-agent business simulator. Their extension of an existing multi-agent business simulator to generate data, conversion into event logs and analysis using process mining methods differs from our approach. We gather domain-specific observational event data (from the JER) which is processed into a format that allows it to be synthesized with digital data to create enriched, hybrid event logs. These hybrid logs are used to identify roles and generate simulation agents.

The idea of agent-based simulation is to compose complex systems from agents with a set of constraints for how to behave. Similarly, recent work in process mining has focused on mining process models that are based on temporal logic constraints [4]. The goal explored in this work is to mine constraints in order to elicit agents to be used in future agent-based simulations.

Agent-based simulation has been used for prediction and workflow analysis in emergency departments, showing that agent-based models can simulate bottlenecks and relationships that comprise the workflow [11,15]. However, these studies offer simulations that though holistic are high-level.

We do not design an agent-based simulation but rather investigate generating one from data. In order to do this we take advantage of generalized formal frameworks from the multi-agent systems community. Development of formal meta-models and frameworks for implementing agent organizations make it feasible to implement systems of agents where agents have roles and collaborative objectives but are capable of acting autonomously. These frameworks are general purpose and thus usable in a range of domains. However, although these frameworks provide a reusable structure for implementing the agents for any domain, significant effort is required to identify the characteristics of a specific domain. Our goal is to use process mining to elicit the agents from historical data in order to reduce the effort needed to create an agent-based simulation and to ameliorate the risks of analytic bias. Our work is further motivated by prominent results of data driven approaches to simulation and decision support in healthcare in the literature [6,17]. Work in process mining often involve domain experts in the production of models, see for example [14]. We believe reducing the potential of domain expert bias in the produces model is more suitable in order to obtain our goal though.

3   Mining Agents Vision

Agent-based simulation is effective for analyzing complex systems composed of many actors where the behavior of each actor can be described and implemented as an agent using a small set of rules or protocols [12]. However, for social-technical systems with human actors that might work autonomously and/or in collaboration, this is not a trivial task. The identification and implementation of agents in the complex socio-technical *melieu* of the JER is complicated by the interplay of their autonomy - arising from their medical specialisms - and their collaboration in multi-disciplinary activities such as diagnosis. Our overall approach is to use process mining to identify roles, task delegations and capabilities from clinical event data and then translate these into agent organization constraints [5] in a formal framework. The BPMN diagram in Fig. 1 shows our agent elicitation pipeline of activities and data artifacts. The figure is adapted from the L^* life-cycle in the process mining manifesto [1,2]. The steps in our pipeline are:

1. Identify the data needed;
2. Gather data from a real-world case study;
3. Convert the data to event logs using Disco;
4. Create a Petri net for the event log using ProM;
5. Convert the Petri net to a process model using BPMN;
6. Create a candidate role set using a swimlane module;
7. Create a handover of work graph;
8. Create a similar-task graph;
9. Translate the candidate role set into role definitions, handover of work into task delegations and similar-task graph into role capabilities.

Clustering activities using the swimlane module identifies candidate roles. The handover of work graph shows the order of resources performing activities.

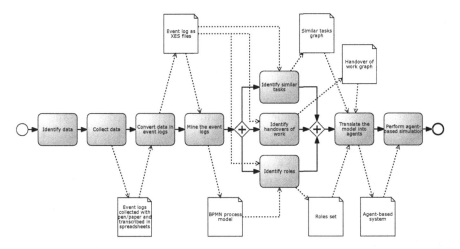

Fig. 1. Agent elicitation pipeline from a real world emergency medicine setting to an agent-based simulation. In order to elicit simulation agents, we use process mining methods and agent-based models for organizations to create a simulation model based on event data. In order to apply process mining, we need an event log that lists a number of instances of the process we want to simulate, and the events with time stamps that took place in each instance. The pipeline also needs to know what staff members are involved in each event and what their position is. Collecting the necessary data requires an understanding of the domain and the scope of available data.

Finally the similar-task graph identifies resources that perform many of the same activities. In the remaining sections we focus on the first two steps but also show how steps 1–8 build cumulatively and discuss their contributions to step 9.

4   Data Collection Method

In this section we present our method for identifying and collecting the data needed to construct the event logs for the agent elicitation pipeline. In order to identify the data, we need to understand the setting from which we gather data and to what extent data is available. For that reason it is critical to involve a clinical expert. We rely on the general guidelines for collecting data for process mining, adapting them for the agent elicitation pipeline.

Long term, we aim to use agent-based simulation for continuously predicting what potential bottlenecks that can occur and help nurses and doctors with preventing the bottlenecks from actually occurring. For this stage of our work we need to identify the contributors to processes enacted in the JER to scope or bound the simulation. Next, we establish which event types we want information about, in particular time stamps and involved agents. Summing up we:

1. Identify which agents to include in the simulation.
2. Identify which activities to include in the simulation.
3. Identify which sorts of events to collection information about.

Next we identify to what extent data is available and how we can collect it. The JER uses a Hospital Information System (HIS) to log data about patient episodes. With guidance from clinical experts, the HIS offers the potential to provide most of the data we need. Alternatively, if data is not available for extraction, we can manually collect data on location. For this we need help from the clinical experts to assess the extent to which we can collect data on location. Collaboratively designing a paper research instrument helped scope and define our data parameters, showing the activities that we are interested in with fields for entering information about staff members and locations, gaining consensus as we worked with the clinical expert.

Having identified the data we need, we decide on how to collect it. Given that we can not extract the data we need from the HIS, we use the form we designed. We discuss with the clinical expert how to fill out the form so that the data collection does not disrupt the workflow of the organization. We also need to establish how much data is necessary to collect. Generally we want as much data as possible but we also need it in a good quality, and if registrations are not done automatically we require help from clinical experts to correct for errors. Doing the data collection over a longer period of time with some time in between each collection allows us to make corrections and improvements, which may not be apparent at first. The overall aim of this aspect of our research design is to achieve a cohesive and comprehensive log for each specific episode, rather than trying to achieve a census of all (undiscriminated) events in the JER. This adds both meaning and validity to the empirical data collection at the case level.

5   Event Log Construction

Having collected the data we need, we convert it into event logs and process models that we can use for creating agent organization constraints. In this section we describe how we processed the data.

To clarify what an event log is, it is a file containing a list of enumerated process instances, i.e. *cases*, where each case contains a series of events with start and end times, plus additional information for each case. An event log is the input for process mining tools such as ProM, and so processing the data to suit the tools we are going to use is critical in order to get meaningful results. In our pipeline we show that the event log is used for (1) generating roles by using a swimlane analysis module, (2) generating task delegations between roles by creating a handover of work graph, and (3) generating capabilities by creating a similar-task graph.

A major point to address in the processing is that existing process mining methods assume that only one resource is involved in each event [1]. In JER however multiple staff members are often involved simultaneously in an event, for example when doctors discuss the course of action for a trauma patient after they have been stabilized. Rather than extending process mining methods to support multiple resources for one event, our solution at this stage of the evolution of our work is to assign names to such groupings. This has the disadvantage that we may lose information about exactly which agents are involved in an event. For the purpose of creating agent organization constraints, we use a naming scheme that combines the positions of the staff members involved.

6 Evaluation

The primary contribution of this phase of our work is the hybrid event log created using the methods outlined above. In this section we evaluate to what extent this type of event log can be used to construct simulation agents for making predictions about patient flow. We do this in two parts. First we discuss the process of applying the method: collecting data and creating an event log, and the content of the log itself. Next we show and discuss preliminary results of applying the three process mining methods for generating roles, task delegations and capabilities.

6.1 Application in a Real-World Setting

Our first step is to identify what parts of the JER that we aim to simulate eventually. In doing so we determine who the agents are and what activities to include in the simulation. The national plan for Danish healthcare services [7] identifies the JER as the common entry point for all acute patients at a hospital. Acute patients are received and treated in JER until they are cleared to either go home or are moved to another department in the hospital. Our scope for agents and activities is limited to those involved from when a patient enters the JER until they leave. We want to collect information about clinical events, which include medical events involving a patient, such as taking samples, and administrative events such as a nurse conveying information to a doctor. We want to include administrative events since they can tell us how tasks are handed over between roles. Next we identify the extent to which data is available and how we can collect it. Here we use the work of Mans et al. [13] as a starting point, as they describe what kind of data is available in most modern HIS and what kind of data is typically required for different forms of analysis in process mining. We identified the data we needed and met with the hospital management in order to clarify to what extent the data was available in their systems. To that end we designed a paper form that showed what activities, kind of events and information we were interested in. The event form contained fields for the following:

1. Arrival time.
2. Triage time, location and outcome.
3. Additional clinical events.
4. What resources were involved in events, both human and material.
5. What decisions were made and who made them.
6. Information about the finish from JER i.e. when the treatment finished in JER and what the outcome was.

We also wanted data about the staff members who were involved. We designed a separate form to collect this information, so we could simply refer to their ID in the event forms and thus simplify the collection process. The form contained an anonymized date field and the following fields for a staff member:

1. Their ID.
2. Their shift.
3. The department were assigned to in their shift.
4. Their position.
5. Their clinical specialties.

With help from the hospital management we found that some, but not all, of the data we needed were present in the HIS. We determined that the system contained the following information of interest to us:

1. Arrival events.
2. Triage events.
3. Treatments, measurements and sampling events.
4. CT, XRAY, blood sample etc. reservations.
5. Times for results from samples and scans.
6. Bed assignments.
7. Movement events within the JER.
8. Request events for additional specialists.
9. What employees were involved in each event.
10. Assignments and reassignments of nurses and doctors.
11. JER departure events.

While the system contained registrations about which nurses and doctors were assigned to patients over time, the potential communication events that could have lead to these registrations were not. We would not be able to tell if or when nurses and doctors communicated in some way beforehand. Such events are of interest in determining the handover of work. In addition, due to data policies, we could not extract data dumps from the HIS. Thus we agreed on having a researcher collect data on location with assistance from a nurse. The researcher could ensure the uniformity of the collected data and the nurse could assist with interpreting the data and identify potentially incorrect or misleading data.

We scheduled data collection in the JER over 9 workdays in order to fully capture the work distribution: patients in the JER are assigned (via triage) to one of three 'tracks', so we planned 3 days in each track: for each of the three tracks, our agenda was as follows

- Day 1: Record data from patients one by one until they leave, following the staff members to register communication events between employees but without interfering with the clinical work.
- Day 2: Copying data from the HIS by hand.
- Day 3: A customized method for collecting data about known parts that have not been covered well during day 1 and day 2.

The reason for using this method is that we could not know beforehand what patients would arrive so having two days with a fixed plan and a third day with an open plan gave us some flexibility.

While collecting the data we observed factors that could be relevant when trying to identify bottlenecks in the JER. We observed 4 different factors: (1) The JER is sometimes used for training nurse students as part of their education, and depending on their level of education more or less time needs to be allocated for the training, (2) some patients have complex treatment paths which require multiple specialties, which involves a doctor handing over the patient to another specialty, (3) preparation and cleanup of beds is done manually and, depending on the circumstances, requires a nurse or cleanup specialist, (4) trauma patients require a team of nurses and doctors from the moment they enter the JER. Having collected the data using paper forms, the next step is to create an event log. As we noted in the description of our method, most events involve multiple staff members. Since the process mining methods assume that one event involve only one resource, we combined the names of the positions of the involved staff members. For example a team of nurses become the resource NURSES.

With that approach we made a spreadsheet manually with the following columns and filled it with information in the paper forms:

process id One number for each paper which corresponds to one case.
event Description of the event.
start Start time.
end End time.
team Team name.
place Room ID.
extra Additional information such as what form of communication was used in communication events, what the color code of a triage was, or what treatment was given.
day The day the data was collected. This is a reference to the staff member paper forms, which we used to create the team names.

In doing so we also adjusted some of the labels from what was stated in the paper forms since we changed our labelling scheme over the period we collected the data. This ensures that the quality of the data in the spreadsheet is the same for all days. For start and end times, we only use the time of day, not the date.

The spreadsheet contains 66 complete cases and 838 events. Having the spreadsheet it was straightforward to convert it to an event log using Disco and ProM, annotating the columns with meta-data. The result is an event log with real-world data about staff activity. The log describes in detail the events that happen for a broad variety of episodes in the JER. The event log also does not contain any personal information but contains detailed information for each event about what roles have been involved in the individual events, which is the organizational information we need to generate agent organization constraints for the JER.

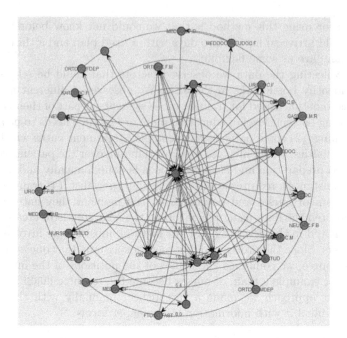

Fig. 2. Handover of work graph. Nodes represent teams and there is an edge between teams if they involved in two sequential events.

6.2   Creating Agent Organization Constraints

Our vision is to use event logs for creating agent organization constraints that can be used for agent-based simulation. In this paper we have focused on creating a real-world event log for this purpose. Although the amount of collected data is low, we would like to discuss preliminary work on the remaining parts of the pipeline showing the potential for eliciting agent organization constraints.

Task Delegation. We can generate a handover of work graph as shown in Fig. 2 directly from the event log using ProM. The graph shows how the teams take turns being used as a resource for an event in a case. Given that nurses are involved in all cases, it is no surprise that there many edges connect to the NURSES node and the centrality of the nurses suggests that addressing nurse shortage would ameliorate bottlenecks. The labelling makes it possible to identify the individual roles involved, for example a nurse is included in both NURSES and NURSEDOC. The tasks are not shown very clearly in the graph though. Our idea for creating delegation constraints is to translate the graph into logical constraints, which requires identifying the tasks for which the handovers occur.

Capabilities. We can generate a similar-task graph as shown in Fig. 3 directly from the event log using ProM. The graph shows which teams work on the same

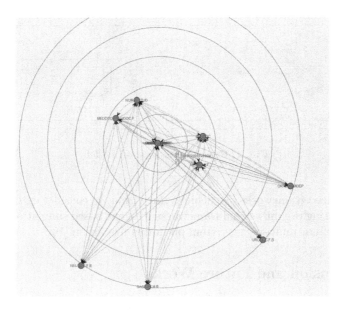

Fig. 3. Similar-task graph. Nodes represent teams and there is an edge between teams if they involved in the same events.

tasks to a high degree. We note that many of the nodes are connected to each other, which is not surprising since we have a small number of event descriptions that are fairly generic for the domain, for example `info` and `treatment`. Creating competence constraints enables us to delineate sets of roles and map the sets to the tasks they can carry out, which requires identifying the tasks that make the roles similar.

Roles. We can generate a role-specific business process model (BPM) as shown in Fig. 4 by generating a Petri net, BPM and a role set from the event log in ProM. The BPM shows the flow of control in the mined process, and is annotated with swimlanes which splits the BPM into role sets. In our case we only identified one role set, which contains the parts concerning ordering of CT scans. Our idea for creating role definitions is to use the generated role sets as a base. Thus our model originates from unadulterated event data, and our pipeline is agnostic to the *de jure* organizational structures. Our role definitions emerge from the set of objectives that agents enacting that role should achieve.

Simulating an Agent Organization. In order to get the best performance, agent-based simulation is often coded from scratch in a general purpose programming language. There are however also agent-based simulation platforms that offer generalized frameworks for coding agents in a high level language which removes the need to handle synchronization issues at a low level. Our vision is to implement the organizational constraints we identify using process mining in

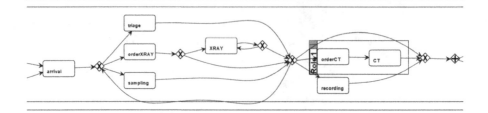

Fig. 4. Role specific process model

such generalized frameworks. To achieve this vision we consider extending work on implementing organizational frameworks in agent-based simulation platforms like GAMA [9] or multi-agent system platforms like Jason [10].

7   Conclusion and Future Work

We have presented our vision for eliciting simulation agents from real-world event logs by using process mining methods. Figure 1 illustrates our vision as a pipeline of activities and emergent products. In this paper we have focused on the creation of a hybrid event log and how it can be used as the basis to develop agent-based prediction models which are agnostic to organizational structures. We created a log that gives a snapshot of reality which is not statistically significant but still offers meaningful insight into reality. We gathered and analyzed data using the protocols summarized in Fig. 1 to offer a proof-of-concept for our approach. These preliminary data also offered meaningful insights into the location of handovers that have the potential to ameliorate bottlenecks in the clinical setting.

For each event in the event log we have assigned a team resource that identifies what roles are involved in an event. To do this we preprocess the data, combining the positions of staff members into a team name so that we can recover the individual roles later when creating agent organization constraints. Alternatively, if the process mining tools supported multiple resources for one event, we could omit this part of the preprocessing and potentially also simplify the translation into agent organization constraints. Future work includes investigating such options.

Time constraints limited the scope of our observations, analysis and modeling. Future work will extend the scale of our models, offering further validation of the methods set out in this paper as the pipeline is more fully loaded. This will enable further exploration of translation of process mining output to agent organization constraints, the preliminary steps for which have been set out in this paper. We also plan to implement these constraints in agent-based simulation, building further on the philosophy of structural agnosticism by using generalized agent organization frameworks, and thus generate an agent-based prediction model from data.

Acknowledgements. This work is supported by the PhD project between PDC A/S and Technical University of Denmark (DTU). We thank the institute Region H, which manages the hospitals in the Danish capital region, for providing input for the use case. Christopher Davis is Otto Mønsted Visiting Professor at DTU Compute, supported by the Otto Mønsteds Fond. Supported by the Process-Oriented Data Science for Healthcare Alliance (PODS4H Alliance).

References

1. van der Aalst, W.: Process Mining: Data Science in Action, 2nd edn. Springer, Heidelberg (2016). https://doi.org/10.1007/978-3-662-49851-4
2. van der Aalst, W., et al.: Process mining manifesto. In: Daniel, F., Barkaoui, K., Dustdar, S. (eds.) Business Process Management Workshops. BPM 2011. LNBIP, vol. 99, pp. 169–194. Springer, Heidelberg (2012). https://doi.org/10.1007/978-3-642-28108-2_19
3. Almagooshi, S.: Simulation modelling in healthcare: challenges and trends. Procedia Manuf. **3**, 301–307 (2015)
4. Burattin, A., Maggi, F.M., Sperduti, A.: Conformance checking based on multi-perspective declarative process models. Expert Syst. Appl. **65**, 194–211 (2016)
5. Dignum, V., Padget, J.: Multiagent organizations. In: Weiss, G. (ed.) Multiagent Systems, pp. 51–98. MIT Press (2013)
6. Goienetxea Uriarte, A., Ruiz Zúñiga, E., Urenda Moris, M., Ng, A.H.: How can decision makers be supported in the improvement of an emergency department? A simulation, optimization and data mining approach. Oper. Res. Health Care **15**, 102–122 (2017)
7. Hansen-Nord, M., Steensen, J.P., Holm, S.: Process driven patient tracks in FAM. Scand. J. Trauma. Resusc. Emerg. Med. **18**(Suppl. 1), 27 (2010)
8. Ito, S., Vymětal, D., Šperka, R., Halaška, M.: Process mining of a multi-agent business simulator. Comput. Math. Organ. Theory **24**(4), 1–32 (2018)
9. Larsen, J.B.: Adding organizational reasoning to agent-based simulations in GAMA. In: Weyns, D., Mascardi, V., Ricci, A. (eds.) EMAS 2018. LNCS (LNAI), vol. 11375, pp. 242–262. Springer, Cham (2019). https://doi.org/10.1007/978-3-030-25693-7_13
10. Larsen, J.B.: Going beyond BDI for agent-based simulation. J. Inf. Telecommun. (TJIT), 1–19 (2019). https://doi.org/10.1080/24751839.2019.1620024
11. Liu, Z., Rexachs, D., Epelde, F., Luque, E.: An agent-based model for quantitatively analyzing and predicting the complex behavior of emergency departments. J. Comput. Sci. **21**, 11–23 (2017)
12. Macal, C.M., North, M.J., Samuelson, D.A.: Agent-based simulation. In: Gass, S.I., Fu, M.C. (eds.) Encyclopedia of Operations Research and Management Science, pp. 8–16. Springer, Boston, MA (2013). https://doi.org/10.1007/978-1-4419-1153-7
13. Mans, R.S., van der Aalst, W.M.P., Vanwersch, R.J.B.: Process Mining in Healthcare: Evaluating and Exploiting Operational Healthcare Processes. Springer International Publishing, Heidelberg (2015)
14. Rovani, M., Maggi, F.M., De Leoni, M., Van Der Aalst, W.M.: Declarative process mining in healthcare. Expert Syst. Appl. **42**(23), 9236–9251 (2015)
15. Wang, L.: An agent-based simulation for workflow in emergency department. In: IEEE Systems and Information Engineering Design Symposium, pp. 19–23 (2009)

16. Wooldridge, M.: Intelligent agents. In: Weiss, G. (ed.) Multiagent Systems, pp. 3–50. MIT Press (2013)
17. Zhou, Z., Wang, Y., Li, L.: Process mining based modeling and analysis of workflows in clinical care. In: Proceedings of the 11th IEEE ICNSC, pp. 590–595 (2014)

A Solution Framework Based on Process Mining, Optimization, and Discrete-Event Simulation to Improve Queue Performance in an Emergency Department

Bianca B. P. Antunes[(✉)], Adrian Manresa, Leonardo S. L. Bastos,
Janaina F. Marchesi, and Silvio Hamacher

Pontifical Catholic University/PUC-Rio, Marquês de São Vicente, 225,
Rio de Janeiro, Gávea 22451-900, Brazil
{biancabrandao,adrianperez,lslbastos,
janaina.marchesi}@tecgraf.puc-rio.br,
hamacher@puc-rio.br

Abstract. Long waiting lines are a frequent problem in hospitals' Emergency Departments and can be critical to the patient's health and experience. This study proposes a three-stage solution framework to address this issue: Process Identification, Process Optimization, and Process Simulation. In the first, we use descriptive statistics to understand the data and obtain indicators as well as Process Mining techniques to identify the main process flow; the Optimization phase is composed of a mathematical model to provide an optimal physician schedule that reduces waiting times; and, finally, Simulation is performed to compare the original process flow and scheduling with the optimized solution. We applied the proposed solution framework to a case study in a Brazilian private hospital. Final data comprised of 65,407 emergency cases which corresponded to 399,631 event log registries in a 13-month period. The main metrics observed were the waiting time before the First General Assessment of a physician and the volume of patients within the system per hour and day of the week. When simulated, the optimal physician scheduling resulted in more than 40% reduction in waiting times and queue length, a 29.3% decrease of queue occurrences, and 54.2% less frequency of large queues.

Keywords: Emergency Department · Process discovery · Physician scheduling

1 Introduction

A hospital's Emergency Department (ED) is the main gateway for a great part of its patients. One of the most frequent problems found in EDs is long waiting times, which result in negative experiences and can worsen the patient's condition. As the use of EDs has a random nature, patient arrival is difficult to predict, which makes tactical and operational hospital planning more complicated. Literature has shown that excessive workload to healthcare staff is prejudicial to patient's safety [1]. Also, in busy hours,

© Springer Nature Switzerland AG 2019
C. Di Francescomarino et al. (Eds.): BPM 2019 Workshops, LNBIP 362, pp. 583–594, 2019.
https://doi.org/10.1007/978-3-030-37453-2_47

patients that are not in severe conditions can end up going through long lines [2], as they do not have priority on getting serviced.

Therefore, it is important to understand and deal with the flow of patients, minimizing waiting times. For that purpose, several studies use discrete-event simulation as an auxiliary tool for operations management in healthcare. One of the most frequent types of work is to simulate the flow of patients, whether concerning clinics [3], surgery divisions [4] or emergency departments [5].

Among the studies regarding simulation, a few also consider Process Mining as an auxiliary tool to better represent the agents' behavior [6]. Liu et al. [7] highlight the difficulties found in the construction of the simulation model that can be overcome through the use of Process Mining software, which automatically creates the process flow; Kovalchuk et al. [8] use process mining, machine learning, discrete-event simulation and queuing theory to simulate the flow of patients; Wang et al. [9] use fuzzy logic to get to the simulation model along with Process Mining; and Rojas et al. [10] implement a performance analysis for emergency rooms through Process Mining.

One way to deal with long lines and improve patient experience in EDs is to raise the number of physicians on duty (increase capacity). However, it can be unsustainable for hospitals due to possible increases in costs, and constraints of workload and law. Therefore, tools to try to find optimal staffing levels can be applied. Savage et al. [11], for example, adopted a mathematical programming model while Green et al. [12] used queuing theory and Kuo [13] proposed a simulation-optimization approach, exploring solutions iteratively, which were evaluated with simulations at each step.

Hence, this paper proposes a solution framework that incorporates Process Mining, optimization, and discrete-event simulation to improve staff schedule and reduce waiting times. Firstly, descriptive statistics and a Process Mining tool provide the identification of the "as is" process and indicators of the system, which are the basis to the simulation model. Then, physician levels are proposed by an optimization model which aims to reduce lines. Lastly, these results are applied in the simulation model to test the optimization results and to represent potential gains with the optimized schedule. We applied this methodology using data from a large sized private hospital in Brazil.

The remainder of this paper is structured as follows: Sect. 2 presents the problem definition; Sect. 3 describes the solution framework; Sect. 4 shows the numerical experiments performed; Sect. 5 presents the conclusions.

2 Problem Definition

This paper is driven by a hospital case study, and it aims to present a framework to analyze the current status of an ED, determine the optimal levels of its staffing at each hour and day of the week and to develop a simulation model that can accurately represent current and future situations.

The problem consists of a hospital's ED, where patients go to the screening process with a nurse, who determines whether they need vital medical care. If the patients do need vital care, they go straight to the physician assessment; else, they stay in a waiting room until a physician is available. After that, the patient can be directed to receive medication, stay in observation or be examined, and, afterward, is evaluated again, preferably by the same physician. These steps can be repeated until the patient is admitted to the hospital or discharged. Patient's arrival might depend on the day of the week (d) and the hour of the day (h); (t) stands for the time between patients' assessments. Figure 1 represents a basic patient flow, which can be adapted to have other activities and, therefore, lines.

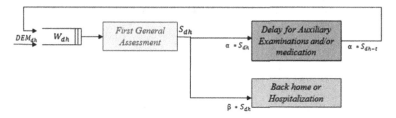

Fig. 1. Basic ED patient flow

In this study, we evaluated indicators of performance regarding the queue for the First General Assessment activity - the average waiting time, the frequency of queue, the frequency of queues with more than ten patients (large queue) and the average number of patients in the queue (queue length). Those indicators belong to the door-to-doctor stage of the process, and their evaluation corresponds to an important measure of ED performance [14].

The problem defined here consists, basically, in addressing physician staffing to meet patient arrival and volume, considering that there would be no increase in the available number of physicians, to minimize patient's waiting time. Furthermore, even though the proposed solution framework was applied in a specific ED in this paper, it can be adapted to other cases.

3   Solution Framework

The framework proposed in this article consists of three general phases: Process Identification, that uses Process Mining and descriptive statistics to define and understand the "as is" process and its current indicators, especially waiting time; Process optimization, in which we defined the best physician staffing to the ED using a Mixed-Integer Programming (MIP) model; and Process simulation, that evaluated the MIP proposed solution and its expected benefits, considering the "as is" process and service times. The proposed framework is represented in Fig. 2, and each phase is described in the following sections.

3.1   Process Identification

In this first phase, we considered Process Mining (PM) techniques to identify the main flows within activities from data of each patient. Therefore, we obtained the event log

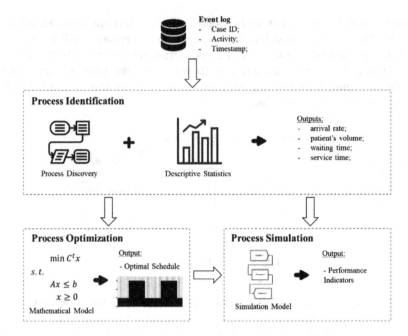

Fig. 2. Diagram of the proposed solution framework.

from the ED and performed process discovery to identify the "as is" flow of activities. For this purpose, we considered the variants that accounted for 80% of cases as the main process flow to be studied [15]. Furthermore, we obtained and analyzed indicators from the process: the arrival rate of patients, average service time, number of patients in the system, and the current staff scheduling.

3.2   Process Optimization

The optimization model is based on the process model identified with the Process Mining software. It represents the basic ED flow described in Sect. 2, but it can be adapted to have more activities and lines, that can be obtained in the Process Identification phase. Its objective is to determine the number of physicians that should be in service during each hour and day of the week, in accordance with service requirements. The objective of the model is to minimize the total number of patients waiting per hour in the ED. Table 1 presents the domains in which our model attributes are defined, while Tables 2 and 3 provide a complete listing of parameter and variable definitions.

The proposed mathematical formulation is as follows:

$$Min\, Z = \sum_{d}\sum_{h} wait_{dh} \tag{1}$$

$$\sum_{d}\sum_{h}\sum_{k} x_{dhk} * DUR_k \leq WL\_TOTAL \tag{2}$$

Table 1. Sets, subsets and corresponding domains

Sets	Indexes	Domain	Description
Days (D)	d	{1, ..., PLAN_HORIZON}	Days of the planning horizon
Hours (H)	h, h'	{0, ..., 23}	Hours of the day (24-h clock format)
Shift type (K)	k	{1, ..., |K|}	Type of shifts

Table 2. Model parameters

Parameters	Description	Unit
PLAN_HORIZON	Number of days in the planning horizon	–
DUR_k	Duration of each shift k	Hour
WL_TOTAL	Total workload available for the planning horizon	Hour
CAP_MIN_{dh}	Minimum number of physicians required per hour h and day d	–
I_QUEUE	Initial queue at the beginning of the planning horizon	–
$START\_COMB_{h,h'k}$	Indicates the hour h' covered if shift type k start in hour h	{0, 1}
a	Rate of patients seen who need ancillary exams and/or medication	–
DEM_{dh}	Demand (arrival of patients for each day d and hour h)	–
CAP_{dh}	Capacity (service rate - number of patients seen per physician each day d and hour h)	–

Table 3. Model variables

Variables	Description	Domain
$wait_{dh}$	Number of patients waiting in each hour h and each day d	$\mathbb{R}_+$
x_{dhk}	Number of physicians assigned in the shift type k starting in hour h in the day d	$\mathbb{Z}_+$
ic_{dh}	Idle capacity in each hour h and day d	$\mathbb{R}_+$
$served_{dh}$	Number of patients seen in each hour h and day d	$\mathbb{R}_+$
n_{dh}	Number of physicians required for each hour h and each day d	$\mathbb{Z}_+$

$$\sum_h \sum_k x_{dhk} * START\_COMB_{h,h'k} \geq n_{dh} + CAP_{MIN_{dh}} \quad \forall_h, \forall_d \qquad (3)$$

$$wait_{dh} = I\_QUEUE + DEM_{dh} - served_{dh} \quad \forall_{d|d=1}, \forall_{h|h=0} \qquad (4)$$

$$wait_{dh} = wait_{d-1,'23'} + DEM_{dh} + \alpha * served_{d-1,'22'} - served_{dh} \qquad (5)$$

$$wait_{dh} = wait_{d,h-1} + DEM_{dh} + \alpha * served_{d,h-t} - served_{dh} \, \forall_d, \forall_{h|h>0} \qquad (6)$$

$$served_{dh} = CAP_{dh} * n_{dh} - ic_{dh} \, \forall_d, \forall_h \qquad (7)$$

$$wait_{dh} \in R_+, \forall h \in H, \forall d \in D \tag{8}$$

$$x_{dhk} \in Z_+, \forall d \in D, \forall h \in H, \forall k \in K \tag{9}$$

$$ic_{dh} \in R_+, \forall h \in H, \forall d \in D \tag{10}$$

$$served_{dh} \in R_+, \forall h \in H, \forall d \in D \tag{11}$$

$$n_{dh} \in Z_+, \forall h \in H, \forall d \in D \tag{12}$$

The objective function in (1) consists of minimizing the total number of patients waiting each day and hour. Constraint (2) guarantees that the sum of the shift durations assigned is less than or equal to the total allowable workload. Constraint (3) enforces that the total of physicians assigned in each shift type (x_{dhk}) meets hourly the staffing level required (n_{dh}). Queue, number of patients served and idle capacity are computed through (4)–(7) based on the number of physicians required per hour to meet demand. Finally, constraints (8)–(12) define the domain of the decision variables.

3.3   Process Simulation

In this framework, we apply Discrete Event Simulation (DES) because of its flexibility, versatility, and ability to model processes in a great level of detail [16]. We use it to evaluate the changes that would be caused if the schedule derived from the optimization phase was applied in the ED. The design of the model is formulated based on the process identification phase. To validate the model and evaluate the optimized schedule, we considered queue-related metrics such as the average number of patients in the queue, average waiting time, frequency of queues and frequency of queues with more than ten patients (large queues).

4   Numerical Experiments

4.1   Case Study and Data

This study was conducted in a large sized private hospital located in the city of Rio de Janeiro, Brazil. Data comprised of 81,736 emergency cases over 13 months for patients aged over 14 years old and with a length-of-stay up to 24 h in the general emergency, resulting in an event log of 544,226 registries. For each record, there are timestamps for their different stages in the emergency flow.

In the original schedule, physicians and nurses work in 12-h shifts. The number of physicians varies during the hour and day of the week: 5 physicians in the first shift (8 am to 8 pm) and 3 in the following shift from Monday to Wednesday, and 4 physicians in the first shift and 2 in the following shift, from Thursday to Sunday. The doctors' entry time into the ED could be modified to meet the demand better. However, this is a difficult task without the support of a decision-making tool. The number of nurses and receptionists were considered as invariable during the whole day.

We emphasize that no information that could identify the patients from each emergency case was provided.

In the Process Identification stage, we used *Disco* for process discovery and *R* 3.5.2 (*tidy verse* package) for data analysis. In Process Optimization, the mathematical programming model was implemented and solved using AIMMS 4.64 and GUROBI 8.1 with default settings in Intel i5-7200U 2.5 GHz 20 GB RAM computer. The optimal solution was obtained within 19 s (548 constraints and 883 variables). In Process Simulation, the model was implemented using Arena Simulation software version 14.7.

4.2 Results

To analyze the current status of the process, we estimated the hourly demand and capacity of each weekday, in terms of patient arrival and volume, as well as service time and waiting time for the First General Assessment. In Fig. 3, we show the proportions regarding the average volume of patients before the first assessment (low blue bars) and during or after the first assessment (top green bars), and the average arrival rate (red line), for each hour of the day.

One can observe that the arrival rate of patients has an uprising trend until its peak at hour 11 (17 patients/hour), which also increases the proportion of patients waiting for the first assessment at this time. The total number of patients in the ED is the highest at around hour 20 (56 patients), which is also the time of shift change, and some of these patients stay in the ED until the next morning (hour 6).

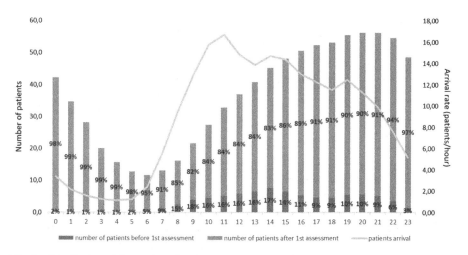

Fig. 3. Distribution of the number of patients and patient arrival rate in the ED (Color figure online)

For the process discovery, we obtained the process variants that accounted for 80% of the emergency cases from the original event log. The comparison of the data from the original and the selected event log is shown in Table 4.

Table 4. Comparison between original and selected event log data

Data Information	Complete	Selected Variants	%
Emergency Cases	81,736	65,407	80
Event Log size	544,226	399,631	73
No. of Variants	1,731	22	1.3

We observed that the selection resulted in 22 out of 1,731 of all the identified process variants, which corresponded to 65,407 emergency cases and composed 73% of the events. Thus, there is a large variability of flows in the ED, which may be associated to different situations and high dynamics in this department due to the severity of patients, complexity of treatments and organizational procedures.

When exploring the 22 selected variants, we observed two main types of flows: either a patient gets discharged right after the first contact with a physician or the patient needs some extra assistance. The former comprehends the most frequent variant, while in the latter type the patient may need nursing assistance for medication, undergo in laboratory or image tests (e.g., x-ray, MRI, etc.), or stay in a dedicated bed for observation. In Table 5, we summarized the activities, events, and time from Variants 1 to 4, which corresponded to 70% of the emergency cases within the previously selected variants.

Table 5. Activities and proportions of Variants 1–4

Activity	Variant 1	Variant 2	Variant 3	Variant 4
1	Door	Door	Door	Door
2	Screening	Screening	Screening	Screening
3	Registration	Registration	Registration	Registration
4	FGA	FGA	FGA	FGA
5	ED discharge	In observation	Lab Tests	Lab tests
6		ED discharge	In observation	Image tests
7			ED discharge	In observation
8				ED discharge
No of events	143,280	62,676	24,710	24,976
No of cases	28,656 (43.8%)	10,446 (16%)	3,530 (5.4%)	3,122 (4.8%)

FGA - First General Assessment

For all variants, the Door-To-Doctor flow was similar. Variant 1 represents the patients that were discharged right after the First General Assessment (about 44%). "In observation" was common among Variants 2 to 4. Therefore, we considered the "as is" process as the set of activities from the 22 selected variants, considering the two main flows observed in Variant 1 and Variants 2–22.

In Table 6, we display the average number of cases, service time, and capacity per day of the week, regarding the First General Assessment activity. Monday has been the day with the largest demand, with an average of 197.4 cases, and it also presented the lowest average service time, 13.5 min, being the highest on Sunday, 16.3 min.

Saturday presented the smallest demand (average of 141.1 patients). In overall, the service times presented a high variability, with standard deviations greater than the average times. We suggest that large patient volume could influence in physician performance by speeding up the First General Assessment. Estimated capacity also shows a slight decrease throughout the week, as the demand is also reduced, although it fluctuates around the average of 4.1 patients per hour.

Table 6. Demand, performance, and capacity of the emergency department

Days of the week	Number of cases		Service time (min)		Estimated capacity per hour	
	Average	SD	Average	SD	Average	SD
Monday	197.4	31.4	13.5	22.8	4.5	2.6
Tuesday	180.3	25.5	14.5	23.2	4.1	2.6
Wednesday	179.7	19.8	15.0	21.5	4.0	2.8
Thursday	162.0	20.2	14.6	35.6	4.1	1.7
Friday	156.4	20.2	15.4	59.6	3.9	1.0
Saturday	141.1	19.3	13.8	28.8	4.4	2.1
Sunday	141.6	19.0	16.3	52.2	3.7	1.2
Total	165.2	29.9	14.7	36.4	4.1	1.6

These statistics were used as input to the Process Optimization and the Simulation. The optimization model provided a schedule considered optimal for this problem to improve the waiting time. The optimal and original schedules are shown in Fig. 4.

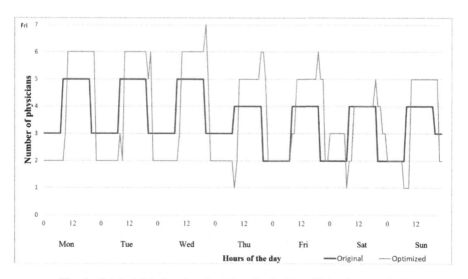

Fig. 4. Original (blue) and optimal (gray) schedules (Color figure online)

In the optimized schedule, there were usually more physicians during the day, between 10 am and 7 pm, which is the period of the day that generally has the highest volume of patients in the system. In the original schedule, there is a maximum of 5 physicians at a given time, while in the optimized schedule, the maximum is 7.

Therefore, using the results from the previous two phases, the distribution of times and arrival rates, the process flow, and the optimal scheduling, we obtained a simulation model that represents the studied ED. We simulated one week of the ED process, starting on Monday, and considered one week for a warm-up period. Patients' inter-arrivals times followed an exponential distribution and service times for each activity in the model were fitted to the most adequate distribution.

To validate the simulation model, we compared its results to data contained in the hospital's information system, such as the average number of patients that arrive in the ED and the service time for the First General Assessment in each day of the week, considering the original physician scheduling. The simulation presented an overall average of 160.7 people arriving in the ED per day (SD = 27), which is a difference of 2.7% to the overall average from the ED's data in Table 6. Service time also had similar results (average of 14.4 min, SD = 8.3, a difference of 2%), being the largest difference on Monday, 11%. These results have a satisfactory accuracy level for our research interest, which also remarks on the credibility of the process knowledge derived from the results provided by the process discovery phase.

Then, we simulated the ED flow with the schedule provided by the optimization model. In Fig. 5, we compare the distribution of waiting times at the First General Assessment obtained from simulating the two schedules for each day of the week.

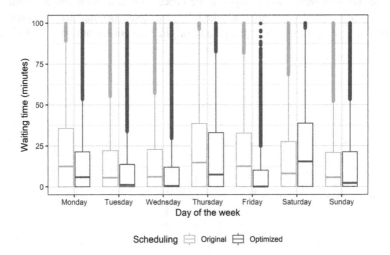

Fig. 5. Waiting times for the original and optimal schedules

The simulation model showed that the optimized schedule could provide reductions in the waiting time for every day of the week, except for Saturday, in which there was an increase of 3% in the average waiting time (less than a minute), which could be influenced by the assignment of fewer physicians in this day as shown in Fig. 4. Results regarding the other indicators of queue performance are in Table 7.

Table 7. Queue indicators for original and optimal schedules (simulation)

Indicators	Original Schedule	Optimized Schedule	Δ%
Average number of patients in queue	3.0	1.6	−46.7
Average waiting time (min)	26.5	14.6	−44.9
Frequency of queues (%)	48.8	34.5	−29.3
Frequency of queues > 10 patients (%)	8.3	3.8	−54.2

The proposed schedule could lead to considerable reductions in queue performance indicators, according to the Simulation model. Both the overall average waiting time and the number of patients in the queue for the FGA activity were decreased by more than 40%, considering all the simulation runs. Also, the frequency of queues was 29.3% lower, and the presence of longer lines decreased by more than 50%. Hence, those results indicate improvements in the process that can provide better patient's experience and care.

5   Conclusion

In this study, we proposed a solution framework to improve waiting times in an emergency department. It is comprised of descriptive statistics to analyze data, process mining to discover the process flow, optimization to find the optimal schedule, and discrete-event simulation to test the proposed schedule. The application of the methodology was performed using data from a large sized private hospital. According to the Simulation, our solution framework could provide an average reduction of 45% in the waiting times and 47% in the number of patients in queue. Moreover, the simulation showed that the frequency of lines could decrease by 29%, with 54% less large queues (more than 10 patients). We emphasize that the proposed schedule has not yet been applied to the ED, owever, the simulation results show promising improvements. Our solution framework could easily be adapted to different contexts of ED's process evaluation and improvement. Suggestions for future works include analyzing ED indicators other than Door-to-Doctor time and developing a more thorough model that includes other lines, and the severity of cases (priorities).

References

1. Carayon, P., Gürses, A.P.: A human factors engineering conceptual framework of nursing workload and patient safety in intensive care units. Intensive Crit. Care Nurs. **21**(5), 284–301 (2005). https://doi.org/10.1016/j.iccn.2004.12.003
2. Wong, Z.S.-Y., Lit, A.C.-H., Leung, S.-Y., Tsui, K.-L., Chin, K.-S.: A discrete-event simulation study for emergency room capacity management in a Hong Kong hospital. In: Proceedings of the 2016 Winter Simulation Conference, pp. 1970–1981. IEEE, Washington (2016). https://doi.org/10.1109/wsc.2016.7822242

3. Côté, M.J.: Patient flow and resource utilization in an outpatient clinic. Soc. Econ. Plann. Sci. **33**(3), 231–245 (1999). https://doi.org/10.1016/S0038-0121(99)00007-5
4. VanBerkel, P.T., Blake, J.T.: A comprehensive simulation for wait time reduction and capacity planning applied in general surgery. Health Care Manag. Sci. **10**(4), 373–385 (2007)
5. Ahmad, N., Ghani, N.A., Kamil, A.A., Tahar, R.M., Teo, A.H.: Evaluating emergency department resource capacity using simulation. Mod. Appl. Sci. **6**(11), 9–19 (2012). https://doi.org/10.5539/mas.v6n11p9
6. Szimanski, F., Ralha, C.G., Wagner, G., Ferreira, D.R.: Improving business process models with agent-based simulation and process mining. In: Nurcan, S., et al. (eds.) BPMDS/EMMSAD -2013. LNBIP, vol. 147, pp. 124–138. Springer, Heidelberg (2013). https://doi.org/10.1007/978-3-642-38484-4_10
7. Liu, Y., Zhang, H., Li, C., Jiao, R.J.: Workflow simulation -for operational decision support using event graph through process mining. Decis. Support Syst. **52**(3), 685–697 (2012). https://doi.org/10.1016/j.dss.2011.11.003
8. Kovalchuk, S.V., Funkner, A.A., Metsker, O.G., Yakovlev, A.N.: Simulation of patient flow in multiple healthcare units using process and data mining techniques for model identification. J. Biomed. Inf. **82**, 128–142 (2018). https://doi.org/10.1016/j.jbi.2018.05.004
9. Wang, Y., Zacharewicz, G., Traoré, M.K., Chen, D.: An integrative approach to simulation model discovery: combining system theory, process mining and fuzzy logic. J. Intell. Fuzzy Syst. **34**(1), 477–490 (2018). https://doi.org/10.3233/JIFS-17403
10. Rojas, E., Cifuentes, A., Burattin, A., Munoz-Gama, J., Sepúlveda, M., Capurro, D.: Analysis of emergency room episodes duration through process mining. In: Daniel, F., Sheng, Quan Z., Motahari, H. (eds.) BPM 2018. LNBIP, vol. 342, pp. 251–263. Springer, Cham (2019). https://doi.org/10.1007/978-3-030-11641-5_20
11. Savage, D.W., Woolford, D.G., Weaver, B., Wood, D.: Developing emergency department physician shift schedules optimized to meet patient demand. Can. J. Emerg. Med. **17**(1), 3–12 (2015). https://doi.org/10.2310/8000.2013.131224
12. Green, L.V., Soares, J., Giglio, J.F., Green, R.A.: Using queueing theory to increase the effectiveness of emergency department provider staffing. Acad. Emerg. Med. **13**(1), 61–68 (2006). https://doi.org/10.1197/j.aem.2005.07.034
13. Kuo, Y.H.: Integrating simulation with simulated annealing for scheduling physicians in an understaffed emergency department. HKIE Trans. **21**(4), 253–261 (2014). https://doi.org/10.1080/1023697X.2014.970748
14. Welch, S., Augustine, J., Camargo Jr., C.A., Reese, C.: Emergency department performance measures and benchmarking summit. Acad. Emerg. Med. **13**(10), 1074–1080 (2006). https://doi.org/10.1197/j.aem.2006.05.026
15. Abohamad, W., Ramy, A., Arisha, A.: A hybrid process-mining approach for simulation modeling. In: Proceedings of the 2017 Winter Simulation Conference, pp. 1527–1538. IEEE, Las Vegas (2017). https://doi.org/10.1109/wsc.2017.8247894
16. Kolker, A.: Process modeling of emergency department patient flow: effect of patient length of stay on ED diversion. J. Med. Syst. **32**(5), 389–401 (2008). https://doi.org/10.1007/s10916-008-9144-x

A Multi-level Approach for Identifying Process Change in Cancer Pathways

Angelina Prima Kurniati[1,2]($\boxtimes$) ⓘ, Ciarán McInerney[1] ⓘ,
Kieran Zucker[3,4] ⓘ, Geoff Hall[3,4] ⓘ, David Hogg[1] ⓘ,
and Owen Johnson[1] ⓘ

[1] School of Computing, University of Leeds, Leeds LS2 9JT, UK
scapk@leeds.ac.uk
[2] School of Computing, Telkom University, 40257 Bandung, Indonesia
[3] School of Medicine, University of Leeds, Leeds LS9 7TF, UK
[4] Leeds Teaching Hospitals Trust, Leeds LS9 7TF, UK

Abstract. An understudied challenge within process mining is the area of process change over time. This is a particular concern in healthcare, where patterns of care emerge and evolve in response to individual patient needs and through complex interactions between people, process, technology and changing organisational structure. We propose a structured approach to analyse process change over time suitable for the complex domain of healthcare. Our approach applies a qualitative process comparison at three levels of abstraction: a holistic perspective summarizing patient pathways (*process model* level), a middle level perspective based on activity sequences for individuals (*trace* level), and a fine-grained detail focus on activities (*activity* level). Our aim is to identify points in time where a process changed (*detection*), to localise and characterise the change (*localisation* and *characterisation*), and to understand process evolution (*unravelling*). We illustrate the approach using a case study of cancer pathways in Leeds Cancer Centre where we found evidence of agreement in process change identified at the process model and activity levels, but not at the trace level. In the experiment we show that this qualitative approach provides a useful understanding of process change over time. Examining change at the three levels provides confirmatory evidence of process change where perspectives agree, while contradictory evidence can lead to focused discussions with domain experts. The approach should be of interest to others dealing with processes that undergo complex change over time.

Keywords: Process mining · Cancer pathways · Process change · Concept drift · Multi-level process comparison

1 Introduction

In general, process-mining research projects work with data collected over months or years and start with the assumption that the processes are largely unchanged during the period of study. In reality, there are many reasons why both the process and the data about that process might change over time. This is a particular concern in healthcare, where patterns of care emerge and evolve in response to individual patient needs and

© Springer Nature Switzerland AG 2019
C. Di Francescomarino et al. (Eds.): BPM 2019 Workshops, LNBIP 362, pp. 595–607, 2019.
https://doi.org/10.1007/978-3-030-37453-2_48

through complex interactions between people, processes, technology and changing organisational structures. Arguably many healthcare processes are in a constant state of flux and evolution. A better understanding of how to identify and model process change over time is important if process mining is to be applied effectively within healthcare.

The changing nature of processes over time has been termed 'concept drift' by the machine learning community [1]. It has been adopted in the process mining community and there is a growing literature exploring potential new approaches [2–4]. Three challenges when dealing with concept drift are: (1) change point detection, which aims to detect that a process change has taken place and the time where this occurred; (2) change localisation and characterisation, which aims to characterise the nature of a change and identify the changed elements of a process, and (3) change process evolution, which aims to unravel the more gradual evolution of a process over time periods.

The standard approach to analysing process changes over time is to construct process models from different time periods in a large dataset and compare them to identify changes. A common application for comparing processes is for conformance checking, where a reference model is compared to the real behaviour recorded in the event log [5]. Partington et al. [6] proposed an approach for comparison by defining points for comparison with various metrics. However, Partington et al. was concerned with differences in process between different clinical settings so their methods are not directly applicable to analysing concept drift. Furthermore, Partington et al.'s approach required the selection of specific clinical metrics that vary between different clinical domains. Bolt et al. [7] suggested another process-comparison approach by comparing frequencies and percentages of the activities in the logs. This facilitated detailed comparison between each activity in two logs, but not between processes. Both of Partington et al. and Bolt et al. works are not directly related to process change analysis, but can be used for analysing concept drift or process change over time.

In our exploratory study, we developed an approach to discovering and analysing changes over time in complex longitudinal healthcare data. Our case study examined process data related to the treatment of endometrial cancer over a 15 year period (2003–2017) in one of the UK's largest cancer centres (Leeds Cancer Centre) with a specific focus on the routes to diagnosis. Process mining has been used and shown promising results to support process analytics in Oncology [8]. Our event logs were drawn from the Electronic Healthcare Record (EHR) system of Leeds Teaching Hospital NHS Trust. In earlier work with this data [9, 10], we had assumed the processes were largely unchanged during the time period but had not explicitly tested this. We were aware that the time period was long, that the system, people and organisation had evolved and changed over time but we were not aware of specific process changes before we commenced the investigation. Our experience with applying process mining on MIMIC-III, an open-access database [11, 12] showed that a change in the system affected the discovered process. The Leeds EHR system, Patient Pathway Manager Plus (PPM+) [13, 14], was developed in-house and we were fortunate in having access to the software developers of the system, the training team and clinical staff and senior clinicians involved in the process. Changes detected through our multi-level process change analysis method could therefore be evaluated and explored with staff who help identify in potential causal links between changes detected in the data and in practice.

2   Method

2.1   General Method

Our structured approach combined the well-established PM^2 process mining method [15] with concept drift analysis [4]. Bose et al. [4] proposed to analyse process change by process change detection, localisation and characterisation, and unravelling process evolution. Our proposed approach applies process comparison at three different levels: process model, trace, and activity levels. We applied our approach to analyse the route to diagnosis of endometrial cancer. Figure 1 shows our general methodology.

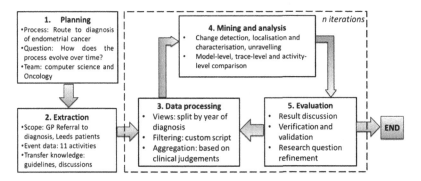

Fig. 1. General methodology for multi-level process change analysis

Planning was done by defining business process, research questions, and team members in this study. The following definitions describe the main components of our study.

Definition 1 (Event logs and traces). An event log E is a set of events *(c, a, t)*. An event refers to an activity a which happened at a timestamp t and is related to a particular case c. A trace T is a sequence of events that happened to a case c ordered by timestamp t, *where* $T \in E$. In this study, a case represents a patient having a set of events that happened between the timestamps of referral and diagnosis of endometrial cancer.

Definition 2 (Sub-logs). A sub-log S is a subset of an event log E based on a subsetting criteria. The subsetting should be done such that a trace is grouped in to a sub-log with no duplication in other sub-logs. For this study, the subsetting was done based on the year of timestamp t where the activity label a is *Diagnosis*. The event log was split into sub-logs based on the year of diagnosis of each patient. There are clearly many subsetting options that could be adopted.

Definition 3 (Process models). A process model M is a directed graph modelling the traces T in the event log E. Process model M draws activities a as nodes and the possible paths p between nodes as arcs from one node to another. Standard process mining algorithms can be used to discover process models with additional components, such as frequency of nodes and arcs as the occurrence of a and p in E, respectively.

2.2 Process Change Analytics

Sub-logs were analysed at the process-model, trace, and activity levels to describe the behaviour of the processes of interest. A summary of the metrics used at each level is presented in Table 1.

Table 1. Metrics for describing the behaviour of processes at multiple levels

Level	Metrics	Description
Model	Replay fitness	The proportion of traces in the log that can be reproduced in the process model
	Precision	The proportion of the sequence of events allowed by the model which is not seen in the event log
	Generalisation	The proportion of the model to reproduce future sequence of events of the process
Trace	Duration	The number of days of the pathway from *Referral* to *Diagnosis*
	Variant proportion	The proportion of variants in the sub-log that were one of the most frequent variants in the complete log
Activity	Frequency	The number of patients having a specific event within one year
	Percentage	The percentage of patients having a specific event out of all patients within a year

Our general process model for the model-level comparison, was built using interactive Data-Aware Heuristics Miner (iDHM) in ProM 6.8 [16] from the complete event log from 2003 to 2017. The iDHM enables an exploration of the parameter space and several heuristics, and focuses on the general pattern based on the most frequent activities. Our model level behaviour was described by the replay fitness, precision and generalisation [5] of each sub-log to the general model. Our trace-level behaviour was described by durations and variants of the traces in the sub-logs using bupaR [17]. Our activity-level behaviour was described by activity frequency and its percentage in the sub-logs. We then adopted a visual-analytic approach to identifying possible changes by quantitatively comparing visualisations of these aforementioned descriptions of process behaviour.

2.3 Case Study: Route to Diagnosis of Leeds Patients Diagnosed with Endometrial Cancer During 2003–2017

We examined the route to diagnosis of Leeds patients diagnosed with endometrial cancer over an extended fifteen year time period. We used an anonymised event log from the PPM+ EHR that had been through significant data cleaning (documented in full in [9]) and extracted events related to the route to diagnosis for endometrial cancer.

The route to diagnosis [18] or cancer waiting time [19] is seen as one of the important performance indicators in cancer treatment. In the UK, it is monitored by Public Health England and forms a key benchmark for high quality cancer care.

Endometrial cancer is a type of cancer that affects the female reproductive system and is more common in women who have been through the menopause. The most common symptom is unusual vaginal bleeding, which is usually followed by a GP consultation and referral to an Oncology specialist (Gynaecologist). The specialist conducts some tests, such as an ultrasound scan, a hysteroscopy or a biopsy. If diagnosed with endometrial cancer, further investigations might be needed to determine the cancer stage [20]. At the Leeds Cancer Centre, all these events have been coded in the PPM EHR since 2003 and should therefore be present in the extracted event log giving a rich data source for process mining of the pathway from GP referral to diagnosis. The PPM EHR includes nine tables relevant to our study and these tables contain broad categories of activity such as *Referrals, Admission,* and *Surgery.*

3 Results

3.1 Data Extraction and Processing

We applied four selection criteria to create the study cohort. Patients were selected if they had (i) a legitimate care relationship with Leeds Teaching Hospital NHS Trust (LTHT), (ii) a primary diagnosis of endometrial cancer (ICD-10 code C54 or C55), (iii) a diagnosis between 2003 and 2017, and (iv) a diagnosis of endometrial cancer at maximum 120 days after referral to an oncology specialty. The last criteria was based on discussion with clinical experts, because a longer time period is implausibly long for the events to be related. Based on those criteria, 943 out of 1126 endometrial cancer patients (84%) were selected in this study. In total, there are 65,200 events selected or 58 events per patient on average.

From those selected patients, we extracted all time-stamped events recorded between GP *Referral* and *Diagnosis,* which resulted in 339 different activity types. For the purpose of our study, we focus simply on the broad categories of activity represented by the nine tables. We split *Admission* and *Discharge* that were from the same table and split out *Diagnostic Surgery* from the *Surgery* table. The resulting 11 activities were agreed with clinical co-authors, they are: *Referral, Outpatient, Consultation, Admission, Discharge, Investigation, Pathology, Diagnosis Surgery, Multi-Disciplinary Team Review (MDT Review), Diagnosis,* and *Surgery.* The event log contained 7,967 events with the minimum 2, median 6, mean 6.3, and max 12 events per patient.

3.2 Process Model Comparison

Figure 2 shows the general process model based on process mining of the complete log for the full 15 years. For simplicity, the process model shows the eight most frequent activities and the most-frequent paths between them. The *Outpatient, Consultation,* and *MDT Review* activities appeared infrequently and were omitted from the process model to produce the simple diagram as shown in Fig. 2. The *Outpatient* activity appeared in 149 out of 943 patients (16%), *Consultation* appeared in 152 out of 943 patients (16%), and *MDT Review* appeared in 231 out of 943 patients (24%). The general process model was highly representative of the complete event log (replay fitness = 0.81, precision = 0.83, generalisation = 0.99).

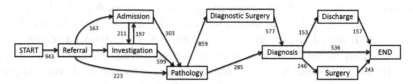

Fig. 2. The directly-follows graph, originally produced by the iDHM plugin in ProM 6.8, shows process model of the pathway from referral to diagnosis. The pathway flows from left to right, with rectangles represent activities and arrows as flows from one activity to the other. Numbers on the arrows show the number of patients having the activity flows to the other activities.

We split the complete event log into 15 sub-logs covering one calendar year each and tested each yearly sub-log for conformance against this general process model. The general process model remained reasonably representative of each yearly sub-log (median [inter-quartile range]: replay fitness = 0.86 [0.10], precision = 0.78 [0.03], and generalisation = 0.93 [0.06]). All measures were similar across years (see Fig. 3). Our qualitative assessment suggest possible changes in 2004 where the replay fitness and generalisation are increased and precision was dropped from the previous year, in 2011 where the trace fitness was dropped while the generalisation increased, and in 2016 where both trace fitness and precision were started to increase. We thus identified three periods of potentially significant change in 2003–2004, 2010–2011 and in 2015–2016.

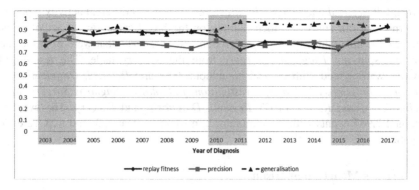

Fig. 3. Conformance to the general process model over years. The shaded areas show the periods where change might have occurred at the process model level.

3.3 Trace Comparison

We examined the profile of trace duration from Referral to Diagnosis for each yearly sub-log to produce the box plots in Fig. 4. There is no obvious qualitative pattern based on the distribution of the duration, except on the inter-quartile range (IQR). The IQR is generally decreasing across year, with exceptions on 2005 where the IQR increased from 42 to 71 days (68%), on 2008 where the IQR increased from 32 to 45 (39%), on 2010 where the IQR increased from 34 to 49 days (44%), on 2011 where the IQR increased from 49 to 50 days (2%), and on 2015 where the IQR increased from 41 to 48 days (18%). Based on those analysis, we identified five periods of potentially change.

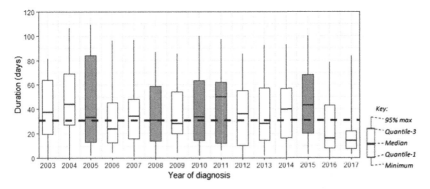

Fig. 4. Boxplot of number of days from GP Referral to Diagnosis by year of diagnosis. Dashed line shows target duration (31 days). The shaded boxes show the periods where change might have occurred.

Figure 5 shows the top ten trace variants (representing 52%) from the general model and the presence of those variants over the years. Those top ten trace variants show only seven activities, excluding *Surgery*. The first variant is the common pathway of patients who got the sequence of *Referral* (R), *Investigation* (I), *Pathology* (P), *Diagnostic Surgery* (DS), and finally got a *Diagnosis* (D) of endometrial cancer. The second variant is similar to the first one, but with no record of *Investigation*. The third variant is similar to the first, except that the patients were admitted after an *Investigation*. The three most common variants (median [IQR]) are R-I-P-DS-D (19[9]%), R-P-DS-D (10[7]%), and R-I-A-P-DS-D-Di (6[3]%). The qualitative distinction between years is the waving trend of the first variant and the decreasing trend of the other variants.

3.4 Activity Comparison

We plotted the total number of patients having each of the main activities over years. There is a sudden increase in almost all activities in 2003–2004. There was also a sudden increase in almost all activities in 2010–2011 except on *Discharge*, and a sudden decrease in all activities in 2015–2016. These three periods of change (see Fig. 6) were also suggested in Sect. 3.2 when reviewing the model level.

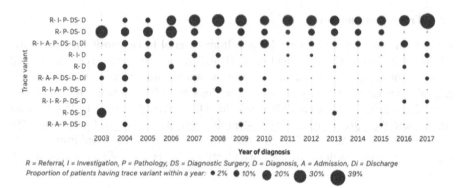

Fig. 5. Summary of the trace variant comparison over years. Size represents the most frequent variants in percentage over the number of patients diagnosed in each year.

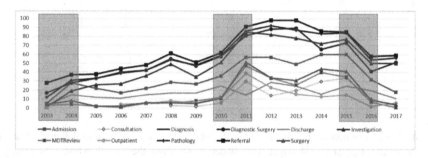

Fig. 6. The total number of patients having each of the main activities over years. The shaded areas show the periods where change might have occurred at the activity level.

We plotted the percentage of each activity for the number of patients each year (see Fig. 7). We grouped the activities into frequent activities occurring in at least 60% patients, and infrequent activities occurring in less than 60% patients. The three most-frequent activities in all years (median [inter-quartile range]) are *Pathology* (93[7]%), *Diagnostic Surgery* (87.5[9]%), and *Investigation* (80[16]%). The four most-infrequent activities in all years (median [inter-quartile range]) are *MDT Review* (12[20]%), *Outpatient* (13[7]%), *Surgery* (15.5[23]%), and *Consultation* (16[24]%). Qualitatively, the period of 2010–2011 was marked by a change in the frequency of the four infrequent activities, while *Discharge* was decreased to be lower than the four infrequent activities. In 2013, the frequency of the infrequent activities were increased except for *Outpatient*.

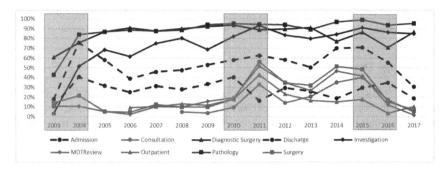

Fig. 7. Percentage of activity presence by the number of patients each year. The frequent activities are presented in solid black lines, the infrequent activities in solid grey lines, and the high-varied activities in dashed grey lines. The shaded areas show the periods where change might have occurred at the activity level. The referral and diagnosis activities are not presented, but both had 100% activity presence in all year groups.

4   Discussion

For our case study of endometrial cancer care in Leeds Cancer Centre, we followed the PM^2 methodology, with the exception of Process Improvement and Support. It is in these blind cases where the presence or characterisations of changes are not known in advance that an exploratory approach is appropriate. Our approach supports this exploration through a qualitative, multi-level perspective for detecting, localising and characterising change in processes. Graphical data visualisations supported discussions about process evolution and change with domain experts and makes use of humans' natural pattern-seeking capacities. We found that such discussions provoked deep reflection on the changing nature of the process over time and generated hypotheses about potential causal links between changes detected in the data and changes in practice.

These hypotheses can then be tested by further, more tightly focused process and data analysis. In that respect, our approach is perhaps just a starting point for further exploratory studies. For example, in this case study, we split a fifteen year event log into 15 yearly sub-log but further division of the log into months, weeks, days or hours can be used to isolate potential change events of interest for further study.

Our approach is based on Bose et al.'s process change analysis: detection, localisation and characterisation, and unravelling [4]. We have improved on Bose et al.'s approach by considering the detection of changes from multiple levels, rather than just at the activity level. Bose et al. suggest that the best place to start analysing process change is by detecting that a process changes has taken place. If that is the case, the next steps are to identify the time periods at which changes happened and characterise the nature of change. The alternative approach described by Partington et al. requires the domain expert to pre-characterise the expected differences that they want to detect and localise. In contrast, our approach did not require prior specification of the changes and instead uses domain experts in the later stage of what Bose et al. describe as

"Unravelling". This supports initial exploration without over-burdening collaborators or when the collaborators are not able to pre-specify the expected differences.

4.1   Change Detection

Changes were detected at all three levels: five at the trace level and three at model and activity levels. From our blinded exploration of our case-study process, we cannot attest to the validity of these detections nor can we know about any true process changes that were not detected. Future work using simulations could attempt to determine the sensitivity of our suggested approach to changes in the magnitude and characteristics of process changes.

4.2   Change Localisation and Characterisation

Assuming that the detected changes were true changes, we have localised those changes to the 2010–2011 and 2015–2016 periods. These localisations were supported by agreement between changes in metric values across multiple levels. The usefulness of our multi-level approach is that the changes detected at one level can guide focused investigation of the same period at other levels. For our case study, no change was independently detected at the trace level in 2003–2004 despite detection at the model and activity level. The duration analysis also shows possible changes in 2005 and 2008, where the IQR of durations were increased. A review of the trace durations led us to consider that the median duration in 2016 was the first example of a year's median duration lying outside the interquartile range of the previous year. Nevertheless, agreement in the localisation of changes lends support to the validity of the detected changes.

The aforementioned, post-hoc review of trace durations is an example of how our approach facilitates characterising change at levels where there were not found when investigated in isolation. The different median duration of traces was the only trace-level characterisation of change in 2016. The strongest evidence for a 2016 change came from the sharp decline in trace frequencies at the activity level, and was supported by an unexpected rise in replay fitness at the model level. This multi-dimensional perspective of a suspected change event informs a more-rounded, complex picture of changes that can be taken to domain experts for discussion.

4.3   Unravelling Process Evolution

The rich characterisation forms the substance for discussion with clinical experts as we collaborate to explain the mechanism and consequences of the detected changes. Our discussion with clinical experts to unravel process evolution found several findings based on three different levels of process comparison. The process model discovered (see Fig. 2) has been agreed to reflect the general pathways of referral to the diagnosis of endometrial cancer. There was no significant change of the duration and sequence from referral to the diagnosis that the clinical experts were aware of, which confirmed the perspective in the trace-level comparison (see Figs. 4 and 5). A concern was raised by clinical experts about some trace variants (see Fig. 5) that have an *Admission*

(A) without a *Discharge* (Di) (variant 7 and 10). Further analysis on these traces revealed that the discharge event happened after diagnosis, hence not captured in the trace variants. In the activity level comparison, clinical experts highlighted their concern on the finding that *MDT Review* is one of the infrequent activities, while in fact, all patients would need to be discussed in the MDT review at some point of their cancer treatment. Further discussion revealed that the MDT reviews usually happened after diagnosis.

Another important discussion about the activity-level comparison is that the system from which the data was collected is evolving. For example, it is shown in Fig. 6 that the *Outpatient* activity started to appear for patients diagnosed in 2006 and the *Consultation* activity started to appear in patients diagnosed in 2008. Further discussion with the PPM development team confirmed that the system was modified in these years to start recording these activities. There were also improvements to the PPM system that introduced automatic imports from other systems (pathology, for example) that were previously captured manually with the effect that data volumes and the reliability of the data increased when the system was improved. There are opportunities for further analysis to examine the effect of these changes in the system on the process over time.

4.4   Reflection on Our Proposed Approach

We discovered our general process model using interactive Data-Aware Heuristics Miner (iDHM) plug-in [16] but other options, like Inductive Miner and Fuzzy Miner or being entirely informed by clinical guidance, might have produced different process models. We used iDHM because its heuristic approach allowed us to get a process model at the desired level of detail. The visualisations provided in the plug-in made it possible to explore the directly-follows graph, dependency graph, causal net, or Petri net. Future work will look at the sensitivity of conclusions to the choice of plug-in.

Further flexibility evident in our approach is the rule for splitting the logs into sub-logs. We split sub-logs based on the calendar year of diagnosis but an equally reasonable method would be to split the log to enforce a uniform number of traces in each sub-log. The consequence of the first method is that the number of traces in each sub-log varies and the consequence of the second is that the duration of the sub-log varies. We addressed the consequence of our choice by analysing the frequency in percentage instead of the number of occurrences.

5   Conclusion

This paper presented an approach to analyse process changes over time and provided a real-life case study. The case study examined the route from GP Referral to Diagnosis of endometrial cancer in Leeds Cancer Centre patients. The general method followed in this study is the PM^2, with a focus on process analytics stage to analyse process changes over time. Process change detection, localisation, and characterisation were carried out at three different levels of comparison: model, trace and activity. This approach allows to detect changes when comparing one year with another. One important limitation of the proposed approach is that it is not able to detect the exact point in time when the change actually occurred. Moreover, this approach could not detect changes back and forth during the same year however, repeating the method with a finer grained time interval would allow the change point to be more accurately detected.

Future work could review the splitting method, the comparison metrics, and the reference model discovery. The comparison metrics used in this study are defined to represent three different level of details but further work might examine other metrics for comparisons. The reference model discovery can be improved by considering clinical guideline as the reference model, or by including only valid traces in the discovery step.

Acknowledgment. This research is supported by ClearPath Connected Cities Project, the Yorkshire and the Humber NIHR Patient Safety Translational Research Centre and the Indonesia Endowment Fund for Education (LPDP). Access to data used in this study is under the Health Research Authority (HRA) Approval Number 206843.

References

1. Schlimmer, J.C., Granger, R.H.: Beyond incremental processing: tracking concept drift. In: AAAI, pp. 502–507 (1986)
2. Carmona, J., Gavaldà, R.: Online techniques for dealing with concept drift in process mining. In: Hollmén, J., Klawonn, F., Tucker, A. (eds.) IDA 2012. LNCS, vol. 7619, pp. 90–102. Springer, Heidelberg (2012). https://doi.org/10.1007/978-3-642-34156-4_10
3. Li, T., He, T., Wang, Z., Zhang, Y., Chu, D.: Unraveling process evolution by handling concept drifts in process mining. In: Proceedings of the IEEE 14th International Conference on Services Computing, SCC 2017, pp. 442–449 (2017)
4. Bose, R.P.J.C., van der Aalst, W.M.P., Žliobaitė, I., Pechenizkiy, M.: Handling concept drift in process mining. In: Mouratidis, H., Rolland, C. (eds.) CAiSE 2011. LNCS, vol. 6741, pp. 391–405. Springer, Heidelberg (2011). https://doi.org/10.1007/978-3-642-21640-4_30
5. Buijs, J.C.A.M., van Dongen, B.F., van der Aalst, W.M.P.: On the role of fitness, precision, generalization and simplicity in process discovery. In: Meersman, R., et al. (eds.) OTM 2012. LNCS, vol. 7565, pp. 305–322. Springer, Heidelberg (2012). https://doi.org/10.1007/978-3-642-33606-5_19
6. Partington, A., Wynn, M., Suriadi, S., Ouyang, C., Karnon, J.: Process mining for clinical processes: a comparative analysis of four australian hospitals. ACM Trans. Manag. Inf. Syst. **5**, 19 (2015)
7. Bolt, A., de Leoni, M., van der Aalst, W.M.P.: A visual approach to spot statistically-significant differences in event logs based on process metrics. In: Nurcan, S., Soffer, P., Bajec, M., Eder, J. (eds.) CAiSE 2016. LNCS, vol. 9694, pp. 151–166. Springer, Cham (2016). https://doi.org/10.1007/978-3-319-39696-5_10
8. Kurniati, A.P., Johnson, O., Hogg, D., Hall, G.: Process mining in oncology: a literature review. In: Proceedings of the 6th International Conference on Information Communication and Management, ICICM 2016, pp. 291–297 (2016)
9. Baker, K., Dunwoodie, E., et al.: Process mining routinely collected electronic health records to define real-life clinical pathways during chemotherapy. Int. J. Med. Inform. **103**, 32–41 (2017)

10. Johnson, O., Hall, P.S., Hulme, C.: NETIMIS: dynamic simulation of health economics outcomes using big data. Pharmacoeconomics **34**(2), 107–114 (2015)
11. Kurniati, A.P., Hall, G., Hogg, D., Johnson, O.: Process mining in oncology using the MIMIC-III dataset (accepted version). In: IOP Journal of Physics: Conference Series, vol. 971, no. 012008, pp. 1–10 (2018)
12. Kurniati, A.P., Rojas, E., Hogg, D., Johnson, O.: The assessment of data quality issues for process mining in healthcare using MIMIC-III, a publicly available e-health record database, no. 2 (2017)
13. Newsham, A., Johnston, C., Hall, G.: Development of an advanced database for clinical trials integrated with an electronic patient record system. Comput. Biol. Med. **41**(8), 575–586 (2011)
14. Johnson, O.A., Abiodun, S.E.: Understanding what success in health information systems looks like: the patient pathway management system at Leeds. UK Academy IS Conference Proceedings, no. 22 (2011)
15. van Eck, M.L., Lu, X., Leemans, S.J.J., van der Aalst, W.M.P.: PM^2: a process mining project methodology. In: Zdravkovic, J., Kirikova, M., Johannesson, P. (eds.) CAiSE 2015. LNCS, vol. 9097, pp. 297–313. Springer, Cham (2015). https://doi.org/10.1007/978-3-319-19069-3_19
16. Mannhart, F., de Leoni, M., Reijers, H.A.: Heuristic mining revamped : an interactive, data-aware, and conformance-aware miner. In: Proceedings of the BPM Demo and Dissertation Award (CEUR Workshop Proceedings), no. 1920, pp. 1–5 (2017)
17. Janssenswillen, G., Depaire, B.: BupaR: business process analysis in R. In: Business Process Management (BPM) Demo (2017)
18. Elliss-Brookes, L., et al.: Routes to diagnosis for cancer - determining the patient journey using multiple routine data sets. Br. J. Cancer **107**(8), 1220–1226 (2012)
19. National Health Service: Delivering Cancer Waiting Times: A Good Practice Guide, pp. 0–67 (2015)
20. National Cancer Institute: Uterine Cancer - Patient Version (2018). https://www.cancer.gov/types/uterine/patient/endometrial-treatment-pdq. Accessed 20 Aug 2007

Adopting Standard Clinical Descriptors for Process Mining Case Studies in Healthcare

Emmanuel Helm[1,3](✉)(iD), Anna M. Lin[1](iD), David Baumgartner[1](iD),
Alvin C. Lin[2], and Josef Küng[3]

[1] Research Department Advanced Information Systems and Technology,
University of Applied Sciences Upper Austria, 4232 Hagenberg, Austria
{emmanuel.helm,anna.lin,david.baumgartner}@fh-hagenberg.at
[2] Faculty of Medicine, University of Toronto, Toronto, Canada
alvin.lin@mail.utoronto.ca
[3] Institute for Applied Knowledge Processing, Johannes Kepler University,
4040 Linz, Austria
josef.kueng@jku.at

Abstract. Process mining can provide greater insight into medical treatment processes and organizational processes in healthcare. A review of the case studies in the literature has identified several different common aspects for comparison, which include methodologies, algorithms or techniques, medical fields and healthcare specialty. However, from a medical perspective, the clinical terms are not reported in a uniform way and do not follow a standard clinical coding scheme. Further, the characteristics of the event log data are not always described. In this paper, we identified 38 clinically-relevant case studies of process mining in healthcare published from 2016 to 2018 that described the tools, algorithms and techniques utilized, and details on the event log data. We then assigned the clinical aspects of patient encounter environment, clinical specialty and medical diagnoses using the standard clinical coding schemes SNOMED CT and ICD-10. The potential outcomes of adopting a standard approach for describing event log data and classifying medical terminology using standard clinical coding schemes are discussed.

Keywords: Process mining · Healthcare · Terminology · ICD · SNOMED

1   Introduction

Process mining is a discipline that allows for greater understanding into real-life processes of recorded systems behaviour. Through process mining techniques, numerous case studies and successful companies have demonstrated valuable insights into quality improvement, compliance, and optimization of processes.

In recent years, several review papers provide an overview on the state of process mining in healthcare. Rojas et al. in 2016 identified classifiers of eleven common aspects across 74 case studies in healthcare [35]. These aspects include

© Springer Nature Switzerland AG 2019
C. Di Francescomarino et al. (Eds.): BPM 2019 Workshops, LNBIP 362, pp. 608–619, 2019.
https://doi.org/10.1007/978-3-030-37453-2_49

methodologies, techniques or algorithms, medical fields and healthcare specialty. In 2018, Erdogan and Tarhan conducted a systematic mapping of 172 case studies with mostly the same metrics and aspects [14]. These papers are very specific as to *how* these case studies were conducted, which enhances comparison between different process mining techniques in different settings. However, from a medical perspective, the terms and categories listed under *medical fields* and *healthcare specialty* are not structured in a uniform way, and do not follow a standardized clinical coding scheme. Further, basic characteristics of the event log data (timeframe, number of cases or patients, healthcare facility/organization) are not always clearly reported.

The number of case studies on process mining in healthcare continues to increase steadily. As such, a standard approach of reporting event log data, clinical specialties and medical diagnoses would provide greater clarity and enhance comparability between treatments of specific diseases across different heathcare settings.

In this paper, further to the studies examined by Rojas et al., we conducted a forward search of processing mining case studies in healthcare for the three-year period from January 2016 to December 2018. We identified case studies that described basic characteristics of the event log data, and where information on the patient encounter environment, clinical specialty and medical diagnoses could be assigned under a standard clinical coding scheme. Section 2 describes how the forward search was conducted and which criteria we applied to filter the results. In addition, the methods describe standard clinical coding systems and terminologies that were used. In Sect. 3, the results of our analysis are presented. Section 4 discusses the benefits and gives an outlook on the potential clinical insights gained by reporting and classifying clinical terms, clinical specialties and medical diagnoses using a standard clinical coding scheme.

2 Methods

Our paper focused on answering three questions: (1) Which clinically-relevant case studies of process mining in healthcare will be selected for this study? (2) What were the technical aspects identified? (3) How can we improve the clarity and comparability of the clinical terms and aspects described?

2.1 Selection of Clinically-Relevant Case Studies

Our starting point was the review paper by Rojas et al. [35] that identified 74 case studies where process mining tools, techniques or algorithms were applied in the healthcare domain. We then performed a forward search using Google Scholar, in reference to the 74 identified articles and the review paper itself. The inclusion criteria (IC) were applied in the Google Scholar search and the exclusion criteria (EC) were applied manually afterwards (see Fig. 1).

IC1: All articles that reference either the review paper by Rojas et al. [35] or any of the 74 articles identified in their review were included.

IC2: All articles published between 01.01.2016 and 31.12.2018 were included.

IC3: All articles published in English were included.

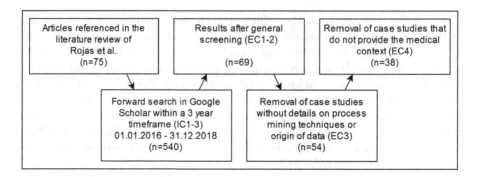

Fig. 1. Flowchart on the case study selection strategy.

EC1: Articles that do not include evidence of a clinically-relevant case study of process mining in healthcare were excluded.

EC2: Articles that present a case study based on data that was already used for an earlier case study were excluded.

EC3: Articles that do not describe the characteritics of the event log data (e.g. timeframe, number of cases or patients, healthcare facility) or do not describe which process mining technique or algorithm was applied were excluded.

EC4: Articles that did not describe any clinical context (i.e. clinical specialty or medical diagnosis) were excluded.

2.2 Technical Aspects

A detailed account of the tools, techniques or algorithms used in process mining case studies in healthcare have been previously described [35]. Also, other technical descriptors such as the data type and geographical analysis have been used to describe the event log data [34]. For the technical scope of our paper, our focus was on (1) the *tools* used in the case studies, (2) the *techniques or algorithms* used, and (3) the *process mining perspectives*.

2.3 Clinical Aspects and Standard Coding Schemes

Medical language is full of homonyms, synonyms, eponyms, acronyms and abbreviations; and each healthcare specialty comes with their own sub-terminology [7]. To improve the clarity and comparability of the clinical aspects described in our selected papers, we adopted the use of standard clinical coding schemes of SNOMED CT and ICD-10. Namely, the clinical terms were matched to their best corresponding standard clinical descriptor, with respect to three clinical categories: (1) the type of patient encounter environment (2) clinical speciality and (3) medical diagnosis (i.e. disease or health problem).

SNOMED CT. The Systematized Nomenclature of Medicine – Clinical Terms is an internationally recognized standard that classifies clinically-relevant terminology and concepts, along with their synonyms and relationships, into numeric coded values. Available in multiple languages and maintained by SNOMED International, there are currently over 340,000 numerically coded concepts that can be combined grammatically to create an expression. We used SNOMED CT international browser[1] in version *v20190131* for clinical descriptors on the *Patient encounter environment* and *Clinical specialty*.

ICD-10. For classification of clinical diagnoses and health problems, the commonly accepted system is the International Classification of Diseases or ICD, which is maintained by the World Health Organization (WHO). The most current version is ICD-10 and it utilizes an alphanumeric coding scheme with more than 14.000 single clinical codes of medical terms organized hierarchically into 22 chapters. We used the WHO ICD-10 browser in the *2016* version[2] for clinical descriptors on *medical diagnoses*.

3 Results

3.1 Selection of Clinically-Relevant Case Studies

Our forward search yielded initially a total of 540 papers, and after our inclusion and exclusion criteria were applied, 38 articles were selected (cf. Figure 1). For all 38 papers, basic characteristics of the event log data were retrieved (e.g. origin of data, number of cases or patients, healthcare facility, timeframe of the study). The results of the technical and clinical aspects are described below.

3.2 Technical Aspects

Tools. ProM[3] is the most used tool in our 38 papers [1,3,4,9,16,17,23,25–28,33,37,39,40,43] (n = 16). Additionally, Disco[4] is also frequently used [2,13,16,24,26,29,33,34,36,38,44] (n = 11). Tools like PALIA [10,15], pMineR [18] and others [4,5,8,10,11,15,31,32,42] (n = 9) are less frequently used. Four papers presented new self-developed solutions [20–22,31].

Techniques or Algorithms. Fuzzy miner (which is also implemented in Disco in a next-generation version) is the most utilized algorithm [2,13,16,24,26,29,33,34,36,38,44] (n = 11). Many papers also presented self-developed approaches [1,9,11,12,19–21,27,43,44] (n = 10), with most of the approaches implemented within the ProM framework. Four studies used the Trace Clustering technique [8,17,24,31]. While the Heuristic Miner algorithm was frequented as per previous reviews, [14,35], it was only used in one of our 38 selected papers [25].

---

[1] https://browser.ihtsdotools.org/.
[2] https://icd.who.int/browse10/2016/en.
[3] https://www.promtools.org.
[4] https://fluxicon.com/disco/.

Process Mining Perspectives. The analysis showed that 30 of the total 38 case studies mainly aimed for the *Control Flow* perspective in their dataset [1, 3–5, 8, 9, 11, 12, 15–17, 19–21, 23, 24, 27–29, 31, 32, 34, 36–40, 42–44]. Five papers analyzed the *Conformance* perspective [18, 22, 25, 26, 33], two focused on *Organizational* [2, 10] and one on *Performance* [13].

3.3   Clinical Aspects Using Standard Clinical Descriptors

Encounter Environment. From the patient's perspective, we considered five clinical settings or encounter environments: (1) Inpatient, (2) Outpatient, (3) Accident and Emergency department or AED, (4) General practitioner or GP practice site (5), and Pharmacy. All five encounter environments could be coded by SNOMED CT. For each paper, at least one of these five encounter environments was retrieved. Most of the papers examined events within the Inpatient environment, followed by AED environment (cf. Table 1).

Table 1. Papers with their corresponding SNOMED CT encounter environment.

SNOMED CT	Environment	Papers
440654001	Inpatient	[1, 4, 5, 8, 9, 11, 13, 17, 19–25, 27–29, 31–34, 36–40, 42–44]
440655000	Outpatient	[3, 5, 16, 18, 26]
225728007	AED	[2, 3, 12, 15, 27, 28, 32, 36, 40, 42–44]
394761003	GP practice site	[10]
264372000	Pharmacy	[34]

Clinical Specialty. SNOMED CT offers the code of 394658006 for *Clinical specialty*, which further contains 18 high-level specialties. Table 2 shows 11 of the 18 high-level clinical specialties were identified in our selected papers. The most identified clinical specialty was *Medical specialty*, followed by *Surgical specialty* and *Emergency medicine*. Of note, some of 18 high-level specialties in SNOMED CT are further divided into sub-specialties of greater clinical specificity. For example, *Medical specialty* has 44 sub-specialties that include e.g. *Dermatology*, *Neurology* and *Cardiology*. In this paper, we identified and assigned sub-specialities to their corresponding high-level *Clinical specialty*. Also, for example, if several different medical sub-specialities were described in one paper, we counted these sub-specialities together as *Medical specialty*.

Medical Diagnosis. For each paper, we focused on identifying the medical diagnosis (i.e. disease or health problem) or description of a medical diagnosis. We then assigned these terms to their corresponding highest chapter or block category in ICD-10. Table 3 shows a total of 15 out of the 22 ICD-10 chapter categories for disease and health related problems were covered amongst the papers. The category with the most papers listed was *Diseases of the circulatory*

Table 2. Papers with their corresponding SNOMED CT clinical specialty.

SNOMED CT	Clinical specialty	Papers
394592004	Clinical oncology	[5, 11, 39]
394581000	Community medicine	[10]
722163006	Dentistry	[16, 26]
722164000	Dietetics and nutrition	[10]
773568002	Emergency medicine	[2–4, 12, 15, 27, 28, 32, 36, 40, 42, 43]
394814009	General practice	[10, 16, 26]
408446006	Gynecological oncology	[21]
394733009	Medical specialty	[1, 4, 8, 9, 17–20, 22, 25, 26, 29, 31, 33, 34, 38, 40, 42]
722165004	Nursing	[2, 10, 26]
394585009	Obstetrics and gynecology	[9, 21]
394732004	Surgical specialty	[4, 11, 14, 18, 21, 23, 24, 26, 37, 39, 43]

system followed by *Neoplasms*. Two papers [9, 25] were not included in Table 3, since several hundred diseases and health problems were cited and classified using ICD-9. Of the remaining 36 case studies, ICD-10 was already used in 8 papers to code the diagnosis [4, 5, 8, 22, 25, 31, 39, 40].

4   Discussion

Whether for process discovery, conformance checking, or enhancement, process mining case studies are influenced by the quality of the labeled data. The benefits of high-quality, labeled data include improved accuracy, efficiency and predictability of processes, not only for the study itself but also for comparability across studies. Further, high-quality, labeled data can make other kinds of future analyses and even machine learning techniques (e.g. supervised learning, trend estimation, clustering, ...) easier and more efficient to achieve. In process mining case studies in healthcare, labeled data often encompasses clinical aspects and terms. As such, our aim was to examine clinically-relevant case studies since Rojas et al. [34] and determine how to improve upon the clarity and comparability of clinical aspects and terms described.

4.1   Reporting Basic Characteristics of the Event Log Data

For our analysis, we selected papers that described basic characteristics of the event log data. These characteristics included the origin or source of the data, the healthcare facility, the number of cases or patients, and the timeframe of the study. For example, in Rinner et al. [33], event logs were extracted for a total of 1023 patients starting melanoma surveillance between January 2010 to June 2017, from a local melanoma registry in a medical university and Hospital Information System (HIS) in Austria. In papers where these characteristics were not

Table 3. Papers with their corresponding ICD-10 medical diagnosis.

ICD-10	Diagnosis	Papers
A00 - B99	Certain Infectious and parasitic diseases	[4, 27, 28, 32, 34, 40]
C00 - D48	Neoplasms	[4, 5, 11, 21, 29, 33, 37, 39]
E00 - E90	Endocrine, nutritional and metabolic diseases	[1, 4, 10, 18, 19, 26]
F00 - F99	Mental and behavioural disorders	[4, 22]
G00 - G99	Diseases of the nervous system	[4]
H60 - H95	Diseases of the ear and mastoid process	[4]
I00 - I99	Diseases of the circulatory system	[1, 4, 8, 13, 15, 17, 19, 20, 29, 31, 42]
J00 - J99	Diseases of the respiratory system	[2, 4]
K00 - K93	Diseases of the digestive system	[2, 4, 16, 36]
M00 - M99	Diseases of the musculoskeletal system and connective tissue	[2, 4, 22]
N00 - N99	Diseases of the genitourinary system	[4]
O00 - O99	Pregnancy, childbirth and the puerperium	[4]
R00 - R99	Symptoms, signs and abnormal clinical and laboratory findings, not elsewhere classified	[2, 4, 12]
S00 - T98	Injury, poisoning and certain other consequences of external causes	[2–4, 12, 38, 43, 44]
Z00 - Z99	Factors influencing health status and contact with health services	[4, 11, 18, 23, 24]

clearly reported, the retrieval process was time-consuming. Several papers provided additional details (e.g. patient age, data from private insurance or public health records). Presumably for reasons of privacy and anonymity, specifics on the healthcare facility (e.g. hospital name) were not always provided, however, the country of origin was always reported. While variations exist in the style of reporting, we recommend case studies include these aforementioned basic characteristics when reporting the event log data.

4.2   Adopting the Use of Standard Clinical Descriptors

Encounter Environment. A patient can have vastly different experiences within the healthcare system depending on the clinical setting or encounter environment. For example, a patient with heart failure who presents to the AED may require admission as a hospital inpatient, follow-up at their GP practice site or outpatient clinic, and prescription drugs at a pharmacy. As such, in our analysis of the selected papers, we focused on five patient encounter environments: Inpatient, Outpatient, AED, GP practice site, and Pharmacy. All five encounter types can be coded by SNOMED CT. While further details can be provided (e.g. Outpatient Clinic for Thyroid Disease [18]), we recommend case studies report at least the patient encounter environment using standard clinical codes e.g. SNOMED CT.

Clinical Specialty. Different clinical specialties are often involved in the care of a patient. For example, for a patient diagnosed with cancer, a multidisciplinary care plan can encompass input from a medical specialty, a surgical specialty and clinical oncology. As each specialty offers their own unique set of knowledge and expertise, it is important to identify which clinical specialty is involved.

For each of our selected papers, we identified at least one of the 18 high-level clinical specialties coded by SNOMED CT. For greater specificity, SNOMED CT offers further standard clinical codes for sub-specialities. In fact, Baek et al. list multiple sub-specialities along with their corresponding SNOMED CT codes in their study [4]. Also, instead of *Clinical specialty*, another category of clinical descriptors such as the type of medical practitioner or occupation could have been considered (e.g. mapping to surgeon instead of surgical specialty).

In any event, the task of identifying and assigning such standard clinical codes is time consuming, and beyond the scope of this paper. For future case studies, we recommend reporting the clinical specialty (or similar clinical descriptor such as medical practitioner) by adopting standard clinical codes e.g. SNOMED CT.

Medical Diagnosis. There are literally thousands of medical diagnoses, and each diagnosis comes with its own treatment and management plan. ICD-10 is a standard coding scheme in healthcare that provides specific clinical descriptors and codes for diseases and health conditions. In our analysis, we were able to identify at least one medical diagnosis or description of a medical diagnosis in each paper, which we could map to the corresponding ICD-10 code. Further, over 25% (10 out of 38) of our selected papers utilized either ICD-9 or ICD-10 codes in their study. For broader comparison across studies, we assigned the selected papers to one or more of the 22 ICD-10 chapters or block categories. In Table 3 we only listed the ICD-10 chapters that were covered in the case studies.

It is important to distinguish the difference between a medical diagnosis (i.e. the process of identifying the disease or medical condition that explains a patient's signs and symptoms) versus a patient's signs (e.g. rash) or symptoms (e.g. cough). While the majority of ICD-10 chapters describe a group of medical diagnoses, some cover other clinical descriptors, such as signs and symptoms (R00-R99), external causes of morbidity and mortality (V01-V98), and codes for special purposes (U00-99). ICD-10 also allows for the coding of location, severity, cause, manifestation and type of health problem [41].

Taken together, we recommend adopting use of a standard coding scheme e.g. ICD-10 for clinical terms and aspects relating to medical diagnosis in process mining case studies in healthcare. Recently developed, ICD-11 is not adopted yet but provides backward compatibility, i.e. ICD-10 coded case studies will be comparable to newer ICD-11 coded ones, once the new coding scheme will be taken on by the information system vendors.

4.3   Conclusions and Future Perspectives

In summary, we propose adopting a standard for describing event log data and reporting medical terminology using standard clinical descriptors and coding schemes. In scientific research, the idea of having a set of guidelines, criteria, or standards for peer-reviewed publications is not novel. In fact, journals such as *Nature* are taking initiatives by creating mandatory reporting summary templates[5], in order to improve comparability, transparency, and reproducibility of the work they publish [30]. Other journals and disciplines, including biomedical informatics, are following suit [6]. Thus, as data sets become more transparent and available, consistency in reporting characteristics of the event log data (e.g. origin of data, number of patients or cases, healthcare facility, timeframe of the study) will aid in improving comparability and reproducibility across future studies.

Further to the work by Rojas et al. [36], we identified and described the clinical terms and aspects in our selected papers with respect to three categories: the patient encounter environment, clinical specialty, and medical diagnosis. We then correlated the clinical terms and aspects to their respective standard clinical descriptors and codes found in SNOMED CT and ICD-10. For each of the five types of patient encounter environments, a more detailed description can be achieved through SNOMED CT by using compositional grammar. Similarly, for *Clinical specialty* in SNOMED CT, reporting of sub-specialties under e.g. *Medical speciality* will provide increased specificity for clarity and comparison.

As aforementioned, several case studies have already adopted the use of a standard clinical coding scheme to describe medical diagnoses. However, our consideration of SNOMED CT and ICD-10 serves only as a starting point. In fact, SNOMED CT also provides standard codes for medical diagnoses, which can provide further specificity and clarity. For example, instead of ICD-10, the one of Systematized Nomenclature for Dentistry or SNODENT CT (which is incorporated into SNOMED CT) could have been used to code for the clinical descriptors of missing and filled teeth in one of our selected papers [16].

Finally, when adopting the use of standard clinical descriptors, we recognize other fundamental clinical categories to consider are medical investigations and procedures. As such, the use of standard clinical descriptors is becoming increasingly relevant, not only for clarity and comparability, but efficiency in outcome measurements such as length of stay (LOS) and financial cost. For example, in their paper, Baek et al. utilized process mining techniques and statistical methods to identify the factors associated with LOS in a South Korean hospital [4]. This study is just one use case for a more detailed description of the medical context where process mining case studies could allow for future meta-studies, e.g. benchmarking LOS in different hospitals or countries, based on diagnoses while also considering other important factors like the patient encounter environment.

Acknowledgements. This work was supported by the Process-Oriented Data Science for Healthcare Alliance (PODS4H Alliance).

---

[5] https://www.nature.com/documents/nr-reporting-summary-flat.pdf.

References

1. Alharbi, A., Bulpitt, A., Johnson, O.: Improving pattern detection in healthcare process mining using an interval-based event selection method. In: Carmona, J., Engels, G., Kumar, A. (eds.) BPM 2017. LNBIP, vol. 297, pp. 88–105. Springer, Cham (2017). https://doi.org/10.1007/978-3-319-65015-9_6
2. Alvarez, C., Rojas, E., Arias, M., Munoz-Gama, J., et al.: Discovering role interaction models in the emergency room using process mining. J. Biomed. Inform. **78**, 60–77 (2018)
3. Andrews, R., et al.: Pre-hospital retrieval and transport of road trauma patients in Queensland. In: Daniel, F., Sheng, Q.Z., Motahari, H. (eds.) BPM 2018. LNBIP, vol. 342, pp. 199–213. Springer, Cham (2019). https://doi.org/10.1007/978-3-030-11641-5_16
4. Baek, H., Cho, M., Kim, S., Hwang, H., Song, M., Yoo, S.: Analysis of length of hospital stay using electronic health records: a statistical and datamining approach. PLoS One **13**(4) (2018)
5. Baker, K., Dunwoodie, E., et al.: Process mining routinely collected electronic health records to define real-life clinical pathways during chemotherapy. Int. J. Med. Inf. **103**, 32–41 (2017)
6. Bakken, S.: The journey to transparency, reproducibility, and replicability. J. Am. Med. Inf. Assoc. **26**, 185–187 (2019)
7. Benson, T., Grieve, G.: Principles of Health Interoperability: SNOMED CT, HL7and FHIR. Springer, Heidelberg (2016). https://doi.org/10.1007/978-3-319-30370-3
8. Chen, J., Sun, L., Guo, C., Wei, W., Xie, Y.: A data-driven framework of typical treatment process extraction and evaluation. J. Biomed. Inform. **83**, 178–195 (2018)
9. Chen, Y., et al.: Learning bundled care opportunities from electronic medical records. J. Biomed. Inform. **77**, 1–10 (2018)
10. Conca, T., et al.: Multidisciplinary collaboration in the treatment of patients with type 2 diabetes in primary care: analysis using process mining. J. Med. Internet Res. **20**(4), e127 (2018)
11. Dagliati, A., et al.: Temporal electronic phenotyping by mining careflows of breast cancer patients. J. Biomed. Inform. **66**, 136–147 (2017)
12. Duma, D., Aringhieri, R.: An ad hoc process mining approach to discover patient paths of an Emergency Department. Flex. Serv. Manuf. J. 1–29 (2018)
13. Erdogan, T.G., Tarhan, A.: A goal-driven evaluation method based on process mining for healthcare processes. Appl. Sci. **8**(6), 894 (2018)
14. Erdogan, T.G., Tarhan, A.: Systematic mapping of process mining studies in healthcare. IEEE Access **6**, 24543–24567 (2018)
15. Fernandez-Llatas, C., et al.: Analyzing medical emergency processes with process mining: the stroke case. In: Daniel, F., Sheng, Q.Z., Motahari, H. (eds.) BPM 2018. LNBIP, vol. 342, pp. 214–225. Springer, Cham (2019). https://doi.org/10.1007/978-3-030-11641-5_17
16. Fox, F., Aggarwal, V.R., Whelton, H., Johnson, O.: A data quality framework for process mining of electronic health record data. In: International Conference on Healthcare Informatics (ICHI), pp. 12–21. IEEE (2018)
17. Funkner, A.A., Yakovlev, A.N., Kovalchuk, S.V.: Data-driven modeling of clinical pathways using electronic health records. Proc. Comput. Sci. **121**, 835–842 (2017)

18. Gatta, R., et al.: A framework for event log generation and knowledge representation for process mining in healthcare. In: International Conference on Tools with Artificial Intelligence (ICTAI), pp. 647–654. IEEE (2018)

19. Huang, Z., Dong, W., Ji, L., He, C., Duan, H.: Incorporating comorbidities into latent treatment pattern mining for clinical pathways. J. Biomed. Inform. **59**, 227–239 (2016)

20. Huang, Z., Ge, Z., Dong, W., He, K., Duan, H.: Probabilistic modeling personalized treatment pathways using electronic health records. J. Biomed. Inform. **86**, 33–48 (2018)

21. Jimenez-Ramirez, A., Barba, I., Reichert, M., Weber, B., Del Valle, C.: Clinical processes - the killer application for constraint-based process interactions? In: Krogstie, J., Reijers, H.A. (eds.) CAiSE 2018. LNCS, vol. 10816, pp. 374–390. Springer, Cham (2018). https://doi.org/10.1007/978-3-319-91563-0_23

22. Johnson, O.A., Ba Dhafari, T., Kurniati, A., Fox, F., Rojas, E.: The clearpath method for care pathway process mining and simulation. In: Daniel, F., Sheng, Q.Z., Motahari, H. (eds.) BPM 2018. LNBIP, vol. 342, pp. 239–250. Springer, Cham (2019). https://doi.org/10.1007/978-3-030-11641-5_19

23. Kirchner, K., Marković, P.: Unveiling hidden patterns in flexible medical treatment processes – a process mining case study. In: Dargam, F., Delias, P., Linden, I., Mareschal, B. (eds.) ICDSST 2018. LNBIP, vol. 313, pp. 169–180. Springer, Cham (2018). https://doi.org/10.1007/978-3-319-90315-6_14

24. Kirchner, K., Marković, P., Delias, P.: Automatic creation of clinical pathways - a case study. Data Sci. Bus. Intell. **179**, 188 (2016)

25. Kurniati, A.P., Rojas, E., Hogg, D., Hall, G., Johnson, O.: The assessment of data quality issues for process mining in healthcare using Medical Information Mart for Intensive Care III, a freely available e-health record database. Health Inform. J. **25**(4), 1878–1893 (2018)

26. Lismont, J., Janssens, A.S., Odnoletkova, I., et al.: A guide for the application of analytics on healthcare processes: a dynamic view on patient pathways. Comput. Biol. Med. **77**, 125–134 (2016)

27. Mannhardt, F., Blinde, D.: Analyzing the trajectories of patients with sepsis using process mining. In: CEUR Workshop Proceedings, vol. 1859, pp. 72–80 (2017)

28. Mannhardt, F., Toussaint, P.J.: Revealing work practices in hospitals using process mining. In: Building Continents of Knowledge in Oceans of Data: The Future of Co-Created eHealth (2018)

29. Metsker, O., Yakovlev, A., Bolgova, E., Vasin, A., Koval-chuk, S.: Identification of pathophysiological subclinical variances during complex treatment process of cardiovascular patients. Proc. Comput. Sci. **138**, 161–168 (2018)

30. Munafò, M.R., Nosek, B.A., Bishop, D.V.M., et al.: A manifesto for reproducible science. Nat. Hum. Behav. **1**(1), 21 (2017)

31. Najjar, A., Reinharz, D., Girouard, C., Gagné, C.: A two-step approach for mining patient treatment pathways in administrative healthcare databases. Artif. Intell. Med. **87**, 34–48 (2018)

32. Neira, R.A.Q., de Vries, G.J., Caffarel, J., Stretton, E.: Extraction of data from a hospital information system to perform process mining. In: MedInfo, pp. 554–558 (2017)

33. Rinner, C., Helm, E., Dunkl, R., Kittler, H., Rinderle-Ma, S.: Process mining and conformance checking of long running processes in the context of melanoma surveillance. Int. J. Env. Res. Public Health **15**(12), 2809 (2018)

34. Rojas, E., Capurro, D.: Characterization of drug use patterns using process mining and temporal abstraction digital phenotyping. In: Daniel, F., Sheng, Q.Z., Motahari, H. (eds.) BPM 2018. LNBIP, vol. 342, pp. 187–198. Springer, Cham (2019). https://doi.org/10.1007/978-3-030-11641-5_15

35. Rojas, E., Munoz-Gama, J., Sepúlveda, M., Capurro, D.: Process mining in healthcare: a literature review. J. Biomed. Inform. **61**, 224–236 (2016)

36. Rojas, E., Sepúlveda, M., Munoz-Gama, J., Capurro, D., Traver, V., Fernandez-Llatas, C.: Question-driven methodology for analyzing emergency room processes using process mining. Appl. Sci. **7**(3), 302 (2017)

37. Stefanini, A., Aloini, D., Dulmin, R., Mininno, V.: Service reconfiguration in healthcare systems: the case of a new focused hospital unit. In: Cappanera, P., Li, J., Matta, A., Sahin, E., Vandaele, N., Visintin, F. (eds.) International Conference on Health Care Systems Engineering, vol. 210, pp. 179–188. Springer, Cham (2017). https://doi.org/10.1007/978-3-319-66146-9_16

38. Stell, A., Piper, I., Moss, L.: Automated measurement of adherence to Traumatic Brain Injury (TBI) guidelines using neurological ICU data. In: International Joint Conference on Biomedical Engineering Systems and Technologies (BIOSTEC). SCITEPRESS (2018)

39. Tóth, K., Machalik, K., Fogarassy, G., Vathy-Fogarassy, Á.: Applicability of process mining in the exploration of healthcare sequences. In: 30th Neumann Colloquium (NC), pp. 151–156. IEEE (2017)

40. de Vries, G.J., Neira, R.A.Q., Geleijnse, G., Dixit, P., Mazza, B.F.: Towards process mining of EMR data. In: International Joint Conference on Biomedical Engineering Systems and Technologies (BIOSTEC) (2017)

41. World Health Organization: International statistical classification of diseases and related health problems, vol. 2. World Health Organization (2004)

42. Yan, H., Van Gorp, P., Kaymak, U., et al.: Aligning event logs to task-time matrix clinical pathways in BPMN for variance analysis. J. Biomed. Health Inform. **22**(2), 311–317 (2018)

43. Yang, S., Sarcevic, A., Farneth, R.A., et al.: An approach to automatic process deviation detection in a time-critical clinical process. J. Biomed. Inform. **85**, 155–167 (2018)

44. Yang, S., et al.: Medical workflow modeling using alignment-guided state-splitting HMM. In: International Conference on Healthcare Informatics (ICHI), pp. 144–153. IEEE (2017)

4th International Workshop on Process Querying (PQ)

4th International Workshop on Process Querying (PQ)

Process querying studies (automated) methods, techniques, and tools for managing, e.g., filtering, inquiring, manipulating, or updating, models that describe observed and/or envisioned processes, and relationships between the processes with the ultimate goal of converting process-related information into decision-making capabilities. Process querying research spans a range of topics from theoretical studies of algorithms and the limits of computability of process querying techniques to practical issues of implementing process querying technologies in software. Process querying techniques have broad applications in Business Process Management and Process Mining. Examples of practical problems tackled using process querying include business process compliance management, business process weakness detection, process variance management, process model translation, syntactical correctness checking, process model comparison, infrequent behavior detection, process instance migration, process monitoring, process reuse, and process standardization.

This workshop aims to provide a high-quality forum for researchers and practitioners to exchange research findings and ideas on methods and practices in the area of process analysis, management, and querying.

PQ 2019 is the 4th edition of the workshop. It attracted two high-quality international submissions. Each paper was reviewed by at least three members of the Program Committee.

The invited talk of Han van der Aa, with which the workshop began, is about Complex Event Processing methods for process querying, covering various use cases, and essential techniques for both online and offline scenarios.

In their work, Esser and Fahland propose a novel approach to store and query multi-dimensional event data in a graph database. The authors highlight the suitability of the approach for multi-dimensional process mining, in which events stem from different sources yet pertain to correlated processes.

The invited talk of Eduardo González López de Murillas closed the workshop, with a discussion centered around the support of process data querying from databases through the Data-Aware Process Oriented Query Language (DAPOQ-Lang).

We hope that the reader will benefit from these proceedings and know more about the latest advances in research about process querying.

September 2019

Artem Polyvyanyy
Claudio Di Ciccio
Arthur ter Hofstede

Organization

Workshop Organizers

Artem Polyvyanyy	The University of Melbourne, Australia
Claudio Di Ciccio	WU Vienna, Austria
Arthur ter Hofstede	Queensland University of Technology, Australia

Program Committee

Hyerim Bae	Pusan National University, South Korea
Massimiliano de Leoni	University of Padua, Italy
Jochen De Weerdt	Katholieke Universiteit Leuven, Belgium
Claudio Di Ciccio	WU Vienna, Austria
Amal Elgammal	Cairo University, Egypt
Dirk Fahland	Eindhoven University of Technology, The Netherlands
Luciano García-Bañuelos	Tecnológico de Monterrey, Mexico
María Teresa Gómez-López	Universidad de Sevilla, Spain
David Knuplesch	alphaQuest GmbH, Germany
Agnes Koschmider	Karlsruhe Institute of Technology, Germany
Marcello La Rosa	The University of Melbourne, Australia
Henrik Leopold	Kühne Logistics University, Germany
Fabrizio M. Maggi	University of Tartu, Estonia
Jorge Munoz-Gama	Pontificia Universidad Católica de Chile, Chile
Artem Polyvyanyy	The University of Melbourne, Australia
Maurizio Proietti	CNR-IASI, Italy
Hajo Reijers	Utrecht University, The Netherlands
Stefan Schönig	University of Bayreuth, Germany
Minseok Song	Pohang University of Science and Technology, South Korea
Arthur ter Hofstede	Queensland University of Technology, Australia
Han van der Aa	Humboldt University of Berlin, Germany
Wil van der Aalst	RWTH Aachen University, Germany
Boudewijn van Dongen	Eindhoven University of Technology, The Netherlands
Seppe vanden Broucke	Katholieke Universiteit Leuven, Belgium
Hagen Völzer	IBM Research, Switzerland

Complex Event Processing
for Event-Based Process Querying

Han van der Aa[✉]

Department of Computer Science, Humboldt-Universität zu Berlin, Berlin, Germany
han.van.der.aa@hu-berlin.de

Abstract. Process querying targets the filtering and transformation of business process representations, such as event data recorded by information systems. This paper argues for the application of models and methods developed in the general field of Complex Event Processing (CEP) for process querying. Specifically, if event data is generated continuously during process execution, CEP techniques may help to filter and transform process-related information by evaluating queries over event streams. This paper motivates the use of such event-based process querying, and discuss common challenges and techniques for the application of CEP for process querying. In particular, focusing on event-activity correlation, automated query derivation, and diagnostics for query matches.

Keywords: Complex event processing · Process querying · Query derivation

1 Introduction

Process querying is concerned with models and methods to filter and transform representations of business processes [14]. As such, it supports various use cases, including process modeling support, variation management, performance simulation, and compliance verification. Although process querying is often performed based on model-based process representations, this paper focuses on querying techniques that target the abundance of event data recorded by information systems during the execution of business processes, i.e., *event-based process querying.*

The notion of *event data* relates to process representations that capture the recorded behavior of a process, such that an event denotes that a certain state has been reached (e.g., an order request has been received) or that an activity has been executed as part of a specific process instance [4]. Event data is often formalized as an *event log*, a set of traces, each trace being a finite sequence of events that denotes past behavior for a particular process instance. Process-related data may also be available as an *event stream*, a potentially infinite sequence of events that represent the ongoing behavior of a process.

Han van der Aa is funded by a fellowship from the Alexander von Humboldt Foundation.

© Springer Nature Switzerland AG 2019
C. Di Francescomarino et al. (Eds.): BPM 2019 Workshops, LNBIP 362, pp. 625–631, 2019.
https://doi.org/10.1007/978-3-030-37453-2_50

Complex Event Processing (CEP) defines models and methods to make sense of such event streams [7]. It defines languages to express queries, which are then evaluated over an event stream. CEP methods employ continuous filtering, transformation, and pattern detection, which are closely related to aspects of process querying. Therefore, it suggests itself to adopt event-based process querying through CEP when process-related information is represented by event streams. While CEP is developed for *online* event processing, event-based process querying also enables various use cases for *offline* event analysis. This is achieved by replaying event logs, which renders online event-based techniques applicable to static event data.

This paper outlines how CEP methods can be used for event-based process querying. It serves as accompanying material for the keynote presentation at the 4th International Workshop on Process Querying and comprises a shortened version of a book chapter on event-based process querying using CEP [2]. In the remainder, Sect. 2 motivates the use of event-based process querying, Sect. 3 briefly introduces CEP, Sect. 4 highlights essential challenges and techniques, prior to concluding in Sect. 5.

2   Motivation

Event-based process querying can be seen as a special variant of process querying, where the filtered and transformed process representations assume the form of event data. Given the event data of a process, event-based querying supports a variety of analysis questions, which may relate to qualitative as well as quantitative properties of a process.

Qualitative process properties relate to recorded execution dependencies [6], e.g., whether two activities have been executed in a specific order or a particular number of times. The analysis of such properties is, in particular, relevant to *compliance verification* of a process, in which recorded event data is compared to the expected behavior of a process, cf., [6,10]. Based on a formalization of compliance requirements, event-based process querying helps to identify cases of non-compliant process execution [22]. Such mechanisms are particularly useful if applied to event streams representing the most recent behavior of a process: Detecting a compliance violation shortly after it occurred enables the immediate implementation of mitigation and compensation schemes.

Quantitative process properties may be defined in terms of execution and wait times, or costs assigned to activity executions [17]. The analysis of such properties is part of *performance monitoring*, which recognizes the importance of efficient process execution for many processes. Event-based process querying helps to measure these properties: It selects events that are used as input for the computation of performance indicators [8], e.g., the average activity execution time, the delay with which a particular activity is executed after activation, or the accumulated costs induced by a specific type of process instance. At the same time, outliers that represent process execution with anomalous performance can be extracted [16]. Again, given that immediate detection of performance issues is a prerequisite for effective countermeasures, the analysis of event streams performed by event-based querying can be highly beneficial.

3   Complex Event Processing

Complex Event Processing (CEP) emerged as a computational paradigm to handle streams of event data [7]. The focus of CEP is on the detection of specific event patterns, which are sets of events that are correlated in terms of their ordering, payload data, and context. Such event patterns are detected by applying *event queries* to *event streams*.

Events. An *event* is a recording of some state change that is considered to be of relevance for a particular setting. In the context of process querying, such state changes typically refer to the progress of process execution as, for instance, indicated by the execution of an activity in a process instance. Whereas events can be defined using different formalisms, events should at least be associated with a unique *identifier*, a *timestamp*, and an *event type*. Using a relational model for the payload of events, Table 1 lists three exemplary events for a Lead-to-Quote process. Each event is of type `Act` (representing that an activity has been executed), while an attribute `name` captures milestones (e.g., `QR` for quote request received) and activities (e.g., `ED` for entering details).

Table 1. Three example events for a Lead-to-Quote scenario.

id	timestamp	type	order_id	name	client	price
11	21.09.18,15:12:36	Act	023	QR	Franklin	745,00
12	21.09.18,15:12:36	Act	067	QR	Meyers	282,00
42	24.09.18,09:72:10	Act	023	ED	Franklin	745,00

Event Streams. An *event stream* is defined by a potentially infinite set of events E and an order relation $\prec \subseteq E \times E$ (either partial or total). The fact that a stream is potentially infinite means that, in practice, processing is based on the stream at a specific point in time, i.e., the prefix of the stream up to this time. The identical timestamps of events 11 and 12 in Table 1 illustrate that events may happen concurrently (e.g., a batch of quote requests is received), inducing a partial order over the stream.

Event Query Languages. Numerous models and languages for the definition of event queries have been proposed in recent years (see [13] for an overview). There is currently no common standard for event query languages. Rather, CEP systems define their own languages, differing in syntax and semantics. Nevertheless, it has been noted that many languages for the specification of event queries, share at least a set of common operator types, such as *disjunction and conjunction, sequencing, Kleene closure,* and *data predicates* [23].

4   Challenges and Solutions

The application of CEP models and methods as discussed above in the context of event-based process querying has to cope with several challenges. This section considers three main challenges in this regard and discusses how state-of-the-art techniques address them. In particular, we focus on: event-activity correlation (Sect. 4.1), automatic event query derivation (Sect. 4.2), and diagnostics for event query matches (Sect. 4.3).

4.1   Event-Activity Correlation

A fundamental requirement for analysis techniques involving event data along-side other representations of a process, i.e., process models, is that observed event types can be linked to process model elements, such as activities or decision points. For instance, in compliance verification, the expected behavior of a process may be formalized by a process model, against which the recorded events are compared. Typically, however, such required event-activity correlation is not readily available [12]. *Manually* establishing correlation is often unfeasible because analysts rarely possess the necessary knowledge on the details of a process implementation [21]. Consequently, it is highly beneficial to establish event-activity correlation in an *automated* fashion. However, to reliably achieve this, challenges including noisy and non-compliant behavior, as well as complex event-activity relations must be taken into account [3].

Several techniques have been developed that aim to overcome such challenges, cf., [5,18]. These techniques consider various aspects to establish a proper correlation, most prominently analyzing label and behavioral information. The labels assigned to recorded events and process activities represent valuable information for the establishment of event-activity correlation, e.g., an event with the label *project information submitted* may correlate to an activity labeled *enter project details*. Correlation techniques employ similarity measures that quantify the similarity between different labels, considering both syntactic and semantic similarity. Behavioral information can, furthermore, be highly relevant when establishing event-activity correlation. Correlation techniques consider behavioral properties in different ways, for instance by quantifying similarity based on the average position in which events or activities occur in process instances [1] or comparing behavioral relations among events and activities [5].

4.2   Event Query Derivation

Establishing relevant event queries is a crucial aspect of event-based process querying to enable, for instance, compliance verification of a process. The actual translation of process properties into event queries is cumbersome, since it requires the formalization of the requirements in a (commonly declarative) query model. To overcome this problem, automatic techniques for query derivation have been developed. Two general directions are followed by such approaches: model-driven and data-driven derivation.

Model-Driven Query Derivation. If a specification of the expected behavior of a process is available in terms of a process model (and an event-activity correlation has been obtained), event queries may be formulated to detect any deviation of the recorded from the modeled behavior. These queries can be derived by first computing behavioral relations for the process model, e.g., using (causal) behavioral profile relations [20] or the relations of the 4C spectrum [15]. Such relations define constraints that should hold between process activities according to a process model. Given a set of behavioral relations, these can then be turned into monitoring queries, where each query identifies when the behavioral relation is not satisfied, i.e., when a compliance violation occurs.

Data-Driven Query Derivation. In case a suitable process model is not available, historic event data may serve as the basis for process querying when such data has been annotated with the situation of interest, e.g., a compliance violation or the attainment of a milestone. The necessary annotations can be obtained by retrospectively recording when such a situation of interest occurred in a process. While the annotations then identify the point in time at which the situation occurred, the actual event pattern leading to the situation is not necessarily known. The problem of event query discovery then becomes a supervised learning problem, which aims to construct a query that matches whenever an annotation indicates that the situation of interest occurred.

For sequential query patterns, query discovery can be framed as frequent sequence mining [19], detecting the subsequences that are shared among all annotated sequences. A few tailored algorithms have been proposed for more complex event query discovery, e.g., iCEP [11] and the IL-Miner [9], which discover queries built of sequence operators, data predicates, and time windows.

4.3 Diagnostics for Event Query Matches

The interpretation of results obtained by event-based process querying can be challenging. Query matches need to be interpreted as a violation of some normative behavior. Here, diagnostics are important, as fine-granular queries may lead to an overload of monitoring alerts for certain behavioral anomalies: e.g., a single out-of-order event may trigger a large amount of order violations, even though these all stem from the same source. To avoid such an overload, monitoring alerts can be filtered by identifying the earliest indicator of non-compliant behavior in a set of compliance violations, i.e., the *trigger* of the violations. Feedback on compliance violations can, furthermore, be filtered by recognizing violations that logically follow from the violations already observed, so-called *consecutive violations* [22].

Next to the identification of particular events that resulted in compliance violations, it is possible to assess whether there are dependencies between violations and their occurrence context as reflected in data attributes associated with process instances. That is, the goal is to check for attribute values that differentiate cases with the violation from cases without the violation. In this way, it is possible to discover that certain violations occur in a specific context, e.g., purchase orders originating from a particular *country* and which are related to a specific *order type* may be more likely to be delivered too late. The automatic detection of such context-related issues can be achieved by applying *classification* techniques on an annotated set of process instances. Classifiers that produce human interpretable output, such as *decision trees*, are particularly useful for this setting. These techniques can produce clear rules indicating in which contexts monitoring violations have been observed.

5   Discussion

This paper outlined how Complex Event Processing methods can be leveraged for process querying that works on event-based representations of processes. In particular, event-based process querying can be used both for online querying, through the analysis of event streams, as well as for offline querying, by replaying event logs containing static event data. We argued that using CEP methods for event-based process querying faces several challenges, which may be addressed by three essential techniques: event-activity correlation, automatic event query derivation, and diagnostics for event query matches.

Open research directions in relation to the challenges discussed in this paper include: dealing with uncertainty in event-activity correlations, identifying the semantically most relevant monitoring queries, and optimizations of event query discovery algorithms in order to overcome scalability issues.

References

1. van der Aa, H., Gal, A., Leopold, H., Reijers, H.A., Sagi, T., Shraga, R.: Instance-based process matching using event-log information. In: Dubois, E., Pohl, K. (eds.) CAiSE 2017. LNCS, vol. 10253, pp. 283–297. Springer, Cham (2017). https://doi.org/10.1007/978-3-319-59536-8_18
2. Van der Aa, H., Artikis, A., Weidlich, M.: Complex event processing methods for process querying. In: Process Querying Methods (2019, in press)
3. Van der Aa, H., Leopold, H., Reijers, H.: Efficient process conformance checking on the basis of uncertain event-to-activity mappings. IEEE TKDE (2019, in press)
4. Van der Aalst, W.M.P.: Process Mining - Data Science in Action. Springer, Heidelberg (2016). https://doi.org/10.1007/978-3-662-49851-4
5. Baier, T., Di Ciccio, C., Mendling, J., Weske, M.: Matching events and activities by integrating behavioral aspects and label analysis. Softw. Syst. Model. **17**(2), 1–26 (2017)
6. Carmona, J., van Dongen, B.F., Solti, A., Weidlich, M.: Conformance Checking-Relating Processes and Models. Springer, Cham (2018). https://doi.org/10.1007/978-3-319-99414-7
7. Cugola, G., Margara, A.: Processing flows of information: from data stream to complex event processing. ACM Comput. Surv. **44**(3), 15:1–15:62 (2012)
8. Del-Río-Ortega, A., Resinas, M., Cabanillas, C., Cortés, A.R.: On the definition and design-time analysis of process performance indicators. Inf. Syst. **38**(4), 470–490 (2013)
9. George, L., Cadonna, B., Weidlich, M.: IL-Miner: instance-level discovery of complex event patterns. PVLDB **10**(1), 25–36 (2016)

10. Ly, L.T., Maggi, F.M., Montali, M., Rinderle-Ma, S., van der Aalst, W.M.P.: Compliance monitoring in business processes: functionalities, application, and tool-support. Inf. Syst. **54**, 209–234 (2015)

11. Margara, A., Cugola, G., Tamburrelli, G.: Learning from the past: automated rule generation for complex event processing. In: DEBS, pp. 47–58. ACM (2014)

12. Oliner, A., Ganapathi, A., Xu, W.: Advances and challenges in log analysis. Commun. ACM **55**(2), 55–61 (2012)

13. Polyvyanyy, A.: Business Process Querying. In: Sakr, S., Zomaya, A. (eds.) Encyclopedia of Big Data Technologies. Springer, Cham (2018). https://doi.org/10.1007/978-3-319-63962-8

14. Polyvyanyy, A., Ouyang, C., Barros, A., van der Aalst, W.M.P.: Process querying: enabling business intelligence through query-based process analytics. Decis. Support Syst. **100**, 41–56 (2017)

15. Polyvyanyy, A., Weidlich, M., Conforti, R., La Rosa, M., ter Hofstede, A.H.M.: The 4C spectrum of fundamental behavioral relations for concurrent systems. In: Ciardo, G., Kindler, E. (eds.) PETRI NETS 2014. LNCS, vol. 8489, pp. 210–232. Springer, Cham (2014). https://doi.org/10.1007/978-3-319-07734-5_12

16. Rogge-Solti, A., Kasneci, G.: Temporal anomaly detection in business processes. In: Sadiq, S., Soffer, P., Völzer, H. (eds.) BPM 2014. LNCS, vol. 8659, pp. 234–249. Springer, Cham (2014). https://doi.org/10.1007/978-3-319-10172-9_15

17. Sadiq, S., Governatori, G., Namiri, K.: Modeling control objectives for business process compliance. In: Alonso, G., Dadam, P., Rosemann, M. (eds.) BPM 2007. LNCS, vol. 4714, pp. 149–164. Springer, Heidelberg (2007). https://doi.org/10.1007/978-3-540-75183-0_12

18. Senderovich, A., Rogge-Solti, A., Gal, A., Mendling, J., Mandelbaum, A.: The ROAD from sensor data to process instances via interaction mining. In: Nurcan, S., Soffer, P., Bajec, M., Eder, J. (eds.) CAiSE 2016. LNCS, vol. 9694, pp. 257–273. Springer, Cham (2016). https://doi.org/10.1007/978-3-319-39696-5_16

19. Wang, J., Han, J.: BIDE: efficient mining of frequent closed sequences. In: ICDE, pp. 79–90. IEEE Computer Society (2004)

20. Weidlich, M., Polyvyanyy, A., Mendling, J., Weske, M.: Causal behavioural profiles-efficient computation, applications, and evaluation. Fund. Informaticae **113**(3–4), 399–435 (2011)

21. Weidlich, M., Ziekow, H., Gal, A., Mendling, J., Weske, M.: Optimizing event pattern matching using business process models. IEEE TKDE **26**(11), 2759–2773 (2014)

22. Weidlich, M., Ziekow, H., Mendling, J., Günther, O., Weske, M., Desai, N.: Event-based monitoring of process execution violations. In: Rinderle-Ma, S., Toumani, F., Wolf, K. (eds.) BPM 2011. LNCS, vol. 6896, pp. 182–198. Springer, Heidelberg (2011). https://doi.org/10.1007/978-3-642-23059-2_16

23. Zhang, H., Diao, Y., Immerman, N.: On complexity and optimization of expensive queries in complex event processing. In: SIGMOD, pp. 217–228. ACM (2014)

Storing and Querying Multi-dimensional Process Event Logs Using Graph Databases

Stefan Esser and Dirk Fahland$^{(\boxtimes)}$

Eindhoven University of Technology, Eindhoven, The Netherlands
s.esser@student.tue.nl, d.fahland@tue.nl

Abstract. Process event data is usually stored either in a sequential process event log or in a relational database. While the sequential, single-dimensional nature of event logs aids querying for (sub)sequences of events based on *temporal relations* such as "directly/eventually-follows", it does not support querying *multi-dimensional* event data of multiple related entities. Relational databases allow storing multi-dimensional event data but existing query languages do not support querying for sequences or paths of events in terms of temporal relations. In this paper, we report on an exploratory case study to store multi-dimensional event data in labeled property graphs and to query the graphs for structural and temporal relations combined. Our main finding is that event data over multiple entities and identifiers with complex relationships can be stored in graph databases in a systematic way. Typical and advanced queries over such multi-dimensional event data can be formulated in the query language Cypher and can be executed efficiently, giving rise to several new research questions.

Keywords: Process mining · Event log · Multi-dimensional processes · Querying · Labeled property graphs · Graph databases

1 Introduction

Retrieving subsets of event data of a particular characteristic is a recurring activity in process analysis and process mining [1]. Each *event* is thereby defined by an *activity*, a *case identifier* referring to the object or case where the activity was carried out, and a *timestamp* or *ordering* attribute defining the order of events. If all events use the same, single case identifier attribute, the event data is *single-dimensional* and can be stored in an *event log* as one *sequence of events* per case. Such sequences can be easily queried for *behavioral properties* such as *event (sub-)sequences* or *temporal relations* such as "directly/eventually-follows" in combination with other data attributes [4, 6, 13, 17, 19, 20].

Most processes in practice however involve multiple inter-related entities which results in *multi-dimensional* event data in which each event is directly or indirectly linked to multiple different case identifiers; sequential event logs cannot represent such multi-dimensional event data [14]. Relational databases (RDBs) can store *1:n and n:m relations between events and case identifiers* and among case identifiers—but the explicit behavioral information of sequences (of arbitrary length) is lost. Reconstructing event sequences of arbitrary length from an RDB requires an arbitrary number of

© Springer Nature Switzerland AG 2019
C. Di Francescomarino et al. (Eds.): BPM 2019 Workshops, LNBIP 362, pp. 632–644, 2019.
https://doi.org/10.1007/978-3-030-37453-2_51

(self-)joins, can only be done under (severe) information loss in the presence of multiple case identifiers [12, 14], and the required queries are large and non-intuitive [16]. Thus, querying and analyzing behavioral properties of multi-dimensional event data in an intuitive way is an open problem.

State of the Art. A recent literature survey of 95 studies [10, Ch.7, pp.133][15] established requirements for querying event data. Focusing on *querying for structure and behavior in multi-dimensional event data* we derive from [10, pp.133] the requirements to **(R1)** query and analyze event data, and to **(R2)** consider relations between multiple data entities. The technique shall support **(R3)** storing and querying business process-oriented concepts (such as activities, cases, resources) and **(R4)** capture information about how events are related to different entities. Queries should **(R5)** be expressed as graphs to specify the behavior of interest in a natural way, **(R6)** allow to query paths (or sequences) of events (connected by some relation), **(R7)** allow to select individual cases based on partial patterns, **(R8)** allow to query temporal properties (such as directly/eventually-follows), **(R9)** correlate events related to the same entity, **(R10)** allow querying aspects related to several processes at the same time on the same dataset, and **(R11)** allow to query multiple event logs and combine results. Altogether, *a user shall be able to query for individual events (and their properties), for different entities/case notions, for behavioral and structural relations, and for patterns of multiple events (within and across entities).*

Of the 95 works surveyed [10, pp.133], several approaches exist to retrieve cases from event logs for temporal properties [13, 19], for most frequent behavior [6], for sequences of activities [4] or algebraic expressions of sequence, choice, and parallelism over activities [20], or to check whether a temporal-logic property holds [17]. Several techniques support graph-based queries [5, 11, 13]. These techniques satisfy R7 and R8. However, they only support a single fixed case notion and thus fail R2, R10, R11. Few techniques support querying over multiple entities or processes: the technique in [3] supports graph-based queries over event data from multiple entities, but does not allow to select individual cases or querying for behavioral properties (R7, R8). The language in [2] allows querying data from different processes, but cannot express properties of events or relations between events (R7, R8). DAPOQ [10, Ch.7] generalizes these approaches to query events in the context of their relational data model for behavior properties, but does not support retrieving individual cases (R7) or specifying behavioral and structural patterns (R8). *No existing query language on sequential event logs or RDBs satisfies R1–R11.*

Hypotheses. The above observations led us to the following hypotheses: (1) A *graph-based* event data storage format allows to explicitly represent both multi-dimensional *structural* relations between events and case identifiers and *behavioral* information of event sequences (as paths in the graph). (2) Queries can easily be formulated over such explicit representation of multi-dimensional event data satisfying R1–R11.

Method. To test the validity of these hypotheses, we conducted an exploratory case study. We wanted to test the ability of *labeled property graphs* (LPGs), the data format of *graph databases* (GDBs) [18], to store structural and behavioral information together. And we wanted to test whether existing declarative query languages on labeled property graphs, such as *Cypher* [9] are expressive enough to satisfy R–R11; Sect. 2 introduces GDBs and Cypher. Thus, we first formulated types of typical and advanced query operations over multi-dimensional event data that address R1–R11 in precise English language (see Sect. 4). To test R1–R4, we had to represent an existing event dataset with multiple case identifiers with 1:1 and 1:n relations as an LPG; the dataset had to be representative of a real-life analysis. To test R5–R11, we had to define for each query type a corresponding query over LPGs and independently solved the same analysis questions through procedural programs. A query was valid when it led to the same result as the program. Further, we evaluated the size of the queries to assess how natural they allow solving each question.

Results. We selected the BPIC'17 [7] dataset of a loan application business process for this case study. It contains 3 entity types *Application* (31,509 cases, 239,595 events), *Workflow* (31,509 cases, 768,823 events), *Offer* (42,995 cases, 193,840 events) each with their own identifier; Application is in a 1:1 relation with Workflow, and in a 1:n relation with Offer; further a common *Case* identifier subsumes all events of/related to the same *Application* into a single case. We chose the graph database system Neo4j (neo4j.com) for LPG storage and querying due to off-the-shelf availability, Cypher support, and suitable performance. We iteratively developed mappings from the BPIC'17 event data to LPG concepts and formulated Cypher queries for all classes, until being able to answer all queries. In our mapping, each event and entity identifier in the BPIC'17 dataset becomes a different node in the LPG, and all structural relations (between entities), correlation (between entities and events), and temporal relations (between events) become labeled relationships in the LPG, satisfying R1–R4; see Sect. 3. We were able to answer each analysis question with a Cypher query that was valid compared to the procedural analysis. The query answer time always satisfied practical requirements on the full BPIC'17 dataset (answers within fractions of a second to at most a few seconds), and significantly outperformed our procedurally implemented single-pass search algorithms for more complex queries. Section 4 provides a summary of the queries and results; full details including implementation are available in a technical report [8]. We discuss limitations and alleys for future work in Sect. 5.

2   Information Systems, Event Logs and Graph Databases

Event Data in Information Systems. Information Systems (IS) created and update information records in structured transactions or *activities*. Each update is linked to one or more *entities* with unique *entity identifiers*, for example a specific order and related invoices. Each update can be recorded as an *event* with attributes for the activity, the entity identifiers, and the *timestamp* (or *ordering* of updates). Events are implicitly related to each other via the structural 1:1, 1:n, and n:m relations between the entities on which the updates occurred [14].

Event Logs. An *event log* is a *view* on the recorded events of an IS assuming a process working on a specific entity, e.g., handling an order. All events related to the same entity identifier form a *case* (or execution) of the process. A log stores each case as a sequence of events (ordered by timestamp or ordering attribute). An IS usually hosts multiple uniquely identifiable entities, e.g., orders and invoices. Each (combination of) entity ids can be used as a *case identifier*, giving a different view on the event data in the IS [1]. The creator of an event log is forced to choose under which perspective (case identifier) the process shall be analyzed, e.g. the invoice handling or the ordering aspects of the procurement process, and *flattens* the data accordingly [12]. The more different entities and potential case identifiers are included, the more "remotely" related events are included, making the analysis of the flattened data harder and erroneous [14]. The event data in the BPIC'17 log used in our case study (see Sect. 1) had been flattened under the case identifier of an Application so events from all 3 entities are interleaved over time.

Graph Databases. *Labeled Property Graphs* (LPGs) are one of the data structures used in graph databases (GDBs) [18]. An LPG consists of nodes (vertices) and relationships (edges), each with an arbitrary number of key-value pairs, called properties.

We explain LPGs on the example of Fig. 1 which shows the relationships between a professor and 2 students. The example contains nodes with the labels *:Person*, *:Professor*, *:Student* and *:Document*. The document you are currently reading is authored by Stefan, a student supervised by Dirk who co-authors this document and say Miro is another student contributing to this paper. The

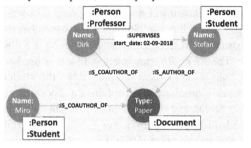

Fig. 1. Labeled property graph

"Name" of each person is a property of the *:Person* nodes; "Type" is a property of *:Document* nodes. The described relationships between the nodes can also hold properties like the starting date of a supervision. Figure 1 shows the LPG structure showing all components of the example. In the GDB system Neo4j used in our case study, a node of an LPG can have multiple labels while relationships have exactly one label and are directed.

Querying Graph Databases. Cypher is a language for querying LPGs [9] and supported by Neo4j. Cypher queries use pattern matching to select sub-graphs of interest. We explain the central Cypher query concepts used in the case study in a single (albeit inefficient) example query. Nodes are denoted in a query by round parentheses and can be used with variables, labels and properties like $(s : Student \{Name : "Miro"\})$ where s is the variable name, *:Student* the node label and *Name* : *"Miro"* the key-value pair of the student's *Name* property. Relationships must have a start and a destination node and may also be specified with variables, a name and properties. $(n1) - [:SUPERVISES]->(n2)$ for example represents the relationship node $n1$ supervises $n2$ with $n1$ and $n2$ being any two nodes in the graph. Cypher also allows to query the directed relationships in an undirected way: $(n1)-[:SUPERVISES]-(n2)$. For the example graph, we want define a query to return documents that Dirk co-authors with students, other than Stefan, working on these documents. We also want to retrieve the longest path that connects the matched students with Dirk. The query may be defined as follows:

```
1 MATCH path = (s:Student)-[*]-(p:Professor {Name: ''Dirk''})
2 WHERE NOT s.Name = ''Stefan''
```

```
3 WITH s AS student, p AS professor, path AS paths
4 MATCH (d:Document)<-[:IS_COAUTHOR_OF]-(professor:Professor)
5 WHERE (student:Student)--(d)
6 RETURN student, d, paths, length(paths) AS pLength
7 ORDER BY pLength DESC
8 LIMIT 1
```

The *MATCH* clause defines the pattern we want to retrieve from the graph. The pattern in line 1 includes students that have any type of relation to Dirk. The *-operator defines that the relationship between a student and Dirk may be direct or indirect over paths of arbitrary length and relationship types between Dirk and the student. The *WHERE* clause in line 2 restricts the pattern such that the student's name cannot be "Stefan". By defining the professors' name property to be "Dirk" in line 1 we also restrict the patterns. *WITH* in line 3 allows to process the matched patterns. In our example we only rename the variables, e.g. from *s*, to *student* and chain queries together: variable *student* in lines 4–6 may only take values retrieved for variable *s* in lines 1–2, e.g., "Miro" but not "Stefan." Line 4 matches the documents Dirk coauthors and line 5 restricts the results to documents that have a direct relationship to a student.

The *RETURN* statement in line 6 is used to define the output of the nested query in lines 4–5. In the example graph, the student is "Miro" and the document is the "paper". Variable *paths* contains the 2 possible paths between Miro and Dirk. One walks over Stefan and one does not. "Length()" is a function of Cypher that returns the hops needed to walk a path. Lines 7–8 sort the results by path lengths (*ORDER BY* clause) in descending order (*DESC*) and return only the first path of this ordered list (*LIMIT* 1). See [8] for more detailed examples of the Cypher concepts used in the case study.

3   Representing Multi-Dimensional Event Data in a Graph DB

To represent the multi-dimensional event log data of BPIC'17 in an LPG, we developed the following mapping.

Entities and Events. For each entity/case type in the data (Application, Workflow, Offer, the log's case notion) and each entity/case id, we created a separate *case node*. We created relationships between case nodes to encode the structural relations among entities/cases, e.g., *OFFER_TO_CASE* in Fig. 2. Each event became an *event node*. Attributes of entities, cases, and events became properties of their respective nodes. We created the relationship *:EVENT_TO_CASE* from event to case

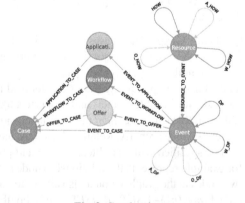

Fig. 2. Graph schema

nodes to describe the *multi-dimensional correlation* of an event to multiple entities, e.g., an event related to an Offer and to the case.

Event Sequences. The temporal order of events within each entity/case is expressed by *one dedicated directly-follows relation per entity/case type*. For BPIC'17 this resulted in

4 different relations *:A_DF*, *:W_DF*, *:O_DF* (for the 3 entities) and *:DF* for the general case, see Fig. 2. For example, the *:O_DF* relations connects any two events that are correlated to the same Offer and directly succeed each other. Thus, each path of *:O_DF* relations forms one event sequence of one concrete Offer. We defined for each *:DF* relation a *duration* property stating the time difference between the 2 events. This way, a single path query can include behavioral relations "along" each entity and structural relations "between" entities in the graph.

Resource Dynamics. As BPIC'17 contains resources, we also created *Resource* nodes. To enable querying for social network dynamics of resources [1, Sect. 9.3.1], we introduced a handover of work relationship *:HOW* between resources; as for *:DF* we created an entity-specific *:HOW* for each entity/case type. Figure 2 shows the schema of the resulting LPG that we obtained. A technical report [8] provides full details for all nodes and relationships created in the case study.

Implementation. To import the BPIC'17 data into Neo4j, we used Cypher's *LOAD CSV* clause which imports a given CSV file row by row such that clauses like *MATCH* or *WITH* can be used to filter or select the values and columns for creating nodes and relationships with the *CREATE* and *MERGE* clauses; the latter only creates graph patterns that do not exist already. We explored two approaches to create the graph. In the first approach, each *CREATE/ MERGE* query used the *LOAD CSV* clause to selectively load only those columns required for the query. This resulted in multiple passes over the CSV files and took about 75 min to complete on an Intel i7 CPU @ 2.8 GHz with 16 GB of memory. The second approach imported each CSV row as an event nodes (with *CREATE* clause storing all columns as event attributes). Subsequent *CREATE/MERGE* queries constructed the other nodes and relationships from the event nodes. This required only a single pass over the CSV file leading to a cumulative execution time of 2:37 min. We refer to [8] for a detailed description of the queries and to https://github.com/multi-dimensional-process-mining/graphdb-eventlogs for the implementation.

Discussion. With the above encoding we could represent events and first-class business process concepts (activity, case, resource) as nodes in a graph which we can query, satisfying (R1, R3). The semi-structured nature of graphs allowed us to represent multiple different, related entities (R2), the relations between entities and events (R4), the correlation of events to the same entity (R9), and the correlation of multiple cases and entities to the same event (e.g., several events are correlated to both *Case* and *Offer*). Thus, the graph database can be seen as a *multi-dimensional event log*, where events of each entity are ordered by "their" directly-follows relation leading to a *partially ordered event log*; the classical directly-follows relation connects events of different entities.

4   Querying Multi-Dimensional Event Data from a Graph DB

In the following we present 7 classes of analysis questions that we formulated to evaluate requirements R5-R11 of Sect. 1 for querying multi-dimensional event data on the LPGs of Sect. 3. For each question we provide a Cypher query and report results and the query processing times (measured on an Intel i7 CPU @ 2.8 GHz machine with 16 GB of memory with Neo4j Browser). The queries are available at https://github.com/multi-dimensional-process-mining/graphdb-eventlogs.

Q1. Query Attributes of Events/Cases. We want to query for the first-class concepts of event logs: query for a case and an event based on event/case attributes by using a partial patterns to satisfy R7. The following query returns the event attribute "completetime" and the case attribute "loangoal" of Case "Application_681547497".

```
1 MATCH (c:Case) <-[:EVENT_TO_CASE]- (e:Event)
2 WHERE c.name = ''Application_681547497" AND e.activity = ''A_Submitted"
3 RETURN e.completetime, c.loangoal
```

The query has been processed in 0.083 s. After modifying the query to consider all cases, i.e. remove the condition for a specific case in line 2, the query completed in 0.944 s. We could also query for properties of an entity related to an event, e.g., for an *Offer* over a specific *offeredAmount*.

Q2. Query Directly-Follows Relations. Q2 is focused on temporal aspects. Here we want a query that satisfies R8 by considering 2 consecutive events. Directly-follows relations of events in a case are an important characteristic of event logs as they represent the case internal temporal order of events and many of today's process mining techniques rely on these relations. Event x directly follows event y if there is no event in between them in the temporal order of the case. The next query returns the event directly following the node with the activity property "O_Created" of a given offer entity by matching the *:O_DF* relationship.

```
1 MATCH (o:Offer) <-[:EVENT_TO_OFFER]- (e1:Event) <-[:O_DF]- (e2:Event)
2 WHERE o.name = ''Offer_716078829" AND e1.activity = ''O_Created"
3 RETURN e1, e2
```

The query execution time for one specific offer was 0.174 s whereas querying the *:O_DF* relations with destination node "O_Created" for all 42,995 offers took 11.291 s. Directly-follows relations of other entities (Application and Workflow) or across entities (Case) can be queried by adjusting the query in the *MATCH* and *WHERE* clauses accordingly.

Q3. Query Eventually-Follows Relations. We want a query that satisfies R8 by considering the temporal relationship of any 2 events of a case. Eventually-follows relations are also related to the case internal order of events. Event y *eventually follows* event x if y occurs after x in the same case, that is, if x and y are connected through a path of directly-follows relations of arbitrary length. We query the offer specific eventually-follows relationship between "O_Created" and "O_Cancelled" for a given offer as follows:

```
1 MATCH (o:Offer) <-[:EVENT_TO_OFFER]- (e1:Event) <-[:O_DF*]- (e2:Event)
2 WHERE o.name = ''Offer_716078829" AND e1.activity = ''O_Created" AND
        e2.activity = ''O_Cancelled"
3 RETURN e1, e2
```

Even though the *MATCH* clause looks similar to the one of the directly-follows query, the *-Operator changes the pattern from a direct relationship to a path of arbitrary length. Since we want to find the eventually-follows relationship of two specific activities we also added condition *e2.activity = "O_Cancelled"* to the *WHERE* clause to define the endpoint of the paths to match in the graph. For the given offer the query took 0.182 s. For all 20,898 offers where "O_Created" is eventually followed by

"O_Cancelled" we removed the condition for "Offer_716078829" from the query which then took 4.469 s.

Q4. Case Variants. We want a query to return a case variant as path in the graph to satisfy R6. A *case variant* is the sequence of activities of a case. Case variants are for example used to detect frequent behaviour of a process. We can query the graph to retain the path of events of a case by walking over all of its *:DF* relationships from the first to the last event. For a given case this can be done as follows:

```
1 MATCH (c:Case) <-[:EVENT_TO_CASE]- (e:Event) <-[:DF*]- (e2:Event)
2 WHERE NOT () <-[:DF]-(e) AND NOT (e2)<-[:DF]-() AND c.name =
          'Application_681547497'
3 RETURN (e:Event) <-[:DF*]- (e2:Event) AS paths
```

The pattern of the match clause follows the same logic as the eventually-follows match pattern. For variants we limit the output to the first and last event of a case, i.e. the events that have no incoming or no outgoing ":DF" relationship. The query completed in 0.023 s. Similarly, we can query the graph for variants of another entity such as Offer. The paths of events returned by the above query can be turned in a list of activity sequences by Cypher's list operators: *UNWIND* processes each path in the *paths* variable iteratively, function *nodes()* translates the path into a list of nodes, and list comprehension maps each event node to its activity property. The resulting list of activities can be compared for equality with other lists, etc.

Q5. Query Handover of Work. We want to show that, next to events, further event log concepts like resources can be correlated to the same process entity to satisfy R9. The LPG created in Sect. 3 already contains resource nodes and one handover-of-work relation *:HOW* for each entity. But the *:HOW* relationships only account for the information that there has been work handed over at least once. We now query the graph for *:DF* relationships to aggregate them along the *:HOW* relationships into a social network including counting frequencies as follows:

```
1 MATCH (r1:Resource) -[:RESOURCE_TO_EVENT]-> (e1:Event) <-[:DF]- (e2:Event)
        <-[:RESOURCE_TO_EVENT]- (r2:Resource)
2 RETURN (r1)-[:HOW]-(r2) as p, count(*) AS frequency
```

This query derives the handover of work network by aggregating *:DF* relationships such that we can count the *:DF* relationships between consecutive events and thus retrieve the number of handovers between their resources in the graph. Note that $r1$ and $r2$ can refer to the same node and thus self loops are also included in the network. The query with frequency and path output, as shown above, had an execution time of 17.517 s. A quicker version of the query only matching on *:HOW* relations with-

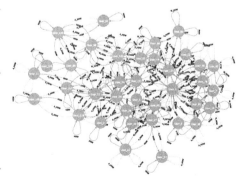

Fig. 3. Handover of work network

out deriving the frequency of handovers took 1.066 s to complete. Figure 3 shows the Neo4j graph output of the query above on a sample of 20 cases. Note that Neo4j by default visualizes all relationships between nodes in the result, even if they are not part of the returned subgraph, i.e. Fig. 3 shows also the entity specific handover of work relationships even though we explicitly specified *:HOW* relationships. This is only a matter of graphical representation, the data output of the query only contains the specified output format.

With traditional event logs, creating a handover of work network typically requires the use of a tool or programming language whereas Neo4j is capable of creating them by in-DB processing only. In the same way an aggregated directly-follows graph can be obtained in-DB by aggregating the *:DF* relations.

Q6. Query Duration/Distance Between Two Specific Activities. The information on how much time or how many activities were needed to get an Offer from "O_Created" to "O_Accepted" for example can be used to measure process performance. For Q6 we want to query temporal relations in the form of durations and path lengths to satisfy R8. Say we are interested in the offer entity that took the longest time to get accepted. We can query the eventually-follows relation of two given activities and use their timestamps to calculate the elapsed time between them:

```
1 MATCH ( start : Event )  -[:EVENT_TO_OFFER]-> ( o : Offer )  <-[:EVENT_TO_OFFER]-
      ( end : Event )
2 WHERE start . activity =  ''O_Created" and end . activity =  ''O_Accepted"
3 WITH start , end , duration . between ( start . completetime , end . starttime ) AS time , o
4 ORDER BY time  DESC
5 LIMIT 1
6 RETURN start , end , time . days AS days , ( toFloat ( time . minutes )/60) AS hours , o
```

The query matches all *:EVENT_TO_OFFER* relationships, filters for the given activities and then uses Cypher's duration function to calculate the time spans. Only the result with the longest duration is returned. In case we want to retrieve the distance wrt. the number of activities, we can aggregate over the nodes along the path between the two events with eventually-follows relation and count the hops with the "Length()" function as shown in [8]. The query for the elapsed time completed in 0.643 s. Querying for the longest path took 2.744 s.

Q7. Query for Behavior across Multi-Instance Relations. Event logs such as BPIC'17 can contain multiple case identifiers. A case identifier may be a single entity, e.g. Offer, or any combination of entities such as the Case notion of BPIC'17 combining Application, Workflow and Offer entities. Querying the behavior across different instances of these entities typically requires multiple steps with traditional event logs such as custom scripts to be able to select, project, aggregate and combine the results accordingly. For Q7 we want to satisfy R10 by combining the results from different (sub)logs and to satisfy R11 by querying 2 (sub)processes in a single query. We defined a query that returns all paths from "A_Create Application" to "O_Cancelled" of the BPIC'17 Cases for Offers that have "O_Created" directly followed by "O_Cancelled" on entity level, but only for those Cases that have more than one Offer with "O_Created" directly followed by "O_Cancelled".

```
1 MATCH ( e1 : Event { activity :  ''O_Created" }) <-[:O_DF]-( e2 : Event { activity :
      ''O_Cancelled" })  -[:EVENT_TO_OFFER]-> ( o : Offer )  -[ rel : OFFER_TO_CASE]->
      ( c : Case )
2 WITH c AS c , count ( o ) AS ct
3 WHERE ct > 1
```

```
4 MATCH (: Event { activity :  ''O_Created''}) <-[:O_DF]-(e : Event { activity :
    ''O_Cancelled''})  -[:EVENT_TO_OFFER]-> (o : Offer)  -[rel : OFFER_TO_CASE]-> (c)
5 WITH  e AS O_Cancelled
6 MATCH p = (A_Created : Event { activity :  ''A_Create Application''})
    <-[:DF*]-(O_Cancelled : Event { activity :  ''O_Cancelled''})
7 RETURN p
```

The query demonstrates several central properties of querying multi-dimensional event data in labeled property graphs.

The first *MATCH* clause (lines 1–3) returns all case nodes of Cases structurally related to more than one Offer (via *:OFFER_TO_CASE*) where "O_Created" is directly followed by "O_Cancelled" in this offer (via *:O_DF*). Note that the case itself typically has several other events not related to the offer in between the two events, i.e., they *only* directly follow each other according to *:O_DF* but not according to *:DF*.

The second *MATCH* clause (lines 4–5) returns all "O_Cancelled" events that directly succeed "O_Created" (via *:O_DF*) in an Offer that is correlated to one of the cases with multiple offers (found in Lines 1–3). The returned "O_Cancelled" events are used in the last *MATCH* clause (lines 6-7) to returns paths from some "A_Create Application" event to one of the "O_Cancelled" events. This way we get a unique path for every Offer that meets the criteria. The query's execution time was 0.453 s in Neo4j Browser. Figure 4 shows 4 of the 218 paths of the query's output in Neo4j's graphical representation; in all 4 cases the "O_Created" and "O_Cancelled" events of one offer are interleaved with events from the Application or other offers.

Discussion. With all above queries, we could demonstrate the ability of Cypher to express queries and results as graphs, satisfying (R5). Q4 and Q7 retrieve entire paths of events (R6) allowing to analyse the sequences. Q1–Q4 and Q7 select individual cases based on partial patterns (R7) allowing to "query by example". Q2, Q3, Q6 and Q7 query for temporal properties (R8) where Q6 specifically considers time; all queries correlate events related to a common entity, Q5 shows that also process concepts such as resources can be used for event correlation (R9); Q7 queries aspects of multiple processes in the same query (R10) and to query multiple logs and combine results (R11) by querying events of Applications and Offer events together. Altogether, we could demonstrate the queries over labeled property graphs satisfy R5–R11.

We validated the correctness of our queries against an independent baseline implementation. The results of Q1–Q6 were obtained by processing the event log with man-

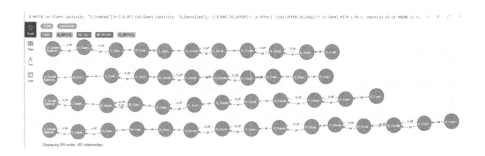

Fig. 4. Q7 output

ual filtering in Disco and social network mining algorithms in ProM. Q7 required a manual procedural algorithm using a single-pass search over the data as the evaluation with existing tools was not possible. Our Cypher queries obtained the same result as the baseline implementations [8]. The graph analysis for Q1–Q7 required only Cypher queries with clauses and functions as described in [9] (except for a typecasts which are not part of Cypher but provided by Neo4j). Notably, the single-pass baseline algorithm for Q7 required 15 min compared to the 0.453 s needed to obtain the same results using Neo4j. Further details on the evaluation of Q1–Q7 regarding time and baselines can be found in [8].

5   Conclusion

We demonstrated the abilities of labeled property graphs and the query language Cypher to store and query multi-dimensional event data of multiple related entities and multiple case notions. We found that we can represent event data of multiple related entities and processes in a single graph database by encoding all first-level concepts of processes (entities, events, resources) as labeled nodes, and the established relations (directly-follows, executes, foreign-key relations between entities) as labeled relationships. This renders our mapping as a candidate for a *multi-dimensional event log* with a partial order between events of different entities. Further, we found that through the chosen relationships, Cypher allows querying paths and subgraphs of the event data based on patterns of nodes and relations. This allows to query cases of individual entities as well as paths across different, related entities, and even to construct aggregates such as the handover of work network. Compared to all prior works discussed in Sect. 1, encoding event data and process concepts alike in a graph database, and querying the graph database with the Cypher query language allowed us to address all requirements (R1–11). The latter suggests that graph databases are a suitable candidate for enabling *multi-dimensional process mining* on event data with multiple identifiers [14].

Limitations and Future Work. We obtained these results through carefully analyzing entities, events, and process concepts from a single event log using domain-knowledge about the studied event log [7]. While we believe the explored ideas to be generalizable, doing so raises a few research questions: How to systematically encode event data (related to multiple entities) into a graph database? How can a graph data schema for such a dataset be specified? A further limitation of this study is that we took a sequential event log as input, from which we reconstructed the relational structure. How to automatically translate event data from a relational database into a graph database in a systematic way is an open problem. Further, a broader understanding of analysis use cases on multi-dimensional event data and the required concepts for an useful and easy-to-use query language have to be established to generalize our findings beyond the concrete case and technology used.

References

1. Van der Aalst, W.M.: Process Mining: Data Science in Action. Springer, Cham (2016). https://doi.org/10.1007/978-3-662-49851-4
2. Baquero, A.V., Molloy, O.: Integration of event data from heterogeneous systems to support business process analysis. In: Fred, A., Dietz, J.L.G., Liu, K., Filipe, J. (eds.) IC3K 2012. CCIS, vol. 415, pp. 440–454. Springer, Heidelberg (2013). https://doi.org/10.1007/978-3-642-54105-6_29
3. Beheshti, S.-M.-R., Benatallah, B., Motahari-Nezhad, H.R., Sakr, S.: A query language for analyzing business processes execution. In: Rinderle-Ma, S., Toumani, F., Wolf, K. (eds.) BPM 2011. LNCS, vol. 6896, pp. 281–297. Springer, Heidelberg (2011). https://doi.org/10.1007/978-3-642-23059-2_22
4. Bottrighi, A., Canensi, L., Leonardi, G., Montani, S., Terenziani, P.: Trace retrieval for business process operational support. Expert Syst. Appl. **55**, 212–221 (2016)
5. Cuevas-Vicenttín, V., Dey, S.C., Wang, M.L.Y., Song, T., Ludäscher, B.: Modeling and querying scientific workflow provenance in the D-OPM. In: 2012 SC Companion, pp. 119–128. IEEE Computer Society (2012)
6. Deutch, D., Milo, T.: TOP-K projection queries for probabilistic business processes. In: ICDT 2009, ACM International Conference Proceeding Series, vol. 361, pp. 239–251. ACM (2009)
7. van Dongen, B.: BPI Challenge Dataset (2017). https://doi.org/10.4121/uuid:5f3067df-f10b-45da-b98b-86ae4c7a310b
8. Esser, S.: Using graph data structures for event logs. Capita selecta research project., Eindhoven University of Technology (2019). https://doi.org/10.5281/zenodo.3333831
9. Francis, N., et al.: Cypher: an evolving query language for property graphs. In: Management of Data, pp. 1433–1445. ACM (2018)
10. Gonzalez Lopez de Murillas, E.: Process mining on databases: extracting event data from real-life data sources. Ph.D. thesis, Department of Mathematics and Computer Science, proefschrift, February 2019
11. Huang, X., Bao, Z., Davidson, S.B., Milo, T., Yuan, X.: Answering regular path queries on workflow provenance. In: ICDE 2015, pp. 375–386. IEEE Computer Society (2015)
12. Jans, M., Soffer, P.: From relational database to event log: decisions with quality impact. In: Teniente, E., Weidlich, M. (eds.) BPM 2017. LNBIP, vol. 308, pp. 588–599. Springer, Cham (2018). https://doi.org/10.1007/978-3-319-74030-0_46
13. Liu, D., Pedrinaci, C., Domingue, J.: Semantic enabled complex event language for business process monitoring. In: 4th International Workshop on Semantic Business Process Management, pp. 31–34 (2009)
14. Lu, X., Nagelkerke, M., van de Wiel, D., Fahland, D.: Discovering interacting artifacts from ERP systems. IEEE Trans. Serv. Comput. **8**(6), 861–873 (2015)
15. González López de Murillas, E., Reijers, H.A., van der Aalst, W.M.P.: Everything you always wanted to know about your process, but did not know how to ask. In: Dumas, M., Fantinato, M. (eds.) BPM 2016. LNBIP, vol. 281, pp. 296–309. Springer, Cham (2017). https://doi.org/10.1007/978-3-319-58457-7_22
16. de Murillas, E.G.L., Reijers, H.A., van der Aalst, W.M.P.: Connecting databases with process mining: a meta model and toolset. Softw. Syst. Model. **18**(2), 1209–1247 (2019)
17. Räim, M., Di Ciccio, C., Maggi, F.M., Mecella, M., Mendling, J.: Log-based understanding of business processes through temporal logic query checking. In: Meersman, R., et al. (eds.) OTM 2014. LNCS, vol. 8841, pp. 75–92. Springer, Heidelberg (2014). https://doi.org/10.1007/978-3-662-45563-0_5
18. Robinson, I., Webber, J., Eifrem, E.: Graph Databases. O'Reilly Media (2013)

19. Song, L., Wang, J., Wen, L., Wang, W., Tan, S., Kong, H.: Querying process models based on the temporal relations between tasks. In: EDOCW 2011, pp. 213–222. IEEE Computer Society (2011)
20. Tang, Y., Mackey, I., Su, J.: Querying workflow logs. Information 9(2), 25 (2018)

Second International Workshop on Security and Privacy-Enhanced Business Process Management (SPBP)

Second International Workshop on Security and Privacy-Enhanced Business Process Management (SPBP)

Despite the growing demand for business processes that comply with security and privacy policies, security and privacy incidents caused by erroneous workflow specifications are regrettably common. This is because business process management (BPM) and security and privacy are seldom addressed together, thereby hindering the development of trustworthy, privacy and security-compliant business processes. The goal of the second edition of the International Workshop on Security and Privacy-Enhanced Business Process Management (SPBP 2019) is to establish a venue to discuss business process privacy and integrity management. The workshop attracted ten submissions. Each paper was reviewed by three members of the Program Committee. The top five submissions are accepted for publication at the workshop and for the inclusion in the proceedings.

All the papers presented in this workshop illustrate the interplay between the BPM and security and privacy management using the emerging technologies (e.g., blockchain, smart contracts, privacy-aware and privacy enhancing technologies, distributed process mining, and other). The paper by Ladleif and Weske shows how the choreography models could impose legal contracts, their states, and legal relationships. In the next paper, Stage and Karastoyanova introduce some architecture components to ensure trust among various system stakeholders. Next, the approach for discovering roles (and their security actions) from event logs is presented by Rafiei and van der Aalst. The authors assess their proposal by estimating the effect on the privacy and accuracy. The process mining topic is continued in another paper, where Mühlberger *et al.* present the approach to retrieve the process data from the Ethereum blockchain ledger in the standardized form. Dasaklis *et al.* present a blockchain token-based framework for tracing the products across the supply chain, thus, guaranteeing auditability and establishing tamper-proof mechanism.

The workshop program also included two invited talks. The first one given by Mayer, presents a risk management framework for compliance of the regulated services. The second invited talk is given by Norta who discusses a secured and flexible blockchain-based identity authentication mechanism for the sociotechnical system applications.

We wish to thank all those who contribute to making SPBP 2019 a success: the authors who submitted papers, the members of the Program Committee who carefully reviewed and discussed the submissions, and the speakers who presented their work at the workshop. We also express our gratitude to the BPM 2019 workshop chairs for their support in preparing the workshop.

July 2019

Raimundas Matulevičius
Nicolas Mayer

Organization

Workshop Chairs

Raimundas Matulevičius   University of Tartu, Estonia
Nicolas Mayer   Luxembourg Institute of Science and Technology,
                          Luxembourg

Program Committee

Massimo Bartoletti   University of Cagliari, Italy
Achim D. Brucker   University of Sheffield, UK
Tiziana Cimoli   University of Cagliari, Italy
Claudio Di Ciccio   Vienna University of Economics and Business, Austria
Vladislav Fomin   Vytautas Magnus University and Vilnius University,
                          Lithuania
Virginia Franqueira   University of Derby, UK
Michael Henke   TU Dortmund University, Germany
Ralph Holz   University of Sydney, Australia
Christos Kalloniatis   University of the Aegean, Greece
Andreas L. Opdahl   University of Bergen, Norway
Constantinos Patsakis   University of Piraeus, Greece
Günther Pernul   Universität Regensburg, Germany
Guttorm Sindre   Norwegian University of Science and Technology, Norway
Mark Strembeck   Vienna University of Economics and Business, Austria
Xiwei Xu   CSIRO, Australia

Additional Reviewers

Wjatscheslav Baumung
Sebastian Groll
Tan Guerpinar
Benedikt Putz
Staros Simou
John Troumpis
Shermin Voshmgir

Keynote Abstracts

A Risk Management Framework for Compliance of Regulated Services

Nicolas Mayer

Luxembourg Institute of Science and Technology,
5, avenue des Hauts-Fourneaux, 4362 Esch-sur-Alzette, Luxembourg
nicolas.mayer@list.lu

Keywords: Information security • Regulatory framework • Regtech

A strong emphasis is placed today on the security of business processes and on the management of information security risks. This tendency can be seen in numerous regulations imposing a risk-based approach for some processes of critical sectors. In the Telecommunications sector, the EU Directive 2009/140/EC introduces Article 13a about security and integrity of networks and services. This article states that Member States shall ensure that providers of public communications networks "take appropriate technical and organizational measures to appropriately manage the risks posed to security of networks and services". The same approach applies for processes managed by the so-called operators of essential services as defined in the EU Directive 2016/1148.

As part of the adoption of these two EU Directives at the national level in Luxembourg, we have developed a project aiming at adapting and facilitating security risk management in regulated sectors. To do so, the project is composed of two parts. The first one consists in the development of a model-based approach and a tool to support the adoption of these regulations by regulated entities at the national level. We have especially extended ArchiMate, an Enterprise Architecture modelling language, with the appropriate concepts coming from the risk management domain. We have then integrated all of the different models into TISRIM, a risk management tool developed in-house. TISRIM is the tool recommended to the regulated entities of the Telecommunications sector by our National Regulatory Authority (NRA) and is currently extended to support additional regulations and standards.

The second one is the development of a framework to analyse the data collected by the NRA through this approach. The framework is currently composed of 10 measurements for regulated entities and 11 measurements to analyse the sector as a whole. This set of measurements is currently extended to take into account systemic risks, i.e. the cascading effect of security risks caused by dependencies between processes.

Acknowledgements. Supported by the National Research Fund (FNR), Luxembourg, and the Luxembourg Regulatory Institute, and financed by the RegTech4ILR project (PUBLIC2-17/IS/11816300).

Secured and Flexible Blockchain-Based Non-governmental Identity-Authentication for Sociotechnical Systems Applications

Alex Norta

Blockchain Technology Group, Tallinn University of Technology,
12618 Akadeemia tee 15A, Tallinn, Estonia
alex.norta.phd@ieee.org

In his talk, the author aims to give an overview of his curious way into security research that culminates in experiencing the Estonian eID system with all its pros and cons. Realizing that government-based identity authentication is potentially a threat to the freedoms of individual citizens, the keynote speech focuses on ongoing research about the non-governmental blockchain-based Authcoin system that is developed formally using Colored Petri Nets (CPN) and further security checked with a set of security risk-oriented patterns (SRP). The initial formal model of Authcoin facilitates the detection and elimination of design flaws, missing specifications as well as security- and privacy issues. The additional risk- and threat analysis based on the Information Systems Security Risk Management (ISSRM) domain model, we perform on the formal CPN models of the protocol. The identified risks are mitigated by applying security risk patterns (SRP) to the formal model of the Authcoin protocol. SRPs are a means to mitigate common security- and privacy risks in a business-process context by applying thoroughly tested and proven best-practice solutions. Thus, by applying such a security test on the untypical domain of the highly distributed CPN-formalized Authcoin protocol, we perform a stress test for the ISSRM and existing set of SRPs that yields limitations, open issues and scope for future work. Since Authcoin is implemented as a first feasibility prototype with the blockchain-based Qtum smart-contracts system for which Alex wrote the ICO-whitepaper, he presents also the planned technical realization path for Authcoin.

A Legal Interpretation of Choreography Models

Jan Ladleif$^{(\boxtimes)}$ and Mathias Weske

Hasso Plattner Institute, University of Potsdam, Potsdam, Germany
{jan.ladleif,mathias.weske}@hpi.de

Abstract. Model-driven smart contract development approaches are gaining in importance since one of the most popular realizations, blockchain-based smart contracts, are prone to coding errors. However, these modeling approaches predominantly focus on operational aspects of smart contracts, neglecting the legal perspective as manifested by deontic concepts such as obligations or permissions. In this paper, we explore an approach at connecting existing models to Legal Ontologies (LOs) on the example of choreography models, effectively interpreting them as legal contracts. We show how the execution of a choreography imposes sequences of legal states, and discuss consequences and limitations.

Keywords: Smart contracts · Choreography · Legal ontologies

1 Introduction

The core idea of smart contracts is that parts of legal contracts are operational and can be automated [5]. Especially since second-generation blockchain platforms have emerged, blockchain-based smart contract realizations are gaining popularity. At the same time, model-driven approaches to smart contract development are being explored, an example being Business Process Model and Notation (BPMN) choreography diagrams [15,20]. Abstracting from internal orchestration details of each participant, choreography models specify their interactions on an interface level and are regarded to represent "a type of business contract between two or more organizations" [18, p. 315]. This has elevated choreography diagrams to be a prime contender for smart contract modeling [16].

However, there is a foundational debate about the role of smart contracts in law, and on which level smart contracts can replace or augment legal contracts [7,14]. Apart from legislation issues, the core problem is that current smart contract realizations and modeling methods ostensibly lack certain aspects that give distinction to legal contracts—namely a normative layer specifying legal relations like obligations and a direct link to laws and regulations. Choreography models are no exception and fail to explicitly consider legal aspects.

In this paper, we propose a novel approach to solving these problems. Instead of extending choreography diagrams with additional domain-specific elements for legal concepts, we introduce deduction rules to derive legal relations from existing models. To this end, we create a conceptual link between choreography models and Legal Ontologies (LOs) by identifying patterns that imply certain legal

© Springer Nature Switzerland AG 2019
C. Di Francescomarino et al. (Eds.): BPM 2019 Workshops, LNBIP 362, pp. 651–663, 2019.
https://doi.org/10.1007/978-3-030-37453-2_52

relations between the participants. In a sense, our approach establishes an alternative interpretation of choreography diagrams, which may serve to broaden the general understanding as to their role in Business Process Management (BPM) and also hint at future research directions towards the specification of legal smart contracts.

We will use a simple rental agreement as an example throughout the paper. Figure 1 shows a Business Process Model and Notation (BPMN) choreography diagram specifying the interactions between a landlord and a tenant. In two first steps, the tenant has to advance some rent and pay a bond, two standard practices in rental law. After that, the tenant regularly pays their rent until they decide to terminate the rental agreement. The landlord at this point has the choice to either demand the bond for repairs on the property, or release the bond back to the tenant.

The paper is structured as follows. We first introduce the reader to legal ontologies in Sect. 2. We then provide a formal specification of choreographies and give an execution semantics in Sect. 3. Using the formal specification, we show how a set of legal positions can be deduced from execution states of a choreography in Sect. 4. We discuss our results in Sect. 5, before comparing them to related work in Sect. 6. We conclude in Sect. 7.

2   Legal Ontologies

LOs aim at the formal and structured specification of legal concepts [4]. The focus of LOs might vary considerably depending on their purpose—from multi-level knowledge-base approaches such as the Legal Knowledge Interchange Format (LKIF) Core ontology [10] to more specialized approaches such as the Unified Foundational Ontology for Legal Relations (UFO-L) [9], which provides a take on bilateral legal relations. LOs are interesting from a model-driven smart contract development perspective as they present a possible solution to a central issue— i.e., how the legal implications of a contract can be lifted from their traditionally prosaic form to a formal, machine-readable form.

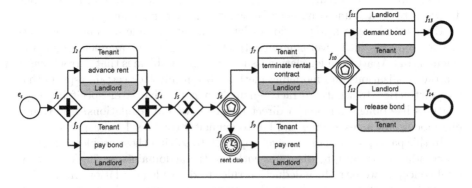

Fig. 1. Example choreography describing the interactions during a tenancy

In the context of this paper, we only consider basic legal relations and do not require a fully-specified LO. The formalization of legal relations in UFO-L and some other ontologies is based on foundational research in law, notably the seminal categorizations of norms and powers by Hohfeld [11]. Figure 2 shows the four basic norms of conduct and their interrelation as identified by Hohfeld; namely rights, obligations, no-rights and permissions. Hohfeld states that there are two fundamental interrelations: correlative and opposite jural positions.

An obligation $O_{A,B}(\phi)$ of A in face of B to perform an action ϕ is a legal position and correlates with the right $R_{B,A}(\phi)$ of B that A performs said action. For example, an obligation/right-relation

$$O_{\text{Tenant,Landlord}}(\text{pay rent}) \xleftrightarrow{\text{correlative}} R_{\text{Landlord,Tenant}}(\text{pay rent})$$

specifies that the tenant has the obligation to pay rent to the landlord, and correlates with the right of the landlord to receive rent from the tenant. Similarly, a permission $P_{A,B}(\phi)$ of A to an action ϕ against B correlates with the no-right $N_{B,A}(\phi)$ of B to demand A to perform ϕ—it is the choice of A. For example, if the tenant has the permission to terminate the rental contract, the landlord may not force them to do so. Additionally, UFO-L adopts the concept of liberties [9]. A liberty $L_{A,B}(\phi) \equiv P_{A,B}(\phi) \wedge P_{A,B}(\neg\phi)$ of A in face of B consists of both the permissions of A to perform and to omit to perform action ϕ.

Lastly, there are four basic norms of power: powers, liabilities, disabilities and immunities [11]. Norms of power concern the alteration and creation of legal positions and are correlative as well: If A has power over a legal position in face of B, then B has a correlative liability. For example, landlords usually have the power to evict a tenant in case of gross negligence, allowing them to effectively extinguish the legal relationship. Similarly, if A has a disability to change a legal position in face of B, then B has an immunity against A.

3   Choreography Modeling and Semantics

A choreography describes the interactions between different participants towards a common business goal, for example in the course of a contract [21]. They capture timing constraints or causal dependencies between actions and choices. In this section, we provide a generic formal definition of choreography models

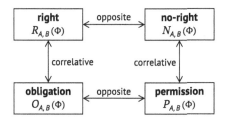

Fig. 2. Hohfeld's norms of conduct and the jural relationships between them

inspired by the BPMN choreography diagram standard [18]. We then provide a formal execution semantics built on a labeled transition system. We later connect these models to concrete legal positions (see Sect. 4).

3.1 Formal Definition

Choreographies are defined as follows:

Definition 1 (Business Process Choreography). *A choreography is a tuple* $q = (T, E, G, P, e_s, \delta, \iota, \rho)$ *with*

- *T a set of choreography tasks,*
- *E a set of events,*
- *G a set of gateways partitioned into disjunct sets G^X, G^E and G^P of exclusive, event-based and parallel gateways, respectively,*
- *P a set of participant roles,*
- *$e_s \in E$ the singular initial start event,*
- *$\delta \subseteq (T \cup E \cup G) \times (T \cup E \cup G)$ a sequence flow relation,*
- *$\iota, \rho : T \to P$ a function assigning an initiating (ι) as well as a responding (ρ) participant to each choreography task.*

For convenience reasons, let $\bullet f := \{(e, f) \mid (e, f) \in \delta\}$ be the incoming sequence flows and let $f\bullet := \{(f, g) \mid (f, g) \in \delta\}$ be the outgoing sequence flows of a flow element $f \in T \cup E \cup G$.

As is apparent from our definition, we do not cover all aspects of choreographies and BPMN choreography diagrams in particular, but rather aim for a minimal baseline to illustrate our approach. For one, we fix choreography tasks to one message only, and do not consider hierarchies as the ones imposed by sub-choreographies. We do not model data as provided through messages and the decision logic on which exclusive gateways operate. For simplicity reasons, we also assume that all choreography models are well-formed as well as sound [21, Ch. 6], and that all structural constraints posed in the standard are fulfilled [18].

3.2 Execution Semantics

The execution semantics of choreography models are defined in terms of a labeled transition system akin to the structural operational semantics approach by Plotkin [19].

Definition 2 (Transition System). *A transition system for a choreography* $q = (T, E, G, P, e_s, \delta, \iota, \rho)$ *is a tuple* $(S, A, \to, s_0, F)$ *with*

- *a set of execution states $S := 2^\delta$, i.e., all markings of the sequence flows,*
- *a set of actions $A := T$ equivalent to the set of tasks,*
- *a set of transitions $\to \subseteq S \times (A \cup \{\epsilon\}) \times S$,*
- *an initial execution state $s_0 := e_s\bullet$, and*
- *final execution states $F := \{\emptyset\}$ consisting entirely of the empty set.*

The transition system specifies the execution states $\mathcal{S}$ of a choreography and transitions between them. An execution state $s \in \mathcal{S}$ is a marking of the sequence flows. A marking in this context is an arbitrary subset of sequence flows assumed to have a kind of token each. Transitions between execution states are triggered by actions $\mathcal{A}$, which are restricted to the sending of messages to execute choreography tasks. Additionally, transitions might not depend on an action, but may be triggered automatically (ϵ). A choreography starts in an initial execution state s_0 in which each outgoing sequence flow of the singular start event e_s is marked, and ends when no markings remain.

A flow element may only be executed once it is enabled in a specific execution state, which depends on the markings:

Definition 3 (Enablement). *Let* $q = (T, E, G, P, e_s, \delta, \iota, \rho)$ *be a choreography with an accompanying transition system* $(\mathcal{S}, \mathcal{A}, \rightarrow, s_0, F)$. *For all execution states* $s \in \mathcal{S}$ *and flow elements* $f \in T \cup E \cup G$ *enablement is defined as follows:*

$$\mathsf{enabled}_s(f) :\Longleftrightarrow \underbrace{\left(f \in G^P \wedge \bullet f \subseteq s\right)}_{\text{parallel gateways}} \vee \underbrace{\left(f \in (T \cup E \cup G^X \cup G^E) \wedge \bullet f \cap s \neq \emptyset\right)}_{\text{other elements}}$$

Parallel gateways are enabled as soon as all incoming sequence flows are marked (join semantics), whereas other elements are enabled when an arbitrary incoming sequence flow is marked. Finally, we define a set of transition rules:

Definition 4 (Transition Rules). *Let* $q = (T, E, G, P, e_s, \delta, \iota, \rho)$ *be a choreography with an accompanying transition system* $(\mathcal{S}, \mathcal{A}, \rightarrow, s_0, F)$. *Then, for all execution states* $s \in \mathcal{S}$, *the following transition rules apply:*

(i) par. gateways, events

$$\forall f \in G^P \cup E :$$
$$\frac{\mathsf{enabled}_s(f)}{s \xrightarrow{\epsilon} s \setminus \bullet f \cup f \bullet}$$

(ii) task, direct

$$\forall f \in T :$$
$$\frac{\mathsf{enabled}_s(f)}{s \xrightarrow{f} s \setminus \bullet f \cup f \bullet}$$

(iii) exclusive gateway

$$\forall f \in G^X : \forall d \in f \bullet :$$
$$\frac{\mathsf{enabled}_s(f)}{s \xrightarrow{\epsilon} s \setminus \bullet f \cup \{d\}}$$

(iv) task after event-based gateway

$$\forall e \in G^E : \forall (e, f) \in e \bullet :$$
$$\frac{f \in T \wedge \mathsf{enabled}_s(e)}{s \xrightarrow{f} s \setminus \bullet e \cup f \bullet}$$

(v) event after event-based gateway

$$\forall e \in G^E : \forall (e, f) \in e \bullet :$$
$$\frac{f \in E \wedge \mathsf{enabled}_s(e)}{s \xrightarrow{\epsilon} s \setminus \bullet e \cup f \bullet}$$

In short, the transition rules specify the token propagation through the choreography model. In rule (i), all tokens from the incoming sequence flows are consumed and one token is produced for each outgoing sequence flow, respectively. The other rules contain variations for different elements, e.g., rule (ii) applies to tasks that are enabled directly and can be executed by performing the respective action. Rule (iii) places a token on one outgoing sequence flow for exclusive gateways, modeling different execution paths. Rules (iv) and (v) specify the execution semantics of event-based gateways, for which exactly one of the following events or tasks is executed.

4 Deduction of a Legal State

The execution states of a choreography model specify which actions may or must be performed. In a sense, this already comes close to legal positions; for example, if a participant must send a message to advance the execution state, they have the obligation to perform the analogue action. In the following, we will introduce patterns in choreography models that imply certain legal positions. We assume that we are targeting an arbitrary choreography q in an execution state s.

4.1 Patterns of Legal Positions

We identified a set of three modeling patterns based on the core legal positions of rights/obligations, permissions/no-rights and liberties. Figure 3 shows these patterns, with circles illustrating the required marking for the pattern to apply.

Obligation. Obligations are modeled by choreography tasks in unconstrained sequence flow (see Fig. 3a). A task $t \in T$ represents an action performed by the sender $\iota(t)$ towards the respondent $\rho(t)$—i.e., A has to send a message to B to advance the choreography. In legal terms, this corresponds to an obligation of A to perform the associated action towards B, or formally $O_{\iota(t),\rho(t)}(\text{action})$.

In the rental agreement example (see Fig. 1), this pattern appears twice right after the parallel gateway f_1. The tenant has both the obligation to advance the rent (f_2), as well as the obligation to pay the bond (f_3).

Permission. A permission to an action can be modeled using an event-based gateway (see Fig. 3b). The action follows the gateway in form of a choreography task $t \in T$. The core aspect of a permission, i.e., the absence of an opposite obligation, is established by an alternative flow through an arbitrary intermediate catch event. Formally, this is equivalent to $P_{\iota(t),\rho(t)}(\text{action})$.

In the tenancy agreement example, a permission is modeled following the event-based gateway f_6 (see Fig. 1). The tenant may terminate the rental contract (f_7), but a timer intermediate catch event (f_8) regularly triggers the rent payment (f_9), which in turn manifests an obligation per the obligation pattern.

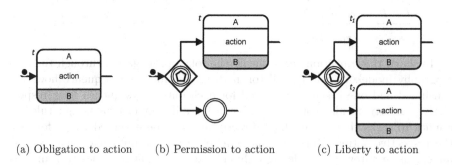

(a) Obligation to action   (b) Permission to action   (c) Liberty to action

Fig. 3. Patterns of legal positions

We did not choose the exclusive gateway for this pattern as permissions—and liberties, for that matter—entail a choice by a participant, which is represented by event-based gateways in choreography diagrams. Exclusive gateways, on the other hand, model decisions based on previously communicated data known to all involved participants independently.

Liberty. A liberty is a stronger form of a permission in that the owner has both the permission to perform an action as well as a permission to omit performing the action, resulting in stronger correlative positions. This relation can be modeled with an event-based gateway followed by two choreography tasks $t_1, t_2 \in T$ representing the two opposing permissions. Note that the choice of t_1 and t_2 is arbitrary. Figure 3c shows the pattern, which implies a liberty $L_{\iota(t_1),\rho(t_1)}(\text{action})$.

In contrast to the other patterns, the liberty pattern requires a semantic match on top of the structural match. The actions specified in t_1 and t_2 need to specify opposite choices, which depends on the interpretation of the natural language label. In our example (see Fig. 1), this pattern appears after the event-based gateway f_{10}. The landlord may choose between demanding (f_{11}) and releasing (f_{12}) the bond, which are somewhat opposite choices. Thus, a liberty of the landlord in face of the tenant to demand the bond is implied.

4.2   Legal States

When a choreography is executed using the transition system introduced in Sect. 3.2, it runs through a sequence of execution states from instantiation to termination. The execution states are markings of the sequence flows. For example, a sequence of execution states in the example choreography could be

$$\{(e_s, f_1)\} = s_0 \xrightarrow{\epsilon(f_1)} s_1 \xrightarrow{f_2} s_2 \xrightarrow{f_3} s_3 \xrightarrow{\epsilon(f_4)} s_4 \xrightarrow{\epsilon(f_5)} s_5 \xrightarrow{f_7} s_6 \xrightarrow{f_{12}} s_7 = \emptyset \in F,$$

i.e., starting in the initial state s_0, the tenant first advances rent (f_2), then pays the bond (f_3) and terminates the rental contract (f_7), following which the landlord releases the bond (f_{12}). There are three ϵ-transitions that are responsible for forking the sequence flow at the parallel gateway f_1, joining it at the parallel gateway f_4 as well as merging it at the exclusive gateway f_5. In the final state s_7, there are no markings left, indicating it is final.

Using the patterns devised earlier in this section, an analogous *legal state* can be derived from every execution state:

Definition 5 (Legal State). *Let q be a choreography with an accompanying transition system $(S, A, \rightarrow, s_0, F)$ and $s \in S$ be an arbitrary execution state. Then the analogous legal state $L(s)$ is defined as a set containing all legal positions deducible from the execution state s.*

Figure 4 shows the deduction of the legal states for two execution states as introduced in the trace given above. The execution state s_0 (see Fig. 4a) contains only one marking which does not imply a legal positions, resulting in an empty legal state $L(s_0)$. Conversely, s_1 (see Fig. 4b) implies both

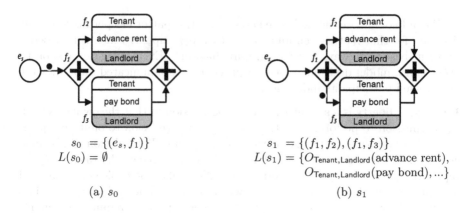

$$s_0 = \{(e_s, f_1)\}$$
$$L(s_0) = \emptyset$$

(a) s_0

$$s_1 = \{(f_1, f_2), (f_1, f_3)\}$$
$$L(s_1) = \{O_{\text{Tenant,Landlord}}(\text{advance rent}),$$
$$O_{\text{Tenant,Landlord}}(\text{pay bond}), ...\}$$

(b) s_1

Fig. 4. Deduction of legal states from execution states

the positions $O_{\text{Tenant,Landlord}}(\text{advance rent})$ and $O_{\text{Tenant,Landlord}}(\text{pay bond})$. This results in an non-empty legal state $L(s_1)$ containing all correlative positions as well.

In some instances, the legal state derived from an execution state might be empty. That is, (i) there are no markings left or (ii) the markings do not imply any legal positions. In the former case, the execution state is final and signifies that the choreography has finished. In a legal sense, this is equivalent to the contract having been terminated or canceled, i.e., no parties are affected by any legal relation stemming from that specific contract anymore. In the second case, intermediate execution states might imply empty legal states. For example, in the trace given above, the state s_0 enables the parallel gateway f_1, but does not yet yield any legal position. For a legal interpretation, these states are irrelevant since they are purely operational and are immediately left using ϵ-transitions.

Figure 5 shows the resulting legal state diagram of the tenancy agreement example (see Fig. 1), with irrelevant empty states omitted. For simplicity reasons, correlative positions are also skipped. The graph starts in state $L(s_1)$ (see Fig. 4b) in which the tenant has both the obligation to pay the bond to the landlord, and the obligation to advance their rent. Both of these tasks may be executed in any order, leading to different legal states. The state transitions continue until we reach a marking in which no element is enabled anymore.

5   Discussion

The connection between execution states and legal states constitutes a novel interpretation of choreography models. Instead of being a pure coordination protocol, the choreography model constitutes a contract with a legal meaning. There are open questions, though, which we will discuss in this section.

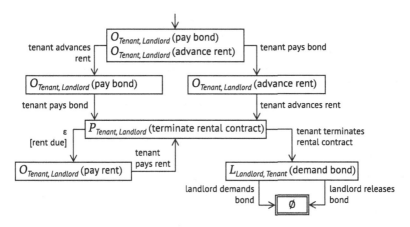

Fig. 5. Legal state diagram of the tenancy agreement example

5.1 Norms of Power

From a foundational perspective, we only target so-called norms of conduct in our deduction, i.e., rights, obligations, permissions and no-rights [9]. That is, the legal positions deduced from the models are behavioral of nature, which directly follows from the fact that the models themselves focus on the behavior of participants. Nevertheless, norms of power—namely powers, liabilities, disabilities and immunities—are inherently present as well.

In general, norms of power are considered to exist on a secondary level above norms of conduct. They are related to the creation and change of other legal positions. Thus, in our case, norms of power have to be considered when the legal state is changed during a state transition. This is the case in two situations: when an action is performed by a participant (i.e., a message is sent), or an ϵ-transition is fired during execution.

In the first case, the power is inherent in the participant performing the action. For example, if the tenant has the obligation to pay the bond, they also have the power to change the legal state by doing so. The act of delivering on the obligation "removes" the obligation and correlative positions from the state, and creates new legal positions based on the subsequent sequence flow. For example, if the tenant chooses to terminate the rental contract, a liberty of the landlord to demand the bond is created (see Fig. 5). The legal relations created are not limited to the parties of the original choreography tasks, but may also include third parties depending on the participants of the following elements.

In the second case, i.e., ϵ-transitions, it is not immediately clear who the power inheres in. These transitions are supposed to happen automatically, and are acted out by the smart contract itself. The contract thus becomes an acting entity on its own, with the power to change the legal state in ways previously agreed upon in the model, underlining its contractual character.

Additionally, there may be further powers and liabilities defined in legal contracts. For example, tenancy agreements usually contain a clause that gives landlords the power to cancel the agreement in case of gross negligence of the tenant, e.g., involving damage to the property. These powers and liabilities often relate to specific cases which stray off the happy path. Choreography models as per our definition—and, as a direct consequence, BPMN choreography diagrams as well—lack the ability to specify such powers. In future work, we want to evaluate whether this limitation may only be alleviated by adding entirely new model elements, or whether sophisticated deduction rules might suffice.

5.2 Coverage of Laws and Regulations

The legal states derived from the execution states are only incomplete snapshots of reality. In particular, jurisdictional aspects are not covered. These include laws, regulations, and customs that govern how contracts are drafted and interpreted. For example, countries might have laws protecting tenants from being evicted within some period of time, while others do not. These governing rules are not explicitly represented in the choreography model.

In a wider sense, this issue is related to the area of compliance checking. In compliance checking, processes are checked for their adherence to compliance rules—i.e., regulations, policies and other constraints that are in effect, usually within an organization. These compliance rules can be interpreted as organizational norms, which describe actual rules governing the execution of intra-organizational workflows. Further, Governatori et al. have shown that contractual obligations can be formalized and the compliance of orchestration processes to these obligations can be automatically checked [8].

This concept could be applied to governing rules as well. A contract specification must adhere to laws and regulations within the target jurisdiction. Thus, a contract specified as a choreography model would have to be checked against a number of formally specified rules that are required to apply.

5.3 Limitation of Expressiveness

As discussed above, choreography models do not cover all elements of LOs and vice-versa. The issue exists in both directions: there are legal positions like powers and immunities which do not have an explicit counterpart in choreography models, and modeling constructs like event-based gateways with more than two involved participants which are not directly translated to legal positions by our deduction rules. These issues have multiple reasons. For one, the deduction rules defined in this paper are by no means exhaustive. Further rules could be defined, based on general or domain-specific understandings of certain elements.

The range of different legal relations deducible from the model is limited by the modeling standard as well. For example, no-rights can not be expressed directly—i.e., we have no element that explicitly models that a participant may *not* perform an action. New modeling elements would have to be introduced to represent such legal relations. These are not the only aspects in which choreography diagrams seem to be lacking when it comes to contract specifications [1], evidently warranting a more general discussion of the standard's shortcomings.

6   Related Work

Historically, deriving semantic models of legal relations from existing artifacts is mainly focused on legal texts like laws and policies. For example, Breaux et al. provide a methodology to extract concrete rights and obligations from legal text through semantic parameterization [3]. In the two-step approach, the legal text statements are first transcribed into a set of restricted natural language statements, which are then transformed into formal semantic models.

Similar research in the area of model-driven smart contract development is sparse, however. Work has been done towards connecting behavioral models and LOs, somewhat enhancing their legal expressiveness: Kabilan uses BPMN collaboration diagrams and annotates message and sequence flows with legal concepts from their own ontology [12]. These flows thus create or extinguish legal positions. In another paper, Kabilan and Johannesson conversely augment a LO with behavioral components [13]. However, these approaches include manual annotation steps, which the proposal in this paper avoids.

Whereas we motivate deducing legal relations from choreography models, there has been previous research in the opposite direction from the contract compliance area—i.e., on how to generate, augment or check models from contract specifications. For example, Milosevic et al. provide a method to derive compliant collaboration processes from contracts through several processing steps including formal contract specifications [17]. Goedertier and Vanthienen focus on temporal deontic assignments extracted from contracts to derive collaborations [6]. Berry and Milosevic use business contract specifications to augment coordination models with enactment and monitoring constraints [2]. These approaches might give valuable insights into future extensions of our approach.

7   Conclusion

The advent of blockchain-based smart contracts has expedited research towards full specifications of contracts, though mostly focused on behavioral and disregarding legal aspects. In this paper, we proposed a method of covering the latter aspects as well on the example of choreography models. To this end, we introduced a set of modeling patterns that correspond to concrete legal concepts founded in LOs, providing deduction rules to transform a state of the choreography execution to a set of legal relations. We thus gave a novel interpretation of choreography models, taking a step back from purely behavioral interface specifications towards a more holistic legal picture which may extend the understanding of choreographies in general. We also gave pointers towards future work, especially regarding the completeness and legal coverage of our approach.

References

1. Adamo, G., Borgo, S., Francescomarino, C.D., Ghidini, C., Rospocher, M.: BPMN 2.0 choreography language: interface or business contract? In: Proceedings of the Joint Ontology Workshops 2017 (2017)
2. Berry, A., Milosevic, Z.: Extending choreography with business contract constraints. Int. J. Coop. Inf. Syst. **14**, 131–179 (2005)
3. Breaux, T.D., Vail, M.W., Antón, A.I.: Towards regulatory compliance: extracting rights and obligations to align requirements with regulations. In: IEEE International Conference on Requirements Engineering (RE), pp. 46–55 (2006). https://doi.org/10.1109/RE.2006.68
4. Casellas, N.: Legal Ontology Engineering: Methodologies, Modelling Trends, and the Ontology of Professional Judicial Knowledge, Law, Governance and Technology, vol. 3. Springer, Cham (2011). https://doi.org/10.1007/978-94-007-1497-7
5. Clack, C.D., Bakshi, V.A., Braine, L.: Smart contract templates: foundations, design landscape and research directions. CoRR (2016). http://arxiv.org/abs/1608.00771
6. Goedertier, S., Vanthienen, J.: Designing compliant business processes with obligations and permissions. In: Eder, J., Dustdar, S. (eds.) BPM 2006. LNCS, vol. 4103, pp. 5–14. Springer, Heidelberg (2006). https://doi.org/10.1007/11837862_2
7. Governatori, G., Idelberger, F., Milosevic, Z., Riveret, R., Sartor, G., Xu, X.: On legal contracts, imperative and declarative smart contracts, and blockchain systems. Artif. Intell. Law **26**(4), 377–409 (2018). https://doi.org/10.1007/s10506-018-9223-3. ISSN 1572-8382
8. Governatori, G., Milosevic, Z., Sadiq, S.: Compliance checking between business processes and business contracts. In: IEEE International Enterprise Distributed Object Computing Conference, EDOC, pp. 221–232 (2006)
9. Griffo, C., Almeida, J.P.A., Guizzardi, G.: Conceptual modeling of legal relations. In: Trujillo, J.C., et al. (eds.) ER 2018. LNCS, vol. 11157, pp. 169–183. Springer, Cham (2018). https://doi.org/10.1007/978-3-030-00847-5_14
10. Hoekstra, R., Breuker, J., Di Bello, M., Boer, A.: The LKIF core ontology of basic legal concepts. Int. J. High Perform. Comput. Appl. **321**, 43–63 (2007)
11. Hohfeld, W.N.: Fundamental legal conceptions as applied in judicial reasoning. Yale Law J. **26**(8), 710–770 (1917)
12. Kabilan, V.: Contract workflow model patterns using BPMN. In: Proceedings of CAiSE 2005 Workshops, CEUR-WS.org, vol. 363 (2005)
13. Kabilan, V., Johannesson, P.: Semantic representation of contract knowledge using multi/tier ontology. In: First International Conference on Semantic Web and Databases, pp. 378–397, CEUR-WS.org (2003)
14. Kõlvart, M., Poola, M., Rull, A.: Smart contracts. In: Kerikmäe, T., Rull, A. (eds.) The Future of Law and eTechnologies, pp. 133–147. Springer, Cham (2016). https://doi.org/10.1007/978-3-319-26896-5_7
15. Ladleif, J., Weske, M., Weber, I.: Modeling and enforcing blockchain-based choreographies. In: Hildebrandt, T., van Dongen, B.F., Röglinger, M., Mendling, J. (eds.) BPM 2019. LNCS, vol. 11675, pp. 69–85. Springer, Cham (2019). https://doi.org/10.1007/978-3-030-26619-6_7
16. Mendling, J., Weber, I., et al.: Blockchains for business process management - challenges and opportunities. ACM Trans. Manag. Inf. Syst. (TMIS) **9**(1), 4:1–4:16 (2018). https://doi.org/10.1145/3183367. ISSN 2158-656X

17. Milosevic, Z., Sadiq, S., Orlowska, M.: Translating business contract into compliant business processes. In: IEEE International Enterprise Distributed Object Computing Conference, EDOC, pp. 211–220 (2006)
18. OMG: Business Process Model and Notation (BPMN), Version 2.0.2 (December 2013). http://www.omg.org/spec/BPMN/2.0.2/
19. Plotkin, G.D.: A Structural Approach to Operational Semantics. Aarhus University Denmark, Computer Science Department (1981)
20. Weber, I., Xu, X., Riveret, R., Governatori, G., Ponomarev, A., Mendling, J.: Untrusted business process monitoring and execution using blockchain. In: La Rosa, M., Loos, P., Pastor, O. (eds.) BPM 2016. LNCS, vol. 9850, pp. 329–347. Springer, Cham (2016). https://doi.org/10.1007/978-3-319-45348-4_19
21. Weske, M.: Business Process Management, 2nd edn. Springer, Cham (2012). https://doi.org/10.1007/978-3-642-28616-2

Provenance Holder: Bringing Provenance, Reproducibility and Trust to Flexible Scientific Workflows and Choreographies

Ludwig Stage$^{(\boxtimes)}$ and Dimka Karastoyanova

Information Systems Group, University of Groningen, Groningen, The Netherlands
bpm@ludwig-stage.de, d.karastoyanova@rug.nl

Abstract. Process adaptation has been in the focus of the BPM community also in the scope of interdisciplinary research on the interface with eScience. In eScience, scientists, software developers and analysts across a range of disciplines model, create and execute different types of data-driven experiments in an exploratory, trial-and-error style. Supporting this style of experimenting by software systems has been enabled in existing work by interactive workflow management systems. The main principles and techniques used are related to the adaptation of workflows and choreographies, as well as to systems integration and software architecture improvement. With the increasing need for collaboration in scientific explorations and adaptation remaining an essential part of the techniques used, the trust in the reproducibility of scientific research gains a most significant importance. To ensure trust in the provenance and reproducibility of collaborative and adaptable scientific experiments, in this paper we introduce a system architecture with focus on a new component that is responsible for ensuring trust in a generic and non-intrusive manner, independent of different process-aware technologies and storage platforms, including blockchain. We contribute the architecture of the Provenance Holder, its components, functionality and two main operations: recording provenance data and retrieving the trusted provenance information. We incorporate the Provenance Holder into the architecture of the interactive ChorSystem and identify the potential directions for future research.

Keywords: Flexible scientific choreographies · Reproducibility · Trust · Provenance · Collaboration · Blockchain

1 Introduction

Blockchain technology has had and still has a significant impact on Business Process Management (BPM) research. Many of these impact factors have been identified in [8]. In this paper, we focus on investigating one unexplored aspect, namely the interplay with and potential impact on the field of adaptable processes, both workflows and choreographies, in particular for the purposes of enabling trust and improving provenance and reproducibility of in-silico

© Springer Nature Switzerland AG 2019
C. Di Francescomarino et al. (Eds.): BPM 2019 Workshops, LNBIP 362, pp. 664–675, 2019.
https://doi.org/10.1007/978-3-030-37453-2_53

scientific experiments, like e.g. simulations, scientific workflows and business analytics and data scientific experiments, that need to be adapted very often. Keeping an immutable trail of transactions without the need of a trusted third party and thus enabling of collaborations between "mutually untrusted parties" [7] is surely a promise of the blockchain technology that needs to be evaluated and exploited. In our previous work [6] we have suggested an approach that uses blockchain to allow for trusted collaborations among adaptable scientific experiments. The approach comprised an architecture of an enabling system and a number of alternative realizations depending on the role blockchain platforms might play. The most prominent and promising alternative realizations we suggested are (a) the use of blockchain platforms for storing purely workflow auditing/historical information or (b) the use of blockchain platforms and smart contracts (and potential extensions) to execute adaptive workflows directly on a blockchain platform. While we deem these alternatives appropriate and worth investigating, in this work we follow a more generic approach to fulfill the requirements of scientists and analysts. We mapped these requirements to properties that an enabling system has to provide in [6] and provide the list here as well:

- provenance to enable Findable Accessible Interoperable Reusable (FAIR) results [9],
- reproducibility for Robust Accountable Reproducible Explained (RARE) experiments [4],
- trust among collaborating parties and
- adaptability of the experiments.

Provenance is an important topic in the scientific community and extensively discussed in publications like [2,5]. There are Scientific Workflow Management Systems (sWfMSs) [2], which already support provenance to a certain extent [5]. The sWfMS Apache Taverna [14], for instance, supports provenance of workflow execution by recording input and output of service calls during execution and provenance of data [1] which is done through annotating the workflow description at design time and generating origin-annotations at run time. However, changes to workflow definitions are not captured and have to be tracked by other means and reproducing changes is not at all in the focus of this or other works.

Since *changes* on workflow and choreography level are essential for collaborative scientific workflows, which represent in-silico experiments, as we have shown in e.g. [6] and [13], we need to establish a way to record information about the execution of choreographies, workflows and their activities, as well as their adaptations so that it is possible to reproduce every step taken and every adaptation made, and at the same time allows storage on a possibly untrusted medium. One prerequisite here is to ensure that all *adaptation* steps are coordinated among all participating partners (as in e.g. [13]). Given that it would be possible to rerun/reproduce choreographies on the basis of the recorded data. Note that runtime information about some activities of involved workflows or even whole parts of workflows might not be explicitly available because they are e.g. calls to service interfaces that may not be providing this information.

Despite that, the need of scientists is to be able to *retrace and reproduce* the executed flexible choreographies since they stand for the explorative way of experimenting. Therefore, in this paper we focus on introducing a system that combines these properties in a generic manner and thus allows for use of different types of technologies and realizations present and at the same time ensures that the collaboratively produced scientific results can be trusted by all participating parties. For this purpose we will mainly focus on introducing the functionality and architecture of a new component in the architecture of an overall sWfMS that brings these properties together and thus enables collaborative, adaptable and reproducible in-silico experiments. We call this new architectural component Provenance Holder (see Fig. 1).

As a basis for the work presented here we use the ChorSystem [12] that supports flexible choreographies of scientific workflows and which is based on a conventional Workflow Management Systems (WfMSs) rather than a system developed by the typical eScience software like the above mentioned Taverna [14] or Kepler[1]. Please note that the work presented here can be applied with other WfMSs as long as they support the Model-as-you-go-approach [10] and are able to provide the required provenance data (cf. Sect. 2.3).

The rest of the paper is structured as follows: in Sect. 2 we introduce the architecture of our system, the *Provenance Holder*, discuss its capabilities, properties and requirements. In Sect. 3 we identify open research questions to be addressed in future work to fully implement and evaluate the new component. We conclude the paper in Sect. 4.

2   Provenance Holder

In this section we will introduce the Provenance Holder as an extension to the ChorSystem [12] we have presented in our previously published work. The ChorSystem allows for modelling, execution and adaptation of scientific workflows and choreographies and supports the trial-and-error manner of exploration scientists prefer. The system is our realization of the Model-as-You-Go approach [13]. The ChorSystem architecture is shown in Fig. 1 and consists of (i) a modeling and monitoring environment (ChorDesigner, Transformer and Process Designer) to model collaborative experiments, generate visible interfaces of participating workflows, refine internal workflow logic of each participating workflows and serve as monitoring tool, (ii) a special middleware (ChorMiddleware) which coordinates choreography adaptations and (iii) interactive WfMSs that support the execution, adaptations and monitoring of the workflows and choreographies together with the rest of the system components. All system components are provided as services and interact with each other using a service middleware, the Enterprise Service Bus (ESB).

In the current system, the adaptation steps on the choreography level or on individual workflows are recorded by the ChorSystem and the records are merely stored into a log file, since trust and provenance requirements were not in the

---

[1] https://kepler-project.org/.

primary focus of our research. As mentioned above, in [6] we introduced two possible approaches as alternative solutions towards enabling the required properties of provenance and trusted reproducibility of collaborative adaptive workflows. The first approach suggested the use of trusted decentralised ledger technology, e.g. blockchain, as the audit trail that can be used to prove the provenance of the adaptations in collaborative in-silico experiments. The second approach was aiming at integrating the properties into an adaptive, blockchain-based Business Process Management System (BPMS)/WfMS. As a natural continuation of work on these approaches, in this paper we contribute an architecture of a system that supports the collaborative adaptations in experiments and at the same time allows for the interchangeable use of different technologies to support trust without third parties through an additional component, the Provenance Holder (see Fig. 1).

The Provenance Holder component is responsible for collecting all information necessary to ensure provenance and reproducibility of and trust in the collaborative adaptations. We argue that this solution is generic and reusable across different scenarios and is less intrusive due to the separation of concerns principle we follow. As depicted in the high-level architecture of the system in Fig. 1, the Provenance Holder is not part of any other functional component of the architecture, but rather is a separate service whose responsibility is to collect all relevant information necessary for reproducibility and provenance in such a manner that the trust in the adaptive collaborations is ensured. Externalizing the functionality needed to support the requirements we identified in a separate service opens up the research questions of what the architecture of such a component is, what functions it has and how it should interact with the rest of the components of the ChorSystem.

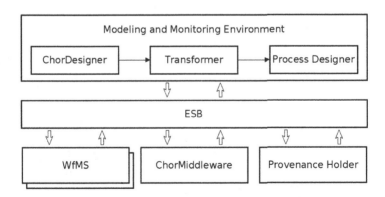

Fig. 1. Architectural overview of the ChorSystem and the Provenance Holder

The *two main operations* the Provenance Holder needs to provide on its interface are: *collect provenance data* and *retrieve provenance information*. These operations can be realized by four *interaction scenarios* carried out by its components, which we also call *methods* and explain in Sect. 2.1. The *components*

of Provenance Holder are the controller, the adapter and one or more provenance providers - they are discussed in Sect. 2.2 and shown in Fig. 2. The impact
on the components comprising the overall architecture, i.e. the Workflow Management Systems (WfMS), the ChorMiddleware and Modelling and Monitoring
Environment will be identified in Sects. 2.3 and 2.4.

2.1   Provenance Holder Methods

The Provenance Holder performs four main methods that realize in combination
the two main operations (see Fig. 2), namely collect provenance data and retrieve
provenance information. The methods are:

1. Record
2. Retrieve (provenance information)
3. Validate
4. Migrate (to new type or instance of storage)

During the execution and adaptation of workflows and choreographies the
Provenance Holder collects provenance data constantly. During normal execution of the experiments (or their runtime) only three out of the four methods
are used regularly. The fourth one, the migration, is used when there is a change
or addition of storage instances or storage technology is necessary or desired.
The data related to the experiment execution and changes also need to be
authenticated. This can be done, for instance, by the executing WfMS environment by signing input, workflow version, and output (see Sect. 2.3 for more
details). Here the input are the parameters handed over to the workflow upon
invocation, the output is the returned message or the result of the workflow, and
such information is made available on the ESB (Fig. 1) as it is the communication
backbone of this service-oriented system.

The *Record* method selects the appropriate provider components (provider a,
b, c and so on in Fig. 2) for a certain workflow type out of the available providers
and uses them to store the provenance information. For the execution of workflows, this might be input data, the executed workflow version and its output
data. The actual provenance information is paired with information about the
corresponding choreography instance, such as a reference to the corresponding
choreography instance, for later ease of retrieval. For an adaptation on workflow
or choreography level, this information might consist of the description of the
actual change performed, the new version and a reference to the preceding one.
Depending on the technology employed to implement a provider, this data is
then stored with the help of the selected provider(s).

The *Retrieve* method is used to fetch the desired provenance information
from the provider components via their interfaces. The information is identified
and retrieved using the corresponding choreography instance ID and/or workflow
IDs. The actual data retrieval is done by each provider itself and returned to
the *retrieve* method. After retrieval, the information is again validated before it
is handed over to the adapter component, i.e. the Provenance Holder interface

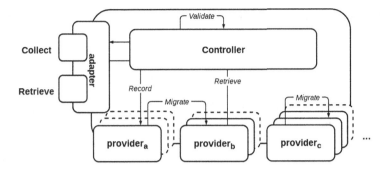

Fig. 2. Provenance Holder components and methods

implementation. The validation is used to rule out storing errors or tampering on the data storage and to guarantee the validity of the data and the freshness of the validity check. The retrieved information should then be presented to the user, in our case by the ChorDesigner, however any other system will have to have its own way of presenting the information to its users.

During *Validation* the provided signature is verified in the controller component. The signed data is validated and the signee is identified. When the record method is called, the signature gets verified before the data is "recorded". If the signature can not be verified or the signee not be identified, the information is rejected and not considered further, hence is not "recorded". The failed signature verification is also communicated to the user, i.e. the experiment modeller, for instance in the ChorDesigner. When calling the retrieve method, the provenance information is fetched from the provenance provider and then validated. If two or more providers are present in a Provenance Holder, retrieved data is not only validated but the data of the different providers and each validation status is compared to one another to identify possible discrepancies. The status of the validation is again, communicated to the user. Especially if the data storage is not situated locally (relative to the Provenance Holder) an adversary might be able to change stored data. If this is done by a participant then he is also able to produce a "valid" signature of the changed data. To be able to validate signatures, the participants' keys need to be exchanged beforehand and stored in the Provenance Holder.

The *Migrate* method is only used, if stored information has to be transferred to a new type or instance of storage, in case an addition or change of instances is desired or needed. It provides the ability to retrieve all stored provenance information from a provider component at once.

Addition of storage instances means that storage is expanded by a new instance of an already existing technology or a new instance of a not yet used storage technology, e.g. data is stored in an SQL database and now will be copied to a second one or will now also be stored in a flat file. Change of storage on the other hand means that one storage instance replaces another one. This can be done within a particular storage technology or done by replacing one technology with another, e.g. data is stored in a flat file and will now be stored in an SQL database, hence it will be migrated. The employed technology has also implications on the complexity of such a migration because of the difference in features and of their characteristics. Migrations can be triggered both automatically or manually by an administrator; the actual procedure for migration is out of the scope of this paper as related work [11] is available.

It is important to note though that the cost of this method must be considered, especially when it comes to blockchain technology where the needed information can be spread virtually over the whole ledger. The migration step might also involve purging data from the source after a successful data transfer. Here, blockchain technology might also pose additional challenges. There are implications on the cost and complexity of such a migration (see Sects. 3.3 and 3.4 for more details).

2.2 Provenance Holder Components

The Provenance Holder consists of the controller, adapter and one or more provider components (see Fig. 2).

The *adapter* is the component ensuring the integration of the Provenance Holder into an existing system. In our case, as we want to enhance the ChorSystem [12], which is a service-based one, the component will enable the communication with the ESB and has two main operations to provide. The first one is to collect all necessary provenance information while collaborative experiments are being executed and changed, whereas the second one is to retrieve the provenance information from the Provenance Holder and relay it for presentation to the user. Since the employed WfMSs publish all necessary data to the ESB and the ChorDesigner communicates with the ChorMiddleware through the ESB, the Provenance Holder adapter ensures the communication with all components of the overall infrastructure.

Data collection and retrieval, which scientists commonly call publication, however, are done differently. The Provenance Holder subscribes to the relevant messages of the WfMS, ChorDesigner and ChorMiddleware and hence collects the needed information autonomously and continuously every time data is published by these systems. Collected provenance information on the other hand is only published/returned upon request to the users, for instance by the ChorDesigner and a user friendly presentation is required. The adapter component serves the request with only the collected provenance information, though. The adapter also captures change events of both choreography and workflow, so every time a change is made or deployed it is recorded by the Provenance Holder. To enable both provenance data collection and retrieval imposes additional requirements, especially on the WfMS (see Sect. 2.3) and the Modeling and Monitoring Environment (Sect. 2.4).

Providers or *provenance providers* have certain properties to fulfil and methods to implement. To be able to provide provenance support a provider has to use appropriate technology to fulfil the requirements. The implementation complexity can be arbitrarily high and is highly dependent on the employed technology:

writing into a log file, for example, is of low complexity, whereas employing blockchain technology is more on the high end of the complexity spectrum. The needs of different workflow types also come into play when deciding which technology to use. The provider components have to implement three out of the four methods explained in Sect. 2.1, namely *record, retrieve and migrate*.

The *controller* is in charge of enabling the interaction between the adapter and the provenance providers so that the Provenance Holder can provide the provenance service operations to each workflow and choreography. The controller combines the four methods we presented already into the realization of the two operations the Provenance Holder provides. For the *collect provenance data operation* the controller receives, validates and relays the provenance information to the providers. For the *operation retrieve provenance data*, it combines the methods retrieve and validate. In both cases the validate method is a crucial step as described in the previous section. The controller has a key storage integrated to be able to verify signatures and identify participants.

2.3   Requirements on Workflow Management Systems

The Provenance Holder as introduced above requires that the WfMSs are able to sign, for example, the provenance data they produce the executed workflow version, the input parameters given at invocation and the calculated result/output of the run workflow. The time when a workflow was started/executed might also be provenance data. The signature of changes made to the workflow model is also necessary. To produce a signature, a private/public key pair has to be generated beforehand. The public key also has to be made available to the Provenance Holder. All created signatures have to be published to the ESB, otherwise the Provenance Holder is not able to consider them.

The idea is that if a workflow is made available, hence its interface is published, the public key of the workflow or WfMS owner is also made available. Every time the workflow is run, the private key to this public key is used to sign the produced provenance data. By doing so the WfMS attests that a workflow instance of a certain version of the workflow model was successfully executed with a certain input and produced a certain output as explained above.

Every WfMS or its owner is interested in producing valid and reproducible results, since he can be held accountable because he signs them and everyone can verify the signature with the published public key. The incentive to produce valid and reproducible results here is, if a not reproducible result is published or a workflow/WfMS owner repudiates a result, he loses his credibility because the signature proves otherwise.

The key exchange and participant identification has to be done before participants can engage in collaborative scientific workflows. Since the key is made public alongside the workflow, the exchange is trivial. The participant identification on the other hand is not. Here principles such as trust on first use (TOFU) or trust upon first use (TUFU) and the web of trust are best applied. An identification through a different channel or even a personal identification on e.g. a conference is also possible. Every participant has to decide himself which method suites best.

2.4   Requirements on Modeling and Monitoring Environment

There are two requirements on the Modeling and Monitoring Environment, specifically the ChorDesigner, which need to be addressed to make use of the Provenance Holder in the ChorSystem. First, provenance data and the corresponding signatures need to be displayed to the users. Especially the validity of a signature is important for a user. A signature is valid if the signee is known, i.e. the public key, and the signature can be verified with help of said public key against the provided provenance data. In terms of visualization to the user, there are two possibilities, either only showing the provenance data to the user if the signature is valid or displaying alongside information about the validity. Pros and cons for both possibilities have to be identified and considered in future. Second, changes made to the workflows and choreographies need to be signed before they are enacted or deployed. Otherwise changes can not be attributed to a participant and authenticated properly later on.

3   Discussion of Open Issues

In this section we discuss the open issues we identify and that need to be addressed separately as part of the future research work and realization of the presented generic architecture.

3.1   Provenance Information

To be able to provide reproducibility of and trust in workflows and choreographies, certain information and data has to be collected and recorded during their execution. For instance, on workflow level input parameters, output/result and the used workflow version naturally come to mind. In practice, depending on many factors, more information may be needed in order to ensure the provenance and reproducibility of the experiments and their results. We will investigate what kind of information should actually be collected and recorded, what information may be omitted, and if and to what extent data and information needs to be preprocessed or enhanced to deliver the desired outcome. Factors influencing this work will be the type of experiments to be carried out, their realization (workflow vs. choreography and the technology used), the provenance type desired, the technology used for storing the provenance information, and others.

3.2   Workflow Types

There are different kinds of workflows, with different requirements, e.g., confidentiality or computational power. These might impose additional demands on the ChorSystem and especially on the Provenance Holder and the provenance providers used. In future work, we will investigate the different types of workflows, classify the distinguishing types for the ChorSystem in particular and the Provenance Holder and identify the technology and storage needs. We will work towards providing a classification that maps the different types of experiments from different domains (business analytics, coupled simulations, data science pipelines, multi-scale and multi-physics simulations, etc.) and different technologies, paradigms and techniques towards their realization.

3.3 Complexity Implications

The available storage technologies are quite diverse, therefore switching between them or exchanging them, as the *migrate method* requires, is certainly a non-trivial task to be addressed. The complexity and multi-dimensionality of this kind of task is discussed for migrating enterprise applications to the cloud and a solution in form of a methodology is presented in [11]. The authors also emphasize on databases as part of such applications, and we consider this approach applicable here as well.

Especially when distributed ledger technology, such as blockchain, is involved, a change or switch from or to it not only brings a different feature-set but also a different set of characteristics. This difference might be explicitly desired and be the reason of the change in the first place. The actual migration might be costly though, for example because the implementation of a new provider is complex or retrieving data from one provider or initially storing a lot of data on the other might be expensive. Therefore, we will investigate in future what implications on complexity and costs might be involved, how they can be addressed and if and when they actually need to be addressed.

3.4 Blockchain Technology as Provenance Provider

Blockchain technology has characteristics that limit its applicability in our case and need to be addressed. The first characteristic is the eventual consistency property of blockchain and the second is that the access to entries is not efficient. These characteristics and issues arising from them are also briefly discussed in [3].

The eventual consistency could be addressed by, for example, adding a cache to a provider which implements the storage capabilities with blockchain. The cache holds the to-be-written data until it is written to the ledger or is part of the main chain and therefore distributed. This means on the other hand, if data is not written, or is not included in the chain, it will be resubmitted for storage until it is written.

To be able to retrieve data efficiently a lookup table could be created in the provider component. This table enables accessing data objects in a more direct manner without searching the whole ledger for individual entries. In future we will investigate what other approaches are possible and evaluate their feasibility.

Another property of blockchain technology is its immutability. This makes it impossible to remove once published data. We also will investigate if and when this is a problem and how to address it in the scope of collaborative and adaptive choreographies of experiments.

4   Conclusions

The focus of this paper is on addressing the requirements of scientists in the field of eScience. The need for adaptation of scientific workflows and choreographies due to the dynamic trial-and-error process of experiment modeling that (natural, data and business analytics) scientists follow and the need for reproducible research to establish trust in research results calls for innovative software systems. In our previous work towards supporting collaborative scientific experiments with exploratory nature, we have created an approach called "Model-as-You-Go for Choreographies" and a corresponding interactive software system, based on the workflow technology and runtime adaptation that allows for interleaving the modeling and execution life cycle phases of workflows and choreographies of scientific workflows. We also identified the additional requirements on these choreographies of scientific workflows such as provenance, reproducibility, trust and adaptability.

In this work, we presented a generic architecture of a software system which enables trusted and coordinated execution and adaption of scientific workflows, choreographies and in-silico experiments with particular focus on how provenance, reproducibility and trust in the collaborative adaptation can be enabled. The architecture is drafted in a generic way allowing implementations which fit the needs of scientists, and their experiments likewise, best. The resulting software system features a separate component, the Provenance Holder that is designed so that it is responsible for and capable of storing provenance data in a trusted manner and able to support the reproducibility of experiments based on the provenance data.

There are more open questions for future research. Some still stand from [6]. These are the user friendliness and technology transparency of the new component, performance characteristics and the access control mechanisms that will satisfy the scientists needs of minimum information disclosure and still allow for reproducibility. Performance characteristics need to be both analyzed in general and evaluated in particular for different implementations. We need to investigate which data and information actually needs to be recorded to be able to provide the desired properties (provenance, reproducibility, trust and adaptability). We also need to explore which technology suits which workflow types and requirements best. Migration of one storage technology to another poses complexity implications and in particular what concerns the blockchain technology and its characteristics that need to be paid special attention when it is employed in a provenance provider component.

References

1. Alper, P., et al.: Enhancing and abstracting scientific workflow provenance for data publishing. In: Proceedings of the Joint EDBT/ICDT 2013 Workshops, pp. 313–318. EDBT 2013. ACM, New York (2013). https://doi.org/10.1145/2457317. 2457370

2. Atkinson, M., et al.: Scientific workflows: past, present and future. Future Gener. Comput. Syst. **75**, 216–227 (2017). https://doi.org/10.1016/j.future.2017.05.041. https://hal.archives-ouvertes.fr/hal-01544818

3. Dumas, M., et al.: Blockchain technology for collaborative information systems (dagstuhl seminar 18332). Dagstuhl Rep. **8**(8), 67–129 (2019). https://doi.org/10. 4230/DagRep.8.8.67. http://drops.dagstuhl.de/opus/volltexte/2019/10236

4. Goble, C.: Results vary: the pragmatics of reproducibility and research object frameworks, keynote. In: iConference (2015). https://www.slideshare.net/carolegoble/i-conference2015-goblefinalupload

5. Herschel, M., et al.: A survey on provenance - what for? what form? what from? Int. J. Very Large Data Bases (VLDB Journal) **26**(6), 881–906 (2017)

6. Karastoyanova, D., Stage, L.: Towards collaborative and reproducible scientific experiments on blockchain. In: Matulevičius, R., Dijkman, R. (eds.) CAiSE 2018. LNBIP, vol. 316, pp. 144–149. Springer, Cham (2018). https://doi.org/10.1007/978-3-319-92898-2_12

7. López-Pintado, O., et al.: Caterpillar: a blockchain-based business process management system. In: Clarisó, et al. (eds.) Proceedings of the BPM Demo Track at (BPM 2017), Barcelona, Spain, September 13, 2017. CEUR Workshop Proceedings, vol. 1920. CEUR-WS.org(2017). http://ceur-ws.org/Vol-1920/BPM_2017_paper_199.pdf

8. Mendling, J., et al.: Blockchains for business process management - challenges and opportunities. ACM Trans. Manag. Inf. Syst. **9**(1), 4:1–4:16 (2018). https://doi.org/10.1145/3183367

9. Mesirov, J.P.: Accessible reproducible research. Science **327**(5964), 415–416 (2010). https://doi.org/10.1126/science.1179653. http://science.sciencemag.org/content/327/5964/415

10. Sonntag, M., Karastoyanova, D.: Model-as-you-go: an approach for an advanced infrastructure for scientific workflows. J. Grid Comput. **11**(3), 553–583 (2013). https://doi.org/10.1007/s10723-013-9268-1

11. Strauch, S., et al.: Migrating enterprise applications to the cloud: methodology and evaluation. Int. J. Big Data Intell. **1**(3), 127–140 (2014)

12. Weiß, A., et al.: ChorSystem: a message-based system for the life cycle management of choreographies. In: CoopIS 2016, pp. 503–521. Springer, Cham (2016). https://doi.org/10.1007/978-3-319-48472-3_30

13. Weiß, A., et al.: Model-as-you-go for choreographies: rewinding and repeating scientific choreographies. IEEE Trans. Serv. Comput. **PP**(99), 1 (2017). https://doi.org/10.1109/TSC.2017.2732988. ISSN 1939-1374

14. Wolstencroft, K.: The Taverna workflow suite: designing and executing workflows of web services on the desktop, web or in the cloud. Nucleic Acids Res. **41**(W1), W557–W561 (2013). https://doi.org/10.1093/nar/gkt328

Mining Roles from Event Logs While Preserving Privacy

Majid Rafiei[(✉)] and Wil M. P. van der Aalst

Chair of Process and Data Science, RWTH Aachen University, Aachen, Germany
majid.rafiei@pads.rwth-aachen.de

Abstract. Process mining aims to provide insights into the actual processes based on event data. These data are widely available and often contain private information about individuals. On the one hand, knowing which individuals (known as resources) performed specific activities can be used for resource behavior analyses like *role mining* and is indispensable for bottleneck analysis. On the other hand, event data with resource information are highly *sensitive*. Process mining should reveal insights in the form of annotated models, but should not reveal sensitive information about individuals. In this paper, we show that the problem cannot be solved by naïve approaches like encrypting data, and an anonymized person can still be identified based on a few well-chosen events. We, therefore, introduce a *decomposition* method and a collection of techniques that preserve the privacy of the individuals, yet, at the same time, roles can be discovered and used for further bottleneck analyses without revealing sensitive information about individuals. To evaluate our approach, we have implemented an interactive environment and applied our approach to several real-life and artificial event logs.

Keywords: Responsible process mining · Privacy preserving · Social network discovery · Role mining · Process mining

1 Introduction

In recent years, process mining has emerged as a field which bridges the gap between data science and process science [1]. Event logs are used by process mining algorithms to extract and analyze the real processes. An event log is a collection of events and such information is widely available in current information systems [3]. Each event is described by its attributes and some of them may refer to individuals, i.e., human actors. The *resource* attribute may refer to the person performing the corresponding activities [1]. Organizational process mining is a sub-discipline of process mining focusing on resource behavior using the resource attributes of events. This form of process mining can be used to extract the roles in a process or organization [4]. A simple example is when two resources perform the same set of activities, the same role can be assigned to them. Moreover, resource information is essential for bottleneck analysis and for finding the root causes of performance degradation.

© Springer Nature Switzerland AG 2019
C. Di Francescomarino et al. (Eds.): BPM 2019 Workshops, LNBIP 362, pp. 676–689, 2019.
https://doi.org/10.1007/978-3-030-37453-2_54

Event data contain highly sensitive information and when the individuals' data are included, privacy issues become more challenging. As discussed in [9], event data may lead to privacy breaches. In addition, data protection regulations like the European General Data Protection Regulation (GDPR) impose many challenges and concerns regarding processing of personal data. In this paper, we show that preserving privacy in process mining cannot be provided by naïve approaches like encryption/anonymization and presence of some implicit information together with background knowledge can be exploited to deduce sensitive data even from minimized encrypted data.

We present a privacy-aware approach to discover roles from event logs. A *decomposition* method along with some techniques are introduced to protect the private information of the individuals in event data against frequency-based attacks in this specific context. The discovered roles can be replaced by the resources and utilized for bottleneck analyses while personal identifiers do not need to be processed anymore. We evaluate our approach w.r.t. the typical trade-off between privacy guarantees and loss of accuracy. To this end, the approach is evaluated on multiple real-life and synthetic event logs.

The rest of the paper is organized as follows. Section 2 outlines related work. In Sect. 3, the main concepts are briefly described. In Sect. 4, the problem is explored in detail. We explain our approach in Sect. 5. In Sect. 6, the implementation and evaluation are described, and Sect. 7 concludes the paper.

2   Related Work

During the last decade, confidentiality and privacy-preserving challenges have received increasing attention. In data science, many privacy algorithms have been presented which cover topics ranging from *privacy quantification* to *downgrading the results* [5]. These algorithms aim to provide privacy guarantees by different methods, e.g., k-anonymity, l-diversity, and t-closeness [8] are series of algorithms having been presented with the initial idea that *each individual should not be distinguished from at least $k - 1$ other individuals*.

Recently, there have been lots of breakthroughs in process mining ranging from *process discovery* and *conformance checking* to *performance analysis*. However, the research field confidentiality and privacy has received rather little attention, although the *Process Mining Manifesto* [3] also points out the importance of privacy. *Responsible Process Mining* (RPM) [2] is the sub-discipline focusing on possible negative side-effects of applying process mining. RPM addresses concerns related to Fairness, Accuracy, Confidentiality, and Transparency (FACT). In [9], the aim is to provide an overview of privacy challenges in process mining in human-centered industrial environments. A method for securing event logs to conduct process discovery by Alpha algorithm has been proposed by [11]. In [6], a possible approach toward a solution, allowing the outsourcing of process mining while ensuring the confidentiality of dataset and processes, has been presented. In [7], the aim is to apply k-anonymity and t-closeness on event data while the assumed background knowledge is a prefix of the sequence of activities. In [10], a framework has been introduced, which provides a generic scheme for confidentiality in process mining. In this paper, for the first time, we focus on the organizational perspective of event data.

3 Preliminaries: Process Mining and Role Mining

In this section, we define basic concepts regarding process mining and discovering social networks from event logs which in turn are used for role mining.

3.1 Process Mining

An event log is a collection of traces, each represented by a sequence of events. For a given set A. A^* is the set of all finite sequences over A, and $\mathcal{B}(A^*)$ is the set of all multisets over the set A^*. A finite sequence over A of length n is a mapping $\sigma \in \{1, ..., n\} \to A$, represented by a string, i.e., $\sigma = \langle a_1, a_2, ..., a_n \rangle$ where $\sigma_i = a_i$ for any $1 \leq i \leq n$. $|\sigma|$ denotes the length of the sequence. Also, $set(\sigma) = \{a \mid a \in \sigma\}$, e.g., $set(\langle a, b, c, c, b \rangle) = \{a, b, c\}$, and $multiset(\sigma) = [a \mid a \in \sigma]$, e.g., $multiset(\langle a, b, c, c, b \rangle) = [a, b^2, c^2]$.

Definition 1 (Event). *An event is a tuple $e = (a, r, c, t, d_1, ..., d_m)$, where $a \in \mathcal{A}$ is the activity associated with the event, $r \in \mathcal{R}$ is the resource, who is performing the activity, $c \in \mathcal{C}$ is the case id, $t \in \mathcal{T}$ is the event timestamp, and $d_1, ..., d_m$ is a list of additional attributes values, where for any $1 \leq i \leq m$, $d_i \in \mathcal{D}_i$ (domain of attributes). We call $\xi = \mathcal{A} \times \mathcal{R} \times \mathcal{C} \times \mathcal{T} \times \mathcal{D}_1 \times ... \times \mathcal{D}_m$ the event universe. An event log is a subset of ξ where each event can appear only once, and events are uniquely identifiable by their attributes.*

Definition 2 (Simple Event Log). *A simple event log $EL \in \mathcal{B}((\mathcal{R} \times \mathcal{A})^*)$ is a multiset of traces. A trace $\sigma \in EL$ is a sequence of events $\sigma = \langle (r_1, a_1), (r_2, a_2), ..., (r_n, a_n) \rangle$ where each event is represented by a resource r_i and activity a_i. Also, $set(EL) = \{set(\sigma) \mid \sigma \in EL\}$, and $multiset(EL) = [multiset(\sigma) \mid \sigma \in EL]$.*

Definition 3 (Activities and Resources of Event Log). *Let $EL \in \mathcal{B}((\mathcal{R} \times \mathcal{A})^*)$ be an event log, $act(EL) = \{a \in \mathcal{A} \mid \exists_{\sigma \in EL} \exists_{r \in \mathcal{R}} (r, a) \in \sigma\}$ is the set of activities in the event log, and $res(EL) = \{r \in \mathcal{R} \mid \exists_{\sigma \in EL} \exists_{a \in \mathcal{A}} (r, a) \in \sigma\}$ is the set of resources in the event log.*

Table 1 shows an event log, where *Case ID*, *Timestamp*, *Activity*, *Resource*, and *Cost* are the attributes. Each row represents an event, e.g., the first row shows that activity "Register" was done by resource "Frank" at time "01-01-2018:08.00" for case "1" with cost "1000". In the remainder, we will refer to the *activities* and the *resources* of Table 1 with their abbreviations.

Definition 4 (Frequencies). *Let $EL \in \mathcal{B}((\mathcal{R} \times \mathcal{A})^*)$ be an event log. The frequency of an activity a is $\#_a(EL) = \sum_{\sigma \in EL} |[(r, a') \in \sigma \mid a' = a]|$, the set of the activity frequencies is $frq(EL) = \{(a, \#_a(EL)) \mid a \in act(EL)\}$. $\#_{most}(EL)$ is the highest frequency, $\#_{least}(EL)$ is the lowest frequency, $\#_{median}(EL)$ is the median of frequencies, and $\#_{sum}(EL)$ is the sum of frequencies.*

Table 1. Sample event log (each row represents an event).

Case ID	Timestamp	Activity	Resource	Cost
1	01-01-2018:08.00	Register (R)	Frank (F)	1000
2	01-01-2018:10.00	Register (R)	Frank (F)	1000
3	01-01-2018:12.10	Register (R)	Joey (J)	1000
3	01-01-2018:13.00	Verify-Documents (V)	Monica (M)	50
1	01-01-2018:13.55	Verify-Documents (V)	Paolo (P)	50
1	01-01-2018:14.57	Check-Vacancies (C)	Frank (F)	100
2	01-01-2018:15.20	Check-Vacancies (C)	Paolo (P)	100
4	01-01-2018:15.22	Register (R)	Joey (J)	1000
2	01-01-2018:16.00	Verify-Documents (V)	Frank (F)	50
2	01-01-2018:16.10	Decision (D)	Alex (A)	500
5	01-01-2018:16.30	Register (R)	Joey (J)	1000
4	01-01-2018:16.55	Check-Vacancies (C)	Monica (M)	100
1	01-01-2018:17.57	Decision (D)	Alex (A)	500
3	01-01-2018:18.20	Check-Vacancies (C)	Joey (J)	50
3	01-01-2018:19.00	Decision (D)	Alex (A)	500
4	01-01-2018:19.20	Verify-Documents (V)	Joey (J)	50
5	01-01-2018:20.00	Special-Case (S)	Katy (K)	800
5	01-01-2018:20.10	Decision (D)	Katy (K)	500
4	01-01-2018:20.55	Decision (D)	Alex (A)	500

In the following, we define the sensitive frequencies on the basis of the box plot of the frequencies in such a way that not only the outliers but also all the other unusual frequencies are classified as sensitive. The activities having the sensitive frequencies are more likely to be identified by an adversary.

Definition 5 (Bounds of Frequencies). *Let $EL \in \mathcal{B}((\mathcal{R} \times \mathcal{A})^*)$ be an event log. We define $upper(EL) = \langle \#_a(EL) \mid \#_a(EL) > upper\_quartile \rangle$ and $lower(EL) = \langle \#_a(EL) \mid \#_a(EL) < lower\_quarile \rangle$ as the bounds of frequencies on the basis of the box plot of the frequencies such that for any $1 \leq i \leq |upper(EL)|-1$, $upper_i(EL) \geq upper_{i+1}(EL)$, and for any $1 \leq i \leq |lower(EL)|-1$, $lower_i(EL) \leq lower_{i+1}(EL)$.*

Definition 6 (Gaps). *Let $EL \in \mathcal{B}((\mathcal{R} \times \mathcal{A})^*)$ be an event log. For each bound of the frequencies, $gap_{bound}(EL) = [|bound_i(EL) - bound_{i+1}(EL)| \mid 1 \leq i \leq |bound(EL)|-1]$, and $mean(gap_{bound}(EL))$ is the mean of the gaps.*

Definition 7 (Sensitive Frequencies). *Let $EL \in \mathcal{B}((\mathcal{R} \times \mathcal{A})^*)$ be an event log. For each bound of the frequencies, $sstv_{bound}(EL) = [bound_i(EL) \mid \forall_{1 \leq i \leq |bound(EL)|-1} |bound_i(EL) - bound_{i+1}(EL)| \leq mean(gap_{bound}(EL))]$. If*

$|sstv_{bound}(EL)|= |bound(EL)|-1$, $sstv_{bound}(EL) = \emptyset$, *i.e., there is no gap greater than the mean of the gaps. Also,* $act(sstv_{bound}(EL)) = \{a \in act(EL) \mid \#_a(EL) \in sstv_{bound}(EL)\}$.

3.2 Role Mining

When discovering a process model from an event log, the focus is on the process activities and their dependencies. When deriving roles and other organizational entities, the focus is on the relation between individuals based on their activities. The metrics based on *joint activities*, used for discovering roles and organization structures, consider each individual as a vector of activity frequencies performed by the individual and use a similarity measure to calculate the similarity between two vectors. A social network is constructed between individuals such that if the similarity is greater than a minimum threshold (Θ), the corresponding individuals are connected with an undirected edge. The individuals in the same connected part are supposed to play the same role [4].

Consider Table 1 and let us assume that the order of the activities in each vector is D, V, C, R, S. Then, Paolo's vector is $P = (0, 1, 1, 0, 0)$, and Monica's vector is $M = (0, 1, 1, 0, 0)$. Therefore, the similarity between these vectors is 1. In this paper, we use a *Resource-Activity Matrix (RAM)*, which is defined as follows, as a basis for extracting the vectors and deriving roles.

Definition 8 (Resource-Activity Matrix (RAM)). *Let* $EL \in \mathcal{B}((\mathcal{R} \times \mathcal{A})^*)$ *be an event log,* $a \in act(EL)$, *and* $r \in res(EL)$: $RAM_{EL}(r, a) = \sum_{\sigma \in EL}|[x \in \sigma \mid x = (r, a)]|$, *and* $RAM_{EL}(r) = (RAM_{EL}(r, a_1), RAM_{EL}(r, a_2), ..., RAM_{EL}(r, a_n))$, *where* n *is the number of unique activities.*

Table 2 shows the *RAM* derived from Table 1. Given the *RAM*, the *joint-activities* social network can be obtained as follows.

Definition 9 (Joint-Activities Social Network (JSN)). *Let* $EL \in \mathcal{B}((\mathcal{R} \times \mathcal{A})^*)$ *be an event log,* RAM_{EL} *be a resource-activity matrix resulting from the* EL, *and* $sim(r_1, r_2)$ *be a similarity relation based on the vectors* $RAM_{EL}(r_1)$ *and* $RAM_{EL}(r_2)$, $JSN_{EL} = (res(EL), E)$ *is the joint-activities social network, where* $E = \{(r_1, r_2) \in res(EL) \times res(EL) \mid sim(r_1, r_2) > \Theta\}$ *is the set of undirected edges between resources, and* Θ *is the threshold of similarities.*

Note that various similarity measures are applicable, e.g., Euclidean, Jaccard, Pearson, etc. Figure 1 shows the network and roles having been obtained by applying threshold 0.1 when using *Pearson* as the similarity measure.

4   The Problem (Attack Analysis)

Here, we discuss the general problem of confidentiality/privacy in process mining, then we focus on the specific problem and the attack model w.r.t. this research.

Table 2. The RAM from Table 1

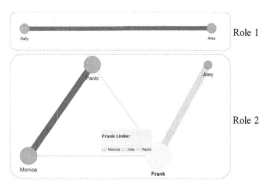

	D	V	C	R	S
Frank	0	1	1	2	0
Joey	0	1	1	3	0
Alex	4	0	0	0	0
Katy	1	0	0	0	1
Paolo	0	1	1	0	0
Monica	0	1	1	0	0

Fig. 1. The network resulting from Table 2 for Pearson similarity 0.1

4.1 General Problem

Consider Table 3 as an entirely encrypted event log with information about surgeries. The standard attributes (*Case ID*, *Activity*, *Resource*, and *Timestamp*) are included. Process mining techniques need to preserve differences. Hence, *Case ID*, *Activity*, and *Resource* are encrypted based on a deterministic encryption method.[1] Numerical data (i.e., *Timestamp*) are encrypted using a homomorphic encryption method so that the basic mathematical computations can be applied. Although the fully encrypted event log seems secure, it is not.

One can find the most or the least frequent activities and given background knowledge, the encrypted values can be simply replaced with the real values. In addition, the position of activities can also be used to infer sensitive information, e.g., when an activity is always the first/last activity, given domain knowledge the real activity can be deduced. These kinds of attacks are considered as *frequency-based*. Note that the corresponding performers are most likely identifiable, after inferring the actual activity names.

In addition to the above-mentioned attacks, other attributes are also exploitable to identify the actual activities and resources. For example, when *timestamp* is encrypted by a deterministic homomorphic encryption method, then the duration between two events is derivable. Based on background knowledge, one can infer that the longest/shortest duration belongs to specific events. When there are more attributes, it is more likely that one can combine these to infer other attributes.

Table 3. An encrypted event log.

Case ID	Activity	Resource	Timestamp
rt!@45	kl56^*	lo09(kl	3125
rt!@45	bn,.^q	lo09(kl	3256
)@!1yt	kl56^*	lo09(kl	4879
)@!1yt	bvS(op	/.,ldf	5214
)@!1yt	jhg!676	nb][,b]	6231
er^7*	kl56^*	lo09(kl	6534
er^7*	2ws34S	v,[]df	7230

---

[1] A deterministic cryptosystem produces the same ciphertext for a given plaintext and key.

These examples clarify that given domain knowledge, data leakage is possible even from a basic event log which is totally encrypted. Moreover, if the mining techniques are applied to encrypted event logs, the results are also encrypted, and data analyst is not able to interpret them without decryption [10].

4.2   Attack Analysis

Now, let us focus on our specific context where the aim is to extract roles without revealing *who performed what?* As described in Sect. 3, roles can be derived from a simple event log, and the *activity* is considered as the *sensitive attribute* in this setting. Therefore, activities get hashed, and we define $H(\mathcal{A})$ as universe of hashed activities $(H(X) = \{H(x) \mid x \in X\})$.[2]

We assume the frequencies of activities as background knowledge (bk) which can be formalized as $bk \in \mathcal{P}_{NE}(\mathbb{U}_{frq}) \times \mathcal{P}_{NE}(H(\mathcal{A})) \rightarrow \mathcal{P}(\mathcal{A})$, where $\mathbb{U}_{frq} = H(\mathcal{A}) \times \mathbb{N}$ is the universe of the hashed activity frequencies, and $\mathcal{P}_{NE}(X)$ is the set of all non-empty sets over the set X. Therefore, the actual activities can be revealed based on the assumed background knowledge. For example, in the event log Table 1, the least frequent activity is "Special-Case" which can be revealed based on background knowledge regarding the frequencies. We consider this information disclosure as *activity disclosure* (kind of *attribute disclosure*). Note that resources are usually not the unique identifiers in event logs. Nevertheless, they could get encrypted or hashed. Here, our focus is on activities, and the challenge is to eliminate the frequency of activities, while they are necessary to measure the similarity of resources and deriving roles. Our approach also improves privacy when background knowledge is about traces, e.g., length of traces and the position of activities in traces.

5   Approach

The idea is to decompose activities into other activities such that the *frequency* and *position* of activities get perturbed. However, at the same time, the similarities between resources should remain as similar as possible. To this end, we need to determine the number of substitutions for each activity, and the way of distributing the frequency of the main activity among its substitutions. We consider $D(H(\mathcal{A}))$ as the universe of hashed activities after the decomposition, and the sanitized event logs are obtained as follows.

Definition 10 (Sanitized Event Logs (EL''_t, EL''_{ms}, and EL''_s)). *Let $EL' \in \mathcal{B}((\mathcal{R} \times H(\mathcal{A}))^*)$ be an event log where activity names are hashed, and Decom $\in H(\mathcal{A}) \rightarrow D(H(\mathcal{A}))$ be a decomposition method. $EL''_t \in \mathcal{B}((\mathcal{R} \times D(H(\mathcal{A})))^*)$ is a trace-based sanitized event log. A multiset-based sanitized event log is $EL''_{ms} = multiset(EL''_t)$, and a set-based sanitized event log is $EL''_s = set(EL')$.*

---

[2] H is a one-way hash function, here we use *SHA-256*.

EL_s'' is used when the similarity measure is binary (Jaccard, hamming, etc.). In this case, the frequencies could be simply ignored, since these measures do not consider the absolute frequency but only whether it is 0 or not. EL_{ms}'' is employed when traces are not needed to be reconstructed from the sanitized event log. In this case, the sanitized event log entirely preserves privacy of the individuals against attribute disclosure when background knowledge is trace-based. Moreover, it is clear that resource-activity matrices and the corresponding joint-activities social networks can be simply derived from the sanitized event logs. In the remainder, we use EL' for the event log where activity names are hashed and EL'' for the sanitized event logs made by applying the decomposition method, i.e., EL_t'' and EL_{ms}''.

5.1 Decomposition Method

The Number of Substitutions for each activity a (NS_a) should be specified in such a way that the activities having the sensitive frequencies are not certainly identifiable anymore. In the following, we introduce some techniques.

- *Fixed-value:* A fixed value is considered as the number of substitutions for each activity such that for any $a \in act(EL')$, $NS_a = n$ where $n \in \mathbb{N}_{>1}$.
- *Selective:* By this technique only the sensitive frequencies are targeted to get perturbed. Hence, only some of the activities having the sensitive frequencies are decomposed. Here, we allocate the substitutions such that for any $a \in act(EL')$: $NS_a = \lceil \#_a(EL')/\#_{median}(EL') \rceil$ if $\#_a(EL') = \#_{most}(EL')$, and for any $a \in act(EL')$: $NS_a = \lceil \#_a(EL')/\#_{least}(EL') \rceil$ if $\#_a(EL') \in sstv_{lower}(EL') \setminus \#_{least}(EL')$. Note that we aim to perturb the bounds of frequencies with the minimum number of activities after the decomposition.
- *Frequency-based:* The substitutions are allocated based on the relative frequencies of the main activities. Here, we allocate the substitutions in such a way that for any $a \in act(EL')$, $NS_a = \lceil \#_a(EL')/\#_{sum}(EL') \times 100 \rceil$.

After specifying the number of substitutions for activity a, we make a substitution set $Sub_a = \{sa_1, sa_2, ..., sa_{NS_a}\}$ such that for any $a_1, a_2 \in act(EL')$: $Sub_{a_1} \cap Sub_{a_2} = \emptyset$ if $a_1 \neq a_2$.[3] Note that $Decom(act(EL')) = \{sa \in D(H(\mathcal{A})) \mid \exists_{a \in act(EL')} sa \in Sub_a\}$. To preserve the main feature of the vectors, we distribute the frequency of the main activity uniformly among its substitutions. To this end, while going through the event log, for each resource, the ith occurrence of the activity $a \in act(EL')$ is replaced by the $sa_i \in Sub_a$, and when $i > NS_a$, i is reset to 1 (*round-robin* manner). Thereby, we guarantee that if the frequency of performing an activity by a resource is greater than or equal to the other resources, the frequency of performing the corresponding substitutions will also be greater or equal to the others.[4]

---

[3] Note that the substitution sets should not be revealed.

[4] We consider a dummy resource in case there is an activity without resource.

684 M. Rafiei and W. M. P. van der Aalst

5.2 Privacy Analysis

To analyze the privacy, we measure the disclosure risk of the original event log, and the sanitized event logs. Two factors are considered to measure the disclosure risk including; the *number of activities* having the sensitive frequencies, and the *presence* of the actual activities having the sensitive frequencies. The presence for each bound of the frequencies before applying the decomposition method is $prs_{bound}(EL) = 1$ if $sstv_{bound}(EL) \neq \emptyset$. Otherwise, $prs_{bound}(EL) = 0$. For the sanitized event logs the presence is obtained as follows.

$$prs_{bound}(EL'') = \frac{|act(sstv_{bound}(EL'')) \cap \{sa \in Decom(act(EL')) \mid \#_a(EL') \in sstv_{bound}(EL')\}|}{|\{sa \in Decom(act(EL')) \mid \#_a(EL') \in sstv_{bound}(EL')\}|}$$

Also for each bound of the frequencies, $PR_{bound}(EL) = 1/|act(sstv_{bound}(EL))|$ is the raw probability of activity disclosure based on the number of activities hav-

Table 4. Similarity between JSN and JSN'' for the *fixed-value* technique

Threshold	Dataset	NS = 2		NS = 4		NS = 8		NS = 16	
		CN	UC	CN	UC	CN	UC	CN	UC
$\Theta = 0.1$	BPIC 2012	1.0	1.0	1.0	1.0	0.99	1.0	0.99	1.0
	BPIC 2017	1.0	1.0	1.0	1.0	0.99	1.0	0.98	1.0
$\Theta = 0.2$	BPIC 2012	1.0	1.0	0.99	1.0	0.98	1.0	0.95	1.0
	BPIC 2017	1.0	1.0	1.0	1.0	0.99	1.0	0.97	1.0
$\Theta = 0.3$	BPIC 2012	1.0	1.0	0.98	1.0	0.95	1.0	0.90	1.0
	BPIC 2017	1.0	1.0	1.0	1.0	0.97	1.0	0.95	1.0
$\Theta = 0.4$	BPIC 2012	1.0	1.0	0.97	1.0	0.92	1.0	0.88	1.0
	BPIC 2017	1.0	1.0	0.99	1.0	0.97	1.0	0.93	1.0
$\Theta = 0.5$	BPIC 2012	1.0	1.0	0.94	1.0	0.91	1.0	0.87	1.0
	BPIC 2017	1.0	1.0	0.99	1.0	0.96	1.0	0.93	1.0
$\Theta = 0.6$	BPIC 2012	1.0	1.0	0.94	1.0	0.90	1.0	0.85	1.0
	BPIC 2017	1.0	1.0	0.98	1.0	0.95	1.0	0.94	1.0
$\Theta = 0.7$	BPIC 2012	1.0	1.0	0.95	1.0	0.91	1.0	0.87	1.0
	BPIC 2017	1.0	1.0	0.99	1.0	0.97	1.0	0.96	1.0
$\Theta = 0.8$	BPIC 2012	1.0	1.0	0.96	1.0	0.95	1.0	0.93	1.0
	BPIC 2017	1.0	1.0	0.99	1.0	0.98	1.0	0.93	1.0
$\Theta = 0.9$	BPIC 2012	1.0	1.0	0.99	1.0	0.96	1.0	0.95	1.0
	BPIC 2017	1.0	1.0	0.99	1.0	0.96	1.0	0.92	1.0
Average	BPIC 2012	1.0	1.0	0.96	1.0	0.94	1.0	0.91	1.0
	BPIC 2017	1.0	1.0	0.99	1.0	0.97	1.0	0.94	1.0
Total average	BPIC 2012	1.0		0.98		0.97		0.955	
	BPIC 2017	1.0		0.995		0.985		0.97	

ing the sensitive frequencies, and $DR_{bound}(EL) = prs_{bound}(EL)/|act(sstv_{bound}(EL))|$ is the disclosure risk. The whole disclosure risk w.r.t. the assumed background knowledge is measured as follows.

$$DR(EL) = \frac{\alpha \times prs_{upper}(EL)}{|act(sstv_{upper}(EL))|} + \frac{(1 - \alpha) \times prs_{lower}(EL)}{|act(sstv_{lower}(EL))|}$$

If $prs_{bound}(EL) = 0$ or $|act(sstv_{upper}(EL))| = 0$, $DR_{bound}(EL) = 0$. Also, α is utilized to set the importance of each bound of the frequencies.

6   Evaluation

To evaluate our approach, we show the effect on the accuracy and privacy for two real life event logs (BPIC 2012 and 2017). To this end, we have implemented an interactive environment in Python. Figure 1 shows an output of our tool.[5]

6.1   Accuracy

To examine the accuracy of our approach, we measure the similarity of joint -activities social networks from the original event log (JSN) and the sanitized event log (JSN''). To this end, we compare the similarity of their *connected (CN)* and *unconnected (UC)* parts. Note that $JSN = (res(EL), E)$, $JSN'' = (res(EL''), E'')$, and $res(EL) = res(EL'')$. Here, we use *Pearson* as the measure of similarity between vectors, which is one of the best measures according to [4].

$$CN = \frac{|E \cap E''|}{|E|} \qquad UC = \frac{|(res(EL) \times res(EL) \backslash E) \cap (res(EL) \times res(EL) \backslash E'')|}{|res(EL) \times res(EL) \backslash E|}$$

Table 4 shows the similarities when the *fixed-value* technique is used to identify the number of substitutions. As can be seen, the networks are almost the same and the accuracy is acceptable. When the number of substitutions increases, the average of similarities decreases, showing the typical trade-off between accuracy and privacy. Moreover, the networks in the unconnected parts are identical, i.e., if two resources are not connected in the JSN, there are not connected in the JSN'' as well.

Figure 2 shows the similarities w.r.t. various thresholds when using the *selective* or *frequency-based* technique. As can be seen, on average the *selective* technique leads to more accurate results. However, in the unconnected parts the *frequency-based* technique has better results. Note that BPIC 2017 is larger than BPIC 2012 in terms of both resources and activities (Table 5).

---

[5] https://github.com/m4jidRafiei/privacyAware-roleMining.

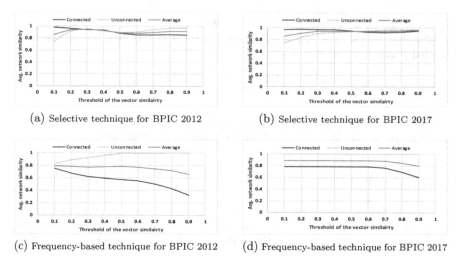

(a) Selective technique for BPIC 2012   (b) Selective technique for BPIC 2017

(c) Frequency-based technique for BPIC 2012   (d) Frequency-based technique for BPIC 2017

Fig. 2. The similarities between JSN and JSN'' when using the *selective* or *frequency-based* technique to identify the number of substitutions.

6.2   Privacy

To evaluate the effect on privacy, we calculate the disclosure risk on the original event logs and the sanitized event logs after applying the decomposition method with different techniques. Tables 6 and 7 show the parameters regarding the disclosure risk for BPIC 2012 and 2017 respectively. As can be seen, when the *fixed-value* technique is used, DR is lower for the larger values as the number of substitutions in both event logs. Moreover, since the relative frequency of the least frequent activities is very low, the *frequency-based* technique does not affect the lower bound of sensitive frequencies. This weakness can be mitigated by combining this technique with the *fixed-value* such that the number of substitutions would be the relative frequency plus a fixed value.

Table 5. Statistics regarding frequencies in BPIC 2012 and BPIC 2017

	BPIC 2012	BPIC 2017
No. resources	69	145
No. unique activities	24	26
No. activities	262200	1202267
$|upper(EL))|$	5	5
Frequency of the most frequent activities ($\#_{most}(EL)$)	54850	209496
Relative frequency for any a: $\#_a(EL) \in \#_{most}(EL)$	0.20	0.17
$|lower(EL))|$	4	6
Frequency of the least frequent activities ($\#_{least}(EL)$)	12	22
Relative frequency for any a: $\#_a(EL) \in \#_{least}(EL)$	4×10^{-5}	1×10^{-5}

Table 6. The DRs before and after applying the method on BPIC 2012

	PR_{upper}	PR_{lower}	prs_{upper}	prs_{lower}	$DR(\alpha = 0.5)$
BPIC 2012	0.5	0	1	0	0.25
Fixed-value NS = 2	0.25	0	1	0	0.12
Fixed-value NS = 4	0.25	0	0.5	0	0.06
Fixed-value NS = 8	0.12	0	0.5	0	0.03
Fixed-value NS = 16	0.06	0	0.5	0	0.01
Selective	1	0	0.09	0	0.04
Frequency-based	0.5	0	0.04	0	0.01

Table 7. The DRs before and after applying the method on BPIC 2017

	PR_{upper}	PR_{lower}	prs_{upper}	prs_{lower}	$DR(\alpha = 0.5)$
BPIC 2017	0.25	0.5	1	1	0.37
Fixed-value NS = 2	0.5	0.25	0.25	1	0.18
Fixed-value NS = 4	0.25	0.12	0.25	1	0.09
Fixed-value NS = 8	0.12	0.07	0.25	1	0.05
Fixed-value NS = 16	0.06	0.04	0.25	1	0.03
Selective	1	0.2	0.09	0.41	0.08
Frequency-based	0.33	0.5	0.04	1	0.25

To compare the introduced techniques, we consider the minimal disclosure risk which can be supplied by all the techniques as the basis of comparison and evaluate the *accuracy* and *complexity* provided by the different techniques for the same disclosure risk. The accuracy is the average similarity between the networks, and the complexity is considered as the number of unique activities. Note that for the *fixed-value* technique, we inspect the event log which has the minimum NS providing the basis disclosure risk. Tables 8 and 9 show the results of this experiment for BPIC 2012 and 2017 respectively. As one can see, in both event logs, the *fixed-value* technique provides more accurate results and the *selective* technique imposes less complexity.

All the above-mentioned explanations and our experiments demonstrate that the decomposition method provides accurate and highly flexible protection for mining roles from event logs, e.g., the decomposition method with the *frequency-based* technique can be used when the upper bound of frequencies is more sensitive and the accuracy of the unconnected parts is more important.

Table 8. Comparison of techniques in BPIC 2012

	DR($\alpha = 0.5$)	Accuracy	Complexity
Fixed_value NS = 8	0.04	0.97	188
Selective	0.04	0.9	87
Frequency-based	0.04	0.75	108

Table 9. Comparison of techniques in BPIC 2017

	DR ($\alpha = 0.5$)	Accuracy	Complexity
Fixed_value NS = 2	0.25	1	52
Selective	0.25	0.93	43
Frequency-based	0.25	0.87	113

7 Conclusions

In this paper, for the first time, we focused on privacy issues in the organizational perspective of process mining. We proposed an approach for discovering joint-activities social networks and mining roles w.r.t. privacy. We introduced the *decomposition* method along with a collection of techniques by which the private information about the individuals would be protected against frequency-based attacks. The discovered roles can be replaced with individuals in the event data for further performance and bottleneck analyses.

The approach was evaluated on BPIC 2012 and 2017, and the effects on accuracy and privacy were demonstrated. To evaluate the accuracy, we measured the similarity between the connected and unconnected parts of two networks separately while different thresholds were considered. Moreover, we introduced three different techniques to identify the number of substitutions in the decomposition method, and we showed their effect on the accuracy and privacy, when the frequencies of activities are assumed as background knowledge. In the future, other techniques or combination of the introduced ones could be explored with respect to the characteristics of the event logs.

References

1. van der Aalst, W.M.P.: Process Mining: Data Science in Action. Springer, Heidelberg (2016). https://doi.org/10.1007/978-3-662-49851-4
2. van der Aalst, W.M.P.: Responsible data science: using event data in a "people friendly" manner. In: Hammoudi, S., Maciaszek, L.A., Missikoff, M.M., Camp, O., Cordeiro, J. (eds.) ICEIS 2016. LNBIP, vol. 291, pp. 3–28. Springer, Cham (2017). https://doi.org/10.1007/978-3-319-62386-3_1
3. van der Aalst, W.M.P., et al.: Process mining manifesto. In: Daniel, F., Barkaoui, K., Dustdar, S. (eds.) BPM 2011. LNBIP, vol. 99, pp. 169–194. Springer, Heidelberg (2012). https://doi.org/10.1007/978-3-642-28108-2_19
4. van der Aalst, W.M.P., Reijers, H.A., Song, M.: Discovering social networks from event logs. Comput. Support. Coop. Work (CSCW) **14**(6), 549–593 (2005)
5. Agrawal, R., Srikant, R.: Privacy-preserving data mining, vol. 29. ACM (2000)
6. Burattin, A., Conti, M., Turato, D.: Toward an anonymous process mining. In: 2015 3rd International Conference on Future Internet of Things and Cloud (FiCloud), pp. 58–63. IEEE (2015)

7. Fahrenkrog-Petersen, S.A., van der Aa, H., Weidlich, M.: Pretsa: event log sanitization for privacy-aware process discovery. In: 1st IEEE International Conference on Process Mining (2019)
8. Li, N., Li, T., Venkatasubramanian, S.: t-closeness: Privacy beyond k-anonymity and l-diversity. In: 2007 IEEE 23rd International Conference on Data Engineering, pp. 106–115. IEEE (2007)
9. Mannhardt, F., Petersen, S.A., Oliveira, M.F.: Privacy challenges for process mining in human-centered industrial environments. In: 2018 14th International Conference on Intelligent Environments (IE), pp. 64–71. IEEE (2018)
10. Rafiei, M., von Waldthausen, L., van der Aalst, W.M.P.: Ensuring confidentiality in process mining. In: Proceedings of the 8th International Symposium on Data-driven Process Discovery and Analysis (SIMPDA 2018), Seville, Spain, 13–14 December 2018, pp. 3–17 (2018)
11. Tillem, G., Erkin, Z., Lagendijk, R.L.: Privacy-preserving alpha algorithm for software analysis. In: 37th WIC Symposium on Information Theory in the Benelux/6th WIC/IEEE SP

Extracting Event Logs for Process Mining from Data Stored on the Blockchain

Roman Mühlberger[1] ⓘ, Stefan Bachhofner[1] ⓘ, Claudio Di Ciccio[1(✉)] ⓘ,
Luciano García-Bañuelos[2,3] ⓘ, and Orlenys López-Pintado[2] ⓘ

[1] Vienna University of Economics and Business, Vienna, Austria
roman@muehlberger.eu.com, {stefan.bachhofner,claudio.di.ciccio}@wu.ac.at
[2] University of Tartu, Tartu, Estonia
orlenyslp@ut.ee
[3] Tecnológico de Monterrey, Monterrey, Mexico
luciano.garcia@tec.mx

Abstract. The integration of business process management with blockchains across organisational borders provides a means to establish transparency of execution and auditing capabilities. To enable process analytics, though, non-trivial extraction and transformation tasks are necessary on the raw data stored in the ledger. In this paper, we describe our approach to retrieve process data from an Ethereum blockchain ledger and subsequently convert those data into an event log formatted according to the IEEE Extensible Event Stream (XES) standard. We show a proof-of-concept software artefact and its application on a data set produced by the smart contracts of a process execution engine stored on the public Ethereum blockchain network.

Keywords: Ethereum · Process discovery · Process monitoring · Process conformance

1 Introduction

Blockchain offers a new prospective environment for the execution of inter-organisational processes [24]. On top of a distributed execution environment, blockchains collate transactions onto a push-only stack (ledger) within backward-linked blocks [25]. Participants are provided with an up-to-date version of the entire blockchain via broadcast after a new block has been added following a consensus algorithm. The utilisation of the blockchain in processes covering a number of participants prevents the necessity of a central orchestrator and provides trust among mutually untrusted parties [31]. Additionally, the decentralised database stores transactional information triggered by smart contracts containing the state and execution details of distinct cases [24]. Data can be retrieved that contains, inter alia, interacting process participants, (discrete) time and duration of the interaction, execution costs on the blockchain as well as the state details of a limited, recurring set of activities. This information enables process monitoring [7], process compliance [32], and process conformance checking [23,26] via process mining [29,30]. In addition, the information is

ⓒ Springer Nature Switzerland AG 2019
C. Di Francescomarino et al. (Eds.): BPM 2019 Workshops, LNBIP 362, pp. 690–703, 2019.
https://doi.org/10.1007/978-3-030-37453-2_55

also valuable for the analysis of the efficiency and effectiveness of process executions, i.e., for performance mining [4, 11, 15, 19].

In order to make proper use of that information, manual effort is required to convert the data from individual blocks into an appropriate format, which causes a variety of challenges [27]. To mention but a few, the information retrieved from the blockchain is represented in hexadecimal and numeric formats; timestamps are approximate as the finest granular unit of time in blockchains is at the level of blocks; the structure of data payloads attached to transactions is for the most part arbitrary. Process mining tools, on the other hand, present clear constraints on data types and their required representation. In order to localise and convert the data, a profound understanding of the data model represented on the blockchain is essential.

In this work-in-progress paper, we follow the Design Science research methodology to devise a framework that retrieves process data from blocks and transforms the process data into an event log, formatted according to the IEEE Extensible Event Stream (XES) standard. The software artefact is reflected in an implemented prototype, which is applied on a real-world case study. The XES output file is used as an input for existing tools in the *ProM* toolkit. Additionally, we discuss the challenges and opportunities pertaining to data extraction, transformation and interpretation of blockchain data to drive future work in this area. To the best of our knowledge, this work is the first one retrieving and transforming blockchain data into a standardised format that can be used in process mining tools. Consequently, our work is the first that enables process monitoring, process conformance and process compliance checking with process mining on blockchain-based process executions.

The remainder of this paper is structured as follows. In Sect. 2, the notions backing our research work are provided. Section 3 describes our approach. Section 4 describes the experiment conducted with a real-world case study. The paper concludes with Sect. 5 including implications for further research.

2   Background

A blockchain consists of a distributed ledger of transactions [25]. Blockchains operate on top of a peer-to-peer network, where each peer stores a local copy of the ledger's transactional data. The transactions are collated in blocks, which are sequentialised as an append-only chain. Compact identification and storage of data, as well as the backward-linking of blocks in the chain, are based upon hashing algorithms such as the SHA-3 compliant algorithm KECCAK [6]. Appending new blocks in a trustworthy manner is enabled by a combination of consensus-making, cryptography, and market mechanisms [25]. A central authority is thus not necessary to that extent, making it advantageous for processes containing a set of untrusted and unknown participants [31]. More recent blockchain protocols such as Ethereum [33] enable programmability of the platform through the *smart contracts*. Ethereum distributions come endowed with dedicated smart-contract programming languages such as Solidity [10], an object-oriented, high-level programming language. Ethereum smart contracts are executed in a decentralised environment named Ethereum Virtual Machine (EVM). Their function signatures and public attributes are exposed through a human- and machine-readable interface called Application Binary Interface (ABI). The execution of smart contract operations is associated to a price expressed in terms of *gas*, exchanged at a variable rate with the Ether cryptocurrency. The invocation of smart contracts' functions occurs via transactions that are stored on the ledger.

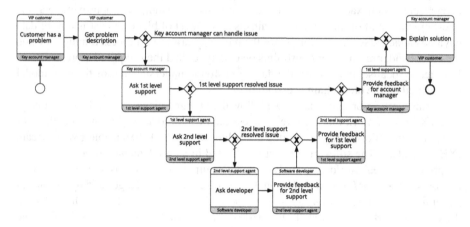

Fig. 1. The choreography Business Process Model and Notation (BPMN) model of an incident management process [31].

Blockchain for Business Process Execution. Smart contracts allow for the codification of business process logic on the blockchain [12], as shown in the seminal work of Weber et al. [31] and later on with tools like Caterpillar [22] and Lorikeet [28]. As several modern Business Process Management Systems (BPMSs) do, Caterpillar and Lorikeet adopt a Model-Driven Engineering (MDE) approach to let the process analysts provide graphical representations of the process and turn it into executable code enacting it. They take as input models that are specified in the Business Process Model and Notation (BPMN) language [15]. Figure 1, e.g., illustrates the BPMN choreography model of the incident management process introduced by Weber et al. [31] to evaluate their execution engine for inter-organisational processes.[1] The process begins with an issue raised by a customer. The key account manager registers the problem description. Until a solution is not found, the ticket escalates to the 1st level support, then to the 2nd level support, and finally to the developer. After the solution is found, a feedback is given and the solution explained to the customer.

The workflow routing of blockchain-based process execution engines is performed by smart contracts generated by compilers that translate BPMN diagrams into smart contract code (e.g., Solidity). Once deployed on the blockchain, the contract instances are identified by the hexadecimal address of their account, which serve as a key for querying the state of the process instances.

---

[1] Diagram in Fig. 1 courtesy of Ingo Weber.

Table 1. An event log excerpt.

Case ID	Event ID	Activity name	Timestamp	Resource	...
1	1	Customer has a problem	2016-03-22T14:06:22.000Z	Key account manager	...
1	2	Get problem description	2016-03-22T14:06:35.000Z	Key account manager	...
1	3	Ask 1st level support	2016-03-22T14:07:20.000Z	Key account manager	...
1	4	Ask 2nd level support	2016-03-22T14:08:38.000Z	1st level support	...
1	5	Provide feedback for 1st level support	2016-03-22T14:09:04.000Z	2nd level support	...
1	6	Provide feedback for account manager	2016-03-22T14:09:14.000Z	1st level support	...
1	7	Explain solution	2016-03-22T14:09:22.000Z	Key account manager	...
2	8	Customer has a problem	2016-03-22T14:37:45.000Z	Key account manager	...
2	9	Get problem description	2016-03-22T14:38:00.000Z	Key account manager	...
...	...	...	...	...	
32	250	Customer has a problem	2016-03-22T12:46:01.000Z	Key account manager	...
32	251	Get problem description	2016-03-22T12:46:21.000Z	Key account manager	...
32	252	Ask 1st level support	2016-03-22T12:47:16.000Z	Key account manager	...
32	253	Ask 2nd level support	2016-03-22T12:47:53.000Z	1st level support	...
32	254	Provide feedback for 1st level support	2016-03-22T12:48:09.000Z	2nd level support	...
32	255	Provide feedback for account manager	2016-03-22T12:48:40.000Z	1st level support	...
32	256	Explain solution	2016-03-22T12:49:17.000Z	Key account manager	...

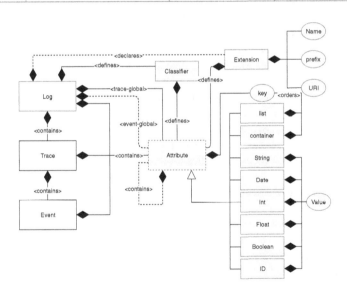

Fig. 2. The meta model class diagram representation of the XES standard [17].

The transactional storage of interactions among the actors operating on the blockchain and with smart contracts enables traceability of processes [13]. Our aim is to extract the process data from the ledger and turn it into a readily processable format for process mining tools, in order to enable analysis and auditing of processes run on the blokchain.

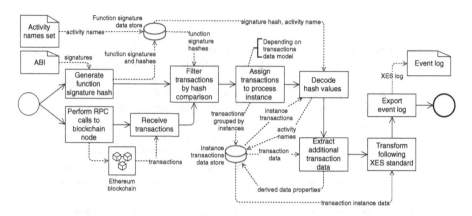

Fig. 3. Approach for the extraction of process data from a blockchain and transformation into an event log.

Event Logs for Process Mining. Process mining makes use of the sequential data stored by BPMS and other process-aware information systems such as Enterprise Resource Plannings (ERPs) or Customer Relationship Managements (CRMs) in order to discover, analyse and enhance existing processes [1]. As such, process mining is a valuable means to conduct auditing and forensics on existing process data [18]. The standard data structure with which those data are stored is named *event log*, or log for short. An event log consists of a collection of traces, where every *trace* represents a process run (or process instance) and activity executions are represented by *events* (each mapped to an activity name). Event logs may contain additional information in the form of event attributes, such as the timestamp or the resource carrying out the related activity. Table 1 shows an excerpt of event log that may originate from the execution of the process in Fig. 1. Process analytics toolkits such as ProM [2] or Apromore [20] take as input event logs that are formatted according to the XML-based IEEE standard XES [17]. Figure 2 illustrates the original version of the standard. The top level element of XES is a log containing trace nodes (one per process instance). Traces incorporate event nodes. Data are stored in *attributes*, which are represented as key-value nodes. The XES standard defines attribute data types and semantics through an extension mechanism. Every extension has a name, a prefix, and a reference Uniform Resource Identifier (URI). For example, extension Concept, with prefix concept:, and URI http://www.xes-standard.org/concept.xesext, is used to name elements such as event (typically with the corresponding activity label) and traces. Other extensions we make use of are, e.g., identity, to denote the unique identifiers of elements, organizational (prefix: org) for resources, time for event time-stamping, lifecycle for the status of activities, and cost for execution costs.

3 Approach

Figure 3 illustrates our approach for the extraction and transformation to XES of process data from the blockchain. We remark that we use only information that is stored on the blockchain.

We begin with the analysis of the smart contract implementing the business process. With this analysis, we aim at producing information to identify transactions that are associated with the execution of process tasks. To that end, we rely on the fact that the payload of an Ethereum transaction stores the *function selector*, i.e., the first four bytes in the SHA-3 hash of a string encoding the function signature. For instance, let us consider task *Customer has a problem*, which is implemented by a function with signature `Customer_Has_a_Problem()`. The selector for such function is `0xefe73dcb`, because the SHA-3 hash for `Customer_Has_a_Problem()` is `0xefe73dcb348c11a7ab31ce1620102e63c94e84ab393a78f187d1485c8a2c72cc`. For convenience, we keep a hash-table to associate task names with their corresponding function selectors, as we use it to generate and enrich the event log.

Blockchain-based BPMSs such as Caterpillar and Lorikeet use the factory pattern to control the creation of process instances: a factory contract deploys, for each new process instance, a smart contract that enacts the business process. We observe that the address of the factory contract is included in the metadata of the transaction associated with the creation of a new process instance as the originator of such transaction, i.e., within the attribute `Transaction.from`. Therefore, if we assume that the address of the factory contract is provided, we can identify the address of each of the process instances created by the factory contract and later each one of the transactions associated with a process instance. We inspect the blockchain data model of blocks and transactions as illustrated in Fig. 4. We identify the hash as a unique identifier for every block and every transaction.

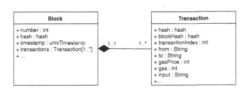

Fig. 4. The data model of blocks and transactions. We depicted the attributes used for the extraction of blockchain data.

The transactions included in a block can be obtained in the Block.transactions field of a block. The UNIX timestamp represented in the block can be used as a basis to derive the timestamps of the activities therein. The transactions are allocated to an enumeration denoting the order of the process executions and saved into a data structure. In order to provide completeness of the dataset, we decode the hash values and save both the function signatures and activity names in a human-readable format. In order to add the activity name, we compare the data from the transaction input field with the signature hash saved in the data structure, which is denoted as *Function signature*

Table 2. Mapping between event log elements and blockchain data fields in our case study.

Event log attribute	Blockchain data field
Case ID	Transaction.to
Activity ID	Transaction.input[2:10] (function selector)
Event ID	Transaction.hash
Activity label	Reverse-engineered from contract ABI and function selector in Transaction.input
Event timestamp	Block.timestamp (plus sorting by Transaction.index)
Event cost	Transaction.gas
Event resource	Transaction.from

```
{   blockHash:        0x1eca7ae74de59dff4f553b052a0e8346b1fc1587cf76ca5ee5b22da84f87822b ,
    blockNumber:      1196772,
    from:             0x1387e74982055e3e1d235aad579350813b329b2b ,
    gas:              1000000,
    gasPrice:         20000000000,
    hash:             0x656252f3ecee102d981520ca9e0ca0f7048bce99e4f6fead89d358cdbedd6156 ,
    input:            0xefe73dcb,
    nonce:            227,
    r:                0xf26831a097ee1a3cf64364a07ff80fa816dd8604461482921a81be74276b5e7b ,
    s:                0x60a41b999ccd10461565d4604cd063c299a5b1006a9ed8b0fcc84e5cfa9960f8d ,
    to:               0x0E6e0313dBe1Ba7A8bCb622EE7A77EaCBc9eF73f ,
    transactionIndex: 3,
    v:                28,
    value:            0 }
```

Listing 1: Key-value representation of the transaction data denoting the execution of an activity.

data store in Fig. 3. For consistency reasons, we use the transaction data from the data structure denoted as *Instance transactions data store* to add additional information such as the costs of the process instances in Ether or fiat currency like US dollars. Finally, the data extraction and processing is finalised and the result can be transformed into an event log that is compliant with the XES format.

To transform the instance transactions data store into a log that complies with XES standard, we map the attributes in the data store to XES event and trace attributes. Whenever suitable, we append XES extension declarations to our file to enrich events with additional information, such as cost or time, declared into the header of the XES document. For example, we use the mentioned transaction hash 0x656252f3ecee102d981520ca9e0ca0f7048bce99e4f6fead89d358cdbedd6156 as the id node of the event. We use the function selector, i.e., the first four bytes of the Transaction.input field (0xefe73dcb) as a custom *activity_id* attribute. Timestamp *2016-03-22T13:44:22.000Z* is mapped to the event with the key time:timestamp.

Table 3. The activity names, their associated smart contract's function signatures, and the first four bytes of the KECCAK hash of those signatures, for the process in Fig. 1.

Activity name	Function signature	Function selector
Customer has a problem	`Customer_Has_a_Problem()`	0xefe73dcb
Get problem description	`Get_problem_description(int32 x)`	0x92ed10ef
Ask 1st level support	`Ask_1st_level_support(int32 y)`	0x82b06df7
Explain solution	`Explain_solution()`	0x95c07f19
Ask 2nd level support	`Ask_2nd_level_support()`	0x63ad6b81
Provide feedback for account manager	`Provide_feedback_for_account_manager()`	0x58a66413
Ask developer	`Ask_developer()`	0xecb07b8c
Provide feedback for 1st level support	`Provide_feedback_for_1st_level_support()`	0x3b26a0eaP
Provide feedback for 2nd level support	`Provide_feedback_for_2nd_level_support()`	0x9ec3200a

```
<event>
  <string key="id" value="i0x656252f3ecee102d981520ca9e0ca0f7048bce99e4f6fead89d358cdbedd6156" />
    <string key="concept:name" value="Customer␣has␣a␣problem" />
    <string key="activity_id" value="e0xefe73dcb" />
    <string key="cost:currency" value="USD"/>
    <date key="time:timestamp" value="2016-03-22T13:44:22.000Z"/>
    <string key="lifecycle:transition" value="complete"/>
    <float key="cost:total" value="0.1754522"/>
    <int key="blockNo" value="1196772"/>
    <string key="from" value="0x1387e74982055e3e1d235aad579350813b329b2b"/>
    <float key="feeEth" value="0.0011393"/>
    <float key="pricePerEth" value="10.96"/>
    <string key="org:resource" value="0x1387e74982055e3e1d235aadb79350813b329b2b"/>
</event>
```

Listing 2: XES event corresponding to the transaction of Listing 1.

4 Case Study

In this section, we present a case study for our approach. We extract process data from a public blockchain and transform the process data into a XES log. The software project, including the Jupyter notebook for our case study,[2] is openly available as a public GitLab repository.[3]

4.1 Extraction of an Event Log from the Public Ethereum Blockchain

To evaluate our approach, we analyse the transactions stored on the public Ethereum network by the execution engine described by Weber et al. [31]. Figure 1 illustrates the BPMN choreography model of the incident management process they implemented. The process is enacted by a smart contract to be run on the blockchain, which we henceforth refer to as process contract. Every activity corresponds to a function, whose signature is exposed by the contract ABI.

---

[2] https://gitlab.com/MacOS/extracting-event-logs-from-process-data-on-the-blockchain/tree/paper/incident_management_process.ipynb.

[3] https://gitlab.com/MacOS/extracting-event-logs-from-process-data-on-the-blockchain/tree/paper.

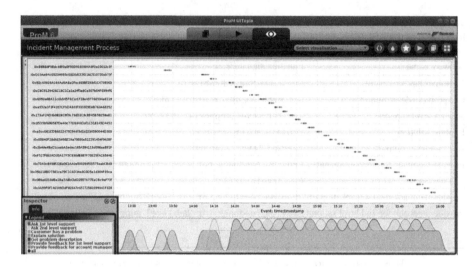

Fig. 5. Dotted Chart visualisation of the incident management log in ProM.

Fig. 6. The incident management process mined via Inductive Visual Miner on ProM.

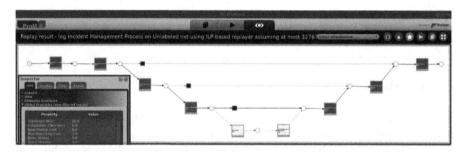

Fig. 7. Conformance checking of the incident management process on the log. (Color figure online)

The process instances implemented on the blockchain are initiated by a factory smart contract at address 0x09890f52cdd5d0743c7d13abe481e705a2706384. The factory contract deploys a new instance of process contract for every new run. To retrieve the process data, we access the Etherscan Ethereum blockchain explorer and analyse the list of transactions having the factory smart contract as the sender, i.e., such that the factory address occurs in their Transaction.from field. Those transactions amount to 32 in our case.[4]

Table 2 summarises the mapping between the event log elements and the blockchain data. The Transaction.to field of each of the transactions from the factory contract identifies a process instance, thus 32 process instances were run on the blockchain. Using

---

[4] https://etherscan.io/address/0x09890f52cdd5d0743c7d13abe481e705a2706384.

again Etherscan, we retrieve the transactions directed to every such process instance contract. Listing 1 shows one of those transactions. To identify which of those correspond to an activity enactment, we operate as follows. Considering the set of activities of the process, we find the corresponding function names in the ABI of the called contract. Thereupon, we compute the KECCAK hash for each of the function signatures of interest and save the hexadecimal representation of its first four bytes, as shown in Table 3. We thus select the transactions denoting the activity enactment by selecting the ones that have those four bytes as the prefix of their Transaction.input field. The transaction of Listing 1, e.g., corresponds to activity *Customer has a problem.*

Finally, we turn the extracted transactions into a XES document. Every process instance contract corresponds to a trace. Therefore, we use its hash as the trace identifier (id). Every (activity-related) transaction towards that contract maps to an event therein, so its hash represents the id of the event. As an additional attribute we add XES-defined standard extensions such as the event time:timestamp, which we approximate by considering the Block.timestamp. We use the Transaction.index to sort events, which belong to the same block. In addition, we include the activity cost:total attribute, on the basis of the consumed gas, plus a correction factor provided by Etherscan for the conversion in US dollars. Furthermore, we include the address of the sender account of the transaction as the org:resource attribute. Listing 2 shows the event extracted from the transaction in Listing 1. The entire generated XES file is available online.[5]

4.2 Analysis of the Event Log with ProM

We used the generated event log as an input for ProM to test its usage for process mining. A full investigation and analysis of the event log goes beyond the scope of this paper. This preliminary experiment shows the possibilities opened up by our approach, which extracts and transforms data stored on the blockchain to make it readily available for process mining. Figure 5 illustrates the event log imported in ProM and visualised through the Dotted Chart plug-in. The id of traces is on the y axis and on the x axis we see the timestamp of events. Traces are sorted by the timestamp of their first event. The colour of dots corresponds to the name (activity label) of events. It can be noticed that the process instances were run sequentially at close distance in time. Figure 6 depicts the output of the Inductive Visual Miner discovery plug-in using the standard set-up [21]. Considering the original model in Fig. 1, we can observe that all instances were such that the 2nd-level support resolved the issue as activities *Ask developer* and *Provide feedback for 2nd level support* were never executed. Our observation is confirmed by the application of the conformance checking plug-in of Adriansyah et al. [3] on the log and on a Workflow net that simulates the behaviour of the original process [1], as illustrated in Fig. 7. As it can be noticed in the information panel on the left-hand side, the model is 100% fitting with the recorded traces. Indeed, the border of all activities (transitions, graphically depicted as boxes) are surrounded by a green line. The background colour of transitions indicates whether the corresponding tasks were executed (blue) or not

---

[5] https://gitlab.com/MacOS/extracting-event-logs-from-process-data-on-the-blockchain/tree/paper/incident_management_process.xes.

(white). Notice that, as expected, the two activities at the bottom of the diagram (*Ask developer* and *Provide feedback for 2nd level support*) did not occur.

A full-fledged quantitative analysis of the process is beyond the scope of this paper. However, the presented outcome hints at the possibilities that the creation of event logs out of transaction data on the blockchain opens up.

4.3 Discussion and Limitations

The described approach shows promising results. However, we acknowledge limitations that are inherently bound to the data analysis we conduct. For process mining, the information found on the blockchain is required to include at least unique identifiers for process instances, timestamps of executions as well as identifiable activity denotations. In our case study, we detect unique identifiers and activity names by considering a smart contract function call as the unit of execution, i.e., the event. However, we understand that smart contract functions may not match process activities. This calls for more advanced techniques to relate smart contract transactions to process task executions [5]. Also, we associate process instances to smart contract life-cycles. This assumption holds when every process run corresponds to one and only one contract. Blockchain BPMSs that adopt other architectural patterns than the factory one could make our assumption not valid any longer, thus requiring more sophisticated reference reconciliation mechanisms, or necessitating transaction payloads to bear a unique identifier for instances. Furthermore, we approximate event timestamps with the block time, although the level of granularity may not be sufficient and fully reliable. Information stemming from off-chain sources should be retrieved from certified sources such as the so-called *oracles* [34]. This opens up new challenges for future work, aimed at mixed on-chain/off-chain information retrieval approaches. In general, our work depends on the data model retrieved from the transactions. In fact, for different cases an adjustment to the corresponding data model may be necessary. In future research endeavours, algorithms from semantic technologies could be leveraged to circumvent manual adjustments [8, 16]. Also, we observe that our solution stores the generated event logs off-chain. This could make their synchronisation and update with new process runs harder and potentially hinder run-time monitoring. To circumvent this issue, we envisage solutions that lie at the core blockchain architecture or leverage the support of additional data stores: an extension of blockchain protocol implementations with on-node storage of state and event information [9], or the adoption of hash-based links connecting transaction data with distributed file systems such as InterPlanetary File System (IPFS).[6]

5 Conclusion

In this paper, we described an approach to generate XES-compliant event logs out of process data stored on the blockchain. Our approach provides a blueprint to retrieve process data from the transactions ledger. As a proof-of-concept, we implemented a software prototype applying our approach on a case study, based on a blockchain-enabled process run on the public Ethereum network.

---

[6] https://ipfs.io.

Our approach shows promising preliminary results, on the basis of which we envision a number of future research avenues. From a practical perspective, we are developing further software artefacts to apply our approach on other blockchain-based BPMSs such as Caterpillar [22]. Furthermore, we aim at investigating how to extract and process information stemming from other Distributed Ledger Technologies (DLTs) such as Hyperledger Fabric [14].

We argue that the main challenge in mining processes from the blockchain is the mapping between the process-specific data and their model and the concrete representations on the blockchain. Challenges arise if that mapping does not ensure traceability. Future research should thus be devoted to the creation of a modelling language describing how blockchain-based process interactions are mapped onto blockchain data, so as to automate the manual adjustments required by injecting such knowledge in the extraction algorithms themselves. Furthermore, an interesting problem to tackle is the analysis of process data that are partially on-chain and partially off-chain, to include also information sources beyond the reach of blockchains. Novel solutions could build upon the ontology-based data access approach of [8].

Acknowledgements. The authors thank Zhivka Dangarska, Dominik Haas, Jan Mendling, and Ingo Weber for their valuable support and fruitful discussions. The work of Claudio Di Ciccio was partially funded by the Austrian FFG grant 861213 (CitySPIN).

References

1. van der Aalst, W.: Process Mining: Data Science in Action, 2nd edn. Springer, Heidelberg (2016). https://doi.org/10.1007/978-3-662-49851-4
2. van der Aalst, W.M., van Dongen, B.F., Günther, C.W., Rozinat, A., Verbeek, E., Weijters, T.: ProM: the process mining toolkit. In: BPM Demos (2009)
3. Adriansyah, A., Munoz-Gama, J., Carmona, J., van Dongen, B.F., van der Aalst, W.M.P.: Measuring precision of modeled behavior. Inf. Syst. E-Bus. Manage. 13(1), 37–67 (2015)
4. Bachhofner, S., Kis, I., Di Ciccio, C., Mendling, J.: Towards a multi-parametric visualisation approach for business process analytics. In: Metzger, A., Persson, A. (eds.) CAiSE 2017. LNBIP, vol. 286, pp. 85–91. Springer, Cham (2017). https://doi.org/10.1007/978-3-319-60048-2_8
5. Baier, T., Di Ciccio, C., Mendling, J., Weske, M.: Matching events and activities by integrating behavioral aspects and label analysis. Softw. Syst. Model. 17(2), 573–598 (2018)
6. Bertoni, G., Daemen, J., Peeters, M., Assche, G.V.: The making of KECCAK. Cryptologia 38(1), 26–60 (2014)
7. Cabanillas, C., Di Ciccio, C., Mendling, J., Baumgrass, A.: Predictive task monitoring for business processes. In: Sadiq, S., Soffer, P., Völzer, H. (eds.) BPM 2014. LNCS, vol. 8659, pp. 424–432. Springer, Cham (2014). https://doi.org/10.1007/978-3-319-10172-9_31
8. Calvanese, D., Kalayci, T.E., Montali, M., Tinella, S.: Ontology-based data access for extracting event logs from legacy data: the onprom tool and methodology. In: Abramowicz, W. (ed.) BIS 2017. LNBIP, vol. 288, pp. 220–236. Springer, Cham (2017). https://doi.org/10.1007/978-3-319-59336-4_16
9. Casino, F., Dasaklis, T.K., Patsakis, C.: A systematic literature review of blockchain-based applications: current status, classification and open issues. Telematics Inform. 36, 55–81 (2019)

10. Dannen, C.: Introducing Ethereum and Solidity: Foundations of Cryptocurrency and Blockchain Programming for Beginners. Apress, Berkely (2017)
11. Del-Río-Ortega, A., Resinas, M., Cabanillas, C., Ruiz-Cortés, A.: On the definition and design-time analysis of process performance indicators. Inf. Syst. **38**(4), 470–490 (2013)
12. Di Ciccio, C., et al.: Blockchain support for collaborative business processes. Informatik Spektrum **42**, 182–190 (2019)
13. Di Ciccio, C., et al.: Blockchain-based traceability of inter-organisational business processes. In: Shishkov, B. (ed.) BMSD 2018. LNBIP, vol. 319, pp. 56–68. Springer, Cham (2018). https://doi.org/10.1007/978-3-319-94214-8_4
14. Duchmann, F., Koschmider, A.: Validation of smart contracts using process mining. In: ZEUS. CEUR Workshop Proceedings, vol. 2339, pp. 13–16 (2019)
15. Dumas, M., La Rosa, M., Mendling, J., Reijers, H.A.: Fundamentals of Business Process Management, 2nd edn. Springer, Heidelberg (2018). https://doi.org/10.1007/978-3-662-56509-4
16. Governatori, G., Hoffmann, J., Sadiq, S., Weber, I.: Detecting regulatory compliance for business process models through semantic annotations. In: Ardagna, D., Mecella, M., Yang, J. (eds.) BPM 2008. LNBIP, vol. 17, pp. 5–17. Springer, Heidelberg (2009). https://doi.org/10.1007/978-3-642-00328-8_2
17. Günther, C.W., Verbeek, E.: XES standard definition. Fluxicon Process Lab. **13**, 14 (2009)
18. Jans, M., Hosseinpour, M.: How active learning and process mining can act as continuous auditing catalyst. Int. J. Acc. Inf. Syst. **32**, 44–58 (2019)
19. Kis, I., Bachhofner, S., Di Ciccio, C., Mendling, J.: Towards a data-driven framework for measuring process performance. In: Reinhartz-Berger, I., Gulden, J., Nurcan, S., Guédria, W., Bera, P. (eds.) BPMDS/EMMSAD -2017. LNBIP, vol. 287, pp. 3–18. Springer, Cham (2017). https://doi.org/10.1007/978-3-319-59466-8_1
20. La Rosa, M., et al.: APROMORE: an advanced process model repository. Expert Syst. Appl. **38**(6), 7029–7040 (2011)
21. Leemans, S.J.J., Fahland, D., van der Aalst, W.M.P.: Scalable process discovery and conformance checking. Softw. Syst. Model. **17**(2), 599–631 (2018)
22. López-Pintado, O., García-Bañuelos, L., Dumas, M., Weber, I., Ponomarev, A.: Caterpillar: a business process execution engine on the ethereum blockchain. Softw. Pract. Exp. **49**(7), 1162–1193 (2019)
23. Mannhardt, F., de Leoni, M., Reijers, H.A., van der Aalst, W.M.P.: Balanced multi-perspective checking of process conformance. Computing **98**(4), 407–437 (2016)
24. Mendling, J., et al.: Blockchains for business process management-challenges and opportunities. ACM Trans. Manage. Inf. Syst. **9**(1), 4:1–4:16 (2018)
25. Nakamoto, S.: Bitcoin: a peer-to-peer electronic cash system (2008)
26. Rozinat, A., van der Aalst, W.M.P.: Conformance testing: measuring the fit and appropriateness of event logs and process models. In: Bussler, C.J., Haller, A. (eds.) BPM 2005. LNCS, vol. 3812, pp. 163–176. Springer, Heidelberg (2006). https://doi.org/10.1007/11678564_15
27. Soffer, P., Hinze, A., Koschmider, A., Ziekow, H., et al.: From event streams to process models and back: challenges and opportunities. Inf. Syst. **81**, 181–200 (2019)
28. Tran, A.B., Lu, Q., Weber, I.: Lorikeet: a model-driven engineering tool for blockchain-based business process execution and asset management. In: BPM Demos, pp. 56–60 (2018)
29. van der Aalst, W.: Process mining: overview and opportunities. ACM Trans. Manage. Inf. Syst. **3**(2), 7 (2012)
30. van der Aalst, W., et al.: Process mining manifesto. In: Daniel, F., Barkaoui, K., Dustdar, S. (eds.) BPM 2011. LNBIP, vol. 99, pp. 169–194. Springer, Heidelberg (2012). https://doi.org/10.1007/978-3-642-28108-2_19

31. Weber, I., Xu, X., Riveret, R., Governatori, G., Ponomarev, A., Mendling, J.: Untrusted business process monitoring and execution using blockchain. In: La Rosa, M., Loos, P., Pastor, O. (eds.) BPM 2016. LNCS, vol. 9850, pp. 329–347. Springer, Cham (2016). https://doi.org/10.1007/978-3-319-45348-4_19

32. Weidlich, M., Polyvyanyy, A., Desai, N., Mendling, J., Weske, M.: Process compliance analysis based on behavioural profiles. Inf. Syst. **36**(7), 1009–1025 (2011)

33. Wood, G.: Ethereum: a secure decentralised generalised transaction ledger (2014)

34. Xu, X., et al.: The blockchain as a software connector. In: WICSA, pp. 182–191 (2016)

A Framework for Supply Chain Traceability Based on Blockchain Tokens

Thomas K. Dasaklis[✉], Fran Casino, Costas Patsakis, and Christos Douligeris

Department of Informatics, University of Piraeus, Piraeus, Greece
{dasaklis,francasino,kpatsak,cdoulig}@unipi.gr

Abstract. Tracing products and processes across complex supply chain networks has become an integral part of current supply chain management practices. However, the effectiveness and efficiency of existing supply chain traceability mechanisms are hindered by several barriers including lack of data interoperability and information sharing, opportunistic behaviour, lack of transparency and visibility and cyber-physical threats, to name a few. In this paper, we propose a forensics-by-design supply chain traceability framework with audit trails for integrity and provenance guarantees based on malleable blockchain tokens. This framework also provides the establishment of different granularity levels for tracing products across the entire supply chain based on their unique characteristics, supply chain processes and stakeholders engagement. To showcase the applicability of our proposal, we develop a functional set of smart contracts and a local private blockchain. The benefits of our framework are further discussed, along with fruitful areas for future research.

Keywords: Supply chain · Traceability · Blockchain tokens · Smart contracts

1 Introduction

The Fourth Industrial Revolution is characterized by the convergence of various technologies, such as the Internet of Things (IoT) and blockchain, which are blurring the lines between the physical and the digital world. Such technologies may transform modern supply chain (SC) networks into complete digital ecosystems. SC digitization offers outstanding business speed, agility, and the development of traceability mechanisms (TM) that allow for an almost complete identification and recording of products and processes. It is worth noting that blockchain-enabled SC approaches coupled with IoT applications could improve the communication and the selective export of SC traceability data, enabling additional benefits to the logistics sector for data management and analytics [2].

SC traceability has attracted considerable attention in the last decade, particularly in safety-sensitive sectors, like food, pharmaceuticals and perishable agri-food products [1,11]. Traceability can be useful in product recalls, may improve process control and production optimization and reduce costs of liability claims and lawsuits. Moreover, traceability mechanisms build trust and foster the establishment of long-term relationships among disparate SC partners [27].

© Springer Nature Switzerland AG 2019
C. Di Francescomarino et al. (Eds.): BPM 2019 Workshops, LNBIP 362, pp. 704–716, 2019.
https://doi.org/10.1007/978-3-030-37453-2_56

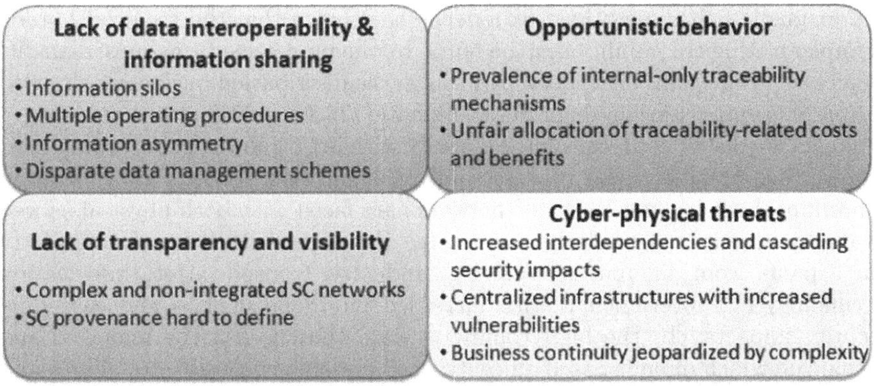

Fig. 1. Barriers hindering the development of TM in modern SC networks.

Blockchain is considered a foundational technology of the Fourth Industrial Revolution and is expected to play a crucial role in SC management [7]. In a nutshell, blockchain is a distributed and immutable data ledger, which enables the transfer of a range of assets among non-trusted parties securely and inexpensively without third-party intermediaries [7]. By providing trust in distributed environments, blockchain has the potential to offer full transparency and visibility within the SC, build confidence in legitimate operations, safeguard products quality and reduce administrative costs. Further benefits from the adoption of blockchain applications in SC traceability may relate to data interoperability, greater access to finance, auditability, integrity, and authenticity. Two features within the blockchain technology are of paramount importance: smart contracts (SmCs) and tokens. SmCs are agreements among mutually distrusting participants that may be executed in multi-peer environments and are automatically enforced by the consensus mechanism of the blockchain without relying on a trusted third party [10]. SmCs enable computations within the blockchain; thus, they operate as a decentralized virtual machine. Tokens are digital entities that may be used as a digital representation of physical assets (ingredients, sub-products, etc.) for tracing these assets individually. SmCs and tokens pave the way for the development of multiple new application scenarios in SC traceability, like product certification, deep traceability and cross-business tracing of products and services [7].

Establishing sound TM at a system level, however, remains a challenging task. As illustrated in Fig. 1, there are several barriers hindering the establishment of robust TM in SC networks. In particular, current SC networks are overwhelmed by information asymmetry, multiple operating procedures, and disparate data management schemes. As a consequence, traceability-related information lives in most of the cases in silos, where each participant has its internal TM and, inevitably, stores its unique traceability records. In addition, not all the collaboration models in SC are based on a "win-win" philosophy, and the prevalence of internal-only TM across SC networks unavoidably gives rise to opportunistic behaviours. Past experience has shown that the financial burden of implementing traceability may be borne by the processing firms (upstream SC members), while gains are reaped by firms in the distribution businesses closer to the end customer (downstream SC members) [22]. In addition, vast amounts of products are distributed/consumed globally with little knowledge regarding their origins, their manufacturing processes involved, and their storage/transportation conditions. Last but not least, SC networks are faced with both physical as well as digital threats, particularly due to: (a) the high and disparate number of participants from different sectors (e.g., industry, transportation, Information Technology-IT, government), with increased interdependencies and cascading security impacts; (b) the high volume of data that need to be managed and exchanged which often is based on centralized systems vulnerable to failures and attacks [28].

The literature related to blockchain tokens for SC traceability is extremely limited [18,30,33]. In addition, essential aspects of SC traceability, like the adoption of different granularity levels have received limited attention so far [13]. The primary goal of this paper is to address such issues. In particular, we propose a structured approach based on blockchain tokens and the usage of SmCs to define various granularity levels in SC traceability. The overall framework captures the multiple needs for precision in SC traceability (depth and breadth) by taking into account the products' unique characteristics, the various SC processes and the various stakeholders involved. Due to the adaptability and customized nature of the blockchain tokens, the proposed framework allows us to represent the entire granularity scales easily and elegantly across the entire SC and, therefore, create an accurate digital representation of the overall SC ecosystem. Moreover, the mutable characteristics of the blockchain tokens enable us to create practically infinite product definitions within the same structure, thus enhancing scalability and performance, a technical aspect which hasn't been addressed so far. Since the proposed framework is based upon the distributed and immutable nature of the blockchain, strong integrity guarantees and provenance regarding the SC traceability processes and the relevant data are provided.

The remainder of the paper is organized as follows. In Sect. 2, we provide an overview of the literature relevant to blockchain-enabled TM, with a particular focus on token-based approaches in SC traceability. In Sect. 3, we describe the proposed blockchain management framework for SC traceability services based on tokens. In Sect. 4, we showcase the applicability of the proposed framework based on a set of experiments. Finally, in Sect. 5, we review the main results and discuss future research directions.

2   Literature Review

The lockchain technology has been recently used for the development of SC traceability applications in various sectors [21]. For example, blockchain-enabled applications have been proposed for tracking gems [6], in the mining industry [24], for tracking medicines and pharmaceuticals [34], in animal product

SC traceability systems [25], for tracking electronic products [17] and in textile industry [14]. The main focus of the available blockchain-based traceability literature has been on the food supply chain [4,15], particularly, in agri-food SC management [5,8,12,32], for ensuring food provenance [23], in meat traceability [29] and in the coffee SC [31]. Apart from specific sectors, certain studies address issues of food security and safety [19] or define different granularity levels in food tracking [13]. It is worth noting that the blockchain-related traceability literature is extremely limited in scope, and the available frameworks present limited applicability (mainly because they do not take into account the invited scepticism related to blockchain's scalability and the high-energy use).

The use of blockchain tokens in business-oriented applications has received limited attention so far. For example, blockchain tokens have been mainly used as a financial engineering instrument [26], particularly for raising funds and engaging stakeholders [9]. Some token-based applications in the health care domain have also been proposed in the literature [20]. Regarding SC traceability, very few studies make use of blockchain tokens for tracking and tracing products [18] or for establishing ingredient certification schemes for commingled foods [30]. Arguably, the most relevant research is the one presented in [33]. The authors propose a blockchain-based approach for tracking manufactured goods based on tokens. The amount of tokenized goods that are required for minting a new token is defined by specific "token recipes", which resemble the bill of materials. It is worth noting that our token-based approach is significantly different from the approach presented in [33], particularly in the way tokens are defined (we make use of a set of adaptable/mutable tokens). In addition, our approach enables the development of TM with a high level of granularity.

3 The Proposed Framework

In this section, we provide the details of our framework for the blockchain-based SC. First, we present the details and equivalences of our framework in terms of physical to digital world transformation. Second, we define the actors/resources of our framework and define a set of bill of materials containing their possible values. Third, we provide a formal definition of elemental and compound tokens, both enabling mutable characteristics.

3.1 Tokenization of the Bill of Materials

For developing the token-based traceability framework we apply a bill-of-materials approach. A *bill of materials* (BOM) is a hierarchical list of raw materials, components, and assemblies required to manufacture a product [16]. By implementing a BOM checker we ensure that each component (at the lowest level of analysis) of a product is digitally represented by a token. The procedure is depicted in Fig. 2a. First, the bill structure checker receives a set of input tokens and checks their pedigree structure as well as their ID, to avoid token reuse. Note that the BOM of tokens information and the IDs that have been used are stored off-chain (by using decentralized permanent storage solutions, such as IPFS [3]). Next, the bill structure checker controls whether the pedigree tokens are included in the BOM of the token that we want to create. Finally, the checker creates the new token if and only if all the pedigree tokens are valid and in the bill structure.

In our approach, we apply a relaxed policy in terms of the BOM structure. We consider a product to be valid only if it contains at least one product of the corresponding BOM (and none out of it). Nevertheless, we can apply restrictions at the BOM structure level, in terms of minimum materials and quantities needed to pass the checker. This also enables the possibility to implement quality thresholds in the token design.

3.2   Main Actors and Resources

In what follows, we describe the characteristics of the main actors/resources of the framework, namely the products, the stakeholders and the processes. Moreover, we provide a generic BOM structure for each of them and describe their main properties.

Product: A product can be defined as an article or substance that is a result of a process. Moreover, a product can be quantified and shared (in a physical and/or digital manner). An example of a BOM structure of a product (with each corresponding pedigree, represented as a child leaf in the tree structure) is depicted in Fig. 2b. Note that a BOM structure may be adapted/modified according to each product specifications and characteristics. We further classify products into *primary* or *secondary* products. The latter are the result of the combination of different raw materials, which could belong to diverse categories or subcategories. In addition, a secondary product p_i may contain or be the result of a set of sub-products (semi-finished etc) so that $p_i = (p_0, p_1, ...p_n)$. Note that, for the sake of clarity, we use p to refer to any kind of product. Depending on the product specification and its BOM structure (see Sect. 3.1), a secondary product p_i may contain zero, one or n different products (i.e. a specific sauce may not contain all the ingredients if a customer is allergic).

Process: A process can be defined as a series of physical, chemical, mechanical or digital operations that aim to transform or preserve a physical or digital element. For example, we may define a set of processes pr such that $pr = (pr_0, pr_1, ..pr_n)$ where pr_0 may correspond to a mixing process, pr_1 a grilling process, etc. In this context, we can define different bills of processes considering their characteristics and the activity that we are performing, such as transportation processes, packaging processes, chemical processes, etc.

Stakeholder: We define a stakeholder as an independent user or party involved in a process. Similarly to the processes, we can define a bill of stakeholders if necessary, depending on the level of analysis needed by our system.

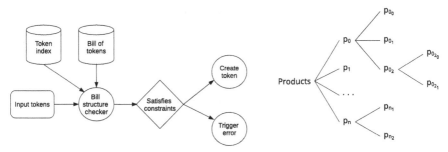

(a) Workflow of the bill structure checker.   (b) Generic bill of products

Fig. 2. Bill of materials checker and a generic example.

3.3 Mutable Tokens

Our framework considers two types of tokens, namely elemental and compound ones. Elemental tokens store the information of basic primary products and their corresponding processes and stakeholders. The information of the primary products corresponds to a unique and integral element, which does not contain other tokens or products in its pedigree structure. The compound tokens inherit the properties and features that elemental tokens have and an additional pedigree structure, which contains information of the pedigree tokens used to create them. More concretely, the pedigree set contains a collection of 2-dimensional tuples in the form $(id, quantity)$, which store a token ID and the quantity of such a token, which is a physical subset of the corresponding product (e.g. one tomato or 0.5 L of milk). Therefore, compound tokens may contain a combination of compound or elemental tokens that contain the pedigree information not only of each inner product but also of all the processes and the stakeholders involved. Without loss of generality, a token t_i is defined as follows:

$$t_i = [(id, token\{(t_0, q_0), (t_1, q_1), ..., (t_n, q_n)\}, processes(pr_0, pr_1, ..., pr_n),$$
$$stakeholders(s_0, s_1, ...s_n), data]$$

Note that for elemental tokens, the pedigree structure corresponds to the source materials (i.e. primary products, as defined in Sect. 3.2). The contents of both types of tokens are described in Table 1. There is an additional field of information, where one can store the timestamp of the token creation, as well as supplementary descriptions.

Figure 3 shows an example of a pedigree structure. This structure follows the bill requirements in a cascade fashion (i.e. each corresponding token follows its own bill of tokens, processes and stakeholders), enabling quality inspections and auditability. The term *mutable* indicates that the contents of the pedigree structure can be different between the same type of tokens, enabling different combinations provided that each BOM is respected. Therefore, we do not need to define a new type of token for each possible combination during the design phase, as discussed in Sect. 3.1. An example is depicted in Fig. 4. A vegetarian salad

Table 1. Contents of each type of token. Note that the information presented in each field relates to the possible values of the bill of each category, as defined in Sect. 3.2. For instance, a compound token can only be created using the set of tokens defined in its bill structure.

Type	elemental/compound
Identifier	a unique alphanumerical identifier
Pedigree	source materials/pedigree structure
Processes	list of processes involved
Stakeholders	list of stakeholders involved
Other	Additional information, timestamps

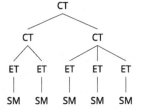

Fig. 3. High-level abstraction of a token pedigree tree. CT stands for compound token, ET for elemental token and SM for source material.

Fig. 4. Simple example of a vegetarian salad's BOM.

may contain lettuce, tomatoes and corn or any combination of such elements. However, it cannot contain chicken since it is not in its BOM. Note that for the existing mechanisms, the more the possible combinations, the less the resulting scalability, since the system needs to store variations (e.g. possible combinations and quantities of tokens to create a vegetarian salad) as a new token with a specific BOM. Nevertheless, our token definition solves this drawback by design, enhancing the scalability and performance of the system. Moreover, every token is linked to its pedigree using unique IDs.

4   Experiments

To showcase our method, we provide a use case scenario. Our example is based on the case presented in Fig. 4, which we now extend. Therefore, our final product is a vegetarian salad, consisting of multiple products, as seen in Fig. 5.

In this case, we consider that tomatoes, lettuce, onion, dried fruits and dressing (including oil) have a specific BOM. For example, the BOM of onions contains two possible onion varieties (vidalia and shallot). In this regard, the set of tokens and compound tokens needed to create the selected vegetarian salad is depicted in Fig. 6a. In the case of source materials, we specify the name and the ID of a product in the pedigree field. In addition, we specify three units of $et1$.

Fig. 5. The complete BOM of a vegetarian salad. Note that each component is digitalised as a token.

(a) Set of tokens of a vegetarian salad.

Type	ID	Pedigree (ID,quantity)	Proc.	Stake.
SM	213	cherry	1,2	1
SM	111	corn	1,2	1
SM	576	nuts	1,2	2
SM	4676	pistachio	1,2	2
SM	2667	balsamic vinegar	2,4	3
ET	et_1	$(213,1)$	-	1,5
ET	et_2	$(111,100)$	-	1,5
ET	et_3	$(576,5)$	-	2,6
ET	et_4	$(4676,10)$	-	2,6
ET	et_5	$(2667,1)$	-	2,5
CT	ct_1	$\{(et3,1),(et4,1)\}$	5	4,7
CT	ct_2	$\{(et_1,3),(et_2,1),(ct_1,1),(et_5,1)\}$	5	8

(b) Processes.

ID	Description
1	Collection
2	Sanitisation
3	Dehydration
4	Aging
5	Mixing

(c) Stakeholders.

ID	Description
1	Manufacturer A
2	Manufacturer B
3	Manufacturer C
4	Manufacturer D
5	Deliverer A
6	Deliverer B
7	Deliverer C
8	Manufacturer E

Fig. 6. An example list of processes and stakeholders.

Thus, this salad will contain 3 cherry tomatoes. Note that the elemental tokens may contain not only a unit of a source material but a set of them. Moreover, a compound token can contain several units of an elemental token, as in the case of the cherry tomato (see the pedigree of ct_2 in Fig. 6a). Lists of possible processes and stakeholders are described in Figs. 6b and c. Finally, all the tokens contain their creation time, as well as additional info, if required. Note that further information of the processes can be specified in the smart contract, such as temperature or delivery time requirements in a transportation process. Nevertheless, the addition of other parties such as insurance and quality inspection organizations that may trigger self-executing penalties for misbehaving is left for future work.

An overview of the framework, from the creation of the primary goods until the generation of the final product is depicted in Fig. 7. First, we create the set of SmCs and fill them with the corresponding information (e.g. stakeholders, processes). Next, all the primary products are collected and transformed into an elemental token. Thereafter, compound tokens are created which contain all the pedigree information. Finally, the vegetarian salad is created using a combination of such tokens. At each step, the operations are translated into the blockchain by using specific functions. All the procedures are checked by the bill structure, which guarantees the validity of each token's contents, enabling features such as quality, provenance and auditability. As previously mentioned, this framework also enables mutable tokens so that similar products can be defined without the

712   T. K. Dasaklis et al.

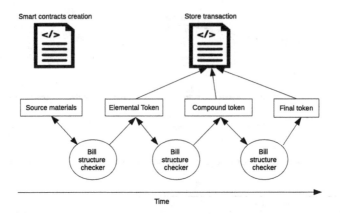

Fig. 7. An overview of the operations performed in our framework.

need to create a specific token in each case. Note that this is critical in contexts where each product may contain a wide variety of combinations, such as in the food industry. Further restrictions can be implemented in terms of quantities so that strict quantity policies can be implemented. It should be noted that we kept the example simple to ease its adoption, considering this functionality as a product-specific feature.

We implemented a set of three SmCs, each one devoted to managing the different resources of our framework (see Sect. 3.2). We use a smart contract to define the stakeholders' functions, another one to manage the processes and their characteristics and another one devoted to product and token creation and management, which checks the information of the other two (utilizing the *call* function) to ensure that the operations are legitimate. More concretely, we use `truffle`[1] and `ganache-cli`[2] to create a local Ethereum blockchain and deploy the SmCs, for which the code is shared in GitHub[3]. The interactions between the different actors can also be stored. Secure access policies manage everything, defining what information is visible by whom using the *require* clause of `solidity`. For the sake of clarity, an excerpt of the most relevant functions implemented in the token management SmCs, as well as a description, is provided in Table 2. In our implementation, we consider that only the creator of a token has permission to add contents to it (e.g. pedigree or processes). Nevertheless, this constraint can be relaxed so that a subset of stakeholders may have permissions over each token. In terms of costs, although writing in the blockchain needs the use of GAS, reading operations are free. Therefore, to minimize the costs, other solutions based on private blockchains will be explored in the future (such as Hyperledger).

---

[1] http://truffleframework.com.
[2] https://github.com/trufflesuite/ganache-cli.
[3] https://github.com/francasino/tokens.

Table 2. Functions available in the tokens' SmCs and their characteristics.

Function	Input	Output	Description
constructor	-	True/False	Creates and initializes structures
addToken	id_token, name, timestamp, pedigree [], processes [], stakeholders[]	True/False	Adds a token to the system
addPedigree	id_token, id_pedigree, quantity, timestamp,	True/False	Adds a pedigree tuple to a token
addPedigreeElemental	id_token, id_product, quantity, timestamp	True/False	Adds product tuple to an elemental token
addProduct	id_product, name, timestamp,	True/False	Adds a product to the system
addProcess	id_token, id_process	True/False	Adds a process to a token
addStakeholder	id_token, id_stakeholder	True/False	Adds a stakeholder to a token
changeStatusToken	id_token, boolean	True/False	Change status of a token
getToken	id_token	Object	Returns a token object and its contents
getPedigreeToken	id_token	List	Returns a token pedigree structure with the corresponding ids
getProcessToken	id_token	List	Returns list of processes used in a token
getStakeholderToken	id_token	List	Returns list of stakeholders involved in a token
getNumberOfTokens	-	Integer	Returns the number of available tokens, for statistics
updateNumOfProcesses	address	Integer	External update to avoid non-existing ids
updateNumOfStakeholders	address	Integer	External update to avoid non-existing ids
retrieveHash	id_token	Hash	The hash of the information of a token
triggerFunctions	-	Alert	Set of functions called after specific operations

5   Discussion and Conclusions

In this paper we have proposed a new blockchain-based framework for SC traceability. Based on a BOM approach we have created a digital representation of all the materials, sub-assemblies and intermediate assemblies of a product by using tokens. This digital representation is complemented by the provision of various granularity levels in the SC traceability processes. Contrary to the existing token-based traceability literature, we use mutable tokens so that similar products can be defined without the need to create a specific token in each case. To achieve our objectives we have used blockchain and SmCs as the building blocks of our framework. In particular, we have used blockchain as a distributed tamper-proof chain-of-custody mechanism and SmCs as an automation mechanism for managing SC stakeholders and processes as well as for token creation.

The proposed framework offers several benefits in terms of improved SC process management, security and resilience. The decentralized and secure nature of blockchain safeguards the accuracy, trustworthiness, timeliness, and usability of the exchanged traceability records. Keeping permanent traceability records on the blockchain creates an undeniable (and potentially unavoidable) transparency in SC where blockchain-certified traceability of products from source to store may be achieved. Checking for compliance by external stakeholders is further enhanced by the usage of SmCs that (a) improve the creation of highly robust audit trails and (b) render the overall process fully automated. By using tokens, each logistics unit obtains a digital representation on the blockchain and it may be individually traced by any stakeholder as it goes through the various SC processes. Moreover, to the best of our knowledge, this is the first work that proposes the use of malleable/mutable tokens, which enable better scalability and avoid asset redundancy, since one token may contain different pedigrees if defined in its BOM. Despite the benefits of our approach, our solution needs to be further refined to enhance its possibilities towards more complex scenarios. Therefore, various issues still need to be addressed such as: (a) large scale application of the proposed framework (with a multitude of SC stakeholders, processes and relevant products) for assessing its scalability and (b) combination of the proposed framework with various SC optimization approaches.

Acknowledgments. This work was supported by the European Commission under the Horizon 2020 Programme, as part of the project *LOCARD* (Grant Agreement no. 832735).

References

1. Badia-Melis, R., Mishra, P., Ruiz-García, L.: Food traceability: new trends and recent advances. A review. Food Control **57**, 393–401 (2015)
2. Banafa, A.: IoT and blockchain convergence: benefits and challenges. IEEE Internet of Things (2017)
3. Benet, J.: IPFS-content addressed, versioned, P2P file system. arXiv preprint: arXiv:1407.3561 (2014)
4. Bettín-Díaz, R., Rojas, A.E., Mejía-Moncayo, C.: Methodological approach to the definition of a blockchain system for the food industry supply chain traceability. In: Gervasi, O., et al. (eds.) ICCSA 2018, Part II. LNCS, vol. 10961, pp. 19–33. Springer, Cham (2018). https://doi.org/10.1007/978-3-319-95165-2_2
5. Caro, M.P., et al.: Blockchain-based traceability in agri-food supply chain management: a practical implementation. In: 2018 IoT Vertical and Topical Summit on Agriculture - Tuscany, IOT Tuscany 2018, pp. 1–4 (2018)

6. Cartier, L.E., Ali, S.H., Krzemnicki, M.S.: Blockchain, chain of custody and trace elements: an overview of tracking and traceability opportunities in the gem industry. J. Gemmol. **36**(3), 212–227 (2018)
7. Casino, F., Dasaklis, T.K., Patsakis, C.: A systematic literature review of blockchain-based applications: current status, classification and open issues. Telemat. Inform. **36**, 55–81 (2019)
8. Casino, F., et al.: Modeling food supply chain traceability based on blockchain technology. In: Manufacturing Modelling, Management and Control - 9th MIM 2019 (2019)
9. Chen, Y.: Blockchain tokens and the potential democratization of entrepreneurship and innovation. Bus. Horiz. **61**(4), 567–575 (2018)
10. Christidis, K., Devetsikiotis, M.: Blockchains and smart contracts for the Internet of Things. IEEE Access **4**, 2292–2303 (2016)
11. Dabbene, F., Gay, P., Tortia, C.: Traceability issues in food supply chain management: a review. Biosyst. Eng. **120**, 65–80 (2014)
12. Dasaklis, T., Casino, F.: Improving vendor-managed inventory strategy based on Internet of Things (IoT) applications and blockchain technology. In: 2019 IEEE International Conference on Blockchain and Cryptocurrency (ICBC), pp. 50–55, May 2019
13. Dasaklis, T.K., et al.: Defining granularity levels for supply chain traceability based on IoT and blockchain. In: Proceedings of the International Conference on Omni-Layer Intelligent Systems, COINS 2019, pp. 184–190. ACM (2019)
14. ElMessiry, M., ElMessiry, A.: Blockchain framework for textile supply chain management: improving Transparency, Traceability, and Quality. In: Chen, S., Wang, H., Zhang, L.-J. (eds.) ICBC 2018. LNCS, vol. 10974, pp. 213–227. Springer, Cham (2018). https://doi.org/10.1007/978-3-319-94478-4_15
15. Galvez, J.F., et al.: Future challenges on the use of blockchain for food traceability analysis. TrAC Trends Anal. Chem. **107**, 222–232 (2018)
16. Hegge, H., Wortmann, J.: Generic bill-of-material: a new product model. Int. J. Prod. Econ. **23**(1–3), 117–128 (1991)
17. Islam, M.N.N., et al.: On IC traceability via blockchain. In: 2018 International Symposium on VLSI Design, Automation and Test, VLSI-DAT 2018, pp. 1–4 (2018)
18. Kim, M., et al.: Integrating blockchain, smart contract-tokens, and IoT to design a food traceability solution. In: IEEE 9th IEMCON 2018, pp. 335–340 (2019)
19. Lin, Q., Wang, H., Pei, X., Wang, J.: Food safety traceability system based on blockchain and EPCIS. IEEE Access **7**, 20698–20707 (2019)
20. Liu, P.T.S.: Medical record system using blockchain, big data and tokenization. In: Lam, K.-Y., Chi, C.-H., Qing, S. (eds.) ICICS 2016. LNCS, vol. 9977, pp. 254–261. Springer, Cham (2016). https://doi.org/10.1007/978-3-319-50011-9_20
21. Lu, Q., Xu, X.: Adaptable blockchain-based systems: a case study for product traceability. IEEE Softw. **34**(6), 21–27 (2017)
22. Mai, N., et al.: Benefits of traceability in fish supply chains-case studies. Br. Food J. **112**(9), 976–1002 (2010)
23. Malik, S., et al.: ProductChain: scalable blockchain framework to support provenance in supply chains. In: IEEE 17th NCA 2018 (2018)
24. Mann, S., et al.: Blockchain technology for supply chain traceability, transparency and data provenance. In: ACM International Conference Proceeding Series, pp. 22–25 (2018)
25. Marinello, F., et al.: Development of a traceability system for the animal product supply chain based on blockchain technology. In: 8th ECPLF, pp. 258–268 (2017)

26. Matsuura, K.: Token model and interpretation function for blockchain-based Fintech applications. IEICE Trans. Fundam. Electron. Commun. Comput. Sci. **1**, 3–10 (2019)
27. Memon, M., et al.: Analysis of traceability optimization and shareholder's profit for efficient supply chain operation under product recall crisis. Math. Probl. Eng. **2015**, 8 (2015)
28. Patsakis, C., Casino, F.: Hydras and IPFS: a decentralised playground for malware. Int. J. Inf. Secur. **18**(6), 787–799 (2019)
29. Sander, F., Semeijn, J., Mahr, D.: The acceptance of blockchain technology in meat traceability and transparency. Br. Food J. **120**(9), 2066–2079 (2018)
30. dos Santos, R.B., et al.: IGR token-raw material and ingredient certification of recipe based foods using smart contracts. Informatics **6**(1), 11 (2019)
31. Thiruchelvam, V., et al.: Blockchain-based technology in the coffee supply chain trade: case of Burundi coffee. J. Telecommun. Electron. Comput. Eng. **10**(3–2), 121–125 (2018)
32. Tian, F.: A supply chain traceability system for food safety based on HACCP, blockchain & Internet of Things. In: 14th ICSSSM (2017)
33. Westerkamp, M., et al.: Tracing manufacturing processes using blockchain-based token compositions. Digit. Commun. Netw. (2019, in press). https://www.sciencedirect.com/science/article/pii/S235286481830244X
34. Zhuang, C., Li, Y., Dai, Q., Liu, H.: A pharmaceutical supply chain traceability system based on blockchain and smart contract. In: Proceedings of International Conference on Computers and Industrial Engineering, CIE, vol. 2018, December 2018

First International Workshop on the Value and Quality of Enterprise Modelling (VEnMo)

First International Workshop on the Value and Quality of Enterprise Modelling (VEnMo)

In computer science and information systems development, Enterprise Modeling (EM) is used for different purposes, such as representing requirements, visualizing established work processes, specifying system design, expressing information structures, formalizing the relationships between organizational structures and IT landscapes, and much more. Despite this large spectrum of modeling purposes and use cases, the value of modeling in general and of Enterprise Models in particular is still the subject of research.

The First International Workshop on the Value and Quality of Enterprise Modelling (VEnMo 2019) invited researchers and practitioners with strong interest in the value of EM to discuss approaches and experiences in measuring, assessing, evaluating, and improving the value of enterprise related models like Enterprise Models, Enterprise Architecture Models, or Business Process Models. While it is common sense that EM adds value to organizations, existing research is still vague regarding how this value manifests. Therefore, the workshop aimed to shed light on existing as well as new theories, concepts, and frameworks for describing, understanding, and measuring the added value of EM.

One important aspect in this context may be the quality of EM, as only models of an adequate quality with respect to their intended purpose allow organizations to exploit the full value of EM. Therefore, the workshop further encourages researchers and practitioners to report on insights or approaches on how to assess the quality of enterprise models. Contributions may span from generic methods and appropriate quantifiable quality metrics to studies that investigate what makes an Enterprise Model of high quality from the perspective of relevant stakeholders.

VEnMo 2019 attracted six international submissions. At least three members of the Program Committee reviewed each paper. From these submissions, the top three were accepted as full papers for presentation at the workshop. The papers presented at the workshop provide a mix of an experiment on EM, measurement of process model reuse, and a survey on BPM anti-patterns tool support.

We hope that the reader will find this selection of papers useful to keep track of the latest direction on the value and quality of EM.

September 2019

Simon Hacks
Felix Timm
Kurt Sandkuhl
Michael Fellmann

Organization

Workshop Chairs

Simon Hacks (Co-chair)
Felix Timm (Co-chair)
Michael Fellmann
Kurt Sandkuhl

Program Committee

Mathias Ekstedt	KTH, Sweden
Peter Fettke	DFKI, Saarland University, Germany
Janis Grabis	Riga Technical University, Latvia
Marite Kirikova	Riga Technical University, Latvia
Agnes Koschmider	Karlsruhe Institute of Technology, Germany
Robert Lagerström	KTH, Sweden
Birger Lantow	Rostock University, Germany
Michael Leyer	Rostock University, Germany
Monika Malinova	Vienna University of Economics and Business, Austria
Judith Michael	RWTH Aachen University, Germany
Alexander Nolte	University of Tartu, Estonia
Erik Proper	Luxembourg Institute for Science and Technology, Luxembourg
Alixandre Santana	UFRPE, Brazil
Rainer Schmidt	Munich University, Germany
Ulf Seigerroth	Jönköping University, Sweden
Nuno Silva	UTL, Portugal
Janis Stirna	Stockholm University, Sweden
Margus Välja	Scania AB, Sweden

Enterprise Modelling of Digital Innovation in Strategies, Services and Processes

Geert Poels[(✉)] [iD]

Faculty of Economics and Business Administration, Ghent University,
Ghent, Belgium
geert.poels@ugent.be

Abstract. We report upon a study performed on 65 cases of digital innovation where graduate business students applied enterprise modelling to analyze and demonstrate the impact and value of implementing digital technologies. As students could freely choose which enterprise modelling techniques to apply, these cases provide insight into which enterprise modelling approaches and which types of enterprise models they preferred to use for the analysis. The study relates those preferences to type of digital technology implemented and the focus area of the digital innovation, i.e., strategy, services (internal and external) and processes. The preliminary insights from this study help directing further research on how enterprise modelling can have value for managerial decision-making on digital innovation.

Keywords: Enterprise modelling · Digital innovation · Impact analysis

1 Introduction

Questions about the value of enterprise modelling and how this value is manifested are existential questions for the discipline. They follow the trend in other modelling disciplines, which are distinct though overlapping with, encompassing or subsumed in enterprise modelling, like business process modelling [1], conceptual modelling [2] and enterprise architecture modelling [3].

A common treat in these discussions is the rapid change of information technologies that organizations currently (need to) implement to innovate, manage and run their business. While modelling fits the enterprise engineering paradigm of planned change, it is less clear what the relevancy (and hence value) of modelling is when organizations are transforming because of emergent change. Such change is often brought about by newcomers that disrupt industries with radically different technology-based business models (e.g., the rise of platform economies enabled through cloud technologies, the pressure of alternative blockchain-based payment and credit systems currently felt by banks and intermediaries in the financial sector). Sometimes, modelling is even seen as a time-consuming, non-value adding step – "Let the data speak for itself" is an often-heard slogan in relation to schema-less databases, Big Data, and NoSQL.

Whether organizations are disruptors or are being disrupted, we see, apart from the pace of change, no essential difference in the basic questions that managers have

© Springer Nature Switzerland AG 2019
C. Di Francescomarino et al. (Eds.): BPM 2019 Workshops, LNBIP 362, pp. 721–732, 2019.
https://doi.org/10.1007/978-3-030-37453-2_57

regarding the impact and value of the implementation of traditional information technologies (e.g., ERP systems, Web applications, BPM systems) and novel digital technologies (e.g., cloud-based platforms, distributed applications based on smart contracts, IoT applications). As such, the current wave of digital innovation does not invalidate a priori the relevancy that enterprise modelling has for assisting managers in finding answers to these questions. It is our position, however, that it might not be clear *how* enterprise modelling should be used to investigate the impact and value of digital innovation. What insights can be gained through enterprise modelling, by means of which modelling languages and model analysis techniques, and how can managers use these insights for more effective and efficient governance of digital innovation initiatives?

To investigate these research questions, an appropriate research design is needed. We felt that the design of our research process and methodology could gain from a pre-study in which we explore the topic to get initial insights that could direct our research. This paper reports on such pre-study with graduate business students, which was performed at Ghent University in the fall semesters of 2017–2018 and 2018–2019. The idea was to have these students apply enterprise modelling to self-chosen cases of digital innovation to find out which enterprise modelling approaches and types of enterprise model they considered most useful to analyze and to demonstrate how technology impacts strategy, services and processes.

Section 2 provides background information on the participating students and the courses in which context the study was performed. Also, the enterprise modelling approaches taught in these courses are reviewed. Further, preliminary findings from a quantitative analysis of the case deliverables are presented and discussed. Section 3 concludes with lessons learned for our research.

2   Preliminary Findings from a Pre-study

2.1   Context

In the study participated Master students of Business Engineering (BE) that were enrolled in the 45-h course Enterprise Architecture (175 in 2017 and 152 in 2018). A further 87 Master students of Business Administration (BA) majoring in 'Management and IT' that were enrolled in the 30-h course IT Management, participated in the fall of 2018. The typical student had no working experience and the mode of the age is 22 years for the BA students and 23 years for the BE students.

The structure of both courses is similar, only IT management doesn't have the third module:

1. Theoretical models of business/IT alignment, digital innovation, technology disruption, IT governance, and enterprise engineering;
2. Practice of enterprise modelling – Analysis of the enterprise as a system;
3. Practice of enterprise architecture – Design of the enterprise as a system.

Twenty percent of course credits are earned with a group assignment (6 to 7 students per group) in which students analyze the impact and value of implementing IT in an organization. Each group chooses its own case, with a preference for an emerging digital technology and a real organization. The task that is of interest to the study was formulated as *Apply the enterprise modelling techniques that, according to you, are best suited to analyze and demonstrate the impact and business value of the IT implementation.* The students of the Enterprise Architecture course were also required to develop an enterprise architecture model (in ArchiMate). The ArchiMate model further elaborates the developed enterprise models. We opt to keep ArchiMate outside the scope of the study as there was no 'free choice' here. The deliverable was a presentation delivered to the professor and selected groups of other students.

2.2   Catalogue of Enterprise Modelling Approaches

Table 1 provides an overview of the enterprise modelling approaches taught. Although the students were free to employ any approach they deemed suitable for the assignment, these approaches were more likely to get chosen.

The classification according to type, focus, view and purpose is the one used in the courses. Enterprise modelling type refers to the form of the enterprise model, which can be a description (i.e., text), a canvas (i.e., a fixed-form template with placeholders), a map (i.e., a table with cells) or a diagrammatic model. The Business Motivation Model (BMM) [4] offers concepts to describe strategic planning concepts, without prescribing a concrete syntax or any other kind of format. The Business Model Canvas (BMC) [5] is a canvassing technique for applying the Business Model Ontology (BMO) [6] where users instantiate concepts by filling in pre-defined slots on the canvas. This technique is also incorporated in the Continuous Business Model Planning method [7]. The Component Business Model (CBM) [8] allows developing business component maps as tabular (i.e., two-dimensional) representations. All other approaches (i.e., i* [9] following iStar 2.0 notation [10], e3-value [11], Value Delivery Modeling Language (VDML) [12], Capability-Driven Development (CDD) [13], which borrows heavily from 4EM [14], and Capability-Actor-Resource-Service (CARS) [15]) require enterprise models to be constructed by directly or indirectly (e.g., by means of a story-telling technique in the CBMP method for applying VDML) instantiating meta-models following a concrete syntax consisting of a notation and often also diagram types.

The focus of the enterprise modelling approaches is derived from the conceptual framework for enterprise modelling that is used in both courses (Fig. 1). In simple terms, the framework distinguishes between approaches that aim at providing an answer to why an organization does what it does (i.e., goals and strategy) and approaches that focus on how an organization choses to do so (i.e., the business model elements), where the latter question can be refined into what an organization offers (i.e., value propositions) and how this is done in terms of resources and activities (i.e., business capabilities, value streams, business processes). The question of who an organization needs for these resources and activities (or with whom value is exchanged) is addressed by enterprise modelling approaches that take a business ecosystem view, whereas other approaches take a predominantly business or (IT) system's perspective of the focal organization.

The final column in Table 1 clarifies the main purpose of using an approach, as it was taught to the students.

Table 1. Enterprise Modelling (EM) approaches taught to the study participants.

EM approach	EM type	EM focus	EM view	EM purpose: developing and analyzing
Business motivation model [4]	Describing	Goals and strategy	Business	Strategic plans
i* [9] /iStar 2.0 [10]	Modelling	Goals and strategy (implicit)	System	System goals and their dependencies
Business model ontology [6]/ Business model canvas [5]	Canvassing	Business model elements	Business	Business models
E3-value [11]	Modelling	Value propositions	Business ecosystem	Value networks
Value delivery modeling language [12]/Continuous business model planning [7]	Canvassing Modelling	Business model elements Value propositions Value streams Business capabilities	Business ecosystem Business	Value networks Business models Strategy maps
Component business model [8]	Mapping	Business components[a]	Business	Business capability (heat) maps
Capability-driven development [13]	Modelling	Goals System context Business processes	System	Capability models (via goals, context and process models)
Capability-actor-resource-service [15]	Modelling	Business capabilities	Business ecosystem Business	Capability models

[a]Similar to business capabilities, though CBM emphasizes that business components should be able to operate as independent units, hence a notion of organizational unit is involved.

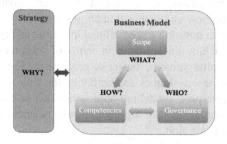

Fig. 1. Enterprise modelling conceptual framework used in the involved courses.

2.3   Quantitative Analysis

The deliverables of a total of 65 groups (26 groups BE 2017, 25 groups BE 2018, 14 groups BA 2018) were analyzed.

Figure 2 shows for each approach in Table 1 the absolute and relative number of groups that used the approach. The median number of approaches used was 2. The 'other' category comprises only BPMN process diagrams.

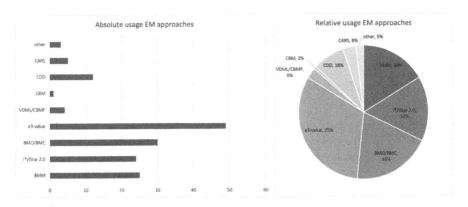

Fig. 2. Usage of Enterprise Modelling (EM) approaches.

Figure 3 repeats the analysis for types of enterprise model used, according to the classification below. The median number of enterprise model types used was also 2.

- Goal models (BMM strategic plans, i*/iStar 2.0 strategic dependency diagrams & strategic rationale diagrams, VDML/CBMP strategy maps, CDD goal models);
- Business models (BMO/BMC business model canvas, VDML/CBMP business model canvas & business model cube);
- Value models (e3-value model, VDML/CBMP business ecosystem model);
- Capability models (CBM capability map, CARS capability model)[1];
- Context models (CDD context model);
- Process models (CDD process model, BPMN process diagram).

We also related the use of enterprise modelling approaches and enterprise model types to the particularities of the chosen cases, more specifically the type of digital technology implemented and the focus area of the digital innovation.

---

[1] CDD does not provide in a specific capability model, but models capabilities via goal models, context models and process models.

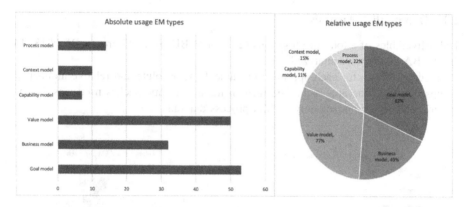

Fig. 3. Usage of types of Enterprise Model (EM)

Regarding the type of digital technology, we started from the SMAC/BRAID[2] classification of well adopted (SMAC) and novel (BRAID) digital technologies, that is often used by consultants [17]. Based on the student groups' cases, we merged the Robotics and Automation of knowledge work classes into an Intelligent Automation (IA) class, added Artificial Intelligence (AI) and Extended Reality (XR) classes for types of digital technology not explicitly covered by SMAC/BRAID, and added Information Systems (IS) and Computer Technology (Tech) classes to cover cases that discussed the implementation of more conventional information and communication technologies. For each case we looked for the dominant technology that enabled the digital innovation, though 6 cases were classified twice as no choice could be made between two technologies that were assessed as equally important for enabling the digital innovation (e.g., self-driving cars at DHL, for which both IoT and Analytics are crucial). Table 2 lists the number of cases for each class of digital/information/computer technology. Classes with less than 5 cases were excluded from further analysis.

Table 2. Types of digital/ICT technologies covered by the cases

Social media	Mobile	Analytics	Cloud	Blockchain	IA	IoT	Digital fabrication	AI	XR	IS	Tech
0	3	4	4	9	5	12	1	16	6	6	5

Figure 4 shows for each class of technology, the percentage of cases that used a particular enterprise modelling approach for analyzing and demonstrating the impact and value of the implemented technology. Figure 5 repeats this analysis for enterprise model types. The results for the IA, XR, IS and Tech classes should be interpreted with caution as these classes were populated by only 5 or 6 cases.

---

[2] Social media, Mobile technologies, Big Data Analytics, Cloud services/Blockchain, Robotics, Automation of knowledge work, Internet of Things, Digital fabrication (e.g., 3-D printing).

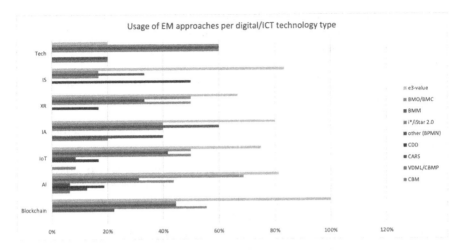

Fig. 4. Usage of Enterprise Modelling (EM) approaches per type of digital/ICT technology.

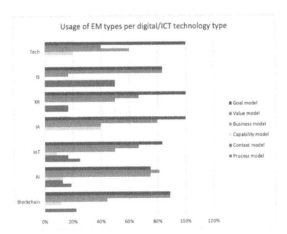

Fig. 5. Usage of Enterprise Model (EM) type per type of digital/ICT technology.

Regarding the focus area of digital innovation, we distinguished, based on the student groups' cases, five classes:

- Strategy (5 groups): Digital innovation aimed at transforming an organization on technological grounds [16], causing disruption or being caused by disruption (e.g., streaming on demand enabled by Artificial Intelligence (AI) and Cloud – case of Netflix);
- Services – external (15 groups): Digital innovation aimed at providing new or improved services to customers, changing the organization's value proposition(s) (e.g., BRUce Pepper, a traveler assistance robot enabled by Intelligent Automation – case Brussels airport);

- Services – internal (15 groups): Digital innovation aimed at providing new or improved services to employees/functions within the organization (e.g., Digital Car Pass, a Blockchain and Internet of Things (IoT) solution for automatic gathering of car performance data – case Volkswagen);
- Process (27 groups): Digital innovation aimed at process improvements (e.g., a digital interviewing system enabled by AI for improving the recruitment process – case UniLever);
- Infrastructure (3 groups): Innovation in an organization's technology infrastructure (e.g., quantum computing – case NASA).

We acknowledge that the boundaries between these classes are not crisp as new services to customers may eventually disrupt an industry and process improvements may result in better internal services, which then may affect services provided to customers. As a pragmatic heuristic we classified a case as process type of digital innovation if one or a few processes could be identified where efficiency improvements were targeted. If no such processes could be identified, then the case was assigned to the internal or external services classes depending on whether the innovation will benefit employees/functions or customers. Only if the digital innovation would radically and irreversibly alter the way of doing business, a case was classified as strategic. The infrastructure class, with only 3 cases, is excluded from further analysis.

Figure 6 shows for each class of digital innovation, the percentage of cases that used a particular enterprise modelling approach for analyzing and demonstrating the impact and value of the implemented technology. Figure 7 repeats this analysis for enterprise model types. Note that the strategy class was populated by only 5 cases, which is probably not sufficient to obtain reliable insights from its results.

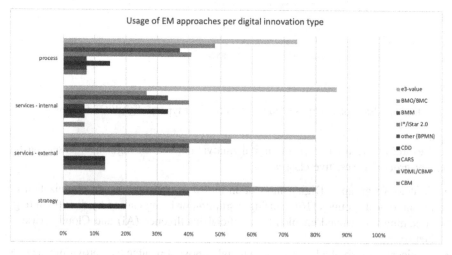

Fig. 6. Usage of Enterprise Modelling (EM) approaches per digital innovation type.

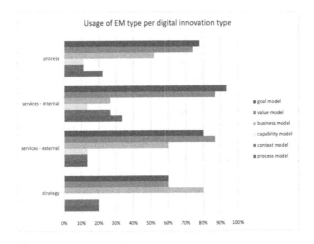

Fig. 7. Usage of Enterprise Model (EM) type per digital innovation type.

2.4 Discussion

The data on the usage of enterprise modelling approach (Fig. 2) reveals e3-value as a clear winner, used by 75% of the groups. Figures 4 and 6 confirm that this observation is independent from the type of digital technology or the type of digital innovation investigated. The reasons for these results require further research, however, it is to be noted that e3-value is one of the few approaches with a business ecosystem view (see Table 1). In our further research, we will investigate if such view is inherently useful or even required for analyzing digital innovation, given that organizations implementing digital technologies usually need to partner up with technology providers, consultants or other intermediaries, and thus issues of sourcing and value co-creation arise.

Three other approaches, BMO/BMC, BMM and i*/iStar 2.0, follow at some distance, being used by respectively 46%, 38% and 37% of the student groups. The first two approaches are relatively simple to apply as they don't require graphical modelling as a meta-model instantiation process. This could explain their relative popularity with our business students. Nevertheless, more than one third of groups used i*/iStar 2.0, which is arguably an easy modelling technique. Like with e3-value, we observe from Figs. 4 and 6 that these results on usage are robust for type of technology and type of innovation considered. Noteworthy is that BMO/BMC was used by more than two-thirds of the groups that investigated AI cases (see Fig. 4), while it was used by less than one-third of groups that investigated cases of digital innovation impacting on internal services (see Fig. 6). This might indicate that AI projects are expected to affect the business model of an organization, which can be shown using a business model canvas, while the impact of digital innovation on services provided to employees/ functions is harder to analyze or demonstrate using a business model canvas.

All other approaches were used by relatively small numbers of student groups, though CDD with 18% was clearly more attractive to the students than VDML/CBMP, CBM and CARS. CDD was used in half of the IS cases (see Fig. 4) and in one-third of the internal services cases (see Fig. 6), which might be explained by the systems view of this enterprise modelling approach.

While the usage of particular enterprise modelling approaches plausibly reflects also other reasons than usefulness (e.g., perceived difficulty – including that of the modelling tool), the usage of enterprise model types might provide a clearer picture of which enterprise models are best suited to analyze and demonstrate the impact and value of implementing digital technologies. Figure 3 shows that goal models (82%) and value models (77%) stand out, followed by business model representations (49%). These results are confirmed in Figs. 5 and 7, hence do not seem to depend on type of technology and type of digital innovation.

The result for value models follows the observation regarding the preference for e3-value. More than four/fifths of groups used goal models, which is also in line with the results of Fig. 2, given that BMM and i*/iStar 2.0 are the predominant approaches employing this type of model and only 5 projects used both approaches. These results indicate that the student groups attached importance to investigating the motivation behind digital innovation.

The usage of business model representations follows the usage of BMO/BMC. Almost half the groups used a business model canvas to clarify which business model elements are impacted by digital innovation. Figure 5 shows that a business model representation was used by three-fourths of the groups investigating AI, which confirms the earlier observation on BMO/BMC for this type of technology. Also, in three-fifths of the cases on digital innovation aimed at services targeted at customers, a business model representation was used (see Fig. 7), probably because of its ability to show how the innovation alters value propositions, channels, customer relationships and customer segments.

Some 22% of groups employed BPMN process diagrams (see Fig. 3), mostly as part of CDD. Remarkable is that the same percentage of usage was observed for cases that investigated the process class of digital innovation (see Fig. 7). As a business process model is clearly less abstract than a goal model, value model or business model canvas, we need to investigate if the level of abstraction of process models is appropriate for the type of analysis envisioned.

3   Conclusion

We recall from sub-Sect. 2.1 that students could choose the enterprise modelling techniques that were, according to them, best suited to analyze and demonstrate the impact and value of digital innovation for their chosen case. The quantitative analysis thus reflects the preferences of the involved students and is not an evaluation in an authentic setting of the effectiveness of these techniques in investigating digital innovation. Further analyses can be conducted with the case data already obtained. For instance, relating type of digital technology with type of digital innovation and investigating whether particular combinations have led to certain preferences for enterprise modelling techniques. Also, contingency factors like size of the organization or industry sector can be included in the analysis. Despite the immaturity of our analysis so far, the results provide insights into which techniques were preferred and when they were used by the students.

As our goal is to clarify *how* enterprise modelling should be used to investigate the impact and value of digital innovation, the main lessons learned from the pre-study can be summarized as:

- Goal models seem to be important for the type of analysis performed as they can be used to examine the motivation behind digital innovation;
- A business ecosystem view such as used by e3-value, seems to be useful for analyzing digital innovation – for blockchain projects this might even be a requirement;
- Apart from a goal model and a value model, a business model representation seems to be useful if digital innovation is aimed at new or improved services towards customers, given that this has an impact on the organization's business model – at least for AI projects this seems to be the case;
- Process models were seldom used, even in cases where digital innovation targeted process improvement – this result is non-intuitive and needs further research;
- Capability models were clearly not the enterprise model of choice of the business students participating in the study – maybe the concept of capability is too abstract to be of practical use for investigating real cases of digital innovation?

Our research will thus be directed towards further investigating these preliminary insights and based on that understanding conducting a Design Science research on a method for enterprise modelling that assists managers in decision-making regarding digital innovation. Based on informal feedback from the students, we are confident that enterprise modelling is still relevant in the digital era and we look forward to a new round of case-studies in the fall semester of 2019–2020.

References

1. Has BPM Missed the IoT Revolution? Panel discussion at BPM 2018, Sidney, Australia. http://bpm2018.web.cse.unsw.edu.au/BPM2018_Panel.pdf
2. Delcambre, Lois M.L., Liddle, Stephen W., Pastor, O., Storey, Veda C.: A reference framework for conceptual modeling. In: Trujillo, Juan C., et al. (eds.) ER 2018. LNCS, vol. 11157, pp. 27–42. Springer, Cham (2018). https://doi.org/10.1007/978-3-030-00847-5_4
3. Babar, Z., Yu, E.: Enterprise architecture in the age of digital transformation. In: Persson, A., Stirna, J. (eds.) CAiSE 2015. LNBIP, vol. 215, pp. 438–443. Springer, Cham (2015). https://doi.org/10.1007/978-3-319-19243-7_40
4. Object Management Group (OMG): Business Motivation Model, version 1.3. May 2015. https://www.omg.org/spec/BMM/1.3/PDF
5. Osterwalder, A., Pigneur, Y.: Business Model Generation. Wiley, Hoboken (2011)
6. Osterwalder, A., Pigneur, Y., Tucci, C.L.: Clarifying business models: origins, present, and future of the concept. Commun. Assoc. Inf. Syst. **16**(1), 1 (2005)
7. Poels, G., Roelens, B., de Man, H., van Donge, T.: Designing value co-creation with the value management platform. In: Satzger, G., Patrício, L., Zaki, M., Kühl, N., Hottum, P. (eds.) IESS 2018. LNBIP, vol. 331, pp. 399–413. Springer, Cham (2018). https://doi.org/10.1007/978-3-030-00713-3_30

8. Cherbakov, L., Galambos, G., Harishankar, R., Kalyana, S., Rackham, G.: Impact of service orientation at the business level. IBM Syst. J. **44**, 653–668 (2005)
9. Yu, E.: Towards modelling and reasoning support for early-phase requirements engineering. In: Proceedings of 3rd International Symposium on Requirements Engineering, RE 1997, pp. 226–235. IEEE (1997)
10. Dalpiaz, F., Franch, X., Horkoff, J.: iStar 2.0 Language Guide. arXiv:1605.07767 (2016)
11. Gordijn, J., Akkermans, H.: Designing and evaluating e-business models. IEEE Intell. Syst. **16**(4), 11–17 (2001)
12. Object Management Group (OMG): Value Delivery Modeling Language (VDML), version 1.1. October 2018. https://www.omg.org/spec/VDML/1.1/PDF
13. Berzisa, S., et al.: Capability driven development: an approach to designing digital enterprises. Bus. Inf. Syst. Eng. **57**(1), 15–25 (2015)
14. Sandkuhl, K., Stirna, J., Persson, A., Wißotzki, M.: Enterprise Modeling. TEES. Springer, Heidelberg (2014). https://doi.org/10.1007/978-3-662-43725-4
15. Rafati, L., Poels, G.: Value-driven strategic sourcing based on service-dominant logic. Serv. Sci. **9**(4), 275–287 (2017)
16. Henderson, J.C., Venkatraman, N.: Strategic alignment: leveraging information technology for transforming organizations. IBM Syst. J. **38**(2/3), 472–484 (1999)
17. Willcocks, L., Hindle, J., Lacity, M.: Keys to RPA Success. February 2019. https://static1.squarespace.com/static/58eceda617bffc97d03b69da/t/5c7d0ab2ec212d1ac3057e42/1551698611602/KCP_Report_Path_to_Maturity_.pdf

Measuring Business Process Model Reuse in a Process Repository

Ross S. Veitch[✉] and Lisa F. Seymour

University of Cape Town, Cape Town, South Africa
vtcros002@myuct.ac.za, lisa.seymour@uct.ac.za

Abstract. The value of process modeling increases with process model reuse. Previous research into process model reuse has focused on behavioral aspects of reuse such as the intention to reuse, the repeated reuse of a process model over time, and the identification of elements of process models which could be reused. However, process model reuse can also be considered from the perspective of the reuse of process models by other process models in the same repository. Such a measure would be a direct measure of whether process modelers are creating bespoke versions of existing process models or are indeed reusing existing process models and reaping some of the purported benefits of reuse. Furthermore, it would provide a measure of reuse which can be automated. Organizations which operate in a multi-channel, multi-product environment have business processes which frequently share functionality (consider authentication for example) and which may be used in different organizational units. While the reuse of complete process models in a process repository is one of the benefits of using a process repository, no research could be found relating to the measurement of the reuse of complete process models by other process models within such a repository. We believe that this paper is the first to propose and validate a measure of complete process model reuse by other process models in the same process repository. The measure is then applied to a real-world process repository of a large financial services organization, illustrating the applicability and potential usefulness of the measure.

Keywords: Reuse measurement · Business process modeling · Business process model reuse

1 Introduction

Complicating enterprise process modeling is the proliferation of new channels. Organizations are now having to design similar processes covering multiple channels (consider bricks and mortar, call centers, the internet, mobile devices, email, social media, instant messaging, and more), while ensuring consistency, as customers are expecting seamless, improved, and consistent experiences when dealing with the organisation [1]. In practice, these organizations will have typically documented their business processes in a process repository. To improve the economic value of these models, organizations wish to reuse them where possible, and the question of how to achieve and measure process model reuse arises. A possible measure of process model reuse is the amount of reuse of models in a repository by other process models in the

© Springer Nature Switzerland AG 2019
C. Di Francescomarino et al. (Eds.): BPM 2019 Workshops, LNBIP 362, pp. 733–744, 2019.
https://doi.org/10.1007/978-3-030-37453-2_58

same repository. Measuring the level of process model reuse by other models in the process repository would be an important indicator of model reuse. For example, it enables the level of process model reuse to be measured directly and automated, the modeling practices within an organization to be managed, and even the direct comparison of modeling methods regarding the reuse of process models. While process model reuse is a frequent topic of research, no research could be found related to measuring the reuse of models within a process repository by other models within the repository.

In this paper, we extend the field of process model reuse by formally defining a measure of process model reuse by other process models in a process repository. Reuse measurement is well established in software engineering and we draw on these principles in the development of a measure of the reuse level of process models. Accordingly the research question for this paper is "How can the reuse of complete process models by other models in a process repository be measured?"

The rest of this paper is structured as follows: in Sect. 2 we review the literature relating to process model and software reuse. In Sect. 3 we describe the method used and in Sect. 4 we evaluate it in a real-world instantiation. The measure is discussed in Sect. 5.

2 Literature Review

Most large organizations try to use a centralized process repository to store process models [2]. Examples of process modeling tools in use are ARIS Business Architect and iGrafx, while ERP systems from SAP and Oracle, and Business Process Management Suites from IBM, Software AG, and Tibco Software also include a process modeling tool [3].

Much research has been conducted into aspects of the reuse of process models: reuse of process fragments, similarity searches, reference models, management of process variants, and patterns. Several literature surveys relating to reuse in business process modeling have also been published [4, 5]. However, while the reuse of software and the measurement thereof has been studied [6], no research into the measurement of reuse of complete business process models by other process models could be found.

Typical benefits of process model reuse are claimed to be: a reduction in modeling time, an increase in productivity, leveraging of existing resources and skills, and an increase in the quality of the process models [5, 7]. The use of the word "claimed" is deliberate because although these benefits are stated in most studies about the reuse of process models, validation of these claims is almost completely missing [7]. One research study showed that only 10.5% of respondents reused complete process models, 55.1% reused individual process elements and that the search for reusable model assets was conducted manually by 40.5% of respondents [5]. It could be argued that the positive effects of process model reuse cannot be realized because process model reuse is not being practiced sufficiently [5]. Related to this dearth of empirical data on process model reuse by other models is the question of how process model reuse within a repository should be measured.

2.1 Software Development and Business Process Modeling (BPM)

Diagrams are used to represent conceptual solutions to problems, and this is particularly prevalent in the fields of software development and BPM [8]. Visual programming languages (VPL's) have been developed which take a graphical representation of program logic and can directly generate the corresponding textual version which can be compiled into an application and executed [8]. BPM notations such as EPC and BPMN are commonly used to represent business process control-flow logic while BPEL, which can be converted directly into an executable application by many workflow systems, is essentially BPMN with additional constraints enforced [9]. Accordingly, business process models can be considered as a programming abstraction for business users where the commands are not executable but syntactical in nature [4, 10].

2.2 Software Reuse

The expected benefits of software reuse are well documented: improvements in productivity and quality, savings during the maintenance phase due to fewer errors [6], quicker time to market for the product [6, 11] and improved interoperability of components [6, 11]. A high correlation between reuse rate, reduced development time, and a decrease in the number of errors has also been identified [11]. However, reuse does not come for free [12], and for software reuse to make economic sense, the quantified benefits of reuse need to exceed the cost of developing for reuse. While this has resulted in various economic models which can be used to quantify the benefits of reuse [6, 11], all economic models depend on some measure of reuse as an input to the model. Measurement models for software reuse are commonly based on counting either lines of code (excluding comment lines), data items (e.g., business objects) or Function points [13].

Reuse Level. The de facto metric for measuring software reuse is the reuse % or level of reuse [6] which measures how much of the product can be attributed to reuse [11]. Their measurement of software reuse was based on the assumption that the software is composed of items at different levels of abstraction. An item could be a module (.c file) which in turn is composed of functions which contain lines of code. A lower level item is used by a higher-level item. The equations of Terry [14] for software reuse can be written as:

$$\text{External Reuse Level} = \text{EU}/\text{T} \tag{1}$$

$$\text{Internal Reuse Level} = \text{IU}/\text{T} \tag{2}$$

$$\text{Total Reuse Level} = \text{Internal Reuse Level} + \text{External Reuse Level} = (\text{IU} + \text{EU})/\text{T} \tag{3}$$

Where:

ITL = internal threshold level, the minimum number of times an internal item must be used before it is counted as reused
ETL = external threshold level, the minimum number of times an external item must be used before it is counted as reused
EU = number of external lower level items that are used more than ETL
IU = number of internal lower level items that are used more than ITL
T = total number of lower level items in the higher-level item, both internal and external

It is important to recognize that these counts are of unique items and not references. In other words, an item is only counted once even if it is reused ten times.

3   Method

The development of the measure of process model reuse by other models in a repository followed a design science research methodology consisting of the following stages: problem identification and motivation, objectives of a solution, design and development, demonstration, evaluation, and communication [15]. To guide the design and evaluation of the measure, we use a real-world process repository. The model is then evaluated in its context using the validation framework of Fenton, Kitchenham, and Pfleeger [16]. Theoretical validation was conducted by assessing each measurement item against criteria required by the framework. Empirical validation was conducted by confirming that the model generated the same results for attribute values which could be confirmed by manually inspecting the process repository.

3.1   Terminology

Before proceeding, the terminology used in this paper must be explained as it forms the basis of the discussion and approach to measuring the reuse of business process models:

- A function can be an individual activity or task at a granular level (e.g., capture customer name) or an abstraction of another process (e.g., authenticate customer). A Function exists as an object of type "Function" in the process repository. If a function is an abstraction of another process model, that process model is said to be "assigned to" that Function (Fig. 1a). Alternately it can be said that Function F2 decomposes into M11.
- A model represents a process and consists of functions which can be individual activities/tasks at a granular level or abstractions of other processes.
- An object (e.g., a function) contains all the attributes and attribute values for a specific object. For example, the attribute "Name" contains the name of the object.
- An object (e.g. a Function), appears in a model as an "Occurrence" (Fig. 1b). The occurrence in a model is a pointer to the object in the repository. It is essential to understand that a specific object can occur multiple times in the same model or even in multiple models.

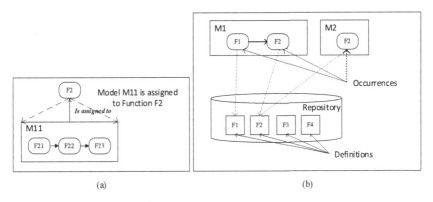

(a)   (b)

Fig. 1. Objects and occurrences

- All occurrences of Functions in a model are represented by a symbol which was selected by the modeler. The same Function can occur as different symbols in different models.
- The parent object of a model is the object to which a model is assigned. In Fig. 1a, object F2 is the parent of model M11.

Significant reuse opportunities arise in repositories which have this structure as updating a process model M which is assigned to a function F will result in all other process models in which function F occurs, also adopting that change. In Fig. 2, by making a change to the function F5 in model M1, the object F5 in the repository is changed. It can be seen that model M2 also has an occurrence of the object F5 and therefore the change made to F5 also reflects in the model M2.

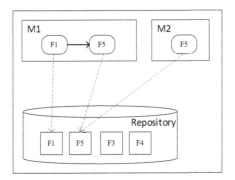

Fig. 2. Object reuse in models

The Business Process Architecture (BPA) typically defines the high-level process grouping found in the organization and provides guidelines regarding the decomposition of such processes into levels of increasing granularity [17–19]. The number of levels of decomposition adopted depends on each organization, but typically 3–5 levels of decomposition are used as shown in Fig. 3 for three levels of decomposition [20]. Level 0 is the highest level (most abstract) while Level 5 would contain the most granular process models.

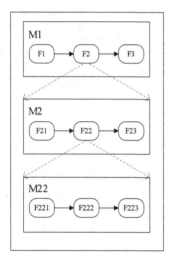 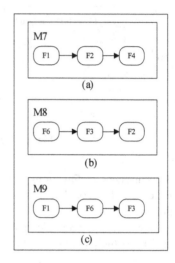

Fig. 3. Business process architecture hierarchy   **Fig. 4.** Complete process model reuse

3.2   Reuse of Complete Process Models

Using Figs. 3 and 4, the concept of complete process model reuse is illustrated. While function F2 is an abstraction of process model M2, F2 is used in both models M7 and M8. Similarly, F1 is used in M7 and M9. A change to the underlying model M2 will change the meaning of F2 and therefore the models M7 and M8 will also be impacted. It stands to reason that the number of times M2 has been used can be calculated by determining the number of occurrences of F2 in models in the repository.

3.3   Developing a Process Model Reuse Measure

To determine how often a process model is reused, it is first necessary to determine the number of times each model is referenced from another process model. This is a three-step process:

1. Determine the parent object of each process (Fig. 1).
2. Determine in which other process models the parent object occurs (Fig. 2)
3. Check the symbol of the parent object in each model that it occurs. If it is NOT a link to the previous/next model at the same architecture level, then it must be counted.

3.4   Defining Complete Process Model Reuse

We define the use of a complete process model in another model to be the occurrence of the parent function of the model in another model such as the occurrence of F2 in

M1 (Fig. 3). This means that M2 has been used in M1. However, the first use of a model cannot be classified as reuse and accordingly a model can only count as reused if it is used more than once. Therefore, we define complete process model reuse as follows:

Complete process model reuse is the occurrence of the parent object of a model in more than one model in the repository and such occurrence is not acting as a 'cross-page connector' to the next process model.

3.5 Developing an Equation for Complete Process Model Reuse

Let:

k = process architecture level

m_k be the process model m at architecture level k

t_k be the number of models in the repository at architecture level k

t_{m_k} be the number of activities in model m_k

p_{tot} be the total number of parent objects in the repository

$A_{m_k} = \left\{ a_{m_k,1}, \ldots, a_{m_k,t_{m_k}} \right\}$ is the set of activities occurring in model m at architecture level k (i.e. in model m_k)

t_{m_R} = number of process models in the repository

Complete process model reuse occurs when an entire process model (m_k) is abstracted into a single activity ($a_{m_{k-1}}$) in another process model $m_{k-1,i}$, one architecture level higher.

Referring to parent-child terminology which is easier to understand:

$a_{m_{k-1,i}}$ is the parent activity of model $m_{k,i}$

We can say that $a_{m_{k-1,i}}$ occurs in model m if $a_{m_{k-1,i}} \in A_{m_i}$

And the count of process models which contain a reference to model $m_{k,i}$ will be:

$$C\left(m_{k,i}\right) = \sum_{k=1}^{k=k_{max}} \sum_{p=1}^{p=p_{tot}} \sum_{t=1}^{t=t_k} \left| \{a_{m_k,p}\} \cap A_{m_{k-1,t}} \right| \tag{4}$$

Where $\{a_{m_{k,p}}\}$ is a single element set consisting of the element $\{a_{m_{k,p}}\}$ only.

$\{a_{m_{k,p}}\} \cap A_{m_{k-1,t}}$ is the intersection of $\{a_{m_{k,p}}\}$ with the set of activities occurring in model $m_{k-1,t}$. Then $||$ denotes cardinality and will evaluate to 1 if $a_{m_{k,p}}$ occurs in $m_{k-1,t}$.

3.6 Defining What to Measure

Reuse Level. For the measurement of the reuse of process models in a repository, we only measure the internal reuse by ensuring that only process models created by the organization are considered. Drawing on the de-facto standard for measuring software reuse [6] we accordingly define the reuse level as follows:

$$\text{Process Reuse Level} = \text{M/L} \tag{5}$$

Where:

> M = the number of models in the repository that are used more than once. Using Eq. (4), the number of times each model is used is calculated. Referring to Eq. (2), we set the threshold value for process model reuse (ITL) to be at least 2. A process model can only be considered as reused when it is used at least twice.
> L = the total number of models in the repository.

How Do We Handle Complexity? While in the measurement of software reuse it has been proposed that the lower level item is weighted based on its size (e.g., number of lines of code) [11], we argue that this approach should not be used to weight process models when measuring reuse as frequently two processes with the same purpose may vary significantly in terms of size. For example, obtaining the approval of a customer can be both a simple process (customer taps on "I approve" in a mobile app) while it could also be a complex process (the company must send a driver to visit the customer to obtain an approval signature on a piece of paper). Two such processes should be equally weighted. Accordingly, we do not provide for complexity when measuring process model reuse.

What Do We Count? An activity in a process model can be considered to be the equivalent of a line of code (LOC) in software [21]. At the lowest level of granularity in a process model (e.g., in a level 5 process model in a 5-level process architecture), an activity does not decompose into any further detail. In ARIS, activities are represented using function objects which, when combined with other object types and connections, make up a process model. Every process model can be (but does not have to be) linked (assigned) to another function object which in turn can form part of a different process model, and in this manner, the process hierarchy is constructed (Figs. 3 and 5). Only functions with a process model assigned to them (Fig. 6) can contribute to the reuse count. A process model which is not assigned to a function (e.g., Y in Fig. 6) cannot be reused because no mechanism then exists by which to reference that process model.

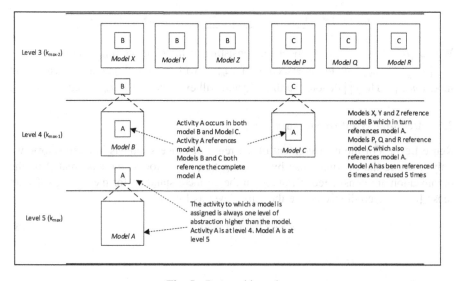

Fig. 5. Process hierarchy

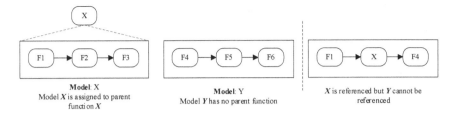

Fig. 6. Counting process model references

We only consider models which depict process flows (EPCs, rowEPCs, column EPCs, BPMN Process diagrams, BPMN Collaboration diagrams) to measure reuse in the repository used for the proof of concept. The number of references to a model must be greater than 1: A model which has only been referenced once has not been reused and therefore must not be counted - referring to Eq. (5), we set the threshold level (ITL) at 2 when calculating M. We only count a process model as reused once: whether a model is referenced twice or ten times, it is counted as reused only once.

4 Evaluation

The proposed measure of process model reuse was tested on the ARIS process repository of a large South African financial services company which has used ARIS as the process modeling tool used for the past 23 years. A five level hierarchial process architecture is used and the modeling method used within the organization is maintained by an organizational Center of Excellence and enforced using the functionality provided by ARIS. Between 100 and 200 modelers are active at any one time. The size of the ARIS process repository (35000+ models) and the duration of use within the organization make this a suitable repository to use as a proof of concept of the proposed measure. Using the ARIS scripting functionality, details of the models, objects, relationships, and symbols were exported into spreadsheets and then imported into Microsoft Access for data analysis. The steps followed to determine the overall level of reuse within the repository were as follows:

1. Determine the number of times each process model is referenced by other process models. Then, using the reuse threshold level, determine how many times the process model has been reused.
2. Determine the frequency distribution of the number of models and the corresponding number of uses in the repository (Table 1). It can be seen that 90.2% of process models in this repository are referenced once or less. This can be restated by saying that 90.2% of models in this repository are not reused at all.
3. Using the frequency distribution table, the levels of reuse for different threshold values (ITL) can be determined.

Table 1. Model reference frequency

Reference frequency	Freq %	Cum %
0	55.5	55.5%
1	34.7	90.2%
2	5.5	95.7%
3	1.8	97.5%

Validating a measure is the process of ensuring that the measure is a proper numerical characterization of the claimed attribute [12]. We adopted the "Framework for software measurement validation" of Fenton, Kitchenham and Pfleeger [16] to validate the measure proposed in this paper. Their framework consists of theoretical and empirical validation tests which are applied to the elements (attributes, measurement units, instruments, and protocols) that make up each measure. For the theoretical validation we validated the measure by confirming the validity of each attribute used in terms of the elements described. We also reviewed and confirmed the direct and indirect attribute requirements that must be satisfied for the measure to be valid. The measure was empirically validated by manually identifying and counting process model references in the repository and comparing the results with those obtained using the measurement protocol. A sample of three models was taken which, based on the results of the measurement protocol, were referenced 2, 3 and 5 times respectively. These results were confirmed by identifying the three models in the repository and manually confirming that the number of references reported for each model by the measurement protocol was correct. A further sample of 50 models were randomly selected and the number of references to each model was determined directly in the repository programmatically and compared to the results based on the data used in the analysis to determine the level of reuse. No differences were found.

5   Discussion

We believe that this paper is the first to quantitatively define a measure of reuse of complete business process models by other process models in a process repository as well as the empirical measurement of such reuse based on the proposed measure. The measure was tested on a real-world process repository containing more than 35000 process models.

Although ARIS terminology has been used in this paper, the measure and approach are not dependent on the repository being an ARIS repository, but rather on the structure and relationships of the models and objects. These are dependent on both the modeling tool functionality and the modeling method used by the organization. The proposed measure was tested on an ARIS repository which is structured as described in Sect. 3.1. Furthermore, the modeling method and process architecture adopted by the organization where the measure was tested supports the method described in this paper.

While we believe that the modeling tool, modeling method, and architecture used are typical, the measure should be tested in other organizations to confirm the generalizability and usefulness of the measure.

6 Conclusions

As the number of channels and devices that are used to interact with customers increases, the principle of reuse of complete process models within a repository will likely increase in importance. The current lack of reported measures for complete process model reuse will need to be adequately addressed for new modeling methods to be quantitatively assessed in this regard. Similarly, economic models of process model reuse will need a suitable measure of reuse if such models are to be compared.

For practitioners, the measure provides a tool which enables the reuse of process models in a repository to be measured and managed. The measure can also be applied to different sets of models in a repository to determine, for example, reuse level per organizational unit carrying out the modeling activities. Accordingly, we identify the following opportunities for further research in this area:

- Further analysis of factors impacting the reuse of complete process models.
- Methods to increase the levels of process model reuse
- Economic models of the reuse of complete process models in a repository

References

1. Seck, A.M., Philippe, J.: Service encounter in multi-channel distribution context: virtual and face-to-face interactions and consumer satisfaction. Serv. Ind. J. **33**, 1–15 (2011)
2. Recker, J.C.: Process modeling in the 21st century. BPTrends (5), 1–6 (2006)
3. Recker, J.C., Rosemann, M., Indulska, M., Green, P.: Business process modeling: a comparative analysis. J. Assoc. Inf. Syst. **10**(4), 333–363 (2009)
4. Fantinato, M., de Toledo, M., Thom, L.H., Gimenes, I.M., Rocha, R., Garcia, D.: A survey on reuse in the business process management domain. Int. J. Bus. Process. Integr. Manag. **6**(1), 52–76 (2012)
5. Koschmider, A., Fellmann, M., Schoknecht, A., Oberweis, A.: Analysis of process model reuse: where are we now, where should we go from here? Decis. Support Syst. **66**, 9–19 (2014)
6. Poulin, J.S.: Measuring Software Reuse: Principles, Practices, and Economic Models, p. 195. Addison-Wesley, Reading (1997)
7. Fellmann, M., Koschmider, A., Schoknecht, A.: Analysis of business process model reuse literature: are research concepts empirically validated? Modellierung, pp. 185–192 (2014)
8. Ferrucci, F., Tortota, G., Vitiello, G.: Exploiting visual languages in software engineering. In: Chang, S.-K. (ed.) Handbook of Software Engineering and Knowledge Engineering, pp. 53–76. World Scientific Publishing Co., Inc., New Jersey (2002)
9. Ouyang, C., Dumas, M., ter Hofstede, A.H., van der Aalst, W.M.: Pattern-based translation of BPMN process models to BPEL web services. Int. J. Web Serv. Res. **5**, 42–61 (2008)

10. Zhao, W., Hauser, R., Bhattacharya, K., Bryant, B.R., Cao, F.: Compiling business processes: untangling unstructured loops in irreducible flow graphs. Int. J. Web Grid Serv. **2**(1), 68–91 (2006)
11. Frakes, W.B., Tech, V., Terry, C.: Software reuse: metrics and models. ACM Comput. Surv. **28**(2), 415–435 (1996)
12. Fenton, N., Pfleeger, S.L.: Software Metrics: A Rigorous and Practical Approach, 2nd edn. (1997)
13. Daneva, M.: Measuring reuse in SAP requirements: a model-based approach. In: Proceedings of the 1999 Symposium on Software Reusability - SSR 1999, pp. 141–150. ACM Press, New York (1999)
14. Terry, C.: Analysis and implementation of software reuse measurement. Virginia Polytechnic Institute and State University, Master's Project and Report (1993)
15. Peffers, K., Tuure, T., Rothenberger, M.A., Chatterjee, S.: A design science research methodology for information systems research. J. Manag. Inf. Syst. **24**(3), 45–77 (2007)
16. Fenton, N.E., Kitchenham, B., Pfleeger, S.L.: Towards a framework for software measurement validation. IEEE Trans. Softw. Eng. **21**(12), 929–944 (1995)
17. Malinova, M., Leopold, H., Mendling, J.: An empirical investigation on the design of process architectures. In: Wirtschaftsinformatik, Leipzig (2013)
18. zur Muehlen, M., Wisnosky, D.E., Kindrick, J.: Primitives: design guidelines and architecture for BPMN models. In: Proceedings of the 21st Australasian Conference on Information Systems, ACIS 2010, Brisbane, Australia, 1–3 December 2010
19. Dijkman, R., Vanderfeesten, I., Reijers, H.A.: Business process architectures: overview, comparison and framework. Enterp. Inf. Syst. **10**(2), 129–158 (2016)
20. Van Nuffel, D., De Backer, M.: Multi-abstraction layered business process modeling. Comput. Ind. **63**(2), 131–147 (2012)
21. Gruhn, V., Laue, R.: Complexity metrics for business process models. In: Abramowicz, W., Mayer, H.C. (eds.) 9th International Conference on Business Information Systems (BIS 2006), Klagenfurt, Austria, pp. 1–12 (2006)

Anti-patterns for Process Modeling Problems: An Analysis of BPMN 2.0-Based Tools Behavior

Clemilson Luís de Brito Dias[1], Vinicius Stein Dani[1], Jan Mendling[2], and Lucineia Heloisa Thom[1(✉)]

[1] Institute of Informatics, Postgraduate Program in Computer Science, Federal University of Rio Grande do Sul, Porto Alegre, Brazil
{clbdias,vsdani,lucineia}@inf.ufrgs.br
[2] Institute for Information Business, Vienna University of Economics and Business, Vienna, Austria
jan.mendling@wu.ac.at

Abstract. Process modeling is increasingly conducted in business by non-expert modelers. For this reason, the increasing uptake of notations like BPMN has been accompanied by problems such syntactic and semantic errors. These modeling problems may hamper process understanding and cause unexpected behavior during process execution. A set of common modeling problems has been classified as *anti-patterns* in the literature. Up until now, it is not clear to which extent these anti-patterns can be spotted during modeling. In this paper, we investigate anti-pattern support based on a selection of prominent BPMN tools. The research contribution is two-fold: we demonstrate the importance of qualification of the analyst for the task of process modeling; and, we identify the need for process modeling tools to detect the use of anti-patterns, feedbacking the user more actively and explicitly about problems to be corrected in the process models.

Keywords: BPM tools analysis · Business process models · BPMN 2.0 · Anti-patterns · Process modeling problems

1 Introduction

Business Process Management (BPM) refers to specification, implementation, and monitoring of executable business processes [3].

By adopting BPM, organizations strive to reduce cost and cycle time while improving the quality of their products and services [11]. Such improvements are achieved by modeling business processes in an understandable way as a basis for analysis and redesign. A key concern is that the resulting process models have to be correct.

Modeling problems may lead to errors during process execution, which entails quality issues and additional costs for the organizations [4].

Various notations have been proposed for process modeling [7] (e.g., Event-Driven Process Chain, UML Activity Diagram, Business Process Model and

© Springer Nature Switzerland AG 2019
C. Di Francescomarino et al. (Eds.): BPM 2019 Workshops, LNBIP 362, pp. 745–757, 2019.
https://doi.org/10.1007/978-3-030-37453-2_59

Notation (BPMN), etc.) and a variety of process modeling tools [4] (e.g., Bonita, Camunda) is available. Recent years have seen a convergence on BPMN [9], and its version 2.0 has been adopted as an ISO standard.

Quality assurance has been less pervasive. Some tools provide functionality for detecting subclasses of problems and many problems go undetected depending on the tool [4]. Rozman et al. [10] identified problems being repeated constantly across a couple of thousands of processes modeled by novice modelers and defined a set of *anti-patterns* of business process modeling. Even professionally created process models such as the SAP Reference Model [6] contain errors.

Anti-patterns are partially useful not only for spotting such errors, but also by providing feedback to the modelers regarding which scenarios should be avoided in the future [10].

In this paper, we investigate the support of anti-patterns by BPMN-based process modeling and execution tools, in order to understand to which extent these tools automatically detect anti-patterns and help the user to correct them. We found that most common anti-patterns are detected by one or the other tool. Our analysis illustrates gaps in detection support. To achieve this analysis, we selected a set of anti-patterns and a set of BPMN 2.0-based process modeling and execution tools to be analyzed. To the best of our knowledge, there is no other paper examining process modeling tools from this point of view.

The organization of our paper is as follows. Section 2 presents the background, while Sect. 3 discusses related work. In Sect. 4 we discuss the methodology we adopted. Section 5 holds our result analysis report. Conclusions and future research are presented in Sect. 6.

2   Background

In this section, we present necessary concepts for understanding the contribution. First, we introduce BPMN 2.0 basic elements. Second, we present examples of anti-patterns selected for our tools analysis.

2.1   Business Process Model and Notation

One of the main purposes of process modeling is to ensure the repeatability of processes in organizations. The notation used to the modeling task of BPM is often the BPMN. There are several different elements in the BPMN to enable the process modeling task. For this paper understanding purposes, we present the process model example in Fig. 1, containing a key subset of core BPMN 2.0 elements needed for further discussion on this paper. This model sample represents the process of a store that sends an invoice to a client and waits for the client to realize the payment. If no payment is received up until 30 days, the main process flow is interrupted and the purchase is cancelled.

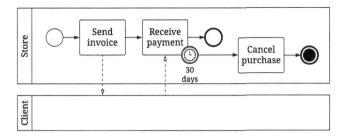

Fig. 1. Process model example illustrating a key subset of core BPMN 2.0 elements.

2.2 Process Modeling Anti-patterns

For the process modeling task there are studies that aims at defining process modeling patterns to support process modeling reuse (e.g., [14]). Based on that, we wanted to identify a set of common problems encountered in the process modeling task and, thus, we conducted a literature research aiming to find works proposing anti-patterns in this area. As a result, we obtained the study of [10] which defines a set of fifteen anti-patterns based on the analysis of a couple of thousands of process models modeled by novice modelers along six years. The advantage of anti-patterns is that typically they come with a proposed solution to fix it, which was the case.

According to Rozman et al. [10], with the adoption of BPMN as the main notation for process modeling, the main commercial tools available for process modeling started adopting BPMN as standard notation. Although BPMN is the most complete notation available, it does not prevent the design of process models with problems which may, for example, prevent the achievement of cost reduction through process execution errors [4]. The process modeling problems, identified by Rozman et al. [10] within their process modeling anti-patterns, are basically of three types: (i) syntactic, regarding the misuse of BPMN notation elements; (ii) pragmatic, concerning the comprehensibility of the process models; and, (iii) semantic, characterized by the compliance of the process modeled and its real-world representative. In many cases, the process can be syntactically valid, however, its semantics may be dubious not only for the user who analyzes the process, but also for the engine that will execute it.

Following, we present two examples of anti-patterns modeled with BPMN 2.0 and based on Rozman et al [10]. For each example, we consider: (i) a description of the *problem* formulated as an anti-pattern; (ii) the *impact* caused by the anti-pattern being exposed; and, finally, (iii) a *proposed solution* represented by a correct process modeling pattern. These examples will be presented in Sect. 5.

Figure 2(a) represents the first anti-pattern example, concerning *activities in a pool are not connected*. This *problem* is characterized by no sequence flow connecting two activities inside the same pool, and its cause is a misinterpretation of modelers who consider multiple pools as a single process; or, consider that there exists a dependency between processes and that the use of the message flows con-

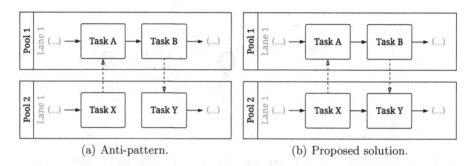

(a) Anti-pattern.   (b) Proposed solution.

Fig. 2. Anti-pattern *activities in a pool not connected.*

necting pools could dismiss the use of sequence flow in this case. The *impact* is that the relationship of dependence between the activities in Pool 2 (cf., Fig. 2(a)) is not defined, which could lead to the non-execution of "Task Y", compromising the execution of the process, since this task will be unreachable. As a *proposed solution* (Fig. 2(b)), the process modeling should be done in such a way that the pools are independently modeled, that is, without taking into account the connections that exist between them. All process elements inserted inside a pool must be fully connected by sequence flows. Although pragmatically incorrect, reproducing this anti-pattern in the process modeling task is syntactically correct, according to *BPMN 2.0*; if a task does not have an output sequence flow (cf., "Task X" in Fig. 2(a)), this represents the end of one of the flow paths [8].

Figure 3 represents the second anti-pattern example, concerning *exception flow is not connected to the exception.* The problem is characterized by missing explicit exception flow. This is a semantic problem, since if any action was expected to be executed after the exception raises, nothing would be performed in this modeling. This type of modeling problem may cause the process to be interrupted with no compensation task executed after the exception. Besides, it may change the semantics of the process, especially when read by third-parties, hampering its comprehension, since it is not known if a sequence flow is missing, if it has been wrongly directly connected to the task to which it is attached, or if the process merely stops right after the exception is raised. As a proposed solution, the exception flow should be modeled to make explicit what the modeler desires to represent (e.g., Fig. 3(a)).

3   Related Work

We conducted a preliminary research of the literature, aiming at identifying approaches that deal with the same subject explored in our paper. Regarding *anti-patterns*, beyond the study presented by Rozman et al. [10], addressing the identification of common mistakes made in the process modeling task, there is the study conducted by [1]. In this paper, the authors focus on the analysis and verification of 14 anti-patterns in collaborative business processes between organizations. The models were implemented using the UML Profile for Collaborative

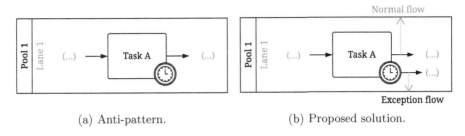

(a) Anti-pattern.   (b) Proposed solution.

Fig. 3. Anti-pattern *exception flow not connected to the exception.*

Business Processes based on Interaction Protocols (UP-ColBPIP), a language for collaborative process modeling that extends UML2 semantics. A set of 10 anti-patterns for control flow of the UP-ColBPIP language were defined, and a tool for checking the process models to identify these anti-patterns was developed. Although, there were no practical application for comparison of the behavior of different process modeling tools concerning any of these anti-patterns, which differs these studies from ours.

Regarding the comparison of process modeling tools, we found approaches on the topic of good practices in process modeling, also known as guidelines, focusing on the use of BPMN. However, these approaches aim at defining good practices in process modeling and how a set of available tools on the market encourage these good practices; or, how these tools react if the user does not follow their good practices recommendation on the process modeling task. The approach presented by Snoeck et al. [12] is an example of this kind of study. The authors explore to what extent BPMN-based tools support a set of process modeling guidelines. Another study that compares process modeling tools is the one by [2]. In this work, the authors compare the performance of the tools focusing on the interoperability among these applications. These studies differ from ours, since we compare the behavior of tools regarding *anti-patterns*, not good practices for the process modeling task or interoperability among the tools.

In [13] the authors analyze the presence of "syntactic anomalies" and "structural anomalies" in business processes modeled with BPMN. Syntactic anomalies were set into three groups: incorrect use of flow objects, misuse of connection objects, and incorrect use of swimming pools. While the structural anomalies were set into four groups: deadlock, lack of synchronization, dead activity, and infinite loop. However, the study does not apply these process models anomalies to process modeling tools nor process execution simulation tools to verify their effects, which differs this study from ours too.

4   Research Methodology

The aims of our investigation on the behavior of a subset of leading tools regarding a subset of process modeling problems is to identify to which extent these tools, which handle processes modeled using the BPMN 2.0 standard, automatically detect anti-patterns and help the user to correct them. Besides, we want to identify if these tools behave similarly or differently to each other when dealing with the same process modeling problems. Since BPMN 2.0 is an ISO standard, we expect that the tools behave similarly. In this section, we describe the steps we follow to perform the tool analyses regarding their behavior in relation to anti-patterns for process modeling problems.

4.1 Selection of the Set of Process Modeling Anti-patterns

After conducting the literature research to search works dealing with common problems encountered in the process modeling task, we identified some studies [5,10,13,15]. They point out to common problems encountered in process modeling tasks, some of them are the same across multiple studies (e.g., *activities in a pool are not connected*). But only the works in [5,10,15] explicitly define anti-patterns. Since [10] presents a more extensive set of anti-patterns, we mainly based our selection in their study, choosing the anti-patterns that also appear in the other papers that also list common problems encountered in process models [5,13,15]. Thus, we selected the ten anti-patterns: (1) *Activities in one pool are not connected*; (2) *The event does not contain an end event*; (3) *Sequence flow crosses sub-process boundary*; (4) *Sequence flow crosses pool boundary*; (5) *Gateway receives, evaluates and sends a message*; (6) *Intermediate events are placed on the edge of the pool*; (7) *Hanging intermediate events or activities in the process model*; (8) *Each swimlane in the pool contains start event*; (9) *Exception flow is not connected to the exception*; (10) *Message flow misuse across swimlanes*.

4.2 Selection of the Set of Process Modeling Tools

After selecting the anti-patterns subset, we selected a set of process modeling and execution tools to submit each of the selected anti-patterns and, thereafter, analyze each tool behavior regarding each process modeling problem. We defined the following inclusion criteria to conduct our tools selection process: (i) the application must have an open source version or a free trial version; (ii) the application must be recognized in the market as one of the references in BPM, considering results from literature and the tools' main clients as parameters; (iii) the application should have recently updated version; (iv) the application, besides providing the necessary resources for the modeling of the processes, should enable execution simulation, so that it is possible to analyze its behavior when executing a process with process modeling anti-pattern; and, finally, (v) the application must support BPMN 2.0.

Moreover, we performed a literature research on modeling and execution tools. We found a variety of papers [4,12] and based our selection on their results. The study [12], for example, which is a related work as presented in Sect. 3, made an exhaustive modeling tools analysis considering 117 tools. Thus, we selected the following process modeling and execution tools, also considering their acceptability in industry and academia communities: *Bizagi (version 3.1)*, *Bonita (version 7.5.4)*, *Camunda, (version 1.10)*, *Signavio, (version 11.9.1)*.

4.3   Tool Analysis Method Definition

In order to answer how the selected tools for process modeling and execution behave regarding process modeling anti-patterns, the following procedure was defined. First, each anti-pattern was modeled in BPMN 2.0 using each one of the selected tools, following an identical process of process modeling on each tool. All of the modeled and analyzed anti-patterns in our study are available at GitHub[1]. Second, we analyzed how the tool behaves regarding the process model containing the anti-pattern. In particular, we wanted to know if: the tools emit any warning or error message to the user in the process modeling task, if the tools emit any warning message pre-execution simulation, and if the tools simulate the execution of the process model containing the anti-pattern. Third, we reported whenever a warning or error message was given to the user (independent of how the visual feedback was composed) regarding any problem detected by the tool in the anti-pattern; we also report whenever an explanation about the modeling problem or correction suggestion was given. Fourth, we verified whether the tool allowed the execution of the process model, how was the behavior of the tool when we tried to execute a process model containing the anti-pattern, and if any warning message was given to the user at runtime. Finally, the data collected from the analysis of each tool was collected and reported in a comprehensive manner, further presented in Sect. 5.

5   Tool Behavior Analysis Concerning Anti-patterns

In this section, we provide an overview of the overall results (Sect. 5.1), followed by a more detailed discussion of two illustrative examples (Sects. 5.2 and 5.3, respectively). Finally, we discuss the overall anti-pattern detection support in the four tools (Sect. 5.4).

5.1   Overall Results

After the analysis of the behavior of every tool regarding each anti-pattern chosen, we were able to build Table 1, which summarizes our work and enables us to point out our findings.

On the process modeling phase, Bonita was the tool that most issued warnings (in 40% of the cases), followed by Signavio (30%), and Bizagi (10%). Camunda was the only tool that did not issue any warning regarding any of the anti-patterns analyzed. Considering all tools, warnings were issued to the user in 60% of the cases, more precisely, in processes modeled containing the anti-patterns 1, 2, 5, 6, 7, and 8. Regarding error messages in the process modeling

---

[1] https://www.github.com/clemilsondias/bpmn-anti-patterns-process-models.

Table 1. Results of the tool behavior analysis concerning the modeling and execution tasks regarding each anti-pattern. For the modeling task we analyzed if the tool emitted (+) or not (−) any warning or error message to the user; while for the simulation task we analyzed if the tool emitted (+) or not (−) any warning, and if the tool executed (+), executed partially (±) or did not execute (−) the anti-pattern.

| Anti-pattern | Modeling | | | | | | | | Simulation | | | | | | | |
| | Warning | | | | Error | | | | Warning | | | | Behavior | | | |
	Bizagi	Bonita	Camunda	Signavio	Bizagi	Bonita	Camunda	Signavio	Bizagi	Bonita	Camunda	Signavio	Bizagi	Bonita	Camunda	Signavio
1	−	+	−	−	−	−	−	+	−	−	−	−	±	+	±	−
2	−	+	−	−	−	−	−	+	−	−	−	−	+	+	+	−
3	−	−	−	−	−	−	−	−	−	−	−	−	−	−	−	−
4	−	−	−	−	−	+	−	+	−	−	−	−	±	−	−	−
5	+	−	−	+	−	+	+	−	−	−	−	−	±	−	−	±
6	−	+	−	−	−	−	−	+	−	−	−	−	±	±	±	−
7	−	+	−	+	−	−	−	+	−	−	−	−	±	+	±	−
8	−	−	−	+	+	−	−	−	−	−	+	−	+	+	−	+
9	−	−	−	−	+	+	−	+	−	−	−	−	−	−	±	−
10	−	−	−	−	+	−	−	−	−	−	−	−	±	+	+	+
Total	1	4	0	3	3	3	1	6	0	0	1	0	N/A	N/A	N/A	N/A
%	10	40	0	30	30	30	10	60	0	0	10	0	N/A	N/A	N/A	N/A

phase, Signavio was the tool that most emitted error messages to user, preventing the modeling of processes containing anti-pattern in 60% of cases. Followed by Bizagi and Bonita (30%), and Camunda (10%). Considering all tools, error messages were issued for 90% of the anti-patterns analyzed (i.e., only one anti-pattern did not generate error messages: anti-pattern 3). The only anti-pattern we could not model was anti-pattern 5. Although, some tools emitted warning or error in the modeling phase.

In general, tools that issued a warning regarding a modeling problem did not issue an error about the same problem, and vice-versa. Signavio was an exception, since it issued a warning and an error to anti-pattern 7. Considering this, Signavio was the tool that most reacted to process models containing anti-patterns, issuing warning or error in 80% of the cases analyzed, followed by Bonita (70%), Bizagi (40%), and Camunda (10%) (cf., Fig. 4).

On the simulation execution phase, only Camunda issued an error, precisely when it simulated a process model containing the anti-pattern 8. Moreover, Table 1 also shows that it was possible to simulate the execution of processes comprising anti-patterns in 90% of the cases (i.e., only one anti-pattern did

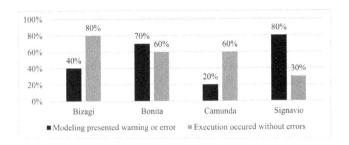

Fig. 4. Comparison of the behavior of each tool regarding the anti-patterns.

not enable the execution simulation: anti-pattern 3). Considering all the anti-patterns that allowed the execution simulation to occur, in 44% of cases the tools performed this procedure showing distinct semantics, that is, the processes were interpreted differently across one or more tools. In 70% of the cases, more than one tool allowed the simulation of the execution of the processes.

Figure 4 shows a comparison of the behavior of each tool analyzed along our study. Bizagi was the tool that most allowed the simulation of the execution of processes containing anti-pattern, executing the simulation in 80% of the cases, followed by Bonita and Camunda (60%), and Signavio (30%).

5.2   Illustrative Example: Anti-pattern 1

As the first illustrative example of anti-pattern analyzed, we present the anti-pattern regarding *activities in one pool are not connected* (cf., Fig. 2(a)). The behavior of the tools are described below.

Bizagi did not issue any warning to the user regarding the anti-pattern in the process modeling task. When executing the process validation function, the tool stated that the process was validated "without warnings or errors". Bizagi allowed the simulation of the execution of the process, and when the execution was initiated in Pool 1, the process passed through the pool activities and ended normally. However, upon initialization of the execution in Pool 2, the process terminated abruptly, right after the end of Task X, and did not perform Task Y.

Bonita issued an information-level warning to the user in the process modeling task, stating that there was no sequence flow on the input of Task Y and, for this reason, the activity would be used as a "possible" initialization activity. Upon the initialization of the process execution simulation through Pool 1, the process was executed and finalized normally. However, when starting through Pool 2, only Task X was executed. Task Y stayed pending, waiting to receive a message from Pool 1, since this task has no input sequence flow. When the message from Pool 1 is received, the process in Pool 2 ends normally.

Camunda has not issued any warning or error to the user about the anti-pattern present during the process modeling task. This tool allowed the simulation of the execution of the process. When beginning the execution simulation in Pool 1, the process passed through the activities of the pool and ended normally. However, upon initialization in Pool 2, the process terminated abruptly, after the end of Task X, and did not perform Task Y.

Signavio pointed to two errors when validating the process model in the process modeling task. One of them, referring to Task X, since it did not contain an output sequence flow; and, another error referring to Task Y, since it did not contain an input sequence flow. Due to these errors, the tool did not allow the execution of the process containing this anti-pattern.

5.3   Illustrative Example: Anti-pattern 9

As second illustrative example, we present the anti-pattern regarding *Exception flow is not connected to the exception* (cf., Fig. 3(a)). The exception flow, in this case, is the flow that should be executed right after the triggering of the interruptive timer intermediate event, attached to the activity, subsequently to the passage of a certain amount of time. The behavior of the tools regarding this anti-patterns are described next.

Bizagi did not allow the process to be modeled without a sequence flow connected to the intermediate event. When attaching an intermediate event to an activity, Bizagi automatically inserts a new sequence flow starting from the intermediate event. If nothing else is modeled from this intermediate event when performing the process validation, Bizagi issue an error message informing that "transition is not connected". In this scenario, it is not possible to simulate the process execution.

Bonita did not allow the modeling of the process containing the anti-pattern. When executing the automatic validation of the model, function provided by the tool, which searches for syntactic errors within the model, Bonita issues an error message informing that "the edge event must have an exception transition". Even so, the process execution simulation option is available. However, when trying to simulate the execution, the tool re-evaluates the model and prevents its simulation.

Camunda allowed the modeling of the process containing the anti-pattern and did not issue any warning or error message to the user. It also allowed the execution simulation without the emission of any warning.

Signavio did not allow the process containing the anti-pattern to be modeled. Upon performing the automatic validation of the model, provided by the tool, it issues an error message informing that "attached intermediate events must have exactly one output stream". Although this tool has several levels of validation, the process with this anti-pattern was not even approved in the most basic validation. Thus, Signavio did not allow the execution simulation to be executed.

5.4   Discussion on the Overall Anti-pattern Detection Support in the Selected Tools

After conducting our tool analysis, our main finding is that distinct process modeling and execution tools behave differently regarding the same anti-pattern modeled with BPMN 2.0 (cf., Table 1), which is a standard notation and, therefore, should trigger similar responses by the tools concerning the same process models.

We point out to the need for the manufacturers of process modeling tools to invest in features to visually feedback and guide modelers not only to the adoption of best practices in the process modeling task but, also, in the warning of known anti-patterns found in processes. Although some tools are already investing in this issue (e.g., Camunda, Signavio), none of the tools analyzed in this study was efficient in this area to avoid the occurrence of the most well-known process modeling problems. Therefore, we believe our findings will inspire tool developers to give more attention to this issue.

Because BPMN 2.0 is consolidated as a standard notation for the process modeling task, it is usual for many users to model their processes in one tool and execute them in another. In this scenario, an analyst could validate a process modeled on one tool but, when executing the same process in another tool, get a different result than expected. As the results of our study point out, this kind of scenario may be very problematic for process execution. Another possible problematic scenario could be an organization changing its BPMN 2.0-based tool supplier. In this case, BPMN 2.0 being a standard induces the stakeholders to think that there is full compatibility across different process modeling and execution tools. However, our results shows that this is not always the case.

6   Conclusions and Future Research

In this paper, we presented an analysis of the behavior of process modeling tools during the modeling and execution (whenever possible) of processes modeled containing some of the most common anti-patterns practiced by users. After modeling and executing these processes on the four selected BPMN 2.0-based tools, we found out that the way manufacturers interpret the BPMN 2.0 notation and visually feedback users about process modeling problems differs from each other. Currently, these tools present simple and passive validation mechanisms based almost entirely on labeling good practice and basic checking (i.e., missing start or end events), being Signavio the tool that presented the highest levels of process validations, although this case study shows that there is still room for the tool to evolve further.

As future research we intend to increment the selected set of process modeling anti-patterns, and even design patterns, to analyze over the toolset; and, we intend to apply a survey with participants from different countries aiming at the identification of more BPMN 2.0-based tools used for modeling and executing of processes. As a limitation of our work, it is important to highlight that the processes modeled in our study were extremely simple, once they consisted of few tasks (i.e., 5 or less in most cases). In real-world situations, the complexity of the models is much higher, what makes this difference of behavior between the tools even more laborious to analyze.

Acknowledgments. Vinicius Stein Dani is a CAPES scholarship holder; Lucineia Heloisa Thom is a CAPES scholarship holder, Program Professor Visitante no Exterior, grant 88881.172071/2018-01; this study was financed (Code 001) in part by CAPES.

References

1. Chiotti, O., Roa, J., Villareal, P.: Detection of anti-patterns in the control flow of collaborative business processes. Simposio Argentino de Ingeniería de Software (2015)
2. Dirndorfer, M., Fischer, H., Sneed, S.: Case study on the interoperability of business process management software. In: Fischer, H., Schneeberger, J. (eds.) S-BPM ONE 2013. CCIS, vol. 360, pp. 229–234. Springer, Heidelberg (2013). https://doi.org/10.1007/978-3-642-36754-0_14
3. Dumas, M., La Rosa, M., Mendling, J., Reijers, H.A., et al.: Fundamentals of Business Process Management, vol. 1. Springer, Heidelberg (2013). https://doi.org/10.1007/978-3-642-33143-5
4. Geiger, M., Harrer, S., Lenhard, J., Wirtz, G.: BPMN 2.0: the state of support and implementation. Future Gener. Comput. Syst. (2017). https://doi.org/10.1016/j.future.2017.01.006
5. Koschmider, A., Laue, R., Fellmann, M.: Business process model anti-patterns: a bibliography and taxonomy of published work. In: vom Brocke, J., Gregor, S., Müller, O. (eds.) 27th European Conference on Information Systems - Information Systems for a Sharing Society, ECIS 2019, Stockholm and Uppsala, Sweden, 8–14 June 2019 (2019). https://aisel.aisnet.org/ecis2019_rp/157
6. Mendling, J.: Empirical studies in process model verification. In: Jensen, K., van der Aalst, W.M.P. (eds.) Transactions on Petri Nets and Other Models of Concurrency II. LNCS, vol. 5460, pp. 208–224. Springer, Heidelberg (2009). https://doi.org/10.1007/978-3-642-00899-3_12
7. Mili, H., et al.: Business process modeling languages: sorting through the alphabet soup. ACM Comput. Surv. **43**(4) (2010). https://doi.org/10.1145/1824795.1824799
8. (Object Management Group) OMG: Business process model and notation (BPMN) version 2.0. Technical report, Object Management Group (OMG), January 2011
9. Recker, J., Indulska, M., Green, P.: How good is BPMN really? Insights from theory and practice. Technical report, Association for Information Systems AIS Electronic Library (AISeL) (2006). www.bpmn.org
10. Rozman, T., Polancic, G., Horvat, R.V.: Analysis of most common process modeling mistakes in bpmn process models. In: 2008 BPM and Workflow Handbook. University of Maribor Slovenia (2008)
11. Rudden, J.: Making the case for BPM making the case for BPM: a benefits checklist. Technical report, BPTrends (2007). www.bptrends.com
12. Snoeck, M., Moreno-Montes de Oca, I., Haegemans, T., Scheldeman, B., Hoste, T.: Testing a selection of BPMN tools for their support of modelling guidelines. In: Ralyté, J., España, S., Pastor, Ó. (eds.) PoEM 2015. LNBIP, vol. 235, pp. 111–125. Springer, Cham (2015). https://doi.org/10.1007/978-3-319-25897-3_8
13. Suchenia, A., Ligeza, A.: Event anomalies in modeling with BPMN. Int. J. Comput. Technol. Appl. **6**(5), 789–797 (2015)

14. Thom, L.H., Reichert, M., Chiao, C.M., Iochpe, C., Hess, G.N.: Inventing less, reusing more, and adding intelligence to business process modeling. In: Bhowmick, S.S., Küng, J., Wagner, R. (eds.) DEXA 2008. LNCS, vol. 5181, pp. 837–850. Springer, Heidelberg (2008). https://doi.org/10.1007/978-3-540-85654-2_75
15. Vidacic, T., Strahonja, V.: Taxonomy of anomalies in business process models. In: Escalona, M.J., Aragón, G., Linger, H., Lang, M., Barry, C., Schneider, C. (eds.) Information System Development, pp. 283–294. Springer, Cham (2014). https://doi.org/10.1007/978-3-319-07215-9_23

Author Index

Agostinelli, Simone   12
Aloini, Davide   326, 508
Antunes, Bianca B. P.   583
Aßfalg, Nico   93
Atzmueller, Martin   417
Azaiz, Imen   186

Bachhofner, Stefan   690
Bastos, Leonardo S. L.   583
Baumgartner, David   608
Beerepoot, Iris   338
Benevento, Elisabetta   508
Bergmann, Ralph   32
Bloemheuvel, Stefan   417
Boltenhagen, Mathilde   160
Borgida, Alexander   19
Budiono, Stephanus   483
Buijs, Joos C. A. M.   459
Bukhsh, Faiza Allah   496
Burattin, Andrea   570

Caballero, Ismael   362
Cabanillas, Cristina   81
Carmona, Josep   160
Casino, Fran   704
Castellano, Maurizio   545
Chatain, Thomas   160
Cho, Minsu   520
Choi, Byung-Kwan   520
Combi, Carlo   68
Comuzzi, Marco   557
Creemers, Mathijs   212
Cuendet, Michel   545

Dasaklis, Thomas K.   704
Davis, Christopher John   570
de Bari, Berardino   545
de Brito Dias, Clemilson Luís   745
de la Fuente, Rene   471
de Vries, J. Gert-Jan   459
De Weerdt, Jochen   250
Di Ciccio, Claudio   690
Dixit, Prabhakar M.   508
Douligeris, Christos   704

Emrich, Andreas   130
Eshuis, Rik   375
Esser, Stefan   632
Exler, Anna-Maria   301

Fagalde, Gonzalo   471
Fahland, Dirk   172, 632
Fernández-Llatas, Carlos   532, 545
Ferreira, Diogo R.   263
Fettke, Peter   56, 130
Folino, Francesco   5
Fuentes, Ricardo   471

Gal, Avigdor   106
Galvez, Victor   471
García-Bañuelos, Luciano   690
Gatta, Roberto   545
Geerdink, Jeroen   496
Gloor, Peter   326
Gómez-López, María Teresa   362
Grumbach, Lisa   32
Gunnarsson, Björn Rafn   250
Gutiérrez Fernández, Antonio Manuel   388

Haarmann, Stephan   400
Hake, Philip   56
Hall, Geoff   595
Hamacher, Silvio   459, 583
Haselböck, Alois   81
Havur, Giray   81
Helm, Emmanuel   608
Hjardem-Hansen, Rasmus   570
Hogg, David   595
Holfter, Adrian   400
Hompes, Bart Franciscus Antonius   459

Imgrund, Florian   288

Janiesch, Christian   288
Jans, Mieke   212
Jochum, Benjamin   93
Johnson, Owen   595
Jooken, Leen   212

Käfer, Tobias   93
Kalokyri, Varvara   19
Karastoyanova, Dimka   664
Keates, Owen   119
Kianoush, Sanaz   430
Klijn, Eva L.   172
Kloepper, Benjamin   417
Ko, Jonghyeon   557
Kogler, Klaus   388
Krofak, Ivan   388
Kuhn, Norbert   32
Küng, Josef   608
Kurniati, Angelina Prima   595

Ladleif, Jan   651
Lambusch, Fabienne   317
Lantow, Birger   317
Larsen, John Bruntse   570
Laue, Ralf   281
Lee, Suhwan   557
Lenkowicz, Jacopo   545
Li, Chiao-Yun   199
Lin, Alvin C.   608
Lin, Anna M.   608
Lopes, Iezalde F.   263
López-Pintado, Orlenys   690

Maletzki, Carsten   32
Mannel, Lisa Luise   224
Manresa, Adrian   583
Marazza, Francesca   496
Marchesi, Janaina F.   583
Marco-Ruiz, Luis   545
Marcos, Mar   545
Marian, Amélie   19
Marrella, Andrea   12
Martin, Niels   532
Martinez-Millana, Antonio   532, 545
Mazza, Bruno F.   459
McInerney, Ciarán   595
Mecella, Massimo   12
Mendling, Jan   301, 745
Meneses, Cesar   471
Montali, Marco   355
Mühlberger, Roman   690
Munoz-Gama, Jorge   471, 545

Nürnberg, Leonard   93

Orini, Stefania   545

Palipana, Sameera   430
Parody, Luisa   362
Pathak, Shreyasi   496
Patsakis, Costas   704
Pegoraro, Marco   238
Peinl, René   441
Perak, Ornella   441
Pika, Anastasiia   483
Pochiero, Federica   326
Poels, Geert   721
Pontieri, Luigi   5
Pufahl, Luise   400

Quintano Neira, Ricardo Alfredo   459

Rafiei, Majid   676
Rebmann, Adrian   130
Rehse, Jana-Rebecca   56
Reijers, Hajo A.   338, 483
Reinhartz-Berger, Iris   147
Richter, Florian   186
Rietzke, Eric   32
Ruhsam, Christoph   388

Sacchi, Lucia   545
Sani, M. F.   508
Savazzi, Stefano   430
Schmitt, Julian   317
Seidl, Thomas   186
Seifert, Christin   496
Sepúlveda, Marcos   471
Seymour, Lisa F.   733
Shadrina, Anna   388
Shraga, Roee   106
Sigg, Stephan   430
Simões de Almeida, Samantha L.   459
Smit, Erik   375
Soffer, Pnina   147
Song, Minseok   520
Stage, Ludwig   664
Stefanini, Alessandro   326, 545
Stein Dani, Vinicius   745
Stretton, Erin   459

Taudes, Alfred   301
ter Hofstede, Arthur H. M.   483

Thom, Lucineia Heloisa   745
Tsoury, Arava   147

Uysal, Merih Seran   238

Valdivieso, Bernardo   532
Valencia-Parra, Álvaro   362
Vallati, Mauro   545
van de Weerd, Inge   338
van der Aa, Han   625
van der Aalst, Wil M. P.   199, 224, 238, 483,
    508, 676
van Keulen, Maurice   496
Van Rijswijk, Freddie   388
van Zelst, Sebastiaan J.   199
vanden Broucke, Seppe K. L. M.   250
Varela-Vaca, Ángel Jesús   362

Veitch, Ross S.   733
Viganò, Luca   68
Vijlbrief, Onno   496
Villadsen, Jørgen   570

Wang, Il-Jae   520
Weske, Mathias   400, 651
Winkel, David   186
Wynn, Moe T.   483

Yeom, Seok-Ran   520

Zasada, Andrea   45
Zavatteri, Matteo   68
Zellner, Ludwig   186
Zucker, Gerhard   388
Zucker, Kieran   595